AFFINE LIE ALGEBRAS, WEIGHT MULTIPLICITIES, AND BRANCHING RULES

PUBLISHED TITLES IN THE *LOS ALAMOS SERIES IN BASIC AND APPLIED SCIENCES*, edited by David H. Sharp and L. M. Simmons, Jr.

1. Wildon Fickett & William C. Davis, *Detonation*
2. Charles L. Mader, *Numerical Modeling of Detonation*
3. Robert D. Cowan, *The Theory of Atomic Structure and Spectra*
4. Ben R. Finney & Eric M. Jones, eds., *Interstellar Migration and the Human Experience*
5. Wildon Ficket, *Introduction to Detonation Theory*
6. Grant Heiken & Kenneth Wohletz, *Volcanic Ash*
7. N. Metropolis, D. H. Sharp, W. J. Worlton, & K. R. Ames, eds. *Frontiers of Supercomputing*
8. Charles L. Mader, *Numerical Modeling of Water Waves*
9. S. Kass, J. Patera, R. Moody, R. Slansky, *Affine Lie Algebras, Weight Multiplicities, and Branching Rules*
10. S. M. Ulam, *Analogies between Analogies. The Mathematical Reports of S. M. Ulam and his Los Alamos Collaborators*

AFFINE LIE ALGEBRAS, WEIGHT MULTIPLICITIES, AND BRANCHING RULES

VOLUME TWO

S. Kass, R. V. Moody, J. Patera, R. Slansky

UNIVERSITY OF CALIFORNIA PRESS
Berkeley • Los Angeles • Oxford

University of California Press
Berkeley and Los Angeles, California

University of California Press
Oxford, England

Library of Congress Cataloging-in-Publication Data

Affine lie algebras, weight multiplicities, and branching rules
/ S. Kass . . . [et. al.]. 2 vols.
p. cm.—(Los Alamos series in basic and applied sciences)
Bibliography: p.
Includes index.
ISBN 0–520–06768-1 (alk. paper)
1. Kac-Moody algebras. 2. Representations of algebras. I. Kass, Sam. II. Series.
QA252.3.A36 1990
512′.55—dc20 89–4977
CIP

Printed in the United States of America

1 2 3 4 5 6 7 8 9

The paper used in this publication meets the minimum requirements of American National Standard for Information Sciences—Permanence of Paper for Printed Library Materials, ANSI Z39.48–1984 ∞

CONTENTS

VOLUME ONE

VOLUME TWO

Part IV—The Tables

Part IV

The Tables

27. INTRODUCTION TO THE TABLES

27.1 After briefly reviewing the root systems of the direct affine algebras and the weight systems of their representations, we summarize the basic formulas for the multiplicities and the strategy for the computation of branching rules. This section is intended to be a guide for using the tables, so many of the basic definitions and concepts discussed in Volume One are not repeated here.

27.2 The root system of a direct affine algebra $\hat{\mathfrak{g}}$ is easily described in terms of the root system of an underlying finite dimensional algebra $\overline{\mathfrak{g}}$. Let $\hat{\Delta}$ and $\overline{\Delta}$ be the root systems of $\hat{\mathfrak{g}}$ and $\overline{\mathfrak{g}}$, respectively. As in Volume One, we include the zero root in $\overline{\Delta}$ and $\hat{\Delta}$. The null root δ of $\hat{\mathfrak{g}}$ is defined as

$$\delta = \sum_{i=0}^{d} c_i \alpha_i, \tag{27.1}$$

where the integers c_i are the marks of $\hat{\mathfrak{g}}$, tabulated in Fig. 2-10. The root system of $\hat{\mathfrak{g}}$ is

$$\hat{\Delta} = \{k\delta + \alpha \mid k \in \mathbb{Z} \text{ and } \alpha \in \overline{\Delta}\}. \tag{27.2}$$

The set of positive roots $\hat{\Delta}^+$ of $\hat{\mathfrak{g}}$ are those elements of $\hat{\Delta}$ with $k \geq 0$ for α positive and $k \geq 1$ if α is negative or zero. The root multiplicities are unity with the exception of the roots $k\delta$, which have multiplicity $d = \text{rank}(\overline{\mathfrak{g}})$ for $k \neq 0$, and multiplicity $d+2$ for $k = 0$. For computations, the subset of the roots

$$R = \{\alpha \in \hat{\Delta} \mid 0 < \alpha \leq \delta\}$$

is stored in a linked list, δ being stored d times, reflecting its multiplicity. Thus, the set of positive roots is $\hat{\Delta}^+ = \cup_{k \geq 0} (R + k\delta)$. The root systems of the direct affine algebras are summarized by the Coxeter-Dynkin diagrams in Fig. 2-10.

27.3 The extended weight lattice $\hat{P}$ of $\hat{\mathfrak{g}}$ is a $(d+2)$-dimensional lattice, and a weight γ in $\hat{P}$ is designated by its **complete weight label**, the $(d+2)$-tuple $(n; a_0, a_1, \ldots, a_d)$, where n is called the **null depth** of γ and $a_i = \gamma(h_i) = (\gamma \mid \breve{\alpha}_i)$. The **fundamental weights** $\omega_0, \ldots, \omega_d$ are defined by the labels

$$\omega_i = (0; 0, \ldots, 1, \ldots, 0),$$

where 1 is in position i. A weight $(n; a_0, \ldots, a_d)$ is **dominant** if $a_0, \ldots, a_d \geq 0$. The set of dominant weights is denoted $\hat{P}^+$. The $(d+1)$-tuple $(a_0, a_1, \ldots, a_d)$ is called the **incomplete weight label** of γ; it is often useful because many properties of weights are independent of null depth. In particular, the structure of the highest weight module $V(\Lambda)$ is independent of the null depth of Λ, so we may adopt the convention that the null depth of Λ is zero. Then the null depth n of a weight γ is the number of times the simple root α_0 is subtracted from Λ in reaching γ.

A highest weight representation of $\hat{\mathfrak{g}}$ is uniquely identified by a highest weight Λ. The weight system of $V(\Lambda)$ is $\Omega(\Lambda)$. One of the two invariants of the weights in $\Omega(\Lambda)$ is the **level**,

$$\text{level}(\gamma) = \gamma(\not{c}) = (\gamma \mid \delta) = \sum_{i=0}^{d} \check{c}_i a_i \tag{27.3}$$

where $\check{c}_i$ are the **comarks** defined in §3.7, and shown in the tables of §28. The level is constant on $\Omega(\Lambda)$. The marks and comarks of an ADE algebra are equal.

The other invariant on the set $\Omega(\Lambda)$ is the **congruence class**; if γ is in $\Omega(\Lambda)$, then $\text{class}(\gamma) := \text{class}(a_1, \ldots, a_d) = \text{class}(\Lambda)$. The class of a weight is an integer (or pair of integers) defined in terms of the (b_i) identified in the Tables of §28 and the equations

$$\begin{aligned} &\text{class}(\gamma) = \sum_{i=1}^{d} b_i a_i \pmod{n}, \qquad \hat{\mathfrak{g}} \text{ not } D_d,\ d \text{ even} \\ &\Omega^+(\Lambda) = \{\gamma \in P_+ \mid \gamma(\not{c}) = \Lambda(\not{c}) \quad \text{and} \quad \text{class}(\gamma) = \text{class}(\Lambda)\}. \end{aligned} \tag{27.4}$$

An important subset of $\Omega^+(\Lambda)$ is $\Omega^+_{\text{high}}(\Lambda)$: for each set of weights in $\Omega^+(\Lambda)$ with the same incomplete weight label, $\Omega^+_{\text{high}}(\Lambda)$ contains the weight with the smallest value of n in its complete weight label. For D_d, d even there are two vectors $(b_i), (b_i')$ to be used. One computes $(\sum b_i a_i, \sum b_i' a_i)$ and interprets the first component modulo 2 and the second modulo 4 thus obtaining the congruence class in $\mathbb{Z}/2\mathbb{Z} \times \mathbb{Z}/4\mathbb{Z}$.

The class and level of a weight are independent of its null depth. Some of the above properties of weights and roots are special to the direct affine algebras, so the twisted and nonaffine Kac-Moody algebras present different challenges from a computational viewpoint.

27.4 The computational tools used to produce the branching rules tables of §32 and multiplicity tables of §31 do not differ greatly from those used in [BMP–85] and [MP–81] for the finite dimensional simple Lie algebras (see end of Volume One for references). However, whereas in the finite dimensional case the entire root system of an algebra and the entire set of dominant weights of a module may be calculated, for the affine case, we must find a suitable stopping point that does not affect the computation. The fact that the affine algebras and their modules have infinite dimension presents some practical problems.

Let $V(\Lambda)$ be a highest weight module for the direct affine algebra $\hat{\mathfrak{g}}$. **Freudenthal's multiplicity formula**, derived in §22, provides a recursive calculation of the multiplicity $\text{mult}_\Lambda(\gamma)$ of the weight $\gamma \in \Omega(\Lambda)$:

$$\begin{aligned} &((\Lambda + \rho \mid \Lambda + \rho) - (\gamma + \rho \mid \gamma + \rho))\, \text{mult}_\Lambda(\gamma) \\ &\quad = 2 \sum_{\alpha \in \hat{\Delta}+} \text{mult}_{\hat{\mathfrak{g}}}(\alpha) \sum_{j \geq 1} (\gamma + j\alpha \mid \alpha)\, \text{mult}_\Lambda(\gamma + j\alpha), \end{aligned} \tag{27.5}$$

where $\text{mult}_{\hat{\mathfrak{g}}}(\alpha)$ is the multiplicity of the root α. There is no point in computing multiplicities of nondominant weights since multiplicities are constant on Weyl orbits and every Weyl orbit has a (unique) dominant weight on it. Note that if γ is a weight of $V(\Lambda)$, the computation of $\text{mult}_\Lambda(\gamma)$ involves only those positive roots α for which $\gamma + j\alpha$ is indeed a weight of $V(\Lambda)$ for some $\alpha \in \hat{\Delta}^+$ and $j \geq 1$. Thus, only a finite number of nonzero terms contribute to the double sum in (27.5).

Denote by $\Omega(\Lambda)$ the set of weights of $V(\Lambda)$, and by $\Omega^+(\Lambda)$ the set of dominant weights of $V(\Lambda)$. Call a set of dominant weights $\tilde{\Pi}$ of $V(\Lambda)$ **saturated** if $\gamma \in \tilde{\Pi}$ implies $\lambda \in \tilde{\Pi}$ for all $\lambda > \gamma$. The multiplicities of the weights in such a set can then be calculated recursively by Freudenthal's multiplicity formula. Computing $\text{mult}_\Lambda(\gamma)$ may require the multiplicity of a nondominant weight $\gamma + j\alpha$. This weight is on a unique Weyl orbit with a unique dominant weight, which must be higher than $\gamma + j\alpha$, hence higher than γ, and hence in $\tilde{\Pi}$. Thus, if the multiplicities of the dominant weights higher than γ have been computed, the multiplicity of $\text{mult}_\Lambda(\gamma)$ is calculable recursively from (27.5). We need to compute the multiplicities of dominant weights only.

Significant economy results by using the improvement to (27.5) due to [MPa–82]; the details of applying this modification to the direct affine case are not difficult. All the computations in this book were performed on an Apple Macintosh II.

27.5 To help explain the use of the multiplicity tables, we recall the geometric structure of a highest weight module by means of the example $V(3,2)$ of affine-A_1. Its first few weights and their multiplicities are shown in Figure 27-1.

The multiplicities of the dominant weights determine all multiplicities. Every other weight is on a Weyl orbit of some dominant weight, and the multiplicity is invariant on a Weyl orbit. For example, in affine-A_1, the $(n;3,2)$ weight is Weyl reflected by r_1 to $(n;7,-2)$, as shown in Fig. 27-2. The multiplicity is unchanged by the reflection. Given the multiplicities of the dominant weights, Weyl reflections are used to find the multiplicities of all remaining weights.

In Figure 27-3 we have excised from the Tables of §31 the multiplicity table labeled "Affine A1 - Level 5 - Class 0," which contains the multiplicities for $V(3,2)$. Detailed comments on reading the table are included in the figure caption.

The basic representations of the ADE affine algebras have weight multiplicities given by classical partition functions of §29. These are included in the Tables of §31, but more values are given in §29.

27.6 The string functions are obtained directly from the multiplicity table. For the example above,

$$
\begin{aligned}
\sigma^{(3,2)}_{(1,4)}(q) &= q(1 + 3q + 8q^2 + 17q^3 + 36q^4 + \cdots), \\
\sigma^{(3,2)}_{(3,2)}(q) &= 1 + 2q + 6q^2 + 13q^3 + 28q^4 + \cdots, \\
\sigma^{(3,2)}_{(5,0)}(q) &= 1 + 3q + 7q^2 + 16q^3 + 32q^4 + \cdots.
\end{aligned}
\tag{27.6}
$$

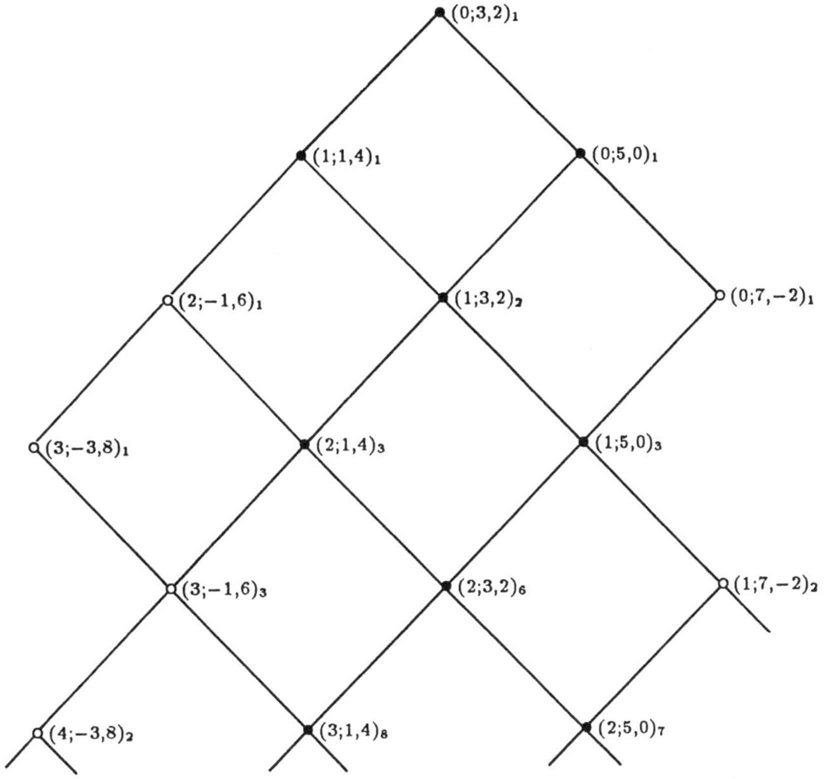

Fig. 27-1. The first few weights and their multiplicities of the weight system of $V(3,2)$ of affine-A_1. The first entry of the complete weight label is the null depth n; a_0 and a_1 follow the semicolon. The subscript is the weight multiplicity, which is the dimension of the space $V^\gamma(\Lambda)$. The dominant (positive) weights are shown in black.

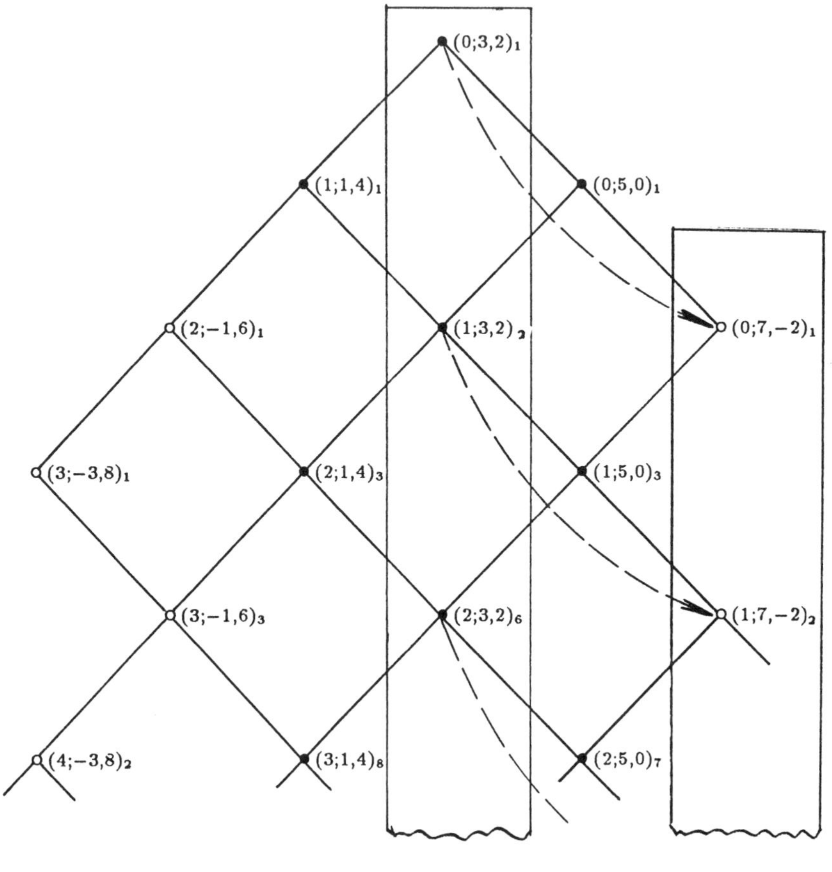

Fig. 27-2. Weyl reflection of $(n, 3, 2)$ to $(n, 7, -2)$ for $n = 0, 1, 2$.

Affine A1 - Level 5 - Class 0

	Rp 1			Rp 2			Rp 3		
		1			3			5	
		4			2			0	
Wt 1	1	148	4638	0	129	4475	0	46	1749
	2	259	7101	1	229	6919	0	86	2706
	5	439	10753	3	399	10544	1	148	4155
	11	730	16093	8	671	15905	2	256	6277
	23	1186	23862	17	1111	23712	6	423	9407
	44	1900	35032	36	1795	35033	12	694	13922
	83	2987	51009	69	2860	51256	25	1106	20453
Wt 2	1	203	5965	1	184	5836	0	71	2294
	3	349	9084	2	324	8934	1	123	3534
	7	589	13663	6	549	13534	2	213	5357
	16	967	20349	13	916	20255	5	353	8057
	32	1562	30000	28	1489	30038	21	930	17623
	114	3873	63529	103	3755	64211	38	1477	25695
Wt 3	1	223	6478	1	209	6364	1	80	2522
	3	385	9836	3	361	9732	1	142	3855
	8	644	14782	7	613	14689	3	238	5846
	17	1060	21965	16	1012	21956	6	398	8742
	36	1704	32345	32	1645	32468	13	643	12977
	68	2701	47188	63	2619	47604	24	1034	19042
	126	4209	68295	116	4116	69163	46	1620	27752

Fig. 27-3. This copy of the dominant weight multiplicity table for affine-A_1, Level 5, Class 0 is a truncated version of the corresponding table in §31. The table includes the multiplicities for all three Level 5, Class 0 modules. The $V(3,2)$ weight multiplicities are in the center column. The null depth is the number of α_0's subtracted from the highest weight $(0;3,2)$. The weight (1; 1, 4) is reached by subtraction of α_0; (0; 1, 4) is not in $\Omega(3,2)$ and the first entry in the Wt 1, Rp 2 box is 0. The weight (0; 5, 0) is reached by subtracting α_1 from the highest weight, and the first entry in the Wt 3, Rp 2 box is nonzero. Thus, the weights in $\Omega^+_{\text{high}}(3,2)$ are (0; 3, 2), (1; 1, 4) and (0; 5, 0). The weight label components are given only at the top of the columns; the "Wt n" has the same incomplete weight label as "Rp n." Note that the Level 5, Class 1 multiplicities are the same up to the number of leading zeros and the interchange of the weight components a_0 and a_1.

Affine A1 - Level 5 - Class 0

	Rp 1	Rp 2	Rp 3
	14	32	50
Wt 1:	-9	-169	-249
Wt 2:	159	-1	-81
Wt 3:	215	55	-25

Den = 280

Fig. 27-4. The modular characteristics of $V(3,2)$ are in the second column, with the layout similar to the multiplicity tables. (The Coxeter-Dynkin diagram has been written in horizontal format.) The modular forms associated with this representation are given in (27.8).

Each string function $\sigma_\gamma^\Lambda(q)$ is related to a modular function $c_\gamma^\Lambda(q)$ by a factor of q to a power called the **modular characteristic** $s(\Lambda,\gamma)$. As in §23,

$$c_\gamma^\Lambda(q) := q^{s(\Lambda,\gamma)}\sigma_\gamma^\Lambda(a),$$

where $s(\Lambda,\gamma)$ is given by the formula

$$s(\Lambda,\gamma) = \frac{(\Lambda+\rho\mid\Lambda+\rho)}{2(\Lambda(\not{c})+g)} - \frac{(\rho\mid\rho)}{2g} - \frac{(\gamma\mid\gamma)}{2\Lambda(\not{c})}. \tag{27.7}$$

Here, as usual, $\Lambda(\not{c})$ is the level of Λ and g is the dual Coxeter number (sum of comarks) of $\hat{\mathfrak{g}}$. While $\sigma_\gamma^\Lambda(q)$ depends on the null depths of Λ and γ (taken to be zero in the examples above), a simple computation shows that the modular function $c_\gamma^\Lambda(q)$ depends only on the incomplete weight labels of Λ and γ.

The modular characteristic $s(\Lambda,\gamma)$ is a rational number and is given for Λ and γ of null depth zero in §30; the appropriate table is reproduced in Figure 27-4. Thus, the modular functions $c_\gamma^\Lambda(q)$ for $V(3,2)$ are

$$\begin{aligned} c_{(1,4)}^{(3,2)}(q) &= q^{-169/280}q(1+3q+8q^2+17q^3+36q^4+\cdots), \\ c_{(3,2)}^{(3,2)}(q) &= q^{-1/280}(1+2q+6q^2+13q^3+28q^4+\cdots), \\ c_{(5,0)}^{(3,2)}(q) &= q^{55/280}(1+3q+7q^2+16q^3+32q^4+\cdots). \end{aligned} \tag{27.8}$$

When the weights of a particular congruence class are related to those of another by a diagram automorphism φ, only one multiplicity (and modular characteristic) table is given. If the weights of the given table are λ_i, the weights of the "missing" table are $\varphi(\lambda_i)$. The multiplicities and modular characteristics are almost, but not exactly, the same. The operator that measures null depth L_0 is not invariant under diagram automorphisms, and the result is that the multiplicities in the missing table are the same as in the given table up to leading zeros in the string functions. The modular characteristics are equal up to the addition of an integer. These two effects cancel each other when we consider the modular function $c_\gamma^\Lambda(q)$, so $c_\gamma^\Lambda(q) = c_{\varphi(\gamma)}^{\varphi(\Lambda)}(q)$.

It is not difficult to reconstruct a missing table if necessary. It suffices to show how to find the number of leading zeros in a string function not found in the tables, say $\sigma_\gamma^\Lambda(q)$. Let $\Lambda = (0; a_0, \ldots, a_d)$ and $\gamma = (0; b_0, \ldots, b_d)$. It is a consequence of the fact that γ is dominant that $\gamma - j\delta$ is a weight of $V(\Lambda)$ if and only if $\gamma - j\delta \leq \Lambda$. Thus the number of leading zeros of $\sigma_\gamma^\Lambda(q)$ is the smallest value of j for which $\gamma - j\delta \leq \Lambda$.

To find j, we can write $\Lambda - (\gamma - j\delta) = (-j; a_0 - b_0, \ldots, a_d - b_d)$ in the basis of simple roots and choose j to be the smallest possible integer to make this a sum of positive roots, i.e., make it lie in $\hat{Q}_+$, $(\delta = (-1; 0, \ldots, 0))$. To give a specific example, let us reconstruct the weight multiplicity table for the $A_1^{(1)}$- representation $V(2,3)$ (level 5, class 1). The diagram automorphism of $A_1^{(1)}$ takes the weight label $(2,3)$ to $(3,2)$, and so we begin with the table containing the weight multiplicities for $V(3,2)$. There are three dominant weight labels for this representation, $(1,4), (3,2)$ and $(5,0)$, with string functions

$$\sigma_{(1,4)}^{(3,2)}(q) = q(1 + 3q + 8q^2 + 17q^3 + 36q^4 + \cdots),$$
$$\sigma_{(3,2)}^{(3,2)}(q) = 1 + 2q + 6q^2 + 13q^3 + 28q^4 + \cdots,$$
$$\sigma_{(5,0)}^{(3,2)}(q) = 1 + 3q + 7q^2 + 16q^3 + 32q^4 + \cdots.$$

The nonzero coefficients of $\sigma_{(4,1)}^{(2,3)}(q)$, $\sigma_{(2,3)}^{(2,3)}(q)$ and $\sigma_{(0,5)}^{(2,3)}(q)$ match these, resepectively, but we must compute the number of leading zeros.

As mentioned above, the number of leading zeros in $\sigma_\gamma^\Lambda(q)$ is the smallest value of j for which $\Lambda - (\gamma - j\delta) \in Q_+ \cup \{0\}$. Recall that the complete weight label of α_i is $-\delta_{i0}$; (the ith row of the Cartan matrix of $\hat{\mathfrak{g}}$). For the first of our string functions, we write

$$(0; 2, 3) - ((0; 4, 1) - j\delta) = (-j; -2, 2)$$

and observe that for $j = 0$, this is $(0; -2, 2) = \alpha_1 \in A_+ \cup \{0\}$. Thus

$$\sigma_{(4,1)}^{(2,3)}(q) = 1 + 3q + \cdots,$$

i.e. it has no leading zeros. For $\sigma_{(2,3)}^{(2,3)}(q)$, we see that

$$(0; 2, 3) - ((0; 2, 3) - j\delta) = 0 \quad \text{if } j = 0,$$

hence

$$\sigma^{(2,3)}_{(2,3)}(q) = 1 + 2q + \cdots .$$

Finally, we consider

$$(0;3,2) - ((0;5,0) - j\delta) = (-j;2,-2).$$

For $j = 0$ this is $\alpha_0 - \delta$, which is not a sum of positive roots, but for $j = 1$, this is α_0, hence $\sigma^{(3,2)}_{(5,0)}(q)$ has one leading 0, and is thus $q + 3q^2 + \cdots$.

Except in the simplest cases, this computation may require the change of basis matrix in $\mathfrak{h}^{e*}$, $\hat{G}^{-1}$ in (5.12), from the basis of fundamental weights to the basis of simple coroots. In that basis we choose j to make $\Lambda - (\gamma - j\delta)$ a sum of positive coroots. To write out $\hat{G}^{-1}$ explicitly requires the matrix of the bilinear form, $(DA)^{-1}$.

27.7 The final computational problem is that of the branching rules. One might anticipate that the multiplicities are a good starting point for computing the branching rules. This is correct if the depth of the calculation is sufficient. Unfortunately, the most effective choice of weights is quite different for the two computations. The choice for the multiplicity tables described in §27.4 has the property that the string functions of one module are computed to roughly the same null depth. There is no particular bias toward any simple root, which is an inconvenience for branching rule computations.

For the branching rule computations, we must choose a set of weights so that the "slices" into which we are decomposing the module are complete before beginning. Figure 27-5 shows the same part of the weight diagram of $V(3,2)$ as in Figure 27-1. The α_1 slicing breaks up $V(3,2)$ into representations of finite dimensional A_1. The only complete slice is slice #0. Thus, this set of weights is not efficient for the branching rule calculations; there are many superfluous weights that do not further the computation. (While the three slices shown can indeed be identified from the weight multiplicities, this is partly due to the simplicity of this example.)

For the branching rule computations, it is better to start with a set of weights that is more skewed, so it includes all the dominant weights in a given number of slices. To understand the problem better, we outline the finite dimensional case, where $\mathfrak{g}'$ is a subalgebra of finite dimensional $\mathfrak{g}$. After the computation of the suitable saturated set of dominant weights in $\mathfrak{g}$, one computes the elements of the Weyl group orbit of each of these $\mathfrak{g}$-dominant weights of V, keeping only those elements dominant in $\mathfrak{g}'$. If one keeps track of a "slice number" in the computation (defined like the null depth, but with reference to the simple root removed from the diagram of $\mathfrak{g}$ instead of with reference to α_0), the set of dominant weights and multiplicities in each slice of V viewed as a $\mathfrak{g}'$-module is evident.

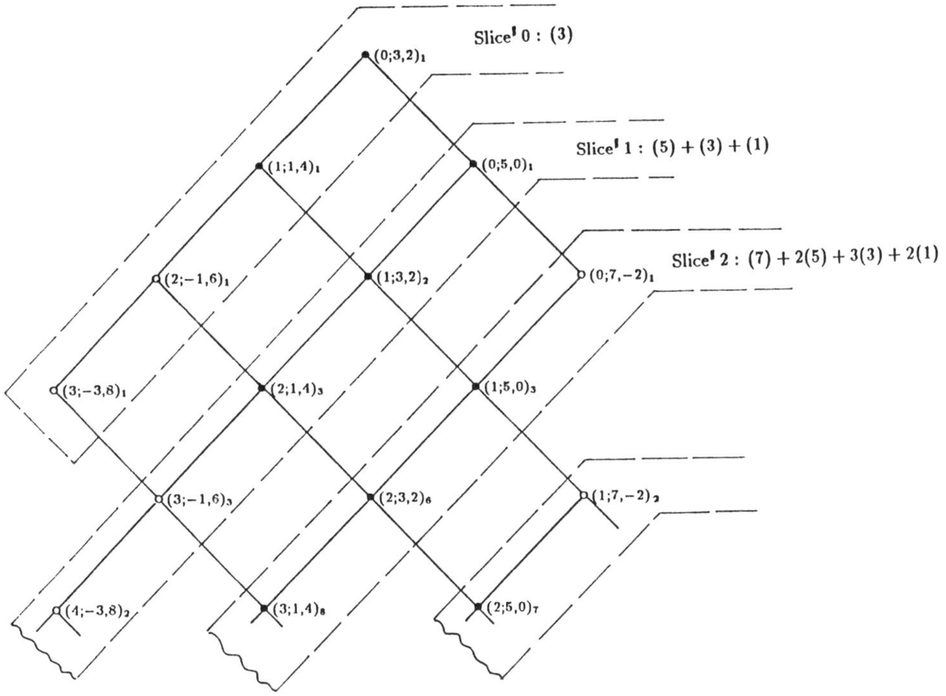

Fig. 27-5. The module $V(3,2)$ of affine-A_1, the slice number being the number of α_1's subtracted from Λ. Thus, α_0 is the simple root system of the finite dimensional A_1 subalgebra. At slice$^{\#}0$, we find the representation (3) of A_1, and so on.

Unfortunately, this computation is extremely complicated for the affine case when the rank is large. Literally millions of elements of the Weyl group orbits must be computed, most of which are discarded because when projected they are not dominant with respect to the subalgebra. Fortunately, another method is possible in the affine case. We start with a dominant weight γ' of the subalgebra, and, working backwards, find the element of $\Omega^+(\Lambda)$ in whose orbit the weight projecting to it lies. Recall that, given the level of a module V, only d of the $d+1$ coordinates of the incomplete weight label of γ can be freely chosen, the equation $\gamma(\not{c}) = \Lambda(\not{c})$ giving the remaining coordinate. Thus we can find the weight $\gamma \in \Omega(\Lambda)$ that projects onto the weight γ' of the subalgebra by filling in the missing coordinate by the equation $\gamma(\not{c}) = \Lambda(\not{c})$. We can then find the unique dominant weight in the Weyl orbit of γ. We repeat this construction for all dominant weights of the subalgebra up to a certain point to obtain the structure of each slice of V as a $\mathfrak{g}'$-module. In this method, we do end up considering a number of weights of $\mathfrak{g}'$ that do not appear in the eventual decomposition, but even with this extra computation the process is more efficient than in the direct computation described above.

To rewrite the slice, now given by its dominant weights and their multiplicities, as a sum of irreducible modules, one merely lists the multiplicities as a column vector according to some ordering of the dominant weights $\mathfrak{g}'$ which respects the usual parital order, then multiplies this by the inverse of the multiplicity matrix of $\mathfrak{g}'$. The multiplicity matrix transforms a linear combination of representations into a set of dominant weights and their multiplicities. These matrices are tabulated in [BMP–85] for simple Lie algebras. Therefore its inverse takes a combination of dominant weights and their multiplicities to the corresponding sum of highest weight representations. Since the multiplicity matrix can be put in upper triangular form, it is easily inverted.

27.8 We look at a branching rule of the module $V(3,2)$ discussed above, where the simple root associated with the 0th node in Table 28 is used to define the slicing. In Figure 27-6 we have copied the first few lines of the "Affine $A_1 \supset A_1$ - Level 5 - Class 1" table of §32.

Symmetries again have been deleted from these sets of tables. There are no asymmetries here as there are for leading zeros in the multiplicity tables.

Affine A1 ⊃ A1 - Level 5 - Class 1

```
         .                         2
Sl 0:   (3)   Highest Weight      (3)

         .          .          .
Sl 1:   (5)  +     (3)  +     (1)

         .          .          .          .
Sl 2:   (7)  +    2(5)  +    3(3)  +    2(1)

         .          .          .          .
Sl 3:  2(7)  +    5(5)  +    6(3)  +    4(1)

         .          .          .          .          .
Sl 4:   (9)  +    5(7)  +   10(5)  +   12(3)  +    8(1)

         .          .          .          .          .
Sl 5:  3(9)  +   10(7)  +   19(5)  +   22(3)  +   15(1)

         .          .          .          .          .          .
Sl 6:  (11)  +    7(9)  +   20(7)  +   35(5)  +   40(3)  +   26(1)
```

Fig. 27-6. Branching Rules of $V(3,2)$ for affine-A_1 to A_1.

28. BIOGRAPHIC TABLES OF SIMPLE LIE ALGEBRAS

For the convenience of the reader, we reproduce here the tables of [MPR–90a, MPR–90b]. They refer mostly to the finite simple Lie algebras. Nevertheless, this information is often needed when working with affine Kac-Moody algebras.

Lie algebra A_n, $n \geq 1$

Algebra dimension	n^2+2n
Coxeter/Dual Coxeter number	$n+1$
Number of positive roots	$n(n+1)/2$
Weyl group order	$(n+1)!$
Cartan matrix determinant / Order of group center	$n+1$
Exponents	$1, 2, \cdots, n$

Numbering of simple roots with the extension

$n>1$

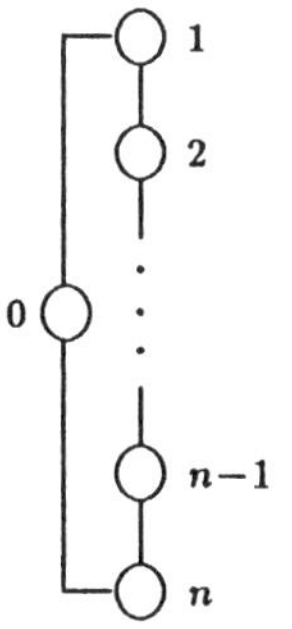

$n=1$

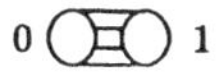

Extended Cartan Matrix

$n>1$

$$\begin{bmatrix} 2 & -1 & \cdot & \cdot & \cdots & \cdot & -1 \\ -1 & 2 & -1 & \cdot & \cdots & \cdot & \cdot \\ \cdot & -1 & 2 & -1 & \cdots & \cdot & \cdot \\ \cdot & \cdot & -1 & 2 & \cdots & \cdot & \cdot \\ \cdots & \cdots & \cdots & \cdots & \cdots & \cdots & \cdots \\ \cdot & \cdot & \cdot & \cdot & \cdots & 2 & -1 \\ -1 & \cdot & \cdot & \cdot & \cdots & -1 & 2 \end{bmatrix}$$

$n=1$

$$\begin{bmatrix} 2 & -2 \\ -2 & 2 \end{bmatrix}$$

Inverse Cartan Matrix

$$\frac{1}{(n+1)} \begin{bmatrix} 1\cdot n & 1\cdot(n-1) & 1\cdot(n-2) & \cdots & 1\cdot 2 & 1\cdot 1 \\ 1\cdot(n-1) & 2\cdot(n-1) & 2\cdot(n-2) & \cdots & 2\cdot 2 & 2\cdot 1 \\ 1\cdot(n-2) & 2\cdot(n-2) & 3\cdot(n-2) & \cdots & 3\cdot 2 & 3\cdot 1 \\ \cdots & \cdots & \cdots & \cdots & \cdots & \cdots \\ 1\cdot 2 & 2\cdot 2 & 3\cdot 2 & \cdots & (n-1)\cdot 2 & (n-1)\cdot 1 \\ 1\cdot 1 & 2\cdot 1 & 3\cdot 1 & \cdots & (n-1)\cdot 1 & n\cdot 1 \end{bmatrix}$$

Marks/Comarks

$n>1$

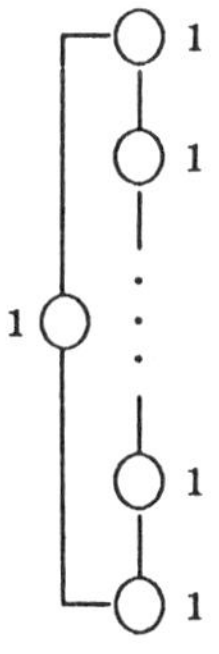

$n=1$

Dimensions of fundamental representations

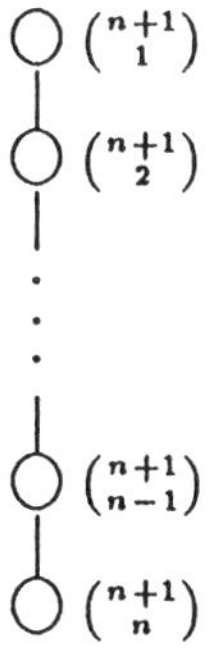

2nd indices of fundamental representations

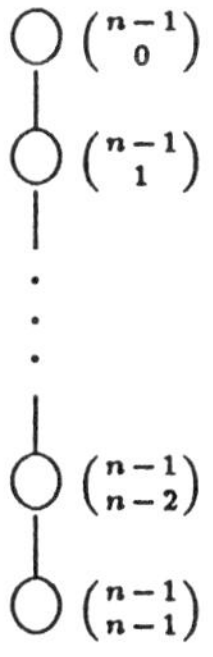

Level vector/Sum of positive roots

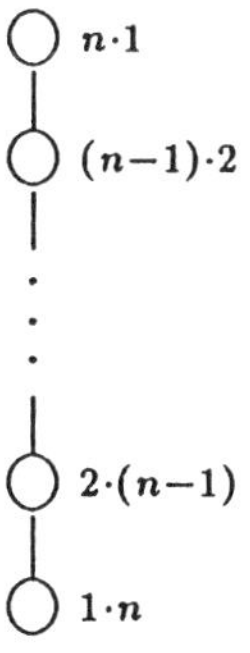

Congruence vector

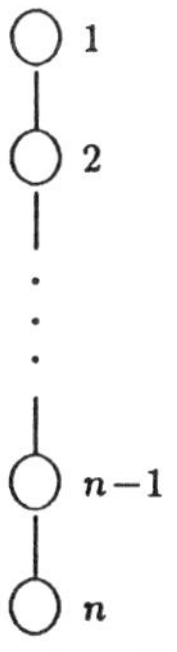

mod$(n+1)$

Table 28-1. Biographic Table of Lie Algebra A_n

Lie algebra B_n, $n \geq 2$

Numbering of simple roots with the extension

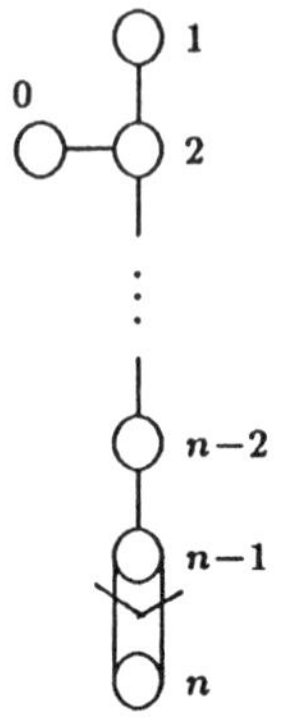

Algebra dimension	$2n^2+n$
Coxeter number	$2n$
Dual Coxeter number	$2n-1$
Number of positive roots	n^2
Weyl group order	$n!\,2^n$
Cartan matrix determinant / Order of group center	2
Exponents	$1, 3, 5, \cdots, 2n-1$

Extended Cartan Matrix

$$\begin{bmatrix} 2 & \cdot & -1 & \cdot & \cdots & \cdot & \cdot \\ \cdot & 2 & -1 & \cdot & \cdots & \cdot & \cdot \\ -1 & -1 & 2 & -1 & \cdots & \cdot & \cdot \\ \cdot & \cdot & -1 & 2 & \cdots & \cdot & \cdot \\ \cdots & \cdots & \cdots & \cdots & \cdots & \cdots & \cdots \\ \cdot & \cdot & \cdot & \cdot & \cdots & 2 & -2 \\ \cdot & \cdot & \cdot & \cdot & \cdots & -1 & 2 \end{bmatrix}$$

Inverse Cartan Matrix

$$\frac{1}{2}\begin{bmatrix} 2 & 2 & 2 & \cdots & 2 & 2 \\ 2 & 4 & 4 & \cdots & 4 & 4 \\ 2 & 4 & 6 & \cdots & 6 & 6 \\ \cdots & \cdots & \cdots & \cdots & \cdots & \cdots \\ 2 & 4 & 6 & \cdots & 2(n-1) & 2(n-1) \\ 1 & 2 & 3 & \cdots & n-1 & n \end{bmatrix}$$

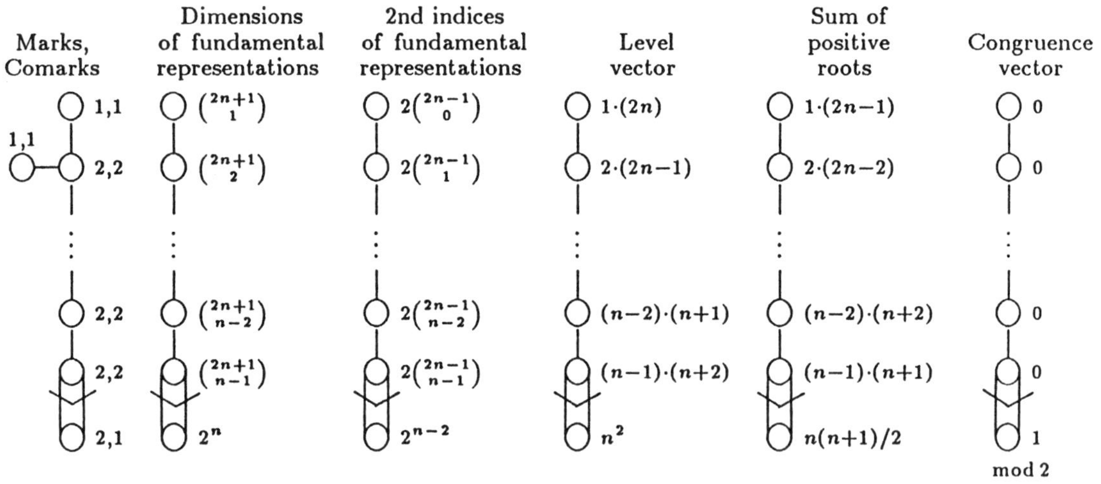

Table 28-2. Biographic Table of Lie Algebra B_n

Lie algebra C_n, $n \geq 2$

Numbering of simple roots with the extension

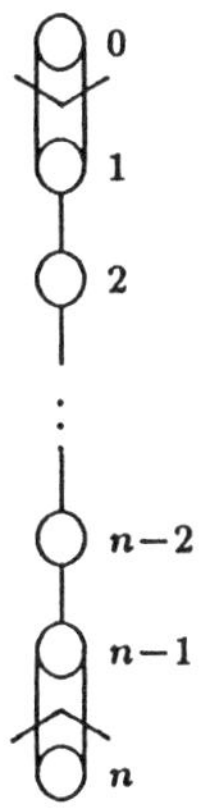

Algebra dimension	$2n^2+n$
Coxeter number	$2n$
Dual Coxeter number	$n+1$
Number of positive roots	n^2
Weyl group order	$n!\,2^n$
Cartan matrix determinant, Order of group center	2
Exponents	$1,3,5,\cdots,2n-1$

Extended Cartan Matrix

$$\begin{bmatrix} 2 & -2 & \cdot & \cdot & \cdots & \cdot & \cdot \\ -1 & 2 & -1 & \cdot & \cdots & \cdot & \cdot \\ \cdot & -1 & 2 & -1 & \cdots & \cdot & \cdot \\ \cdot & \cdot & -1 & 2 & \cdots & \cdot & \cdot \\ \cdots & \cdots & \cdots & \cdots & \cdots & \cdots & \cdots \\ \cdot & \cdot & \cdot & \cdot & \cdots & 2 & -1 \\ \cdot & \cdot & \cdot & \cdot & \cdots & -2 & 2 \end{bmatrix}$$

Inverse Cartan Matrix

$$\frac{1}{2}\begin{bmatrix} 1 & 1 & 1 & \cdots & 1 & 1 \\ 1 & 2 & 2 & \cdots & 2 & 2 \\ 1 & 2 & 3 & \cdots & 3 & 3 \\ \cdots & \cdots & \cdots & \cdots & \cdots & \cdots \\ 1 & 2 & 3 & \cdots & n-1 & n-1 \\ 1 & 2 & 3 & \cdots & n-1 & n \end{bmatrix}$$

Marks, Comarks	Dimensions of fundamental representations	2nd indices of fundamental representations	Level vector	Sum of positive roots	Congruence vector
1,1					
2,1	$\binom{2n}{1}$	$\binom{2n-1}{0}$	$1\cdot(2n-1)$	$1\cdot(2n)$	1
2,1	$\binom{2n}{2}-\binom{2n}{0}$	$\binom{2n-1}{1}-\binom{2n-1}{0}$	$2\cdot(2n-2)$	$2\cdot(2n-1)$	2
⋮	⋮	⋮	⋮	⋮	⋮
2,1	$\binom{2n}{n-2}-\binom{2n}{n-4}$	$\binom{2n-1}{n-3}-\binom{2n-1}{n-4}$	$(n-2)\cdot(n+2)$	$(n-2)\cdot(n+3)$	$n-2$
2,1	$\binom{2n}{n-1}-\binom{2n}{n-3}$	$\binom{2n-1}{n-2}-\binom{2n-1}{n-3}$	$(n-1)\cdot(n+1)$	$(n-1)\cdot(n+2)$	$n-1$
1,1	$\binom{2n}{n}-\binom{2n}{n-2}$	$\binom{2n-1}{n-1}-\binom{2n-1}{n-2}$	n^2	$n(n+1)/2$	n
					mod 2

Table 28-3. Biographic Table of Lie Algebra C_n

Lie algebra D_n, $n \geq 4$

Algebra dimension	$2n^2-n$
Coxeter number / Dual Coxeter number	$2n-2$
Number of positive roots	$n(n-1)$
Weyl group order	$n!\,2^{n-1}$
Cartan matrix determinant / Order of group center	4
Exponents	$1,3,5,\cdots,2n-5,2n-3,n-1$

Numbering of simple roots with the extension

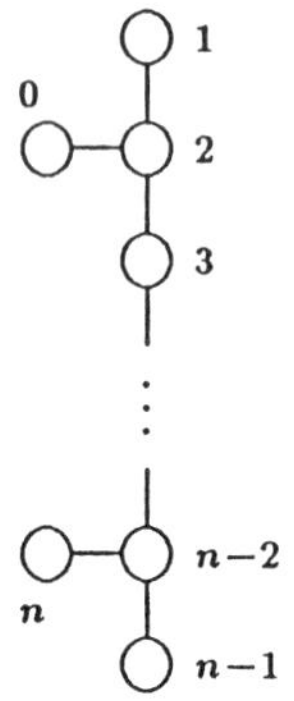

Extended Cartan Matrix

$$\begin{bmatrix} 2 & \cdot & -1 & \cdot & \cdots & \cdot & \cdot & \cdot \\ \cdot & 2 & -1 & \cdot & \cdots & \cdot & \cdot & \cdot \\ -1 & -1 & 2 & -1 & \cdots & \cdot & \cdot & \cdot \\ \cdot & \cdot & -1 & 2 & \cdots & \cdot & \cdot & \cdot \\ \cdots & & & & & & & \cdots \\ \cdot & \cdot & \cdot & \cdot & \cdots & 2 & -1 & -1 \\ \cdot & \cdot & \cdot & \cdot & \cdots & -1 & 2 & \cdot \\ \cdot & \cdot & \cdot & \cdot & \cdots & -1 & \cdot & 2 \end{bmatrix}$$

Inverse Cartan Matrix

$$\frac{1}{2}\begin{bmatrix} 2 & 2 & 2 & \cdots & 2 & 1 & 1 \\ 2 & 4 & 4 & \cdots & 4 & 2 & 2 \\ 2 & 4 & 6 & \cdots & 6 & 3 & 3 \\ \cdots & & & & & & \cdots \\ 2 & 4 & 6 & \cdots & 2(n-2) & n-2 & n-2 \\ 1 & 2 & 3 & \cdots & (n-2) & n/2 & (n-2)/2 \\ 1 & 2 & 3 & \cdots & (n-2) & (n-2)/2 & n/2 \end{bmatrix}$$

Marks/ Comarks | Dimensions of fundamental representations | 2nd indices of fundamental representations | Level vector/ Sum of positive roots | Congruence vectors

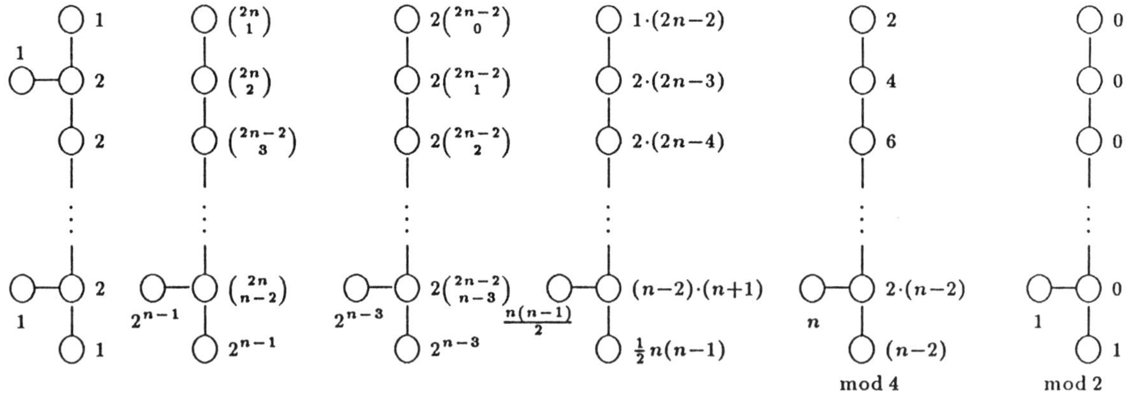

Table 28-4. Biographic Table of Lie Algebra D_n

Lie algebra E_6

Algebra dimension	78
Coxeter/Dual Coxeter number	12
Number of positive roots	36
Weyl group order	$51840 = 2^7 \cdot 3^4 \cdot 5$
Cartan matrix determinant / Order of group center	3
Exponents	1, 4, 5, 7, 8, 11

Numbering of simple roots with the extension

(Dynkin)

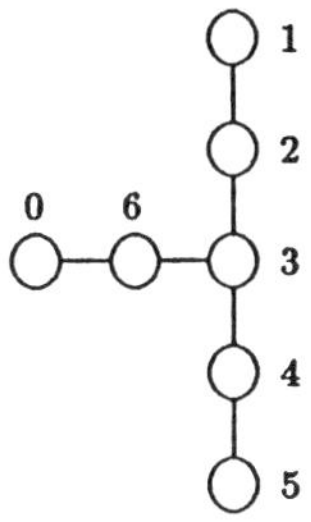

(Bourbaki)

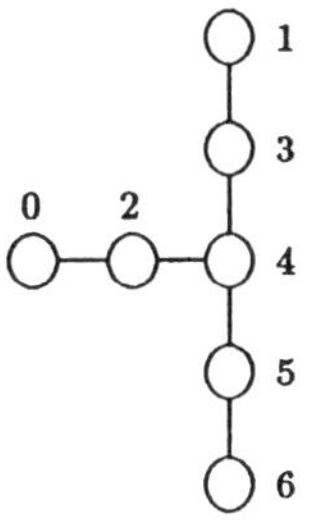

Extended Cartan Matrix

$$\begin{bmatrix} 2 & \cdot & \cdot & \cdot & \cdot & \cdot & -1 \\ \cdot & 2 & -1 & \cdot & \cdot & \cdot & \cdot \\ \cdot & -1 & 2 & -1 & \cdot & \cdot & \cdot \\ \cdot & \cdot & -1 & 2 & -1 & \cdot & -1 \\ \cdot & \cdot & \cdot & -1 & 2 & -1 & \cdot \\ \cdot & \cdot & \cdot & \cdot & -1 & 2 & \cdot \\ -1 & \cdot & \cdot & -1 & \cdot & \cdot & 2 \end{bmatrix}$$

Inverse Cartan Matrix

$$\frac{1}{3}\begin{bmatrix} 4 & 5 & 6 & 4 & 2 & 3 \\ 5 & 10 & 12 & 8 & 4 & 6 \\ 6 & 12 & 18 & 12 & 6 & 9 \\ 4 & 8 & 12 & 10 & 5 & 6 \\ 2 & 4 & 6 & 5 & 4 & 3 \\ 3 & 6 & 9 & 6 & 3 & 6 \end{bmatrix}$$

Marks/ Comarks

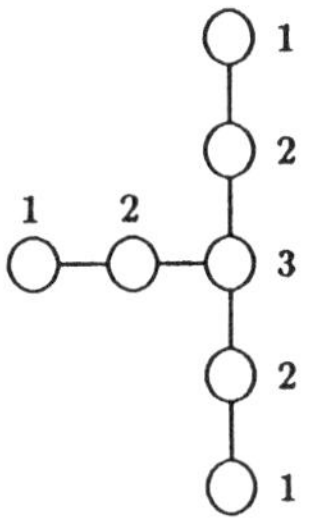

Dimensions of fundamental representations

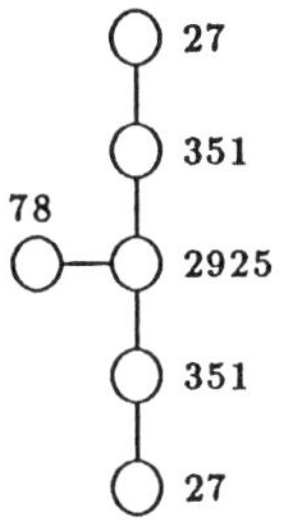

2nd indices of fundamental representations

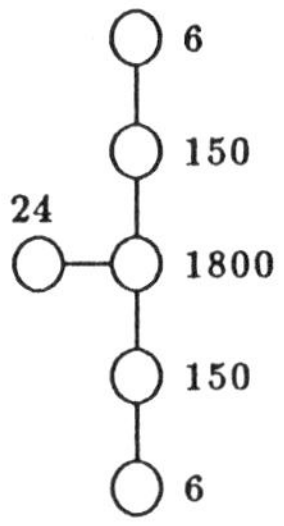

Level vector/ Sum of positive roots

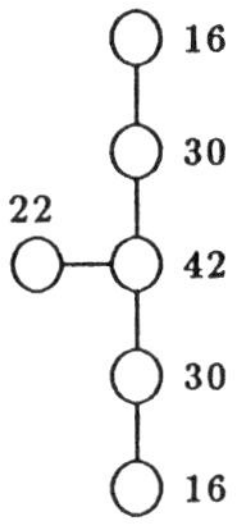

Congruence vector

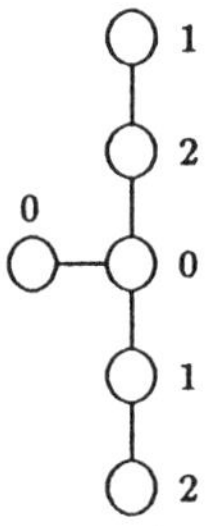

mod 3

Table 28-5. Biographic Table of Lie Algebra E_6

Lie algebra E_7

Numbering of simple roots with the extension

(Dynkin) (Bourbaki)

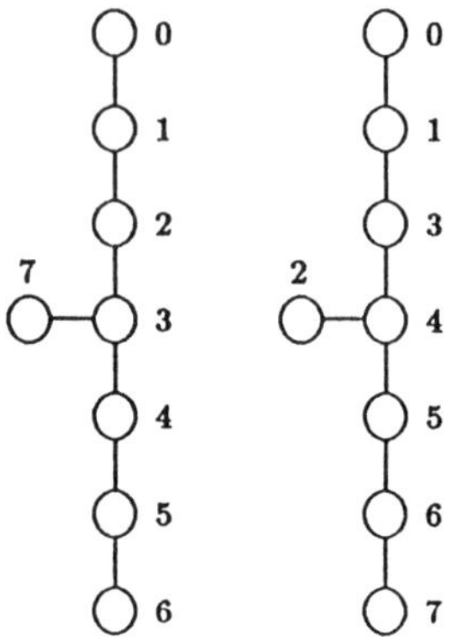

Algebra dimension	133
Coxeter/Dual Coxeter number	18
Number of positive roots	63
Weyl group order	$2903040 = 2^{10}\cdot 3^4\cdot 5\cdot 7$
Cartan matrix determinant } Order of group center	2
Exponents	1,5 7,9,11,13,17

Extended Cartan Matrix

$$\begin{bmatrix} 2 & -1 & \cdot & \cdot & \cdot & \cdot & \cdot & \cdot \\ -1 & 2 & -1 & \cdot & \cdot & \cdot & \cdot & \cdot \\ \cdot & -1 & 2 & -1 & \cdot & \cdot & \cdot & \cdot \\ \cdot & \cdot & -1 & 2 & -1 & \cdot & \cdot & -1 \\ \cdot & \cdot & \cdot & -1 & 2 & -1 & \cdot & \cdot \\ \cdot & \cdot & \cdot & \cdot & -1 & 2 & -1 & \cdot \\ \cdot & \cdot & \cdot & \cdot & \cdot & -1 & 2 & \cdot \\ \cdot & \cdot & \cdot & -1 & \cdot & \cdot & \cdot & 2 \end{bmatrix}$$

Inverse Cartan Matrix

$$\frac{1}{2}\begin{bmatrix} 4 & 6 & 8 & 6 & 4 & 2 & 4 \\ 6 & 12 & 16 & 12 & 8 & 4 & 8 \\ 8 & 16 & 24 & 18 & 12 & 6 & 12 \\ 6 & 12 & 18 & 15 & 10 & 5 & 9 \\ 4 & 8 & 12 & 10 & 8 & 4 & 6 \\ 2 & 4 & 6 & 5 & 4 & 3 & 3 \\ 4 & 8 & 12 & 9 & 6 & 3 & 7 \end{bmatrix}$$

Marks/ Comarks

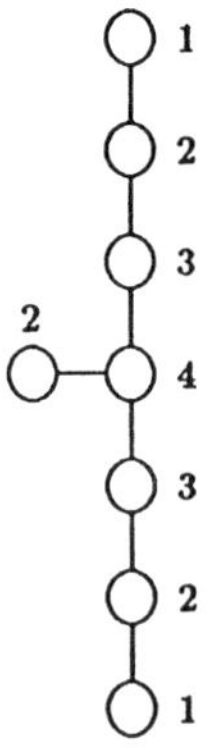

Dimensions of fundamental representations

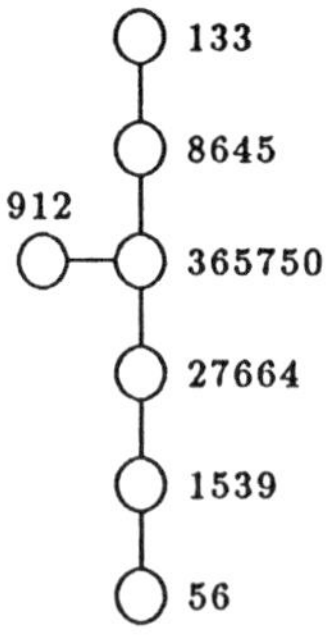

2nd indices of fundamental representations

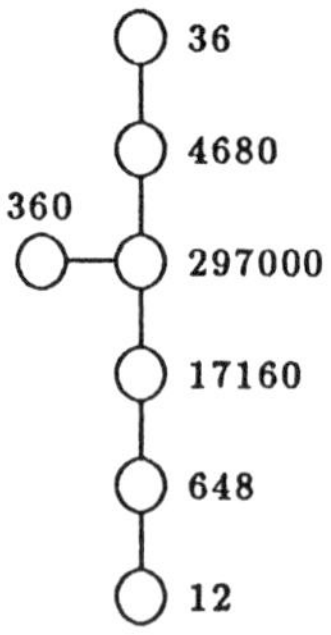

Level vector/ Sum of positive roots

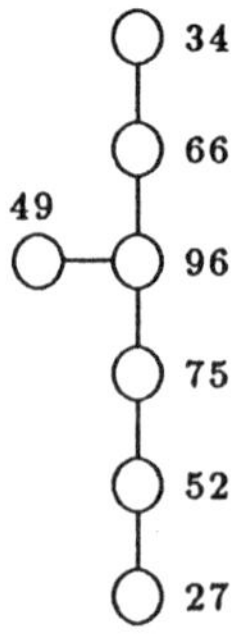

Congruence vector

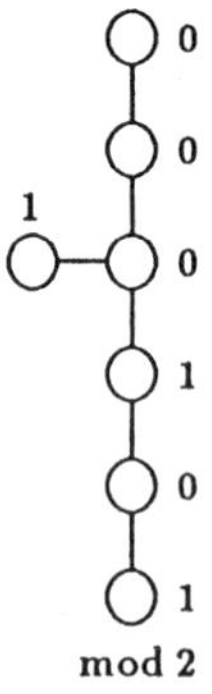

Table 28-6. Biographic Table of Lie Algebra E_7

Lie algebra E_8

Numbering of simple roots with the extension
(Dynkin) (Bourbaki)

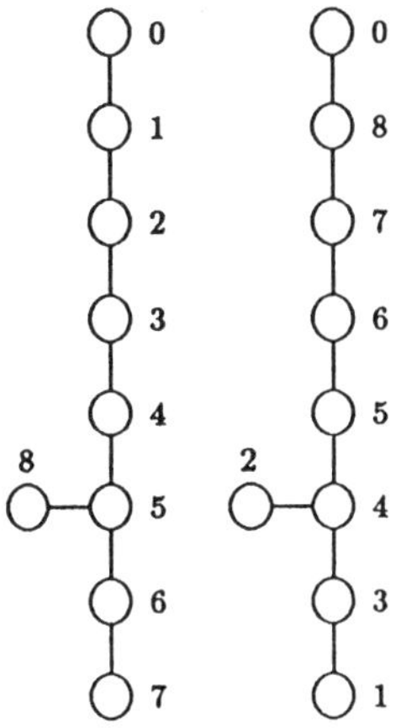

Algebra dimension	248
Coxeter/Dual Coxeter number	30
Number of positive roots	120
Weyl group order	$696729600 = 2^{14} \cdot 3^5 \cdot 5^2 \cdot 7$
Cartan matrix determinant } Order of group center	1
Exponents	1, 7, 11, 13, 17, 19, 23, 29

Extended Cartan Matrix

2	−1	·	·	·	·	·	·	·
−1	2	−1	·	·	·	·	·	·
·	−1	2	−1	·	·	·	·	·
·	·	−1	2	−1	·	·	·	·
·	·	·	−1	2	−1	·	·	·
·	·	·	·	−1	2	−1	·	−1
·	·	·	·	·	−1	2	−1	·
·	·	·	·	·	·	−1	2	·
·	·	·	·	·	−1	·	·	2

Inverse Cartan Matrix

2	3	4	5	6	4	2	3
3	6	8	10	12	8	4	6
4	8	12	15	18	12	6	9
5	10	15	20	24	16	18	12
6	12	18	24	30	20	10	15
4	8	12	16	20	14	7	10
2	4	6	8	10	7	4	5
3	6	9	12	15	10	5	8

Marks/
Comarks

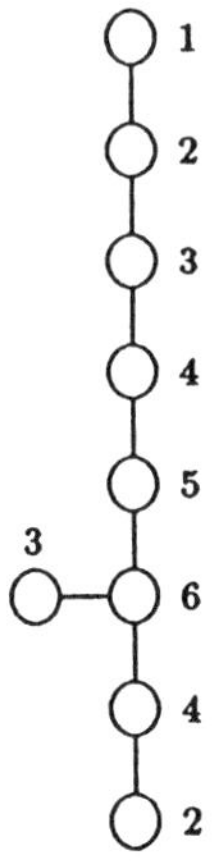

Dimensions
of fundamental
representations

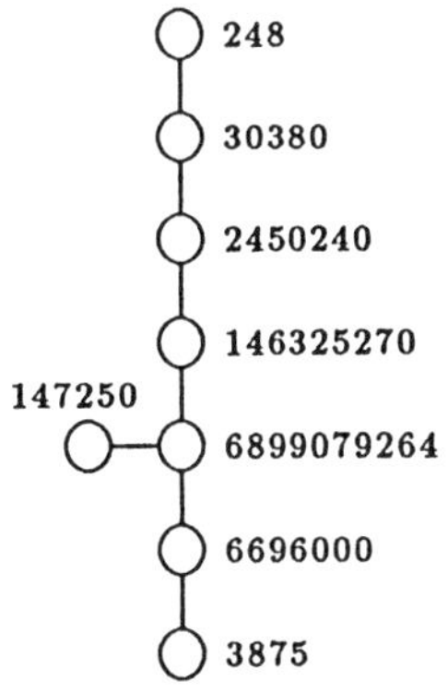

2nd indices
of fundamental
representations

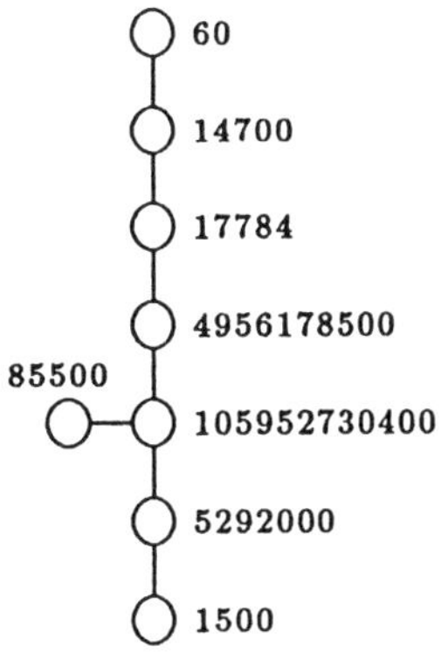

Level vector/
Sum of
positive roots

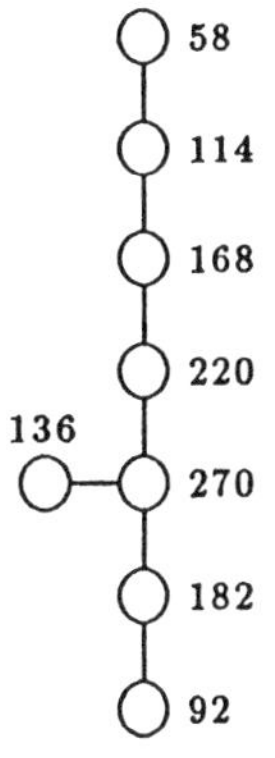

Table 28-7. Biographic Table of Lie Algebra E_8

Lie algebra F_4

Numbering of simple roots with the extension

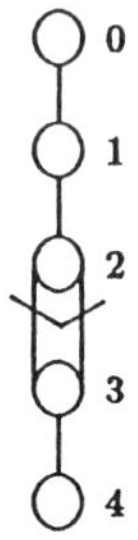

Algebra dimension	52
Coxeter number	12
Dual Coxeter number	9
Number of positive roots	24
Weyl group order	$1152 = 2^7 \cdot 3^2$
Cartan matrix determinant, Order of group center	2
Exponents	1, 5, 7, 11

Extended Cartan Matrix

$$\begin{bmatrix} 2 & -1 & \cdot & \cdot & \cdot \\ -1 & 2 & -1 & \cdot & \cdot \\ \cdot & -1 & 2 & -2 & \cdot \\ \cdot & \cdot & -1 & 2 & -1 \\ \cdot & \cdot & \cdot & -1 & 2 \end{bmatrix}$$

Inverse Cartan Matrix

$$\begin{bmatrix} 2 & 3 & 4 & 2 \\ 3 & 6 & 8 & 4 \\ 2 & 4 & 6 & 3 \\ 1 & 2 & 3 & 2 \end{bmatrix}$$

Marks, Comarks

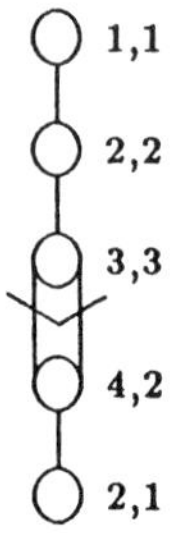

Dimensions of fundamental representations

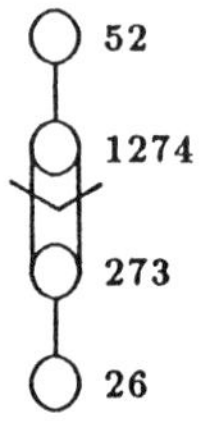

2nd indices of fundamental representations

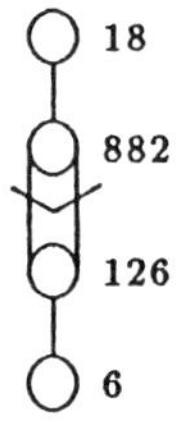

Level vector

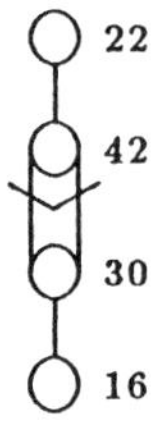

Sum of positive positive roots

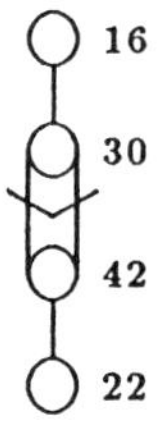

Table 28-8. Biographic Table of Lie Algebra F_4

Lie algebra G_2

Algebra dimension	14
Coxeter number	6
Dual Coxeter number	4
Number of positive roots	6
Weyl group order	$12 = 2^2 \cdot 3$
Cartan matrix determinant, Order of group center	1
Exponents	1, 5

Numbering of simple roots with the extension

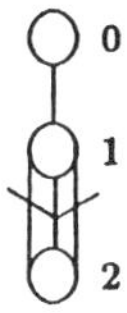

Extended Cartan Matrix

$$\begin{bmatrix} 2 & -1 & \cdot \\ -1 & 2 & -3 \\ \cdot & -1 & 2 \end{bmatrix}$$

Inverse Cartan Matrix

$$\begin{bmatrix} 2 & 3 \\ 1 & 2 \end{bmatrix}$$

Marks, Comarks

Dimensions of fundamental representations

2nd indices of fundamental representations

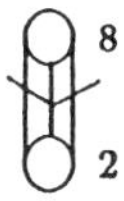

Level vector

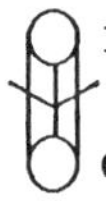

Sum of positive positive roots

Table 28-9. Biographic Table of Lie Algebra G_2

29. EXPANSIONS OF PARTITION FUNCTIONS

29.1 Values of the Partition Functions p_n for $n \leq 8$.

These tables contain the exact values of the first eight powers of the generating function of the classical partition function for arguments ≤ 80. The values of p_d are the weight multiplicities of the basic representation of the algebras A_d, D_d and E_d, or equivalently, the multiplicities of the weights of the unique irreducible representation of the d-fold Heisenberg algebra. The combinatorial interpretation of $p_d(i)$ is the number of partitions of the integer i into integers of d different colors.

Euler's φ-function is given by

$$\varphi(q) = \prod_{j \geq 1} (1 - q^j),$$

and

$$\varphi(q)^{-1} = \prod_{j \geq 1} (1 - q^j)^{-1} = \sum_{k \geq 0} p_1(k)\, q^k,$$

where $p_1(k)$ is the classical partition function of number theory, the number of different ways of writing k as a sum of positive integers (ignoring the order). For example, $p_1(4) = 5$, the 5 ways of writing 4 being 4, $3+1$, $2+2$, $2+1+1$ and $1+1+1+1$. For $d > 1$,

$$\varphi(q)^{-d} = \sum_{k \geq 0} p_d(k)\, q^k.$$

For example, $p_2(2) = 5$, with partitions 2, $\underline{2}$, $1+1$, $\underline{1}+1$, and $\underline{1}+\underline{1}$.

$p_1(0)$ =	1
$p_1(1)$ =	1
$p_1(2)$ =	2
$p_1(3)$ =	3
$p_1(4)$ =	5
$p_1(5)$ =	7
$p_1(6)$ =	11
$p_1(7)$ =	15
$p_1(8)$ =	22
$p_1(9)$ =	30
$p_1(10)$ =	42
$p_1(11)$ =	56
$p_1(12)$ =	77
$p_1(13)$ =	101
$p_1(14)$ =	135
$p_1(15)$ =	176
$p_1(16)$ =	231
$p_1(17)$ =	297
$p_1(18)$ =	385
$p_1(19)$ =	490
$p_1(20)$ =	627
$p_1(21)$ =	792
$p_1(22)$ =	1002
$p_1(23)$ =	1255
$p_1(24)$ =	1575
$p_1(25)$ =	1958
$p_1(26)$ =	2436
$p_1(27)$ =	3010
$p_1(28)$ =	3718
$p_1(29)$ =	4565
$p_1(30)$ =	5604
$p_1(31)$ =	6842
$p_1(32)$ =	8349
$p_1(33)$ =	10143
$p_1(34)$ =	12310
$p_1(35)$ =	14883
$p_1(36)$ =	17977
$p_1(37)$ =	21637
$p_1(38)$ =	26015
$p_1(39)$ =	31185
$p_1(40)$ =	37338
$p_1(41)$ =	44583
$p_1(42)$ =	53174
$p_1(43)$ =	63261
$p_1(44)$ =	75175
$p_1(45)$ =	89134
$p_1(46)$ =	105558
$p_1(47)$ =	124754
$p_1(48)$ =	147273
$p_1(49)$ =	173525
$p_1(50)$ =	204226
$p_1(51)$ =	239943
$p_1(52)$ =	281589
$p_1(53)$ =	329931
$p_1(54)$ =	386155
$p_1(55)$ =	451276
$p_1(56)$ =	526823
$p_1(57)$ =	614154
$p_1(58)$ =	715220
$p_1(59)$ =	831820
$p_1(60)$ =	966467
$p_1(61)$ =	1121505
$p_1(62)$ =	1300156
$p_1(63)$ =	1505499
$p_1(64)$ =	1741630
$p_1(65)$ =	2012558
$p_1(66)$ =	2323520
$p_1(67)$ =	2679689
$p_1(68)$ =	3087735
$p_1(69)$ =	3554345
$p_1(70)$ =	4087968
$p_1(71)$ =	4697205
$p_1(72)$ =	5392783
$p_1(73)$ =	6185689
$p_1(74)$ =	7089500
$p_1(75)$ =	8118264
$p_1(76)$ =	9289091
$p_1(77)$ =	10619863
$p_1(78)$ =	12132164
$p_1(79)$ =	13848650
$p_1(80)$ =	15796476

$p_2(0)$ = 1
$p_2(1)$ = 2
$p_2(2)$ = 5
$p_2(3)$ = 10
$p_2(4)$ = 20
$p_2(5)$ = 36
$p_2(6)$ = 65
$p_2(7)$ = 110
$p_2(8)$ = 185
$p_2(9)$ = 300
$p_2(10)$ = 481
$p_2(11)$ = 752
$p_2(12)$ = 1165
$p_2(13)$ = 1770
$p_2(14)$ = 2665
$p_2(15)$ = 3956
$p_2(16)$ = 5822
$p_2(17)$ = 8470
$p_2(18)$ = 12230
$p_2(19)$ = 17490
$p_2(20)$ = 24842
$p_2(21)$ = 35002
$p_2(22)$ = 49010
$p_2(23)$ = 68150
$p_2(24)$ = 94235
$p_2(25)$ = 129512
$p_2(26)$ = 177087
$p_2(27)$ = 240840
$p_2(28)$ = 326015
$p_2(29)$ = 439190
$p_2(30)$ = 589128
$p_2(31)$ = 786814
$p_2(32)$ = 1046705
$p_2(33)$ = 1386930
$p_2(34)$ = 1831065
$p_2(35)$ = 2408658
$p_2(36)$ = 3157789
$p_2(37)$ = 4126070
$p_2(38)$ = 5374390
$p_2(39)$ = 6978730
$p_2(40)$ = 9035539
$p_2(41)$ = 11664896
$p_2(42)$ = 15018300
$p_2(43)$ = 19283830
$p_2(44)$ = 24697480
$p_2(45)$ = 31551450
$p_2(46)$ = 40210481
$p_2(47)$ = 51124970
$p_2(48)$ = 64854575
$p_2(49)$ = 82088400
$p_2(50)$ = 103679156
$p_2(51)$ = 130673928
$p_2(52)$ = 164363280
$p_2(53)$ = 206327710
$p_2(54)$ = 258508230
$p_2(55)$ = 323275512
$p_2(56)$ = 403531208
$p_2(57)$ = 502810130
$p_2(58)$ = 625425005
$p_2(59)$ = 776616430
$p_2(60)$ = 962759294
$p_2(61)$ = 1191580872
$p_2(62)$ = 1472454540
$p_2(63)$ = 1816715170
$p_2(64)$ = 2238075315
$p_2(65)$ = 2753078840
$p_2(66)$ = 3381689157
$p_2(67)$ = 4147937540
$p_2(68)$ = 5080752250
$p_2(69)$ = 6214880700
$p_2(70)$ = 7592053897
$p_2(71)$ = 9262292216
$p_2(72)$ = 11285536125
$p_2(73)$ = 13733486100
$p_2(74)$ = 16691879795
$p_2(75)$ = 20263074134
$p_2(76)$ = 24569214653
$p_2(77)$ = 29755845120
$p_2(78)$ = 35996306025
$p_2(79)$ = 43496760380
$p_2(80)$ = 52502280642

$p_3(0)$ =	1
$p_3(1)$ =	3
$p_3(2)$ =	9
$p_3(3)$ =	22
$p_3(4)$ =	51
$p_3(5)$ =	108
$p_3(6)$ =	221
$p_3(7)$ =	429
$p_3(8)$ =	810
$p_3(9)$ =	1479
$p_3(10)$ =	2640
$p_3(11)$ =	4599
$p_3(12)$ =	7868
$p_3(13)$ =	13209
$p_3(14)$ =	21843
$p_3(15)$ =	35581
$p_3(16)$ =	57222
$p_3(17)$ =	90882
$p_3(18)$ =	142769
$p_3(19)$ =	221910
$p_3(20)$ =	341649
$p_3(21)$ =	521196
$p_3(22)$ =	788460
$p_3(23)$ =	1183221
$p_3(24)$ =	1762462
$p_3(25)$ =	2606604
$p_3(26)$ =	3829437
$p_3(27)$ =	5590110
$p_3(28)$ =	8111346
$p_3(29)$ =	11701998
$p_3(30)$ =	16790136
$p_3(31)$ =	23964594
$p_3(32)$ =	34034391
$p_3(33)$ =	48104069
$p_3(34)$ =	67679109
$p_3(35)$ =	94800537
$p_3(36)$ =	132230021
$p_3(37)$ =	183686994
$p_3(38)$ =	254170332
$p_3(39)$ =	350371088
$p_3(40)$ =	481225800
$p_3(41)$ =	658626822
$p_3(42)$ =	898365204
$p_3(43)$ =	1221340197
$p_3(44)$ =	1655151489
$p_3(45)$ =	2236138826
$p_3(46)$ =	3012047298
$p_3(47)$ =	4045435515
$p_3(48)$ =	5418100581
$p_3(49)$ =	7236723897
$p_3(50)$ =	9640159893
$p_3(51)$ =	12808713258
$p_3(52)$ =	16976054058
$p_3(53)$ =	22444345566
$p_3(54)$ =	29603582728
$p_3(55)$ =	38956080441
$p_3(56)$ =	51147644598
$p_3(57)$ =	67006939237
$p_3(58)$ =	87595398657
$p_3(59)$ =	114270097032
$p_3(60)$ =	148763165694
$p_3(61)$ =	193281563784
$p_3(62)$ =	250632677790
$p_3(63)$ =	324381699990
$p_3(64)$ =	419049110067
$p_3(65)$ =	540357492807
$p_3(66)$ =	695540306883
$p_3(67)$ =	893726827236
$p_3(68)$ =	1146422320191
$p_3(69)$ =	1468105216993
$p_3(70)$ =	1876969980867
$p_3(71)$ =	2395848783159
$p_3(72)$ =	3053355042465
$p_3(73)$ =	3885298927059
$p_3(74)$ =	4936439201418
$p_3(75)$ =	6262646818988
$p_3(76)$ =	7933576194687
$p_3(77)$ =	10035957093903
$p_3(78)$ =	12677649592650
$p_3(79)$ =	15992630479809
$p_3(80)$ =	20147121920736

$p_4(0)$	1
$p_4(1)$	4
$p_4(2)$	14
$p_4(3)$	40
$p_4(4)$	105
$p_4(5)$	252
$p_4(6)$	574
$p_4(7)$	1240
$p_4(8)$	2580
$p_4(9)$	5180
$p_4(10)$	10108
$p_4(11)$	19208
$p_4(12)$	35693
$p_4(13)$	64960
$p_4(14)$	116090
$p_4(15)$	203984
$p_4(16)$	353017
$p_4(17)$	602348
$p_4(18)$	1014580
$p_4(19)$	1688400
$p_4(20)$	2778517
$p_4(21)$	4524760
$p_4(22)$	7296752
$p_4(23)$	11658920
$p_4(24)$	18468245
$p_4(25)$	29015700
$p_4(26)$	45235414
$p_4(27)$	70005376
$p_4(28)$	107585845
$p_4(29)$	164245380
$p_4(30)$	249162620
$p_4(31)$	375704920
$p_4(32)$	563251038
$p_4(33)$	839757660
$p_4(34)$	1245383930
$p_4(35)$	1837560960
$p_4(36)$	2698091422
$p_4(37)$	3943014992
$p_4(38)$	5736319830
$p_4(39)$	8308877360
$p_4(40)$	11984575498
$p_4(41)$	17216193316
$p_4(42)$	24634595782
$p_4(43)$	35115884320
$p_4(44)$	49872929610
$p_4(45)$	70579640748
$p_4(46)$	99539408850
$p_4(47)$	139912622304
$p_4(48)$	196023438555
$p_4(49)$	273772115260
$p_4(50)$	381188229432
$p_4(51)$	529170799416
$p_4(52)$	732476650347
$p_4(53)$	1011036830300
$p_4(54)$	1391706766520
$p_4(55)$	1910587448448
$p_4(56)$	2616098398791
$p_4(57)$	3573036775872
$p_4(58)$	4867929566570
$p_4(59)$	6616075955920
$p_4(60)$	8970797573976
$p_4(61)$	12135564764120
$p_4(62)$	16379866259050
$p_4(63)$	22059938995960
$p_4(64)$	29645802130705
$p_4(65)$	39756449836628
$p_4(66)$	53205592346174
$p_4(67)$	71061006335032
$p_4(68)$	94721425187725
$p_4(69)$	126015991964800
$p_4(70)$	167332703917878
$p_4(71)$	221784043689400
$p_4(72)$	293420254726949
$p_4(73)$	387503559146240
$p_4(74)$	510860238900970
$p_4(75)$	672332046038856
$p_4(76)$	883354180960398
$p_4(77)$	1158694312894228
$p_4(78)$	1517396274581230
$p_4(79)$	1983983526116480
$p_4(80)$	2589991944446707

$p_5(0)$ = 1
$p_5(1)$ = 5
$p_5(2)$ = 20
$p_5(3)$ = 65
$p_5(4)$ = 190
$p_5(5)$ = 506
$p_5(6)$ = 1265
$p_5(7)$ = 2990
$p_5(8)$ = 6765
$p_5(9)$ = 14725
$p_5(10)$ = 31027
$p_5(11)$ = 63505
$p_5(12)$ = 126730
$p_5(13)$ = 247170
$p_5(14)$ = 472295
$p_5(15)$ = 885723
$p_5(16)$ = 1633000
$p_5(17)$ = 2963840
$p_5(18)$ = 5302075
$p_5(19)$ = 9358470
$p_5(20)$ = 16313440
$p_5(21)$ = 28107365
$p_5(22)$ = 47902010
$p_5(23)$ = 80803485
$p_5(24)$ = 134992865
$p_5(25)$ = 223474667
$p_5(26)$ = 366772720
$p_5(27)$ = 597049255
$p_5(28)$ = 964375855
$p_5(29)$ = 1546208695
$p_5(30)$ = 2461649861
$p_5(31)$ = 3892774130
$p_5(32)$ = 6116396095
$p_5(33)$ = 9551122255
$p_5(34)$ = 14826771315
$p_5(35)$ = 22886283845
$p_5(36)$ = 35134760670
$p_5(37)$ = 53656441020
$p_5(38)$ = 81529441675
$p_5(39)$ = 123280423945
$p_5(40)$ = 185539111677

$p_5(41)$ = 277976935445
$p_5(42)$ = 414648673725
$p_5(43)$ = 615903376755
$p_5(44)$ = 911097530745
$p_5(45)$ = 1342434636590
$p_5(46)$ = 1970382520705
$p_5(47)$ = 2881293221825
$p_5(48)$ = 4198090331015
$p_5(49)$ = 6095215310245
$p_5(50)$ = 8819473130897
$p_5(51)$ = 12719026386865
$p_5(52)$ = 18283618530375
$p_5(53)$ = 26200230998780
$p_5(54)$ = 37429905920025
$p_5(55)$ = 53313523261796
$p_5(56)$ = 75717101668375
$p_5(57)$ = 107230925161695
$p_5(58)$ = 151441819854255
$p_5(59)$ = 213304624055315
$p_5(60)$ = 299647895750902
$p_5(61)$ = 419860902953965
$p_5(62)$ = 586824953908055
$p_5(63)$ = 818173402728110
$p_5(64)$ = 1137992946226480
$p_5(65)$ = 1579116288160506
$p_5(66)$ = 2186205854034750
$p_5(67)$ = 3019893740872790
$p_5(68)$ = 4162329528356680
$p_5(69)$ = 5724601363094080
$p_5(70)$ = 7856645398062380
$p_5(71)$ = 10760455091110555
$p_5(72)$ = 14707659423971630
$p_5(73)$ = 20062876126065510
$p_5(74)$ = 27314686577501345
$p_5(75)$ = 37116653960457726
$p_5(76)$ = 50341555613708600
$p_5(77)$ = 68152975679478020
$p_5(78)$ = 92099671694824475
$p_5(79)$ = 124239773795576365
$p_5(80)$ = 167304007785834396

$p_6(0) = 1$
$p_6(1) = 6$
$p_6(2) = 27$
$p_6(3) = 98$
$p_6(4) = 315$
$p_6(5) = 918$
$p_6(6) = 2492$
$p_6(7) = 6372$
$p_6(8) = 15525$
$p_6(9) = 36280$
$p_6(10) = 81816$
$p_6(11) = 178794$
$p_6(12) = 380051$
$p_6(13) = 788004$
$p_6(14) = 1597725$
$p_6(15) = 3174210$
$p_6(16) = 6190182$
$p_6(17) = 11867310$
$p_6(18) = 22395359$
$p_6(19) = 41650050$
$p_6(20) = 76413078$
$p_6(21) = 138421358$
$p_6(22) = 247783113$
$p_6(23) = 438616728$
$p_6(24) = 768291650$
$p_6(25) = 1332444330$
$p_6(26) = 2289213495$
$p_6(27) = 3898064226$
$p_6(28) = 6581591157$
$p_6(29) = 11023247880$
$p_6(30) = 18320981569$
$p_6(31) = 30227301468$
$p_6(32) = 49522366656$
$p_6(33) = 80590617278$
$p_6(34) = 130307480730$
$p_6(35) = 209396014650$
$p_6(36) = 334492740175$
$p_6(37) = 531278722086$
$p_6(38) = 839204245278$
$p_6(39) = 1318587244800$
$p_6(40) = 2061234013149$
$p_6(41) = 3206265612102$
$p_6(42) = 4963610011174$
$p_6(43) = 7648740634158$
$p_6(44) = 11733857681850$
$p_6(45) = 17923026108974$
$p_6(46) = 27262103713383$
$p_6(47) = 41299026921900$
$p_6(48) = 62316761884412$
$p_6(49) = 93670787579400$
$p_6(50) = 140276509484724$
$p_6(51) = 209311092717496$
$p_6(52) = 311221082489247$
$p_6(53) = 461164908282750$
$p_6(54) = 681072218736560$
$p_6(55) = 1002575819281092$
$p_6(56) = 1471174898831406$
$p_6(57) = 2152131323752182$
$p_6(58) = 3138799331684016$
$p_6(59) = 4564363803456240$
$p_6(60) = 6618341988415724$
$p_6(61) = 9569726912844018$
$p_6(62) = 13799370667455450$
$p_6(63) = 19845194082605218$
$p_6(64) = 28465163415636075$
$p_6(65) = 40724826208003224$
$p_6(66) = 58118725494437321$
$p_6(67) = 82738453811650308$
$p_6(68) = 117504789092547003$
$p_6(69) = 166487707267085750$
$p_6(70) = 235346673673043703$
$p_6(71) = 331935256651796730$
$p_6(72) = 467129825087621531$
$p_6(73) = 655963278944277444$
$p_6(74) = 919173274601444100$
$p_6(75) = 1285312714955089064$
$p_6(76) = 1793621662235878482$
$p_6(77) = 2497928656596353478$
$p_6(78) = 3471941462435113530$
$p_6(79) = 4816410159748817880$
$p_6(80) = 6668809352618868378$

$p_7(0)$ = 1
$p_7(1)$ = 7
$p_7(2)$ = 35
$p_7(3)$ = 140
$p_7(4)$ = 490
$p_7(5)$ = 1547
$p_7(6)$ = 4522
$p_7(7)$ = 12405
$p_7(8)$ = 32305
$p_7(9)$ = 80465
$p_7(10)$ = 192899
$p_7(11)$ = 447146
$p_7(12)$ = 1006145
$p_7(13)$ = 2204475
$p_7(14)$ = 4715510
$p_7(15)$ = 9869132
$p_7(16)$ = 20247710
$p_7(17)$ = 40786690
$p_7(18)$ = 80782800
$p_7(19)$ = 157510780
$p_7(20)$ = 302666903
$p_7(21)$ = 573720808
$p_7(22)$ = 1073720305
$p_7(23)$ = 1985506775
$p_7(24)$ = 3630307835
$p_7(25)$ = 6567206471
$p_7(26)$ = 11760658378
$p_7(27)$ = 20860415590
$p_7(28)$ = 36665885170
$p_7(29)$ = 63891010155
$p_7(30)$ = 110415782785
$p_7(31)$ = 189320804673
$p_7(32)$ = 322174588225
$p_7(33)$ = 544313510685
$p_7(34)$ = 913274346530
$p_7(35)$ = 1522188092649
$p_7(36)$ = 2520942484767
$p_7(37)$ = 4149449539025
$p_7(38)$ = 6789686865150
$p_7(39)$ = 11046722446970
$p_7(40)$ = 17874289316814
$p_7(41)$ = 28768544837739
$p_7(42)$ = 46065759830255
$p_7(43)$ = 73397788167025
$p_7(44)$ = 116386144869365
$p_7(45)$ = 183695569433761
$p_7(46)$ = 288627213359933
$p_7(47)$ = 451518983340670
$p_7(48)$ = 703349049433800
$p_7(49)$ = 1091126798413415
$p_7(50)$ = 1685930588557978
$p_7(51)$ = 2594852348869557
$p_7(52)$ = 3978691071349855
$p_7(53)$ = 6078080199178295
$p_7(54)$ = 9251951511540605
$p_7(55)$ = 14033992077506758
$p_7(56)$ = 21215270934140557
$p_7(57)$ = 31964823610360830
$p_7(58)$ = 48005145278868680
$p_7(59)$ = 71866905262741890
$p_7(60)$ = 107257668780789891
$p_7(61)$ = 159594275946114163
$p_7(62)$ = 236769575980276185
$p_7(63)$ = 350253951850294200
$p_7(64)$ = 516673992315272715
$p_7(65)$ = 760069639569275287
$p_7(66)$ = 1115113920667866707
$p_7(67)$ = 1631695331336874545
$p_7(68)$ = 2381425052399170575
$p_7(69)$ = 3466857358141543450
$p_7(70)$ = 5034526527400420694
$p_7(71)$ = 7293341292995563757
$p_7(72)$ = 10540485080207226305
$p_7(73)$ = 15197811061879210800
$p_7(74)$ = 21862883126302198155
$p_7(75)$ = 31380417129736719506
$p_7(76)$ = 44942084936061587008
$p_7(77)$ = 64225679689056244555
$p_7(78)$ = 91588807809231769120
$p_7(79)$ = 130337983266497806590
$p_7(80)$ = 185101811335892977043

$p_8(0)$ =	1
$p_8(1)$ =	8
$p_8(2)$ =	44
$p_8(3)$ =	192
$p_8(4)$ =	726
$p_8(5)$ =	2464
$p_8(6)$ =	7704
$p_8(7)$ =	22528
$p_8(8)$ =	62337
$p_8(9)$ =	164560
$p_8(10)$ =	417140
$p_8(11)$ =	1020416
$p_8(12)$ =	2418710
$p_8(13)$ =	5573568
$p_8(14)$ =	12520744
$p_8(15)$ =	27484160
$p_8(16)$ =	59068372
$p_8(17)$ =	124505880
$p_8(18)$ =	257770964
$p_8(19)$ =	524871424
$p_8(20)$ =	1052316364
$p_8(21)$ =	2079491744
$p_8(22)$ =	4053978040
$p_8(23)$ =	7803219968
$p_8(24)$ =	14840711765
$p_8(25)$ =	27907041392
$p_8(26)$ =	51917588800
$p_8(27)$ =	95608651776
$p_8(28)$ =	174374048226
$p_8(29)$ =	315116024256
$p_8(30)$ =	564480785968
$p_8(31)$ =	1002743087104
$p_8(32)$ =	1767059165529
$p_8(33)$ =	3090160346056
$p_8(34)$ =	5364345550500
$p_8(35)$ =	9246688953216
$p_8(36)$ =	15831025142756
$p_8(37)$ =	26927556198400
$p_8(38)$ =	45515024095024
$p_8(39)$ =	76468217309184
$p_8(40)$ =	127722692222125
$p_8(41)$ =	212130564940976
$p_8(42)$ =	350403749917320
$p_8(43)$ =	575760966987776
$p_8(44)$ =	941232951872944
$p_8(45)$ =	1531098085437568
$p_8(46)$ =	2478710665057880
$p_8(47)$ =	3994185055723520
$p_8(48)$ =	6407218457844325
$p_8(49)$ =	10233082112753456
$p_8(50)$ =	16273944632919964
$p_8(51)$ =	25773891757087232
$p_8(52)$ =	40655278297166302
$p_8(53)$ =	63877868631171232
$p_8(54)$ =	99982762342805888
$p_8(55)$ =	155913539423358976
$p_8(56)$ =	242252035659499837
$p_8(57)$ =	375072432344924160
$p_8(58)$ =	578714734503739980
$p_8(59)$ =	889921447128006656
$p_8(60)$ =	1363989908490835692
$p_8(61)$ =	2083896965430862400
$p_8(62)$ =	3173795153766576680
$p_8(63)$ =	4818921535153127424
$p_8(64)$ =	7294889573349093342
$p_8(65)$ =	11010676281531958896
$p_8(66)$ =	16571550116078875124
$p_8(67)$ =	24870964118368774720
$p_8(68)$ =	37224424840547394450
$p_8(69)$ =	55564052517808446432
$p_8(70)$ =	82720695702227851136
$p_8(71)$ =	122832075958896601088
$p_8(72)$ =	181931954719477114990
$p_8(73)$ =	268798759007819843792
$p_8(74)$ =	396175316146196643056
$p_8(75)$ =	582518309188160694272
$p_8(76)$ =	854502339558676285326
$p_8(77)$ =	1250596839296642580480
$p_8(78)$ =	1826165337213412528080
$p_8(79)$ =	2660720809824362250240
$p_8(80)$ =	3868228952300778857578

30. MODULAR CHARACTERISTICS

Each table gives the modular characteristics associated with every representation of a given algebra, level and congruence class. The modular characteristics are rational numbers: the numerators are given in the table, and the common denominator is given alongside or just above the table. The configuration of the weights in these tables is a reflection in the line $y = -x$ of the configuration used in §31 and §32. In all tables the node corresponding to α_0 is in the upper left-hand corner of the diagram. The modular characteristics of the representation "Rp n" lie in the column below the "Rp n" label, and "Wt k" refers to the same weight as "Rp k."

These tables are in one-to-one correspondence with the multiplicity tables in §31, except that we have provided a few extra modular characteristic tables of type E to aid with the "leading zeros" problem discussed in §28.

Modular Characteristics for Affine A Algebras

Affine A1 - Level 1 - Class 0

	Rp 1
	10
Wt 1:	-1

Denominator = 24

Affine A1 - Level 2 - Class 0

	Rp 1	Rp 2
	02	20
Wt 1:	-1	-9
Wt 2:	7	-1

Den = 16

Affine A1 - Level 2 - Class 1

	Rp 1
	11
Wt 1:	0

Den = 1

Affine A1 - Level 3 - Class 0

	Rp 1	Rp 2
	12	30
Wt 1:	-1	-49
Wt 2:	39	-9

Den = 120

Affine A1 - Level 4 - Class 0

	Rp 1	Rp 2	Rp 3
	04	22	40
Wt 1:	-1	-9	-13
Wt 2:	8	0	-4
Wt 3:	11	3	-1

Den = 12

Affine A1 - Level 4 - Class 1

Den = 48

	Rp 1 (13)	Rp 2 (31)
Wt 1:	-1	-25
Wt 2:	23	-1

Affine A1 - Level 5 - Class 0

Den = 280

	Rp 1 (14)	Rp 2 (32)	Rp 3 (50)
Wt 1:	-9	-169	-249
Wt 2:	159	-1	-81
Wt 3:	215	55	-25

Affine A1 - Level 6 - Class 0

Den = 96

	Rp 1 (06)	Rp 2 (24)	Rp 3 (42)	Rp 4 (60)
Wt 1:	-9	-81	-129	-153
Wt 2:	71	-1	-49	-73
Wt 3:	119	47	-1	-25
Wt 4:	135	63	15	-9

Affine A1 - Level 6 - Class 1

Den = 24

	Rp 1 (15)	Rp 2 (33)	Rp 3 (51)
Wt 1:	-1	-16	-25
Wt 2:	15	0	-9
Wt 3:	23	8	-1

Affine A1 - Level 7 - Class 0

Den = 504

	Rp 1	Rp 2	Rp 3	Rp 4
	16	34	52	70
Wt 1:	-25	-361	-585	-697
Wt 2:	335	-1	-225	-337
Wt 3:	551	215	-9	-121
Wt 4:	623	287	63	-49

Affine A1 - Level 8 - Class 0

Den = 40

	Rp 1	Rp 2	Rp 3	Rp 4	Rp 5
	08	26	44	62	80
Wt 1:	-4	-36	-60	-76	-84
Wt 2:	31	-1	-25	-41	-49
Wt 3:	56	24	0	-16	-24
Wt 4:	71	39	15	-1	-9
Wt 5:	76	44	20	4	-4

Affine A1 - Level 8 - Class 1

Den = 160

	Rp 1	Rp 2	Rp 3	Rp 4
	17	35	53	71
Wt 1:	-9	-121	-201	-249
Wt 2:	111	-1	-81	-129
Wt 3:	191	79	-1	-49
Wt 4:	231	119	39	-9

Affine A1 - Level 9 - Class 0

Den = 792

	Rp 1	Rp 2	Rp 3	Rp 4	Rp 5
	18	36	54	72	90
Wt 1:	-49	-625	-1057	-1345	-1489
Wt 2:	567	-9	-441	-729	-873
Wt 3:	1007	431	-1	-289	-433
Wt 4:	1271	695	263	-25	-169
Wt 5:	1359	783	351	63	-81

Affine A2 - Level 1 - Class 0

Den = 12

	Rp 1
	10
	0
Wt 1:	-1

Affine A2 - Level 2 - Class 0

Den = 30

	Rp 1	Rp 2
	01	20
	1	0
Wt 1:	-1	-19
Wt 2:	14	-4

Affine A2 - Level 3 - Class 0

Den = 6

	Rp 1	Rp 2	Rp 3	Rp 4
	00	03	11	30
	3	0	1	0
Wt 1:	-1	-1	-4	-7
Wt 2:	-1	-1	-4	-7
Wt 3:	3	3	0	-3
Wt 4:	5	5	2	-1

Affine A2 - Level 3 - Class 1

Den = 18

	Rp 1	Rp 2	Rp 3
	02	10	21
	1	2	0
Wt 1:	-1	-7	-13
Wt 2:	5	-1	-7
Wt 3:	11	5	-1

Affine A2 - Level 4 - Class 0

Den = 84

	Rp 1	Rp 2	Rp 3	Rp 4	Rp 5
	02	10	13	21	40
	2	3	0	1	0
Wt 1:	-4	-28	-28	-64	-100
Wt 2:	17	-7	-7	-43	-79
Wt 3:	17	-7	-7	-43	-79
Wt 4:	59	35	35	-1	-37
Wt 5:	80	56	56	20	-16

Affine A2 - Level 5 - Class 0

Den = 120

	Rp 1	Rp 2	Rp 3	Rp 4	Rp 5	Rp 6	Rp 7
	01	04	12	20	23	31	50
	4	1	2	3	0	1	0
Wt 1:	-13	-13	-73	-103	-103	-148	-193
Wt 2:	-13	-13	-73	-103	-103	-148	-193
Wt 3:	59	59	-1	-31	-31	-76	-121
Wt 4:	83	83	23	-7	-7	-52	-97
Wt 5:	83	83	23	-7	-7	-52	-97
Wt 6:	131	131	71	41	41	-4	-49
Wt 7:	155	155	95	65	65	20	-25

Affine A2 - Level 6 - Class 0

Den = 18

	Rp 1	Rp 2	Rp 3	Rp 4	Rp 5	Rp 6	Rp 7	Rp 8	Rp 9	Rp10
	00	03	06	11	14	22	30	33	41	60
	6	3	0	4	1	2	3	0	1	0
Wt 1:	-4	-10	-4	-16	-16	-24	-28	-28	-34	-40
Wt 2:	5	-1	5	-7	-7	-15	-19	-19	-25	-31
Wt 3:	-4	-10	-4	-16	-16	-24	-28	-28	-34	-40
Wt 4:	11	5	11	-1	-1	-9	-13	-13	-19	-25
Wt 5:	11	5	11	-1	-1	-9	-13	-13	-19	-25
Wt 6:	20	14	20	8	8	0	-4	-4	-10	-16
Wt 7:	23	17	23	11	11	3	-1	-1	-7	-13
Wt 8:	23	17	23	11	11	3	-1	-1	-7	-13
Wt 9:	29	23	29	17	17	9	5	5	-1	-7
Wt10:	32	26	32	20	20	12	8	8	2	-4

Affine A2 - Level 6 - Class 1

Den = 54

	Rp 1	Rp 2	Rp 3	Rp 4	Rp 5	Rp 6	Rp 7	Rp 8	Rp 9
	02 4	05 1	10 5	13 2	21 3	24 0	32 1	40 2	51 0
Wt 1:	-4	2	-16	-28	-46	-40	-64	-76	-88
Wt 2:	-13	-7	-25	-37	-55	-49	-73	-85	-97
Wt 3:	5	11	-7	-19	-37	-31	-55	-67	-79
Wt 4:	23	29	11	-1	-19	-13	-37	-49	-61
Wt 5:	41	47	29	17	-1	5	-19	-31	-43
Wt 6:	32	38	20	8	-10	-4	-28	-40	-52
Wt 7:	59	65	47	35	17	23	-1	-13	-25
Wt 8:	68	74	56	44	26	32	8	-4	-16
Wt 9:	77	83	65	53	35	41	17	5	-7

Affine A2 - Level 7 - Class 0

Den = 210

	Rp 1	Rp 2	Rp 3	Rp 4	Rp 5	Rp 6	Rp 7	Rp 8	Rp 9	Rp10	Rp11	Rp12
	02 5	05 2	10 6	13 3	16 0	21 4	24 1	32 2	40 3	43 0	51 1	70 0
Wt 1:	-19	-19	-61	-124	-61	-187	-187	-271	-313	-313	-376	-439
Wt 2:	-19	-19	-61	-124	-61	-187	-187	-271	-313	-313	-376	-439
Wt 3:	11	11	-31	-94	-31	-157	-157	-241	-283	-283	-346	-409
Wt 4:	101	101	59	-4	59	-67	-67	-151	-193	-193	-256	-319
Wt 5:	11	11	-31	-94	-31	-157	-157	-241	-283	-283	-346	-409
Wt 6:	161	161	119	56	119	-7	-7	-91	-133	-133	-196	-259
Wt 7:	161	161	119	56	119	-7	-7	-91	-133	-133	-196	-259
Wt 8:	251	251	209	146	209	83	83	-1	-43	-43	-106	-169
Wt 9:	281	281	239	176	239	113	113	29	-13	-13	-76	-139
Wt10:	281	281	239	176	239	113	113	29	-13	-13	-76	-139
Wt11:	341	341	299	236	299	173	173	89	47	47	-16	-79
Wt12:	371	371	329	266	329	203	203	119	77	77	14	-49

Affine A3 - Level 1 - Class 0

Den = 8

	Rp 1
	10 0 0
Wt 1:	-1

Affine A3 - Level 2 - Class 0

	Rp 1	Rp 2	Rp 3
	00 2 0	01 0 1	20 0 0
Wt 1:	-5	-13	-29
Wt 2:	7	-1	-17
Wt 3:	19	11	-5

Den = 24

Affine A3 - Level 2 - Class 1

	Rp 1	Rp 2
	00 1 1	11 0 0
Wt 1:	-1	-7
Wt 2:	5	-1

Den = 12

Affine A3 - Level 3 - Class 0

	Rp 1	Rp 2	Rp 3	Rp 4	Rp 5
	00 1 2	02 1 0	10 2 0	11 0 1	30 0 0
Wt 1:	-21	-21	-69	-117	-213
Wt 2:	-21	-21	-69	-117	-213
Wt 3:	35	35	-13	-61	-157
Wt 4:	91	91	43	-5	-101
Wt 5:	147	147	99	51	-45

Den = 168

Affine A3 - Level 4 - Class 0

Den = 16

	Rp 1	Rp 2	Rp 3	Rp 4	Rp 5	Rp 6	Rp 7	Rp 8	Rp 9	Rp10
	00	00	01	02	04	10	12	20	21	40
	0	4	2	0	0	1	1	2	0	0
	4	0	1	2	0	2	0	0	1	0
Wt 1:	-5	3	-5	-9	-5	-13	-13	-17	-21	-29
Wt 2:	-13	-5	-13	-17	-13	-21	-21	-25	-29	-37
Wt 3:	-1	7	-1	-5	-1	-9	-9	-13	-17	-25
Wt 4:	3	11	3	-1	3	-5	-5	-9	-13	-21
Wt 5:	-5	3	-5	-9	-5	-13	-13	-17	-21	-29
Wt 6:	7	15	7	3	7	-1	-1	-5	-9	-17
Wt 7:	7	15	7	3	7	-1	-1	-5	-9	-17
Wt 8:	11	19	11	7	11	3	3	-1	-5	-13
Wt 9:	15	23	15	11	15	7	7	3	-1	-9
Wt10:	19	27	19	15	19	11	11	7	3	-5

Affine A3 - Level 4 - Class 1

Den = 64

	Rp 1	Rp 2	Rp 3	Rp 4	Rp 5	Rp 6	Rp 7	Rp 8
	00	01	03	10	11	12	20	31
	3	1	1	0	2	0	1	0
	1	2	0	3	0	1	1	0
Wt 1:	-11	-35	-27	-59	-51	-67	-83	-107
Wt 2:	21	-3	5	-27	-19	-35	-51	-75
Wt 3:	5	-19	-11	-43	-35	-51	-67	-91
Wt 4:	37	13	21	-11	-3	-19	-35	-59
Wt 5:	37	13	21	-11	-3	-19	-35	-59
Wt 6:	53	29	37	5	13	-3	-19	-43
Wt 7:	69	45	53	21	29	13	-3	-27
Wt 8:	85	61	69	37	45	29	13	-11

Affine A3 - Level 4 - Class 2

Den = 8

	Rp 1	Rp 2	Rp 3	Rp 4	Rp 5	Rp 6	Rp 7	Rp 8	Rp 9
	00 2 2	01 0 3	02 2 0	03 0 1	10 3 0	11 1 1	20 0 2	22 0 0	30 1 0
Wt 1:	-1	-3	-1	-3	-3	-6	-9	-9	-11
Wt 2:	1	-1	1	-1	-1	-4	-7	-7	-9
Wt 3:	-1	-3	-1	-3	-3	-6	-9	-9	-11
Wt 4:	1	-1	1	-1	-1	-4	-7	-7	-9
Wt 5:	1	-1	1	-1	-1	-4	-7	-7	-9
Wt 6:	5	3	5	3	3	0	-3	-3	-5
Wt 7:	7	5	7	5	5	2	-1	-1	-3
Wt 8:	7	5	7	5	5	2	-1	-1	-3
Wt 9:	9	7	9	7	7	4	1	1	-1

Affine A3 - Level 5 - Class 0

Den = 360

	Rp 1	Rp 2	Rp 3	Rp 4	Rp 5	Rp 6	Rp 7	Rp 8	Rp 9	Rp10	Rp11	Rp12	Rp13	Rp14
	00 3 2	01 1 3	02 3 0	03 1 1	10 0 4	10 4 0	11 2 1	12 0 2	14 0 0	20 1 2	22 1 0	30 2 0	31 0 1	50 0 0
Wt 1:	-53	-173	-53	-173	-293	-133	-293	-373	-293	-453	-453	-533	-613	-773
Wt 2:	91	-29	91	-29	-149	11	-149	-229	-149	-309	-309	-389	-469	-629
Wt 3:	-53	-173	-53	-173	-293	-133	-293	-373	-293	-453	-453	-533	-613	-773
Wt 4:	91	-29	91	-29	-149	11	-149	-229	-149	-309	-309	-389	-469	-629
Wt 5:	163	43	163	43	-77	83	-77	-157	-77	-237	-237	-317	-397	-557
Wt 6:	19	-101	19	-101	-221	-61	-221	-301	-221	-381	-381	-461	-541	-701
Wt 7:	235	115	235	115	-5	155	-5	-85	-5	-165	-165	-245	-325	-485
Wt 8:	307	187	307	187	67	227	67	-13	67	-93	-93	-173	-253	-413
Wt 9:	163	43	163	43	-77	83	-77	-157	-77	-237	-237	-317	-397	-557
Wt10:	379	259	379	259	139	299	139	59	139	-21	-21	-101	-181	-341
Wt11:	379	259	379	259	139	299	139	59	139	-21	-21	-101	-181	-341
Wt12:	451	331	451	331	211	371	211	131	211	51	51	-29	-109	-269
Wt13:	523	403	523	403	283	443	283	203	283	123	123	43	-37	-197
Wt14:	595	475	595	475	355	515	355	275	355	195	195	115	35	-125

Affine A4 - Level 1 - Class 0

Den = 6

	Rp 1
	10
	0
	0
	0
Wt 1:	-1

Affine A4 - Level 2 - Class 0

Den = 14

	Rp 1	Rp 2	Rp 3
	00	01	20
	1	0	0
	1	0	0
	0	1	0
Wt 1:	-2	-8	-18
Wt 2:	5	-1	-11
Wt 3:	12	6	-4

Affine A4 - Level 3 - Class 0

Den = 24

	Rp 1	Rp 2	Rp 3	Rp 4	Rp 5	Rp 6	Rp 7
	00	00	01	02	10	11	30
	0	1	2	0	1	0	0
	2	0	0	1	1	0	0
	1	2	0	0	0	1	0
Wt 1:	-5	-11	-5	-11	-17	-26	-41
Wt 2:	3	-3	3	-3	-9	-18	-33
Wt 3:	-5	-11	-5	-11	-17	-26	-41
Wt 4:	3	-3	3	-3	-9	-18	-33
Wt 5:	11	5	11	5	-1	-10	-25
Wt 6:	19	13	19	13	7	-2	-17
Wt 7:	27	21	27	21	15	6	-9

Affine A4 - Level 4 - Class 0

Den = 36

	Rp 1	Rp 2	Rp 3	Rp 4	Rp 5	Rp 6	Rp 7	Rp 8	Rp 9	Rp10	Rp11	Rp12	Rp13	Rp14
	00 0 1 3	00 2 2 0	00 3 0 1	01 0 3 0	01 1 1 1	02 0 0 2	03 1 0 0	10 0 2 1	10 1 0 2	11 2 0 0	12 0 1 0	20 1 1 0	21 0 0 1	40 0 0 0
Wt 1:	-10	10	2	2	-10	-22	-10	-22	-30	-22	-30	-38	-50	-70
Wt 2:	-28	-8	-16	-16	-28	-40	-28	-40	-48	-40	-48	-56	-68	-88
Wt 3:	-19	1	-7	-7	-19	-31	-19	-31	-39	-31	-39	-47	-59	-79
Wt 4:	-19	1	-7	-7	-19	-31	-19	-31	-39	-31	-39	-47	-59	-79
Wt 5:	-1	19	11	11	-1	-13	-1	-13	-21	-13	-21	-29	-41	-61
Wt 6:	8	28	20	20	8	-4	8	-4	-12	-4	-12	-20	-32	-52
Wt 7:	-10	10	2	2	-10	-22	-10	-22	-30	-22	-30	-38	-50	-70
Wt 8:	8	28	20	20	8	-4	8	-4	-12	-4	-12	-20	-32	-52
Wt 9:	17	37	29	29	17	5	17	5	-3	5	-3	-11	-23	-43
Wt10:	8	28	20	20	8	-4	8	-4	-12	-4	-12	-20	-32	-52
Wt11:	17	37	29	29	17	5	17	5	-3	5	-3	-11	-23	-43
Wt12:	26	46	38	38	26	14	26	14	6	14	6	-2	-14	-34
Wt13:	35	55	47	47	35	23	35	23	15	23	15	7	-5	-25
Wt14:	44	64	56	56	44	32	44	32	24	32	24	16	4	-16

Affine A5 - Level 1 - Class 0

Den = 24

	Rp 1
	10 0 0 0 0
Wt 1:	-5

Affine A5 - Level 2 - Class 0

Den = 96

	Rp 1	Rp 2	Rp 3	Rp 4
	00	00	01	20
	0	1	0	0
	2	0	0	0
	0	1	0	0
	0	0	1	0
Wt 1:	-35	-59	-107	-179
Wt 2:	13	-11	-59	-131
Wt 3:	61	37	-11	-83
Wt 4:	109	85	37	-35

Affine A5 - Level 2 - Class 1

Den = 24

	Rp 1	Rp 2	Rp 3
	00	00	11
	0	1	0
	1	0	0
	1	0	0
	0	1	0
Wt 1:	-5	-14	-29
Wt 2:	7	-2	-17
Wt 3:	19	10	-5

Affine A5 - Level 3 - Class 0

Den = 72

	Rp 1	Rp 2	Rp 3	Rp 4	Rp 5	Rp 6	Rp 7	Rp 8	Rp 9	Rp10
	00	00	00	00	01	02	10	10	11	30
	0	0	1	3	1	0	0	1	0	0
	0	1	0	0	1	0	2	0	0	0
	3	1	0	0	0	1	0	1	0	0
	0	1	2	0	0	0	0	0	1	0
Wt 1:	-35	-59	-83	-35	-59	-83	-83	-99	-131	-179
Wt 2:	13	-11	-35	13	-11	-35	-35	-51	-83	-131
Wt 3:	37	13	-11	37	13	-11	-11	-27	-59	-107
Wt 4:	-35	-59	-83	-35	-59	-83	-83	-99	-131	-179
Wt 5:	13	-11	-35	13	-11	-35	-35	-51	-83	-131
Wt 6:	37	13	-11	37	13	-11	-11	-27	-59	-107
Wt 7:	37	13	-11	37	13	-11	-11	-27	-59	-107
Wt 8:	61	37	13	61	37	13	13	-3	-35	-83
Wt 9:	85	61	37	85	61	37	37	21	-11	-59
Wt10:	109	85	61	109	85	61	61	45	13	-35

Affine A5 - Level 3 - Class 1

Den = 216

	Rp 1	Rp 2	Rp 3	Rp 4	Rp 5	Rp 6	Rp 7	Rp 8	Rp 9
	00	00	00	01	01	02	10	10	21
	0	0	2	0	1	0	0	1	0
	0	1	1	2	0	0	1	0	0
	2	0	0	0	1	0	1	0	0
	1	2	0	0	0	1	0	1	0
Wt 1:	-65	-113	7	-41	-89	-185	-161	-233	-353
Wt 2:	7	-41	79	31	-17	-113	-89	-161	-281
Wt 3:	-137	-185	-65	-113	-161	-257	-233	-305	-425
Wt 4:	-65	-113	7	-41	-89	-185	-161	-233	-353
Wt 5:	7	-41	79	31	-17	-113	-89	-161	-281
Wt 6:	79	31	151	103	55	-41	-17	-89	-209
Wt 7:	79	31	151	103	55	-41	-17	-89	-209
Wt 8:	151	103	223	175	127	31	55	-17	-137
Wt 9:	223	175	295	247	199	103	127	55	-65

Affine A6 - Level 1 - Class 0

Den = 4

	Rp 1
	10
	0
	0
	0
	0
	0
Wt 1:	-1

Affine A6 - Level 2 - Class 0

Den = 18

	Rp 1	Rp 2	Rp 3	Rp 4
	00	00	01	20
	0	1	0	0
	1	0	0	0
	1	0	0	0
	0	1	0	0
	0	0	1	0
Wt 1:	-5	-11	-21	-35
Wt 2:	4	-2	-12	-26
Wt 3:	13	7	-3	-17
Wt 4:	22	16	6	-8

Affine A6 - Level 3 - Class 0

Den = 30

	Rp 1	Rp 2	Rp 3	Rp 4	Rp 5	Rp 6	Rp 7	Rp 8	Rp 9	Rp10	Rp11	Rp12
	00	00	00	00	00	01	01	02	10	10	11	30
	0	0	0	1	2	0	1	0	0	1	0	0
	0	0	1	0	1	2	0	0	1	0	0	0
	1	2	0	0	0	0	1	0	1	0	0	0
	2	0	1	0	0	0	0	1	0	1	0	0
	0	1	1	2	0	0	0	0	0	0	1	0
Wt 1:	-12	-18	-24	-36	-12	-18	-24	-36	-33	-42	-57	-78
Wt 2:	-2	-8	-14	-26	-2	-8	-14	-26	-23	-32	-47	-68
Wt 3:	8	2	-4	-16	8	2	-4	-16	-13	-22	-37	-58
Wt 4:	18	12	6	-6	18	12	6	-6	-3	-12	-27	-48
Wt 5:	-12	-18	-24	-36	-12	-18	-24	-36	-33	-42	-57	-78
Wt 6:	-2	-8	-14	-26	-2	-8	-14	-26	-23	-32	-47	-68
Wt 7:	8	2	-4	-16	8	2	-4	-16	-13	-22	-37	-58
Wt 8:	18	12	6	-6	18	12	6	-6	-3	-12	-27	-48
Wt 9:	18	12	6	-6	18	12	6	-6	-3	-12	-27	-48
Wt10:	28	22	16	4	28	22	16	4	7	-2	-17	-38
Wt11:	38	32	26	14	38	32	26	14	17	8	-7	-28
Wt12:	48	42	36	24	48	42	36	24	27	18	3	-18

Affine A7 - Level 1 - Class 0

Den = 24

	Rp 1
	10
	0
	0
	0
	0
	0
	0
Wt 1:	-7

Affine A7 - Level 2 - Class 0

	Rp 1	Rp 2	Rp 3	Rp 4	Rp 5
	00	00	00	01	20
	0	0	1	0	0
	0	1	0	0	0
	2	0	0	0	0
	0	1	0	0	0
	0	0	1	0	0
	0	0	0	1	0
Wt 1:	-21	-29	-45	-69	-101
Wt 2:	-1	-9	-25	-49	-81
Wt 3:	19	11	-5	-29	-61
Wt 4:	39	31	15	-9	-41
Wt 5:	59	51	35	11	-21

Den = 40

Affine A7 - Level 2 - Class 1

	Rp 1	Rp 2	Rp 3	Rp 4
	00	00	00	11
	0	0	1	0
	0	1	0	0
	1	0	0	0
	1	0	0	0
	0	1	0	0
	0	0	1	0
Wt 1:	-7	-13	-23	-37
Wt 2:	3	-3	-13	-27
Wt 3:	13	7	-3	-17
Wt 4:	23	17	7	-7

Den = 20

Affine A7 - Level 3 - Class 0

Den = 264

	Rp 1	Rp 2	Rp 3	Rp 4	Rp 5	Rp 6	Rp 7	Rp 8	Rp 9	Rp10	Rp11	Rp12	Rp13	Rp14	Rp15
	00	00	00	00	00	00	00	01	01	02	10	10	10	11	30
	0	0	0	0	1	1	2	0	1	0	0	0	1	0	0
	0	0	0	1	0	2	0	1	0	0	0	1	0	0	0
	0	1	1	0	0	0	1	1	0	0	2	0	0	0	0
	2	0	1	0	0	0	0	0	1	0	0	1	0	0	0
	1	2	0	1	0	0	0	0	0	1	0	0	1	0	0
	0	0	1	1	2	0	0	0	0	0	0	0	0	1	0
Wt 1:	-133	-181	-229	-301	-421	-133	-181	-229	-301	-421	-325	-373	-469	-613	-805
Wt 2:	-45	-93	-141	-213	-333	-45	-93	-141	-213	-333	-237	-285	-381	-525	-717
Wt 3:	43	-5	-53	-125	-245	43	-5	-53	-125	-245	-149	-197	-293	-437	-629
Wt 4:	131	83	35	-37	-157	131	83	35	-37	-157	-61	-109	-205	-349	-541
Wt 5:	219	171	123	51	-69	219	171	123	51	-69	27	-21	-117	-261	-453
Wt 6:	-133	-181	-229	-301	-421	-133	-181	-229	-301	-421	-325	-373	-469	-613	-805
Wt 7:	-45	-93	-141	-213	-333	-45	-93	-141	-213	-333	-237	-285	-381	-525	-717
Wt 8:	43	-5	-53	-125	-245	43	-5	-53	-125	-245	-149	-197	-293	-437	-629
Wt 9:	131	83	35	-37	-157	131	83	35	-37	-157	-61	-109	-205	-349	-541
Wt10:	219	171	123	51	-69	219	171	123	51	-69	27	-21	-117	-261	-453
Wt11:	131	83	35	-37	-157	131	83	35	-37	-157	-61	-109	-205	-349	-541
Wt12:	219	171	123	51	-69	219	171	123	51	-69	27	-21	-117	-261	-453
Wt13:	307	259	211	139	19	307	259	211	139	19	115	67	-29	-173	-365
Wt14:	395	347	299	227	107	395	347	299	227	107	203	155	59	-85	-277
Wt15:	483	435	387	315	195	483	435	387	315	195	291	243	147	3	-189

Affine A8 - Level 1 - Class 0

Den = 3

	Rp 1
	10
	0
	0
	0
	0
	0
	0
	0
Wt 1:	-1

Affine A8 - Level 2 - Class 0

	Rp 1	Rp 2	Rp 3	Rp 4	Rp 5
	00	00	00	01	20
	0	0	1	0	0
	0	1	0	0	0
	1	0	0	0	0
	1	0	0	0	0
	0	1	0	0	0
	0	0	1	0	0
	0	0	0	1	0
Wt 1:	-28	-46	-76	-118	-172
Wt 2:	5	-13	-43	-85	-139
Wt 3:	38	20	-10	-52	-106
Wt 4:	71	53	23	-19	-73
Wt 5:	104	86	56	14	-40

Den = 66

Modular Characteristics for Affine B Algebras

Affine B3 - Level 1 - Class 0

	Rp 1	Rp 2
	0 1	1 0
	0	0
	0	0
Wt 1:	-7	-31
Wt 2:	17	-7

Denominator = 48

Affine B3 - Level 1 - Class 1

	Rp 1
	0 0
	0
	1
Wt 1:	-1

Den = 12

Affine B3 - Level 2 - Class 0

	Rp 1	Rp 2	Rp 3	Rp 4	Rp 5
	0 0	0 0	0 2	1 1	2 0
	0	1	0	0	0
	2	0	0	0	0
Wt 1:	-4	-8	0	-16	-28
Wt 2:	3	-1	7	-9	-21
Wt 3:	-11	-15	-7	-23	-35
Wt 4:	10	6	14	-2	-14
Wt 5:	17	13	21	5	-7

Den = 28

Affine B3 - Level 2 - Class 1

	Rp 1	Rp 2
	0 1	1 0
	0	0
	1	1
Wt 1:	-1	-9
Wt 2:	7	-1

Den = 16

Affine B3 - Level 3 - Class 0

Den = 192

	Rp 1	Rp 2	Rp 3	Rp 4	Rp 5	Rp 6	Rp 7	Rp 8
	0 1	0 1	0 3	1 0	1 0	1 2	2 1	3 0
	0	1	0	0	1	0	0	0
	2	0	0	2	0	0	0	0
Wt 1:	-15	-39	33	-111	-135	-87	-183	-255
Wt 2:	17	-7	65	-79	-103	-55	-151	-223
Wt 3:	-111	-135	-63	-207	-231	-183	-279	-351
Wt 4:	81	57	129	-15	-39	9	-87	-159
Wt 5:	113	89	161	17	-7	41	-55	-127
Wt 6:	49	25	97	-47	-71	-23	-119	-191
Wt 7:	145	121	193	49	25	73	-23	-95
Wt 8:	177	153	225	81	57	105	9	-63

Affine B3 - Level 3 - Class 1

Den = 48

	Rp 1	Rp 2	Rp 3	Rp 4	Rp 5
	0 0	0 0	0 2	1 1	2 0
	0	1	0	0	0
	3	1	1	1	1
Wt 1:	-9	-18	-6	-33	-54
Wt 2:	7	-2	10	-17	-38
Wt 3:	-9	-18	-6	-33	-54
Wt 4:	23	14	26	-1	-22
Wt 5:	39	30	42	15	-6

Affine B3 - Level 4 - Class 0

Den = 72

	Rp 1	Rp 2	Rp 3	Rp 4	Rp 5	Rp 6	Rp 7	Rp 8	Rp 9	Rp10	Rp11	Rp12	Rp13	Rp14
	0 0	0 0	0 0	0 2	0 2	0 4	1 1	1 1	1 3	2 0	2 0	2 2	3 1	4 0
	0	1	2	0	1	0	0	1	0	0	1	0	0	0
	4	2	0	2	0	0	2	0	0	2	0	0	0	0
Wt 1:	-16	-32	-40	-16	-24	8	-56	-64	-40	-88	-96	-80	-112	-136
Wt 2:	11	-5	-13	11	3	35	-29	-37	-13	-61	-69	-53	-85	-109
Wt 3:	20	4	-4	20	12	44	-20	-28	-4	-52	-60	-44	-76	-100
Wt 4:	-7	-23	-31	-7	-15	17	-47	-55	-31	-79	-87	-71	-103	-127
Wt 5:	2	-14	-22	2	-6	26	-38	-46	-22	-70	-78	-62	-94	-118
Wt 6:	-52	-68	-76	-52	-60	-28	-92	-100	-76	-124	-132	-116	-148	-172
Wt 7:	38	22	14	38	30	62	-2	-10	14	-34	-42	-26	-58	-82
Wt 8:	47	31	23	47	39	71	7	-1	23	-25	-33	-17	-49	-73
Wt 9:	11	-5	-13	11	3	35	-29	-37	-13	-61	-69	-53	-85	-109
Wt10:	65	49	41	65	57	89	25	17	41	-7	-15	1	-31	-55
Wt11:	74	58	50	74	66	98	34	26	50	2	-6	10	-22	-46
Wt12:	56	40	32	56	48	80	16	8	32	-16	-24	-8	-40	-64
Wt13:	83	67	59	83	75	107	43	35	59	11	3	19	-13	-37
Wt14:	92	76	68	92	84	116	52	44	68	20	12	28	-4	-28

Affine B3 - Level 4 - Class 1

Den = 288

	Rp 1	Rp 2	Rp 3	Rp 4	Rp 5	Rp 6	Rp 7	Rp 8
	0 1	0 1	0 3	1 0	1 0	1 2	2 1	3 0
	0	1	0	0	1	0	0	0
	3	1	1	3	1	1	1	1
Wt 1:	-31	-79	17	-175	-223	-159	-303	-415
Wt 2:	41	-7	89	-103	-151	-87	-231	-343
Wt 3:	-103	-151	-55	-247	-295	-231	-375	-487
Wt 4:	113	65	161	-31	-79	-15	-159	-271
Wt 5:	185	137	233	41	-7	57	-87	-199
Wt 6:	113	65	161	-31	-79	-15	-159	-271
Wt 7:	257	209	305	113	65	129	-15	-127
Wt 8:	329	281	377	185	137	201	57	-55

Affine B4 - Level 1 - Class 0

Den = 16

	Rp 1	Rp 2
	0 1	1 0
	0	0
	0	0
	0	0
Wt 1:	-3	-11
Wt 2:	5	-3

Affine B4 - Level 1 - Class 1

Den = 8

	Rp 1
	0 0
	0
	0
	1
Wt 1:	-1

Affine B4 - Level 2 - Class 0

Den = 36

	Rp 1	Rp 2	Rp 3	Rp 4	Rp 5	Rp 6
	0 0	0 0	0 0	0 2	1 1	2 0
	0	0	1	0	0	0
	0	1	0	0	0	0
	2	0	0	0	0	0
Wt 1:	-8	-12	-20	-12	-32	-48
Wt 2:	1	-3	-11	-3	-23	-39
Wt 3:	10	6	-2	6	-14	-30
Wt 4:	-8	-12	-20	-12	-32	-48
Wt 5:	19	15	7	15	-5	-21
Wt 6:	28	24	16	24	4	-12

Affine B4 - Level 2 - Class 1

Den = 12

	Rp 1	Rp 2
	0 1	1 0
	0	0
	0	0
	1	1
Wt 1:	-1	-7
Wt 2:	5	-1

Affine B4 - Level 3 - Class 0

Den = 60

	Rp 1	Rp 2	Rp 3	Rp 4	Rp 5	Rp 6	Rp 7	Rp 8	Rp 9	Rp10
	0 1	0 1	0 1	0 3	1 0	1 0	1 0	1 2	2 1	3 0
	0	0	1	0	0	0	1	0	0	0
	0	1	0	0	0	1	0	0	0	0
	2	0	0	0	2	0	0	0	0	0
Wt 1:	-7	-13	-25	-7	-37	-43	-55	-43	-73	-97
Wt 2:	3	-3	-15	3	-27	-33	-45	-33	-63	-87
Wt 3:	13	7	-5	13	-17	-23	-35	-23	-53	-77
Wt 4:	-27	-33	-45	-27	-57	-63	-75	-63	-93	-117
Wt 5:	23	17	5	23	-7	-13	-25	-13	-43	-67
Wt 6:	33	27	15	33	3	-3	-15	-3	-33	-57
Wt 7:	43	37	25	43	13	7	-5	7	-23	-47
Wt 8:	23	17	5	23	-7	-13	-25	-13	-43	-67
Wt 9:	53	47	35	53	23	17	5	17	-13	-37
Wt10:	63	57	45	63	33	27	15	27	-3	-27

Affine B4 - Level 3 - Class 1

Den = 60

	Rp 1	Rp 2	Rp 3	Rp 4	Rp 5	Rp 6
	0 0	0 0	0 0	0 2	1 1	2 0
	0	0	1	0	0	0
	0	1	0	0	0	0
	3	1	1	1	1	1
Wt 1:	-18	-27	-42	-30	-63	-90
Wt 2:	2	-7	-22	-10	-43	-70
Wt 3:	22	13	-2	10	-23	-50
Wt 4:	2	-7	-22	-10	-43	-70
Wt 5:	42	33	18	30	-3	-30
Wt 6:	62	53	38	50	17	-10

Affine B5 - Level 1 - Class 0

Den = 48

	Rp 1	Rp 2
	0 1	1 0
	0	0
	0	0
	0	0
	0	0
Wt 1:	-11	-35
Wt 2:	13	-11

Affine B5 - Level 1 - Class 1

Den = 6

	Rp 1
	0 0
	0
	0
	0
	1
Wt 1:	-1

Affine B5 - Level 2 - Class 0

Den = 132

	Rp 1	Rp 2	Rp 3	Rp 4	Rp 5	Rp 6	Rp 7
	0 0	0 0	0 0	0 0	0 2	1 1	2 0
	0	0	0	1	0	0	0
	0	0	1	0	0	0	0
	0	1	0	0	0	0	0
	2	0	0	0	0	0	0
Wt 1:	-40	-52	-76	-112	-88	-160	-220
Wt 2:	-7	-19	-43	-79	-55	-127	-187
Wt 3:	26	14	-10	-46	-22	-94	-154
Wt 4:	59	47	23	-13	11	-61	-121
Wt 5:	-7	-19	-43	-79	-55	-127	-187
Wt 6:	92	80	56	20	44	-28	-88
Wt 7:	125	113	89	53	77	5	-55

Affine B5 - Level 2 - Class 1

Den = 48

	Rp 1	Rp 2
	0 1	1 0
	0	0
	0	0
	0	0
	1	1
Wt 1:	-5	-29
Wt 2:	19	-5

Affine B5 - Level 3 - Class 0

Den = 96

	Rp 1	Rp 2	Rp 3	Rp 4	Rp 5	Rp 6	Rp 7	Rp 8	Rp 9	Rp10	Rp11	Rp12
	0 1	0 1	0 1	0 1	0 3	1 0	1 0	1 0	1 0	1 2	2 1	3 0
	0	0	0	1	0	0	0	0	1	0	0	0
	0	0	1	0	0	0	0	1	0	0	0	0
	0	1	0	0	0	0	1	0	0	0	0	0
	2	0	0	0	0	2	0	0	0	0	0	0
Wt 1:	-15	-23	-39	-63	-39	-63	-71	-87	-111	-95	-143	-183
Wt 2:	1	-7	-23	-47	-23	-47	-55	-71	-95	-79	-127	-167
Wt 3:	17	9	-7	-31	-7	-31	-39	-55	-79	-63	-111	-151
Wt 4:	33	25	9	-15	9	-15	-23	-39	-63	-47	-95	-135
Wt 5:	-31	-39	-55	-79	-55	-79	-87	-103	-127	-111	-159	-199
Wt 6:	33	25	9	-15	9	-15	-23	-39	-63	-47	-95	-135
Wt 7:	49	41	25	1	25	1	-7	-23	-47	-31	-79	-119
Wt 8:	65	57	41	17	41	17	9	-7	-31	-15	-63	-103
Wt 9:	81	73	57	33	57	33	25	9	-15	1	-47	-87
Wt10:	49	41	25	1	25	1	-7	-23	-47	-31	-79	-119
Wt11:	97	89	73	49	73	49	41	25	1	17	-31	-71
Wt12:	113	105	89	65	89	65	57	41	17	33	-15	-55

Affine B5 - Level 3 - Class 1

Den = 24

	Rp 1	Rp 2	Rp 3	Rp 4	Rp 5	Rp 6	Rp 7
	0 0	0 0	0 0	0 0	0 2	1 1	2 0
	0	0	0	1	0	0	0
	0	0	1	0	0	0	0
	0	1	0	0	0	0	0
	3	1	1	1	1	1	1
Wt 1:	-10	-13	-18	-25	-21	-34	-45
Wt 2:	-2	-5	-10	-17	-13	-26	-37
Wt 3:	6	3	-2	-9	-5	-18	-29
Wt 4:	14	11	6	-1	3	-10	-21
Wt 5:	6	3	-2	-9	-5	-18	-29
Wt 6:	22	19	14	7	11	-2	-13
Wt 7:	30	27	22	15	19	6	-5

Affine B6 - Level 1 - Class 0

Den = 48

	Rp 1	Rp 2
	0 1	1 0
	0	0
	0	0
	0	0
	0	0
	0	0
Wt 1:	-13	-37
Wt 2:	11	-13

Affine B6 - Level 1 - Class 1

Den = 24

	Rp 1
	0 0
	0
	0
	0
	0
	1
Wt 1:	-5

Affine B6 - Level 2 - Class 0

Den = 52

	Rp 1	Rp 2	Rp 3	Rp 4	Rp 5	Rp 6	Rp 7	Rp 8
	0 0	0 0	0 0	0 0	0 0	0 2	1 1	2 0
	0	0	0	0	1	0	0	0
	0	0	0	1	0	0	0	0
	0	0	1	0	0	0	0	0
	0	1	0	0	0	0	0	0
	2	0	0	0	0	0	0	0
Wt 1:	-20	-24	-32	-44	-60	-52	-80	-104
Wt 2:	-7	-11	-19	-31	-47	-39	-67	-91
Wt 3:	6	2	-6	-18	-34	-26	-54	-78
Wt 4:	19	15	7	-5	-21	-13	-41	-65
Wt 5:	32	28	20	8	-8	0	-28	-52
Wt 6:	6	2	-6	-18	-34	-26	-54	-78
Wt 7:	45	41	33	21	5	13	-15	-39
Wt 8:	58	54	46	34	18	26	-2	-26

Affine B6 - Level 2 - Class 1

Den = 8

	Rp 1	Rp 2
	0 1	1 0
	0	0
	0	0
	0	0
	0	0
	1	1
Wt 1:	-1	-5
Wt 2:	3	-1

Affine B6 - Level 3 - Class 0

Den = 168

	Rp 1	Rp 2	Rp 3	Rp 4	Rp 5	Rp 6	Rp 7	Rp 8	Rp 9	Rp10	Rp11	Rp12	Rp13	Rp14
	0 1	0 1	0 1	0 1	0 1	0 3	1 0	1 0	1 0	1 0	1 0	1 2	2 1	3 0
	0	0	0	0	1	0	0	0	0	0	1	0	0	0
	0	0	0	1	0	0	0	0	0	1	0	0	0	0
	0	0	1	0	0	0	0	0	1	0	0	0	0	0
	0	1	0	0	0	0	0	1	0	0	0	0	0	0
	2	0	0	0	0	0	2	0	0	0	0	0	0	0
Wt 1:	-33	-45	-69	-105	-153	-117	-117	-129	-153	-189	-237	-213	-297	-369
Wt 2:	-5	-17	-41	-77	-125	-89	-89	-101	-125	-161	-209	-185	-269	-341
Wt 3:	23	11	-13	-49	-97	-61	-61	-73	-97	-133	-181	-157	-241	-313
Wt 4:	51	39	15	-21	-69	-33	-33	-45	-69	-105	-153	-129	-213	-285
Wt 5:	79	67	43	7	-41	-5	-5	-17	-41	-77	-125	-101	-185	-257
Wt 6:	-33	-45	-69	-105	-153	-117	-117	-129	-153	-189	-237	-213	-297	-369
Wt 7:	51	39	15	-21	-69	-33	-33	-45	-69	-105	-153	-129	-213	-285
Wt 8:	79	67	43	7	-41	-5	-5	-17	-41	-77	-125	-101	-185	-257
Wt 9:	107	95	71	35	-13	23	23	11	-13	-49	-97	-73	-157	-229
Wt10:	135	123	99	63	15	51	51	39	15	-21	-69	-45	-129	-201
Wt11:	163	151	127	91	43	79	79	67	43	7	-41	-17	-101	-173
Wt12:	107	95	71	35	-13	23	23	11	-13	-49	-97	-73	-157	-229
Wt13:	191	179	155	119	71	107	107	95	71	35	-13	11	-73	-145
Wt14:	219	207	183	147	99	135	135	123	99	63	15	39	-45	-117

Affine B6 - Level 3 - Class 1

Den = 84

	Rp 1	Rp 2	Rp 3	Rp 4	Rp 5	Rp 6	Rp 7	Rp 8
	0 0	0 0	0 0	0 0	0 0	0 2	1 1	2 0
	0	0	0	0	1	0	0	0
	0	0	0	1	0	0	0	0
	0	0	1	0	0	0	0	0
	0	1	0	0	0	0	0	0
	3	1	1	1	1	1	1	1
Wt 1:	-45	-54	-69	-90	-117	-105	-150	-189
Wt 2:	-17	-26	-41	-62	-89	-77	-122	-161
Wt 3:	11	2	-13	-34	-61	-49	-94	-133
Wt 4:	39	30	15	-6	-33	-21	-66	-105
Wt 5:	67	58	43	22	-5	7	-38	-77
Wt 6:	39	30	15	-6	-33	-21	-66	-105
Wt 7:	95	86	71	50	23	35	-10	-49
Wt 8:	123	114	99	78	51	63	18	-21

Affine B7 - Level 1 - Class 0

Den = 16

	Rp 1	Rp 2
	0 1	1 0
	0	0
	0	0
	0	0
	0	0
	0	0
	0	0
Wt 1:	-5	-13
Wt 2:	3	-5

Affine B7 - Level 1 - Class 1

Den = 4

	Rp 1
	0 0
	0
	0
	0
	0
	0
	1
Wt 1:	-1

Affine B7 - Level 2 - Class 0

Den = 60

	Rp 1	Rp 2	Rp 3	Rp 4	Rp 5	Rp 6	Rp 7	Rp 8	Rp 9
	0 0	0 0	0 0	0 0	0 0	0 0	0 2	1 1	2 0
	0	0	0	0	0	1	0	0	0
	0	0	0	0	1	0	0	0	0
	0	0	0	1	0	0	0	0	0
	0	0	1	0	0	0	0	0	0
	0	1	0	0	0	0	0	0	0
	2	0	0	0	0	0	0	0	0
Wt 1:	-28	-32	-40	-52	-68	-88	-80	-112	-140
Wt 2:	-13	-17	-25	-37	-53	-73	-65	-97	-125
Wt 3:	2	-2	-10	-22	-38	-58	-50	-82	-110
Wt 4:	17	13	5	-7	-23	-43	-35	-67	-95
Wt 5:	32	28	20	8	-8	-28	-20	-52	-80
Wt 6:	47	43	35	23	7	-13	-5	-37	-65
Wt 7:	17	13	5	-7	-23	-43	-35	-67	-95
Wt 8:	62	58	50	38	22	2	10	-22	-50
Wt 9:	77	73	65	53	37	17	25	-7	-35

Affine B7 - Level 2 - Class 1

```
      |Rp 1|Rp 2|
      |----|----|   Den = 48
      | 0 1| 1 0|
      |  0 |  0 |
      |  0 |  0 |
      |  0 |  0 |
      |  0 |  0 |
      |  0 |  0 |
      |  1 |  1 |
-----------------
Wt 1:   -7  -31
-----------------
Wt 2:   17   -7
-----------------
```

Affine B8 - Level 1 - Class 0

```
      |Rp 1|Rp 2|
      |----|----|   Den = 48
      | 0 1| 1 0|
      |  0 |  0 |
      |  0 |  0 |
      |  0 |  0 |
      |  0 |  0 |
      |  0 |  0 |
      |  0 |  0 |
      |  0 |  0 |
-----------------
Wt 1:  -17  -41
-----------------
Wt 2:    7  -17
-----------------
```

Affine B8 - Level 1 - Class 1

```
      |Rp 1|
      |----|   Den = 24
      | 0 0|
      |  0 |
      |  0 |
      |  0 |
      |  0 |
      |  0 |
      |  0 |
      |  1 |
------------
Wt 1:   -7
------------
```

Affine B8 - Level 2 - Class 0

Den = 204

	Rp 1	Rp 2	Rp 3	Rp 4	Rp 5	Rp 6	Rp 7	Rp 8	Rp 9	Rp10
	0 0	0 0	0 0	0 0	0 0	0 0	0 0	0 2	1 1	2 0
	0	0	0	0	0	0	1	0	0	0
	0	0	0	0	0	1	0	0	0	0
	0	0	0	0	1	0	0	0	0	0
	0	0	0	1	0	0	0	0	0	0
	0	0	1	0	0	0	0	0	0	0
	0	1	0	0	0	0	0	0	0	0
	2	0	0	0	0	0	0	0	0	0
Wt 1:	-112	-124	-148	-184	-232	-292	-364	-340	-448	-544
Wt 2:	-61	-73	-97	-133	-181	-241	-313	-289	-397	-493
Wt 3:	-10	-22	-46	-82	-130	-190	-262	-238	-346	-442
Wt 4:	41	29	5	-31	-79	-139	-211	-187	-295	-391
Wt 5:	92	80	56	20	-28	-88	-160	-136	-244	-340
Wt 6:	143	131	107	71	23	-37	-109	-85	-193	-289
Wt 7:	194	182	158	122	74	14	-58	-34	-142	-238
Wt 8:	92	80	56	20	-28	-88	-160	-136	-244	-340
Wt 9:	245	233	209	173	125	65	-7	17	-91	-187
Wt10:	296	284	260	224	176	116	44	68	-40	-136

Affine B8 - Level 2 - Class 1

Den = 6

	Rp 1	Rp 2
	0 1	1 0
	0	0
	0	0
	0	0
	0	0
	0	0
	0	0
	1	1
Wt 1:	-1	-4
Wt 2:	2	-1

Modular Characteristics for Affine C Algebras

Affine C2 - Level 1 - Class 0

Denominator = 48

	Rp 1	Rp 2
	0	1
	0	0
	1	0
Wt 1:	-5	-29
Wt 2:	19	-5

Affine C2 - Level 1 - Class 1

Den = 24

	Rp 1
	0
	1
	0
Wt 1:	-1

Affine C2 - Level 2 - Class 0

Den = 60

	Rp 1	Rp 2	Rp 3	Rp 4
	0	0	1	2
	0	2	0	0
	2	0	1	0
Wt 1:	-10	-34	-46	-70
Wt 2:	20	-4	-16	-40
Wt 3:	35	11	-1	-25
Wt 4:	50	26	14	-10

Affine C2 - Level 2 - Class 1

Den = 24

	Rp 1	Rp 2
	0	1
	1	1
	1	0
Wt 1:	-1	-13
Wt 2:	11	-1

Affine C2 - Level 3 - Class 0

Den = 24

	Rp 1	Rp 2	Rp 3	Rp 4	Rp 5	Rp 6
	0	0	1	1	2	3
	0	2	0	2	0	0
	3	1	2	0	1	0
Wt 1:	-5	-17	-21	-29	-33	-41
Wt 2:	11	-1	-5	-13	-17	-25
Wt 3:	15	3	-1	-9	-13	-21
Wt 4:	23	11	7	-1	-5	-13
Wt 5:	27	15	11	3	-1	-9
Wt 6:	31	19	15	7	3	-5

Affine C2 - Level 3 - Class 1

Den = 12

	Rp 1	Rp 2	Rp 3	Rp 4
	0	0	1	2
	1	3	1	1
	2	0	1	0
Wt 1:	-1	-5	-8	-13
Wt 2:	3	-1	-4	-9
Wt 3:	7	3	0	-5
Wt 4:	11	7	4	-1

Affine C2 - Level 4 - Class 0

Den = 168

	Rp 1	Rp 2	Rp 3	Rp 4	Rp 5	Rp 6	Rp 7	Rp 8	Rp 9
	0	0	0	1	1	2	2	3	4
	0	2	4	0	2	0	2	0	0
	4	2	0	3	1	2	0	1	0
Wt 1:	-40	-136	-184	-160	-232	-256	-304	-328	-376
Wt 2:	86	-10	-58	-34	-106	-130	-178	-202	-250
Wt 3:	128	32	-16	8	-64	-88	-136	-160	-208
Wt 4:	107	11	-37	-13	-85	-109	-157	-181	-229
Wt 5:	191	95	47	71	-1	-25	-73	-97	-145
Wt 6:	212	116	68	92	20	-4	-52	-76	-124
Wt 7:	254	158	110	134	62	38	-10	-34	-82
Wt 8:	275	179	131	155	83	59	11	-13	-61
Wt 9:	296	200	152	176	104	80	32	8	-40

Affine C2 - Level 4 - Class 1

Den = 336

	Rp 1	Rp 2	Rp 3	Rp 4	Rp 5	Rp 6
	0 1 3	0 3 1	1 1 2	1 3 0	2 1 1	3 1 0
Wt 1:	-41	-185	-257	-353	-425	-545
Wt 2:	127	-17	-89	-185	-257	-377
Wt 3:	211	67	-5	-101	-173	-293
Wt 4:	295	151	79	-17	-89	-209
Wt 5:	379	235	163	67	-5	-125
Wt 6:	463	319	247	151	79	-41

Affine C2 - Level 5 - Class 0

Den = 480

	Rp 1	Rp 2	Rp 3	Rp 4	Rp 5	Rp 6	Rp 7	Rp 8	Rp 9	Rp10	Rp11	Rp12
	0 0 5	0 2 3	0 4 1	1 0 4	1 2 2	1 4 0	2 0 3	2 2 1	3 0 2	3 2 0	4 0 1	5 0 0
Wt 1:	-125	-425	-605	-485	-725	-845	-785	-965	-1025	-1145	-1205	-1325
Wt 2:	259	-41	-221	-101	-341	-461	-401	-581	-641	-761	-821	-941
Wt 3:	451	151	-29	91	-149	-269	-209	-389	-449	-569	-629	-749
Wt 4:	307	7	-173	-53	-293	-413	-353	-533	-593	-713	-773	-893
Wt 5:	595	295	115	235	-5	-125	-65	-245	-305	-425	-485	-605
Wt 6:	691	391	211	331	91	-29	31	-149	-209	-329	-389	-509
Wt 7:	643	343	163	283	43	-77	-17	-197	-257	-377	-437	-557
Wt 8:	835	535	355	475	235	115	175	-5	-65	-185	-245	-365
Wt 9:	883	583	403	523	283	163	223	43	-17	-137	-197	-317
Wt10:	979	679	499	619	379	259	319	139	79	-41	-101	-221
Wt11:	1027	727	547	667	427	307	367	187	127	7	-53	-173
Wt12:	1075	775	595	715	475	355	415	235	175	55	-5	-125

Affine C2 - Level 5 - Class 1

Den = 240

	Rp 1	Rp 2	Rp 3	Rp 4	Rp 5	Rp 6	Rp 7	Rp 8	Rp 9
	0	0	0	1	1	2	2	3	4
	1	3	5	1	3	1	3	1	1
	4	2	0	3	1	2	0	1	0
Wt 1:	-37	-157	-217	-202	-292	-337	-397	-442	-517
Wt 2:	107	-13	-73	-58	-148	-193	-253	-298	-373
Wt 3:	155	35	-25	-10	-100	-145	-205	-250	-325
Wt 4:	155	35	-25	-10	-100	-145	-205	-250	-325
Wt 5:	251	131	71	86	-4	-49	-109	-154	-229
Wt 6:	299	179	119	134	44	-1	-61	-106	-181
Wt 7:	347	227	167	182	92	47	-13	-58	-133
Wt 8:	395	275	215	230	140	95	35	-10	-85
Wt 9:	443	323	263	278	188	143	83	38	-37

Affine C3 - Level 1 - Class 0

Den = 40

	Rp 1	Rp 2
	0	1
	0	0
	1	0
	0	0
Wt 1:	-3	-27
Wt 2:	17	-7

Affine C3 - Level 2 - Class 0

Den = 24

	Rp 1	Rp 2	Rp 3	Rp 4	Rp 5	Rp 6
	0	0	0	0	1	2
	0	0	1	2	0	0
	0	2	0	0	1	0
	2	0	1	0	0	0
Wt 1:	-7	-15	-19	-27	-31	-43
Wt 2:	5	-3	-7	-15	-19	-31
Wt 3:	11	3	-1	-9	-13	-25
Wt 4:	17	9	5	-3	-7	-19
Wt 5:	23	15	11	3	-1	-13
Wt 6:	29	21	17	9	5	-7

Affine C3 - Level 2 - Class 1

Den = 24

	Rp 1	Rp 2	Rp 3	Rp 4
	0	0	1	1
	0	1	0	1
	1	1	0	0
	1	0	1	0
Wt 1:	-3	-13	-19	-27
Wt 2:	9	-1	-7	-15
Wt 3:	15	5	-1	-9
Wt 4:	21	11	5	-3

Affine C3 - Level 3 - Class 0

Den = 168

	Rp 1	Rp 2	Rp 3	Rp 4	Rp 5	Rp 6	Rp 7	Rp 8	Rp 9	Rp10
	0	0	0	0	1	1	1	1	2	3
	0	0	1	2	0	0	1	2	0	0
	1	3	1	1	0	2	0	0	1	0
	2	0	1	0	2	0	1	0	0	0
Wt 1:	-35	-83	-119	-179	-155	-203	-227	-275	-299	-371
Wt 2:	21	-27	-63	-123	-99	-147	-171	-219	-243	-315
Wt 3:	77	29	-7	-67	-43	-91	-115	-163	-187	-259
Wt 4:	133	85	49	-11	13	-35	-59	-107	-131	-203
Wt 5:	105	57	21	-39	-15	-63	-87	-135	-159	-231
Wt 6:	161	113	77	17	41	-7	-31	-79	-103	-175
Wt 7:	189	141	105	45	69	21	-3	-51	-75	-147
Wt 8:	217	169	133	73	97	49	25	-23	-47	-119
Wt 9:	245	197	161	101	125	77	53	5	-19	-91
Wt10:	273	225	189	129	153	105	81	33	9	-63

Affine C4 - Level 1 - Class 0

Den = 12

	Rp 1	Rp 2	Rp 3
	0	0	1
	0	0	0
	0	1	0
	0	0	0
	1	0	0
Wt 1:	-3	-7	-15
Wt 2:	3	-1	-9
Wt 3:	9	5	-3

Affine C4 - Level 1 - Class 1

Den = 8

	Rp 1	Rp 2
	0	0
	0	1
	0	0
	1	0
	0	0
Wt 1:	-1	-5
Wt 2:	3	-1

Affine C4 - Level 2 - Class 0

Den = 28

	Rp 1	Rp 2	Rp 3	Rp 4	Rp 5	Rp 6	Rp 7	Rp 8	Rp 9
	0	0	0	0	0	0	1	1	2
	0	0	0	0	1	2	0	0	0
	0	0	1	2	0	0	0	1	0
	0	2	0	0	1	0	0	0	0
	2	0	1	0	0	0	1	0	0
Wt 1:	-12	-20	-24	-32	-36	-48	-44	-52	-68
Wt 2:	2	-6	-10	-18	-22	-34	-30	-38	-54
Wt 3:	9	1	-3	-11	-15	-27	-23	-31	-47
Wt 4:	16	8	4	-4	-8	-20	-16	-24	-40
Wt 5:	23	15	11	3	-1	-13	-9	-17	-33
Wt 6:	30	22	18	10	6	-6	-2	-10	-26
Wt 7:	30	22	18	10	6	-6	-2	-10	-26
Wt 8:	37	29	25	17	13	1	5	-3	-19
Wt 9:	44	36	32	24	20	8	12	4	-12

Affine C4 - Level 2 - Class 1

Den = 56

	Rp 1	Rp 2	Rp 3	Rp 4	Rp 5	Rp 6
	0	0	0	0	1	1
	0	0	1	1	0	1
	0	1	0	1	0	0
	1	1	0	0	1	0
	1	0	1	0	0	0
Wt 1:	-13	-33	-45	-61	-73	-97
Wt 2:	15	-5	-17	-33	-45	-69
Wt 3:	29	9	-3	-19	-31	-55
Wt 4:	43	23	11	-5	-17	-41
Wt 5:	57	37	25	9	-3	-27
Wt 6:	71	51	39	23	11	-13

Affine C5 - Level 1 - Class 0

Den = 168

	Rp 1	Rp 2	Rp 3
	0	0	1
	0	0	0
	0	1	0
	0	0	0
	1	0	0
	0	0	0
Wt 1:	-31	-103	-223
Wt 2:	53	-19	-139
Wt 3:	137	65	-55

Affine C5 - Level 2 - Class 0

Den = 96

	Rp 1	Rp 2	Rp 3	Rp 4	Rp 5	Rp 6	Rp 7	Rp 8	Rp 9	Rp10	Rp11	Rp12
	0	0	0	0	0	0	0	0	0	1	1	2
	0	0	0	0	0	0	1	1	2	0	0	0
	0	0	0	0	1	2	0	0	0	0	1	0
	0	0	1	2	0	0	0	1	0	0	0	0
	0	2	0	0	1	0	0	0	0	1	0	0
	2	0	1	0	0	0	1	0	0	0	0	0
Wt 1:	-55	-79	-91	-115	-127	-163	-151	-175	-223	-199	-235	-295
Wt 2:	-7	-31	-43	-67	-79	-115	-103	-127	-175	-151	-187	-247
Wt 3:	17	-7	-19	-43	-55	-91	-79	-103	-151	-127	-163	-223
Wt 4:	41	17	5	-19	-31	-67	-55	-79	-127	-103	-139	-199
Wt 5:	65	41	29	5	-7	-43	-31	-55	-103	-79	-115	-175
Wt 6:	89	65	53	29	17	-19	-7	-31	-79	-55	-91	-151
Wt 7:	89	65	53	29	17	-19	-7	-31	-79	-55	-91	-151
Wt 8:	113	89	77	53	41	5	17	-7	-55	-31	-67	-127
Wt 9:	137	113	101	77	65	29	41	17	-31	-7	-43	-103
Wt10:	137	113	101	77	65	29	41	17	-31	-7	-43	-103
Wt11:	161	137	125	101	89	53	65	41	-7	17	-19	-79
Wt12:	185	161	149	125	113	77	89	65	17	41	5	-55

Affine C5 - Level 2 - Class 1

Den = 48

	Rp 1	Rp 2	Rp 3	Rp 4	Rp 5	Rp 6	Rp 7	Rp 8	Rp 9
	0	0	0	0	0	0	1	1	1
	0	0	0	0	1	1	0	0	1
	0	0	1	1	0	1	0	0	0
	0	1	0	1	0	0	0	1	0
	1	1	0	0	1	0	0	0	0
	1	0	1	0	0	0	1	0	0
Wt 1:	-17	-32	-41	-53	-62	-80	-77	-89	-113
Wt 2:	7	-8	-17	-29	-38	-56	-53	-65	-89
Wt 3:	19	4	-5	-17	-26	-44	-41	-53	-77
Wt 4:	31	16	7	-5	-14	-32	-29	-41	-65
Wt 5:	43	28	19	7	-2	-20	-17	-29	-53
Wt 6:	55	40	31	19	10	-8	-5	-17	-41
Wt 7:	55	40	31	19	10	-8	-5	-17	-41
Wt 8:	67	52	43	31	22	4	7	-5	-29
Wt 9:	79	64	55	43	34	16	19	7	-17

Affine C6 - Level 1 - Class 0

Den = 32

	Rp 1	Rp 2	Rp 3	Rp 4
	0	0	0	1
	0	0	0	0
	0	0	1	0
	0	0	0	0
	0	1	0	0
	0	0	0	0
	1	0	0	0
Wt 1:	-13	-21	-37	-61
Wt 2:	3	-5	-21	-45
Wt 3:	19	11	-5	-29
Wt 4:	35	27	11	-13

Affine C6 - Level 1 - Class 1

Den = 8

Rp 1	Rp 2	Rp 3
0	0	0
0	0	1
0	0	0
0	1	0
0	0	0
1	0	0
0	0	0

Wt 1:	-2	-5	-10
Wt 2:	2	-1	-6
Wt 3:	6	3	-2

Affine C7 - Level 1 - Class 0

Den = 72

Rp 1	Rp 2	Rp 3	Rp 4
0	0	0	1
0	0	0	0
0	0	1	0
0	0	0	0
0	1	0	0
0	0	0	0
1	0	0	0
0	0	0	0

Wt 1:	-23	-47	-87	-143
Wt 2:	13	-11	-51	-107
Wt 3:	49	25	-15	-71
Wt 4:	85	61	21	-35

Affine C8 - Level 1 - Class 0

Den = 30

Rp 1	Rp 2	Rp 3	Rp 4	Rp 5
0	0	0	0	1
0	0	0	0	0
0	0	0	1	0
0	0	0	0	0
0	0	1	0	0
0	0	0	0	0
0	1	0	0	0
0	0	0	0	0
1	0	0	0	0

Wt 1:	-17	-23	-35	-53	-77
Wt 2:	-2	-8	-20	-38	-62
Wt 3:	13	7	-5	-23	-47
Wt 4:	28	22	10	-8	-32
Wt 5:	43	37	25	7	-17

Affine C8 - Level 1 - Class 1

Den = 120

	Rp 1	Rp 2	Rp 3	Rp 4
	0	0	0	0
	0	0	0	1
	0	0	0	0
	0	0	1	0
	0	0	0	0
	0	1	0	0
	0	0	0	0
	1	0	0	0
	0	0	0	0
Wt 1:	-47	-83	-143	-227
Wt 2:	13	-23	-83	-167
Wt 3:	73	37	-23	-107
Wt 4:	133	97	37	-47

Modular Characteristics for Affine D algebras

Affine D4 - Level 1 - Class (0,0)

```
        |Rp 1|
        |----|   Denominator = 6
        | 1 0|
        |  0 |
        | 0 0|
-----------
Wt 1:   -1
-----------
```

Affine D4 - Level 2 - Class (0,0)

```
        |Rp 1|Rp 2|Rp 3|Rp 4|Rp 5|
        |----|----|----|----|----|   Den = 24
        | 0 0| 0 0| 0 0| 0 2| 2 0|
        |  0 |  0 |  1 |  0 |  0 |
        | 0 2| 2 0| 0 0| 0 0| 0 0|
--------------------------------
Wt 1:   -7   -7  -13   -7  -31
--------------------------------
Wt 2:   -7   -7  -13   -7  -31
--------------------------------
Wt 3:    5    5   -1    5  -19
--------------------------------
Wt 4:   -7   -7  -13   -7  -31
--------------------------------
Wt 5:   17   17   11   17   -7
--------------------------------
```

Affine D4 - Level 2 - Class (0,1)

```
        |Rp 1|Rp 2|
        |----|----|   Den = 48
        | 0 1| 1 0|
        |  0 |  0 |
        | 1 0| 0 1|
----------------
Wt 1:   -5  -29
----------------
Wt 2:   19   -5
----------------
```

Affine D4 - Level 3 - Class (0,0)

Den = 18

	Rp 1	Rp 2	Rp 3	Rp 4	Rp 5	Rp 6
	0 1 0 1 1	1 0 0 0 2	1 0 0 2 0	1 0 1 0 0	1 2 0 0 0	3 0 0 0 0
Wt 1:	-1	-9	-9	-13	-9	-25
Wt 2:	5	-3	-3	-7	-3	-19
Wt 3:	5	-3	-3	-7	-3	-19
Wt 4:	11	3	3	-1	3	-13
Wt 5:	5	-3	-3	-7	-3	-19
Wt 6:	17	9	9	5	9	-7

Affine D4 - Level 4 - Class (0,0)

Den = 60

	Rp 1	Rp 2	Rp 3	Rp 4	Rp 5	Rp 6	Rp 7	Rp 8	Rp 9	Rp10	Rp11	Rp12	Rp13	Rp14	Rp15	Rp16
	0 0 0 0 4	0 0 0 2 2	0 0 0 4 0	0 0 1 0 2	0 0 1 2 0	0 0 2 0 0	0 2 0 0 2	0 2 0 2 0	0 2 1 0 0	0 4 0 0 0	1 1 0 1 1	2 0 0 0 2	2 0 0 2 0	2 0 1 0 0	2 2 0 0 0	4 0 0 0 0
Wt 1:	-28	-40	-28	-52	-52	-64	-40	-40	-52	-28	-76	-100	-100	-112	-100	-148
Wt 2:	2	-10	2	-22	-22	-34	-10	-10	-22	2	-46	-70	-70	-82	-70	-118
Wt 3:	-28	-40	-28	-52	-52	-64	-40	-40	-52	-28	-76	-100	-100	-112	-100	-148
Wt 4:	17	5	17	-7	-7	-19	5	5	-7	17	-31	-55	-55	-67	-55	-103
Wt 5:	17	5	17	-7	-7	-19	5	5	-7	17	-31	-55	-55	-67	-55	-103
Wt 6:	32	20	32	8	8	-4	20	20	8	32	-16	-40	-40	-52	-40	-88
Wt 7:	2	-10	2	-22	-22	-34	-10	-10	-22	2	-46	-70	-70	-82	-70	-118
Wt 8:	2	-10	2	-22	-22	-34	-10	-10	-22	2	-46	-70	-70	-82	-70	-118
Wt 9:	17	5	17	-7	-7	-19	5	5	-7	17	-31	-55	-55	-67	-55	-103
Wt10:	-28	-40	-28	-52	-52	-64	-40	-40	-52	-28	-76	-100	-100	-112	-100	-148
Wt11:	47	35	47	23	23	11	35	35	23	47	-1	-25	-25	-37	-25	-73
Wt12:	62	50	62	38	38	26	50	50	38	62	14	-10	-10	-22	-10	-58
Wt13:	62	50	62	38	38	26	50	50	38	62	14	-10	-10	-22	-10	-58
Wt14:	77	65	77	53	53	41	65	65	53	77	29	5	5	-7	5	-43
Wt15:	62	50	62	38	38	26	50	50	38	62	14	-10	-10	-22	-10	-58
Wt16:	92	80	92	68	68	56	80	80	68	92	44	20	20	8	20	-28

Affine D4 - Level 4 - Class (0,1)

Den = 120

	Rp 1 0 1 0 1 2	Rp 2 0 1 0 3 0	Rp 3 0 1 1 1 0	Rp 4 0 3 0 1 0	Rp 5 1 0 0 0 3	Rp 6 1 0 0 2 1	Rp 7 1 0 1 0 1	Rp 8 1 2 0 0 1	Rp 9 2 1 0 1 0	Rp10 3 0 0 0 1
Wt 1:	-11	1	-35	1	-59	-71	-95	-71	-131	-179
Wt 2:	-41	-29	-65	-29	-89	-101	-125	-101	-161	-209
Wt 3:	19	31	-5	31	-29	-41	-65	-41	-101	-149
Wt 4:	-41	-29	-65	-29	-89	-101	-125	-101	-161	-209
Wt 5:	19	31	-5	31	-29	-41	-65	-41	-101	-149
Wt 6:	49	61	25	61	1	-11	-35	-11	-71	-119
Wt 7:	79	91	55	91	31	19	-5	19	-41	-89
Wt 8:	49	61	25	61	1	-11	-35	-11	-71	-119
Wt 9:	109	121	85	121	61	49	25	49	-11	-59
Wt10:	139	151	115	151	91	79	55	79	19	-29

Affine D5 - Level 1 - Class 0

Den = 24

	Rp 1 1 0 0 0 0 0
Wt 1:	-5

Affine D5 - Level 2 - Class 0

Den = 40

	Rp 1 0 0 0 0 1 1	Rp 2 0 0 1 0 0 0	Rp 3 0 2 0 0 0 0	Rp 4 2 0 0 0 0 0
Wt 1:	-7	-23	-15	-55
Wt 2:	13	-3	5	-35
Wt 3:	-7	-23	-15	-55
Wt 4:	33	17	25	-15

Affine D5 - Level 2 - Class 1

Den = 8

	Rp 1	Rp 2
	0 1	1 0
	0	0
	0	0
	1 0	0 1
Wt 1:	-1	-5
Wt 2:	3	-1

Affine D5 - Level 3 - Class 0

Den = 264

	Rp 1	Rp 2	Rp 3	Rp 4	Rp 5	Rp 6	Rp 7
	0 1	0 1	0 1	1 0	1 0	1 2	3 0
	0	0	0	0	1	0	0
	0	0	1	0	0	0	0
	0 2	2 0	0 0	1 1	0 0	0 0	0 0
Wt 1:	-55	-55	-103	-199	-295	-247	-487
Wt 2:	-55	-55	-103	-199	-295	-247	-487
Wt 3:	33	33	-15	-111	-207	-159	-399
Wt 4:	121	121	73	-23	-119	-71	-311
Wt 5:	209	209	161	65	-31	17	-223
Wt 6:	121	121	73	-23	-119	-71	-311
Wt 7:	297	297	249	153	57	105	-135

Affine D6 - Level 1 - Class (0,0)

Den = 4

	Rp 1
	1 0
	0
	0
	0
	0 0
Wt 1:	-1

Affine D6 - Level 2 - Class (0,0)

Den = 24

	Rp 1	Rp 2	Rp 3	Rp 4	Rp 5	Rp 6
	0 0	0 0	0 0	0 0	0 2	2 0
	0	0	0	1	0	0
	0	0	0	0	0	0
	0	0	1	0	0	0
	0 2	2 0	0 0	0 0	0 0	0 0
Wt 1:	-11	-11	-15	-27	-23	-47
Wt 2:	-11	-11	-15	-27	-23	-47
Wt 3:	1	1	-3	-15	-11	-35
Wt 4:	13	13	9	-3	1	-23
Wt 5:	1	1	-3	-15	-11	-35
Wt 6:	25	25	21	9	13	-11

Affine D6 - Level 2 - Class (1,0)

Den = 48

	Rp 1	Rp 2
	0 1	1 0
	0	0
	0	0
	0	0
	0 1	1 0
Wt 1:	-7	-31
Wt 2:	17	-7

Affine D6 - Level 2 - Class (1,1)

Den = 12

	Rp 1	Rp 2	Rp 3
	0 0	0 0	1 1
	0	0	0
	0	1	0
	0	0	0
	1 1	0 0	0 0
Wt 1:	-3	-7	-15
Wt 2:	3	-1	-9
Wt 3:	9	5	-3

Affine D6 - Level 3 - Class (0,0)

	Rp 1	Rp 2	Rp 3	Rp 4	Rp 5	Rp 6	Rp 7	Rp 8
	0 1	0 1	1 0	1 0	1 0	1 0	1 2	3 0
	0	0	0	0	0	1	0	0
	0	1	0	0	0	0	0	0
	0	0	0	0	1	0	0	0
	1 1	0 0	0 2	2 0	0 0	0 0	0 0	0 0
Wt 1:	-19	-67	-91	-91	-115	-187	-163	-307
Wt 2:	33	-15	-39	-39	-63	-135	-111	-255
Wt 3:	33	-15	-39	-39	-63	-135	-111	-255
Wt 4:	33	-15	-39	-39	-63	-135	-111	-255
Wt 5:	85	37	13	13	-11	-83	-59	-203
Wt 6:	137	89	65	65	41	-31	-7	-151
Wt 7:	85	37	13	13	-11	-83	-59	-203
Wt 8:	189	141	117	117	93	21	45	-99

Den = 156

Affine D7 - Level 1 - Class 0

	Rp 1
	1 0
	0
	0
	0
	0
	0 0
Wt 1:	-7

Den = 24

Affine D7 - Level 2 - Class 0

	Rp 1	Rp 2	Rp 3	Rp 4	Rp 5
	0 0	0 0	0 0	0 2	2 0
	0	0	1	0	0
	0	0	0	0	0
	0	1	0	0	0
	0	0	0	0	0
	1 1	0 0	0 0	0 0	0 0
Wt 1:	-55	-103	-199	-175	-343
Wt 2:	29	-19	-115	-91	-259
Wt 3:	113	65	-31	-7	-175
Wt 4:	29	-19	-115	-91	-259
Wt 5:	197	149	53	77	-91

Den = 168

Affine D7 - Level 2 - Class 3

Den = 6

	Rp 1	Rp 2
	0 1	1 0
	0	0
	0	0
	0	0
	0	0
	0 1	1 0
Wt 1:	-1	-4
Wt 2:	2	-1

Affine D7 - Level 3 - Class 0

Den = 120

	Rp 1	Rp 2	Rp 3	Rp 4	Rp 5	Rp 6	Rp 7	Rp 8	Rp 9
	0 1	0 1	0 1	0 1	1 0	1 0	1 0	1 2	3 0
	0	0	0	0	0	0	1	0	0
	0	0	0	1	0	0	0	0	0
	0	0	0	0	0	1	0	0	0
	0	0	1	0	0	0	0	0	0
	0 2	2 0	0 0	0 0	1 1	0 0	0 0	0 0	0 0
Wt 1:	-35	-35	-51	-99	-99	-131	-195	-179	-291
Wt 2:	-35	-35	-51	-99	-99	-131	-195	-179	-291
Wt 3:	5	5	-11	-59	-59	-91	-155	-139	-251
Wt 4:	45	45	29	-19	-19	-51	-115	-99	-211
Wt 5:	45	45	29	-19	-19	-51	-115	-99	-211
Wt 6:	85	85	69	21	21	-11	-75	-59	-171
Wt 7:	125	125	109	61	61	29	-35	-19	-131
Wt 8:	85	85	69	21	21	-11	-75	-59	-171
Wt 9:	165	165	149	101	101	69	5	21	-91

Affine D8 - Level 1 - Class (0,0)

Den = 3

	Rp 1
	1 0
	0
	0
	0
	0
	0
	0 0
Wt 1:	-1

Affine D8 - Level 2 - Class (0,0)

Den = 8

	Rp 1	Rp 2	Rp 3	Rp 4	Rp 5	Rp 6	Rp 7
	0 0	0 0	0 0	0 0	0 0	0 2	2 0
	0	0	0	0	1	0	0
	0	0	0	0	0	0	0
	0	0	0	1	0	0	0
	0	0	0	0	0	0	0
	0	0	1	0	0	0	0
	0 2	2 0	0 0	0 0	0 0	0 0	0 0
Wt 1:	-5	-5	-6	-9	-14	-13	-21
Wt 2:	-5	-5	-6	-9	-14	-13	-21
Wt 3:	-1	-1	-2	-5	-10	-9	-17
Wt 4:	3	3	2	-1	-6	-5	-13
Wt 5:	7	7	6	3	-2	-1	-9
Wt 6:	3	3	2	-1	-6	-5	-13
Wt 7:	11	11	10	7	2	3	-5

Affine D8 - Level 2 - Class (0,1)

Den = 16

	Rp 1	Rp 2
	0 1	1 0
	0	0
	0	0
	0	0
	0	0
	0	0
	1 0	0 1
Wt 1:	-3	-11
Wt 2:	5	-3

Affine D8 - Level 2 - Class (1,1)

Den = 32

	Rp 1	Rp 2	Rp 3	Rp 4
	0 0	0 0	0 0	1 1
	0	0	0	0
	0	0	1	0
	0	0	0	0
	0	1	0	0
	0	0	0	0
	1 1	0 0	0 0	0 0
Wt 1:	-13	-21	-37	-61
Wt 2:	3	-5	-21	-45
Wt 3:	19	11	-5	-29
Wt 4:	35	27	11	-13

Affine D8 - Level 3 - Class (0,0)

Den = 51

	Rp 1	Rp 2	Rp 3	Rp 4	Rp 5	Rp 6	Rp 7	Rp 8	Rp 9	Rp10
	0 1	0 1	0 1	1 0	1 0	1 0	1 0	1 0	1 2	3 0
	0	0	0	0	0	0	0	1	0	0
	0	0	1	0	0	0	0	0	0	0
	0	0	0	0	0	0	1	0	0	0
	0	1	0	0	0	0	0	0	0	0
	0	0	0	0	0	1	0	0	0	0
	1 1	0 0	0 0	0 2	2 0	0 0	0 0	0 0	0 0	0 0
Wt 1:	-10	-22	-46	-34	-34	-40	-58	-88	-82	-130
Wt 2:	7	-5	-29	-17	-17	-23	-41	-71	-65	-113
Wt 3:	24	12	-12	0	0	-6	-24	-54	-48	-96
Wt 4:	7	-5	-29	-17	-17	-23	-41	-71	-65	-113
Wt 5:	7	-5	-29	-17	-17	-23	-41	-71	-65	-113
Wt 6:	24	12	-12	0	0	-6	-24	-54	-48	-96
Wt 7:	41	29	5	17	17	11	-7	-37	-31	-79
Wt 8:	58	46	22	34	34	28	10	-20	-14	-62
Wt 9:	41	29	5	17	17	11	-7	-37	-31	-79
Wt10:	75	63	39	51	51	45	27	-3	3	-45

Modular Characteristics for Affine Exceptional Algebras

Affine E6 - Level 1 - Class 0

```
       |Rp 1|
       |----|   Denominator = 4
       |   0|
       |   0|
       | 100|
       |   0|
       |   0|
-----------
Wt 1:   -1
-----------
```

Affine E6 - Level 2 - Class 0

```
       |Rp 1|Rp 2|Rp 3|
       |----|----|----|   Den = 28
       |   0|   1|   0|
       |   0|   0|   0|
       | 010| 000| 200|
       |   0|   0|   0|
       |   0|   1|   0|
---------------------
Wt 1:   -3    9  -27
---------------------
Wt 2: -17   -5  -41
---------------------
Wt 3:  11   23  -13
---------------------
```

Affine E6 - Level 2 - Class 1

```
       |Rp 1|Rp 2|Rp 3|
       |----|----|----|   Den = 28
       |   0|   0|   1|
       |   0|   0|   0|
       | 000| 000| 100|
       |   0|   1|   0|
       |   2|   0|   0|
---------------------
Wt 1: -13  -17  -33
---------------------
Wt 2:   1   -3  -19
---------------------
Wt 3:  15   11   -5
---------------------
```

Affine E6 - Level 3 - Class 0

	Rp 1	Rp 2	Rp 3	Rp 4	Rp 5	Rp 6	Rp 7	Rp 8
	0	0	0	1	3	0	1	0
	0	0	0	1	0	0	0	0
	000	000	001	000	000	110	100	300
	0	1	0	0	0	0	0	0
	3	1	0	0	0	0	1	0
Wt 1:	-39	-51	-63	-51	-39	-111	-87	-159
Wt 2:	1	-11	-23	-11	1	-71	-47	-119
Wt 3:	21	9	-3	9	21	-51	-27	-99
Wt 4:	1	-11	-23	-11	1	-71	-47	-119
Wt 5:	-39	-51	-63	-51	-39	-111	-87	-159
Wt 6:	61	49	37	49	61	-11	13	-59
Wt 7:	41	29	17	29	41	-31	-7	-79
Wt 8:	81	69	57	69	81	9	33	-39

Den = 60

Affine E6 - Level 3 - Class 1

	Rp 1	Rp 2	Rp 3	Rp 4	Rp 5	Rp 6
	0	1	2	0	0	1
	1	0	0	0	0	0
	000	010	000	100	100	200
	0	0	0	0	1	0
	1	0	1	2	0	0
Wt 1:	-17	-77	7	-113	-137	-233
Wt 2:	43	-17	67	-53	-77	-173
Wt 3:	-77	-137	-53	-173	-197	-293
Wt 4:	43	-17	67	-53	-77	-173
Wt 5:	103	43	127	7	-17	-113
Wt 6:	163	103	187	67	43	-53

Den = 180

Affine E6 - Level 4 - Class 0 **Den = 16**

	Rp 1	Rp 2	Rp 3	Rp 4	Rp 5	Rp 6	Rp 7	Rp 8	Rp 9	Rp10	Rp11	Rp12	Rp13	Rp14
	0	0	0	1	2	2	0	0	0	1	3	0	1	0
	1	1	0	0	0	0	0	0	0	1	0	0	0	0
	000	000	020	010	000	000	100	100	101	100	100	210	200	400
	0	1	0	0	0	1	0	1	0	0	0	0	0	0
	2	0	0	1	2	0	3	1	0	0	0	0	1	0
Wt 1:	-3	-5	-15	-9	-1	-3	-11	-14	-17	-14	-11	-29	-23	-41
Wt 2:	1	-1	-11	-5	3	1	-7	-10	-13	-10	-7	-25	-19	-37
Wt 3:	9	7	-3	3	11	9	1	-2	-5	-2	1	-17	-11	-29
Wt 4:	5	3	-7	-1	7	5	-3	-6	-9	-6	-3	-21	-15	-33
Wt 5:	-7	-9	-19	-13	-5	-7	-15	-18	-21	-18	-15	-33	-27	-45
Wt 6:	-3	-5	-15	-9	-1	-3	-11	-14	-17	-14	-11	-29	-23	-41
Wt 7:	1	-1	-11	-5	3	1	-7	-10	-13	-10	-7	-25	-19	-37
Wt 8:	9	7	-3	3	11	9	1	-2	-5	-2	1	-17	-11	-29
Wt 9:	13	11	1	7	15	13	5	2	-1	2	5	-13	-7	-25
Wt10:	9	7	-3	3	11	9	1	-2	-5	-2	1	-17	-11	-29
Wt11:	1	-1	-11	-5	3	1	-7	-10	-13	-10	-7	-25	-19	-37
Wt12:	21	19	9	15	23	21	13	10	7	10	13	-5	1	-17
Wt13:	17	15	5	11	19	17	9	6	3	6	9	-9	-3	-21
Wt14:	25	23	13	19	27	25	17	14	11	14	17	-1	5	-13

Affine E6 - Level 4 - Class 1

Den = 16

	Rp 1	Rp 2	Rp 3	Rp 4	Rp 5	Rp 6	Rp 7	Rp 8	Rp 9	Rp10	Rp11	Rp12	Rp13	Rp14
	0 2 000 0 0	0 0 010 0 2	0 0 010 1 0	1 0 000 0 3	1 0 000 1 1	1 0 001 0 0	2 1 000 0 0	4 0 000 0 0	0 1 100 0 1	1 0 110 0 0	2 0 100 0 1	0 0 200 0 2	0 0 200 1 0	1 0 300 0 0
Wt 1:	-3	-7	-9	1	-2	-5	-1	3	-13	-18	-11	-21	-23	-31
Wt 2:	1	-3	-5	5	2	-1	3	7	-9	-14	-7	-17	-19	-27
Wt 3:	5	1	-1	9	6	3	7	11	-5	-10	-3	-13	-15	-23
Wt 4:	-11	-15	-17	-7	-10	-13	-9	-5	-21	-26	-19	-29	-31	-39
Wt 5:	-3	-7	-9	1	-2	-5	-1	3	-13	-18	-11	-21	-23	-31
Wt 6:	1	-3	-5	5	2	-1	3	7	-9	-14	-7	-17	-19	-27
Wt 7:	-7	-11	-13	-3	-6	-9	-5	-1	-17	-22	-15	-25	-27	-35
Wt 8:	-19	-23	-25	-15	-18	-21	-17	-13	-29	-34	-27	-37	-39	-47
Wt 9:	9	5	3	13	10	7	11	15	-1	-6	1	-9	-11	-19
Wt10:	13	9	7	17	14	11	15	19	3	-2	5	-5	-7	-15
Wt11:	5	1	-1	9	6	3	7	11	-5	-10	-3	-13	-15	-23
Wt12:	13	9	7	17	14	11	15	19	3	-2	5	-5	-7	-15
Wt13:	17	13	11	21	18	15	19	23	7	2	9	-1	-3	-11
Wt14:	21	17	15	25	22	19	23	27	11	6	13	3	1	-7

Affine E7 - Level 1 - Class 0

Den = 24

	Rp 1
	1 0 0 00 0 0 0
Wt 1:	-7

Affine E7 - Level 2 - Class 0

Den = 240

	Rp 1	Rp 2	Rp 3	Rp 4
	0	0	0	2
	0	0	1	0
	0	0	0	0
	00	00	00	00
	0	0	0	0
	0	1	0	0
	2	0	0	0
Wt 1:	-133	-157	-277	-493
Wt 2:	-13	-37	-157	-373
Wt 3:	107	83	-37	-253
Wt 4:	227	203	83	-133

Affine E7 - Level 2 - Class 1

Den = 60

	Rp 1	Rp 2
	0	1
	0	0
	0	0
	10	00
	0	0
	0	0
	0	1
Wt 1:	-7	-43
Wt 2:	23	-13

Affine E7 - Level 3 - Class 0

Den = 168

	Rp 1	Rp 2	Rp 3	Rp 4	Rp 5	Rp 6
	0	0	1	1	1	3
	0	0	0	0	1	0
	1	0	0	0	0	0
	00	10	00	00	00	00
	0	0	0	0	0	0
	0	0	0	1	0	0
	0	1	2	0	0	0
Wt 1:	-13	35	-61	-77	-157	-301
Wt 2:	-69	-21	-117	-133	-213	-357
Wt 3:	-13	35	-61	-77	-157	-301
Wt 4:	43	91	-5	-21	-101	-245
Wt 5:	99	147	51	35	-45	-189
Wt 6:	155	203	107	91	11	-133

Affine E7 - Level 3 - Class 1

	Rp 1	Rp 2	Rp 3	Rp 4	Rp 5	Rp 6
	0	0	0	0	1	2
	0	0	0	1	0	0
	0	0	0	0	0	0
	00	00	00	00	10	00
	0	0	1	0	0	0
	0	1	0	0	0	0
	3	1	0	1	0	1
Wt 1:	-133	-157	-181	-245	-301	-397
Wt 2:	-21	-45	-69	-133	-189	-285
Wt 3:	35	11	-13	-77	-133	-229
Wt 4:	91	67	43	-21	-77	-173
Wt 5:	147	123	99	35	-21	-117
Wt 6:	203	179	155	91	35	-61

Den = 168

Affine E7 - Level 4 - Class 0

	Rp 1	Rp 2	Rp 3	Rp 4	Rp 5	Rp 6	Rp 7	Rp 8	Rp 9	Rp10	Rp11	Rp12	Rp13	Rp14	Rp15
	0	0	0	0	0	0	0	0	0	1	1	2	2	2	4
	0	0	0	0	0	0	1	1	2	0	0	0	0	1	0
	0	0	0	0	0	0	0	0	0	1	0	0	0	0	0
	00	00	00	00	01	20	00	00	00	00	10	00	00	00	00
	0	0	0	1	0	0	0	0	0	0	0	0	0	0	0
	0	1	2	0	0	0	0	1	0	0	0	0	1	0	0
	4	2	0	1	0	0	2	0	0	0	1	2	0	0	0
Wt 1:	-133	-157	-169	-181	-205	-193	-229	-241	-301	-313	-277	-349	-361	-421	-529
Wt 2:	-34	-58	-70	-82	-106	-94	-130	-142	-202	-214	-178	-250	-262	-322	-430
Wt 3:	-1	-25	-37	-49	-73	-61	-97	-109	-169	-181	-145	-217	-229	-289	-397
Wt 4:	32	8	-4	-16	-40	-28	-64	-76	-136	-148	-112	-184	-196	-256	-364
Wt 5:	65	41	29	17	-7	5	-31	-43	-103	-115	-79	-151	-163	-223	-331
Wt 6:	32	8	-4	-16	-40	-28	-64	-76	-136	-148	-112	-184	-196	-256	-364
Wt 7:	65	41	29	17	-7	5	-31	-43	-103	-115	-79	-151	-163	-223	-331
Wt 8:	98	74	62	50	26	38	2	-10	-70	-82	-46	-118	-130	-190	-298
Wt 9:	131	107	95	83	59	71	35	23	-37	-49	-13	-85	-97	-157	-265
Wt10:	164	140	128	116	92	104	68	56	-4	-16	20	-52	-64	-124	-232
Wt11:	131	107	95	83	59	71	35	23	-37	-49	-13	-85	-97	-157	-265
Wt12:	164	140	128	116	92	104	68	56	-4	-16	20	-52	-64	-124	-232
Wt13:	197	173	161	149	125	137	101	89	29	17	53	-19	-31	-91	-199
Wt14:	230	206	194	182	158	170	134	122	62	50	86	14	2	-58	-166
Wt15:	263	239	227	215	191	203	167	155	95	83	119	47	35	-25	-133

Den = 132

Affine E7 - Level 4 - Class 1

Den = 528

	Rp 1	Rp 2	Rp 3	Rp 4	Rp 5	Rp 6	Rp 7	Rp 8	Rp 9	Rp10
	0	0	0	0	1	1	1	1	2	3
	0	0	0	1	0	0	0	1	0	0
	1	0	0	0	0	0	0	0	0	0
	00	10	10	10	00	00	00	00	10	00
	0	0	0	0	0	0	1	0	0	0
	0	0	1	0	0	1	0	0	0	0
	1	2	0	0	3	1	0	1	0	1
Wt 1:	-37	131	83	-181	-157	-229	-301	-493	-661	-949
Wt 2:	-301	-133	-181	-445	-421	-493	-565	-757	-925	-1213
Wt 3:	-169	-1	-49	-313	-289	-361	-433	-625	-793	-1081
Wt 4:	95	263	215	-49	-25	-97	-169	-361	-529	-817
Wt 5:	-169	-1	-49	-313	-289	-361	-433	-625	-793	-1081
Wt 6:	95	263	215	-49	-25	-97	-169	-361	-529	-817
Wt 7:	227	395	347	83	107	35	-37	-229	-397	-685
Wt 8:	359	527	479	215	239	167	95	-97	-265	-553
Wt 9:	491	659	611	347	371	299	227	35	-133	-421
Wt10:	623	791	743	479	503	431	359	167	-1	-289

Affine E8 - Level 1

Den = 3

	Rp 1
	1
	0
	0
	0
	0
	00
	0
	0
Wt 1:	-1

Affine E8 - Level 2

Den = 48

	Rp 1	Rp 2	Rp 3
	0	0	2
	1	0	0
	0	0	0
	0	0	0
	0	0	0
	00	00	00
	0	0	0
	0	1	0
Wt 1:	-10	17	-55
Wt 2:	-34	-7	-79
Wt 3:	14	41	-31

Affine E8 - Level 3

Den = 33

	Rp 1	Rp 2	Rp 3	Rp 4	Rp 5
	0	0	1	1	3
	0	0	1	0	0
	1	0	0	0	0
	0	0	0	0	0
	0	0	0	0	0
	00	10	00	00	00
	0	0	0	0	0
	0	0	0	1	0
Wt 1:	-4	8	-34	-16	-64
Wt 2:	-15	-3	-45	-27	-75
Wt 3:	18	30	-12	6	-42
Wt 4:	7	19	-23	-5	-53
Wt 5:	29	41	-1	17	-31

Affine E8 - Level 4

Den = 204

	Rp 1	Rp 2	Rp 3	Rp 4	Rp 5	Rp 6	Rp 7	Rp 8	Rp 9	Rp10
	0	0	0	0	0	1	1	2	2	4
	2	0	0	1	0	0	0	1	0	0
	0	0	0	0	0	1	0	0	0	0
	0	1	0	0	0	0	0	0	0	0
	0	0	0	0	0	0	0	0	0	0
	00	00	00	00	00	00	10	00	00	00
	0	0	1	0	0	0	0	0	0	0
	0	0	0	1	2	0	0	0	1	0
Wt 1:	-80	88	136	28	148	-92	-20	-272	-164	-452
Wt 2:	-182	-14	34	-74	46	-194	-122	-374	-266	-554
Wt 3:	-233	-65	-17	-125	-5	-245	-173	-425	-317	-605
Wt 4:	-131	37	85	-23	97	-143	-71	-323	-215	-503
Wt 5:	-284	-116	-68	-176	-56	-296	-224	-476	-368	-656
Wt 6:	-29	139	187	79	199	-41	31	-221	-113	-401
Wt 7:	-80	88	136	28	148	-92	-20	-272	-164	-452
Wt 8:	73	241	289	181	301	61	133	-119	-11	-299
Wt 9:	22	190	238	130	250	10	82	-170	-62	-350
Wt10:	124	292	340	232	352	112	184	-68	40	-248

Affine E8 - Level 5

Den = 105

	Rp 1	Rp 2	Rp 3	Rp 4	Rp 5	Rp 6	Rp 7	Rp 8	Rp 9	Rp10	Rp11	Rp12	Rp13	Rp14	Rp15
	0	0	0	0	0	1	1	1	1	1	2	2	3	3	5
	1	0	1	0	0	2	0	0	1	0	0	0	1	0	0
	1	0	0	1	0	0	0	0	0	0	1	0	0	0	0
	0	0	0	0	0	0	1	0	0	0	0	0	0	0	0
	0	1	0	0	0	0	0	0	0	0	0	0	0	0	0
	00	00	10	00	10	00	00	00	00	00	00	10	00	00	00
	0	0	0	0	0	0	0	1	0	0	0	0	0	0	0
	0	0	0	1	1	0	0	0	1	2	0	0	0	1	0
Wt 1:	-23	58	13	34	73	-116	-32	-8	-62	-2	-122	-86	-212	-158	-302
Wt 2:	-86	-5	-50	-29	10	-179	-95	-71	-125	-65	-185	-149	-275	-221	-365
Wt 3:	-44	37	-8	13	52	-137	-53	-29	-83	-23	-143	-107	-233	-179	-323
Wt 4:	-65	16	-29	-8	31	-158	-74	-50	-104	-44	-164	-128	-254	-200	-344
Wt 5:	-107	-26	-71	-50	-11	-200	-116	-92	-146	-86	-206	-170	-296	-242	-386
Wt 6:	40	121	76	97	136	-53	31	55	1	61	-59	-23	-149	-95	-239
Wt 7:	-2	79	34	55	94	-95	-11	13	-41	19	-101	-65	-191	-137	-281
Wt 8:	-23	58	13	34	73	-116	-32	-8	-62	-2	-122	-86	-212	-158	-302
Wt 9:	19	100	55	76	115	-74	10	34	-20	40	-80	-44	-170	-116	-260
Wt10:	-44	37	-8	13	52	-137	-53	-29	-83	-23	-143	-107	-233	-179	-323
Wt11:	61	142	97	118	157	-32	52	76	22	82	-38	-2	-128	-74	-218
Wt12:	40	121	76	97	136	-53	31	55	1	61	-59	-23	-149	-95	-239
Wt13:	103	184	139	160	199	10	94	118	64	124	4	40	-86	-32	-176
Wt14:	82	163	118	139	178	-11	73	97	43	103	-17	19	-107	-53	-197
Wt15:	124	205	160	181	220	31	115	139	85	145	25	61	-65	-11	-155

Affine F4 - Level 1

Den = 60

	Rp 1	Rp 2
	0	1
	0	0
	0	0
	0	0
	1	0
Wt 1:	-7	-43
Wt 2:	23	-13

Affine F4 - Level 2

Den = 132

	Rp 1	Rp 2	Rp 3	Rp 4	Rp 5
	0	0	0	1	2
	0	0	1	0	0
	0	0	0	0	0
	0	1	0	0	0
	2	0	0	1	0
Wt 1:	-28	-40	-76	-112	-184
Wt 2:	5	-7	-43	-79	-151
Wt 3:	38	26	-10	-46	-118
Wt 4:	71	59	23	-13	-85
Wt 5:	104	92	56	20	-52

Affine F4 - Level 3

Den = 24

	Rp 1	Rp 2	Rp 3	Rp 4	Rp 5	Rp 6	Rp 7	Rp 8	Rp 9
	0	0	0	0	1	1	1	2	3
	0	0	0	1	0	0	1	0	0
	0	0	1	0	0	0	0	0	0
	0	1	0	0	0	1	0	0	0
	3	1	0	1	2	0	0	1	0
Wt 1:	-7	-10	-13	-17	-23	-25	-31	-37	-49
Wt 2:	1	-2	-5	-9	-15	-17	-23	-29	-41
Wt 3:	5	2	-1	-5	-11	-13	-19	-25	-37
Wt 4:	9	6	3	-1	-7	-9	-15	-21	-33
Wt 5:	13	10	7	3	-3	-5	-11	-17	-29
Wt 6:	17	14	11	7	1	-1	-7	-13	-25
Wt 7:	21	18	15	11	5	3	-3	-9	-21
Wt 8:	25	22	19	15	9	7	1	-5	-17
Wt 9:	29	26	23	19	13	11	5	-1	-13

Affine F4 - Level 4

Den = 312

	Rp 1	Rp 2	Rp 3	Rp 4	Rp 5	Rp 6	Rp 7	Rp 8	Rp 9	Rp10	Rp11	Rp12	Rp13	Rp14	Rp15	Rp16
	0	0	0	0	0	0	0	1	1	1	1	2	2	2	3	4
	0	0	0	0	1	1	2	0	0	0	1	0	0	1	0	0
	0	0	0	1	0	0	0	0	0	1	0	0	0	0	0	0
	0	1	2	0	0	1	0	0	1	0	0	0	1	0	0	0
	4	2	0	1	2	0	0	3	1	0	1	2	0	0	1	0
Wt 1:	-112	-160	-184	-208	-256	-280	-352	-328	-364	-400	-448	-520	-544	-616	-688	-832
Wt 2:	5	-43	-67	-91	-139	-163	-235	-211	-247	-283	-331	-403	-427	-499	-571	-715
Wt 3:	44	-4	-28	-52	-100	-124	-196	-172	-208	-244	-292	-364	-388	-460	-532	-676
Wt 4:	83	35	11	-13	-61	-85	-157	-133	-169	-205	-253	-325	-349	-421	-493	-637
Wt 5:	122	74	50	26	-22	-46	-118	-94	-130	-166	-214	-286	-310	-382	-454	-598
Wt 6:	161	113	89	65	17	-7	-79	-55	-91	-127	-175	-247	-271	-343	-415	-559
Wt 7:	200	152	128	104	56	32	-40	-16	-52	-88	-136	-208	-232	-304	-376	-520
Wt 8:	161	113	89	65	17	-7	-79	-55	-91	-127	-175	-247	-271	-343	-415	-559
Wt 9:	239	191	167	143	95	71	-1	23	-13	-49	-97	-169	-193	-265	-337	-481
Wt10:	278	230	206	182	134	110	38	62	26	-10	-58	-130	-154	-226	-298	-442
Wt11:	317	269	245	221	173	149	77	101	65	29	-19	-91	-115	-187	-259	-403
Wt12:	356	308	284	260	212	188	116	140	104	68	20	-52	-76	-148	-220	-364
Wt13:	395	347	323	299	251	227	155	179	143	107	59	-13	-37	-109	-181	-325
Wt14:	434	386	362	338	290	266	194	218	182	146	98	26	2	-70	-142	-286
Wt15:	473	425	401	377	329	305	233	257	221	185	137	65	41	-31	-103	-247
Wt16:	512	464	440	416	368	344	272	296	260	224	176	104	80	8	-64	-208

Affine G2 - Level 1

Den = 60

	Rp 1	Rp 2
	0	1
	0	0
	1	0
Wt 1:	-3	-27
Wt 2:	17	-7

Affine G2 - Level 2

Den = 36

	Rp 1	Rp 2	Rp 3	Rp 4
	0	0	1	2
	0	1	0	0
	2	0	1	0
Wt 1:	-3	-7	-19	-31
Wt 2:	3	-1	-13	-25
Wt 3:	15	11	-1	-13
Wt 4:	21	17	5	-7

Affine G2 - Level 3

Den = 252

	Rp 1	Rp 2	Rp 3	Rp 4	Rp 5	Rp 6
	0 0 3	0 1 1	1 0 2	1 1 0	2 0 1	3 0 0
Wt 1:	-27	-63	-147	-171	-243	-315
Wt 2:	29	-7	-91	-115	-187	-259
Wt 3:	113	77	-7	-31	-103	-175
Wt 4:	141	105	21	-3	-75	-147
Wt 5:	197	161	77	53	-19	-91
Wt 6:	225	189	105	81	9	-63

Affine G2 - Level 4

Den = 24

	Rp 1	Rp 2	Rp 3	Rp 4	Rp 5	Rp 6	Rp 7	Rp 8	Rp 9
	0 0 4	0 1 2	0 2 0	1 0 3	1 1 1	2 0 2	2 1 0	3 0 1	4 0 0
Wt 1:	-3	-7	-9	-15	-18	-25	-27	-33	-39
Wt 2:	3	-1	-3	-9	-12	-19	-21	-27	-33
Wt 3:	5	1	-1	-7	-10	-17	-19	-25	-31
Wt 4:	11	7	5	-1	-4	-11	-13	-19	-25
Wt 5:	15	11	9	3	0	-7	-9	-15	-21
Wt 6:	21	17	15	9	6	-1	-3	-9	-15
Wt 7:	23	19	17	11	8	1	-1	-7	-13
Wt 8:	27	23	21	15	12	5	3	-3	-9
Wt 9:	29	25	23	17	14	7	5	-1	-7

Affine G2 - Level 5

Den = 540

	Rp 1 0 0 5	Rp 2 0 1 3	Rp 3 0 2 1	Rp 4 1 0 4	Rp 5 1 1 2	Rp 6 1 2 0	Rp 7 2 0 3	Rp 8 2 1 1	Rp 9 3 0 2	Rp10 3 1 0	Rp11 4 0 1	Rp12 5 0 0
Wt 1:	-75	-175	-235	-355	-435	-475	-595	-655	-795	-835	-955	-1075
Wt 2:	69	-31	-91	-211	-291	-331	-451	-511	-651	-691	-811	-931
Wt 3:	141	41	-19	-139	-219	-259	-379	-439	-579	-619	-739	-859
Wt 4:	249	149	89	-31	-111	-151	-271	-331	-471	-511	-631	-751
Wt 5:	357	257	197	77	-3	-43	-163	-223	-363	-403	-523	-643
Wt 6:	393	293	233	113	33	-7	-127	-187	-327	-367	-487	-607
Wt 7:	501	401	341	221	141	101	-19	-79	-219	-259	-379	-499
Wt 8:	573	473	413	293	213	173	53	-7	-147	-187	-307	-427
Wt 9:	681	581	521	401	321	281	161	101	-39	-79	-199	-319
Wt10:	717	617	557	437	357	317	197	137	-3	-43	-163	-283
Wt11:	789	689	629	509	429	389	269	209	69	29	-91	-211
Wt12:	825	725	665	545	465	425	305	245	105	65	-55	-175

Affine G2 - Level 6

Den = 180

	Rp 1 0 0 6	Rp 2 0 1 4	Rp 3 0 2 2	Rp 4 0 3 0	Rp 5 1 0 5	Rp 6 1 1 3	Rp 7 1 2 1	Rp 8 2 0 4	Rp 9 2 1 2	Rp10 2 2 0	Rp11 3 0 3	Rp12 3 1 1	Rp13 4 0 2	Rp14 4 1 0	Rp15 5 0 1	Rp16 6 0 0
Wt 1:	-27	-63	-87	-99	-123	-153	-171	-207	-231	-243	-279	-297	-339	-351	-387	-423
Wt 2:	23	-13	-37	-49	-73	-103	-121	-157	-181	-193	-229	-247	-289	-301	-337	-373
Wt 3:	53	17	-7	-19	-43	-73	-91	-127	-151	-163	-199	-217	-259	-271	-307	-343
Wt 4:	63	27	3	-9	-33	-63	-81	-117	-141	-153	-189	-207	-249	-261	-297	-333
Wt 5:	83	47	23	11	-13	-43	-61	-97	-121	-133	-169	-187	-229	-241	-277	-313
Wt 6:	123	87	63	51	27	-3	-21	-57	-81	-93	-129	-147	-189	-201	-237	-273
Wt 7:	143	107	83	71	47	17	-1	-37	-61	-73	-109	-127	-169	-181	-217	-253
Wt 8:	173	137	113	101	77	47	29	-7	-31	-43	-79	-97	-139	-151	-187	-223
Wt 9:	203	167	143	131	107	77	59	23	-1	-13	-49	-67	-109	-121	-157	-193
Wt10:	213	177	153	141	117	87	69	33	9	-3	-39	-57	-99	-111	-147	-183
Wt11:	243	207	183	171	147	117	99	63	39	27	-9	-27	-69	-81	-117	-153
Wt12:	263	227	203	191	167	137	119	83	59	47	11	-7	-49	-61	-97	-133
Wt13:	293	257	233	221	197	167	149	113	89	77	41	23	-19	-31	-67	-103
Wt14:	303	267	243	231	207	177	159	123	99	87	51	33	-9	-21	-57	-93
Wt15:	323	287	263	251	227	197	179	143	119	107	71	53	11	-1	-37	-73
Wt16:	333	297	273	261	237	207	189	153	129	117	81	63	21	9	-27	-63

31. TABLES OF DOMINANT WEIGHT MULTIPLICITIES

These are the tables of dominant weight multiplicities for direct affine algebras of rank ≤ 8. Each table gives the weight multiplicites, level and congruence class for every representation of the given algebra. Multiplicities of a particular representation are given in the volume labeled with the Coxeter-Dynkin diagram of the highest weight of that representation, just as in §29. Within a representation, the weight multiplicities of $\gamma - n\delta$ are given in one "box" of the table. The highest weight Λ is labeled "Rp Λ" and the multiplicities of the weight $\gamma - n\delta$ are given in the row labeled "Wt γ." They are listed in the box beginning with the multiplicity of $\gamma - 0\delta$ and reading down. The weights Λ and γ are considered to have null depth zero, so the first entry in the box is zero if $\gamma > \Lambda$.

The coordinates of the weights are presented in the same configuration as the nodes on the Coxeter-Dynkin diagram in the upper right-hand corner of the page. An effort has been made to reduce redundancy; tables have been eliminated that can be reconstructed from the information given here (see §27).

Affine A1 - Level 1 - Class 0

Weight Multiplicities for Affine A1

	Rp 1 1 0				
Wt 1	1	22	231	1575	8349
	1	30	297	1958	10143
	2	42	385	2436	12310
	3	56	490	3010	14883
	5	77	627	3718	17977
	7	101	792	4565	21637
	11	135	1002	5604	26015
	15	176	1255	6842	31185

Affine A1 - Level 2 - Class 0

	Rp 1 0 2					Rp 2 2 0				
Wt 1	1	105	2741	37128	351841	0	86	2331	32490	312677
	1	161	3740	48406	444314	1	130	3206	42411	395537
	3	236	5096	62777	559727	2	196	4368	55159	498852
	5	350	6868	81182	703002	4	287	5922	71416	627509
	10	501	9233	104494		7	420	7967	92152	787171
	16	722	12306	134131		13	602	10670	118434	
	28	1016	16357	171467		21	858	14193	151725	
	43	1431	21581	218607		35	1206	18803	193676	
	70	1981	28394	277691		55	1687	24766	246491	
Wt 2	1	130	3206	42411	395537	1	105	2741	37128	351841
	2	196	4368	55159	498852	1	161	3740	48406	444314
	4	287	5922	71416	627509	3	236	5096	62777	559727
	7	420	7967	92152	787171	5	350	6868	81182	703002
	13	602	10670	118434		10	501	9233	104494	
	21	858	14193	151725		16	722	12306	134131	
	35	1206	18803	193676		28	1016	16357	171467	
	55	1687	24766	246491		43	1431	21581	218607	
	86	2331	32490	312677		70	1981	28394	277691	

Affine A1 - Level 2 - Class 1

	Rp 1 1 1				
Wt 1	1	100	2062	23528	192612
	2	154	2864	31066	246256
	4	232	3948	40824	313808
	8	344	5400	53408	398640
	14	504	7336	69568	504886
	24	728	9904	90248	637592
	40	1040	13288	116624	802936
	64	1472	17728	150144	1008448

Affine A1 - Level 3 - Class 0

	Rp 1 1 2					Rp 2 3 0				
Wt 1	1	289	10765	193609	2328131	0	149	5769	106308	1296468
	2	459	15239	259517	3013742	1	235	8209	142714	1680580
	5	711	21425	346203	3889283	2	371	11558	190791	2171315
	10	1090	29881	459901	5003902	5	567	16187	253818	2797115
	20	1640	41418	608312		9	864	22473	336358	3592332
	36	2444	57015	801504		18	1287	31041	443790	
	64	3591	78049	1051921		31	1908	42561	583425	
	108	5231	106200	1375650		55	2781	58078	763965	
	180	7533	143775	1792571		90	4032	78752	997011	
Wt 2	1	339	12094	213598	2538163	1	173	6503	117316	1414190
	3	531	17091	285810	3282330	1	276	9201	157323	1830958
	6	822	23949	380775	4231416	3	426	12951	209907	2363451
	13	1249	33348	505045	5438971	6	657	18063	278961	3041298
	24	1878	46098	667203		12	985	25057	369040	
	44	2779	63355	877869		21	1473	34494	486411	
	76	4078	86530	1150828		39	2163	47256	638499	
	129	5910	117564	1503096		64	3156	64303	835245	
	210	8496	158841	1956537		108	4542	87105	1088574	

Affine A1 - Level 4 - Class 0

	Rp 1					Rp 2					Rp 3				
	0 4					2 2					4 0				
Wt 1	1	75	2097	30031	298411	0	103	3156	47473	485244	0	41	1368	21055	218612
	1	131	3153	42504	405157	1	180	4800	67473	661122	0	75	2082	30031	298276
	3	215	4712	59669	547732	3	306	7212	95214	896397	1	126	3153	42448	405157
	6	354	6936	83323		7	507	10716	133461		2	215	4690	59669	547556
	13	561	10151	115536		15	822	15759	185892		6	347	6936	83246	
	23	889	14677	159409		30	1308	22956	257379		11	561	10121	115536	
	44	1368	21097	218612		57	2047	33147	354348		23	878	14677	159308	
Wt 2	1	112	2850	38968	375506	1	156	4332	61824	612252	0	66	1880	27512	276142
	2	190	4256	54852	508092	2	270	6516	87432	831024	1	112	2850	38968	375506
	5	310	6303	76656		6	446	9722	122712		2	190	4256	54852	508092
	10	501	9222	106530		12	732	14316	171234		5	310	6303	76656	
	20	788	13396	147112		26	1164	20922	237374		10	501	9222	106530	
	36	1230	19262	202114		48	1836	30252	327324		20	788	13396	147112	
	66	1880	27512	276142		90	2832	43440	448752		36	1230	19262	202114	
Wt 3	1	126	3153	42448	405157	1	180	4800	67473	661122	1	75	2097	30031	298411
	2	215	4690	59669	547556	3	306	7212	95214	896397	1	131	3153	42504	405157
	6	347	6936	83246		7	507	10716	133461		3	215	4712	59669	547732
	11	561	10121	115536		15	822	15759	185892		6	354	6936	83323	
	23	878	14677	159308		30	1308	22956	257379		13	561	10151	115536	
	41	1368	21055	218612		57	2047	33147	354348		23	889	14677	159409	
	75	2082	30031	298276		103	3156	47473	485244		44	1368	21097	218612	

Affine A1 - Level 4 - Class 1

	Rp 1					Rp 2				
	1 3					3 1				
Wt 1	1	385	17977	386155	5392783	0	297	14883	329931	4697205
	2	627	26015	526823	7089500	1	490	21637	451276	6185689
	5	1002	37338	715220	9289091	3	792	31185	614154	8118264
	11	1575	53174	966467	12132164	7	1255	44583	831820	10619863
	22	2436	75175	1300156		15	1958	63261	1121505	13848650
	42	3718	105558	1741630		30	3010	89134	1505499	
	77	5604	147273	2323520		56	4565	124754	2012558	
	135	8349	204226	3087735		101	6842	173525	2679689	
	231	12310	281589	4087968		176	10143	239943	3554345	
Wt 2	1	490	21637	451276	6185689	1	385	17977	386155	5392783
	3	792	31185	614154	8118264	2	627	26015	526823	7089500
	7	1255	44583	831820	10619863	5	1002	37338	715220	9289091
	15	1958	63261	1121505	13848650	11	1575	53174	966467	12132164
	30	3010	89134	1505499		22	2436	75175	1300156	
	56	4565	124754	2012558		42	3718	105558	1741630	
	101	6842	173525	2679689		77	5604	147273	2323520	
	176	10143	239943	3554345		135	8349	204226	3087735	
	297	14883	329931	4697205		231	12310	281589	4087968	

Affine A1 - Level 5 - Class 0

	Rp 1					Rp 2					Rp 3				
			1					3					5		
			4					2					0		
Wt 1	1	148	4638	73651	798350	0	129	4475	74398	827849	0	46	1749	29735	335725
	2	259	7101	105566	1096789	1	229	6919	107088	1140801	0	86	2706	42940	463259
	5	439	10753	150205	1499569	3	399	10544	153059	1564037	1	148	4155	61473	636168
	11	730	16093	212337		8	671	15905	217181		2	256	6277	87455	869178
	23	1186	23862	298227		17	1111	23712	306224		6	423	9407	123501	
	44	1900	35032	416418		36	1795	35033	429025		12	694	13922	173410	
	83	2987	51009	578111		69	2860	51256	597659		25	1106	20453	241915	
Wt 2	1	203	5965	91357	965998	1	184	5836	92701	1004321	0	71	2294	37207	407852
	3	349	9084	130379	1323372	2	324	8934	132872	1379606	1	123	3534	53403	561153
	7	589	13663	184826		6	549	13534	189037		2	213	5357	76167	768072
	16	967	20349	260278		13	916	20255	267218		5	353	8057	107811	
	32	1562	30000	364333		28	1489	30038	375264		10	582	11959	151726	
	62	2477	43847	506981		54	2388	44095	523955		21	930	17623	212123	
	114	3873	63529	701677		103	3755	64211	727317		38	1477	25695	294989	
Wt 3	1	223	6478	98049	1028873	1	209	6364	99731	1070678	1	80	2522	40062	435182
	3	385	9836	139784	1408119	3	361	9732	142684	1469411	1	142	3855	57459	597947
	8	644	14782	197882		7	613	14689	202765		3	238	5846	81754	817798
	17	1060	21965	278377		16	1012	21956	286175		6	398	8742	115633	
	36	1704	32345	389193		32	1645	32468	401463		13	643	12977	162414	
	68	2701	47188	541056		63	2619	47604	559780		24	1034	19042	226882	
	126	4209	68295	748014		116	4116	69163	776281		46	1620	27752	315013	

Affine A1 - Level 6 - Class 0

	Rp 1				Rp 2				Rp 3				Rp 4			
	0 6				2 4				4 2				6 0			
Wt 1	1	82	2784	47455	0	122	4734	86031	0	81	3689	70585	0	27	1335	26430
	1	147	4300	68712	1	222	7412	125304	0	155	5828	103344	0	51	2133	38767
	3	249	6590		3	390	11462	181083	1	278	9105	149991	0	97	3337	56448
	6	421	9954		7	670	17505		3	492	14004		1	170	5172	81498
	13	687	14929		16	1122	26449		9	836	21329		2	297	7895	
	24	1118	22122		33	1845	39549		19	1403	32082		6	499	11938	
	47	1769	32559		65	2978	58593		42	2289	47832		13	827	17835	
Wt 2	1	128	3974	64419	1	197	6863	117574	0	142	5432	97176	0	48	1999	36519
	2	224	6090	92726	2	353	10625	170209	1	257	8490	141230	0	91	3127	53233
	5	376	9251		6	605	16286		3	453	13090	203718	1	159	4854	76904
	10	626	13877		13	1025	24638		8	775	19951		2	279	7409	
	21	1013	20649		29	1686	36942		18	1298	30067		6	466	11221	
	39	1625	30409		56	2741	54797		38	2127	44867		12	775	16771	
	73	2553	44454		109	4361	80632		75	3428	66313		26	1249	24881	
Wt 3	1	159	4854	76904	1	257	8490	141230	1	197	6863	117574	0	73	2553	44454
	2	279	7409		3	453	13090	203718	2	353	10625	170209	1	128	3974	64419
	6	466	11221		8	775	19951		6	605	16286		2	224	6090	92726
	12	775	16771		18	1298	30067		13	1025	24638		5	376	9251	
	26	1249	24881		38	2127	44867		29	1686	36942		10	626	13877	
	48	1999	36519		75	3428	66313		56	2741	54797		21	1013	20649	
	91	3127	53233		142	5432	97176		109	4361	80632		39	1625	30409	
Wt 4	1	170	5172	81498	1	278	9105	149991	1	222	7412	125304	1	82	2784	47455
	2	297	7895		3	492	14004		3	390	11462	181083	1	147	4300	68712
	6	499	11938		9	836	21329		7	670	17505		3	249	6590	
	13	827	17835		19	1403	32082		16	1122	26449		6	421	9954	
	27	1335	26430		42	2289	47832		33	1845	39549		13	687	14929	
	51	2133	38767		81	3689	70585		65	2978	58593		24	1118	22122	
	97	3337	56448		155	5828	103344		122	4734	86031		47	1769	32559	

Affine A1 - Level 6 - Class 1

	Rp 1					Rp 2					Rp 3				
	1 5					3 3					5 1				
Wt 1	1	154	5223	89228	1031184	0	142	5502	99290	1183896	0	74	3227	60559	737259
	2	272	8088	129120	1428896	1	258	8610	144459	1646220	0	139	5088	88436	1027466
	5	467	12376	185486	1970129	3	455	13296	208574	2277246	1	250	7912	128118	1424331
	11	785	18726	264641		8	780	20289	298980		3	437	12145	184231	1964525
	23	1292	28048	375130		18	1308	30618	425680		8	743	18429	263066	
	45	2091	41609	528505		38	2148	45740	602202		18	1236	27663	373172	
	85	3328	61186	740269		75	3466	67686	846788		38	2014	41119	526069	
Wt 2	1	216	6919	113787	1280274	1	212	7428	127600	1475772	0	120	4440	78385	922141
	3	377	10648	163935	1768432	2	378	11520	184692	2044932	1	216	6919	113787	1280274
	7	643	16194	234517		6	654	17646	265368		3	377	10648	163935	1768432
	16	1071	24363	333266		14	1104	26724	378668		7	643	16194	234517	
	33	1749	36294	470619		30	1824	40058	536810		16	1071	24363	333266	
	64	2809	53568	660630		60	2962	59456	756336		33	1749	36294	470619	
	120	4440	78385	922141		116	4724	87468	1059426		64	2809	53568	660630	
Wt 3	1	250	7912	128118	1424331	1	258	8610	144459	1646220	1	154	5223	89228	1031184
	3	437	12145	184231	1964525	3	455	13296	208574	2277246	2	272	8088	129120	1428896
	8	743	18429	263066		8	780	20289	298980		5	467	12376	185486	1970129
	18	1236	27663	373172		18	1308	30618	425680		11	785	18726	264641	
	38	2014	41119	526069		38	2148	45740	602202		23	1292	28048	375130	
	74	3227	60559	737259		75	3466	67686	846788		45	2091	41609	528505	
	139	5088	88436	1027466		142	5502	99290	1183896		85	3328	61186	740269	

Affine A1 - Level 7 - Class 0

	Rp 1				Rp 2				Rp 3				Rp 4			
	1 6				3 4				5 2				7 0			
Wt 1	1	156	5571	100109	0	148	6142	117694	0	87	4260	86061	0	28	1478	30918
	2	278	8689	145883	1	271	9704	172557	0	168	6792	126849	0	53	2383	45663
	5	480	13397		3	484	15114	251073	1	306	10692	185367	0	102	3759	66942
	11	813	20416		8	837	23273		3	546	16584		1	181	5874	97281
	23	1347	30808		18	1420	35412		9	940	25442		2	320	9034	
	45	2198	46026		39	2352	53354		20	1590	38565		6	542	13767	
	86	3522	68172		77	3835	79587		44	2622	57901		13	908	20715	
Wt 2	1	222	7511	130074	1	225	8489	154514	0	148	6045	114322	0	49	2147	41345
	3	390	11651	188687	2	407	13278	225361	1	270	9533	167332	0	93	3387	60681
	7	671	17851		6	710	20526		3	482	14815	243096	1	164	5297	88272
	16	1126	27063		14	1214	31342		8	831	22770		2	290	8149	
	33	1855	40607		31	2024	47380		18	1407	34578		6	489	12430	
	65	3001	60377		62	3322	70883		39	2325	52006		12	819	18720	
	122	4784	88973		122	5346	105118		77	3784	77433		26	1332	27967	
Wt 3	1	263	8873	151260	1	286	10236	181404	1	203	7497	135659	0	74	2703	49462
	3	465	13725		3	509	15949	263735	2	366	11707	197798	1	130	4239	72163
	8	797	20994		8	885	24529		6	634	18083		2	229	6542	104577
	18	1340	31744		19	1497	37322		13	1082	27582		5	387	10010	
	39	2201	47540		40	2486	56178		29	1797	41669		10	649	15120	
	76	3560	70512		81	4047	83773		57	2944	62294		21	1057	22661	
	145	5658	103704		155	6487	123778		111	4727	92340		39	1709	33601	
Wt 4	1	277	9353	158900	1	307	10890	191239	1	228	8052	143692	1	83	2935	52518
	3	489	14470		3	549	16927		3	403	12556	209148	1	149	4566	76551
	8	842	22110		9	947	26017		7	699	19323		3	254	7045	
	19	1412	33428		20	1605	39515		16	1179	29433		6	432	10718	
	40	2324	50020		44	2654	59431		33	1957	44342		13	710	16183	
	80	3754	74162		87	4320	88495		66	3182	66204		24	1162	24152	
	152	5970	108989		168	6904	130652		124	5103	97924		47	1853	35785	

Affine A1 - Level 8 - Class 0

	Rp 1 0 8	Rp 2 2 6	Rp 3 4 4	Rp 4 6 2	Rp 5 8 0
Wt 1	1 47 1185 16957	0 66 2014 31297	0 45 1718 29108	0 21 1044 19171	0 6 343 6576
	1 83 1897 25426	1 125 3294 47414	0 89 2860 44503	0 46 1777 29571	0 13 585 10174
	3 150 3019	3 230 5309 71165	1 174 4695 67394	0 93 2955 45082	0 29 989 15599
	6 256 4717	7 409 8425	3 320 7557	1 181 4831 68037	0 55 1621 23602
	13 437 7313	16 712 13206	9 578 12012	3 334 7757	1 107 2633
	24 721 11177	33 1208 20439	20 1004 18800	9 600 12282	2 192 4181
Wt 2	1 74 1753 23930	1 112 3056 44720	0 83 2691 42244	0 45 1699 28277	0 13 565 9775
	2 131 2787 35662	2 205 4932 67215	1 161 4419 64027	0 89 2820 43110	0 28 952 14979
	5 231 4390	6 372 7869	3 299 7134	1 174 4616 65107	0 54 1561 22671
	10 392 6809	13 647 12354	8 538 11346	3 319 7406	1 104 2531 34019
	21 660 10468	29 1111 19193	19 942 17801	9 575 11738	2 186 4019
	39 1080 15891	57 1854 29434	41 1609 27580	20 996 18316	6 331 6317
Wt 3	1 94 2231 29994	1 150 3987 56800	1 128 3684 54870	0 78 2436 37556	0 26 842 13192
	2 166 3537	3 276 6412 85058	2 238 5972 82612	1 150 3987 56800	0 49 1376 19970
	6 295 5562	8 495 10170	6 436 9560	3 276 6412 85058	1 94 2231 29994
	12 500 8602	18 860 15904	14 766 15058	8 495 10170	2 166 3537
	26 842 13192	39 1464 24583	32 1324 23452	18 860 15904	6 295 5562
	49 1376 19970	78 2436 37556	64 2224 36048	39 1464 24583	12 500 8602
Wt 4	1 104 2531 34019	1 174 4616 65107	1 161 4419 64027	1 112 3056 44720	0 39 1080 15891
	2 186 4019	3 319 7406	3 299 7134	2 205 4932 67215	1 74 1753 23930
	6 331 6317	9 575 11738	8 538 11346	6 372 7869	2 131 2787 35662
	13 565 9775	20 996 18316	19 942 17801	13 647 12354	5 231 4390
	28 952 14979	45 1699 28277	41 1609 27580	29 1111 19193	10 392 6809
	54 1561 22671	89 2820 43110	83 2691 42244	57 1854 29434	21 660 10468
Wt 5	1 107 2633	1 181 4831 68037	1 174 4695 67394	1 125 3294 47414	1 47 1185 16957
	2 192 4181	3 334 7757	3 320 7557	3 230 5309 71165	1 83 1897 25426
	6 343 6576	9 600 12282	9 578 12012	7 409 8425	3 150 3019
	13 585 10174	21 1044 19171	20 1004 18800	16 712 13206	6 256 4717
	29 989 15599	46 1777 29571	45 1718 29108	33 1208 20439	13 437 7313
	55 1621 23602	93 2955 45082	89 2860 44503	66 2014 31297	24 721 11177

Affine A1 - Level 8 - Class 1

	Rp 1				Rp 2				Rp 3				Rp 4			
	1 7				3 5				5 3				7 1			
Wt 1	1	157	5766	107327	0	150	6518	130315	0	93	4902	104817	0	41	2511	56475
	2	280	9040	157175	1	277	10360	192181	0	181	7892	155559	0	82	4087	84182
	5	486	14005		3	497	16242	281194	1	335	12523	228913	0	158	6540	124384
	11	826	21453		8	866	25158		3	603	19596		1	290	10315	182274
	23	1375	32532		18	1477	38524		9	1052	30291		3	517	16053	
	45	2253	48853		39	2465	58382		21	1794	46286		8	896	24682	
	86	3629	72716		78	4042	87617		46	2992	69997		19	1516	37525	
Wt 2	1	224	7860	141187	1	231	9136	173644	0	161	7112	141847	0	77	3752	77388
	3	396	12255	205862	2	420	14388	254691	1	299	11304	209046	0	147	6002	114419
	7	684	18879		6	739	22378		3	538	17723	305642	1	269	9466	167779
	16	1154	28769		14	1271	34393		8	941	27443		3	478	14731	
	33	1910	43401		31	2136	52302		19	1607	42007		8	825	22658	
	65	3108	64861		63	3527	78732		41	2686	63628		18	1395	34460	
	123	4978	96086		124	5718	117437		83	4408	95435		39	2307	51887	
Wt 3	1	269	9466	167779	1	299	11304	209046	1	231	9136	173644	0	123	4978	96086
	3	478	14731		3	538	17723	305642	2	420	14388	254691	1	224	7860	141187
	8	825	22658		8	941	27443		6	739	22378		3	396	12255	205862
	18	1395	34460		19	1607	42007		14	1271	34393		7	684	18879	
	39	2307	51887		41	2686	63628		31	2136	52302		16	1154	28769	
	77	3752	77388		83	4408	95435		63	3527	78732		33	1910	43401	
	147	6002	114419		161	7112	141847		124	5718	117437		65	3108	64861	
Wt 4	1	290	10315	182274	1	335	12523	228913	1	277	10360	192181	1	157	5766	107327
	3	517	16053		3	603	19596		3	497	16242	281194	2	280	9040	157175
	8	896	24682		9	1052	30291		8	866	25158		5	486	14005	
	19	1516	37525		21	1794	46286		18	1477	38524		11	826	21453	
	41	2511	56475		46	2992	69997		39	2465	58382		23	1375	32532	
	82	4087	84182		93	4902	104817		78	4042	87617		45	2253	48853	
	158	6540	124384		181	7892	155559		150	6518	130315		86	3629	72716	

Affine A1 - Level 9 - Class 0

	Rp 1			Rp 2			Rp 3			Rp 4			Rp 5		
		1			3			5			7			9	
		8			6			4			2			0	
Wt 1	1	281	14357	0	279	16905	0	187	13657	0	95	8371	0	29	2781
	2	488	22064	1	503	26302	0	348	21494	0	187	13328	0	56	4441
	5	832	33576	3	879	40443	1	632	33431	0	347	20904	0	109	7019
	11	1388	50593	8	1506	61561	3	1109	51369	1	629	32409	0	197	10915
	23	2281		18	2522	92773	9	1907	78134	3	1100	49634	1	354	16811
	45	3684		39	4155		21	3199		9	1886	75259	2	608	25559
	86	5873		78	6726		47	5279		21	3153		6	1033	
	157	9235		151	10739		95	8550		47	5187		13	1704	
Wt 2	1	398	19488	1	426	23513	0	312	19588	0	176	12375	0	54	4169
	3	690	29809	2	752	36294	1	567	30522	0	325	19406	0	105	6582
	7	1167	45132	6	1300	55448	3	998	46984	1	588	30093	0	188	10228
	16	1938	67704	14	2193	83827	8	1719	71584	3	1025	46095	1	336	15741
	33	3163		31	3640		19	2891		9	1756	69917	2	576	23922
	65	5085		63	5925		42	4781		20	2931		6	975	36049
	123	8055		125	9513		85	7761		45	4819		13	1605	
	225	12606		233	15047		167	12420		90	7773		28	2615	
Wt 3	1	484	23687	1	551	29302	1	449	25320	0	278	16552	0	94	5713
	3	838	36169	3	970	45074	2	795	39133	1	501	25696	0	167	8869
	8	1423	54688	8	1664	68584	6	1381	59859	3	873	39415	1	297	13650
	18	2362		19	2798	103351	14	2336	90589	8	1493	59867	2	505	20742
	39	3859		41	4613		32	3889		18	2493	90013	6	853	31266
	77	6196		84	7484		65	6345		39	4099		12	1399	
	148	9815		163	11952		130	10210		78	6617		26	2275	
	271	15335		305	18836		244	16175		151	10542		49	3621	
Wt 4	1	530	26347	1	632	33212	1	551	29487	1	374	19846	0	131	6960
	3	924	40244	3	1108	50987	3	971	45402	2	653	30557	1	232	10736
	8	1571	60829	9	1904	77481	8	1666	69143	6	1124	46600	2	394	16351
	19	2617		21	3192		19	2804		13	1883	70321	5	665	24707
	41	4279		47	5263		41	4626		29	3113		10	1091	36943
	83	6884		95	8517		84	7512		57	5044		21	1776	
	160	10908		187	13592		163	12006		112	8075		39	2831	
	296	17059		348	21373		305	18939		206	12728		74	4474	
Wt 5	1	545	27255	1	657	34572	1	591	31027	1	411	21093	1	150	7464
	3	950	41625	3	1157	53073	3	1033	47682	3	718	32423	1	257	11445
	8	1620		9	1983	80597	9	1775	72550	7	1221	49300	3	439	17418
	19	2698		22	3330		20	2973		16	2043	74283	6	726	26204
	42	4420		48	5484		45	4903		33	3351		13	1196	
	84	7110		99	8880		90	7936		66	5421		24	1920	
	164	11277		194	14157		176	12675		125	8631		47	3063	
	303	17636		363	22268		326	19944		231	13581		83	4801	

Weight Multiplicities for Affine A2

Affine A2 - Level 1 - Class 0

	Rp 1 1 00				
Wt 1	1	185	5822	94235	1046705
	2	300	8470	129512	1386930
	5	481	12230	177087	1831065
	10	752	17490	240840	2408658
	20	1165	24842	326015	3157789
	36	1770	35002	439190	4126070
	65	2665	49010	589128	5374390
	110	3956	68150	786814	6978730

Affine A2 - Level 2 - Class 0

	Rp 1 0 11					Rp 2 2 00				
Wt 1	1	3444	454745	22231560	628287275	0	1445	210960	10814802	314730044
	4	6410	726820	32959060	888572552	1	2736	339471	16095932	446281663
	13	11636	1149398	48559068	177643663*	4	5041	540232	23802444	630087340
	36	20663	1799740	71118954	681432101*	12	9068	850880	34984060	885897404
	89	35976	2792039	103571096		32	15974	1327296	51119553	166833085*
	204	61539	4293928	150016797		77	27612	2051832	74282584	
	441	103592	6549881	216167864		172	46922	3145056	107368431	
	908	171848	9914232	309946591		365	78516	4782378	154402672	
	1798	281280	14897574	442298212		740	129539	7217416	220959900	
Wt 2	2	4708	575698	27090392	747583188	1	1996	268064	13204604	374986020
	7	8662	915120	40036212	1055034448	2	3736	428816	19589308	530560964
	22	15536	1439992	58808492	409024493*	8	6780	678940	28878188	747507200
	56	27328	2244068	85884068	1001844777*	20	12084	1063960	42319872	1048859008
	136	47136	3466140	124731136		52	21048	1652224	61664078	
	300	79996	5308416	180194154		116	36096	2542916	89365238	
	636	133632	8065910	259001075		256	60810	3882241	128835712	
	1280	220212	12163715	370470976		522	101065	5880454	184818952	
	2498	358139	18214118	527447376		1045	165568	8842864	263860844	

* Multiplicity is given modulo 1073741527

Affine A2 - Level 3 - Class 0

	Rp 1			Rp 2			Rp 3			Rp 4		
	0			0			1			3		
	03			30			11			00		
Wt 1	1	2060	375613	0	2027	375120	0	4144	845472	0	918	210648
	2	4328	658126	1	4278	657467	2	8914	1493222	0	2027	375120
	8	8814	1136846	6	8744	1135954	10	18484	2597892	1	4278	657467
	24	17412	1937628	20	17314	1936458	36	37124	4457600	6	8744	1135954
	67	33567	3262251	61	33424	3260708	112	72512	7551056	20	17314	1936458
	170	63182	5429338	161	62989	5427303	308	138196	12639604	61	33424	3260708
	414	116572	8939790	398	116306	8937148	778	257648	20923146	161	62989	5427303
	940	211016		918	210648		1844	470932	34276596	398	116306	8937148
Wt 2	0	2027	375120	1	2060	375613	0	4144	845472	0	918	210648
	1	4278	657467	2	4328	658126	2	8914	1493222	0	2027	375120
	6	8744	1135954	8	8814	1136846	10	18484	2597892	1	4278	657467
	20	17314	1936458	24	17412	1937628	36	37124	4457600	6	8744	1135954
	61	33424	3260708	67	33567	3262251	112	72512	7551056	20	17314	1936458
	161	62989	5427303	170	63182	5429338	308	138196	12639604	61	33424	3260708
	398	116306	8937148	414	116572	8939790	778	257648	20923146	161	62989	5427303
	918	210648		940	211016		1844	470932	34276596	398	116306	8937148
Wt 3	1	3366	546438	1	3366	546438	1	6936	1237470	0	1572	310172
	4	6932	948360	4	6932	948360	6	14550	2163456	1	3366	546438
	15	13865	1623737	15	13865	1623737	24	29520	3728904	4	6932	948360
	44	26952	2744840	44	26952	2744840	78	58176	6343068	15	13865	1623737
	121	51187	4585920	121	51187	4585920	222	111750	10658880	44	26952	2744840
	300	95080	7577684	300	95080	7577684	576	209820	17708412	121	51187	4585920
	707	173280	12393330	707	173280	12393330	1392	385968	29108880	300	95080	7577684
	1572	310172		1572	310172		3180	696960		707	173280	12393330
Wt 4	1	4278	657467	1	4278	657467	2	8914	1493222	1	2060	375613
	6	8744	1135954	6	8744	1135954	10	18484	2597892	2	4328	658126
	20	17314	1936458	20	17314	1936458	36	37124	4457600	8	8814	1136846
	61	33424	3260708	61	33424	3260708	112	72512	7551056	24	17412	1937628
	161	62989	5427303	161	62989	5427303	308	138196	12639604	67	33567	3262251
	398	116306	8937148	398	116306	8937148	778	257648	20923146	170	63182	5429338
	918	210648		918	210648		1844	470932	34276596	414	116572	8939790
	2027	375120		2027	375120		4144	845472		940	211016	

Affine A2 - Level 3 - Class 1

	Rp 1				Rp 2				Rp 3			
	0 21				1 02				2 10			
Wt 1	1	4448	799629	49599509	0	3422	661038	42332110	0	2642	545820	36096138
	4	9334	1399002	79300458	2	7294	1162572	67879222	1	5694	964813	58050641
	16	18968	2413106	125729560	9	14982	2014637	107917204	6	11839	1680012	92550018
	50	37410	4107501	197765596	33	29868	3444342	170191459	22	23812	2884752	146342066
	143	71953	6906711	308743055	98	57924	5814820	266352012	70	46598	4890200	229603444
	368	135212	11481322	478559132	266	109728	9702826	413824508	193	88913	8190833	357579188
	892	248986	18883586		659	203407	16014218	638511887	493	165991	13567153	552982810
	2035	449967	30748372		1543	369939	26162392		1170	303706	22238296	
Wt 2	1	5694	964813	58050641	1	4448	799629	49599509	0	3422	661038	42332110
	6	11839	1680012	92550018	4	9334	1399002	79300458	2	7294	1162572	67879222
	22	23812	2884752	146342066	16	18968	2413106	125729560	9	14982	2014637	107917204
	70	46598	4890200	229603444	50	37410	4107501	197765596	33	29868	3444342	170191459
	193	88913	8190833	357579188	143	71953	6906711	308743055	98	57924	5814820	266352012
	493	165991	13567153	552982810	368	135212	11481322	478559132	266	109728	9702826	413824508
	1170	303706	22238296		892	248986	18883586		659	203407	16014218	638511887
	2642	545820	36096138		2035	449967	30748372		1543	369939	26162392	
Wt 3	2	7294	1162572	67879222	1	5694	964813	58050641	1	4448	799629	49599509
	9	14982	2014637	107917204	6	11839	1680012	92550018	4	9334	1399002	79300458
	33	29868	3444342	170191459	22	23812	2884752	146342066	16	18968	2413106	125729560
	98	57924	5814820	266352012	70	46598	4890200	229603444	50	37410	4107501	197765596
	266	109728	9702826	413824508	193	88913	8190833	357579188	143	71953	6906711	308743055
	659	203407	16014218	638511887	493	165991	13567153	552982810	368	135212	11481322	478559132
	1543	369939	26162392		1170	303706	22238296		892	248986	18883586	
	3422	661038	42332110		2642	545820	36096138		2035	449967	30748372	

Affine A2 - Level 4 - Class 0

	Rp 1		Rp 2		Rp 3		Rp 4		Rp 5	
	0 22		1 03		1 30		2 11		4 00	
Wt 1	1	82876	0	54024	0	54024	0	70877	0	12640
	4	167868	2	110229	2	110229	1	146512	0	26520
	19	331464	9	219280	9	219280	8	294661	1	53952
	64	640273	36	426123	36	426123	35	578432	4	107095
	202	1211688	115	811172	115	811172	124	1110880	18	207584
	560	2251654	336	1514980	336	1514980	379	2091760	56	394333
	1464	4113692	890	2781198	890	2781198	1052	3868304	167	734832
	3568	7400588	2224	5024851	2224	5024851	2700	7036712	440	1346387
	8315	13123300	5241	8946784	5241	8946784	6536	12606986	1103	2427804
	18512		11840		11840		15047		2588	4315116
	39822		25701		25701		33248		5844	
Wt 2	1	99110	1	65114	0	64362	0	85226	0	15254
	6	199394	4	131816	2	130651	2	174902	0	31735
	25	391540	16	260328	11	258558	12	349554	1	64180
	86	752263	54	502838	44	500173	49	682276	6	126588
	260	1417044	163	951890	143	947934	166	1303640	23	244148
	716	2621653	450	1769185	414	1763363	494	2443270	74	461444
	1833	4770868	1161	3233342	1096	3224872	1340	4499194	212	856244
	4426	8550960	2824	5818382	2714	5806152	3387	8152278	556	1562325
	10183	15111984	6549	10321164	6364	10303674	8086	14553003	1366	2806898
	22488		14572		14272		18415		3184	4971383
	47942		31289		30808		40302		7100	
Wt 3	1	99110	0	64362	1	65114	0	85226	0	15254
	6	199394	2	130651	4	131816	2	174902	0	31735
	25	391540	11	258558	16	260328	12	349554	1	64180
	86	752263	44	500173	54	502838	49	682276	6	126588
	260	1417044	143	947934	163	951890	166	1303640	23	244148
	716	2621653	414	1763363	450	1769185	494	2443270	74	461444
	1833	4770868	1096	3224872	1161	3233342	1340	4499194	212	856244
	4426	8550960	2714	5806152	2824	5818382	3387	8152278	556	1562325
	10183	15111984	6364	10303674	6549	10321164	8086	14553003	1366	2806898
	22488		14272		14572		18415		3184	4971383
	47942		30808		31289		40302		7100	
Wt 4	2	140626	1	92301	1	92301	1	122814	0	22328
	11	279602	6	184795	6	184795	6	248488	1	45656
	44	543200	25	361263	25	361263	27	490414	4	90981
	145	1033526	85	691265	85	691265	96	946528	15	177140
	424	1929558	255	1297255	255	1297255	298	1790348	48	337758
	1133	3540650	695	2391715	695	2391715	836	3324781	138	631878
	2830	6394535	1764	4338431	1764	4338431	2173	6071324	366	1161739
	6688	11380724	4226	7752638	4226	7752638	5310	10916466	913	2102052
	15102		9653		9653		12341	19349382	2156	3747779
	32805		21179		21179		27486		4874	
	68922		44888		44888		59029		10606	
Wt 5	3	166928	1	109858	1	109858	2	147158	1	27040
	14	330168	8	218644	8	218644	10	295682	2	54824
	58	638146	32	425256	32	425256	40	579944	8	108334
	184	1208766	110	809759	110	809759	133	1113192	24	209568
	536	2247048	322	1513056	322	1513056	398	2095120	72	397128
	1408	4107393	872	2778181	872	2778181	1084	3873312	190	739154
	3492	7390976	2183	5020760	2183	5020760	2760	7043874	490	1352414
	8160	13110230	5186	8940568	5186	8940568	6632	12617432	1176	2436860
	18299		11730		11730		15214		2729	4327632
	39428		25552		25552		33508		6048	
	82332		53752		53752		71304		13002	

Affine A2 - Level 5 - Class 0 - Page 1 of 2

	Rp 1		Rp 2		Rp 3		Rp 4		Rp 5		Rp 6		Rp 7	
	0		0		1		2		2		3		5	
	14		41		22		03		30		11		00	
Wt 1	1	7689	0	7109	0	11384	0	5463	0	5107	0	5170	0	721
	4	17721	1	16693	2	27144	1	13089	0	12438	0	12810	0	1847
	16	39425	8	37646	12	62027	6	30093	3	28943	2	30216	0	4460
	54	84927	35	81915	53	136709	25	66667	16	64677	14	68378	1	10305
	168	177886	126	172879	189	291835	90	143084	69	139723	62	149316	6	22864
	476	363159	390	354978	597	605763	282	298427	235	292845	222	316058	26	49095
	1271	724633	1102	711467	1708	1226109	812	606902	716	597803	694	650848	87	102299
	3198		2880		4546		2167		1977		1966		264	207746
Wt 2	0	7109	1	7689	0	11384	0	5107	0	5463	0	5170	0	721
	1	16693	4	17721	2	27144	0	12438	1	13089	0	12810	0	1847
	8	37646	16	39425	12	62027	3	28943	6	30093	2	30216	0	4460
	35	81915	54	84927	53	136709	16	64677	25	66667	14	68378	1	10305
	126	172879	168	177886	189	291835	69	139723	90	143084	62	149316	6	22864
	390	354978	476	363159	597	605763	235	292845	282	298427	222	316058	26	49095
	1102	711467	1271	724633	1708	1226109	716	597803	812	606902	694	650848	87	102299
	2880		3198		4546		1977		2167		1966		264	207746
Wt 3	1	12241	1	12241	1	19220	0	9013	0	9013	0	9064	0	1321
	6	27871	6	27871	6	44660	2	21206	2	21206	1	21725	0	3226
	25	61196	25	61196	30	99870	11	47877	11	47877	8	49852	1	7544
	89	130258	89	130258	114	215902	47	104422	47	104422	38	110207	4	16916
	276	269612	276	269612	379	453189	162	220795	162	220795	142	235848	18	36686
	785	544433	785	544433	1124	926468	500	454429	500	454429	459	490480	60	77136
	2071		2071		3082		1398	912676	1398	912676	1336	994356	184	157974
	5162		5162		7902		3657		3657		3591		508	315760
Wt 4	2	14686	1	14147	1	22763	1	10939	0	10559	0	10904	0	1602
	9	33061	6	32116	8	52515	4	25375	2	24709	2	25854	0	3878
	36	71975	28	70357	38	116637	19	56715	13	55563	12	58798	1	8955
	119	152003	100	149290	144	250708	68	122562	55	120625	52	128994	6	19930
	359	312619	318	308150	468	523442	222	257291	193	254073	186	274236	23	42865
	988	627608	905	620369	1374	1065094	648	526069	591	520835	582	567032	78	89583
	2560		2399		3717		1767	1050791	1654	1042375	1656	1143684	230	182352
	6268		5969		9445		4506		4298		4378		629	362724
Wt 5	1	14147	2	14686	1	22763	0	10559	1	10939	0	10904	0	1602
	6	32116	9	33061	8	52515	2	24709	4	25375	2	25854	0	3878
	28	70357	36	71975	38	116637	13	55563	19	56715	12	58798	1	8955
	100	149290	119	152003	144	250708	55	120625	68	122562	52	128994	6	19930
	318	308150	359	312619	468	523442	193	254073	222	257291	186	274236	23	42865
	905	620369	988	627608	1374	1065094	591	520835	648	526069	582	567032	78	89583
	2399		2560		3717		1654	1042375	1767	1050791	1656	1143684	230	182352
	5969		6268		9445		4298		4506		4378		629	362724
Wt 6	2	19729	2	19729	2	31702	1	15152	1	15152	1	15844	0	2440
	11	44125	11	44125	14	72125	6	34731	6	34731	6	36718	1	5709
	47	95405	47	95405	62	158230	28	76767	28	76767	27	81936	4	12838
	160	200127	160	200127	225	336386	101	164232	101	164232	100	176904	15	27953
	487	408865	487	408865	708	695410	325	341567	325	341567	323	370967	48	59044
	1345	815596	1345	815596	2024	1402373	937	692466	937	692466	946	757930	143	121527
	3476		3476		5358		2514		2514		2567		388	244160
	8472		8472		13365		6326		6326		6544		1001	

continued next page

Affine A2 - Level 5 - Class 0 - Page 2 of 2

	Rp 1	Rp 2	Rp 3	Rp 4	Rp 5	Rp 6	Rp 7
Wt 7	2 22950	2 22950	3 37287	1 17921	1 17921	2 19114	1 3043
	13 51125	13 51125	18 84262	8 40800	8 40800	10 43778	2 6968
	55 110045	55 110045	79 183829	35 89549	35 89549	40 96748	8 15457
	190 229961	190 229961	278 388752	128 190515	128 190515	138 207180	24 33216
	575 467995	575 467995	866 799978	400 394095	400 394095	426 431420	72 69512
	1588	1588	2440	1145 795305	1145 795305	1208 876100	196 141818
	4078	4078	6404	3027	3027	3200	515 282995
	9907	9907	15828	7555	7555	8010	1274

Affine A2 - Level 6 - Class 0 - Page 1 of 2

	Rp 1	Rp 2	Rp 3	Rp 4	Rp 5	Rp 6	Rp 7	Rp 8	Rp 9	Rp10
	0 06	0 33	0 60	1 14	1 41	2 22	3 03	3 30	4 11	6 00
Wt 1	1 522	0 1252	0 317	0 1243	0 906	0 996	0 424	0 332	0 280	0 29
	2 1304	1 3412	0 892	2 3358	0 2646	0 2916	0 1252	0 1044	0 906	0 100
	8 3163	8 8784	1 2356	10 8584	2 7147	3 7956	1 3412	0 2970	0 2646	0 317
	24	35	6	40	16	18 20376	8 8784	4 7892	2 7147	0 892
	72	132	29	138	75	84	35	21	16	1 2356
	196	424	100	432	280	306	132	94	75	6
Wt 2	0 652	1 2077	0 652	0 1778	0 1778	0 1724	0 670	0 670	0 539	0 64
	1 1697	4 5484	1 1697	2 4828	2 4828	1 4862	0 1939	0 1939	0 1624	0 201
	6 4192	19 13740	6 4192	12 12352	12 12352	8 12824	2 5218	2 5218	1 4506	0 576
	23	72	23	52	52	41 32016	13 13230	13 13230	8 11708	1 1541
	78	242	78	190	190	164	58	58	41	4 3876
	235	736	235	609	609	562	210	210	160	18
Wt 3	0 317	0 1252	1 522	0 906	0 1243	0 996	0 332	0 424	0 280	0 29
	0 892	1 3412	2 1304	0 2646	2 3358	0 2916	0 1044	0 1252	0 906	0 100
	1 2356	8 8784	8 3163	2 7147	10 8584	3 7956	0 2970	1 3412	0 2646	0 317
	6	35	24	16	40	18 20376	4 7892	8 8784	2 7147	0 892
	29	132	72	75	138	84	21	35	16	1 2356
	100	424	196	280	432	306	94	132	75	6
Wt 4	1 1028	1 2868	0 795	1 2711	0 2296	0 2454	0 1036	0 903	0 784	0 91
	4 2550	6 7456	1 2090	6 7072	2 6228	2 6760	1 2868	0 2586	0 2296	0 282
	15 6071	28 18411	6 5187	27 17540	14 15885	14 17490	6 7456	3 6893	2 6228	0 795
	48	105	26	100	65	66	28 18411	18 17316	14 15885	1 2090
	143	347	91	328	242	250	105	80	65	6 5187
	394	1036	282	978	784	824	347	286	242	26
Wt 5	0 795	1 2868	1 1028	0 2296	1 2711	0 2454	0 903	0 1036	0 784	0 91
	1 2090	6 7456	4 2550	2 6228	6 7072	2 6760	0 2586	1 2868	0 2296	0 282
	6 5187	28 18411	15 6071	14 15885	27 17540	14 17490	3 6893	6 7456	2 6228	0 795
	26	105	48	65	100	66	18 17316	28 18411	14 15885	1 2090
	91	347	143	242	328	250	80	105	65	6 5187
	282	1036	394	784	978	824	286	347	242	26

-continued next page-

Affine A2 - Level 6 - Class 0 - Page 2 of 2

	Rp 1	Rp 2	Rp 3	Rp 4	Rp 5	Rp 6	Rp 7	Rp 8	Rp 9	Rp10
Wt 6	1 1411	2 4524	1 1411	1 4009	1 4009	1 4119	0 1670	0 1670	0 1450	0 189
	4 3522	11 11515	4 3522	8 10408	8 10408	6 10956	2 4524	2 4524	1 4009	0 530
	18	50	18	38 25634	38 25634	33 27600	11 11515	11 11515	8 10408	1 1411
	60	179	60	146	146	132	50	50	38 25634	4 3522
	189	576	189	484	484	465	179	179	146	18
	530	1670	530	1450	1450	1440	576	576	484	60
Wt 7	1 1697	2 5218	1 1541	2 4828	1 4506	1 4862	1 2077	0 1939	0 1778	0 235
	6 4192	13 13230	4 3876	12 12352	8 11708	8 12824	4 5484	2 5218	2 4828	0 652
	23	58	18	52	41	41 32016	19 13740	13 13230	12 12352	1 1697
	78	210	64	190	160	164	72	58	52	6 4192
	235	670	201	609	539	562	242	210	190	23
	652	1939	576	1778	1624	1724	736	670	609	78
Wt 8	1 1541	2 5218	1 1697	1 4506	2 4828	1 4862	0 1939	1 2077	0 1778	0 235
	4 3876	13 13230	6 4192	8 11708	12 12352	8 12824	2 5218	4 5484	2 4828	0 652
	18	58	23	41	52	41 32016	13 13230	19 13740	12 12352	1 1697
	64	210	78	160	190	164	58	72	52	6 4192
	201	670	235	539	609	562	210	242	190	23
	576	1939	652	1624	1778	1724	670	736	609	78
Wt 9	1 2090	3 6893	1 2090	2 6228	2 6228	2 6760	1 2868	1 2868	1 2711	0 394
	6 5187	18 17316	6 5187	14 15885	14 15885	14 17490	6 7456	6 7456	6 7072	1 1028
	26	80	26	65	65	66	28 18411	28 18411	27 17540	4 2550
	91	286	91	242	242	250	105	105	100	15 6071
	282	903	282	784	784	824	347	347	328	48
	795	2586	795	2296	2296	2454	1036	1036	978	143
Wt10	1 2356	4 7892	1 2356	2 7147	2 7147	3 7956	1 3412	1 3412	2 3358	1 522
	6	21	6	16	16	18 20376	8 8784	8 8784	10 8584	2 1304
	29	94	29	75	75	84	35	35	40	8 3163
	100	332	100	280	280	306	132	132	138	24
	317	1044	317	906	906	996	424	424	432	72
	892	2970	892	2646	2646	2916	1252	1252	1243	196

Affine A2 - Level 6 - Class 1 - Page 1 of 2

	Rp 1	Rp 2	Rp 3	Rp 4	Rp 5	Rp 6	Rp 7	Rp 8	Rp 9
	0	0	1	1	2	2	3	4	5
	24	51	05	32	13	40	21	02	10
Wt 1	1 1884	0 1198	0 1017	0 2043	0 1498	0 796	0 927	0 391	0 177
	4 4920	1 3215	2 2686	2 5628	1 4179	0 2264	0 2738	0 1164	0 560
	19 12256	8 8152	9 6709	12 14553	8 10948	3 6032	2 7468	1 3227	0 1597
	68 29152	35 19638	36 16064	56 35762	38 27184	16 15112	16 19154	6 8348	2 4256
	227	130	119	209	148	72	75 46645	31 20544	11 10664
	674	412	364	688	495	252	284	116	50 25496

--continued next page-

Affine A2 - Level 6 - Class 1 - Page 2 of 2

	Rp 1	Rp 2	Rp 3	Rp 4	Rp 5	Rp 6	Rp 7	Rp 8	Rp 9
Wt 2	0 1392	1 1303	0 650	0 1846	0 1098	0 840	0 808	0 286	0 142
	1 3802	4 3336	0 1850	2 5054	0 3206	1 2292	0 2387	0 900	0 459
	8 9760	16 8168	2 4898	12 13034	3 8679	6 5908	2 6532	0 2570	0 1338
	38 23797	54 19212	13 12225	53 31980	20 22098	25 14505	14 16797	3 6828	1 3608
	144	168	58	194	92	90	66 41046	18 17098	8 9148
	471	482	207	627	340	287	247	80	38 22048
Wt 3	1 2292	0 1338	1 1303	0 2387	0 1846	0 900	0 1098	0 471	0 207
	6 5908	1 3608	4 3336	2 6532	2 5054	0 2570	0 3206	0 1392	0 650
	25 14505	8 9148	16 8168	14 16797	12 13034	3 6828	3 8679	1 3802	0 1850
	90	38 22048	54 19212	66 41046	53 31980	18 17098	20 22098	8 9760	2 4898
	287	142	168	247	194	80	92	38 23797	13 12225
	840	459	482	808	627	286	340	144	58
Wt 4	1 2846	1 2189	0 1446	1 3510	0 2374	0 1503	0 1677	0 668	0 339
	6 7377	6 5574	2 3830	6 9293	2 6503	2 4033	1 4699	0 1928	0 996
	28 18161	25 13528	11 9580	30 23301	14 16740	11 10214	8 12320	2 5172	1 2714
	105	89 31512	47 22868	118 55777	65 40932	47 24667	41 30598	13 13072	6 6946
	346	281	164	404	245	166	162	58 31458	28 16898
	1031	812	510	1240	802	523	551	210	104
Wt 5	2 4033	1 2714	1 2189	1 4699	1 3510	0 1928	0 2374	0 1031	0 510
	11 10214	6 6946	6 5574	8 12320	6 9293	2 5172	2 6503	1 2846	0 1446
	47 24667	28 16898	25 13528	41 30598	30 23301	13 13072	14 16740	6 7377	2 3830
	166	104	89 31512	162	118 55777	58 31458	65 40932	28 18161	11 9580
	523	339	281	551	404	210	245	105	47 22868
	1503	996	812	1677	1240	668	802	346	164
Wt 6	1 3227	2 2686	0 1597	1 4179	0 2738	1 1884	0 2043	0 796	0 412
	6 8348	9 6709	2 4256	8 10948	2 7468	4 4920	2 5628	0 2264	0 1198
	31 20544	36 16064	11 10664	38 27184	16 19154	19 12256	12 14553	3 6032	1 3215
	116	119	50 25496	148	75 46645	68 29152	56 35762	16 15112	8 8152
	391	364	177	495	284	227	209	72	35 19638
	1164	1017	560	1498	927	674	688	252	130
Wt 7	2 5172	2 3830	1 2714	2 6503	1 4699	1 2846	1 3510	0 1503	0 812
	13 13072	11 9580	6 6946	14 16740	8 12320	6 7377	6 9293	2 4033	1 2189
	58 31458	47 22868	28 16898	65 40932	41 30598	28 18161	30 23301	11 10214	6 5574
	210	164	104	245	162	105	118 55777	47 24667	25 13528
	668	510	339	802	551	346	404	166	89 31512
	1928	1446	996	2374	1677	1031	1240	523	281
Wt 8	3 6032	2 4256	1 3215	2 7468	2 5628	1 3227	1 4179	1 1884	0 1017
	16 15112	11 10664	8 8152	16 19154	12 14553	6 8348	8 10948	4 4920	2 2686
	72	50 25496	35 19638	75 46645	56 35762	31 20544	38 27184	19 12256	9 6709
	252	177	130	284	209	116	148	68 29152	36 16064
	796	560	412	927	688	391	495	227	119
	2264	1597	1198	2738	2043	1164	1498	674	364
Wt 9	3 6828	2 4898	1 3608	3 8679	2 6532	1 3802	2 5054	1 2292	1 1303
	18 17098	13 12225	8 9148	20 22098	14 16797	8 9760	12 13034	6 5908	4 3336
	80	58	38 22048	92	66 41046	38 23797	53 31980	25 14505	16 8168
	286	207	142	340	247	144	194	90	54 19212
	900	650	459	1098	808	471	627	287	168
	2570	1850	1338	3206	2387	1392	1846	840	482

Affine A2 - Level 7 - Class 0 - Page 1 of 3

	Rp 1	Rp 2	Rp 3	Rp 4	Rp 5	Rp 6	Rp 7	Rp 8	Rp 9	Rp10	Rp11	Rp12
	0	0	1	1	1	2	2	3	4	4	5	7
	25	52	06	33	60	14	41	22	03	30	11	00
Wt 1	1	0	0	0	0	0	0	0	0	0	0	0
	4	1	2	2	0	1	0	0	0	0	0	0
	19	8	9	12	2	8	3	2	1	0	0	0
	68	38	36	56	13	38	20	16	6	3	2	0
	227	148	119	214	58	148	95	79	31	20	14	1
	680	493	364	718	211	500	360	311	120	91	68	6
	1916	1492	1023	2184	673	1532	1195	1054	413	337	262	26
	5066	4160	2721	6156	1953	4338	3574	3221	1266	1091	874	95
								9073	3598	3205	2628	300
												869
Wt 2	0	1	0	0	0	0	0	0	0	0	0	0
	1	4	0	2	2	0	1	0	0	0	0	0
	8	19	2	12	9	3	8	2	0	1	0	0
	38	68	13	56	36	20	38	16	3	6	2	0
	148	227	58	214	119	95	148	79	20	31	14	1
	493	680	211	718	364	360	500	311	91	120	68	6
	1492	1916	673	2184	1023	1195	1532	1054	337	413	262	26
	4160	5066	1953	6156	2721	3574	4338	3221	1091	1266	874	95
								9073	3205	3598	2628	300
												869
Wt 3	1	0	1	0	0	0	0	0	0	0	0	0
	6	1	4	2	0	2	0	0	0	0	0	0
	25	8	16	14	2	12	3	3	1	0	0	0
	90	41	54	66	13	53	22	20	8	4	2	0
	287	160	168	252	61	194	105	97	38	23	16	1
	846	543	482	840	224	633	402	370	148	105	78	6
	2326	1645	1310	2536	725	1883	1333	1239	495	383	300	29
	6062	4606	3374	7094	2114	5227	3992	3735	1502	1236	998	104
									4204	3607	2988	335
												967
Wt 4	1	1	0	1	0	0	0	0	0	0	0	0
	6	6	2	6	2	2	2	1	0	0	0	0
	28	28	11	30	11	14	14	8	2	2	1	0
	105	105	47	122	47	65	65	44	13	13	8	1
	351	351	164	429	164	249	249	182	61	61	41	4
	1058	1058	515	1356	515	829	829	648	227	227	164	18
	2969	2969	1475	3945	1475	2503	2503	2046	747	747	564	64
	7820	7820	3959	10724	3959	6993	6993	5935	2217	2217	1738	206
									6118	6118	4928	598
												1631
Wt 5	0	1	0	0	1	0	0	0	0	0	0	0
	1	6	0	2	4	0	2	0	0	0	0	0
	8	25	2	14	16	3	12	3	0	1	0	0
	41	90	13	66	54	22	53	20	4	8	2	0
	160	287	61	252	168	105	194	97	23	38	16	1
	543	846	224	840	482	402	633	370	105	148	78	6
	1645	2326	725	2536	1310	1333	1883	1239	383	495	300	29
	4606	6062	2114	7094	3374	3992	5227	3735	1236	1502	998	104
									3607	4204	2988	335
												967

continued next page

Affine A2 - Level 7 - Class 0 - Page 2 of 3

	Rp 1	Rp 2	Rp 3	Rp 4	Rp 5	Rp 6	Rp 7	Rp 8	Rp 9	Rp10	Rp11	Rp12
Wt 6	2	1	1	1	0	1	0	0	0	0	0	0
	11	6	6	8	2	6	2	2	1	0	0	0
	47	31	25	41	11	30	16	14	6	3	2	0
	166	120	89	166	50	118	78	69	28	18	14	1
	528	412	281	578	181	409	304	270	109	83	65	6
	1532	1259	818	1804	583	1272	1021	921	368	303	246	26
	4167	3565	2222	5181	1698	3661	3093	2823	1130	983	811	91
						9861	8643	7996	3204	2881	2420	287
											6692	818
												2186
Wt 7	1	2	0	1	1	0	1	0	0	0	0	0
	6	11	2	8	6	2	6	2	0	1	0	0
	31	47	11	41	25	16	30	14	3	6	2	0
	120	166	50	166	89	78	118	69	18	28	14	1
	412	528	181	578	281	304	409	270	83	109	65	6
	1259	1532	583	1804	818	1021	1272	921	303	368	246	26
	3565	4167	1698	5181	2222	3093	3661	2823	983	1130	811	91
						8643	9861	7996	2881	3204	2420	287
											6692	818
												2186
Wt 8	2	2	1	2	1	1	1	1	0	0	0	0
	13	13	6	14	6	8	8	6	2	2	1	0
	58	58	28	68	28	41	41	33	11	11	8	1
	214	214	104	265	104	166	166	136	50	50	38	4
	691	691	344	896	344	576	576	490	183	183	146	18
	2034	2034	1023	2731	1023	1795	1795	1562	599	599	489	60
	5557	5557	2833	7688	2833	5142	5142	4581	1775	1775	1482	189
								12504	4908	4908	4157	536
												1438
Wt 9	3	2	1	2	1	2	1	1	1	0	0	0
	16	13	8	16	6	12	8	8	4	2	2	0
	72	61	35	78	28	56	44	41	19	13	12	1
	256	227	130	304	108	213	180	168	72	58	52	6
	819	745	417	1024	360	715	634	589	247	214	190	23
	2373	2203	1226	3108	1087	2169	1982	1854	762	693	614	78
	6430	6055	3339	8710	3030	6103	5696	5358	2194	2047	1812	235
									5906	5615	4984	658
												1725
Wt10	2	3	1	2	1	1	2	1	0	1	0	0
	13	16	6	16	8	8	12	8	2	4	2	0
	61	72	28	78	35	44	56	41	13	19	12	1
	227	256	108	304	130	180	213	168	58	72	52	6
	745	819	360	1024	417	634	715	589	214	247	190	23
	2203	2373	1087	3108	1226	1982	2169	1854	693	762	614	78
	6055	6430	3030	8710	3339	5696	6103	5358	2047	2194	1812	235
									5615	5906	4984	658
												1725

continued next page

Affine A2 - Level 7 - Class 0 - Page 3 of 3

	Rp 1	Rp 2	Rp 3	Rp 4	Rp 5	Rp 6	Rp 7	Rp 8	Rp 9	Rp10	Rp11	Rp12
Wt11	3	3	1	3	1	2	2	2	1	1	1	0
	18	18	8	22	8	14	14	14	6	6	6	1
	83	83	38	105	38	69	69	66	28	28	27	4
	303	303	146	402	146	267	267	255	105	105	100	15
	980	980	481	1337	481	901	901	856	352	352	328	48
	2861	2861	1435	4012	1435	2734	2734	2606	1064	1064	984	143
						7672	7672	7336	2996	2996	2750	394
											7250	1035
												2582
Wt12	3	3	1	4	1	2	2	3	1	1	2	1
	20	20	8	26	8	16	16	18	8	8	10	2
	91	91	41	122	41	79	79	84	35	35	40	8
	337	337	158	462	158	307	307	312	132	132	138	24
	1087	1087	529	1524	529	1031	1031	1031	429	429	432	72
	3182	3182	1580	4546	1580	3118	3118	3082	1282	1282	1250	196
									3548	3548	3400	522
												1312

Weight Multiplicities for Affine A3

Affine A3 - Level 1 - Class 0

	Rp 1				
	1				
	000				
Wt 1	1	810	57222	1762462	34034391
	3	1479	90882	2606604	48104069
	9	2640	142769	3829437	67679109
	22	4599	221910	5590110	94800537
	51	7868	341649	8111346	132230021
	108	13209	521196	11701998	183686994
	221	21843	788460	16790136	254170332
	429	35581	1183221	23964594	350371088

Affine A3 - Level 2 - Class 0

	Rp 1				Rp 2				Rp 3			
	0				0				2			
	020				101				000			
Wt 1	1	7295	2348317	232218704	0	10878	3799557	389332644	0	3001	1236639	136347682
	3	16615	4376081	391532317	3	25236	7121793	658569561	0	7186	2345445	232176591
	15	36572	8027381	654031065	15	56211	13132080	29725529*	2	16470	4372257	391477150
	50	77819	14509718	9182660*	63	121089	23853108	758653889*	11	36305	8021522	653953417
	162	161120	25873470	704408691*	210	253035	42724638	869563495*	45	77462	14501944	9081393*
	457	325001	45553571	749099567*	636	515067	75539292	632629760*	148	160502	25861870	704268212*
	1226	641113	79259551	387979032*	1743	1023636	131944368	470147759*	439	324175	45538242	748917172*
	3055	1238454	136377107		4485	1991517	227863977		1185	639754	79237189	387729243*
Wt 2	1	10997	3210622	301868066	1	16660	5212989	506980545	0	4710	1708062	178166949
	6	24658	5935856	506578162	7	37836	9689739	853456334	1	10997	3210622	301868066
	25	53389	10807999	842443984	32	82836	17730751	349778621*	6	24658	5935856	506578162
	87	112114	19402759	315273406*	117	175658	31977358	206214356*	25	53389	10807999	842443984
	266	229151	34377061	124120484*	371	362153	56899365	638322943*	87	112114	19402759	315273406*
	741	457175	60164903	465076418*	1063	728159	99981354	910412274*	266	229151	34377061	124120484*
	1923	892449	104094442		2819	1431447	173632536		741	457175	60164903	465076418*
	4710	1708062	178166949		7029	2757188	298236157		1923	892449	104094442	
Wt 3	2	16470	4372257	391477150	3	25236	7121793	658569561	1	7295	2348317	232218704
	11	36305	8021522	653953417	15	56211	13132080	29725529*	3	16615	4376081	391532317
	45	77462	14501944	9081393*	63	121089	23853108	758653889*	15	36572	8027381	654031065
	148	160502	25861870	704268212*	210	253035	42724638	869563495*	50	77819	14509718	9182660*
	439	324175	45538242	748917172*	636	515067	75539292	632629760*	162	161120	25873470	704408691*
	1185	639754	79237189	387729243*	1743	1023636	131944368	470147759*	457	325001	45553571	749099567*
	3001	1236639	136347682		4485	1991517	227863977		1226	641113	79259551	387979032*
	7186	2345445	232176591		10878	3799557	389332644		3055	1238454	136377107	

* Multiplicity is given modulo 1073741527

Affine A3 - Level 2 - Class 1

	Rp 1					Rp 2				
	0					1				
	011					100				
Wt 1	1	31732	14979446	922719317*	813946996*	0	21082	11088696	480222159*	372204527*
	6	69545	27030558	53629110*	273670429*	2	47072	20153920	412018618*	1022685389*
	27	147620	48122514	1034610626*	1062159859*	12	101540	36120120	960466115*	296956753*
	98	304631	84600770	18564044*	632509975*	50	212516	63896098	337110054*	844288569*
	310	613030	146990096	920433650*		172	433054	111663688	173510614*	1006700266*
	888	1206089	252584870	513445415*		522	861616	192929402	257375748*	
	2357	2324844	429554069	433103214*		1444	1677766	329788336	837631959*	
	5882	4398602	723398612	1022448421*		3724	3203748	558077398	712628412*	
	13957	8181250	133296504*	685542785*		9072	6009390	935452628	802158623*	
Wt 2	2	47072	20153920	412018618*	1022685389*	1	31732	14979446	922719317*	813946996*
	12	101540	36120120	960466115*	296956753*	6	69545	27030558	53629110*	273670429*
	50	212516	63896098	337110054*	844288569*	27	147620	48122514	1034610626*	1062159859*
	172	433054	111663688	173510614*	1006700266*	98	304631	84600770	18564044*	632509975*
	522	861616	192929402	257375748*		310	613030	146990096	920433650*	
	1444	1677766	329788336	837631959*		888	1206089	252584870	513445415*	
	3724	3203748	558077398	712628412*		2357	2324844	429554069	433103214*	
	9072	6009390	935452628	802158623*		5882	4398602	723398612	1022448421*	
	21082	11088696	480222159*	372204527*		13957	8181250	133296504*	685542785*	

* Multiplicity is given modulo 1073741527

Affine A3 - Level 3 - Class 0

	Rp 1		Rp 2		Rp 3		Rp 4		Rp 5	
	0		0		1		1		3	
	012		210		020		101		000	
Wt 1	1	689352	0	684753	0	670225	0	754884	0	112260
	6	1593318	3	1585450	3	1564677	2	1779570	0	270099
	33	3575019	24	3561810	21	3541959	18	4063066	1	627975
	143	7810983	121	7789140	108	7799267	99	9015912	9	1416026
	543	16659774	492	16624193	441	16751871	430	19499562	45	3107286
	1836	34762905	1728	34705683	1577	35176473	1593	41205119	186	6652563
	5702	71095761	5481	71004879	5067	72357318	5274	85246758	657	13928570
	16473	142742464	16044	142599695	15048	146042613	16005	172970511	2103	28572714
	44883	281740494	44073	281518584	41833	289658109	45324	344743842	6194	57525282
	116269		114790		110214		121200		17127	113827399
	288510		285870		277227		308829		44847	
Wt 2	0	684753	1	689352	0	670225	0	754884	0	112260
	3	1585450	6	1593318	3	1564677	2	1779570	0	270099
	24	3561810	33	3575019	21	3541959	18	4063066	1	627975
	121	7789140	143	7810983	108	7799267	99	9015912	9	1416026
	492	16624193	543	16659774	441	16751871	430	19499562	45	3107286
	1728	34705683	1836	34762905	1577	35176473	1593	41205119	186	6652563
	5481	71004879	5702	71095761	5067	72357318	5274	85246758	657	13928570
	16044	142599695	16473	142742464	15048	146042613	16005	172970511	2103	28572714
	44073	281518584	44883	281740494	41833	289658109	45324	344743842	6194	57525282
	114790		116269		110214		121200		17127	113827399
	285870		288510		277227		308829		44847	
Wt 3	1	912440	1	912440	1	893468	0	1007601	0	150437
	9	2088676	9	2088676	7	2063101	4	2349655	0	358088
	49	4645865	49	4645865	41	4624613	31	5312976	2	824311
	214	10069992	214	10069992	182	10094816	161	11687003	14	1842435
	795	21322135	795	21322135	700	21511588	665	25077645	70	4010241
	2646	44193943	2646	44193943	2379	44848808	2380	52611634	276	8522793
	8063	89826487	8063	89826487	7406	91649196	7658	108128350	957	17723232
	22933	179317261	22933	179317261	21407	183868904	22721	218067108	2989	36130610
	61553	352049698	61553	352049698	58317	362654072	63120	432185068	8661	72320576
	157398		157398		150967		166085		23552	142335927
	385936		385936		374297		417295		60858	
Wt 4	2	1205190	2	1205190	1	1182586	1	1343178	0	203195
	15	2732430	15	2732430	11	2704203	10	3097684	1	477076
	79	6025000	79	6025000	62	6008941	60	6935957	6	1085620
	328	12956735	328	12956735	276	13011927	274	15123950	31	2401825
	1184	27237142	1184	27237142	1031	27526401	1056	32196921	126	5181325
	3832	56081874	3832	56081874	3442	57004487	3601	67064452	456	10923429
	11436	113295702	11436	113295702	10493	115771385	11199	136930461	1477	22552168
	31918	224895304	31918	224895304	29839	230934898	32354	274493595	4430	45673278
	84347		84347		80023		88009	540994678	12416	90874391
	212703		212703		204436		227555		32973	
	515307		515307		500731		563390		83512	
Wt 5	3	1586694	3	1586694	2	1560438	3	1783324	1	272973
	25	3563919	25	3563919	18	3534759	21	4069452	3	632892
	123	7792671	123	7792671	99	7787265	107	9026571	15	1424262
	498	16629978	498	16629978	419	16732155	450	19517085	60	3120897
	1742	34715058	1742	34715058	1527	35144553	1638	41233533	219	6674734
	5511	71019858	5511	71019858	4962	72306294	5367	85292241	726	13964199
	16104	142623342	16104	142623342	14835	145962030	16194	173042428	2245	28629291
	44193	281555490	44193	281555490	41424	289532195	45687	344856381	6465	57614178
	115011		115011		109449		121876		17637	113965548
	286275		286275		275841		310056		45778	
	685471		685471		667773		757056		113913	

Affine A3 - Level 4 - Class 0 - Page 1 of 2

	Rp 1	Rp 2	Rp 3	Rp 4	Rp 5	Rp 6	Rp 7	Rp 8	Rp 9	Rp10
	0 004	0 040	0 121	0 202	0 400	1 012	1 210	2 020	2 101	4 000
Wt 1	1	0	0	0	0	0	0	0	0	0
	3	1	3	2	0	3	0	0	0	0
	15	12	33	18	3	21	7	6	3	0
	60	69	198	114	22	117	66	44	33	1
	234	315	942	550	126	537	384	252	198	12
	822	1212	3786	2274	546	2169	1753	1122	942	69
	2741	4194	13582	8310	2074	7866	6828	4380	3786	315
	8529	13302	44481	27722	7011	26214	23778	15270	13582	1212
	25203	39459	135624	85764	21894	81346	75921	49074	44481	4194
									135624	13302
										39459
Wt 2	0	1	0	0	0	0	0	0	0	0
	0	3	3	0	0	0	0	0	0	0
	1	15	21	6	1	3	3	2	0	0
	12	60	117	44	12	33	33	18	7	0
	69	234	537	252	69	198	198	114	66	3
	315	822	2169	1122	315	942	942	550	384	22
	1212	2741	7866	4380	1212	3786	3786	2274	1753	126
	4194	8529	26214	15270	4194	13582	13582	8310	6828	546
	13302	25203	81346	49074	13302	44481	44481	27722	23778	2074
	39459				39459	135624	135624	85764	75921	7011
										21894
Wt 3	0	1	1	0	0	0	0	0	0	0
	1	6	10	3	1	2	2	1	0	0
	9	31	66	28	9	21	21	11	5	0
	51	136	337	167	51	134	134	74	47	2
	231	528	1452	785	231	658	658	378	278	17
	902	1861	5514	3151	902	2727	2727	1615	1288	95
	3156	6062	18989	11274	3156	10001	10001	6069	5098	419
	10144	18468	60433	36905	10144	33391	33391	20669	17999	1594
	30439	53161	180179	112442	30439	103407	103407	65074	58202	5443
								192059	175365	17105
										50275
Wt 4	0	1	1	1	0	0	0	0	0	0
	2	6	13	7	2	4	4	1	1	0
	14	37	88	51	14	34	34	16	13	1
	76	165	454	261	76	204	204	106	88	6
	328	663	1943	1156	328	950	950	535	454	37
	1246	2348	7315	4428	1246	3814	3814	2233	1943	165
	4254	7683	24923	15406	4254	13622	13622	8237	7315	663
	13434	23340	78508	49273	13434	44588	44588	27555	24923	2348
	39684	66983	231743	147570	39684	135782	135782	85477	78508	7683
									231743	23340
										66983
Wt 5	0	0	0	0	1	0	0	0	0	0
	0	1	3	2	3	0	3	0	0	0
	3	12	33	18	15	7	21	6	3	0
	22	69	198	114	60	66	117	44	33	1
	126	315	942	550	234	384	537	252	198	12
	546	1212	3786	2274	822	1753	2169	1122	942	69
	2074	4194	13582	8310	2741	6828	7866	4380	3786	315
	7011	13302	44481	27722	8529	23778	26214	15270	13582	1212
	21894	39459	135624	85764	25203	75921	81346	49074	44481	4194
									135624	13302
										39459

continued next page

Affine A3 - Level 4 - Class 0 - Page 2 of 2

	Rp 1	Rp 2	Rp 3	Rp 4	Rp 5	Rp 6	Rp 7	Rp 8	Rp 9	Rp10
Wt 6	1	1	2	1	0	1	0	0	0	0
	6	9	21	11	2	10	5	3	2	0
	31	51	134	74	17	66	47	28	21	1
	136	231	658	378	95	337	278	167	134	9
	528	902	2727	1615	419	1452	1288	785	658	51
	1861	3156	10001	6069	1594	5514	5098	3151	2727	231
	6062	10144	33391	20669	5443	18989	17999	11274	10001	902
	18468	30439	103407	65074	17105	60433	58202	36905	33391	3156
	53161			192059	50275	180179	175365	112442	103407	10144
										30439
Wt 7	0	1	2	1	1	0	1	0	0	0
	2	9	21	11	6	5	10	3	2	0
	17	51	134	74	31	47	66	28	21	1
	95	231	658	378	136	278	337	167	134	9
	419	902	2727	1615	528	1288	1452	785	658	51
	1594	3156	10001	6069	1861	5098	5514	3151	2727	231
	5443	10144	33391	20669	6062	17999	18989	11274	10001	902
	17105	30439	103407	65074	18468	58202	60433	36905	33391	3156
	50275			192059	53161	175365	180179	112442	103407	10144
										30439
Wt 8	1	2	4	1	1	1	1	1	0	0
	6	14	34	16	6	13	13	7	4	0
	37	76	204	106	37	88	88	51	34	2
	165	328	950	535	165	454	454	261	204	14
	663	1246	3814	2233	663	1943	1943	1156	950	76
	2348	4254	13622	8237	2348	7315	7315	4428	3814	328
	7683	13434	44588	27555	7683	24923	24923	15406	13622	1246
	23340	39684	135782	85477	23340	78508	78508	49273	44588	4254
	66983				66983	231743	231743	147570	135782	13434
										39684
Wt 9	1	2	5	3	1	2	2	1	1	0
	9	17	47	28	9	21	21	11	10	1
	51	95	278	167	51	134	134	74	66	6
	231	419	1288	785	231	658	658	378	337	31
	902	1594	5098	3151	902	2727	2727	1615	1452	136
	3156	5443	17999	11274	3156	10001	10001	6069	5514	528
	10144	17105	58202	36905	10144	33391	33391	20669	18989	1861
	30439	50275	175365	112442	30439	103407	103407	65074	60433	6062
								192059	180179	18468
										53161
Wt10	1	3	7	6	1	3	3	2	3	1
	12	22	66	44	12	33	33	18	21	3
	69	126	384	252	69	198	198	114	117	15
	315	546	1753	1122	315	942	942	550	537	60
	1212	2074	6828	4380	1212	3786	3786	2274	2169	234
	4194	7011	23778	15270	4194	13582	13582	8310	7866	822
	13302	21894	75921	49074	13302	44481	44481	27722	26214	2741
	39459				39459	135624	135624	85764	81346	8529
										25203

Affine A3 - Level 4 - Class 1 - Page 1 of 2

	Rp 1 0 031	Rp 2 0 112	Rp 3 0 310	Rp 4 1 003	Rp 5 1 120	Rp 6 1 201	Rp 7 2 011	Rp 8 3 100
Wt 1	1	0	0	0	0	0	0	0
	6	3	1	0	2	0	0	0
	33	27	12	3	18	6	3	0
	153	161	75	28	111	54	33	4
	627	765	363	159	541	315	207	36
	2316	3111	1479	717	2253	1445	999	207
	7862	11275	5352	2779	8331	5685	4067	933
	24828	37356	17664	9654	28081	19986	14679	3603
	73755	115077	54180	30816	87792	64434	48306	12444
	207811	333663	156393	91883	257841	193704	147681	39462
	559176	918912	428868	258849	718120	549525	424869	116841
				694890			1161216	326880
Wt 2	1	1	0	0	0	0	0	0
	9	10	3	2	4	2	1	0
	55	69	27	15	37	22	14	2
	265	359	155	85	222	143	97	18
	1093	1565	703	389	1043	710	501	108
	4005	5982	2739	1545	4187	2961	2147	503
	13394	20684	9549	5503	14975	10900	8074	2006
	41587	65984	30564	18001	48999	36475	27488	7128
	121399	197019	91309	54890	149214	113113	86494	23186
	336296	556351	257609	157852	428133	329467	255101	70224
	890453	1497726	692309	431760	1167906	910267	712539	200502
								544495
Wt 3	0	0	1	0	0	0	0	0
	3	3	6	0	3	2	0	0
	28	33	33	4	27	18	6	1
	159	207	153	36	161	111	54	12
	717	999	627	207	765	541	315	75
	2779	4067	2316	933	3111	2253	1445	363
	9654	14679	7862	3603	11275	8331	5685	1479
	30816	48306	24828	12444	37356	28081	19986	5352
	91883	147681	73755	39462	115077	87792	64434	17664
	258849	424869	207811	116841	333663	257841	193704	54180
	694890	1161216	559176	326880	918912	718120	549525	156393
								428868
Wt 4	1	2	0	1	0	0	0	0
	12	18	4	6	6	3	3	0
	75	111	36	33	54	33	27	3
	363	541	207	153	315	207	161	28
	1479	2253	933	627	1445	999	765	159
	5352	8331	3603	2316	5685	4067	3111	717
	17664	28081	12444	7862	19986	14679	11275	2779
	54180	87792	39462	24828	64434	48306	37356	9654
	156393	257841	116841	73755	193704	147681	115077	30816
	428868	718120	326880	207811	549525	424869	333663	91883
				559176		1161216	918912	258849
								694890

-continued next page-

Affine A3 - Level 4 - Class 1 - Page 2 of 2

	Rp 1	Rp 2	Rp 3	Rp 4	Rp 5	Rp 6	Rp 7	Rp 8
Wt 5	2 15 85 389 1545 5503 18001 54890 157852 431760	1 14 97 501 2147 8074 27488 86494 255101 712539	1 9 55 265 1093 4005 13394 41587 121399 336296 890453	0 2 18 108 503 2006 7128 23186 70224 200502 544495	1 10 69 359 1565 5982 20684 65984 197019 556351 1497726	0 4 37 222 1043 4187 14975 48999 149214 428133 1167906	0 2 22 143 710 2961 10900 36475 113113 329467 910267	0 0 3 27 155 703 2739 9549 30564 91309 257609 692309
Wt 6	2 18 108 503 2006 7128 23186 70224 200502 544495	2 22 143 710 2961 10900 36475 113113 329467 910267	2 15 85 389 1545 5503 18001 54890 157852 431760	0 3 27 155 703 2739 9549 30564 91309 257609 692309	1 14 97 501 2147 8074 27488 86494 255101 712539	1 10 69 359 1565 5982 20684 65984 197019 556351 1497726	0 4 37 222 1043 4187 14975 48999 149214 428133 1167906	0 1 9 55 265 1093 4005 13394 41587 121399 336296 890453
Wt 7	3 27 155 703 2739 9549 30564 91309 257609 692309	4 37 222 1043 4187 14975 48999 149214 428133 1167906	2 18 108 503 2006 7128 23186 70224 200502 544495	1 9 55 265 1093 4005 13394 41587 121399 336296 890453	2 22 143 710 2961 10900 36475 113113 329467 910267	1 14 97 501 2147 8074 27488 86494 255101 712539	1 10 69 359 1565 5982 20684 65984 197019 556351 1497726	0 2 15 85 389 1545 5503 18001 54890 157852 431760
Wt 8	4 36 207 933 3603 12444 39462 116841 326880	6 54 315 1445 5685 19986 64434 193704 549525	3 28 159 717 2779 9654 30816 91883 258849 694890	1 12 75 363 1479 5352 17664 54180 156393 428868	3 33 207 999 4067 14679 48306 147681 424869 1161216	3 27 161 765 3111 11275 37356 115077 333663 918912	2 18 111 541 2253 8331 28081 87792 257841 718120	1 6 33 153 627 2316 7862 24828 73755 207811 559176

Affine A3 - Level 4 - Class 2 - Page 1 of 2

	Rp 1 0 022	Rp 2 0 103	Rp 3 0 220	Rp 4 0 301	Rp 5 1 030	Rp 6 1 111	Rp 7 2 002	Rp 8 2 200	Rp 9 3 010
Wt 1	1	0	0	0	0	0	0	0	0
	6	3	3	1	3	2	0	0	0
	39	21	27	15	21	24	3	3	1
	188	120	157	96	120	162	27	24	15
	807	541	720	479	541	824	157	148	96
	3036	2145	2835	1971	2145	3504	720	684	479
	10461	7599	9984	7197	7599	13104	2835	2742	1971
	33270	24795	32253	23841	24795	44434	9984	9728	7197
	99354	75375	97173	73341	75375	139368	32253	31665	23841
	280656	216306	276281	211950	216306	409968	97173	95802	73341
							276281	273385	211950
Wt 2	1	1	0	0	0	0	0	0	0
	9	7	4	2	3	4	1	0	0
	55	41	37	22	27	40	9	4	3
	268	198	213	138	157	248	55	37	27
	1117	829	967	662	720	1200	268	213	157
	4140	3102	3764	2676	2835	4924	1117	967	720
	13991	10603	13106	9586	9984	17928	4140	3764	2835
	43863	33617	41885	31318	32253	59528	13991	13106	9984
	129174	100064	124938	95092	97173	183536	43863	41885	32253
		282199		271842	276281	532208	129174	124938	97173
									276281
Wt 3	0	0	1	0	0	0	0	0	0
	3	1	6	3	3	2	0	0	0
	27	15	39	21	21	24	3	3	1
	157	96	188	120	120	162	24	27	15
	720	479	807	541	541	824	148	157	96
	2835	1971	3036	2145	2145	3504	684	720	479
	9984	7197	10461	7599	7599	13104	2742	2835	1971
	32253	23841	33270	24795	24795	44434	9728	9984	7197
	97173	73341	99354	75375	75375	139368	31665	32253	23841
	276281	211950	280656	216306	216306	409968	95802	97173	73341
							273385	276281	211950
Wt 4	0	0	1	1	0	0	0	0	0
	4	2	9	7	3	4	0	1	0
	37	22	55	41	27	40	4	9	3
	213	138	268	198	157	248	37	55	27
	967	662	1117	829	720	1200	213	268	157
	3764	2676	4140	3102	2835	4924	967	1117	720
	13106	9586	13991	10603	9984	17928	3764	4140	2835
	41885	31318	43863	33617	32253	59528	13106	13991	9984
	124938	95092	129174	100064	97173	183536	41885	43863	32253
		271842		282199	276281	532208	124938	129174	97173
									276281
Wt 5	1	0	1	0	1	0	0	0	0
	9	3	9	3	7	4	0	0	0
	55	27	55	27	41	40	4	4	2
	268	157	268	157	198	248	37	37	22
	1117	720	1117	720	829	1200	213	213	138
	4140	2835	4140	2835	3102	4924	967	967	662
	13991	9984	13991	9984	10603	17928	3764	3764	2676
	43863	32253	43863	32253	33617	59528	13106	13106	9586
	129174	97173	129174	97173	100064	183536	41885	41885	31318
		276281		276281	282199	532208	124938	124938	95092
									271842

continued next page

Affine A3 - Level 4 - Class 2 - Page 2 of 2

	Rp 1	Rp 2	Rp 3	Rp 4	Rp 5	Rp 6	Rp 7	Rp 8	Rp 9
Wt 6	2	1	2	1	1	1	0	0	0
	18	11	18	11	11	14	2	2	1
	108	71	108	71	71	104	18	18	11
	510	352	510	352	352	560	108	108	71
	2058	1470	2058	1470	1470	2474	510	510	352
	7398	5423	7398	5423	5423	9520	2058	2058	1470
	24318	18194	24318	18194	18194	33008	7398	7398	5423
	74358	56573	74358	56573	56573	105408	24318	24318	18194
	214128	165215	214128	165215	165215	314744	74358	74358	56573
							214128	214128	165215
Wt 7	3	3	3	1	1	2	1	0	0
	27	21	24	15	15	24	6	3	3
	157	120	148	96	96	162	39	27	21
	720	541	684	479	479	824	188	157	120
	2835	2145	2742	1971	1971	3504	807	720	541
	9984	7599	9728	7197	7197	13104	3036	2835	2145
	32253	24795	31665	23841	23841	44434	10461	9984	7599
	97173	75375	95802	73341	73341	139368	33270	32253	24795
	276281	216306	273385	211950	211950	409968	99354	97173	75375
							280656	276281	216306
Wt 8	3	1	3	3	1	2	0	1	0
	24	15	27	21	15	24	3	6	3
	148	96	157	120	96	162	27	39	21
	684	479	720	541	479	824	157	188	120
	2742	1971	2835	2145	1971	3504	720	807	541
	9728	7197	9984	7599	7197	13104	2835	3036	2145
	31665	23841	32253	24795	23841	44434	9984	10461	7599
	95802	73341	97173	75375	73341	139368	32253	33270	24795
	273385	211950	276281	216306	211950	409968	97173	99354	75375
							276281	280656	216306
Wt 9	4	3	4	3	2	4	1	1	1
	37	27	37	27	22	40	9	9	7
	213	157	213	157	138	248	55	55	41
	967	720	967	720	662	1200	268	268	198
	3764	2835	3764	2835	2676	4924	1117	1117	829
	13106	9984	13106	9984	9586	17928	4140	4140	3102
	41885	32253	41885	32253	31318	59528	13991	13991	10603
	124938	97173	124938	97173	95092	183536	43863	43863	33617
		276281		276281	271842	532208	129174	129174	100064
									282199

Affine A3 - Level 5 - Class 0 - Page 1 of 3

	Rp 1 0 032	Rp 2 0 113	Rp 3 0 230	Rp 4 0 311	Rp 5 1 004	Rp 6 1 040	Rp 7 1 121	Rp 8 1 202	Rp 9 1 400	Rp10 2 012	Rp11 2 210	Rp12 3 020	Rp13 3 101	Rp14 5 000
Wt 1	1 6 39 198 891 3564 13035	0 3 27 173 883 3855 14961	0 3 27 163 791 3308 12399	0 1 18 138 771 3526 14079	0 0 3 28 165 788 3234 11915	0 3 21 120 561 2322 8664	0 2 24 174 960 4425 17886	0 0 6 63 408 2064 8850	0 0 2 21 137 690 2940 11103	0 0 3 36 249 1308 5766 22398	0 0 3 33 234 1245 5559 21780	0 0 1 15 111 618 2826 11314	0 0 0 6 63 405 2019 8529	0 0 0 0 2 20 120 570 2307
Wt 2	1 9 61 319 1433 5680 20478	1 10 69 375 1727 7018 25840	0 4 40 249 1220 5092 18952	0 3 37 256 1345 5908 22858	0 2 15 85 399 1650 6159	0 3 27 169 826 3477 13032	0 4 46 318 1686 7505 29440	0 2 22 158 861 3935 15765	0 0 4 39 241 1167 4815	0 1 14 103 574 2652 10745	0 0 7 70 452 2259 9606	0 0 3 35 230 1177 5070 19370	0 0 2 24 169 898 3993 15612	0 0 0 1 9 57 276 1156 4296
Wt 3	0 3 27 163 791 3308 12399	0 1 18 138 771 3526 14079	1 6 39 198 891 3564 13035	0 3 27 173 883 3855 14961	0 0 2 21 137 690 2940 11103	0 3 21 120 561 2322 8664	0 2 24 174 960 4425 17886	0 0 6 63 408 2064 8850	0 0 3 28 165 788 3234 11915	0 0 3 33 234 1245 5559 21780	0 0 3 36 249 1308 5766 22398	0 0 1 15 111 618 2826 11314	0 0 0 6 63 405 2019 8529	0 0 0 0 2 20 120 570 2307
Wt 4	0 4 40 249 1220 5092 18952	0 3 37 256 1345 5908 22858	1 9 61 319 1433 5680 20478	1 10 69 375 1727 7018 25840	0 0 4 39 241 1167 4815	0 3 27 169 826 3477 13032	0 4 46 318 1686 7505 29440	0 2 22 158 861 3935 15765	0 2 15 85 399 1650 6159	0 0 7 70 452 2259 9606	0 1 14 103 574 2652 10745	0 0 3 35 230 1177 5070 19370	0 0 2 24 169 898 3993 15612	0 0 0 1 9 57 276 1156 4296
Wt 5	1 12 81 426 1881 7359	2 18 111 561 2451 9587	0 5 50 315 1533 6372	0 4 48 330 1716 7468	1 6 33 153 642 2451 8679	0 3 33 210 1041 4365	0 6 66 439 2261 9844	0 3 33 225 1185 5259 20610	0 0 5 48 299 1443 5941	0 3 27 171 864 3768 14643	0 0 9 93 591 2922 12261	0 0 6 53 330 1606 6729	0 0 3 36 240 1233 5328	0 0 0 1 12 75 366 1505
Wt 6	1 9 55 278 1213 4750 17005	0 3 33 219 1126 4901 18888	1 9 55 278 1213 4750 17005	0 3 33 219 1126 4901 18888	0 0 3 31 195 948 3932	1 7 41 198 854 3317 11880	0 4 40 264 1371 6065 23789	0 0 9 87 551 2720 11463	0 0 3 31 195 948 3932	0 0 4 48 325 1687 7342	0 0 4 48 325 1687 7342	0 0 2 22 156 830 3711 14557	0 0 0 8 82 522 2572 10758	0 0 0 0 3 25 151 706 2848

continued next page

Affine A3 - Level 5 - Class 0 - Page 2 of 3

	Rp 1	Rp 2	Rp 3	Rp 4	Rp 5	Rp 6	Rp 7	Rp 8	Rp 9	Rp10	Rp11	Rp12	Rp13	Rp14
Wt 7	2	1	2	1	0	1	1	0	0	0	0	0	0	0
	18	14	18	14	2	11	14	4	2	2	2	1	0	0
	114	106	114	106	18	71	113	44	18	25	25	11	5	0
	571	592	571	592	114	368	666	298	114	178	178	84	50	2
	2461	2737	2461	2737	564	1614	3208	1548	564	965	965	468	321	17
	9434	11045	9434	11045	2394	6302	13383	6785	2394	4354	4354	2181	1602	101
					9039			26259	9039	17256	17256	8838	6808	471
													25691	1904
														6883
Wt 8	2	2	2	2	0	1	1	1	0	0	0	0	0	0
	21	22	21	22	3	11	18	10	3	4	4	1	1	0
	137	152	137	152	27	80	148	79	27	40	40	16	13	1
	700	811	700	811	161	428	869	459	161	265	265	118	94	6
	3027	3633	3027	3633	771	1926	4143	2215	771	1357	1357	648	517	37
	11613	14339	11613	14339	3180	7584	17093	9256	3180	5909	5909	2933	2372	175
					11754			34717	11754	22774	22774	11644	9538	735
														2755
Wt 9	0	0	1	2	0	0	0	0	1	0	0	0	0	0
	5	4	12	18	0	3	6	3	6	0	3	0	0	0
	50	48	81	111	5	33	66	33	33	9	27	6	3	0
	315	330	426	561	48	210	439	225	153	93	171	53	36	1
	1533	1716	1881	2451	299	1041	2261	1185	642	591	864	330	240	12
	6372	7468	7359	9587	1443	4365	9844	5259	2451	2922	3768	1606	1233	75
					5941			20610	8679	12261	14643	6729	5328	366
														1505
Wt10	3	4	3	2	1	1	2	1	0	1	0	0	0	0
	30	37	27	26	9	15	28	14	3	10	5	3	2	0
	191	234	177	187	55	105	213	109	30	75	53	28	21	1
	946	1171	890	1009	275	561	1194	626	189	422	346	182	140	9
	4001	5026	3821	4529	1185	2480	5522	2951	922	1995	1757	922	731	51
	15047	19204	14521	17820	4575	9657	22273	12106	3843	8226	7554	3999	3230	241
					16150				14246	30586	28808	15374	12635	982
														3616
Wt11	3	2	3	4	0	1	2	1	1	0	1	0	0	0
	27	26	30	37	3	15	28	14	9	5	10	3	2	0
	177	187	191	234	30	105	213	109	55	53	75	28	21	1
	890	1009	946	1171	189	561	1194	626	275	346	422	182	140	9
	3821	4529	4001	5026	922	2480	5522	2951	1185	1757	1995	922	731	51
	14521	17820	15047	19204	3843	9657	22273	12106	4575	7554	8226	3999	3230	241
					14246				16150	28808	30586	15374	12635	982
														3616
Wt12	4	4	4	4	1	2	4	1	1	1	1	1	0	0
	40	43	40	43	9	22	44	19	9	13	13	7	4	0
	247	283	247	283	61	147	307	149	61	97	97	51	34	2
	1211	1441	1211	1441	316	751	1641	842	316	548	548	280	210	14
	5057	6210	5057	6210	1400	3245	7353	3899	1400	2562	2562	1322	1034	76
	18842	23715	18842	23715	5467	12384	28984	15720	5467	10469	10469	5451	4400	338
												20322	16720	1335
														4773

---continued next page-

Affine A3 - Level 5 - Class 0 - Page 3 of 3

	Rp 1	Rp 2	Rp 3	Rp 4	Rp 5	Rp 6	Rp 7	Rp 8	Rp 9	Rp10	Rp11	Rp12	Rp13	Rp14
Wt13	5	6	5	6	1	2	5	3	1	2	2	1	1	0
	49	60	49	60	12	26	59	34	12	21	21	11	10	1
	309	383	309	383	81	177	408	230	81	146	146	74	66	6
	1509	1903	1509	1903	420	920	2149	1207	420	778	778	400	347	31
	6290	8050	6290	8050	1830	3979	9492	5336	1830	3506	3506	1816	1564	136
					7053			20852	7053	13913	13913	7301	6263	543
													22835	1976
														6709
Wt14	6	8	6	8	1	3	7	6	1	3	3	2	3	1
	62	81	62	81	15	33	81	54	15	33	33	18	21	3
	389	507	389	507	105	225	547	339	105	213	213	114	117	15
	1897	2481	1897	2481	546	1152	2826	1683	546	1092	1092	576	552	60
	7851	10342	7851	10342	2361	4959	12270	7191	2361	4749	4749	2516	2313	234
												9786	8802	843
														2886

Weight Multiplicities for Affine A4

Affine A4 - Level 1 - Class 0

	Rp 1				
	1				
	0000				
Wt 1	1	2580	353017	18468245	563251038
	4	5180	602348	29015700	839757660
	14	10108	1014580	45235414	171642403*
	40	19208	1688400	70005376	763819433*
	105	35693	2778517	107585845	550608368*
	252	64960	4524760	164245380	721790411*
	574	116090	7296752	249162620	367612195*
	1240	203984	11658920	375704920	792686671*

* Multiplicity is given modulo 1073741527

Affine A4 - Level 2 - Class 0

	Rp 1				Rp 2				Rp 3			
	0				0				2			
	0110				1001				0000			
Wt 1	1	70497	60657750	233217651*	0	56752	54972768	730189923*	0	11948	14256636	346295520*
	8	185072	125676344	555878934*	3	152487	114882194	679698606*	0	33456	30242464	220785566*
	46	465300	255607280	57398675*	24	390944	235507832	668776895*	2	88794	62852934	490235725*
	208	1126544	511034816	85741242*	128	962548	474298212	871028435*	16	225184	128191968	987511204*
	804	2638247	1005582179	402301446*	544	2287424	939629376	916317056*	83	549075	256949659	863799295*
	2768	5998008	875857961*	381374834*	1997	5268093	759498115*	22481316*	344	1293520	506798144	695221515*
	8716	13277896	506580675*	265019671*	6584	11797392	304770899*	424620763*	1236	2955716	984702804	988533931*
	25536	28693152	593393782*		19986	25760042	249286252*		4000	6571904	812893257*	752607458*
Wt 2	2	114150	87461962	652251576*	1	93870	79690460	987091528*	0	20418	20847112	588976002*
	17	293561	179533239	474244552*	10	245960	164901088	341615963*	1	55374	43750193	478042626*
	92	724724	362010336	227713355*	60	617199	334987169	442533602*	8	143148	90037616	541335671*
	396	1726339	717989710	148210467*	276	1491552	668994580	335898116*	43	355062	181986795	173067273*
	1466	3984274	328553933*	346085910*	1073	3486927	241290275*	1060590303*	186	849406	361748478	683238110*
	4865	8939342	552312574*	825159762*	3700	7914382	399567640*	312735110*	693	1968027	708002335	192000764*
	14844	19552916	833530472*		11643	17493090	570713635*		2316	4431332	292023905*	151523035*
	42332	41791871	1030036633*		34064	37747348	585511980*		7115	9724321	451684355*	
Wt 3	5	183075	125523857	550610841*	4	153180	114937568	681666633*	1	34260	30312961	223423813*
	36	461600	255361320	49484293*	26	392180	235596626	671732611*	4	90260	62967084	494219999*
	184	1119960	510643872	73943850*	136	964864	474441360	875459767*	24	227952	128377040	993509212*
	744	2626604	1004964980	384808356*	560	2291424	939854560	922888960*	100	553940	257243220	872738637*
	2640	5978022	874895413*	355614792*	2040	5275208	759853177*	32205637*	390	1302236	507263444	708499411*
	8440	13243832	505089123*		6667	11809340	305319974*	438877399*	1328	2970560	985427528	1008086847*
	24992	28636400	591106358*		20172	25780460	250135658*		4208	6597440	814019801*	
	69424	60563880	229730724*		57096	55006224	731483443*		12344	14298968	348021859*	

* Multiplicity is given modulo 1073741527

Affine A4 - Level 3 - Class 0 - Page 1 of 3

	Rp 1 0 0021	Rp 2 0 0102	Rp 3 0 1200	Rp 4 0 2010	Rp 5 1 0110	Rp 6 1 1001	Rp 7 3 0000
Wt 1	1	0	0	0	0	0	0
	8	4	3	2	3	0	0
	56	38	36	28	36	8	0
	312	254	245	214	264	88	3
	1497	1334	1297	1204	1485	592	33
	6331	5998	5785	5618	6979	3088	213
	24247	23918	22852	22906	28787	13640	1076
	85461	86898	82091	84366	107285	53392	4609
	281075	292438	273298	286448	368555	190360	17581
	871224	923354	853969	909784	1183369	629464	61255
	2565834	2760732	2528827	2731152	3588581	1955256	198539
	7225516	7874096	7148481	7811662	10359626	5759272	605821
	19557966	21547242	19401784	21419220	28648436	16203000	1756504
	51102582	56838468	50793348	56582478	76272042	43787712	4872109
	129354434	145085150	128755089	144584830	196299562	114188448	12999570
							33509480
Wt 2	1	1	0	0	0	0	0
	12	10	6	5	6	2	0
	87	78	62	57	66	26	1
	492	458	407	388	459	198	12
	2328	2255	2073	2044	2461	1134	81
	9688	9657	8991	9081	11149	5392	432
	36415	37194	34639	35719	44595	22410	1940
	126135	131367	121860	127806	162038	83986	7731
	408088	432114	398267	423894	544724	289720	27997
	1246241	1338177	1224556	1319956	1716849	933336	93943
	3620754	3935318	3574476	3896260	5122646	2837966	295650
	10071539	11062917	9975677	10981692	14578883	8211212	881324
	26957563	29889883	26764148	29725341	39808287	22750776	2506420
	69720837	77953130	69339666	77627665	104786559	60671302	6840558
					266935836	156376540	17999545
							45840288
Wt 3	0	0	1	0	0	0	0
	3	2	8	4	3	0	0
	36	28	56	38	36	8	0
	245	214	312	254	264	88	3
	1297	1204	1497	1334	1485	592	33
	5785	5618	6331	5998	6979	3088	213
	22852	22906	24247	23918	28787	13640	1076
	82091	84366	85461	86898	107285	53392	4609
	273298	286448	281075	292438	368555	190360	17581
	853969	909784	871224	923354	1183369	629464	61255
	2528827	2731152	2565834	2760732	3588581	1955256	198539
	7148481	7811662	7225516	7874096	10359626	5759272	605821
	19401784	21419220	19557966	21547242	28648436	16203000	1756504
	50793348	56582478	51102582	56838468	76272042	43787712	4872109
	128755089	144584830	129354434	145085150	196299562	114188448	12999570
							33509480

continued next page

Affine A4 - Level 3 - Class 0 - Page 2 of 3

	Rp 1	Rp 2	Rp 3	Rp 4	Rp 5	Rp 6	Rp 7
Wt 4	0	0	1	1	0	0	0
	6	5	12	10	6	2	0
	62	57	87	78	66	26	1
	407	388	492	458	459	198	12
	2073	2044	2328	2255	2461	1134	81
	8991	9081	9688	9657	11149	5392	432
	34639	35719	36415	37194	44595	22410	1940
	121860	127806	126135	131367	162038	83986	7731
	398267	423894	408088	432114	544724	289720	27997
	1224556	1319956	1246241	1338177	1716849	933336	93943
	3574476	3896260	3620754	3935318	5122646	2837966	295650
	9975677	10981692	10071539	11062917	14578883	8211212	881324
	26764148	29725341	26957563	29889883	39808287	22750776	2506420
	69339666	77627665	69720837	77953130	104786559	60671302	6840558
					266935836	156376540	17999545
							45840288
Wt 5	2	1	2	1	1	0	0
	21	16	21	16	16	4	0
	148	126	148	126	137	48	2
	803	739	803	739	840	344	20
	3689	3553	3689	3553	4192	1880	133
	14934	14890	14934	14890	18052	8612	684
	54861	56108	54861	56108	69573	34720	3003
	186195	194387	186195	194387	245480	126860	11690
	591856	628280	591856	628280	805854	428296	41528
	1779284	1915599	1779284	1915599	2489810	1354408	136886
	5098008	5554938	5098008	5554938	7304120	4052248	424150
	14004710	15420738	14004710	15420738	20483726	11558660	1246648
	37066305	41192105	37066305	41192105	55213913	31622240	3500899
	94893122	106325801	94893122	106325801	143679492	83376884	9445674
						212705848	24597172
							62049544
Wt 6	3	3	3	3	2	1	0
	35	33	35	33	31	14	1
	241	234	241	234	244	113	8
	1282	1269	1282	1269	1420	678	53
	5740	5808	5740	5808	6788	3348	278
	22727	23412	22727	23412	28276	14356	1267
	81775	85632	81775	85632	106000	55229	5120
	272538	289443	272538	289443	365500	194810	18866
	852239	916569	852239	916569	1176414	639745	64310
	2525032	2745942	2525032	2745942	3573350	1978076	205494
	7140456	7842879	7140456	7842879	10327334	5808205	621052
	19385293	21483231	19385293	21483231	28581941	16304808	1788796
	50760369	56710473	50760369	56710473	76138546	43993957	4938604
	128690619	144834990	128690619	144834990	196037662	114596430	13133066
							33771380

continued next page-

Affine A4 - Level 3 - Class 0 - Page 3 of 3

	Rp 1	Rp 2	Rp 3	Rp 4	Rp 5	Rp 6	Rp 7
Wt 7	5	6	5	6	5	4	1
	59	64	59	64	60	36	4
	396	414	396	414	435	236	24
	2041	2130	2041	2130	2380	1260	120
	8905	9330	8905	9330	10911	5760	545
	34425	36378	34425	36378	43954	23396	2244
	121347	129438	121347	129438	160429	86468	8507
	397107	427734	397107	427734	540900	295632	29858
	1222011	1328580	1222011	1328580	1708175	946800	98235
	3569085	3914940	3569085	3914940	5103702	2867500	305170
	9964570	11020858	9964570	11020858	14538879	8273880	901746
	26741853	29805174	26741853	29805174	39726210	22879952	2548962
	69295875	77786370	69295875	77786370	104622440	60930860	6926934
						156886200	18170770

Affine A4 - Level 4 - Class 0 - Page 1 of 3

	Rp 1 0 0013	Rp 2 0 0220	Rp 3 0 0301	Rp 4 0 1030	Rp 5 0 1111	Rp 6 0 2002	Rp 7 0 3100	Rp 8 1 0021	Rp 9 1 0102	Rp10 1 1200	Rp11 1 2010	Rp12 2 0110	Rp13 2 1001	Rp14 4 0000
Wt 1	1	0	0	0	0	0	0	0	0	0	0	0	0	0
	8	6	3	6	8	2	0	4	3	0	0	0	0	0
	56	84	52	76	120	28	10	48	36	16	12	12	3	0
	332	663	458	580	1028	250	124	378	294	209	168	150	48	1
	1737	3960	2932	3428	6583	1664	939	2344	1884	1632	1356	1176	414	16
	8148	19752	15344	17090	34816	9122	5428	12286	10173	9700	8259	7098	2664	128
	34824					43284	26352	56664	48108	48232	41868	36024	14139	772
													65328	3874
														17088
Wt 2	0	1	0	0	0	0	0	0	0	0	0	0	0	0
	0	8	4	4	3	0	0	0	0	0	0	0	0	0
	4	66	38	38	48	6	4	8	3	8	3	2	0	0
	58	416	284	284	438	72	58	100	60	100	60	40	7	0
	456	2295	1678	1678	2968	599	456	794	546	794	546	402	104	3
	2735	11064	8560	8560	16492	3712	2735	4828	3636	4828	3636	2800	868	32
	13660	48170	38619	38619	79224	19292	13660	24660	19629	24660	19629	15686	5416	250
									91668		91668	75184	27903	1456
														7196
Wt 3	0	1	1	0	0	0	0	0	0	0	0	0	0	0
	0	12	10	4	6	0	1	0	0	2	0	0	0	0
	6	97	78	50	84	9	16	11	6	26	9	5	0	0
	82	620	498	390	713	118	137	146	96	222	120	78	13	0
	642	3351	2722	2353	4598	929	880	1150	837	1462	963	683	174	4
	3803	15902	13114	11964	24618	5590	4678	6910	5370	8026	5892	4432	1370	50
	18814	67953	56918	53576	114854	28168	21690	34765	28278	38368	30144	23648	8180	374
												109522	40798	2142
														10313

-continued next page-

Affine A4 - Level 4 - Class 0 - Page 2 of 3

	Rp 1	Rp 2	Rp 3	Rp 4	Rp 5	Rp 6	Rp 7	Rp 8	Rp 9	Rp10	Rp11	Rp12	Rp13	Rp14
Wt 4	0 1 16 137 880 4678 21690	1 12 97 620 3351 15902 67953	0 4 50 390 2353 11964 53576	1 10 78 498 2722 13114 56918	0 6 84 713 4598 24618 114854	0 0 9 118 929 5590 28168	0 0 6 82 642 3803 18814	0 2 26 222 1462 8026 38368	0 0 9 120 963 5892 30144	0 0 11 146 1150 6910 34765	0 0 6 96 837 5370 28278	0 0 5 78 683 4432 23648 109522	0 0 0 13 174 1370 8180 40798	0 0 0 0 4 50 374 2142 10313
Wt 5	0 3 39 309 1877 9554 42694	2 25 204 1284 6772 31267	1 16 144 965 5312 25312	1 16 144 965 5312 25312	1 22 238 1778 10526 52868 234475	0 3 44 382 2478 13250 61538	0 3 39 309 1877 9554 42694	0 4 57 486 3116 16538 76432	0 2 37 353 2422 13454 64277	0 4 57 486 3116 16538 76432	0 2 37 353 2422 13454 64277	0 1 24 253 1840 10632 52298	0 0 5 72 605 3814 19915 90668	0 0 0 2 24 183 1074 5313 23168
Wt 6	0 6 66 493 2854 14049 61092	3 34 286 1786 9384 42914	1 22 201 1362 7436 35106	1 22 201 1362 7436 35106	2 38 382 2723 15584 76212	1 10 99 698 4100 20516 91224	0 6 66 493 2854 14049 61092	0 6 89 740 4626 23964 108448	0 6 75 624 3924 20646 94794	0 6 89 740 4626 23964 108448	0 6 75 624 3924 20646 94794	0 2 45 428 2929 16158 76803	0 1 18 166 1148 6465 31436	0 0 1 8 63 370 1908 8632
Wt 7	0 0 10 124 939 5428 26352	0 6 84 663 3960 19752	0 6 76 580 3428 17090	0 3 52 458 2932 15344	0 8 120 1028 6583 34816	0 2 28 250 1664 9122 43284	1 8 56 332 1737 8148 34824	0 0 16 209 1632 9700 48232	0 0 12 168 1356 8259 41868	0 4 48 378 2344 12286 56664	0 3 36 294 1884 10173 48108	0 0 12 150 1176 7098 36024	0 0 3 48 414 2664 14139 65328	0 0 0 1 16 128 772 3874 17088
Wt 8	1 12 97 614 3283 15398 65065	3 39 312 1917 9870 44595	1 22 206 1387 7558 35541	3 33 258 1583 8226 37602	2 40 397 2800 15897 77321	0 5 72 600 3752 19462 88144	0 4 53 426 2588 13105 58109	1 14 131 916 5260 26024 114602	0 6 78 639 3999 20916 95718	0 6 85 715 4502 23484 106790	0 3 57 531 3561 19359 90744	0 3 48 438 2943 16161 76656	0 0 9 120 954 5772 29190	0 0 0 3 37 275 1576 7623
Wt 9	2 21 158 953 4908 22356	4 54 436 2672 13656 61124	2 34 306 2004 10712 49544	3 42 349 2184 11362 51646	4 68 630 4253 23378 110846	1 16 153 1072 6100 29812	0 7 88 670 3908 19164	1 20 194 1354 7656 37212	1 16 155 1104 6391 31742 140001	0 10 137 1102 6710 34050	0 8 111 914 5684 29390 132814	0 6 86 726 4614 24310 111678	0 2 30 266 1762 9614 45496	0 0 1 12 91 542 2738 12216

-continued next page-

Affine A4 - Level 4 - Class 0 - Page 3 of 3

	Rp 1	Rp 2	Rp 3	Rp 4	Rp 5	Rp 6	Rp 7	Rp 8	Rp 9	Rp10	Rp11	Rp12	Rp13	Rp14
Wt10	0	3	3	1	2	0	1	0	0	1	0	0	0	0
	4	39	33	22	40	5	12	6	3	14	6	3	0	0
	53	312	258	206	397	72	97	85	57	131	78	48	9	0
	426	1917	1583	1387	2800	600	614	715	531	916	639	438	120	3
	2588	9870	8226	7558	15897	3752	3283	4502	3561	5260	3999	2943	954	37
	13105	44595	37602	35541	77321	19462	15398	23484	19359	26024	20916	16161	5772	275
	58109					88144	65065	106790	90744	114602	95718	76656	29190	1576
														7623
Wt11	0	4	3	2	4	1	2	0	0	1	1	0	0	0
	7	54	42	34	68	16	21	10	8	20	16	6	2	0
	88	436	349	306	630	153	158	137	111	194	155	86	30	1
	670	2672	2184	2004	4253	1072	953	1102	914	1354	1104	726	266	12
	3908	13656	11362	10712	23378	6100	4908	6710	5684	7656	6391	4614	1762	91
	19164	61124	51646	49544	110846	29812	22356	34050	29390	37212	31742	24310	9614	542
									132814		140001	111678	45496	2738
														12216
Wt12	2	7	5	5	8	1	2	2	1	2	1	1	0	0
	25	88	66	66	119	24	25	32	22	32	22	16	4	0
	204	678	527	527	1024	234	204	294	223	294	223	166	52	2
	1269	4004	3204	3204	6568	1616	1269	1990	1596	1990	1596	1224	430	20
	6612	19874	16285	16285	34772	8985	6612	11000	9147	11000	9147	7236	2712	143
	30156	86848	72584	72584	160129	42936	30156	52468	44812	52468	44812	36376	14276	816
												161624	65676	3996
														17392
Wt13	3	10	7	7	14	4	3	3	3	3	3	2	1	0
	40	126	95	95	194	53	40	50	45	50	45	31	14	1
	316	962	755	755	1587	430	316	451	396	451	396	288	123	8
	1911	5618	4541	4541	9814	2681	1911	2975	2604	2975	2604	1981	830	53
	9687	27548	22779	22779	50495	13986	9687	16041	14118	16041	14118	11130	4662	298
	43148					63929	43148	74851	66447	74851	66447	53888	22764	1477
													99437	6594
														26948
Wt14	4	16	10	10	24	10	4	5	6	5	6	5	4	1
	61	184	140	140	311	100	61	80	84	80	84	60	36	4
	476	1384	1090	1090	2436	740	476	696	669	696	669	500	256	24
	2830	7920	6464	6464	14560	4280	2830	4444	4140	4444	4140	3184	1505	120
	14020				72920	21290	14020	23334	21420	23334	21420	17016	7740	580
													35516	2560
														10616

Weight Multiplicities for Affine A5

Affine A5 - Level 1 - Class 0

	Rp 1 1 00000				
Wt 1	1	6765	1633000	134992865	747688460*
	5	14725	2963840	223474667	961190039*
	20	31027	5302075	366772720	868131464*
	65	63505	9358470	597049255	337711778*
	190	126730	16313440	964375855	775031806*
	506	247170	28107365	472467168*	1043106197*
	1265	472295	47902010	314166807*	998827150*
	2990	885723	80803485	671549549*	873889867*

* Multiplicity is given modulo 1073741527

Affine A5 - Level 2 - Class 0 - Page 1 of 2

	Rp 1 0 00200			Rp 2 0 01010			Rp 3 0 10001			Rp 4 2 00000		
Wt 1	1	373070	291943058*	0	680004	550407951*	0	384720	726562308*	0	66482	402877180
	5	1062530	848297216*	5	1965145	578992041*	0	1146230	788265744*	0	207765	911582135
	35	2898900	10836596*	40	5429415	1059978048*	9	3248435	168673244*	0	612950	946872413*
	175	7616805	782953461*	250	14417670	533656580*	80	8816927	538984096*	5	1721565	99529652*
	800	19347235	576453033*	1225	36965110	392734709*	490	23038425	998514478*	46	4634320	800012609*
	3192	47675635		5225	91835968		2350	58203355	768526058*	280	12018275	409034239*
	11690	114291024		19840	221769610		9725	142662220		1330	30152435	47289074*
	39440	267204600		69055	521888980		35929	340258695		5435	73435675	
	124820	610506935		223345	125783633*		121800	791617640		19840	174117425	
Wt 2	1	626303	951429299*	1	1157432	792225901*	0	672395	547446514*	0	120504	608592141
	10	1749503	107538728*	13	3272042	175651259*	3	1949000	573732786*	0	363752	288508357*
	66	4691025	813921221*	97	8864956	551136922*	33	5395787	1050768368*	2	1042708	842228708*
	343	12130733	432908951*	541	23134214	1019424420*	221	14350102	517706590*	21	2858459	234683*
	1513	30370273	87840352*	2502	58389425	804131546*	1142	36831672	365440487*	135	7535586	764666832*
	5913	73849657		10117	143017453		4983	91579380	1051881881*	676	19188972	545229940*
	21025	174889349		36915	340922959		19235	221283848		2877	47371847	
	69244	404293556		124092	792841162		67559	520988276		10880	113721190	
	213965	914140250		389916	728778137*		219941	124136610*		37562	266150392	
Wt 3	2	1042708	842228708*	3	1949000	573732786*	1	1157432	792225901*	0	213965	914140250
	21	2858459	234683*	33	5395787	1050768368*	13	3272042	175651259*	1	626303	951429299*
	135	7535586	764666832*	221	14350102	517706590*	97	8864956	551136922*	10	1749503	107538728*
	676	19188972	545229940*	1142	36831672	365440487*	541	23134214	1019424420*	66	4691025	813921221*
	2877	47371847		4983	91579380	1051881881*	2502	58389425	804131546*	343	12130733	432908951*
	10880	113721190		19235	221283848		10117	143017453		1513	30370273	87840352*
	37562	266150392		67559	520988276		36915	340922959		5913	73849657	
	120504	608592141		219941	124136610*		124092	792841162		21025	174889349	
	363752	288508357*		672395	547446514*		389916	728778137*		69244	404293556	

-continued next page-

Affine A5 - Level 2 - Class 0 - Page 2 of 2

	Rp 1			Rp 2			Rp 3			Rp 4		
Wt 4	5	1721565	99529652*	9	3248435	168673244*	5	1965145	578992041*	1	373070	291943058*
	46	4634320	800012609*	80	8816927	538984096*	40	5429415	1059978048*	5	1062530	848297216*
	280	12018275	409034239*	490	23038425	998514478*	250	14417670	533656580*	35	2898900	10836596*
	1330	30152435	47289074*	2350	58203355	768526058*	1225	36965110	392734709*	175	7616805	782953461*
	5435	73435675		9725	142662220		5225	91835968		800	19347235	576453033*
	19840	174117425		35929	340258695		19840	221769610		3192	47675635	
	66482	402877180		121800	791617640		69055	521888980		11690	114291024	
	207765	911582135		384720	726562308*		223345	125783633*		39440	267204600	
	612950	946872413*		1146230	788265744*		680004	550407951*		124820	610506935	

* Multiplicity is given modulo 1073741527

Affine A5 - Level 2 - Class 1

	Rp 1				Rp 2				Rp 3			
	0				0				1			
	00110				01001				10000			
Wt 1	1	271225	561805050	499744676*	0	247680	579227480	404301071*	0	83140	240222105	483666501*
	10	807100	205854763*	921497845*	4	755060	256983413*	1058472195*	0	264460	560172050	364751811*
	70	2287499	706535251*	533617882*	40	2183680	844165506*	10734622*	5	792375	202890923*	698023178*
	380	6213190	874694245*	550996876*	260	6034360	149946998*	317130853*	50	2256472	701233176*	166845162*
	1740	16253650	529418066*	169581436*	1320	16023640	301487429*	897875188*	315	6149685	865335775*	1027689148*
	7012	41118580	424777358*	384096930*	5700	41073020	254077360*	836891249*	1550	16126920	513104626*	278947108*
	25585	100936015	968005182*	762673525*	21848	102008960	81373507*	271975661*	6506	40871410	396669993*	985371289*
	86130	241107828	564469986*		76360	246238700	802130213*		24320	100463720	920103172*	448506718*
Wt 2	2	465894	849081924	787029949*	1	435577	880432734	908359999*	0	151734	368533610	221531344*
	22	1355706	840104985*	1054579387*	14	1291980	926716051*	1037975070*	2	465894	849081924	787029949*
	148	3766436	1006332253*	23345620*	107	3649457	156691870*	615394818*	22	1355706	840104985*	1054579387*
	770	10048634	575310618*	601655865*	602	9879508	53802255*	1034935591*	148	3766436	1006332253*	23345620*
	3382	25865660	199900888*	838125764*	2794	25761571	435279513*	831568543*	770	10048634	575310618*	601655865*
	13134	64482280	132033482*	61612895*	11312	64971806	989944290*	768966297*	3382	25865660	199900888*	838125764*
	46382	156184328	762639782*		41281	159026425	1033331132*		13134	64482280	132033482*	61612895*
	151734	368533610	221531344*		138714	378828804	903700435*		46382	156184328	762639782*	
Wt 3	5	792375	202890923*	698023178*	4	755060	256983413*	1058472195*	1	271225	561805050	499744676*
	50	2256472	701233176*	166845162*	40	2183680	844165506*	10734622*	10	807100	205854763*	921497845*
	315	6149685	865335775*	1027689148*	260	6034360	149946998*	317130853*	70	2287499	706535251*	533617882*
	1550	16126920	513104626*	278947108*	1320	16023640	301487429*	897875188*	380	6213190	874694245*	550996876*
	6506	40871410	396669993*	985371289*	5700	41073020	254077360*	836891249*	1740	16253650	529418066*	169581436*
	24320	100463720	920103172*	448506718*	21848	102008960	81373507*	271975661*	7012	41118580	424777358*	384096930*
	83140	240222105	483666501*		76360	246238700	802130213*		25585	100936015	968005182*	762673525*
	264460	560172050	364751811*		247680	579227480	404301071*		86130	241107828	564469986*	

* Multiplicity is given modulo 1073741527

Affine A5 - Level 3 - Class 0 - Page 1 of 2

	Rp 1	Rp 2	Rp 3	Rp 4	Rp 5	Rp 6	Rp 7	Rp 8	Rp 9	Rp10
	0 00030	0 00111	0 01002	0 03000	0 11100	0 20010	1 00200	1 01010	1 10001	3 00000
Wt 1	1	0	0	0	0	0	0	0	0	0
	5	5	0	0	0	0	0	0	0	0
	35	55	10	5	16	5	10	9	0	0
	210	440	130	75	235	85	130	135	16	0
	1145	2855	1040	620	1975	809	1040	1215	235	5
	5658	15790	6530	3831	12520	5560	6530	8190	1975	75
	25575	77020	34590	19735	66095	31085	34590	45630	12520	620
	106485	339096	161430	89135	305442	150020	161430	220887	66095	3831
	412695	1370455	681385	364045	1273470	647130	681385	958815	305442	19735
			2648975			2552514	2648975	3810510	1273470	89135
										364045
Wt 2	1	1	0	0	0	0	0	0	0	0
	10	18	4	3	8	2	4	3	0	0
	81	183	58	44	124	43	58	60	8	0
	523	1370	522	367	1099	442	522	609	124	3
	2901	8334	3534	2324	7279	3197	3534	4449	1099	44
	14166	43474	19820	12247	39819	18583	19820	26205	7279	367
	62452	201152	96598	56556	189580	92521	96598	132363	39819	2324
	252604	844744	422073	235651	810636	409669	422073	594132	189580	12247
	950010	3273486	1687668	903845	3178732	1652346	1687668	2426991	810636	56556
									3178732	235651
										903845
Wt 3	1	2	1	0	0	0	0	0	0	0
	15	34	14	6	17	7	7	9	2	0
	128	333	143	79	238	105	105	135	34	1
	846	2402	1079	634	1985	916	916	1215	333	15
	4658	14123	6633	3873	12549	6019	6019	8190	2402	128
	22457	71483	34841	19847	66169	32767	32767	45630	14123	846
	97385	322100	162016	89410	305619	155550	155550	220887	71483	4658
	387484	1321602	682681	364687	1273878	663846	663846	958815	322100	22457
	1434175	5017640	2651730	1372533	4887045	2599825	2599825	3810510	1321602	97385
									5017640	387484
										1434175
Wt 4	0	0	0	1	0	0	0	0	0	0
	0	0	0	5	5	0	0	0	0	0
	5	16	5	35	55	10	10	9	0	0
	75	235	85	210	440	130	130	135	16	0
	620	1975	809	1145	2855	1040	1040	1215	235	5
	3831	12520	5560	5658	15790	6530	6530	8190	1975	75
	19735	66095	31085	25575	77020	34590	34590	45630	12520	620
	89135	305442	150020	106485	339096	161430	161430	220887	66095	3831
	364045	1273470	647130	412695	1370455	681385	681385	958815	305442	19735
			2552514			2648975	2648975	3810510	1273470	89135
										364045
Wt 5	0	0	0	1	1	0	0	0	0	0
	3	8	2	10	18	4	4	3	0	0
	44	124	43	81	183	58	58	60	8	0
	367	1099	442	523	1370	522	522	609	124	3
	2324	7279	3197	2901	8334	3534	3534	4449	1099	44
	12247	39819	18583	14166	43474	19820	19820	26205	7279	367
	56556	189580	92521	62452	201152	96598	96598	132363	39819	2324
	235651	810636	409669	252604	844744	422073	422073	594132	189580	12247
	903845	3178732	1652346	950010	3273486	1687668	1687668	2426991	810636	56556
									3178732	235651
										903845

-continued next page-

Affine A5 - Level 3 - Class 0 - Page 2 of 2

	Rp 1	Rp 2	Rp 3	Rp 4	Rp 5	Rp 6	Rp 7	Rp 8	Rp 9	Rp10
Wt 6	0	0	0	1	2	1	0	0	0	0
	6	17	7	15	34	14	7	9	2	0
	79	238	105	128	333	143	105	135	34	1
	634	1985	916	846	2402	1079	916	1215	333	15
	3873	12549	6019	4658	14123	6633	6019	8190	2402	128
	19847	66169	32767	22457	71483	34841	32767	45630	14123	846
	89410	305619	155550	97385	322100	162016	155550	220887	71483	4658
	364687	1273878	663846	387484	1321602	682681	663846	958815	322100	22457
	1372533	4887045	2599825	1434175	5017640	2651730	2599825	3810510	1321602	97385
									5017640	387484
										1434175
Wt 7	1	2	0	1	2	0	1	0	0	0
	15	34	7	15	34	7	14	9	0	0
	128	333	105	128	333	105	143	135	17	0
	846	2402	916	846	2402	916	1079	1215	238	6
	4658	14123	6019	4658	14123	6019	6633	8190	1985	79
	22457	71483	32767	22457	71483	32767	34841	45630	12549	634
	97385	322100	155550	97385	322100	155550	162016	220887	66169	3873
	387484	1321602	663846	387484	1321602	663846	682681	958815	305619	19847
	1434175	5017640	2599825	1434175	5017640	2599825	2651730	3810510	1273878	89410
									4887045	364687
										1372533
Wt 8	2	4	1	2	4	1	1	1	0	0
	26	65	23	26	65	23	23	25	4	0
	219	607	245	219	607	245	245	296	65	2
	1407	4197	1844	1407	4197	1844	1844	2361	607	26
	7579	23804	11082	7579	23804	11082	11082	14804	4197	219
	35732	116826	56825	35732	116826	56825	56825	78292	23804	1407
	151881	512665	258085	151881	512665	258085	258085	364367	116826	7579
	593190	2055779	1064392	593190	2055779	1064392	1064392	1532562	512665	35732
			4054875			4054875	4054875	5934958	2055779	151881
										593190
Wt 9	3	8	4	3	8	4	2	3	1	0
	44	124	58	44	124	58	43	60	18	1
	367	1099	522	367	1099	522	442	609	183	10
	2324	7279	3534	2324	7279	3534	3197	4449	1370	81
	12247	39819	19820	12247	39819	19820	18583	26205	8334	523
	56556	189580	96598	56556	189580	96598	92521	132363	43474	2901
	235651	810636	422073	235651	810636	422073	409669	594132	201152	14166
	903845	3178732	1687668	903845	3178732	1687668	1652346	2426991	844744	62452
									3273486	252604
										950010
Wt10	5	16	10	5	16	10	5	9	5	1
	75	235	130	75	235	130	85	135	55	5
	620	1975	1040	620	1975	1040	809	1215	440	35
	3831	12520	6530	3831	12520	6530	5560	8190	2855	210
	19735	66095	34590	19735	66095	34590	31085	45630	15790	1145
	89135	305442	161430	89135	305442	161430	150020	220887	77020	5658
	364045	1273470	681385	364045	1273470	681385	647130	958815	339096	25575
			2648975			2648975	2552514	3810510	1370455	106485
										412695

Affine A5 - Level 3 - Class 1 - Page 1 of 2

	Rp 1	Rp 2	Rp 3	Rp 4	Rp 5	Rp 6	Rp 7	Rp 8	Rp 9
	0 00021	0 00102	0 02100	0 10200	0 11010	0 20001	1 00110	1 01001	2 10000
Wt 1	1	0	0	0	0	0	0	0	0
	10	5	5	6	5	0	4	0	0
	85	60	80	90	90	10	60	15	0
	580	495	701	775	875	150	550	210	11
	3385	3195	4565	5070	6144	1290	3825	1775	155
	17308	17450	24575	27545	35060	8295	21970	11425	1265
	79435	83735	115305	130722	172350	44230	109443	61340	7850
	332815	362535	486720	558130	755960	205855	487465	288190	40695
	1290860	1441686	1886883	2187700	3026750	862890	1982575	1220159	185238
	4685000	5337350		7984820	11240448	3324545	7474945	4747945	762300
									2891970
Wt 2	1	1	0	0	0	0	0	0	0
	15	13	11	12	13	2	8	3	0
	136	127	153	164	195	38	114	51	3
	951	934	1253	1361	1698	380	1002	501	47
	5515	5668	7794	8583	11109	2714	6681	3610	414
	27807	29569	40484	45252	60208	15695	37025	21164	2753
	125383	137041	184531	209127	284536	77969	178817	106731	15147
	516061	576574	760151	872592	1209246	345032	775584	479088	72633
	1967539	2238611	2885874	3351546	4717371	1391907	3082518	1958389	312825
	7027717	8118770				5203175	11390555	7408956	1235361
									4538246
Wt 3	0	0	1	0	0	0	0	0	0
	0	0	10	5	4	0	0	0	0
	11	10	85	60	60	6	15	5	0
	155	150	580	495	550	90	210	90	5
	1265	1290	3385	3195	3825	775	1775	875	80
	7850	8295	17308	17450	21970	5070	11425	6144	701
	40695	44230	79435	83735	109443	27545	61340	35060	4565
	185238	205855	332815	362535	487465	130722	288190	172350	24575
	762300	862890	1290860	1441686	1982575	558130	1220159	755960	115305
	2891970	3324545	4685000	5337350	7474945	2187700	4747945	3026750	486720
						7984820		11240448	1886883
Wt 4	0	0	1	1	0	0	0	0	0
	3	2	15	13	8	0	3	0	0
	47	38	136	127	114	12	51	13	0
	414	380	951	934	1002	164	501	195	11
	2753	2714	5515	5668	6681	1361	3610	1698	153
	15147	15695	27807	29569	37025	8583	21164	11109	1253
	72633	77969	125383	137041	178817	45252	106731	60208	7794
	312825	345032	516061	576574	775584	209127	479088	284536	40484
	1235361	1391907	1967539	2238611	3082518	872592	1958389	1209246	184531
	4538246	5203175	7027717	8118770	11390555	3351546	7408956	4717371	760151
									2885874
Wt 5	0	0	2	1	1	0	0	0	0
	6	5	27	21	22	3	6	2	0
	85	81	240	214	247	45	99	44	3
	723	728	1626	1581	1933	416	923	464	46
	4644	4872	9179	9412	12016	2862	6351	3455	408
	24834	26828	45095	48051	63328	16230	35823	20594	2724
	116082	128382	198672	217804	294364	79713	174879	104846	15036
	488920	551093	800703	897721	1238089	350298	763665	473324	72253
	1892781	2167857	2995928	3420316	4797218	1406820	3048723	1941887	311649
	6831594	7931807		12194623	17349056	5243247	11299728	7364186	1231965
									4529019

continued next page

Affine A5 - Level 3 - Class 1 - Page 2 of 2

	Rp 1	Rp 2	Rp 3	Rp 4	Rp 5	Rp 6	Rp 7	Rp 8	Rp 9
Wt 6	0	0	3	2	3	1	0	0	0
	11	12	47	38	51	13	13	8	1
	153	164	414	380	501	127	195	114	15
	1253	1361	2753	2714	3610	934	1698	1002	136
	7794	8583	15147	15695	21164	5668	11109	6681	951
	40484	45252	72633	77969	106731	29569	60208	37025	5515
	184531	209127	312825	345032	479088	137041	284536	178817	27807
	760151	872592	1235361	1391907	1958389	576574	1209246	775584	125383
	2885874	3351546	4538246	5203175	7408956	2238611	4717371	3082518	516061
						8118770		11390555	1967539
									7027717
Wt 7	2	1	3	3	2	0	1	0	0
	27	21	46	45	44	5	22	6	0
	240	214	408	416	464	81	247	99	6
	1626	1581	2724	2862	3455	728	1933	923	85
	9179	9412	15036	16230	20594	4872	12016	6351	723
	45095	48051	72253	79713	104846	26828	63328	35823	4644
	198672	217804	311649	350298	473324	128382	294364	174879	24834
	800703	897721	1231965	1406820	1941887	551093	1238089	763665	116082
	2995928	3420316	4529019	5243247	7364186	2167857	4797218	3048723	488920
		12194623				7931807	17349056	11299728	1892781
									6831594
Wt 8	3	3	6	5	6	1	2	1	0
	46	45	85	81	99	21	44	22	2
	408	416	723	728	923	214	464	247	27
	2724	2862	4644	4872	6351	1581	3455	1933	240
	15036	16230	24834	26828	35823	9412	20594	12016	1626
	72253	79713	116082	128382	174879	48051	104846	63328	9179
	311649	350298	488920	551093	763665	217804	473324	294364	45095
	1231965	1406820	1892781	2167857	3048723	897721	1941887	1238089	198672
	4529019	5243247	6831594	7931807	11299728	3420316	7364186	4797218	800703
						12194623		17349056	2995928
Wt 9	5	6	11	10	15	5	5	4	1
	80	90	155	150	210	60	90	60	10
	701	775	1265	1290	1775	495	875	550	85
	4565	5070	7850	8295	11425	3195	6144	3825	580
	24575	27545	40695	44230	61340	17450	35060	21970	3385
	115305	130722	185238	205855	288190	83735	172350	109443	17308
	486720	558130	762300	862890	1220159	362535	755960	487465	79435
	1886883	2187700	2891970	3324545	4747945	1441686	3026750	1982575	332815
		7984820				5337350	11240448	7474945	1290860
									4685000

Weight Multiplicities for Affine A6

Affine A6 - Level 1 - Class 0

	Rp 1 1 000000				
Wt 1	1	15525	6190182	768291650	130256414*
	6	36280	11867310	258702803*	60002753*
	27	81816	22395359	141730441*	384755963*
	98	178794	41650050	676839645*	16416885*
	315	380051	76413078	139141995*	559125278*
	918	788004	138421358	285832610*	850407748*
	2492	1597725	247783113	67375610*	612112691*
	6372	3174210	438616728	162538712*	32649644*

* Multiplicity is given modulo 1073741527

Affine A6 - Level 2 - Class 0

	Rp 1			Rp 2			Rp 3			Rp 4		
	0 001100			0 010010			0 100001			2 000000		
Wt 1	1	2868072	1004462287*	0	2918339	44723757*	0	1282248	449889404*	0	186571	665489298*
	12	9046690	30999916*	5	9397968	889796235*	0	4280769	494539744*	0	657484	486559005*
	99	27233324	1053665820*	60	28792779	57336800*	9	13520808	926663099*	0	2173171	577447919*
	628	78663607	168291136*	460	84440236	736795572*	108	40705830	463776014*	5	6802088	679138473*
	3329	218997248	595091553*	2720	238223384	960297465*	801	117499716	689405723*	60	20310589	746386420*
	15396	589796138		13530	649167912		4572	326739213	945874502*	439	58189368	305266377*
	63964	467667769*		59200	640778874*		21996	878696172		2468	160703360	190965817*
	243512	698367674*		234371	106396764*		93384	145269223*		11701	429459276	
	861971	56262609*		855492	271757694*		359856	451594129*		49012	40332366*	
Wt 2	2	5064359	882456993*	1	5256874	854710426*	0	2384682	1022209792*	0	361908	143900130*
	27	15639304	468156309*	18	16509372	792059159*	3	7715490	133100980*	0	1225730	65266678*
	218	46186310	19221819*	168	49477872	746804363*	42	23728638	512417937*	2	3921044	1027558825*
	1336	131111926	188509710*	1124	142292621	186382016*	342	69810780	140715519*	26	11939477	973400765*
	6816	359278814	748488225*	6091	394464676	443111225*	2094	197484198	198094844*	201	34817412	38402020*
	30409	953678158		28414	1058063962*		10641	539409378	562411255*	1186	97716916	132036150*
	122262	311946134*		118266	607096566*		47238	353804621*		5860	265005832	
	451976	808915423*		449570	536561318*		188940	450063123*		25440	696804811	
	1558344	113810783*		1586270	67329584*		694950	607914976*		99912	707680443*	
Wt 3	5	8859860	961815269*	4	9349815	849980619*	1	4356572	564447407*	0	685260	517284154*
	64	26803105	781115612*	54	28680380	1054765188*	16	13700812	1061722184*	1	2241196	637537410*
	488	77702172	729345730*	432	84186128	592565129*	143	41119124	720942264*	12	6962542	794860940*
	2840	216905543	716680578*	2615	237664868	691255136*	940	118420874	98737595*	94	20676780	966078453*
	13857	585353992		13180	647971094	617560678*	5055	328738384	768107202*	572	59001067	716793200*
	59484	458438682*		58143	638272221*		23508	882932231		2923	162456168	952230552*
	231282	679573255*		231396	101255057*		97765	154051746*		13108	433156243	
	830296	18686538*		847609	261409874*		371764	469444472*		53044	47964753*	
	2789644	930599465*		2898440	24260720*		1312971	485513473*		197428	680941173*	
Wt 4	14	15357671	270816281*	14	16432932	736521331*	6	7832694	230454100*	1	1270470	108493146*
	154	45546534	719136806*	148	49303840	641194642*	57	24002538	699103293*	6	4028192	37631820*
	1092	129699088	563450041*	1056	141906282	1062011148*	414	70429632	493760982*	48	12186496	59834026*
	5984	356239388	532041656*	5858	393630608	76058069*	2352	198845688	856981064*	280	35371708	340270712*
	27868	947288146		27727	1056301786*		11508	542326974		1470	98925232	693741984*
	115018	298795662*		116298	603455794*		49818	359920325*		6728	267580544	
	432628	782366637*		444368	529177618*		196224	462611931*		28000	702165794	
	1509088	61162920*		1572956	52629744*		714144	633188248*		106844	718628768*	
	4944442	779745525*		5224502	825912222*		2433216	1072210795*		380000	165836602*	

* Multiplicity is given modulo 1073741527

Affine A6 - Level 3 - Class 0 - Page 1 of 3

	Rp 1 0 000120	Rp 2 0 000201	Rp 3 0 001011	Rp 4 0 010002	Rp 5 0 021000	Rp 6 0 102000	Rp 7 0 110100	Rp 8 0 200010	Rp 9 1 001100	Rp10 1 010010	Rp11 1 100001	Rp12 3 000000
Wt 1	1 12 120 970 6690 40299 216710 1057779	0 6 87 855 6564 42388 239049 1208217	0 5 90 990 8225 56100 329415 1716770	0 0 10 180 1875 14685 94650 527910	0 0 21 373 3717 27456 166892 880554	0 0 26 447 4425 32820 201036 1070451	0 0 30 555 5815 44895 283575 1546830	0 0 5 120 1470 12560 85245 490995	0 0 24 408 4160 32016 203616 1123224	0 0 9 198 2340 19794 134493 779141	0 0 0 16 312 3336 26112 166656 916896	0 0 0 0 5 96 993 7513 46536 249387
Wt 2	1 18 196 1638 11304 67367 356555 1710578	1 16 188 1666 12087 74834 408198 2005925	0 10 175 1865 14955 98700 562520 2854820	0 0 20 340 3435 26030 162865 884230	0 5 101 1104 8794 56980 317640 1575957	0 6 116 1263 10172 66819 377803 1899611	0 5 120 1490 12970 89800 527160 2726895	0 0 15 290 3110 24385 155755 856895	0 4 84 1012 8844 62132 371320 1956064	0 0 24 447 4769 37626 242693 1349877	0 0 0 36 628 6292 46956 288596 1539128	0 0 0 0 11 187 1815 13170 78926 411642
Wt 3	2 33 355 2900 19565 113898 589534	1 26 326 2924 20921 126983 677892 3262587	1 28 390 3738 28068 176521 968667 4765523	0 4 78 900 7579 51675 301159 1553115	0 11 200 2059 15707 98274 532053	0 12 220 2319 18057 115101 633370 3108764	0 13 270 3028 24613 161880 912853 4569923	0 2 58 765 6871 48505 288572 1507535	0 8 168 1944 16296 110264 637720 3265096	0 3 84 1116 10210 73413 444335 2357332	0 0 8 160 1784 14520 95904 543592 2735856	0 0 0 3 55 568 4374 27660 151436 740677
Wt 4	3 58 630 5091 33632 191442 969285	2 48 597 5222 36411 215373 1122575	3 66 818 7275 51696 311426 1650159	1 18 228 2116 15747 98874 543360 2679963	0 23 387 3792 27783 168156 884994	0 24 423 4265 31998 197475 1056771	0 33 586 6015 45909 287948 1563621	0 9 168 1800 14307 93033 521711 2605554	0 18 346 3756 29952 194626 1088376	0 12 228 2538 20817 139013 796899 4054455	0 2 42 504 4412 31152 187398 994168	0 0 1 18 186 1475 9684 55197 280819
Wt 5	0 0 21 373 3717 27456 166892 880554	0 0 26 447 4425 32820 201036 1070451	0 0 30 555 5815 44895 283575 1546830	0 0 5 120 1470 12560 85245 490995	1 12 120 970 6690 40299 216710 1057779	0 6 87 855 6564 42388 239049 1208217	0 5 90 990 8225 56100 329415 1716770	0 0 10 180 1875 14685 94650 527910	0 0 24 408 4160 32016 203616 1123224	0 0 9 198 2340 19794 134493 779141	0 0 0 16 312 3336 26112 166656 916896	0 0 0 0 5 96 993 7513 46536 249387

-continued next page-

Affine A6 - Level 3 - Class 0 - Page 2 of 3

	Rp 1	Rp 2	Rp 3	Rp 4	Rp 5	Rp 6	Rp 7	Rp 8	Rp 9	Rp10	Rp11	Rp12
Wt 6	0 5 101 1104 8794 56980 317640 1575957	0 6 116 1263 10172 66819 377803 1899611	0 5 120 1490 12970 89800 527160 2726895	0 0 15 290 3110 24385 155755 856895	1 18 196 1639 11304 67367 356555 1710578	1 16 188 1666 12087 74834 408198 2005925	0 10 175 1865 14955 98700 562520 2854820	0 0 20 340 3435 26030 162865 884230	0 4 84 1012 8844 62132 371320 1956064	0 0 24 447 4769 37626 242693 1349877	0 0 0 36 628 6292 46956 288596 1539128	0 0 0 0 11 187 1815 13170 78926 411642
Wt 7	0 11 200 2059 15707 98274 532053	0 12 220 2319 18057 115101 633370 3108764	0 13 270 3028 24613 161880 912853 4569923	0 2 58 765 6871 48505 288572 1507535	2 33 355 2900 19565 113898 589534	1 26 326 2924 20921 126983 677892 3262587	1 28 390 3738 28068 176521 968667 4765523	0 4 78 900 7579 51675 301159 1553115	0 8 168 1944 16296 110264 637720 3265096	0 3 84 1116 10210 73413 444335 2357332	0 0 8 160 1784 14520 95904 543592 2735856	0 0 0 3 55 568 4374 27660 151436 740677
Wt 8	0 23 387 3792 27783 168156 884994	0 24 423 4265 31998 197475 1056771	0 33 586 6015 45909 287948 1563621	0 9 168 1800 14307 93033 521711 2605554	3 58 630 5091 33632 191442 969285	2 48 597 5222 36411 215373 1122575	3 66 818 7275 51696 311426 1650159	1 18 228 2116 15747 98874 543360 2679963	0 18 346 3756 29952 194626 1088376	0 12 228 2538 20817 139013 796899 4054455	0 2 42 504 4412 31152 187398 994168	0 0 1 18 186 1475 9684 55197 280819
Wt 9	3 57 622 5043 33405 190504 965811	3 57 654 5508 37654 220206 1139820	2 57 759 6970 50334 306009 1630454	0 7 147 1659 13573 89787 508980 2560026	3 57 622 5043 33405 190504 965811	3 57 654 5508 37654 220206 1139820	2 57 759 6970 50334 306009 1630454	0 7 147 1659 13573 89787 508980 2560026	1 30 435 4260 32343 204588 1125916 5546154	0 9 198 2339 19788 134466 779042 3990408	0 0 17 318 3363 26212 166983 917868	0 0 0 6 102 1020 7612 46857 250332
Wt10	6 108 1142 8960 57627 319932 1583532	5 105 1190 9792 65142 371224 1876074	6 132 1572 13406 91754 534923 2754961	1 30 406 3779 27568 168992 906536	6 108 1142 8960 57627 319932 1583532	5 105 1190 9792 65142 371224 1876074	6 132 1572 13406 91754 534923 2754961	1 30 406 3779 27568 168992 906536	2 62 858 8010 58314 355872 1899264	1 34 518 5160 39458 250262 1378258 6787332	0 4 82 950 8026 54830 320186 1654316	0 0 2 32 328 2548 16411 91638 457126

continued next page

Affine A6 - Level 3 - Class 0 - Page 3 of 3

	Rp 1	Rp 2	Rp 3	Rp 4	Rp 5	Rp 6	Rp 7	Rp 8	Rp 9	Rp10	Rp11	Rp12
Wt11	11	10	15	5	11	10	15	5	5	4	1	0
	202	201	290	90	202	201	290	90	134	98	22	1
	2076	2198	3160	985	2076	2198	3160	985	1708	1220	270	12
	15808	17452	25290	8040	15808	17452	25290	8040	15034	10795	2430	115
	98747	112504	164845	53775	98747	112504	164845	53775	104569	76144	17660	884
	533974	623428	924390	309600	533974	623428	924390	309600	615002	455529	109224	5808
				1583920				1583920	3182692	2398820	594486	33444
											2914700	172726
Wt12	21	21	35	15	21	21	35	15	14	14	6	1
	378	392	615	230	378	392	615	230	294	252	78	6
	3767	4074	6210	2190	3767	4074	6210	2190	3396	2709	736	48
	27747	31059	46915	16275	27747	31059	46915	16275	28048	21742	5628	336
	168258	193529	292320	101535	168258	193529	292320	101535	186150	143220	36624	2142
		1041942	1580460	554520		1041942	1580460	554520	1055040	813750	209362	12384
											1074696	65282
												315858

Weight Multiplicities for Affine A7

Affine A7 - Level 1 - Class 0

	Rp 1 1 0000000				
Wt 1	1	32305	20247710	409083254*	52130125*
	7	80465	40786690	124757309*	1000298023*
	35	192899	80782800	1023243108*	594048580*
	140	447146	157510780	459326577*	696348890*
	490	1006145	302666903	158673252*	871120898*
	1547	2204475	573720808	540260062*	512278697*
	4522	4715510	1073720305*	894147031*	419189929*
	12405	9869132	911765248*	342295921*	69617194*

* Multiplicity is given modulo 1073741527

Affine A7 - Level 2 - Class 0

	Rp 1		Rp 2		Rp 3		Rp 4		Rp 5	
	0		0		0		0		2	
	0002000		0010100		0100010		1000001		0000000	
Wt 1	1	43412586	0	88710188	0	66514140	0	25133641	0	3398437
	7	137933565	7	284994794	0	218568910	0	85535100	0	12149767
	63	421049188	77	878336277	20	686862540	0	277026939	0	41046460
	420	165972779*	637	460387400*	245	1000660313*	28	858963812	0	132034217
	2492	311259318*	4116	1043824906*	1995	676562820*	358	414627518*	14	406796600
	12873	105409028*	22736	439841860*	12565	951322635*	2898	938485224*	182	132484678*
	60060	515572273*	110579	577559451*	66990	563650229*	17997	203024447*	1470	234558288*
	255873	240407855*	486570	626425187*	314650	86760003*	94157	56499592*	9072	1007478628*
	1010940	640059958*	1968981	1003036165*	1338645	1228411*	433769	576306480*	47068	139426447*
	3739596		7425642		5250740	943050256*	1810634	770532183*	214950	505901907*
	13066550		26353572		19241075		6975360		889497	8250568*
Wt 2	1	76831801	1	158608407	0	121307940	0	47164744	0	6610423
	14	239791825	19	499651955	5	389537560	0	156082518	0	22837917
	127	720024401	198	438942145*	80	125610388*	9	493415672	0	74956716
	887	1014248728*	1501	123855347*	745	335171574*	136	424003497*	5	235191375
	5162	497708898*	9219	685832473*	5180	530358932*	1202	89031707*	74	709066427
	26169	980616890*	48556	1029459761*	29675	416807073*	7993	606635003*	640	988834901*
	118928	689426592*	227059	444558923*	147535	333823553*	44087	853360802*	4174	440180599*
	494229	890782800*	964820	713468107*	657080	852449911*	212176	61015983*	22636	853245831*
	1905524	816651155*	3787124	255681827*	2677615	959928751*	918725	967650679*	107341	413102725*
	6891311		13900170		10134505		3652677	682622582*	458794	302477033*
	23577319		48153769		36025825		13527449		1803268	659502384*
Wt 3	2	134942360	3	280997329	1	218602199	0	87115825	0	12598041
	31	413842492	51	868622997	23	686929668	3	280997329	0	42209124
	281	149072611*	491	437455794*	259	1000789392*	51	868622997	2	134942360
	1916	272616789*	3490	991056274*	2032	676811508*	491	437455794*	31	413842492
	10815	19027334*	20315	321250518*	12681	951785056*	3490	991056274*	281	149072611*
	53146	326508336*	102246	316692551*	67268	564508821*	20315	321250518*	1916	272616789*
	234472	908231482*	459860	63889413*	315371	88311080*	102246	316692551*	10815	19027334*
	948010	857850742*	1889126	885699607*	1340249	4020985*	459860	63889413*	53146	326508336*
	3564341		7198949	952217259*	5254450	947969791*	1889126	885699607*	234472	908231482*
	12598041		25740379		19248859		7198949	952217259*	948010	857850742*
	42209124		87115825		66530780		25740379		3564341	
Wt 4	5	235191375	9	493415672	5	389537560	1	158608407	0	23577319
	74	709066427	136	424003497*	80	125610388*	19	499651955	1	76831801
	640	988834901*	1202	89031707*	745	335171574*	198	438942145*	14	239791825
	4174	440180599*	7993	606635003*	5180	530358932*	1501	123855347*	127	720024401
	22636	853245831*	44087	853360802*	29675	416807073*	9219	685832473*	887	1014248728*
	107341	413102725*	212176	61015983*	147535	333823553*	48556	1029459761*	5162	497708898*
	458794	302477033*	918725	967650679*	657080	852449911*	227059	444558923*	26169	980616890*
	1803268	659502384*	3652677	682622582*	2677615	959928751*	964820	713468107*	118928	689426592*
	6610423		13527449		10134505		3787124	255681827*	494229	890782800*
	22837917		47164744		36025825		13900170		1905524	816651155*
	74956716		156082518		121307940		48153769		6891311	
Wt 5	14	406796600	28	858963812	20	686862540	7	284994794	1	43412586
	182	132484678*	358	414627518*	245	1000660313*	77	878336277	7	137933565
	1470	234558288*	2898	938485224*	1995	676562820*	637	460387400*	63	421049188
	9072	1007478628*	17997	203024447*	12565	951322635*	4116	1043824906*	420	165972779*
	47068	139426447*	94157	56499592*	66990	563650229*	22736	439841860*	2492	311259318*
	214950	505901907*	433769	576306480*	314650	86760003*	110579	577559451*	12873	105409028*
	889497	8250568*	1810634	770532183*	1338645	1228411*	486570	626425187*	60060	515572273*
	3398437		6975360		5250740	943050256*	1968981	1003036165*	255873	240407855*
	12149767		25133641		19241075		7425642		1010940	640059958*
	41046460		85535100		66514140		26353572		3739596	
	132034217		277026939		218568910		88710188		13066550	

* Multiplicity is given modulo 1073741527

Affine A7 - Level 2 - Class 1

	Rp 1			Rp 2			Rp 3			Rp 4		
	0			0			0			1		
	0001100			0010010			0100001			1000000		
Wt 1	1	8742496	1042121128*	0	9442426	686725450*	0	4764984	626869786*	0	1104292	459499225*
	14	30384571	715923851*	6	33518912	224690223*	0	17546249	154247616*	0	4294640	652616012*
	133	100417884	1003766920*	84	112775222	233972962*	14	60883886	926434890*	0	15597876	214513589*
	966	317403989	133335493*	742	361984672	851430809*	196	200658807	629422389*	14	53446724	653927529*
	5824	964092262	86575015*	5012	40595445*	1038803991*	1673	632078608	371307731*	196	174133162	297400187*
	30436	677729775*		28252	83117355*		10878	838915800*	34874844*	1646	542768240	281358943*
	142009	498086043*		139104	882837128*		59087	214220373*		10500	552816231*	747636197*
	603794	595026199*		615874	565029044*		281162	742022883*		55958	410590132*	
	2375051	90328299*		2500468	318252430*		1206870	320105474*		261548	301006800*	
Wt 2	2	16133272	223725472*	1	17796690	690858555*	0	9293808	319699996*	0	2255634	815065884*
	32	54848864	766156237*	22	61545205	977530453*	4	33116532	522023288*	0	8410382	294867083*
	302	177672064	642467291*	243	202340840	801638739*	68	111729532	766726994*	5	29500530	217940277*
	2136	551412806	3525806*	1890	636217083	787787517*	652	359363390	767210214*	78	98153356	973396569*
	12446	573323461*	140876573*	11747	848799609*	878509234*	4610	34231225*	814262713*	703	311797038	1025722453*
	62864	457971506*		62198	237200679*		26692	68102271*	901379912*	4746	950624962	745642489*
	284040	407885872*		291396	794165666*		133670	848317108*		26502	646256126*	
	1172624	695361209*		1238260	435810546*		598432	487513838*		128872	426335736*	
	4490642	88995901*		4855868	878413179*		2448122	147900874*		562985	435125860*	
Wt 3	5	29500530	217940277*	4	33116532	522023288*	1	17796690	690858555*	0	4490642	88995901*
	78	98153356	973396569*	68	111729532	766726994*	22	61545205	977530453*	2	16133272	223725472*
	703	311797038	1025722453*	652	359363390	767210214*	243	202340840	801638739*	32	54848864	766156237*
	4746	950624962	745642489*	4610	34231225*	814262713*	1890	636217083	787787517*	302	177672064	642467291*
	26502	646256126*		26692	68102271*	901379912*	11747	848799609*	878509234*	2136	551412806	3525806*
	128872	426335736*		133670	848317108*		62198	237200679*		12446	573323461*	140876573*
	562985	435125860*		598432	487513838*		291396	794165666*		62864	457971506*	
	2255634	815065884*		2448122	147900874*		1238260	435810546*		284040	407885872*	
	8410382	294867083*		9293808	319699996*		4855868	878413179*		1172624	695361209*	
Wt 4	14	53446724	653927529*	14	60883886	926434890*	6	33518912	224690223*	1	8742496	1042121128*
	196	174133162	297400187*	196	200658807	629422389*	84	112775222	233972962*	14	30384571	715923851*
	1646	542768240	281358943*	1673	632078608	371307731*	742	361984672	851430809*	133	100417884	1003766920*
	10500	552816231*	747636197*	10878	838915800*	34874844*	5012	40595445*	1038803991*	966	317403989	133335493*
	55958	410590132*		59087	214220373*		28252	83117355*		5824	964092262	86575015*
	261548	301006800*		281162	742022883*		139104	882837128*		30436	677729775*	
	1104292	459499225*		1206870	320105474*		615874	565029044*		142009	498086043*	
	4294640	652616012*		4764984	626869786*		2500468	318252430*		603794	595026199*	
	15597876	214513589*		17546249	154247616*		9442426	686725450*		2375051	90328299*	

* Multiplicity is given modulo 1073741527

Affine A7 - Level 3 - Class 0 - Page 1 of 6 (see page guide at right)

	Rp 1	Rp 2	Rp 3	Rp 4	Rp 5	Rp 6	Rp 7	Rp 8
	0 0000210	0 0001020	0 0001101	0 0010011	0 0100002	0 0120000	0 0201000	0 1011000
Wt 1	1	0	0	0	0	0	0	0
	14	7	6	0	0	0	0	0
	161	119	126	35	0	21	21	35
	1505	1358	1617	700	50	476	504	805
	11998	12103	15645	8365	1015	5824	6481	10171
	83426	90461	123893	75110	11830	51604	59675	93359
	516194	588315	840910	554218	102487	370391	441245	692692
				3526026	727664			
Wt 2	1	1	0	0	0	0	0	0
	21	19	12	5	0	0	0	0
	267	261	249	125	10	47	51	80
	2600	2710	3123	1785	230	965	1091	1680
	20868	22948	29348	18390	2985	11127	12997	20001
	144081	165111	225701	151259	28371	94474	113196	175241
	879413	1040914	1489818	1050606	217894	655286	801687	1251639
					1427661			
Wt 3	2	1	1	0	0	0	0	0
	39	31	34	10	0	11	10	15
	493	462	566	255	20	249	252	370
	4728	4890	6440	3535	450	3112	3379	4995
	37271	41053	56963	35130	5695	28356	32273	48372
	252140	290356	418685	279014	52491	209215	246494	374960
	1507441	1794043	2664690	1877239	391341	1321601	1598912	2467057
					2494362			
Wt 4	3	3	2	1	0	0	0	0
	68	69	70	38	4	23	24	33
	883	948	1129	677	98	490	545	771
	8464	9495	12400	8011	1388	5884	6796	9880
	65809	76479	105926	72419	14109	51778	61537	91508
	437156	522421	754099	538752	114330	370625	450504	682973
	2562907	3132980	4663342	3451119	782152	2279242	2822358	4350230
					4686208			
Wt 5	5	6	5	4	1	0	0	0
	122	142	150	108	22	47	56	75
	1605	1878	2280	1622	333	960	1151	1615
	15226	18111	23896	17251	3679	11072	13452	19486
	116248	140722	196311	144768	32307	94084	115861	172131
	756339	930979	1351232	1018698	237402	653136	814887	1235877
				6246747	1514478			

1	2	(Page Guide)
3	4	
5	6	

continued next page

Affine A7 - Level 3 - Class 0 - Page 2 of 6

	Rp 9	Rp10	Rp11	Rp12	Rp13	Rp14	Rp15
	0 1100100	0 2000010	1 0002000	1 0010100	1 0100010	1 1000001	3 0000000
Wt 1	0	0	0	0	0	0	0
	0	0	0	0	0	0	0
	14	0	15	14	0	0	0
	406	35	315	364	63	0	0
	5929	805	3850	5068	1386	77	0
	59780	10150	35210	50274	17157	1630	21
	473662	92260	263585	397908	155232	19222	441
	3154950	675626	1696940	2666573	1139089	166110	5097
						1169322	43085
							296968
Wt 2	0	0	0	0	0	0	0
	0	0	0	0	0	0	0
	44	5	30	38	9	0	0
	1029	155	610	846	261	16	0
	13183	2355	7285	10712	3879	389	5
	121850	24360	64755	99426	40020	5097	117
	906867	196660	471405	747933	324802	47981	1464
		1329201	2955890	4812613	2213635	362802	13223
						2336355	96563
							603801
Wt 3	0	0	0	0	0	0	0
	5	0	5	4	0	0	0
	178	15	120	136	24	0	0
	2856	385	1665	2184	609	36	0
	30652	5170	16675	23840	8233	806	11
	254612	49255	133700	202140	79410	9974	234
	1761444	374605	907620	1427720	612017	89785	2771
		2418317			4002495	654355	24075
						4084570	170323
Wt 4	0	0	0	0	0	0	0
	18	2	9	12	3	0	0
	474	73	233	330	108	8	0
	6529	1183	3187	4754	1785	196	3
	63689	12814	31087	48166	19827	2647	66
	494633	107432	242076	386317	170032	25759	816
	3252885	749860	1598232	2611366	1210526	201014	7451
		4549550			7471277	1332199	55375
							353191
Wt 5	0	0	0	0	0	0	0
	54	11	20	33	15	2	0
	1164	246	479	775	345	50	1
	14275	3135	6234	10124	4590	711	21
	128574	29386	58310	95727	44590	7342	255
	941509	223506	438234	728536	349303	60793	2366
	5915100	1454455	2805759	4723456	2328585	426024	18116
						2615209	119781
							703031

continued next page

Affine A7 - Level 3 - Class 0 - Page 3 of 6

	Rp 1	Rp 2	Rp 3	Rp 4	Rp 5	Rp 6	Rp 7	Rp 8
Wt 6	0 0 21 476 5824 51604 370391	0 0 21 504 6481 59675 441245	0 0 35 805 10171 93359 692692	0 0 14 406 5929 59780 473662 3154950	0 0 0 35 805 10150 92260 675626	1 14 161 1505 11998 83426 516194	0 7 119 1358 12103 90461 588315	0 6 126 1617 15645 123893 840910
Wt 7	0 0 47 965 11127 94474 655286	0 0 51 1091 12997 113196 801687	0 0 80 1680 20001 175241 1251639	0 0 44 1029 13183 121850 906867	0 0 5 155 2355 24360 196660 1329201	1 21 267 2600 20868 144081 879413	1 19 261 2710 22948 165111 1040914	0 12 249 3123 29348 225701 1489818
Wt 8	0 11 249 3112 28356 209215 1321601	0 10 252 3379 32273 246494 1598912	0 15 370 4995 48372 374960 2467057	0 5 178 2856 30652 254612 1761444	0 0 15 385 5170 49255 374605 2418317	2 39 493 4728 37271 252140 1507441	1 31 462 4890 41053 290356 1794043	1 34 566 6440 56963 418685 2664690
Wt 9	0 23 490 5884 51778 370625 2279242	0 24 545 6796 61537 450504 2822358	0 33 771 9880 91508 682973 4350230	0 18 474 6529 63689 494633 3252885	0 2 73 1183 12814 107432 749860 4549550	3 68 883 8464 65809 437156 2562907	3 69 948 9495 76479 522421 3132980	2 70 1129 12400 105926 754099 4663342
Wt10	0 47 960 11072 94084 653136	0 56 1151 13452 115861 814887	0 75 1615 19486 172131 1235877	0 54 1164 14275 128574 941509 5915100	0 11 246 3135 29386 223506 1454455	5 122 1605 15226 116248 756339	6 142 1878 18111 140722 930979	5 150 2280 23896 196311 1351232

continued next page-

Affine A7 - Level 3 - Class 0 - Page 4 of 6

	Rp 9	Rp10	Rp11	Rp12	Rp13	Rp14	Rp15
Wt 6	0	0	0	0	0	0	0
	0	0	0	0	0	0	0
	35	0	15	14	0	0	0
	700	50	315	364	63	0	0
	8365	1015	3850	5068	1386	77	0
	75110	11830	35210	50274	17157	1630	21
	554218	102487	263585	397908	155232	19222	441
	3526026	727664	1696940	2666573	1139089	166110	5097
						1169322	43085
							296968
Wt 7	0	0	0	0	0	0	0
	5	0	0	0	0	0	0
	125	10	30	38	9	0	0
	1785	230	610	846	261	16	0
	18390	2985	7285	10712	3879	389	5
	151259	28371	64755	99426	40020	5097	117
	1050606	217894	471405	747933	324802	47981	1464
		1427661	2955890	4812613	2213635	362802	13223
						2336355	96563
							603801
Wt 8	0	0	0	0	0	0	0
	10	0	5	4	0	0	0
	255	20	120	136	24	0	0
	3535	450	1665	2184	609	36	0
	35130	5695	16675	23840	8233	806	11
	279014	52491	133700	202140	79410	9974	234
	1877239	391341	907620	1427720	612017	89785	2771
		2494362			4002495	654355	24075
						4084570	170323
Wt 9	1	0	0	0	0	0	0
	38	4	9	12	3	0	0
	677	98	233	330	108	8	0
	8011	1388	3187	4754	1785	196	3
	72419	14109	31087	48166	19827	2647	66
	538752	114330	242076	386317	170032	25759	816
	3451119	782152	1598232	2611366	1210526	201014	7451
		4686208			7471277	1332199	55375
							353191
Wt10	4	1	0	0	0	0	0
	108	22	20	33	15	2	0
	1622	333	479	775	345	50	1
	17251	3679	6234	10124	4590	711	21
	144768	32307	58310	95727	44590	7342	255
	1018698	237402	438234	728536	349303	60793	2366
	6246747	1514478	2805759	4723456	2328585	426024	18116
						2615209	119781
							703031

continued next page

Affine A7 - Level 3 - Class 0 - Page 5 of 6

	Rp 1	Rp 2	Rp 3	Rp 4	Rp 5	Rp 6	Rp 7	Rp 8
Wt11	3	2	3	0	0	3	2	3
	69	58	81	23	0	69	58	81
	893	865	1215	541	42	893	865	1215
	8536	9001	12930	7081	906	8536	9001	12930
	66217	73955	108725	67152	10968	66217	73955	108725
	439150	510952	767188	512811	97278	439150	510952	767188
	2571581	3085464	4718923	3336572	701286	2571581	3085464	4718923
Wt12	6	5	6	2	0	6	5	6
	131	129	165	80	7	131	129	165
	1664	1774	2406	1380	189	1664	1774	2406
	15534	17464	24701	15659	2652	15534	17464	24701
	117656	137331	200594	136038	26291	117656	137331	200594
	762142	915314	1371219	976869	207006	762142	915314	1371219
		5361333	8198593	6066390	1377540		5361333	8198593
Wt13	11	12	13	8	1	11	12	13
	247	276	345	222	37	247	276	345
	3090	3565	4793	3238	609	3090	3565	4793
	28195	33389	47118	33149	6794	28195	33389	47118
	208301	252167	368404	268153	58717	208301	252167	368404
	1317181	1624392	2436923	1825502	421884	1317181	1624392	2436923
					2628415			
Wt14	21	26	30	24	6	21	26	30
	471	574	730	556	129	471	574	730
	5759	7036	9551	7205	1665	5759	7036	9551
	51124	63035	89489	67931	15954	51124	63035	89489
	367651	458275	672601	516884	124476	367651	458275	672601
		2857656	4303370	3355258	830682		2857656	4303370
Wt15	42	56	70	64	21	42	56	70
	910	1176	1546	1309	371	910	1176	1546
	10766	13727	18942	15435	4095	10766	13727	18942
	92568	117761	168868	135639	35196	92568	117761	168868
	646639	825237	1219456	977564	253316	646639	825237	1219456

continued next page-

Affine A7 - Level 3 - Class 0 - Page 6 of 6

	Rp 9	Rp10	Rp11	Rp12	Rp13	Rp14	Rp15
Wt11	0	0	1	0	0	0	0
	23	0	23	16	0	0	0
	541	42	359	380	63	0	0
	7081	906	4048	5156	1385	80	0
	67152	10968	35984	50660	17149	1654	23
	512811	97278	266313	399376	155188	19354	457
	3336572	701286	1705808	2671589	1138894	166694	5185
					7160957	1171564	43472
							298444
Wt12	2	0	1	1	0	0	0
	80	7	39	46	9	0	0
	1380	189	663	891	261	17	0
	15659	2652	7531	10913	3878	398	6
	136038	26291	65736	100205	40011	5151	125
	976869	207006	474902	750642	324750	48235	1509
	6066390	1377540	2967329	4821290	2213398	363827	13424
						2340045	97343
							606519
Wt13	8	1	2	3	1	0	0
	222	37	78	117	43	4	0
	3238	609	1303	2003	803	99	2
	33149	6794	14315	22582	9652	1373	38
	268153	58717	120889	195031	87785	13777	460
	1825502	421884	846892	1393018	654372	110812	4210
		2628415	5147262	8609049	4192851	755396	31738
						4520183	206070
							1187190
Wt14	24	6	5	9	5	1	0
	556	129	169	289	145	26	1
	7205	1665	2626	4406	2140	374	14
	67931	15954	27425	45926	22305	3938	155
	516884	124476	222361	374284	184104	33382	1384
	3355258	830682	1505784	2554841	1279484	239759	10541
						1507525	70188
							417394
Wt15	64	21	14	28	20	7	1
	1309	371	378	700	420	105	7
	15435	4095	5342	9513	5250	1141	63
	135639	35196	52556	91931	48930	10052	504
	977564	253316	407673	708799	372918	75348	3682
					2440480	495159	24423
							147490

Weight Multiplicities for Affine A8

Affine A8 - Level 1 - Class 0

	Rp 1 1 00000000				
Wt 1	1	62337	59068372	882071914*	754353614*
	8	164560	124505880	1063503217*	1005972877*
	44	417140	257770964	377995504*	1006623135*
	192	1020416	524871424	45655873*	700664219*
	726	2418710	1052316364	427920852*	853810195*
	2464	5573568	1005750217*	509756845*	266184294*
	7704	12520744	832753459*	766484293*	194507021*
	22528	27484160	287029279*	942242413*	640722352*

* Multiplicity is given modulo 1073741527

Affine A8 - Level 2 - Class 0

	Rp 1		Rp 2		Rp 3		Rp 4		Rp 5	
	0		0		0		0		2	
	00011000		00100100		01000010		10000001		00000000	
Wt 1	1	323323456	0	379856568	0	225618112	0	72682600	0	8681984
	16	36468475*	7	251018217*	0	808098372	0	270689728	0	34112972
	172	432500091*	112	124131616*	20	611153714*	0	955744419	0	126113088
	1408	835592726*	1120	205961045*	320	433603348*	28	1072418130*	0	442253804
	9520	1046135072*	8512	844955712*	3090	489086122*	448	741174093*	14	406848809*
	55520	786542114*	53662	42243402*	22560	478892969*	4249	175474446*	224	461574432*
	287684	394224561*	293888	395452289*	136668	25329349*	30352	823710010*	2108	767205445*
	1352256	662910188*	1440068	2509217*	721088	507159046*	179816	907894600*	14912	66214845*
	5855940	186074508*	6441792	515486984*	3414560	154071835*	928704	289003883*	87455	267883983*
	23640464		26693436		14806208	283452121*	4310832	410863420*	447280	839772196*
	89797116		103601456		59644208		18351328		2057076	1054813486*
Wt 2	2	594294849	1	708509444	0	430986471	0	143402500	0	17824063
	37	928574431*	26	271132641*	5	431210897*	0	517836207	0	67442734
	400	33237827*	332	397186477*	100	728555174*	9	706567643*	0	241527279
	3212	856648370*	2948	226502734*	1105	83983*	164	490981794*	5	824178332
	21082	707996009*	20710	665011394*	8880	404825883*	1693	294793829*	88	546872162*
	119118	676526839*	122930	216504248*	57834	790730456*	12910	846503404*	883	959991563*
	598428	812119035*	641249	397697422*	323144	766058318*	80613	245269476*	6580	1059401894*
	2732408	783974815*	3016496	478995628*	1603533	75753205*	435000	14791790*	40313	665014010*
	11519372	921467779*	13030068	175264095*	7231352	667911894*	2095949	222365536*	214078	1061323480*
	45372483		52375932		30122100	466996322*	9215582	127176700*	1017432	619675004*
	168501056		197902721		117300540		37548999		4420532	910857154*
Wt 3	5	9729557*	4	234269663*	1	812195224	0	277923292	0	35762528
	92	362179681*	82	79586156*	28	622149216*	3	975749864	0	130837118
	960	656172790*	920	91176730*	370	462188980*	60	52061266*	2	455252748
	7400	600760124*	7470	557517952*	3340	561287790*	668	879142941*	36	441373261*
	46679	782122912*	49020	414678081*	23657	656521052*	5420	521889954*	375	550408812*
	254244	1064884100*	275501	872095156*	140988	451901838*	35646	597210475*	2908	989338869*
	1235428	124707551*	1373540	402975812*	736744	435294423*	201024	783371936*	18504	607413157*
	5473816	731448393*	6218200	209739192*	3467440	311737597*	1006116	718775919*	101744	481721994*
	22458772		25986664	877365900*	14974585	118085024*	4572872	485548546*	499073	615626931*
	86315556		101481242		60153552		19184551	3878149*	2231236	375408090*
	313492026		373777784		227091562		75194728		9232314	
Wt 4	14	885516282*	14	243591982*	6	437703438*	1	530012452	0	70312557
	238	995261903*	244	325172527*	111	745816388*	22	739582528*	1	249543166
	2332	575140945*	2455	43744733*	1180	44541414*	262	577551135*	16	845764087
	17047	17239228*	18392	213618559*	9277	516095230*	2252	515047665*	165	603128278*
	102764	95949968*	113337	202631851*	59616	1062059710*	15571	318024377*	1302	28597326*
	538168	158747981*	605270	1057286356*	330237	338377631*	91730	488107837*	8511	336314602*
	2526776	484151149*	2891833	1013354539*	1629264	507018152*	477008	974298057*	48188	434143852*
	10860979	238216218*	12625564	770004485*	7317965	882388336*	2242404	958798780*	243285	894726855*
	43375386		51134419		30396084		9693280	286838987*	1117740	884290898*
	162719617		194270702		118122684		39021758		4743682	468192822*
	578228652		698320636		433343488		147727954		18809992	
Wt 5	42	293239963*	48	34725572*	27	632408642*	8	996018560	1	135756016
	624	479656286*	699	1049468596*	376	489206780*	100	106019899*	8	468684500
	5664	161357648*	6360	268696336*	3440	630212154*	928	1018323723*	80	476827385*
	39000	789765175*	44144	784306856*	24248	827454736*	6720	870825070*	600	641206372*
	224124	685718740*	256456	269386423*	143844	865023461*	41376	375821084*	3972	141771006*
	1127424	706926665*	1305148	792451401*	748272	336908742*	223454	669065964*	22864	83056523*
	5114176	292786147*	5989872	956474956*	3509632	417865791*	1086960	94830548*	118272	711809225*
	21331664		25267968		15116016	927883568*	4843616	598970192*	556720	423284582*
	82962239		99332960		60598588		20039592		2420490	830935387*
	303951432		367633872		228415224		77758168		9819032	
	1057374576*		217376737*		815956448		285275648		37498240	

* Multiplicity is given modulo 1073741527

Weight Multiplicities for Affine B3

Affine B3 - Level 1 - Class 0

	Rp 1 (0 00 1)					Rp 2 (1 00 0)				
Wt 1	1	2560	378031	20713069	651689441	0	1758	292383	16801825	543141339
	3	4823	612014	31087469	932319947	1	3380	476791	25310042	778861824
	10	8849	980446	46363032	254318729*	4	6328	768742	37878658	38228839*
	26	15915	1554943	68729635	810191097*	13	11553	1226589	56336926	507080534*
	65	28042	2443349	101300489		35	20657	1937975	83295013	90700388*
	147	48594	3805581	148488171		87	36218	3034071	122457416	
	321	82833	5878808	216511555		198	62429	4709234	179061377	
	662	139217	9010586	314108015		429	105911	7250321	260475719	
	1326	230802	13709567	453492662		884	177147	11077187	377029942	
Wt 2	1	3380	476791	25310042	778861824	1	2560	378031	20713069	651689441
	4	6328	768742	37878658	38228839*	3	4823	612014	31087469	932319947
	13	11553	1226589	56336926	507080534*	10	8849	980446	46363032	254318729*
	35	20657	1937975	83295013	90700388*	26	15915	1554943	68729635	810191097*
	87	36218	3034071	122457416		65	28042	2443349	101300489	
	198	62429	4709234	179061377		147	48594	3805581	148488171	
	429	105911	7250321	260475719		321	82833	5878808	216511555	
	884	177147	11077187	377029942		662	139217	9010586	314108015	
	1758	292383	16801825	543141339		1326	230802	13709567	453492662	

* Multiplicity is given modulo 1073741527

Affine B3 - Level 1 - Class 1

	Rp 1 (0 01 0)				
Wt 1	1	1901	201822	8519173	214811311
	4	3688	334684	13055208	313149460
	13	6955	548158	19859113	454206821
	36	12792	887500	29998024	655619380
	90	23019	1421602	45012751	941952353
	208	40612	2254460	67116436	273552941*
	455	70395	3541928	99472320	845038371*
	948	120072	5515900	146580028	573870270*

* Multiplicity is given modulo 1073741527

Affine B3 - Level 2 - Class 0

	Rp 1		Rp 2		Rp 3		Rp 4		Rp 5	
	0		0		0		1		2	
	02		10		00		00		00	
	0		0		2		1		0	
Wt 1	1	558840	0	493773	0	311528	0	393768	0	135006
	6	1251911	4	1112172	3	695976	3	892848	0	311528
	36	2725170	24	2433048	18	1511373	18	1965447	3	695976
	152	5782680	117	5184384	84	3199048	88	4212360	18	1511373
	564	11988960	446	10789548	312	6617304	336	8814220	84	3199048
	1836	24338892	1512	21976892	1033	13403568	1149	18044400	312	6617304
	5504	48465666	4608	43895796	3090	26634419	3530	36212202	1033	13403568
	15312	94813992	13058	86108592	8616	51999726	10092	71351698	3090	26634419
	40263	182468160	34704	166132405	22592	99882660	27048	138235896	8616	51999726
	100696		87708		56424		68919		22592	99882660
	241632		212032		135006		167880		56424	
Wt 2	1	683633	1	611813	1	386849	0	478360	0	169653
	9	1520890	8	1365351	7	855738	3	1079217	1	386849
	49	3290423	43	2962549	34	1842288	22	2363920	7	855738
	207	6942564	181	6267906	135	3870421	107	5043034	34	1842288
	746	14319841	652	12961503	467	7952935	420	10505614	135	3870421
	2398	28932498	2101	26251159	1459	16014185	1431	21417934	467	7952935
	7067	57361011	6213	52162807	4213	31652926	4401	42812889	1459	16014185
	19439	111759169	17152	101847437	11410	61501146	12514	84045235	4213	31652926
	50523		44746	195654503	29297		33403	162256399	11410	61501146
	125195		111285		71894		84617		29297	
	297805		265640		169653		205068		71894	
Wt 3	0	453033	0	404819	1	257250	0	314634	0	110916
	3	1023351	3	917343	3	577589	1	721377	0	257250
	21	2244315	19	2017863	18	1259706	9	1602890	3	577368
	102	4793592	90	4322059	70	2678379	51	3463641	15	1259706
	397	9997494	348	9037911	261	5562921	218	7299507	70	2677950
	1350	20404500	1184	18491817	837	11313351	789	15039160	252	5562921
	4149	40830084	3645	37089011	2517	22564836	2541	30353172	837	11312541
	11802	80233254	10401	73041507	6990	44213386	7493	60115869	2495	22564836
	31524	155044997	27878	141438438	18432	85212921	20601	117012151	6990	44211907
	79944		70950		46151		53499		18381	85212921
	193969		172746		111024		132427		46151	
Wt 4	2	839523	1	743912	1	467436	1	595318	0	205993
	15	1853184	10	1650665	8	1029141	8	1329445	1	467436
	74	3981664	55	3562596	41	2205606	42	2886538	8	1029141
	296	8349086	232	7500139	166	4613772	176	6110926	41	2205606
	1026	17124817	833	15438112	578	9441602	633	12644450	166	4613772
	3205	34424196	2666	31132409	1808	18938047	2038	25623590	578	9441602
	9242	67931931	7823	61612926	5209	37294522	6027	50943224	1808	18938047
	24982	131790092	21432	119844459	14054	72210374	16646	99517001	5209	37294522
	64006		55496		35923		43450	191270088	14054	72210374
	156715		137050		87726		108134		35923	
	368980		324990		205993		258299		87726	
Wt 5	3	1023351	3	917343	3	577368	1	721377	1	257250
	21	2244315	19	2017863	15	1259706	9	1602890	3	577589
	102	4793592	90	4322059	70	2677950	51	3463641	18	1259706
	397	9997494	348	9037911	252	5562921	218	7299507	70	2678379
	1350	20404500	1184	18491817	837	11312541	789	15039160	261	5562921
	4149	40830084	3645	37089011	2495	22564836	2541	30353172	837	11313351
	11802	80233254	10401	73041507	6990	44211907	7493	60115869	2517	22564836
	31524	155044997	27878	141438438	18381	85212921	20601	117012151	6990	44213386
	79944		70950		46151		53499		18432	85212921
	193969		172746		110916		132427		46151	
	453033		404819		257250		314634		111024	

Affine B3 - Level 2 - Class 1

	Rp 1 (0 01 1)					Rp 2 (1 01 0)				
Wt 1	1	142769	132230021	612561499*	707881204*	0	90882	94800537	969515026*	331436422*
	9	341649	254170332	681792829*	449401799*	3	221910	183686994	301385469*	502082373*
	51	788460	481225800	622334970*	773793211*	22	521196	350371088	434964563*	586233524*
	221	1762462	898365204	586834968*	1057124888*	108	1183221	658626822	453495170*	768782561*
	810	3829437	581409962*	450901999*		429	2606604	147598670*	8088924*	324176671*
	2640	8111346	864564244*	289914537*		1479	5590110	88655772*	111758836*	
	7868	16790136	49392946*	829538914*		4599	11701998	824210934*	265504726*	
	21843	34034391	1050227677*	740110882*		13209	23964594	794274735*	373876772*	
	57222	67679109	869931153*	69791671*		35581	48104069	997556461*	300549584*	
Wt 2	3	221910	183686994	301385469*	502082373*	1	142769	132230021	612561499*	707881204*
	22	521196	350371088	434964563*	586233524*	9	341649	254170332	681792829*	449401799*
	108	1183221	658626822	453495170*	768782561*	51	788460	481225800	622334970*	773793211*
	429	2606604	147598670*	8088924*	324176671*	221	1762462	898365204	586834968*	1057124888*
	1479	5590110	88655772*	111758836*		810	3829437	581409962*	450901999*	
	4599	11701998	824210934*	265504726*		2640	8111346	864564244*	289914537*	
	13209	23964594	794274735*	373876772*		7868	16790136	49392946*	829538914*	
	35581	48104069	997556461*	300549584*		21843	34034391	1050227677*	740110882*	
	90882	94800537	969515026*	331436422*		57222	67679109	869931153*	69791671*	

* Multiplicity is given modulo 1073741527

Affine B3 - Level 3 - Class 0 - Page 1 of 2

	Rp 1 0 02 1	Rp 2 0 10 1	Rp 3 0 00 3	Rp 4 1 02 0	Rp 5 1 10 0	Rp 6 1 00 2	Rp 7 2 00 1	Rp 8 3 00 0
Wt 1	1	0	0	0	0	0	0	0
	12	6	3	3	2	3	0	0
	96	56	25	33	24	27	6	1
	542	351	147	225	162	171	57	12
	2523	1707	678	1164	846	837	342	72
	10149	7091	2709	5056	3690	3522	1618	354
	36690	26148	9672	19320	14193	13138	6549	1461
	121692	88080	31740	66966	49461	44781	23670	5406
	376476	275602	97123	214618	159396	141624	78353	18225
	1098274	811500	280593	644886	481314	421211	241626	57213
	3047868	2268876	771588	1835001	1375947	1188453	702474	168909
							1942548	473799
Wt 2	1	1	1	0	0	0	0	0
	16	12	7	5	2	3	1	0
	123	91	45	48	28	31	14	1
	689	503	220	308	195	200	95	13
	3149	2306	944	1526	1026	996	504	83
	12510	9188	3568	6445	4467	4196	2212	414
	44674	32982	12340	24103	17101	15647	8570	1725
	146711	108834	39487	82200	59243	53138	30018	6387
	449823	335344	118733	259992	189751	167385	97208	21514
	1302179	975178	338156	772890	569528	495512	294708	67327
		2699188		2179371	1618832	1391730	845580	198076
								553386
Wt 3	0	0	1	0	0	0	0	0
	3	3	3	0	0	1	0	0
	33	27	18	9	5	9	3	0
	225	169	87	88	54	66	27	3
	1165	864	387	522	345	364	171	24
	5058	3729	1530	2478	1692	1668	837	144
	19326	14272	5548	10056	7059	6654	3522	671
	66981	49620	18495	36501	26082	23883	13137	2691
	214650	159699	57720	121329	87945	78762	44778	9633
	644952	481869	169833	375801	275349	242391	141616	31662
	1835136	1376952	475434	1097049	811029	703860	421191	96966
							1188408	280296
Wt 4	3	2	1	1	0	0	0	0
	33	24	12	12	6	6	3	0
	225	162	72	96	56	57	27	3
	1164	846	354	542	351	342	171	25
	5056	3690	1461	2523	1707	1618	837	147
	19320	14193	5406	10149	7091	6549	3522	678
	66966	49461	18225	36690	26148	23670	13138	2709
	214618	159396	57213	121692	88080	78353	44781	9672
	644886	481314	168909	376476	275602	241626	141624	31740
	1835001	1375947	473799	1098274	811500	702474	421211	97123
				3047868	2268876	1942548	1188453	280593
								771588

continued next page

Affine B3 - Level 3 - Class 0 - Page 2 of 2

	Rp 1	Rp 2	Rp 3	Rp 4	Rp 5	Rp 6	Rp 7	Rp 8
Wt 5	5	2	1	1	1	1	0	0
	48	28	13	16	12	14	3	1
	308	195	83	123	91	95	31	7
	1526	1026	414	689	503	504	200	45
	6445	4467	1725	3149	2306	2212	996	220
	24103	17101	6387	12510	9188	8570	4196	944
	82200	59243	21514	44674	32982	30018	15647	3568
	259992	189751	67327	146711	108834	97208	53138	12340
	772890	569528	198076	449823	335344	294708	167385	39487
	2179371	1618832	553386	1302179	975178	845580	495512	118733
					2699188		1391730	338156
Wt 6	2	1	1	0	0	1	0	0
	25	14	8	6	5	8	1	0
	173	110	52	62	46	57	15	3
	920	612	260	390	287	306	110	22
	4057	2801	1118	1913	1400	1399	593	127
	15733	11104	4234	7962	5848	5572	2630	575
	55150	39615	14604	29446	21708	20064	10190	2292
	178545	129883	46592	99357	73616	66412	35638	8176
	541064	397705	139522	311464	231852	205349	114968	26903
	1551275	1149671	395742	918451	686891	599067	347119	82582
				2571632	1931549	1663260	991434	239541
								661347
Wt 7	6	5	3	2	1	1	1	0
	62	46	22	25	14	15	8	1
	390	287	127	173	110	110	57	8
	1913	1400	575	920	612	593	306	52
	7962	5848	2292	4057	2801	2630	1399	260
	29446	21708	8176	15733	11104	10190	5572	1118
	99357	73616	26903	55150	39615	35638	20064	4234
	311464	231852	82582	178545	129883	114968	66412	14604
	918451	686891	239541	541064	397705	347119	205349	46592
	2571632	1931549	661347	1551275	1149671	991434	599067	139522
							1663260	395742
Wt 8	9	5	3	3	3	3	1	1
	88	54	24	33	27	27	9	3
	522	345	144	225	169	171	66	18
	2478	1692	671	1165	864	837	364	87
	10056	7059	2691	5058	3729	3522	1668	387
	36501	26082	9633	19326	14272	13137	6654	1530
	121329	87945	31662	66981	49620	44778	23883	5548
	375801	275349	96966	214650	159699	141616	78762	18495
	1097049	811029	280296	644952	481869	421191	242391	57720
				1835136	1376952	1188408	703860	169833
								475434

Affine B3 - Level 3 - Class 1

	Rp 1 0 03 0		Rp 2 0 11 0		Rp 3 0 01 2		Rp 4 1 01 1		Rp 5 2 01 0	
Wt 1	1	4021227	0	6198258	0	5575798	0	5968899	0	2093418
	6	10327605	6	16030342	6	14290680	3	15622647	0	5575798
	45	25630341	54	40030266	58	35398782	35	39427188	6	14290680
	255	61685994	334	96879000	348	85049044	249	96327141	58	35398782
	1191	144410843	1644	227938770	1662	198789516	1317	228584148	348	85049044
	4842	329683089	6894	522749076	6792	453165378	5803	528311628	1662	198789516
	17664	735562452	25692	97444493*	24768	1009706466	22425	118328579*	6792	453165378
	59064	533130120*	87360	420853272*	82680	55524526*	78402	483735472*	24768	1009706466
	183987	221390910*	275736	153302281*	256938	419379782*	253077	322624873*	82680	55524526*
	540005		818150		752232		765033		256938	419379782*
	1506390		2303490		2093418		2188212		752232	
Wt 2	1	5528226	1	8542864	1	7662073	0	8261725	0	2915543
	11	14031387	13	21827823	15	19406667	7	21352101	1	7662073
	79	34456677	102	53921479	111	47566553	70	53285331	15	19406667
	426	82144202	579	129236275	604	113203200	444	128882513	111	47566553
	1912	190653669	2687	301407328	2707	262328858	2196	303074113	604	113203200
	7504	431841945	10801	685724011	10592	593337377	9224	694718858	2707	262328858
	26584	956561265	38969	451343634*	37385	238806999*	34379	482022340*	10592	593337377
	86777	1002055170*	129067	100663420*	121609	697349210*	116818	189028578*	37385	238806999*
	264843		398652	655220984*	369953		368420	886843705*	121609	697349210*
	763829		1161390		1063890		1092240		369953	
	2098688		3218762		2915543		3072679		1063890	
Wt 3	0	4018695	0	6196599	1	5579544	0	5969610	0	2092947
	4	10323321	5	16027527	9	14297051	3	15623873	0	5574945
	39	25623267	51	40025547	66	35409411	36	39429258	6	14289183
	240	61674444	327	96871242	368	85066503	251	96330594	57	35396202
	1159	144392241	1626	227926221	1707	198817816	1323	228589776	346	85044660
	4776	329653563	6855	522728994	6885	453210648	5817	528320706	1656	198782235
	17529	735516102	25614	97412795*	24957	1009777992	22455	118343018*	6777	453153426
	58812	533058068*	87204	420803839*	83043	55636385*	78462	483758146*	24736	1009687110
	183519	221280204*	275442	153225895*	257613	419552873*	253197	322660081*	82614	55493616*
	539160		817617		753458		765253		256803	419330963*
	1504914		2302535		2095587		2188614		751979	
Wt 4	2	7569279	2	11724878	3	10485571	1	11385375	0	4041461
	21	18994574	26	29611004	31	26257765	17	29069887	3	10485571
	139	46171729	184	72390164	202	63707274	136	71765290	31	26257765
	708	109064543	978	171879692	1017	150231164	777	171902763	202	63707274
	3040	251026922	4318	397460742	4330	345244712	3606	400701257	1017	150231164
	11525	564261537	16718	897247812	16308	774951697	14484	911180251	4330	345244712
	39700	167391795*	58538	907591829*	55884	628610738*	52193	951860820*	16308	774951697
	126624	528449561*	189214	1066137325*	177479	444720861*	172664	115304009*	55884	628610738*
	378978		572630		529273		532721	829599435*	177479	444720861*
	1074851		1639534		1496488		1550438		529273	
	2910564		4476184		4041461		4293172		1496488	
Wt 5	4	10323321	5	16027527	6	14289183	3	15623873	1	5579544
	39	25623267	51	40025547	57	35396202	36	39429258	9	14297051
	240	61674444	327	96871242	346	85044660	251	96330594	66	35409411
	1159	144392241	1626	227926221	1656	198782235	1323	228589776	368	85066503
	4776	329653563	6855	522728994	6777	453153426	5817	528320706	1707	198817816
	17529	735516102	25614	97412795*	24736	1009687110	22455	118343018*	6885	453210648
	58812	533058068*	87204	420803839*	82614	55493616*	78462	483758146*	24957	1009777992
	183519	221280204*	275442	153225895*	256803	419330963*	253197	322660081*	83043	55636385*
	539160		817617		751979		765253		257613	419552873*
	1504914		2302535		2092947		2188614		753458	
	4018695		6196599		5574945		5969610		2095587	

* Multiplicity is given modulo 1073741527

Affine B3 - Level 4 - Class 0 - Page 1 of 3

	Rp 1	Rp 2	Rp 3	Rp 4	Rp 5	Rp 6	Rp 7	Rp 8	Rp 9	Rp10	Rp11	Rp12	Rp13	Rp14
	0 04 0	0 12 0	0 20 0	0 02 2	0 10 2	0 00 4	1 02 1	1 10 1	1 00 3	2 02 0	2 10 0	2 00 2	3 00 1	4 00 0
Wt 1	1	0	0	0	0	0	0	0	0	0	0	0	0	0
	6	6	2	6	3	1	3	0	0	0	0	0	0	0
	48	63	30	76	48	18	42	17	10	6	3	6	0	0
	304	474	232	580	372	124	372	190	109	76	48	60	10	1
	1643	2760	1400	3349	2196	694	2370	1346	744	580	372	447	109	18
	7646	13548	6962	16162	10704	3190	12354	7370	3934	3349	2196	2490	744	124
	31796	58242	30320	68332	45708	13066	55452	34145	17668	16162	10704	11802	3934	694
										68332	45708	49080	17668	3190
														13066
Wt 2	1	1	0	1	0	0	0	0	0	0	0	0	0	0
	11	15	5	19	8	3	7	4	3	1	0	0	0	0
	89	140	59	179	96	33	93	57	37	19	8	10	3	0
	558	943	437	1180	698	224	738	459	275	179	96	114	37	3
	2895	5123	2494	6255	3890	1181	4399	2772	1562	1180	698	797	275	33
	13024	23801	11942	28422	18224	5286	21735	13831	7444	6255	3890	4309	1562	224
	52420	98140	50241				93675	60114	31197	28422	18224	19703	7444	1181
													31197	5286
Wt 3	1	1	1	1	1	1	0	0	0	0	0	0	0	0
	13	18	9	24	16	7	11	4	3	1	1	1	0	0
	110	169	88	220	147	57	127	65	41	24	16	22	3	1
	680	1139	582	1436	950	319	957	536	314	220	147	178	41	7
	3496	6131	3174	7506	5000	1567	5494	3257	1804	1436	950	1111	314	57
	15530	28238	14699	33737	22582	6658	26517	16218	8615	7506	5000	5597	1804	319
							112259	70198	36063	33737	22582	24620	8615	1567
													36063	6658
Wt 4	0	0	0	1	0	0	0	0	0	0	0	0	0	0
	4	6	2	12	6	3	3	2	3	0	0	0	0	0
	46	75	32	117	63	25	51	33	27	9	3	6	1	0
	334	573	267	783	471	165	450	287	196	99	54	72	21	1
	1875	3336	1627	4285	2670	854	2859	1822	1100	734	429	525	169	19
	8861	16215	8136	19936	12816	3851	14748	9437	5307	4117	2559	2937	1029	139
	36902	69054	35329	82445	53949	15431	65616	42216	22520	19522	12474	13806	5089	786
										81328	53133	57159	22001	3650
														14958
Wt 5	0	0	0	1	1	1	0	0	0	0	0	0	0	0
	5	7	4	16	12	7	5	2	3	0	0	1	0	0
	58	91	48	146	102	45	72	37	31	10	7	14	1	0
	413	694	361	967	657	244	591	334	225	121	81	118	22	3
	2279	4007	2085	5182	3494	1160	3603	2139	1281	883	588	745	190	30
	10620	19291	10081	23813	16051	4930	18092	11073	6167	4932	3289	3867	1172	198
	43707	81408	42780				78967	49348	26135	23139	15501	17381	5858	1030
													25337	4595
Wt 6	0	0	0	0	0	1	0	0	0	0	0	0	0	0
	0	0	0	3	3	3	0	0	1	0	0	0	0	0
	8	12	8	33	27	18	9	5	9	0	0	3	0	0
	88	147	79	258	186	87	123	69	66	16	12	27	3	0
	616	1065	567	1553	1080	420	936	553	399	190	129	204	39	6
	3305	5928	3112	7852	5361	1814	5448	3304	2054	1331	897	1173	304	46
	15059	27696	14581	34633	23550	7321	26367	16357	9289	7209	4833	5799	1766	297
										32928	22155	25071	8499	1481
														6474

-continued next page-

Affine B3 - Level 4 - Class 0 - Page 2 of 3

	Rp 1	Rp 2	Rp 3	Rp 4	Rp 5	Rp 6	Rp 7	Rp 8	Rp 9	Rp10	Rp11	Rp12	Rp13	Rp14
Wt 7	2 23 176 1040 5130 22143	2 32 278 1761 9092 40535	1 16 141 902 4702 21144	3 44 363 2211 11075 48200	2 29 239 1463 7375 32307	1 12 86 480 2262 9407	1 22 220 1514 8272 38462	0 9 114 861 4940 23665 99454	0 6 70 495 2702 12451	0 3 44 363 2211 11075 48200	0 2 29 239 1463 7375 32307	0 3 38 290 1688 8208 34988	0 0 6 70 495 2702 12451	0 0 1 12 86 480 2262 9407
Wt 8	2 27 211 1244 6097 26109	3 43 357 2193 11080 48593	1 18 164 1061 5528 24759	5 62 476 2782 13566 57925	2 33 279 1718 8647 37722	1 13 97 549 2600 10820	1 27 270 1840 9935 45699	1 18 174 1175 6356 29372	1 14 114 700 3549 15646	0 5 62 476 2782 13566 57925	0 2 33 279 1718 8647 37722	0 3 42 331 1950 9510 40495	0 1 14 114 700 3549 15646	0 0 1 13 97 549 2600 10820
Wt 9	0 6 70 498 2726 12589	0 10 120 879 4939 23321	0 4 55 423 2453 11832	2 25 202 1256 6475 29007	1 14 121 781 4134 18875	1 8 52 284 1350 5723	0 6 89 723 4355 21614	0 5 61 474 2817 13955 60425	1 8 57 341 1755 7958	0 0 14 159 1115 6065 27917	0 0 7 91 678 3826 18039	0 1 15 133 856 4474 20125	0 0 3 37 274 1557 7424	0 0 0 3 32 219 1163 5230
Wt10	4 46 334 1875 8861 36902	6 75 573 3336 16215 69054	2 32 267 1627 8136 35329	9 99 734 4117 19522 81328	3 54 429 2559 12474 53133	1 19 139 786 3650 14958	3 51 450 2859 14748 65616	2 33 287 1822 9437 42216	1 21 169 1029 5089 22001	1 12 117 783 4285 19936 82445	0 6 63 471 2670 12816 53949	0 6 72 525 2937 13806 57159	0 3 27 196 1100 5307 22520	0 0 3 25 165 854 3851 15431
Wt11	5 58 413 2279 10620 43707	7 91 694 4007 19291 81408	4 48 361 2085 10081 42780	10 121 883 4932 23139	7 81 588 3289 15501	3 30 198 1030 4595	5 72 591 3603 18092 78967	2 37 334 2139 11073 49348	1 22 190 1172 5858 25337	1 16 146 967 5182 23813	1 12 102 657 3494 16051	1 14 118 745 3867 17381	0 3 31 225 1281 6167 26135	0 1 7 45 244 1160 4930
Wt12	3 34 264 1518 7338 31008	3 52 431 2640 13206 57412	2 27 226 1369 6898 30113	6 77 582 3360 16184 68404	5 53 395 2253 10877 46102	3 22 145 739 3336 13370	2 40 360 2343 12260 55248	1 20 202 1378 7465 34372	1 15 129 804 4103 18093	0 6 77 582 3360 16184 68404	0 5 53 395 2253 10877 46102	1 8 75 482 2616 12084 49856	0 1 15 129 804 4103 18093	0 0 3 22 145 739 3336 13370

continued next page

Affine B3 - Level 4 - Class 0 - Page 3 of 3

	Rp 1	Rp 2	Rp 3	Rp 4	Rp 5	Rp 6	Rp 7	Rp 8	Rp 9	Rp10	Rp11	Rp12	Rp13	Rp14
Wt13	6	10	4	14	7	3	6	5	3	2	1	1	1	0
	70	120	55	159	91	32	89	61	37	25	14	15	8	1
	498	879	423	1115	678	219	723	474	274	202	121	133	57	8
	2726	4939	2453	6065	3826	1163	4355	2817	1557	1256	781	856	341	52
	12589	23321	11832	27917	18039	5230	21614	13955	7424	6475	4134	4474	1755	284
								60425		29007	18875	20125	7958	1350
														5723
Wt14	8	12	8	16	12	6	9	5	3	3	3	3	1	1
	88	147	79	190	129	46	123	69	39	33	27	27	9	3
	616	1065	567	1331	897	297	936	553	304	258	186	204	66	18
	3305	5928	3112	7209	4833	1481	5448	3304	1766	1553	1080	1173	399	87
	15059	27696	14581	32928	22155	6474	26367	16357	8499	7852	5361	5799	2054	420
										34633	23550	25071	9289	1814
														7321

Affine B3 - Level 4 - Class 1 - Page 1 of 2

	Rp 1 0 03 1	Rp 2 0 11 1	Rp 3 0 01 3	Rp 4 1 03 0	Rp 5 1 11 0	Rp 6 1 01 2	Rp 7 2 01 1	Rp 8 3 01 0
Wt 1	1	0	0	0	0	0	0	0
	12	9	6	3	2	3	0	0
	112	110	71	37	35	48	12	1
	760	826	505	297	315	414	144	22
	4159	4764	2782	1808	2036	2580	1053	193
	19436	23001	12933	9120	10657	13116	5907	1202
	80520	97467	53109	40031	47995	57675	27852	6064
	303084	373167	198045	157776	192786	227160	115665	26452
	1054837	1316290	683105	570203	707135	819573	435309	103384
	3437894	4337437	2207987	1918138	2407518	2751369	1513062	370419
	10594262	13490526	6753097	6073139	7699342	8692830	4922763	1235827
							15141579	3883052
Wt 2	1	1	1	0	0	0	0	0
	19	19	15	5	4	7	1	0
	176	190	130	60	61	89	26	3
	1156	1310	820	462	507	683	251	41
	6124	7169	4227	2707	3109	3980	1672	318
	27837	33367	18818	13234	15658	19342	8876	1846
	112665	137522	74927	56619	68453	82322	40303	8908
	415866	515016	272895	218453	268526	316191	162736	37632
	1423465	1783950	923707	775436	965972	1118059	599089	143515
	4573358	5789254	2939566	2568803	3235460	3691499	2045160	504189
		17771322		8025886	10203604	11500241	6554243	1654988
								5128903
Wt 3	0	0	1	0	0	0	0	0
	4	5	9	0	0	3	0	0
	57	66	66	13	14	36	6	0
	451	526	403	152	170	291	84	10
	2678	3170	2099	1078	1245	1800	663	109
	13159	15833	9593	5899	6979	9246	3915	751
	56448	68913	39357	27246	32862	41274	19152	4023
	218074	269661	147615	111214	136264	165129	81816	18272
	774655	968625	513478	412497	512057	604659	314928	73556
	2567240	3241396	1675205	1415987	1777294	2057574	1115079	269651
	8022905	10216393	5171406	4557389	5774863	6580848	3684774	916363
							11485647	2923598
Wt 4	3	2	1	1	0	0	0	0
	37	35	22	12	9	12	3	0
	297	315	193	112	110	144	48	6
	1808	2036	1202	760	826	1053	414	71
	9120	10657	6064	4159	4764	5907	2580	505
	40031	47995	26452	19436	23001	27852	13116	2782
	157776	192786	103384	80520	97467	115665	57675	12933
	570203	707135	370419	303084	373167	435309	227160	53109
	1918138	2407518	1235827	1054837	1316290	1513062	819573	198045
	6073139	7699342	3883052	3437894	4337437	4922763	2751369	683105
				10594262	13490526	15141579	8692830	2207987
								6753097

continued next page

Affine B3 - Level 4 - Class 1 - Page 2 of 2

	Rp 1	Rp 2	Rp 3	Rp 4	Rp 5	Rp 6	Rp 7	Rp 8
Wt 5	5	4	3	1	1	1	0	0
	60	61	41	19	19	26	7	1
	462	507	318	176	190	251	89	15
	2707	3109	1846	1156	1310	1672	683	130
	13234	15658	8908	6124	7169	8876	3980	820
	56619	68453	37632	27837	33367	40303	19342	4227
	218453	268526	143515	112665	137522	162736	82322	18818
	775436	965972	504189	415866	515016	599089	316191	74927
	2568803	3235460	1654988	1423465	1783950	2045160	1118059	272895
	8025886	10203604	5128903	4573358	5789254	6554243	3691499	923707
					17771322		11500241	2939566
Wt 6	2	2	3	0	0	1	0	0
	33	35	31	8	8	17	3	0
	282	316	227	96	104	164	49	6
	1763	2041	1307	710	804	1117	417	70
	8996	10676	6358	4015	4694	6092	2590	501
	39721	48053	27204	19062	22805	28338	13140	2766
	157041	192948	105199	79601	96962	116864	57732	12883
	568558	707547	374567	300958	371950	438102	227276	52965
	1914597	2408508	1244926	1050116	1313503	1519291	819800	197669
	6065805	7701596	3902281	3427824	4331332	4936119	2751766	682177
				10573448	13477626	15169303	8693479	2205826
								6748266
Wt 7	8	8	6	2	2	3	1	0
	96	104	70	33	35	49	17	3
	710	804	501	282	316	417	164	31
	4015	4694	2766	1763	2041	2590	1117	227
	19062	22805	12883	8996	10676	13140	6092	1307
	79601	96962	52965	39721	48053	57732	28338	6358
	300958	371950	197669	157041	192948	227276	116864	27204
	1050116	1313503	682177	568558	707547	819800	438102	105199
	3427824	4331332	2205826	1914597	2408508	2751766	1519291	374567
	10573448	13477626	6748266	6065805	7701596	8693479	4936119	1244926
							15169303	3902281
Wt 8	13	14	10	4	5	6	3	1
	152	170	109	57	66	84	36	9
	1078	1245	751	451	526	663	291	66
	5899	6979	4023	2678	3170	3915	1800	403
	27246	32862	18272	13159	15833	19152	9246	2099
	111214	136264	73556	56448	68913	81816	41274	9593
	412497	512057	269651	218074	269661	314928	165129	39357
	1415987	1777294	916363	774655	968625	1115079	604659	147615
	4557389	5774863	2923598	2567240	3241396	3684774	2057574	513478
				8022905	10216393	11485647	6580848	1675205
								5171406

Weight Multiplicities for Affine B4

Affine B4 - Level 1 - Class 0

	Rp 1 0 000 1					Rp 2 1 000 0				
Wt 1	1	7895	2215001	204271475	425798368*	0	4999	1636850	160349560	672218719*
	4	16112	3817255	323267388	95326567*	1	10490	2846862	254926170	501207198*
	15	31973	6500389	507949705	23254275*	5	21337	4888380	402299272	167601747*
	45	61965	10945989	792739584	235662593*	19	42205	8294760	630415265	585462284*
	125	117471	18239855	155460085*		59	81484	13919305	981274880	924548066*
	316	218430	30095327	820446823*		165	153903	23116329	443911883*	
	758	398913	49197790	754162131*		421	285050	38016945	185411695*	
	1720	716821	79721851	124745652*		1010	518588	61950320	343872132*	
	3755	1268790	128116241	252863824*		2295	928260	100078011	48890560*	
Wt 2	1	10490	2846862	254926170	501207198*	1	7895	2215001	204271475	425798368*
	5	21337	4888380	402299272	167601747*	4	16112	3817255	323267388	95326567*
	19	42205	8294760	630415265	585462284*	15	31973	6500389	507949705	23254275*
	59	81484	13919305	981274880	924548066*	45	61965	10945989	792739584	235662593*
	165	153903	23116329	443911883*		125	117471	18239855	155460085*	
	421	285050	38016945	185411695*		316	218430	30095327	820446823*	
	1010	518588	61950320	343872132*		758	398913	49197790	754162131*	
	2295	928260	100078011	48890560*		1720	716821	79721851	124745652*	
	4999	1636850	160349560	672218719*		3755	1268790	128116241	252863824*	

* Multiplicity is given modulo 1073741527

Affine B4 - Level 1 - Class 1

	Rp 1 0 001 0				
Wt 1	1	5310	1035260	71280736	596084625*
	5	11229	1834002	115513040	981062014*
	19	22998	3203735	185637555	1031292812*
	60	45795	5524138	295982048	30231738*
	169	88958	9410657	468384450	647500144*
	436	169005	15851550	735928460	421806960*
	1055	314729	26419990	74704933*	454553540*
	2419	575559	43599545	706842863*	188113279*

* Multiplicity is given modulo 1073741527

Affine B4 - Level 2 - Class 0 - Page 1 of 2

	Rp 1 0 002 0	Rp 2 0 010 0	Rp 3 0 100 0	Rp 4 0 000 2	Rp 5 1 000 1	Rp 6 2 000 0
Wt 1	1	0	0	0	0	0
	8	5	4	3	0	0
	60	40	32	24	10	3
	320	244	195	140	80	24
	1452	1152	920	640	460	140
	5712	4710	3772	2554	2080	640
	20336	17120	13760	9120	8181	2554
	66560	57052	46064	29988	28808	9120
	203731	176800	143440	92000	93404	29988
	588800	516365	421032	266627	282752	92000
	1620964	1433168	1174368	735600	809048	266627
	4276672	3807652	3135245	1945180	2205440	735600
	10870794	9734528	8053088	4954496	5766288	1945180
	26729648	24056072	19990596	12207160	14532000	4954496
	63799364	57666000	48126720	29192112	35449578	12207160
	148237920	134498932	112710196	67954060	83984760	29192112
	336114698	305990720	257420640	154351040	193791552	67954060
						154351040
Wt 2	1	1	0	0	0	0
	12	11	4	3	3	0
	85	77	39	28	28	3
	452	407	241	167	171	28
	1998	1795	1163	785	824	167
	7736	6950	4777	3158	3398	785
	27075	24352	17450	11341	12493	3158
	87435	78777	58234	37313	42013	11341
	264309	238627	180701	114390	131433	37313
	755799	683865	527927	330715	387155	114390
	2060735	1868798	1465583	909719	1083482	330715
	5390769	4899607	3893361	2397145	2900805	909719
	13596867	12385060	9952081	6083185	7470443	2397145
	33199540	30304661	24586927	14930753	18587485	6083185
	78735959	72016726	58921149	35569119	44845070	14930753
	181874177	166677411	137380251	82485209	105230401	35569119
	410158094	376586286	312439813	186664752	240773644	82485209
						186664752
Wt 3	2	1	1	1	0	0
	21	14	12	10	3	1
	136	101	83	63	32	10
	682	536	434	312	205	63
	2890	2356	1898	1314	1012	312
	10839	9062	7300	4914	4221	1314
	37000	31508	25445	16760	15594	4914
	117103	101132	81978	53060	52492	16760
	348142	304008	247510	157898	164012	53060
	981628	864894	707432	445826	481885	157898
	2644424	2347253	1928943	1202951	1344114	445826
	6845640	6114284	5047958	3119476	3585114	1202951
	17108288	15361949	12740050	7809935	9196208	3119476
	41433574	37375724	31131300	18948268	22789128	7809935
	97547612	88349085	73894676	44688995	54760968	18948268
	223845322	203459860	170848536	102726300	127989835	44688995
					291718844	102726300

-continued next page-

Affine B4 - Level 2 - Class 0 - Page 2 of 2

	Rp 1	Rp 2	Rp 3	Rp 4	Rp 5	Rp 6
Wt 4	0	0	0	1	0	0
	6	4	4	4	1	0
	52	38	33	30	12	4
	304	236	196	148	88	26
	1408	1137	926	669	476	148
	5628	4668	3780	2596	2124	655
	20144	17044	13788	9236	8264	2596
	66209	56876	46104	30164	28998	9196
	203012	176492	143548	92413	93748	30164
	587532	515728	421192	267264	283457	92308
	1618556	1432084	1174737	736936	810280	267264
	4272557	3805580	3135800	1947252	2207780	736684
	10863396	9731089	8054236	4958509	5770256	1947252
	26717344	24049864	19992332	12213368	14539132	4957935
	63778164	57655920	48130044	29203432	35461360	12213368
	148203485	134481524	112715216	67971468	84005050	29202192
		305963005	257429704	154381335	193824320	67971468
						154378755
Wt 5	3	3	1	1	1	0
	31	29	14	11	11	1
	195	178	102	74	75	11
	953	862	545	378	392	74
	3953	3563	2403	1617	1719	378
	14561	13114	9259	6094	6636	1617
	48948	44115	32224	20845	23219	6094
	152879	137987	103475	65997	75078	20845
	449275	406269	311090	196079	227441	65997
	1253921	1136235	884982	552165	652080	196079
	3347403	3039727	2401305	1485033	1783040	552165
	8595069	7821784	6253385	3837136	4678198	1485033
	21322650	19445149	15706427	9570551	11834822	3837136
	51295176	46874172	38200588	23131204	28982273	9570551
	120027174	109898403	90266285	54346977	68931727	23131204
	273883440	251244971	207798781	124459949	159669727	54346977
					361046964	124459949
Wt 6	6	4	4	4	1	1
	52	38	33	26	12	4
	304	236	196	148	88	30
	1408	1137	926	655	476	148
	5628	4668	3780	2596	2124	669
	20144	17044	13788	9196	8264	2596
	66209	56876	46104	30164	28998	9236
	203012	176492	143548	92308	93748	30164
	587532	515728	421192	267264	283457	92413
	1618556	1432084	1174737	736684	810280	267264
	4272557	3805580	3135800	1947252	2207780	736936
	10863396	9731089	8054236	4957935	5770256	1947252
	26717344	24049864	19992332	12213368	14539132	4958509
	63778164	57655920	48130044	29202192	35461360	12213368
	148203485	134481524	112715216	67971468	84005050	29203432
		305963005	257429704	154378755	193824320	67971468
						154381335

Affine B4 - Level 2 - Class 1

	Rp 1 0 001 1					Rp 2 1 001 0				
Wt 1	1	1014580	550608368*	137747528*	1055126713*	0	602348	763819433*	646053393*	583804496*
	14	2778517	367612195*	464039019*	863892545*	4	1688400	721790411*	401271915*	979467210*
	105	7296752	173418701*	659224679*	985337349*	40	4524760	792686671*	698715543*	198975590*
	574	18468245	1012282188*	760857418*	854747735*	252	11658920	36328884*	754408073*	503761042*
	2580	45235414	480819368*	1013006192*		1240	29015700	756155456*	138025966*	168174297*
	10108	107585845	755188366*	872311762*		5180	70005376	786441493*	993065272*	
	35693	249162620	602480641*	625941797*		19208	164245380	326223794*	96057926*	
	116090	563251038	9987347*	242641893*		64960	375704920	1041767402*	792078172*	
	353017	171642403*	184928933*	824350198*		203984	839757660	889968132*	612614553*	
Wt 2	4	1688400	721790411*	401271915*	979467210*	1	1014580	550608368*	137747528*	1055126713*
	40	4524760	792686671*	698715543*	198975590*	14	2778517	367612195*	464039019*	863892545*
	252	11658920	36328884*	754408073*	503761042*	105	7296752	173418701*	659224679*	985337349*
	1240	29015700	756155456*	138025966*	168174297*	574	18468245	1012282188*	760857418*	854747735*
	5180	70005376	786441493*	993065272*		2580	45235414	480819368*	1013006192*	
	19208	164245380	326223794*	96057926*		10108	107585845	755188366*	872311762*	
	64960	375704920	1041767402*	792078172*		35693	249162620	602480641*	625941797*	
	203984	839757660	889968132*	612614553*		116090	563251038	9987347*	242641893*	
	602348	763819433*	646053393*	583804496*		353017	171642403*	184928933*	824350198*	

* Multiplicity is given modulo 1073741527

Affine B4 - Level 3 - Class 0 - Page 1 of 2

	Rp 1 0 002 1	Rp 2 0 010 1	Rp 3 0 100 1	Rp 4 0 000 3	Rp 5 1 002 0	Rp 6 1 010 0	Rp 7 1 100 0	Rp 8 1 000 2	Rp 9 2 000 1	Rp10 3 000 0
Wt 1	1	0	0	0	0	0	0	0	0	0
	20	10	6	3	4	3	0	0	0	0
	228	140	84	36	66	52	15	15	6	0
	1760	1200	724	288	626	492	201	189	84	6
	10735	7706	4676	1749	4324	3413	1624	1476	684	78
	55227	41002	25074	8976	24300	19237	9984	8820	4266	600
	250090	189876	117056	40437	117514	93396	51257	44296	22194	3552
	1023427	789934	491002	164862	506442	404050	230608	195671	101248	17658
	3856314	3013744	1888212	619083	1990042	1593900	936450	782550	416598	77370
							3500302		1576928	307182
Wt 2	1	1	0	0	0	0	0	0	0	0
	27	22	6	3	7	3	2	3	0	0
	303	243	96	40	102	62	39	42	6	1
	2312	1837	856	330	903	606	372	369	93	18
	13863	11012	5628	2052	5951	4239	2593	2421	791	153
	70302	55929	30347	10636	32362	23866	14634	13167	5037	997
	314225	250728	141824	48171	152658	115354	71145	62200	26469	5334
	1271481	1017914	593784	196647	645313	496131	308116	263580	121223	24930
	4743512	3810828	2276258	738021	2496503	1944952	1216819	1022574	499080	104575
									1886511	402477
Wt 3	2	1	1	1	0	0	0	0	0	0
	44	26	18	10	9	8	2	3	1	0
	448	299	191	87	137	112	43	46	20	1
	3234	2283	1418	572	1191	957	436	423	191	19
	18663	13686	8448	3177	7731	6166	3103	2846	1347	173
	92070	69223	42784	15290	41384	32987	17684	15667	7686	1157
	402882	308555	191647	65989	192622	153760	86167	74416	37771	6294
	1603074	1244841	778050	260100	804607	644019	372858	315789	165155	29622
			2914251	951265	3080715	2473504	1468397	1224080	657871	124645
										479773
Wt 4	0	0	0	1	0	0	0	0	0	0
	6	4	4	4	0	0	0	1	0	0
	96	64	48	30	20	18	7	12	4	0
	876	612	404	192	276	231	102	118	48	4
	5857	4252	2696	1109	2220	1800	882	878	396	44
	32070	23888	14928	5634	13588	10914	5716	5301	2516	348
	151839	115380	71920	25694	69552	55686	30619	27200	13464	2116
	643176	496140	310052	106402	312312	250147	142578	123110	63036	10832
	2491251	1944836	1221408	406689	1266852	1016592	595738	503682	265772	48720
									1027980	198084
Wt 5	4	3	0	0	1	0	0	0	0	0
	66	52	15	6	20	10	6	6	0	0
	626	492	201	78	228	140	84	84	15	3
	4324	3413	1624	600	1760	1200	724	684	189	36
	24300	19237	9984	3552	10735	7706	4676	4266	1476	288
	117514	93396	51257	17658	55227	41002	25074	22194	8820	1749
	506442	404050	230608	77370	250090	189876	117056	101248	44296	8976
	1990042	1593900	936450	307182	1023427	789934	491002	416598	195671	40437
			3500302		3856314	3013744	1888212	1576928	782550	164862
										619083

continued next page

Affine B4 - Level 3 - Class 0 - Page 2 of 2

	Rp 1	Rp 2	Rp 3	Rp 4	Rp 5	Rp 6	Rp 7	Rp 8	Rp 9	Rp10
Wt 6	7	3	2	1	1	1	0	0	0	0
	102	62	39	18	27	22	6	6	3	0
	903	606	372	153	303	243	96	93	42	3
	5951	4239	2593	997	2312	1837	856	791	369	40
	32362	23866	14634	5334	13863	11012	5628	5037	2421	330
	152658	115354	71145	24930	70302	55929	30347	26469	13167	2052
	645313	496131	308116	104575	314225	250728	141824	121223	62200	10636
	2496503	1944952	1216819	402477	1271481	1017914	593784	499080	263580	48171
					4743512	3810828	2276258	1886511	1022574	196647
										738021
Wt 7	9	8	2	1	2	1	1	1	0	0
	137	112	43	19	44	26	18	20	3	1
	1191	957	436	173	448	299	191	191	46	10
	7731	6166	3103	1157	3234	2283	1418	1347	423	87
	41384	32987	17684	6294	18663	13686	8448	7686	2846	572
	192622	153760	86167	29622	92070	69223	42784	37771	15667	3177
	804607	644019	372858	124645	402882	308555	191647	165155	74416	15290
	3080715	2473504	1468397	479773	1603074	1244841	778050	657871	315789	65989
							2914251		1224080	260100
										951265
Wt 8	3	3	1	1	0	0	0	1	0	0
	60	52	20	11	15	9	7	11	1	0
	600	491	222	98	207	136	90	105	21	4
	4234	3408	1694	665	1689	1185	749	755	215	41
	24023	19217	10195	3743	10519	7661	4762	4483	1567	309
	116742	93331	51834	18170	54630	40876	25335	22796	9101	1819
	504435	403860	232090	78661	248543	189555	117778	102814	45083	9187
	1985124	1593394	940042	310263	1019645	789154	492864	420440	197724	41013
					3847476	3011949	1892748	1585937	787598	166338
										622650
Wt 9	15	9	7	4	3	3	1	1	1	0
	207	136	90	41	60	52	20	21	11	1
	1689	1185	749	309	600	491	222	215	105	11
	10519	7661	4762	1819	4234	3408	1694	1567	755	98
	54630	40876	25335	9187	24023	19217	10195	9101	4483	665
	248543	189555	117778	41013	116742	93331	51834	45083	22796	3743
	1019645	789154	492864	166338	504435	403860	232090	197724	102814	18170
	3847476	3011949	1892748	622650	1985124	1593394	940042	787598	420440	78661
									1585937	310263
Wt10	20	18	7	4	6	4	4	4	1	1
	276	231	102	44	96	64	48	48	12	4
	2220	1800	882	348	876	612	404	396	118	30
	13588	10914	5716	2116	5857	4252	2696	2516	878	192
	69552	55686	30619	10832	32070	23888	14928	13464	5301	1109
	312312	250147	142578	48720	151839	115380	71920	63036	27200	5634
	1266852	1016592	595738	198084	643176	496140	310052	265772	123110	25694
					2491251	1944836	1221408	1027980	503682	106402
										406689

Affine B4 - Level 3 - Class 1 - Page 1 of 2

	Rp 1	Rp 2	Rp 3	Rp 4	Rp 5	Rp 6
	0 003 0	0 011 0	0 101 0	0 001 2	1 001 1	2 001 0
Wt 1	1	0	0	0	0	0
	8	8	4	3	0	0
	78	96	65	56	24	3
	584	792	624	546	320	56
	3563	5104	4366	3756	2576	546
	18552	27456	24772	20898	15888	3756
	85178	128912	120777	100016	81968	20898
	353244	543704	524016	426852	370816	100016
	1347297	2101248	2070958	1662828	1514256	426852
	4789224	7550096	7578564	6009001	5690880	1662828
	16029490	25499344	25991358	20382192	19957104	6009001
	50921508	81633864	84310512	65474290	65981472	20382192
	154519076	249389280	260533080	200579764	207296480	65474290
	450208256	730958688	771375376	589275774	622767744	200579764
	191147309*	990864377*	50999115*	594031257*	724385489*	589275774
	217919943*	271707333*	686338093*	269351008*	715517156*	594031257*
	486597258*	992138037*	62087673*	306198691*	636134004*	269351008*
						306198691*
Wt 2	1	1	0	0	0	0
	15	18	10	9	3	0
	146	194	141	129	66	9
	1054	1484	1217	1080	679	129
	6173	9014	7910	6828	4878	1080
	30950	46348	42548	35854	28031	6828
	137550	209875	199075	164415	137606	35854
	554930	859174	835838	678686	599525	164415
	2067138	3238565	3215410	2573408	2375926	678686
	7199780	11391620	11503024	9092420	8711649	2573408
	23673136	37772007	38693055	30254959	29921440	9092420
	74035942	118991878	123416369	95587160	97171596	30254959
	221558413	358376229	375771306	288588994	300560382	95587160
	637553297	1037131566*	24287385*	836923960	890594638	288588994
	697521875*	748609742*	945307344*	193851713*	392518804*	836923960
	472191726*	314291701*	911727422*	972169839*	557391366*	193851713*
	653864499*	159813699*	812606020*	567912899*	447319059*	972169839*
						567912899*
Wt 3	2	2	1	1	0	0
	30	38	26	26	10	1
	277	380	300	280	154	26
	1894	2720	2322	2072	1362	280
	10596	15672	14086	12149	8958	2072
	51140	77270	72062	60558	48486	12149
	220202	338194	324442	267032	227740	60558
	865040	1346070	1320792	1068701	958674	267032
	3150050	4955016	4953177	3950863	3694920	1068701
	10758854	17079142	17342574	13665282	13237548	3950863
	34776245	55642118	57266778	44649078	44579024	13665282
	107133688	172598368	179737812	138841330	142326306	44649078
	316336334	512747502	539525856	413350976	433698714	138841330
	899403006	392043369*	482931935*	110142385*	194444695*	413350976
	324256097*	826821589*	40018150*	53835331*	353048833*	110142385*
	144580004*	98409068*	954335260*	190136437*	81650827*	53835331*
					14243056*	190136437*

continued next page

* Multiplicity is given modulo 1073741527

Affine B4 - Level 3 - Class 1 - Page 2 of 2

	Rp 1	Rp 2	Rp 3	Rp 4	Rp 5	Rp 6
Wt 4	0	0	0	1	0	0
	9	12	9	14	4	0
	120	168	136	149	72	10
	963	1392	1197	1146	704	135
	5892	8728	7844	7024	4964	1106
	30162	45540	42353	36386	28292	6919
	135492	207752	198548	165771	138328	36134
	549862	853916	834502	681958	601388	165197
	2055240	3226160	3212204	2580965	2380468	680718
	7172966	11363536	11495662	9109218	8722216	2578385
	23614784	37710644	38676784	30291099	29945068	9104038
	73912799	118861900	123381565	95662688	97222652	30280991
	221305480	358108360	375698984	288742853	300667448	95643480
	637046219	1036592884*	24140887*	837230236	890813288	288707160
	696527205*	747550086*	945017334*	194448855*	392954870*	837165276
	470278845*	312248503*	911165064*	973312199*	558242654*	194332765*
		155945311*	811535916*	570060722*	448949017*	973108215*
						569707705*
Wt 5	4	5	3	4	1	0
	60	81	61	63	29	4
	519	735	608	573	339	63
	3361	4913	4314	3848	2642	573
	17976	26894	24620	21174	16087	3848
	83646	127395	120367	100768	82518	21174
	349428	539849	522972	428771	372246	100768
	1338227	2091982	2068448	1667469	1517764	428771
	4768554	7528800	7572772	6019703	5699091	1667469
	15984086	25452180	25978491	20405956	19975600	6019703
	50824852	81532831	84282852	65525325	66021704	20405956
	154318947	249179037	260475308	200686172	207381378	65525325
	449804188	730532170	771257742	589491977	622942205	200686172
	190349463*	990018871*	50765117*	594460490*	724735440*	589491977
	216376242*	270065764*	685882181*	270185400*	716204033*	594460490*
	483665414*	989010380*	61216155*	307790167*	637455989*	270185400*
					8217672*	307790167*
Wt 6	9	12	9	10	4	1
	120	168	136	135	72	14
	963	1392	1197	1106	704	149
	5892	8728	7844	6919	4964	1146
	30162	45540	42353	36134	28292	7024
	135492	207752	198548	165197	138328	36386
	549862	853916	834502	680718	601388	165771
	2055240	3226160	3212204	2578385	2380468	681958
	7172966	11363536	11495662	9104038	8722216	2580965
	23614784	37710644	38676784	30280991	29945068	9109218
	73912799	118861900	123381565	95643480	97222652	30291099
	221305480	358108360	375698984	288707160	300667448	95662688
	637046219	1036592884*	24140887*	837165276	890813288	288742853
	696527205*	747550086*	945017334*	194332765*	392954870*	837230236
	470278845*	312248503*	911165064*	973108215*	558242654*	194448855*
		155945311*	811535916*	569707705*	448949017*	973312199*
						570060722*

* Multiplicity is given modulo 1073741527

Weight Multiplicities for Affine B5

Affine B5 - Level 1 - Class 0

	Rp 1 (0 0000 1)					Rp 2 (1 0000 0)				
Wt 1	1	20657	10155288	404469381*	92720599*	0	12121	7177192	50249914*	1061348411*
	5	45221	18484704	303286759*	1026898512*	1	27443	13209767	800081586*	441654903*
	21	96061	33207954	810364675*	544595520*	6	60017	23972636	951050331*	592607257*
	71	198771	58931405	140428421*	673497798*	26	127304	42938511	789166894*	336660779*
	217	401609	103386620	1008831469*		91	262873	75973989	764539284*	73121159*
	603	794300	179428074	5678618*		282	529886	132895113	507450892*	
	1573	1540547	308246046	650056408*		793	1045252	229974966	38918520*	
	3874	2935172	524479488	761051576*		2079	2021680	393960861	1051734138*	
	9126	5501141	884315809	358728864*		5140	3840733	668455372	780629020*	
Wt 2	1	27443	13209767	800081586*	441654903*	1	20657	10155288	404469381*	92720599*
	6	60017	23972636	951050331*	592607257*	5	45221	18484704	303286759*	1026898512*
	26	127304	42938511	789166894*	336660779*	21	96061	33207954	810364675*	544595520*
	91	262873	75973989	764539284*	73121159*	71	198771	58931405	140428421*	673497798*
	282	529886	132895113	507450892*		217	401609	103386620	1008831469*	
	793	1045252	229974966	38918520*		603	794300	179428074	5678618*	
	2079	2021680	393960861	1051734138*		1573	1540547	308246046	650056408*	
	5140	3840733	668455372	780629020*		3874	2935172	524479488	761051576*	
	12121	7177192	50249914*	1061348411*		9126	5501141	884315809	358728864*	

* Multiplicity is given modulo 1073741527

Affine B5 - Level 1 - Class 1

	Rp 1 (0 0001 0)				
Wt 1	1	12718	4214925	445774999	193901729*
	6	28990	7911566	758759242	91861831*
	26	63800	14623292	205967411*	1031746536*
	92	136148	26645728	1065916857*	190819869*
	287	282737	47911845	326854200*	433243912*
	814	573028	85089164	468855031*	814908580*
	2150	1136174	149370252	942845468*	574051845*
	5356	2208336	259369604	423890274*	780891461*

* Multiplicity is given modulo 1073741527

Affine B5 - Level 2 - Class 0 - Page 1 of 3

	Rp 1 0 0002 0	Rp 2 0 0010 0	Rp 3 0 0100 0	Rp 4 0 1000 0	Rp 5 0 0000 2	Rp 6 1 0000 1	Rp 7 2 0000 0
Wt 1	1	0	0	0	0	0	0
	10	6	5	0	0	0	0
	90	60	50	15	10	10	0
	580	440	366	150	100	100	10
	3120	2480	2060	1030	675	690	100
	14412	11900	9890	5500	3550	3700	675
	59510	50272	41840	25126	15990	17015	3550
	224020	192710	160750	101840	64020	69470	15990
	782215	681820	570292	376600	234190	258950	64020
	2562360	2257570	1894030	1291640	795600	895380	234190
	7948650	7063120	5944670	4161565	2541790	2908506	795600
	23513600	21043625	17769260	12709912	7704680	8954540	2541790
	66714290	60063250	50883420	37062050	22315345	26316600	7704680
	182384740	165044290	140271820	103761440	62095080	74237180	22315345
	482281320	438357900	373741840	280187410	166747506	201927425	62095080
							166747506
Wt 2	1	1	0	0	0	0	0
	15	14	5	4	3	0	0
	131	121	61	47	34	10	3
	843	774	459	349	243	115	34
	4432	4058	2641	1997	1349	820	243
	20125	18408	12734	9621	6345	4519	1349
	81673	74707	53953	40838	26413	21037	6345
	302996	277332	206823	157084	99976	86565	26413
	1044039	956602	731346	557875	350292	323735	99976
	3380552	3101384	2418574	1853857	1150702	1120712	350292
	10377719	9534168	7556521	5821703	3577757	3638376	1150702
	30413601	27982919	22479806	17408810	10605955	11184329	3577757
	85563477	78844334	64066574	49870919	30150301	32794971	10605955
	232123574	214219159	175783591	137529354	82578873	92262394	30150301
			466213858	366561717	218753473	250205617	82578873
							218753473
Wt 3	2	1	1	0	0	0	0
	27	18	16	4	3	3	0
	218	162	139	54	38	38	3
	1325	1044	882	421	285	289	38
	6672	5457	4577	2470	1628	1682	285
	29300	24590	20568	12052	7784	8204	1628
	115768	99007	82794	51474	32712	35166	7784
	420172	364550	305228	198443	124445	136358	32712
	1421518	1247414	1046625	704694	436988	487580	124445
	4531731	4013346	3376156	2338067	1435947	1629814	436988
	13726272	12248526	10333790	7323988	4460432	5144576	1435947
	39759042	35704994	30215330	21834108	13198981	15454416	4460432
	110707050	99958994	84852674	62335410	37433826	44453630	13198981
	297590332	269958206	229869980	171286231	102250495	123043435	37433826
				454859254	270070114	329053956	102250495
							270070114

continued next page-

Affine B5 - Level 2 - Class 0 - Page 2 of 3

	Rp 1	Rp 2	Rp 3	Rp 4	Rp 5	Rp 6	Rp 7
Wt 4	3	3	1	1	1	0	0
	41	39	19	16	13	3	1
	323	301	174	138	102	42	13
	1914	1767	1127	871	610	338	102
	9427	8663	5896	4506	3043	2030	610
	40615	37247	26547	20211	13291	10081	3043
	157848	144675	106724	81265	52360	43655	13291
	564737	517754	392253	299399	189778	170185	52360
	1886701	1731121	1339660	1026300	641835	609925	189778
	5948040	5463620	4301991	3310189	2046825	2039333	641835
	17837836	16406277	13105408	10132052	6204524	6430513	2046825
	51208421	47164362	38135902	29628737	17991118	19280404	6204524
	141437821	130455776	106586831	83219776	50158653	55321093	17991118
	377406071	348607905	287404642	225494814	135017054	152687514	50158653
						407077266	135017054
Wt 5	0	0	0	0	1	0	0
	10	10	5	5	5	1	0
	110	105	60	51	45	15	5
	775	725	455	365	270	135	40
	4230	3915	2626	2045	1440	885	270
	19585	18035	12690	9760	6561	4710	1420
	80310	73775	53830	41195	26995	21547	6561
	299760	275146	206515	157967	101285	87845	26930
	1036665	951650	730605	559920	353390	326770	101285
	3364425	3090625	2416897	1858445	1157300	1127605	353200
	10343535	9511445	7552845	5831565	3592078	3653430	1157300
	30343301	27936365	22472065	17429455	10635025	11216165	3591572
	85422440	78751142	64050690	49912830	30209560	32860355	10635025
	231847215	214036955	175751930	137612585	82694275	92393340	30208295
				366723155		250461890	82694275
Wt 6	6	4	4	1	1	1	0
	72	53	48	18	14	14	1
	525	410	356	164	117	118	14
	2950	2394	2038	1068	726	744	117
	13968	11644	9820	5618	3702	3866	726
	58364	49631	41688	25416	16380	17444	3702
	221324	191102	160356	102586	65002	70572	16380
	775984	678152	569465	378292	236464	261524	65002
	2548778	2249284	1892130	1295527	800759	901307	236464
	7919580	7045485	5940826	4169836	2552884	2921360	800759
	23453851	21006588	17761116	12727496	7728078	8981970	2552884
	66593568	59988582	50867518	37097580	22362878	26372768	7728078
	182148008	164895820	140240052	103832756	62189966	74350391	22362878
	481824256	438071397	373681798	280325484	166931700	202148794	62189966
							166931700

continued next page

Affine B5 - Level 2 - Class 0 - Page 3 of 3

	Rp 1	Rp 2	Rp 3	Rp 4	Rp 5	Rp 6	Rp 7
Wt 7	10	10	5	5	5	1	1
	110	105	60	51	40	15	5
	775	725	455	365	270	135	45
	4230	3915	2626	2045	1420	885	270
	19585	18035	12690	9760	6561	4710	1440
	80310	73775	53830	41195	26930	21547	6561
	299760	275146	206515	157967	101285	87845	26995
	1036665	951650	730605	559920	353200	326770	101285
	3364425	3090625	2416897	1858445	1157300	1127605	353390
	10343535	9511445	7552845	5831565	3591572	3653430	1157300
	30343301	27936365	22472065	17429455	10635025	11216165	3592078
	85422440	78751142	64050690	49912830	30208295	32860355	10635025
	231847215	214036955	175751930	137612585	82694275	92393340	30209560
				366723155		250461890	82694275

Affine B5 - Level 2 - Class 1

	Rp 1 0 0001 1					Rp 2 1 0001 0				
Wt 1	1	5302075	775031806*	350030332*	32308497*	0	2963840	337711778*	937739980*	549165716*
	20	16313440	998827150*	70408916*	284127488*	5	9358470	1043106197*	108963192*	661689132*
	190	47902010	855569033*	241144648*	163874120*	65	28107365	873889867*	653826313*	310372377*
	1265	134992865	184444303*	995294066*	96141746*	506	80803485	951621479*	501009130*	358809626*
	6765	366772720	564715849*	513347434*		2990	223474667	649481784*	48617263*	6351239*
	31027	964375855	66818660*	873733854*		14725	597049255	257727840*	612760069*	
	126730	314166807*	834701972*	459346549*		63505	472467168*	444704884*	57871997*	
	472295	747688460*	833969646*	563703936*		247170	671549549*	658402993*	1064892925*	
	1633000	868131464*	1021550146*	261871909*		885723	961190039*	557999550*	1025228403*	
Wt 2	5	9358470	1043106197*	108963192*	661689132*	1	5302075	775031806*	350030332*	32308497*
	65	28107365	873889867*	653826313*	310372377*	20	16313440	998827150*	70408916*	284127488*
	506	80803485	951621479*	501009130*	358809626*	190	47902010	855569033*	241144648*	163874120*
	2990	223474667	649481784*	48617263*	6351239*	1265	134992865	184444303*	995294066*	96141746*
	14725	597049255	257727840*	612760069*		6765	366772720	564715849*	513347434*	
	63505	472467168*	444704884*	57871997*		31027	964375855	66818660*	873733854*	
	247170	671549549*	658402993*	1064892925*		126730	314166807*	834701972*	459346549*	
	885723	961190039*	557999550*	1025228403*		472295	747688460*	833969646*	563703936*	
	2963840	337711778*	937739980*	549165716*		1633000	868131464*	1021550146*	261871909*	

* Multiplicity is given modulo 1073741527

Affine B5 - Level 3 - Class 0 - Page 1 of 3

	Rp 1 0 0002 1	Rp 2 0 0010 1	Rp 3 0 0100 1	Rp 4 0 1000 1	Rp 5 0 0000 3	Rp 6 1 0002 0	Rp 7 1 0010 0	Rp 8 1 0100 0	Rp 9 1 1000 0	Rp10 1 0000 2	Rp11 2 0000 1	Rp12 3 0000 0
Wt 1	1	0	0	0	0	0	0	0	0	0	0	0
	30	15	10	0	0	5	4	0	0	0	0	0
	460	290	195	40	15	115	95	29	15	15	0	0
	4615	3235	2185	675	240	1440	1190	520	275	255	36	6
	35475	26275	17824	6700	2290	12715	10535	5400	2875	2565	595	105
	225318	172830	117855	49675	16425	89110	73990	41530	22305	19271	5750	1035
	1239020	973185	667395	303600	97685	527640	439268	261460	141700	119470	41720	7685
	6078150	4857215	3350705	1609365	506025	2743885	2290625	1420617	777295	642340	250340	47085
									3801600		1306729	250833
												1197120
Wt 2	1	1	0	0	0	0	0	0	0	0	0	0
	41	35	10	6	3	9	4	3	0	0	0	0
	629	529	224	124	51	185	115	81	15	15	6	0
	6236	5206	2615	1424	537	2161	1495	1031	306	280	120	6
	47137	39295	21737	11792	4177	18178	13359	9157	3344	2936	1300	114
	294818	245891	144757	78803	26702	122985	93756	64252	26539	22609	10366	1176
	1598813	1335732	821261	449950	147254	708822	554153	380853	170563	142075	67240	8952
			4118558	2274350	723978	3608267	2872593	1982775	940737	769585	375237	55672
											1861736	299045
Wt 3	2	1	1	0	0	0	0	0	0	0	0	0
	69	42	32	6	3	12	11	3	2	3	0	0
	967	664	474	136	55	256	220	91	54	57	6	1
	9061	6607	4602	1641	604	2941	2484	1225	687	658	129	24
	65807	49884	34459	13954	4836	24320	20386	11122	6125	5520	1472	270
	399522	310878	214451	94573	31440	161813	135346	78812	43198	37474	12073	2237
	2116923	1678961	1160586	543805	175146	919042	768807	468581	257447	217120	79638	14944
				2756553	865784	4618289	3868660	2438648	1347338	1112151	448534	85572
											2235954	433945
Wt 4	3	3	1	1	1	0	0	0	0	0	0	0
	97	87	36	24	13	21	13	11	2	3	1	0
	1333	1145	559	331	144	404	275	205	58	61	26	1
	12220	10333	5572	3143	1207	4347	3152	2244	782	737	323	25
	87149	73204	42199	23402	8354	34378	25914	18096	7195	6386	2892	299
	520720	436525	263793	145680	49418	221338	171459	118921	51636	44182	20598	2562
	2721516	2281820	1428128	790486	258344	1225829	968515	671122	310520	258872	124255	17491
								3362635	1631734	1333917	658076	101412
												517955
Wt 5	0	0	0	0	1	0	0	0	0	0	0	0
	10	10	5	5	5	0	0	0	0	1	0	0
	230	210	105	75	45	50	35	30	9	15	5	0
	2785	2425	1295	795	360	865	625	479	165	185	75	5
	23600	20114	11420	6600	2580	8602	6435	4660	1805	1740	765	70
	158990	134280	79908	45010	16143	64030	49215	34780	14695	13170	6015	690
	909196	765080	472260	263710	89480	393335	308545	215805	97445	83618	39435	5225
				1367335	445850	2097090	1671450	1165450	553754	461725	224040	32940
												180325

continued next page

Affine B5 - Level 3 - Class 0 - Page 2 of 3

	Rp 1	Rp 2	Rp 3	Rp 4	Rp 5	Rp 6	Rp 7	Rp 8	Rp 9	Rp10	Rp11	Rp12
Wt 6	5	4	0	0	0	1	0	0	0	0	0	0
	115	95	29	15	6	30	15	10	0	0	0	0
	1440	1190	520	275	105	460	290	195	40	36	15	0
	12715	10535	5400	2875	1035	4615	3235	2185	675	595	255	15
	89110	73990	41530	22305	7685	35475	26275	17824	6700	5750	2565	240
	527640	439268	261460	141700	47085	225318	172830	117855	49675	41720	19271	2290
	2743885	2290625	1420617	777295	250833	1239020	973185	667395	303600	250340	119470	16425
				3801600	1197120	6078150	4857215	3350705	1609365	1306729	642340	97685
												506025
Wt 7	9	4	3	0	0	1	1	0	0	0	0	0
	185	115	81	15	6	41	35	10	6	6	0	0
	2161	1495	1031	306	114	629	529	224	124	120	15	3
	18178	13359	9157	3344	1176	6236	5206	2615	1424	1300	280	51
	122985	93756	64252	26539	8952	47137	39295	21737	11792	10366	2936	537
	708822	554153	380853	170563	55672	294818	245891	144757	78803	67240	22609	4177
	3608267	2872593	1982775	940737	299045	1598813	1335732	821261	449950	375237	142075	26702
								4118558	2274350	1861736	769585	147254
												723978
Wt 8	12	11	3	2	1	2	1	1	0	0	0	0
	256	220	91	54	24	69	42	32	6	6	3	0
	2941	2484	1225	687	270	967	664	474	136	129	57	3
	24320	20386	11122	6125	2237	9061	6607	4602	1641	1472	658	55
	161813	135346	78812	43198	14944	65807	49884	34459	13954	12073	5520	604
	919042	768807	468581	257447	85572	399522	310878	214451	94573	79638	37474	4836
	4618289	3868660	2438648	1347338	433945	2116923	1678961	1160586	543805	448534	217120	31440
									2756553	2235954	1112151	175146
												865784
Wt 9	21	13	11	2	1	3	3	1	1	1	0	0
	404	275	205	58	25	97	87	36	24	26	3	1
	4347	3152	2244	782	299	1333	1145	559	331	323	61	13
	34378	25914	18096	7195	2562	12220	10333	5572	3143	2892	737	144
	221338	171459	118921	51636	17491	87149	73204	42199	23402	20598	6386	1207
	1225829	968515	671122	310520	101412	520720	436525	263793	145680	124255	44182	8354
			3362635	1631734	517955	2721516	2281820	1428128	790486	658076	258872	49418
											1333917	258344
Wt10	6	4	4	1	1	0	0	0	0	1	0	0
	160	110	89	26	14	30	28	12	9	14	1	0
	2019	1459	1076	374	159	561	493	239	149	168	27	5
	17542	13172	9359	3671	1378	5909	5050	2698	1566	1527	356	66
	120523	92976	65038	27866	9738	45810	38701	22098	12428	11292	3308	620
	700290	551303	383596	175348	58415	290033	243837	146128	81265	70604	24131	4538
	3581079	2863197	1991580	956473	307850	1583077	1329160	825931	458482	386421	147586	28073
									2301538		787732	151924
												738657

-continued next page-

Affine B5 - Level 3 - Class 0 - Page 3 of 3

	Rp 1	Rp 2	Rp 3	Rp 4	Rp 5	Rp 6	Rp 7	Rp 8	Rp 9	Rp10	Rp11	Rp12
Wt11	30	28	12	9	5	6	4	4	1	1	1	0
	561	493	239	149	66	160	110	89	26	27	14	1
	5909	5050	2698	1566	620	2019	1459	1076	374	356	168	14
	45810	38701	22098	12428	4538	17542	13172	9359	3671	3308	1527	159
	290033	243837	146128	81265	28073	120523	92976	65038	27866	24131	11292	1378
	1583077	1329160	825931	458482	151924	700290	551303	383596	175348	147586	70604	9738
				2301538	738657	3581079	2863197	1991580	956473	787732	386421	58415
												307850
Wt12	50	35	30	9	5	10	10	5	5	5	1	1
	865	625	479	165	70	230	210	105	75	75	15	5
	8602	6435	4660	1805	690	2785	2425	1295	795	765	185	45
	64030	49215	34780	14695	5225	23600	20114	11420	6600	6015	1740	360
	393335	308545	215805	97445	32940	158990	134280	79908	45010	39435	13170	2580
	2097090	1671450	1165450	553754	180325	909196	765080	472260	263710	224040	83618	16143
									1367335		461725	89480
												445850

Affine B5 - Level 3 - Class 1 - Page 1 of 3

	Rp 1	Rp 2	Rp 3	Rp 4	Rp 5	Rp 6	Rp 7
	0 0003 0	0 0011 0	0 0101 0	0 1001 0	0 0001 2	1 0001 1	2 0001 0
Wt 1	1	0	0	0	0	0	0
	10	10	5	0	0	0	0
	120	150	104	40	30	10	0
	1120	1550	1260	680	530	255	30
	8495	12440	11050	6920	5396	3190	530
	54448	82500	77555	53008	40830	27475	5396
	304200	471790	461280	334200	253770	187286	40830
	1518500	2395962	2411557	1822800	1365010	1081045	253770
	6901885	11036220	11361390	8872640	6560190	5496460	1365010
	28979550	46839340	49104585	39367120	28775718	25254825	6560190
	113677509	185382610	197306090	161654064	116960790	106747520	28775718
	420323380	690669530	744584795	621385440	445500750	420504507	116960790
	401717793*	292923272*	512782732*	108221826*	530296593*	485342293*	445500750
	650906452*	711499681*	466748004*	271040311*	128158395*	114119020*	530296593*
	876149942*	830463352*	543575521*	1014859839*	848436602*	148051781*	128158395*
							848436602*

-continued next page-

* Multiplicity is given modulo 1073741527

Affine B5 - Level 3 - Class 1 - Page 2 of 3

	Rp 1	Rp 2	Rp 3	Rp 4	Rp 5	Rp 6	Rp 7
Wt 2	1	1	0	0	0	0	0
	19	23	13	4	3	0	0
	234	316	237	116	93	34	3
	2134	3067	2612	1544	1237	641	93
	15609	23326	21340	14028	11040	6837	1237
	96388	147910	141803	99828	77102	53621	11040
	520710	814621	807536	596884	453101	342863	77102
	2523866	4007366	4074813	3125020	2336411	1887387	453101
	11180690	17963909	18639113	14718564	10858930	9247022	2336411
	45903520	74475299	78567990	63546188	46340526	41228501	10858930
	176546143	288800379	308960500	254972484	184044943	169944600	46340526
	641471061	1056793171*	70480752*	960729216	687228531	655279628	184044943
	69442863*	453699637*	799631795*	206004175*	284388887*	237421410*	687228531
	885595399*	402878709*	603154361*	912941786*	691006554*	735270969*	284388887*
			456734771*	357987883*	782660096*	452475951*	691006554*
							782660096*
Wt 3	2	2	1	0	0	0	0
	39	50	35	14	12	3	0
	466	652	533	290	244	107	12
	4065	5974	5315	3330	2708	1514	244
	28465	43156	40528	27600	21824	14135	2708
	169109	262092	255635	184270	142368	102140	21824
	883550	1392214	1397063	1050192	796044	616681	142368
	4161126	6642908	6816704	5294600	3949684	3250735	796044
	17980795	29012628	30318621	24180920	17795736	15388684	3949684
	72238651	117605836	124785320	101744550	74010316	66704660	17795736
	272587693	447182166	480689415	399365170	287571936	268514129	74010316
	973851422	534515785*	674635700*	402657443*	1053673824*	1014499512*	287571936
	94055989*	138395361*	677901504*	884436242*	446050671*	406214947*	1053673824*
	73713708*	872539444*	669215915*	167192038*	384697087*	544665693*	446050671*
				982637043*	286211952*	579425773*	384697087*
							286211952*
Wt 4	4	5	3	1	1	0	0
	81	111	86	43	40	13	1
	925	1339	1154	688	591	286	40
	7675	11500	10560	6943	5685	3361	591
	51386	78886	75615	53061	41993	28239	5685
	293806	459299	454426	334315	257931	190225	41993
	1485886	2356193	2389371	1823051	1378612	1091166	257931
	6805842	10917890	11294474	8873145	6601549	5528519	1378612
	28711612	46506538	48914225	39368102	28894173	25349779	6601549
	112963986	184490337	196790784	161655956	117283313	107013338	28894173
	418498473	688374621	743248307	621388940	446341458	421214078	117283313
	397213715*	287232061*	509443892*	108228154*	532406189*	487160005*	446341458
	640137073*	697834720*	458678854*	271051624*	133276878*	118610143*	532406189*
	851122446*	798589528*	524643793*	1014879570*	860488023*	158798431*	133276878*
						209119170*	860488023*

-continued next page-

* Multiplicity is given modulo 1073741527

Affine B5 - Level 3 - Class 1 - Page 3 of 3

	Rp 1	Rp 2	Rp 3	Rp 4	Rp 5	Rp 6	Rp 7
Wt 5	0	0	0	0	1	0	0
	21	30	25	14	20	5	0
	360	530	470	290	290	125	15
	3575	5396	5009	3330	2920	1620	270
	26525	40830	39270	27600	22670	14625	2855
	162255	253770	251055	184270	145392	104080	22480
	861364	1365010	1381870	1050192	805965	623535	144886
	4094210	6560190	6769930	5294600	3980020	3272921	804700
	17790435	28775718	30183216	24180920	17883205	15455600	3977030
	71723345	116960790	124413200	101744550	74250170	66895020	17876440
	271251205	445500750	479711450	399365170	288201671	269029435	74235445
	970512582	530296593*	672163788*	402657443*	1055265250*	1015836000*	288170644
	85986839*	128158395*	671864805*	884436242*	449938519*	409553787*	1055201745*
	54781980*	848436602*	654918997*	167192038*	393911483*	552734843*	449811789*
				982637043*		598357501*	393664313*
Wt 6	9	12	9	4	5	1	0
	171	244	205	116	111	44	5
	1828	2708	2436	1544	1339	704	111
	14351	21824	20564	14028	11500	7143	1339
	91808	142368	138839	99828	78886	54879	11500
	505517	796044	797360	596884	459299	347443	78886
	2477092	3949684	4042633	3125020	2356193	1902580	459299
	11045285	17795736	18543937	14718564	10917890	9293796	2356193
	45531400	74010316	78301726	63546188	46506538	41363906	10917890
	175568178	287571936	308250092	254972484	184490337	170316720	46506538
	638999149	1053673824*	68661532*	960729216	688374621	656257593	184490337
	63406164*	446050671*	795137955*	206004175*	287232061*	239893322*	688374621
	871298481*	384697087*	592402885*	912941786*	697834720*	741307668*	287232061*
	693088808*	286211952*	431738251*	357987883*	798589528*	466772869*	697834720*
							798589528*
Wt 7	21	30	25	14	15	5	1
	360	530	470	290	270	125	20
	3575	5396	5009	3330	2855	1620	290
	26525	40830	39270	27600	22480	14625	2920
	162255	253770	251055	184270	144886	104080	22670
	861364	1365010	1381870	1050192	804700	623535	145392
	4094210	6560190	6769930	5294600	3977030	3272921	805965
	17790435	28775718	30183216	24180920	17876440	15455600	3980020
	71723345	116960790	124413200	101744550	74235445	66895020	17883205
	271251205	445500750	479711450	399365170	288170644	269029435	74250170
	970512582	530296593*	672163788*	402657443*	1055201745*	1015836000*	288201671
	85986839*	128158395*	671864805*	884436242*	449811789*	409553787*	1055265250*
	54781980*	848436602*	654918997*	167192038*	393664313*	552734843*	449938519*
				982637043*		598357501*	393911483*

* Multiplicity is given modulo 1073741527

Weight Multiplicities for Affine B6

Affine B6 - Level 1 - Class 0

	Rp 1 0 00000 1					Rp 2 1 00000 0				
Wt 1	1	48020	39003917	1034796530*	296899817*	0	26243	26394948	933727371*	956913017*
	6	111944	74560332	823732941*	690401515*	1	63573	51113111	228345781*	1047807213*
	28	252701	140533482	644513134*	358082053*	7	148329	97482560	658582175*	417608551*
	105	554484	261409970	142908286*	748275859*	34	334866	183302028	132416330*	839081024*
	350	1185961	480279363	780135665*		132	734328	340143228	116373816*	880646907*
	1057	2478877	872198684	490558066*		448	1568952	623413301	57156709*	
	2974	5073682	492927864*	789282527*		1372	3274624	55629769*	899819632*	
	7876	10187366	637624910*	915847851*		3892	6690817	949924290*	807312037*	
	19865	20096846	607887981*	1034974674*		10369	13408190	367569914*	699650814*	
Wt 2	1	63573	51113111	228345781*	1047807213*	1	48020	39003917	1034796530*	296899817*
	7	148329	97482560	658582175*	417608551*	6	111944	74560332	823732941*	690401515*
	34	334866	183302028	132416330*	839081024*	28	252701	140533482	644513134*	358082053*
	132	734328	340143228	116373816*	880646907*	105	554484	261409970	142908286*	748275859*
	448	1568952	623413301	57156709*		350	1185961	480279363	780135665*	
	1372	3274624	55629769*	899819632*		1057	2478877	872198684	490558066*	
	3892	6690817	949924290*	807312037*		2974	5073682	492927864*	789282527*	
	10369	13408190	367569914*	699650814*		7876	10187366	637624910*	915847851*	
	26243	26394948	933727371*	956913017*		19865	20096846	607887981*	1034974674*	

* Multiplicity is given modulo 1073741527

Affine B6 - Level 1 - Class 1

	Rp 1 0 00001 0				
Wt 1	1	27293	14530612	111345981*	153780607*
	7	66513	28725628	797070578*	113195936*
	34	156073	55852375	635434896*	359328192*
	133	354283	106936970	539536572*	51784370*
	454	780976	201833814	959327764*	173411885*
	1400	1677004	375882889	123554541*	184506729*
	3997	3516957	691309052	35140761*	590770669*
	10718	7219065	182818600*	1030945895*	589798908*

* Multiplicity is given modulo 1073741527

Affine B6 - Level 2 - Class 0 - Page 1 of 2

	Rp 1	Rp 2	Rp 3	Rp 4	Rp 5	Rp 6	Rp 7	Rp 8
	0 00002 0	0 00010 0	0 00100 0	0 01000 0	0 10000 0	0 00000 2	1 00000 1	2 00000 0
Wt 1	1	0	0	0	0	0	0	0
	12	7	6	0	0	0	0	0
	126	84	72	21	15	10	0	0
	952	720	616	252	180	120	35	10
	5931	4720	4032	2016	1442	945	420	120
	31500	26040	22236	12432	8904	5740	3276	945
	148204	125496	107200	64821	46560	29556	19712	5740
	631248	544450	465549	297540	214520	134352	100401	29556
	2477514	2166264	1855068	1236984	895860	554456	451332	134352
	9072108	8019585	6880014	4741968	3451296	2113992	1841608	554456
	31302558	27914640	23996928	16988289	12429326	7543956	6943344	2113992
					42249516		24507414	7543956
Wt 2	1	1	0	0	0	0	0	0
	18	17	6	5	0	0	0	0
	187	175	88	71	15	10	10	0
	1415	1317	782	623	206	135	135	10
	8645	8024	5240	4147	1717	1105	1120	135
	45139	41838	29080	22950	10900	6905	7111	1105
	208744	193379	140454	110795	57812	36131	37843	6905
	875704	811285	608879	480819	268734	166008	176837	36131
	3389786	3141776	2418504	1913794	1127413	689452	746641	166008
	12260867	11371593	8932571	7087714	4352824	2638485	2903072	689452
	41838341	38836988	31014957	24686681	15682276	9432025	10536494	2638485
							36058096	9432025
Wt 3	2	1	1	0	0	0	0	0
	33	22	20	5	4	3	0	0
	320	238	210	82	62	44	10	3
	2295	1812	1576	758	559	381	150	44
	13440	11021	9513	5180	3776	2498	1306	381
	67839	57102	49100	29055	21080	13629	8540	2498
	305164	261825	224776	141220	102382	64992	46279	13629
	1251101	1089216	934914	614570	446299	279089	218728	64992
	4749750	4183062	3593062	2447112	1782712	1100890	929912	279089
	16896240	15019598	12918080	9052558	6621690	4045663	3629896	1100890
	56829184	50907047	43858453	31463856	23121616	13996900	13199121	4045663
							45189786	13996900
Wt 4	3	3	1	1	0	0	0	0
	51	49	24	21	4	3	3	0
	486	459	267	222	69	48	48	3
	3406	3189	2048	1661	657	438	442	48
	19515	18185	12465	9981	4577	2967	3036	438
	96600	89787	64461	51291	26025	16528	17184	2967
	427143	396531	294702	233901	127791	79875	84431	16528
	1724925	1600668	1221966	969654	560571	345897	371700	79875
	6461688	5997027	4677270	3716109	2246409	1371341	1497471	345897
	22714418	21091239	16739752	13328536	8353920	5053347	5603938	1371341
			56562282	45159552	29164009	17502435	19697577	5053347
								17502435

continued next page

Affine B6 - Level 2 - Class 0 - Page 2 of 2

	Rp 1	Rp 2	Rp 3	Rp 4	Rp 5	Rp 6	Rp 7	Rp 8
Wt 5	6	4	4	1	1	1	0	0
	92	68	63	24	20	16	3	1
	816	640	572	268	208	151	52	16
	5428	4423	3880	2064	1548	1064	506	151
	29884	25012	21720	12599	9291	6160	3600	1064
	143320	122358	105681	65284	47772	30864	20838	6160
	617605	535380	461220	298872	218144	138189	103860	30864
	2441140	2141861	1843656	1240400	905832	564932	461356	138189
	8980028	7956648	6850868	4750730	3477655	2141342	1868672	564932
	31077804	27759888	23926134	17009296	12495376	7612004	7013281	2141342
					42409610	25600182	24679776	7612004
								25600182
Wt 6	0	0	0	0	0	1	0	0
	20	15	15	6	6	6	1	0
	255	200	186	87	73	63	18	6
	2040	1656	1488	780	612	446	192	57
	12551	10458	9212	5257	3975	2748	1480	446
	65052	55292	48180	29304	21760	14394	9165	2721
	297036	256440	222120	141948	104469	67341	48294	14394
	1228870	1074255	927729	616584	452304	285540	224719	67243
	4691916	4143640	3574464	2452365	1798924	1118223	946584	285540
	16752348	14920482	12872034	9065706	6663456	4089264	3673851	1117908
				31495458	23224587	14103798	13309782	4089264
								14102880
Wt 7	10	10	5	5	1	1	1	0
	145	140	81	72	22	17	17	1
	1240	1176	747	629	241	170	171	17
	8015	7527	5101	4175	1857	1245	1267	170
	43102	40256	28597	23055	11391	7395	7636	1245
	202675	188717	138941	111145	59365	37678	39530	7395
	858777	798391	604490	481876	273284	170530	181849	37678
	3345076	3107931	2406554	1916770	1139923	701857	760593	170530
	12148097	11286641	8901673	7095604	4385479	2670790	2939899	701857
	41564977	38631824	30938524	24706615	15763798	9512490	10629363	2670790
							36283293	9512490
Wt 8	20	15	15	6	6	6	1	1
	255	200	186	87	73	57	18	6
	2040	1656	1488	780	612	446	192	63
	12551	10458	9212	5257	3975	2721	1480	446
	65052	55292	48180	29304	21760	14394	9165	2748
	297036	256440	222120	141948	104469	67243	48294	14394
	1228870	1074255	927729	616584	452304	285540	224719	67341
	4691916	4143640	3574464	2452365	1798924	1117908	946584	285540
	16752348	14920482	12872034	9065706	6663456	4089264	3673851	1118223
				31495458	23224587	14102880	13309782	4089264
								14103798

Affine B6 - Level 2 - Class 1

	Rp 1 0 00001 1	Rp 2 1 00001 0
Wt 1	1 22395359 559125278* 115643514* 768760914*	0 11867310 16416885* 438626939* 364302389*
	27 76413078 612112691* 830510680* 434408061*	6 41650050 850407748* 806949125* 837523838*
	315 247783113 724022836* 519004171* 438327720*	98 138421358 32649644* 42681799*1055384485*
	2492 768291650 776673380* 5653273* 111683918*	918 438616728 73412480* 241298048* 160821221*
	15525 141730441* 10274794*1044638414*	6372 258702803* 479737337* 184748883* 128867103*
	81816 139141995* 880084380* 657035163*	36280 676839645* 132540290* 385480604*
	380051 67375610* 24881913* 35016329*	178794 285832610* 780310426* 521608304*
	1597725 130256414* 768914390*1049386919*	788004 162538712* 797988501* 147676111*
	6190182 384755963* 322112878* 729287993*	3174210 60002753* 214410224* 66708569*
Wt 2	6 41650050 850407748* 806949125* 837523838*	1 22395359 559125278* 115643514* 768760914*
	98 138421358 32649644* 42681799*1055384485*	27 76413078 612112691* 830510680* 434408061*
	918 438616728 73412480* 241298048* 160821221*	315 247783113 724022836* 519004171* 438327720*
	6372 258702803* 479737337* 184748883* 128867103*	2492 768291650 776673380* 5653273* 111683918*
	36280 676839645* 132540290* 385480604*	15525 141730441* 10274794*1044638414*
	178794 285832610* 780310426* 521608304*	81816 139141995* 880084380* 657035163*
	788004 162538712* 797988501* 147676111*	380051 67375610* 24881913* 35016329*
	3174210 60002753* 214410224* 66708569*	1597725 130256414* 768914390*1049386919*
	11867310 16416885* 438626939* 364302389*	6190182 384755963* 322112878* 729287993*

* Multiplicity is given modulo 1073741527

Affine B6 - Level 3 - Class 0 - Page 1 of 3

	Rp 1 0 00002 1	Rp 2 0 00010 1	Rp 3 0 00100 1	Rp 4 0 01000 1	Rp 5 0 10000 1	Rp 6 0 00000 3	Rp 7 1 00002 0	Rp 8 1 00010 0	Rp 9 1 00100 0	Rp10 1 01000 0	Rp11 1 10000 0	Rp12 1 00000 2	Rp13 2 00000 1	Rp14 3 00000 0
Wt 1	1	0	0	0	0	0	0	0	0	0	0	0	0	0
	42	21	15	0	0	0	6	5	0	0	0	0	0	0
	831	532	384	84	40	15	183	156	49	29	0	0	0	0
	10458	7455	5403	1807	870	306	2913	2487	1128	678	105	91	36	0
	98469	74319	54048	22098	10740	3612	31863	27256	14565	8805	2175	1833	762	36
	752988	589139	430071	197889	97095	31467	271133	232311	136304	82909	25665	21162	9120	722
	4916908	3942501	2889924	1438418	712890	224208	1919547	1647828	1027167	628851	223140	180601	80466	8250
										4065516	1582165	1260672	578934	69810
														483646
Wt 2	1	1	0	0	0	0	0	0	0	0	0	0	0	0
	58	51	15	10	0	0	11	5	4	0	0	0	0	0
	1162	1008	444	279	40	15	303	191	144	29	15	15	0	0
	14471	12476	6542	4047	955	331	4509	3177	2352	758	380	346	36	6
	134048	115357	66768	41102	12325	4073	46984	35196	25889	10314	5115	4453	827	141
	1008932	868164	535569	329968	114335	36447	385542	299917	220296	99477	49320	41533	10341	1778
	6493247	5592428	3607207	2230524	852155	264080	2653358	2118054	1557676	764083	380575	312642	93683	16361
											2493920	2008656	685132	121713
														774712

continued next page

Affine B6 - Level 3 - Class 0 - Page 2 of 3

	Rp 1	Rp 2	Rp 3	Rp 4	Rp 5	Rp 6	Rp 7	Rp 8	Rp 9	Rp10	Rp11	Rp12	Rp13	Rp14
Wt 3	2	1	1	0	0	0	0	0	0	0	0	0	0	0
	100	62	50	10	6	3	15	14	4	3	0	0	0	0
	1842	1288	984	308	164	66	429	380	164	110	15	15	6	0
	21692	16137	12064	4702	2404	877	6294	5488	2832	1808	411	371	156	6
	192916	149273	110619	49068	24708	8443	64444	55792	31890	19969	5814	4999	2148	150
	1407604	1118710	826694	399782	200702	65519	519781	448770	274080	170407	57699	48018	21198	[illegible]
		7163045	5295105	2723428	1371214	432315	3522379	3039256	1944438	1208073	453079	368265	167036	[illegible]
											2999777	2393885	1115168	[illegible]
														[illegible]
Wt 4	3	3	1	1	0	0	0	0	0	0	0	0	0	0
	144	132	57	42	6	3	27	17	15	3	2	3	0	0
	2604	2306	1179	789	176	70	699	486	387	120	69	72	6	1
	30013	26205	14825	9509	2706	969	9609	7113	5436	2082	1107	1038	165	30
	261900	227172	137478	86598	28710	9633	93987	72372	54279	23703	12192	10737	2385	[illegible]
	1879233	1625696	1031940	645609	237844	76362	732521	579792	431317	205762	104457	88434	24311	4306
				4133763	1644684	511059	4833222	3902988	2895621	1472352	745464	613435	195690	34384
													1324353	[illegible]
Wt 5	6	4	4	1	1	1	0	0	0	0	0	0	0	0
	244	171	145	46	30	16	40	38	17	14	2	3	1	0
	4068	3004	2368	903	511	216	1001	905	462	327	73	76	32	1
	44453	34110	26014	11228	5958	2216	13427	11852	6659	4407	1233	1142	491	[illegible]
	373298	294381	221023	104193	53773	18543	128552	112199	67368	43168	14052	12219	5371	[illegible]
	2601176	2093582	1561164	785220	400941	131283	983942	854160	538245	339948	123128	103007	46291	[illegible]
										2272972	891081	725662	334073	40379
														[illegible]
Wt 6	0	0	0	0	0	1	0	0	0	0	0	0	0	0
	20	15	15	6	6	6	0	0	0	0	0	1	0	0
	600	450	390	156	108	63	105	100	54	45	11	18	6	0
	8880	6809	5469	2358	1382	606	2295	2094	1184	861	243	267	108	6
	89886	70557	54511	25316	13776	5208	28161	25044	14922	10077	3219	3041	1314	102
	713148	570843	432567	213666	112218	38907	252657	221690	138144	89871	31689	27798	12354	1206
					779146	255405		1602600	1035270	661201	252342	211773	96402	[illegible]
													647934	[illegible]
Wt 7	6	5	0	0	0	0	1	0	0	0	0	0	0	0
	183	156	49	29	0	0	42	21	15	0	0	0	0	0
	2913	2487	1128	678	105	36	831	532	384	84	40	36	0	[illegible]
	31863	27256	14565	8805	2175	722	10458	7455	5403	1807	870	762	91	[illegible]
	271133	232311	136304	82909	25665	8250	98469	74319	54048	22098	10740	9120	1833	[illegible]
	1919547	1647828	1027167	628851	223140	69810	752988	589139	430071	197889	97095	80466	21162	[illegible]
				4065516	1582165	483646	4916908	3942501	2889924	1438418	712890	578934	180601	[illegible]
													1260672	[illegible]
Wt 8	11	5	4	0	0	0	1	1	0	0	0	0	0	0
	303	191	144	29	15	6	58	51	15	10	0	0	0	0
	4509	3177	2352	758	380	141	1162	1008	444	279	40	36	15	0
	46984	35196	25889	10314	5115	1778	14471	12476	6542	4047	955	827	346	15
	385542	299917	220296	99477	49320	16361	134048	115357	66768	41102	12325	10341	4453	331
	2653358	2118054	1557676	764083	380575	121713	1008932	868164	535569	329968	114335	93683	41533	4073
					2493920	774712	6493247	5592428	3607207	2230524	852155	685132	312642	36447
													2008656	264080

—continued next page—

Affine B6 - Level 3 - Class 0 - Page 3 of 3

	Rp 1	Rp 2	Rp 3	Rp 4	Rp 5	Rp 6	Rp 7	Rp 8	Rp 9	Rp10	Rp11	Rp12	Rp13	Rp14
Wt 9	15 429 6294 64444 519781 3522379	14 380 5488 55792 448770 3039256	4 164 2832 31890 274080 1944438	3 110 1808 19969 170407 1208073	0 15 411 5814 57699 453079 2999777	0 6 150 1984 18804 142636 919164	2 100 1842 21692 192916 1407604	1 62 1288 16137 149273 1118710 7163045	1 50 984 12064 110619 826694 5295105	0 10 308 4702 49068 399782 2723428	0 6 164 2404 24708 200702 1371214	0 6 156 2148 21198 167036 1115168	0 0 15 371 4999 48018 368265 2393885	0 0 3 66 877 8443 65519 432315
Wt10	27 699 9609 93987 732521 4833222	17 486 7113 72372 579792 3902988	15 387 5436 54279 431317 2895621	3 120 2082 23703 205762 1472352	2 69 1107 12192 104457 745464	1 30 423 4306 34884 238980	3 144 2604 30013 261900 1879233	3 132 2306 26205 227172 1625696	1 57 1179 14825 137478 1031940	1 42 789 9509 86598 645609 4133763	0 6 176 2706 28710 237844 1644684	0 6 165 2385 24311 195690 1324353	0 3 72 1038 10737 88434 613435	0 0 3 70 969 9633 76362 511059
Wt11	40 1001 13427 128552 983942	38 905 11852 112199 854160	17 462 6659 67368 538245	14 327 4407 43168 339948 2272972	2 73 1233 14052 123128 891081	1 31 461 4861 40379 281203	6 244 4068 44453 373298 2601176	4 171 3004 34110 294381 2093582	4 145 2368 26014 221023 1561164	1 46 903 11228 104193 785220	1 30 511 5958 53773 400941	1 32 491 5371 46291 334073	0 3 76 1142 12219 103007 725662	0 1 16 216 2216 18543 131283
Wt12	10 360 5765 61376 504933	10 335 5169 54022 440429 3004401	5 175 2904 32273 275817 1951509	5 136 2012 21195 176543 1234977	1 32 566 6823 63049 477532	1 17 235 2490 21346 153837	0 70 1600 20216 185467 1374755	0 50 1184 15478 145826 1103111	0 45 969 12053 110795 828123	0 15 371 5167 51706 412355	0 11 223 2843 27228 212898 1423251	1 17 246 2713 24174 180694 1171475	0 1 33 533 6048 53533 393278	0 0 6 97 1096 9647 71171 455841
Wt13	70 1600 20216 185467 1374755	50 1184 15478 145826 1103111	45 969 12053 110795 828123	15 371 5167 51706 412355	11 223 2843 27228 212898 1423251	6 97 1096 9647 71171 455841	10 360 5765 61376 504933	10 335 5169 54022 440429 3004401	5 175 2904 32273 275817 1951509	5 136 2012 21195 176543 1234977	1 32 566 6823 63049 477532	1 33 533 6048 53533 393278	1 17 246 2713 24174 180694 1171475	0 1 17 235 2490 21346 153837
Wt14	105 2295 28161 252657	100 2094 25044 221690 1602600	54 1184 14922 138144 1035270	45 861 10077 89871 661201	11 243 3219 31689 252342	6 102 1206 10978 82782	20 600 8880 89886 713148	15 450 6809 70557 570843	15 390 5469 54511 432567	6 156 2358 25316 213666	6 108 1382 13776 112218 779146	6 108 1314 12354 96402 647934	1 18 267 3041 27798 211773	1 6 63 606 5208 38907 255405

Affine B6 - Level 3 - Class 1 - Page 1 of 2

	Rp 1	Rp 2	Rp 3	Rp 4	Rp 5	Rp 6	Rp 7	Rp 8
	0	0	0	0	0	0	1	2
	00003	00011	00101	01001	10001	00001	00001	00001
	0	0	0	0	0	2	1	0
Wt 1	1	0	0	0	0	0	0	0
	12	12	6	0	0	0	0	0
	171	216	152	60	15	10	0	0
	1914	2684	2226	1240	484	345	120	10
	17415	25872	23544	15300	7413	5370	2640	345
	133278	205116	198050	140928	77244	55811	32780	5370
	882952	1392032	1400019	1058056	628881	450264	297288	55811
	5186250	8322084	8623260	6805608	4286100	3036003	2183988	450264
	27527595	44774832	47470857	38722272	25480608	17856918	13733856	3036003
	134022304	220365652	237947016	199253280	135751740	94184196	76477044	17856918
	605669454	1004780184*	27644990*	942343296	660546906	454076286	385793952	94184196
					829265890*		718189509*	454076286
Wt 2	1	1	0	0	0	0	0	0
	23	28	16	5	0	0	0	0
	343	468	358	180	55	40	10	0
	3793	5524	4818	2950	1289	955	400	40
	33483	50801	47738	32660	16857	12421	6599	955
	247548	386116	380886	279808	159647	116157	71444	12421
	1586656	2524636	2577487	1990464	1215947	872715	594968	116157
	9044257	14610156	15306355	12269834	7882262	5584371	4115812	872715
	46737384	76406933	81689939	67426064	45041858	31536378	24728395	5584371
	222203782	366816976	398710467	336985740	232330277	160953478	132795820	31536378
	983177101	562577463*	729548730*	480791205*	26486254*	755030165	650211148	160953478
							797596566*	755030165
Wt 3	2	2	1	0	0	0	0	0
	48	62	44	18	4	3	0	0
	706	1000	835	470	182	140	44	3
	7544	11252	10274	6700	3228	2451	1132	140
	64058	98718	95389	67840	36967	27510	15470	2451
	456296	719396	722898	544956	321832	235025	150218	27510
	2827992	4534712	4690656	3689206	2308185	1657786	1162720	235025
	15647608	25424132	26896162	21855676	14289618	10116160	7622960	1657786
	78768790	129352988	139342104	116228792	78704227	55025082	43929922	10116160
	365923480	606250220	662937620	565045326	393812678	272342363	228099540	55025082
	512456162*	500267838*	784885581*	397923762*	743926289*	171285937*	12267533*	272342363
							508320008*	171285937*
Wt 4	4	5	3	1	0	0	0	0
	102	141	111	57	18	15	3	0
	1458	2139	1889	1170	516	413	158	15
	14936	22740	21483	14757	7622	5871	2949	413
	121590	189897	187611	137829	78369	58635	34665	5871
	833772	1326735	1353908	1043394	634095	463663	306805	58635
	4997594	8068001	8439954	6744689	4307675	3092082	2225925	463663
	26856954	43862004	46801245	38492616	25562253	18070038	13900221	3092082
	131786952	217297611	235666047	198452226	136038663	94933617	77083603	18070038
	598619358	995037733*	20321192*	939723004	661493972	456546684	387856566	94933617
			442008242*	917072124*	832228258*	963149873*	724801880*	456546684
								963149873*

continued next page

* Multiplicity is given modulo 1073741527

Affine B6 - Level 3 - Class 1 - Page 2 of 2

	Rp 1	Rp 2	Rp 3	Rp 4	Rp 5	Rp 6	Rp 7	Rp 8
Wt 5	9	12	9	4	1	1	0	0
	222	320	274	160	64	57	16	1
	3016	4536	4193	2776	1354	1105	476	57
	29380	45452	44152	31572	17238	13374	7168	1105
	228777	361212	363564	274260	161556	121097	74712	13374
	1509770	2421288	2503742	1965928	1224394	894881	610736	121097
	8756090	14218844	15021742	12172424	7916084	5673506	4182828	894881
	45733674	75032700	80675350	67071044	45166532	31864887	24986560	5673506
	218917578	362286840	395324963	335779696	232759028	162080490	133714492	31864887
	972974996	548431905*	718869385*	476929697*	27875799*	758670132	653271488	162080490
					553739893*	86361552*	807235434*	758670132
								86361552*
Wt 6	0	0	0	0	0	1	0	0
	51	76	69	44	20	27	6	0
	1068	1632	1548	1062	540	510	198	21
	12663	19740	19405	14082	7800	6476	3268	483
	110498	175076	177114	134332	79384	61788	36588	6378
	786207	1262538	1307622	1027632	638967	478002	316452	61473
	4813026	7816794	8256162	6680892	4328292	3150674	2268250	477084
	26196216	42956272	46130361	38255854	25641382	18289509	14067648	3148182
	129574626	214246296	233381952	197634408	136319322	95698314	77692908	18283137
	591620841	985333308	12990024*	937064556	662426217	459052408	389925888	95682789
				908870351*	835156850*	970937219*	731429915*	459016128
								970855403*
Wt 7	21	30	25	14	5	6	1	0
	487	726	659	423	195	178	62	6
	6206	9524	9111	6354	3330	2742	1296	178
	57274	89778	89177	65876	37574	29211	16544	2742
	426275	679253	694342	535550	324848	243353	155920	29211
	2708392	4373006	4573509	3649452	2321319	1693611	1188816	243353
	15209758	24826872	26457080	21703188	14341266	10255577	7729646	1693611
	77274192	127299350	137813861	115688070	78891280	55525496	44328676	10255577
	361115197	599602722	657939532	563249290	394445397	274021851	229484173	55525496
	497757186*	479838724*	769388345*	392281067*	745945981*	176610195*	16786911*	274021851
							522308680*	176610195*
Wt 8	51	76	69	44	20	21	6	1
	1068	1632	1548	1062	540	483	198	27
	12663	19740	19405	14082	7800	6378	3268	510
	110498	175076	177114	134332	79384	61473	36588	6476
	786207	1262538	1307622	1027632	638967	477084	316452	61788
	4813026	7816794	8256162	6680892	4328292	3148182	2268250	478002
	26196216	42956272	46130361	38255854	25641382	18283137	14067648	3150674
	129574626	214246296	233381952	197634408	136319322	95682789	77692908	18289509
	591620841	985333308	12990024*	937064556	662426217	459016128	389925888	95698314
				908870351*	835156850*	970855403*	731429915*	459052408
								970937219*

* Multiplicity is given modulo 1073741527

Weight Multiplicities for Affine B7

Affine B7 - Level 1 - Class 0

	Rp 1 0 000000 1					Rp 2 1 000000 0				
Wt 1	1	102031	130871366	880959794*	678125799*	0	52177	84918117	807671814*	699331403*
	7	251854	261596933	775180750*	679434707*	1	134372	172238856	226555729*	948501226*
	36	600813	515111283	111481490*	25357047*	8	332467	343715926	1066934098*	375114327*
	148	1390598	1000170689	23615698*	795028187*	43	794245	675613280	921276903*	582034530*
	534	3132375	842861329*	421321815*		183	1839474	235629299*	585507907*	20552257*
	1738	6885039	406347125*	962318628*		674	4143716	356793822*	899149478*	
	5239	14799626	343864929*	1011718118*		2228	9104146	435521828*	569076599*	
	14816	31169728	745238396*	852988158*		6786	19554219	241838173*	807986773*	
	39765	64425522	442709001*	871625451*		19339	41138620	201553779*	553758124*	
Wt 2	1	134372	172238856	226555729*	948501226*	1	102031	130871366	880959794*	678125799*
	8	332467	343715926	1066934098*	375114327*	7	251854	261596933	775180750*	679434707*
	43	794245	675613280	921276903*	582034530*	36	600813	515111283	111481490*	25357047*
	183	1839474	235629299*	585507907*	20552257*	148	1390598	1000170689	23615698*	795028187*
	674	4143716	356793822*	899149478*		534	3132375	842861329*	421321815*	
	2228	9104146	435521828*	569076599*		1738	6885039	406347125*	962318628*	
	6786	19554219	241838173*	807986773*		5239	14799626	343864929*	1011718118*	
	19339	41138620	201553779*	553758124*		14816	31169728	745238396*	852988158*	
	52177	84918117	807671814*	699331403*		39765	64425522	442709001*	871625451*	

* Multiplicity is given modulo 1073741527

Affine B7 - Level 1 - Class 1

	Rp 1 0 000001 0				
Wt 1	1	53907	44136666	83681501*	650410511*
	8	139568	91470872	872969816*	365709570*
	43	347100	186244911	872140223*	346014113*
	184	833336	373048408	972685687*	820679483*
	681	1939277	735901872	511885943*	301155995*
	2264	4388784	357415857*	1026692402*	1034119308*
	6934	9685621	598888705*	1071944968*	705224903*
	19872	20892648	909628820*	443321729*	145743654*

* Multiplicity is given modulo 1073741527

Affine B7 - Level 2 - Class 0 - Page 1 of 3

	Rp 1	Rp 2	Rp 3	Rp 4	Rp 5	Rp 6	Rp 7	Rp 8	Rp 9
	0 000002 0	0 000010 0	0 000100 0	0 001000 0	0 010000 0	0 100000 0	0 000000 2	1 000000 1	2 000000 0
Wt 1	1	0	0	0	0	0	0	0	0
	14	8	7	0	0	0	0	0	0
	168	112	98	28	21	0	0	0	0
	1456	1099	960	392	294	56	35	35	0
	10318	8218	7168	3584	2688	784	490	490	35
	61992	51296	44709	25088	18816	6958	4312	4368	490
	327712	277984	242242	147211	110523	47236	29008	29792	4312
	1559392	1347647	1174768	755594	568302	269024	163730	170667	29008
	6800591	5959998	5199264	3491712	2632728	1343776	811048	857990	163730
	27540240	24405752	21313019	14804608	11195856	6058745	3629024	3895360	811048
					44313004	25126038	14945952	16270912	3629024
									14945952
Wt 2	1	1	0	0	0	0	0	0	0
	21	20	7	6	0	0	0	0	0
	253	239	120	100	21	15	10	0	0
	2203	2071	1231	1014	336	236	155	35	10
	15360	14402	9430	7716	3213	2241	1445	546	155
	90802	85011	59303	48352	23148	16098	10207	5075	1445
	472013	441571	322108	262211	138105	96038	60005	35630	10207
	2211807	2068581	1559974	1269606	716967	499486	308136	207781	60005
	9510579	8895306	6883034	5605871	3339114	2333463	1423720	1057316	308136
	38026864	35577784	28099957	22917572	14238773	9989841	6036663	4838272	1423720
						39755320	23820753	20312698	6036663
									23820753
Wt 3	2	1	1	0	0	0	0	0	0
	39	26	24	6	5	0	0	0	0
	442	329	296	116	92	15	10	10	0
	3662	2896	2572	1244	964	262	170	170	10
	24514	20143	17759	9736	7452	2614	1661	1676	170
	140145	118238	103810	61893	47067	19352	12094	12356	1661
	708474	609387	533832	338088	256354	117657	72536	75145	12094
	3242616	2830550	2477200	1642112	1244365	619446	377562	396822	72536
	13665302	12068288	10560548	7255180	5502866	2916524	1760459	1877152	377562
		47869774	41909138	29632109	22518685	12546046	7509470	8121464	1760459
								32614285	7509470
Wt 4	3	3	1	1	0	0	0	0	0
	61	59	29	26	5	4	3	0	0
	684	652	381	327	103	77	54	10	3
	5555	5254	3392	2846	1141	825	554	185	54
	36422	34292	23637	19581	9099	6467	4213	1918	554
	204239	191779	138475	113889	58545	41257	26275	14605	4213
	1014672	951315	711158	582651	322466	226463	141687	90753	26275
	4572414	4283587	3289606	2690651	1575681	1106253	682248	486236	141687
	19003744	17798548	13965590	11419992	6993272	4918254	2997185	2322603	682248
			55163136	45140699	28663086	20219061	12197545	10113235	2997185
								40776699	12197545

continued next page

Affine B7 - Level 2 - Class 0 - Page 2 of 3

	Rp 1	Rp 2	Rp 3	Rp 4	Rp 5	Rp 6	Rp 7	Rp 8	Rp 9
Wt 5	6	4	4	1	1	0	0	0	0
	112	83	78	30	26	4	3	3	0
	1177	926	844	400	326	84	58	58	3
	9094	7434	6662	3584	2828	949	626	630	58
	57324	48132	42708	25041	19408	7694	4928	5012	626
	311360	266662	235230	146822	112676	50176	31469	32417	4928
	1506587	1310112	1151836	753946	575690	279440	172474	180142	31469
	6640014	5844226	5129422	3485428	2656040	1378049	839780	889630	172474
	27078184	24067608	21110864	14784088	11265789	6163730	3716992	3993604	839780
				58337578	44511436	25430970	15200970	16559612	3716992
									15200970
Wt 6	10	10	5	5	1	1	1	0	0
	180	175	102	93	29	24	19	3	1
	1831	1753	1123	977	383	293	210	62	19
	13760	13057	8920	7559	3431	2521	1709	709	210
	84717	79959	57233	47763	24028	17278	11297	5848	1709
	450954	424246	314604	260211	141287	100457	64021	38781	11297
	2144239	2013383	1535024	1263294	727557	514617	321699	219249	64021
	9306925	8729833	6805706	5587142	3372012	2381607	1466382	1095592	321699
	37444654	35106730	27873829	22864736	14335321	10134031	6163182	4957333	1466382
						40165430	24177570	20662082	6163182
									24177570
Wt 7	0	0	0	0	0	0	1	0	0
	35	35	21	21	7	7	7	1	0
	525	511	329	301	119	99	84	21	7
	4802	4606	3136	2744	1232	952	686	259	77
	33341	31689	22547	19201	9507	7042	4802	2296	686
	193011	182427	134379	112623	60144	43554	28455	16233	4767
	977109	920339	697081	578648	328090	234675	149366	96943	28455
	4455381	4187827	3244766	2678782	1593949	1133384	706840	507647	149226
	18659669	17518648	13831216	11386291	7048776	5001941	3072174	2391230	706840
					28822787	20463312	12412869	10319848	3071684
									12412869
Wt 8	20	15	15	6	6	1	1	1	0
	325	256	242	115	101	26	20	20	1
	3096	2525	2315	1234	1018	333	233	234	20
	22172	18554	16706	9676	7708	2968	1970	1996	233
	131504	112222	99950	61612	48110	20845	13384	13717	1970
	679291	588662	520848	336966	260152	123242	77324	80286	13384
	3150858	2764350	2436475	1638116	1257048	638498	393797	414541	77324
	13393417	11869612	10440034	7242142	5542347	2976892	1811642	1933866	393797
				29592446	22634624	12725999	7661391	8292181	1811642
								33102074	7661391

-continued next page-

Affine B7 - Level 2 - Class 0 - Page 3 of 3

	Rp 1	Rp 2	Rp 3	Rp 4	Rp 5	Rp 6	Rp 7	Rp 8	Rp 9
Wt 9	35	35	21	21	7	7	7	1	1
	525	511	329	301	119	99	77	21	7
	4802	4606	3136	2744	1232	952	686	259	84
	33341	31689	22547	19201	9507	7042	4767	2296	686
	193011	182427	134379	112623	60144	43554	28455	16233	4802
	977109	920339	697081	578648	328090	234675	149226	96943	28455
	4455381	4187827	3244766	2678782	1593949	1133384	706840	507647	149366
	18659669	17518648	13831216	11386291	7048776	5001941	3071684	2391230	706840
					28822787	20463312	12412869	10319848	3072174
									12412869

Affine B7 - Level 2 - Class 1

	Rp 1 0 000001 1					Rp 2 1 000001 0				
Wt 1	1	80782800	871120898*	735844174*	957971546*	0	40786690	696348890*	929399637*	380307994*
	35	302666903	419189929*	228428375*	779568364*	7	157510780	512278697*	267366479*	777361393*
	490	1073720305*	787858372*	925934431*	659199652*	140	573720808	69617194*	60427440*	616032167*
	4522	409083254*	100838901*	930346486*	782624144*	1547	911765248*	861846355*	807803587*	422989512*
	32305	1023243108*	79533254*	92287561*		12405	124757309*	38605886*	86492252*	919712623*
	192899	158673252*	122194698*	921807520*		80465	459326577*	942736128*	466174646*	
	1006145	894147031*	30880085*	908987249*		447146	540260062*	1007563427*	439806144*	
	4715510	52130125*	698646563*	563485427*		2204475	342295921*	322349758*	326470886*	
	20247710	594048580*	898835340*	424428086*		9869132	1000298023*	256352642*	745191799*	
Wt 2	7	157510780	512278697*	267366479*	777361393*	1	80782800	871120898*	735844174*	957971546*
	140	573720808	69617194*	60427440*	616032167*	35	302666903	419189929*	228428375*	779568364*
	1547	911765248*	861846355*	807803587*	422989512*	490	1073720305*	787858372*	925934431*	659199652*
	12405	124757309*	38605886*	86492252*	919712623*	4522	409083254*	100838901*	930346486*	782624144*
	80465	459326577*	942736128*	466174646*		32305	1023243108*	79533254*	92287561*	
	447146	540260062*	1007563427*	439806144*		192899	158673252*	122194698*	921807520*	
	2204475	342295921*	322349758*	326470886*		1006145	894147031*	30880085*	908987249*	
	9869132	1000298023*	256352642*	745191799*		4715510	52130125*	698646563*	563485427*	
	40786690	696348890*	929399637*	380307994*		20247710	594048580*	898835340*	424428086*	

* Multiplicity is given modulo 1073741527

Weight Multiplicities for Affine B8

Affine B8 - Level 1 - Class 0

	Rp 1 (0 0000000 1)					Rp 2 (1 0000000 0)				
Wt 1	1	201925	394485218	185180592*	327790882*	0	97038	245784007	649895533*	826681721*
	8	525362	821669323	321708866*	991448619*	1	264307	520274795	1050027908*	648095457*
	45	1318591	610818914*	363761138*	995271426*	9	690123	8803895*	280235420*	339230191*
	201	3205617	181655445*	382485739*	862678633*	53	1736510	69246671*	556438219*	875538746*
	780	7573360	336734785*	985607489*		245	4228742	176942157*	141342455*	247409177*
	2718	17435919	445350060*	479365662*		972	10000746	306119179*	485521909*	
	8729	39209243	121486237*	687801923*		3444	23035458	286164158*	1041070245*	
	26217	86295162	314869082*	478854948*		11193	51803508	584026604*	1026152967*	
	74502	186201711	968234808*	234998353*		33922	113977750	492913699*	251370976*	
Wt 2	1	264307	520274795	1050027908*	648095457*	1	201925	394485218	185180592*	327790882*
	9	690123	8803895*	280235420*	339230191*	8	525362	821669323	321708866*	991448619*
	53	1736510	69246671*	556438219*	875538746*	45	1318591	610818914*	363761138*	995271426*
	245	4228742	176942157*	141342455*	247409177*	201	3205617	181655445*	382485739*	862678633*
	972	10000746	306119179*	485521909*		780	7573360	336734785*	985607489*	
	3444	23035458	286164158*	1041070245*		2718	17435919	445350060*	479365662*	
	11193	51803508	584026604*	1026152967*		8729	39209243	121486237*	687801923*	
	33922	113977750	492913699*	251370976*		26217	86295162	314869082*	478854948*	
	97038	245784007	649895533*	826681721*		74502	186201711	968234808*	234998353*	

* Multiplicity is given modulo 1073741527

Affine B8 - Level 1 - Class 1

	Rp 1 (0 0000001 0)				
Wt 1	1	99747	121286856	520587919*	356641892*
	9	272983	262530048	311109219*	763123334*
	53	716095	557748500	16925424*	245693707*
	246	1810040	90873759*	38515299*	104727166*
	980	4427223	245458829*	441163103*	1030765991*
	3489	10514915	548425838*	196516091*	841892969*
	11394	24320128	2340496*	154063709*	947227840*
	34701	54912096	783269576*	468272329*	544846572*

* Multiplicity is given modulo 1073741527

Affine B8 - Level 2 - Class 0 - Page 1 of 2

	Rp 1 0 0000002 0	Rp 2 0 0000010 0	Rp 3 0 0000100 0	Rp 4 0 0001000 0	Rp 5 0 0010000 0	Rp 6 0 0100000 0	Rp 7 0 1000000 0	Rp 8 0 0000000 2	Rp 9 1 0000000 1	Rp10 2 0000000 0
Wt 1	1 16 216 2112 16784 112608 661296 3479424 16703632	0 9 144 1592 13376 93252 561696 3011616 14665088	0 8 128 1413 11856 82584 497216 2665788 12984864	0 0 36 576 5928 46464 303301 1722832 8767800	0 0 28 448 4608 36096 235644 1339456 6825432 31790208	0 0 0 84 1344 13392 101376 639330 3517344 17384624	0 0 0 56 896 8946 67872 429360 2370944 11768364	0 0 0 35 560 5544 41664 261072 1428896 7034832	0 0 0 0 126 2016 19824 147840 918372 4980864 24293880	0 0 0 0 35 560 5544 41664 261072 1428896 7034832
Wt 2	1 24 329 3242 25426 168050 971115 5032412 23818652	1 23 313 3071 24027 158568 915524 4742183 22441603	0 8 157 1827 15765 110931 670097 3589668 17435222	0 7 134 1543 13230 92744 559057 2992039 14531207	0 0 28 512 5532 44665 296546 1703033 8729046 40783800	0 0 21 378 4053 32586 215967 1240176 6363876 29793290	0 0 0 56 994 10374 81074 522670 2923903 14640570	0 0 0 35 616 6363 49238 314566 1745422 8675793	0 0 0 35 616 6419 50232 324919 1826118 9194410 42328246	0 0 0 0 35 616 6363 49238 314566 1745422 8675793
Wt 3	2 45 584 5496 41466 265269 1491072 7545950 34991388	1 30 435 4352 34128 224218 1285103 6600952 30970000	1 28 397 3926 30578 200060 1143778 5866890 27509152	0 7 156 1907 16852 119981 728950 3914810 19028284	0 6 128 1532 13376 94563 572260 3067569 14903660	0 0 21 420 4740 39312 265704 1545396 7996200 37633288	0 0 15 292 3250 26764 180363 1048772 5435001 25650424	0 0 10 190 2071 16760 111310 639302 3278183 15330368	0 0 0 35 672 7336 59192 390928 2229292 11339517	0 0 0 10 190 2071 16760 111310 639302 3278183 15330368
Wt 4	3 71 917 8487 62815 394405 2178954 10855378	3 69 880 8088 59609 373286 2058894 10247273 46824566	1 34 516 5243 41269 270768 1546341 7906888 36917755	1 31 453 4510 35073 228426 1298769 6624556 30895206	0 6 144 1826 16463 118619 726123 3918994 19112767	0 5 113 1389 12305 87773 534380 2877030 14024685	0 0 15 318 3721 31648 217963 1286483 6735985 32015397	0 0 10 205 2343 19579 132975 775953 4024139 18970156	0 0 10 205 2358 19897 136691 807488 4240713 20244334	0 0 0 10 205 2343 19579 132975 775953 4024139 18970156
Wt 5	6 132 1608 14200 101160 615416 3310937 16126192	4 98 1268 11637 85156 528416 2886480 14229457	4 93 1172 10594 76804 473805 2578732 12684672 57364888	1 36 560 5748 45413 298176 1701606 8687956 40490318	1 32 472 4704 36521 237276 1345648 6847116 31863140	0 5 124 1608 14728 107363 663148 3604728 17682504	0 4 92 1147 10280 74052 454664 2465897 12098484 54611820	0 3 64 762 6620 46614 281224 1503976 7294216	0 0 10 220 2656 23176 163021 979332 5205084 25059420	0 0 3 64 762 6620 46614 281224 1503976 7294216

continued next page

Affine B8 - Level 2 - Class 0 - Page 2 of 2

	Rp 1	Rp 2	Rp 3	Rp 4	Rp 5	Rp 6	Rp 7	Rp 8	Rp 9	Rp10
Wt 6	10	10	5	5	1	1	0	0	0	0
	215	210	123	114	36	31	4	3	3	0
	2548	2456	1583	1409	561	451	99	68	68	3
	21927	20967	14413	12527	5772	4467	1297	849	853	68
	152662	145244	104592	89581	45699	34589	12031	7637	7736	849
	910188	863165	643825	546377	300558	224523	88777	55148	56444	7637
	4810675	4552748	3490677	2945520	1717416	1273366	554391	338841	350841	55148
	23065943	21801521	17089930	14372460	8777277	6482794	3042861	1836197	1924523	338841
				64550190	40936780	30192112	15053988	8990365	9540289	1836197
									43468541	8990365
Wt 7	20	15	15	6	6	1	1	1	0	0
	395	312	298	143	129	34	28	22	3	1
	4404	3604	3352	1808	1544	519	393	279	72	22
	36175	30374	27792	16281	13462	5312	3846	2580	947	279
	242720	207799	188171	117232	95036	42076	29702	19212	8900	2580
	1404051	1220468	1097840	717334	574450	277440	193068	121726	66661	19212
	7235938	6367044	5703026	3870412	3076520	1590505	1098424	679432	422342	121726
	33957920	30180260	26962140	18871711	14936895	8156252	5613848	3421048	2349073	679432
							26252498	15806612	11761552	3421048
										15806612
Wt 8	0	0	0	0	0	0	0	1	0	0
	70	56	56	28	28	8	8	8	1	0
	1176	966	924	504	456	156	129	108	24	8
	11872	9968	9296	5440	4672	1832	1400	1000	336	100
	90440	77308	70918	43928	36512	15937	11628	7846	3368	1000
	571648	495824	450032	291844	237700	112920	80264	51928	26784	7802
	3147872	2763328	2490724	1676984	1348440	685780	480096	302528	178928	51928
	15562652	13798816	12381808	8599558	6860225	3688784	2561144	1581768	1042396	302336
						17972064	12429508	7560917	5434688	1581768
										7560191
Wt 9	35	35	21	21	7	7	1	1	1	0
	651	637	413	385	155	135	30	23	23	1
	6979	6741	4627	4145	1899	1544	440	306	307	23
	55622	53284	38206	33392	16841	13156	4457	2936	2966	306
	364161	346997	257455	221528	120286	91741	35313	22483	22923	2936
	2063600	1959490	1494175	1272899	732674	550779	233883	145440	149890	22483
	10448683	9899136	7718812	6534428	3942553	2938742	1348773	824295	859473	145440
				30600002	19191668	14239910	6960202	4196936	4429275	824295
									20883122	4196936
Wt10	70	56	56	28	28	8	8	8	1	1
	1176	966	924	504	456	156	129	100	24	8
	11872	9968	9296	5440	4672	1832	1400	1000	336	108
	90440	77308	70918	43928	36512	15937	11628	7802	3368	1000
	571648	495824	450032	291844	237700	112920	80264	51928	26784	7846
	3147872	2763328	2490724	1676984	1348440	685780	480096	302336	178928	51928
	15562652	13798816	12381808	8599558	6860225	3688784	2561144	1581768	1042396	302528
						17972064	12429508	7560191	5434688	1581768
										7560917

Affine B8 - Level 2 - Class 1

	Rp 1 0 0000001 1					Rp 2 1 0000001 0				
Wt 1	1	257770964	853810195*	427864205*	27413283*	0	124505880	700664219*	568960973*	800682125*
	44	1052316364*	194507021*	1032882360*	769377353*	8	524871424	266184294*	126182823*	985518320*
	726	832753459*	63843948*	629875865*	106884840*	192	1005750217*	640722352*	1052120647*	300991854*
	7704	882071914*	13737667*	749356204*	229132687*	2464	287029279*	41383802*	638171282*	658189773*
	62337	377995504*	792978487*	535047123*		22528	1063503217*	359121363*	970525679*	645445101*
	417140	427920852*	898550447*	212721456*		164560	45655873*	649978026*	401934820*	
	2418710	766484293*	902828249*	445772448*		1020416	509756845*	792974395*	210226483*	
	12520744	754353614*	222215972*	357606660*		5573568	942242413*	16760775*	870369686*	
	59068372	1006623135*	996958226*	828824828*		27484160	1005972877*	932644781*	661995912*	
Wt 2	8	524871424	266184294*	126182823*	985518320*	1	257770964	853810195*	427864205*	27413283*
	192	1005750217*	640722352*	1052120647*	300991854*	44	1052316364*	194507021*	1032882360*	769377353*
	2464	287029279*	41383802*	638171282*	658189773*	726	832753459*	63843948*	629875865*	106884840*
	22528	1063503217*	359121363*	970525679*	645445101*	7704	882071914*	13737667*	749356204*	229132687*
	164560	45655873*	649978026*	401934820*		62337	377995504*	792978487*	535047123*	
	1020416	509756845*	792974395*	210226483*		417140	427920852*	898550447*	212721456*	
	5573568	942242413*	16760775*	870369686*		2418710	766484293*	902828249*	445772448*	
	27484160	1005972877*	932644781*	661995912*		12520744	754353614*	222215972*	357606660*	
	124505880	700664219*	568960973*	800682125*		59068372	1006623135*	996958226*	828824828*	

* Multiplicity is given modulo 1073741527

Weight Multiplicities for Affine C2

Affine C2 - Level 1 - Class 0

	Rp 1					Rp 2				
	001					100				
Wt 1	1	647	44556	1318113	24371948	0	484	36099	1107140	20913123
	2	1110	66963	1858429	32993387	1	838	54568	1565324	28365342
	6	1852	99803	2605943	44502870	3	1423	81683	2201077	38329786
	13	3051	147413	3635828	59815509	8	2361	121229	3078725	51609716
	29	4922	216111	5047673		18	3857	178411	4284993	69250485
	57	7849	314365	6975598		39	6191	260590	5935312	
	112	12319	454229	9596468		77	9805	377819	8183962	
	205	19131	651835	13146268		148	15314	544117	11235080	
	372	29333	929753	17934770		271	23649	778501	15359416	
Wt 2	1	838	54568	1565324	28365342	1	647	44556	1318113	24371948
	3	1423	81683	2201077	38329786	2	1110	66963	1858429	32993387
	8	2361	121229	3078725	51609716	6	1852	99803	2605943	44502870
	18	3857	178411	4284993	69250485	13	3051	147413	3635828	59815509
	39	6191	260590	5935312		29	4922	216111	5047673	
	77	9805	377819	8183962		57	7849	314365	6975598	
	148	15314	544117	11235080		112	12319	454229	9596468	
	271	23649	778501	15359416		205	19131	651835	13146268	
	484	36099	1107140	20913123		372	29333	929753	17934770	

Affine C2 - Level 1 - Class 1

	Rp 1				
	010				
Wt 1	1	538	27741	656835	10063319
	3	942	42543	942364	13841107
	8	1610	64565	1343596	18956002
	19	2694	97034	1904354	25854967
	41	4427	144519	2684008	35126329
	83	7153	213432	3762667	47542081
	161	11387	312720	5248002	64112357
	299	17884	454803	7284132	86155099

Affine C2 - Level 2 - Class 0

	Rp 1			Rp 2			Rp 3			Rp 4		
	002			020			101			200		
Wt 1	1	2387	436343	0	3477	693847	0	2952	617844	0	1084	245072
	2	5010	764506	2	7448	1223476	1	6386	1093029	0	2367	436158
	9	10227	1320760	9	15386	2125477	6	13298	1904552	2	5010	764506
	26	20198	2251058	32	30804	3642088	24	26805	3272500	7	10191	1320460
	77	38972	3790202	97	60002	6161894	76	52520	5550672	26	20198	2251058
	194	73352	6307916	264	114078	10302346	213	100359	9302291	72	38907	3789721
	477	135396	10386831	661	212223	17035667	546	187542	15415758	194	73352	6307916
	1084	245072		1556	387156		1305	343507	25280471	467	135286	10386079
Wt 2	1	3469	578615	1	5119	923192	0	4360	823368	0	1612	327624
	4	7176	1006504	4	10744	1615428	3	9261	1445499	1	3469	578615
	15	14414	1726971	18	21858	2786966	12	18944	2500656	4	7176	1006504
	44	28124	2925164	56	43120	4744496	44	37656	4268720	15	14414	1726971
	122	53598	4896310	163	82987	7979060	128	72800	7196000	44	28124	2925164
	304	99872	8104704	420	155976	13265484	346	137573	11991409	122	53598	4896310
	721	182543	13277080	1024	287336	21820840	848	254396	19766168	304	99872	8104704
	1612	327624		2336	519360		1978	461716		721	182543	13277080
Wt 3	1	4141	664804	1	6163	1062945	1	5324	949925	0	1941	377835
	5	8523	1152604	6	12849	1853388	5	11165	1661013	1	4141	664804
	18	17005	1971321	23	25930	3186609	19	22648	2863369	5	8523	1152604
	55	33019	3329334	73	50858	5408208	61	44631	4871526	18	17005	1971321
	149	62592	5557286	205	97278	9068586	172	85737	8187383	55	33019	3329334
	372	116127	9175138	525	181952	15036340	444	160994	13604395	149	62592	5557286
	872	211317		1256	333546	24670438	1071	296190	22366021	372	116127	9175138
	1941	377835		2846	600399		2443	534926		872	211317	
Wt 4	2	5010	764506	2	7448	1223476	1	6386	1093029	1	2387	436343
	7	10191	1320460	9	15386	2125477	6	13298	1904552	2	5010	764506
	26	20198	2251058	32	30804	3642088	24	26805	3272500	9	10227	1320760
	72	38907	3789721	97	60002	6161894	76	52520	5550672	26	20198	2251058
	194	73352	6307916	264	114078	10302346	213	100359	9302291	77	38972	3790202
	467	135286	10386079	661	212223	17035667	546	187542	15415758	194	73352	6307916
	1084	245072		1556	387156		1305	343507	25280471	477	135396	10386831
	2367	436158		3477	693847		2952	617844		1084	245072	

Affine C2 - Level 2 - Class 1

	Rp 1					Rp 2				
	011					110				
Wt 1	1	12230	3157789	258508230	548120855*	0	8470	2408658	206327710	672360000*
	5	24842	5374390	403531208	585756890*	2	17490	4126070	323275512	848587776*
	20	49010	9035539	625425005	946901059*	10	35002	6978730	502810130	935726648*
	65	94235	15018300	962759294	562835634*	36	68150	11664896	776616430	764823891*
	185	177087	24697480	398713013*		110	129512	19283830	117839345*	547099300*
	481	326015	40210481	90592261*		300	240840	31551450	742973643*	
	1165	589128	64854575	160464576*		752	439190	51124970	605595786*	
	2665	1046705	103679156	785786142*		1770	786814	82088400	926712959*	
	5822	1831065	164363280	75863208*		3956	1386930	130673928	846173065*	
Wt 2	2	17490	4126070	323275512	848587776*	1	12230	3157789	258508230	548120855*
	10	35002	6978730	502810130	935726648*	5	24842	5374390	403531208	585756890*
	36	68150	11664896	776616430	764823891*	20	49010	9035539	625425005	946901059*
	110	129512	19283830	117839345*	547099300*	65	94235	15018300	962759294	562835634*
	300	240840	31551450	742973643*		185	177087	24697480	398713013*	
	752	439190	51124970	605595786*		481	326015	40210481	90592261*	
	1770	786814	82088400	926712959*		1165	589128	64854575	160464576*	
	3956	1386930	130673928	846173065*		2665	1046705	103679156	785786142*	
	8470	2408658	206327710	672360000*		5822	1831065	164363280	75863208*	

* Multiplicity is given modulo 1073741527

Affine C2 - Level 3 - Class 0

	Rp 1	Rp 2	Rp 3	Rp 4	Rp 5	Rp 6
	003	021	102	120	201	300
Wt 1	1 11150	0 27080	0 17169	0 17438	0 11044	0 2954
	2 25060	2 62178	1 39675	0 40965	0 26136	0 7126
	9 54377	12 137504	6 88326	3 92425	2 59292	0 16448
	30 114675	50 294550	29 190162	21 201328	12 129856	2 36508
	97 235321	184 613094	107 397755	89 425325	54 275562	10 78396
	279 471767	580 1244270	351 810513	313 874423	190 568946	42 163462
	768 925293	1678 2467848	1027 1613731	965 1754670	596 1145892	142 332166
	1958 1779872	4486 4794592	2792 3145633	2709 3444694	1686 2257380	424 659400
	4787	11312	7101	7072	4450 4357932	1156 1281950
Wt 2	1 19076	1 47243	0 30067	0 31039	0 19767	0 5393
	4 41976	6 105817	3 67766	2 70875	1 45424	0 12583
	18 89389	31 229027	16 147576	13 156057	8 100513	1 28233
	60 185282	118 481297	68 311621	59 332804	35 215398	7 61179
	190 374380	396 984842	233 640578	214 690071	131 448463	29 128672
	533 739949	1177 1968195	716 1285116	680 1395320	418 910337	101 263445
	1417 1432603	3241 3849640	1996 2522699	1952 2758372	1217 1805797	306 526712
	3521	8326 7384578	5207 4854515	5200 5342334	3263 3508706	852 1030522
	8381	20301	12804	13025	8246	2203 1977243
Wt 3	1 21771	1 54151	1 34660	0 35745	0 22658	0 6242
	5 47639	8 120551	5 77584	3 81063	1 51799	0 14476
	21 101031	38 259647	24 167848	17 177495	9 114081	2 32225
	72 208507	145 543100	89 352625	75 376593	41 243415	9 69485
	223 419816	474 1106895	295 721397	262 777453	154 504789	38 145345
	626 826841	1398 2203857	871 1441483	819 1565708	490 1020940	123 296382
	1644 1595974	3798 4296364	2389 2819118	2312 3084221	1419 2018423	371 590126
	4068	9684	6123 5407335	6094 5953906	3783 3909731	1006 1150731
	9609	23411	14908	15112	9504	2583 2200650
Wt 4	1 28233	2 70875	1 45424	1 47243	0 30067	0 8381
	7 61179	13 156057	8 100513	6 105817	3 67766	1 19076
	29 128672	59 332804	35 215398	31 229027	16 147576	4 41976
	101 263445	214 690071	131 448463	118 481297	68 311621	18 89389
	306 526712	680 1395320	418 910337	396 984842	233 640578	60 185282
	852 1030522	1952 2758372	1217 1805797	1177 1968195	716 1285116	190 374380
	2203 1977243	5200 5342334	3263 3508706	3241 3849640	1996 2522699	533 739949
	5393	13025	8246	8326 7384578	5207 4854515	1417 1432603
	12583	31039	19767	20301	12804	3521
Wt 5	2 32225	3 81063	1 51799	1 54151	1 34660	0 9609
	9 69485	17 177495	9 114081	8 120551	5 77584	1 21771
	38 145345	75 376593	41 243415	38 259647	24 167848	5 47639
	123 296382	262 777453	154 504789	145 543100	89 352625	21 101031
	371 590126	819 1565708	490 1020940	474 1106895	295 721397	72 208507
	1006 1150731	2312 3084221	1419 2018423	1398 2203857	871 1441483	223 419816
	2583 2200650	6094 5953906	3783 3909731	3798 4296364	2389 2819118	626 826841
	6242	15112	9504	9684	6123 5407335	1644 1595974
	14476	35745	22658	23411	14908	4068
Wt 6	2 36508	3 92425	2 59292	2 62178	1 39675	1 11150
	10 78396	21 201328	12 129856	12 137504	6 88326	2 25060
	42 163462	89 425325	54 275562	50 294550	29 190162	9 54377
	142 332166	313 874423	190 568946	184 613094	107 397755	30 114675
	424 659400	965 1754670	596 1145892	580 1244270	351 810513	97 235321
	1156 1281950	2709 3444694	1686 2257380	1678 2467848	1027 1613731	279 471767
	2954	7072	4450 4357932	4486 4794592	2792 3145633	768 925293
	7126	17438	11044	11312	7101	1958 1779872
	16448	40965	26136	27080	17169	4787

Affine C2 - Level 3 - Class 1

	Rp 1			Rp 2			Rp 3			Rp 4		
	012			030			111			210		
Wt 1	1	14397	5265114	0	10676	4242732	0	14800	6244320	0	5799	2739053
	5	33529	9927106	2	25280	8047050	2	35446	11896354	0	14212	5259292
	23	75124	18411777	12	57428	15006244	14	81312	22275692	3	33229	9918636
	86	162733	33633319	52	125878	27550172	64	179744	41051552	18	74643	18399547
	283	342193	60579039	184	267412	49853348	236	384700	74546380	76	161981	33615829
	839	700764	107690015	574	552574	89006906	756	800212	133529588	263	341028	60554197
	2300	1401401	189105558	1628	1113920	156929232	2190	1622732	236148750	803	698994	107655013
	5909	2743009		4294	2196040		5872	3216336	412676848	2235	1398736	189056548
Wt 2	1	19062	6513612	1	14309	5262242	0	19904	7756544	0	7928	3413798
	7	43873	12213531	4	33386	9922918	4	46952	14690028	1	19062	6513612
	33	97264	22536966	21	74891	18405740	24	106328	27357208	7	43873	12213531
	123	208698	40973931	82	162366	33624648	100	232420	50162880	33	97264	22536966
	398	435089	73474731	274	341629	60566724	352	492544	90667104	123	208698	40973931
	1161	884084	130074971	822	699894	107672656	1088	1015584	161700224	398	435089	73474731
	3131	1755486	227529334	2271	1400093	189081214	3064	2043320	284808560	1161	884084	130074971
	7928	3413798		5856	2741066		8040	4021216		3131	1755486	227529334
Wt 3	2	25280	8047050	1	19062	6513612	1	26632	9615496	0	10676	4242732
	12	57428	15006244	7	43873	12213531	8	61928	18106000	2	25280	8047050
	52	125878	27550172	33	97264	22536966	40	138512	33540232	12	57428	15006244
	184	267412	49853348	123	208698	40973931	156	299520	61198848	52	125878	27550172
	574	552574	89006906	398	435089	73474731	520	628740	110110544	184	267412	49853348
	1628	1113920	156929232	1161	884084	130074971	1552	1285456	195543168	574	552574	89006906
	4294	2196040		3131	1755486	227529334	4260	2566656	343050048	1628	1113920	156929232
	10676	4242732		7928	3413798		10944	5016408		4294	2196040	
Wt 4	3	33229	9918636	2	25280	8047050	2	35446	11896354	1	14397	5265114
	18	74643	18399547	12	57428	15006244	14	81312	22275692	5	33529	9927106
	76	161981	33615829	52	125878	27550172	64	179744	41051552	23	75124	18411777
	263	341028	60554197	184	267412	49853348	236	384700	74546380	86	162733	33633319
	803	698994	107655013	574	552574	89006906	756	800212	133529588	283	342193	60579039
	2235	1398736	189056548	1628	1113920	156929232	2190	1622732	236148750	839	700764	107690015
	5799	2739053		4294	2196040		5872	3216336	412676848	2300	1401401	189105558
	14212	5259292		10676	4242732		14800	6244320		5909	2743009	

Affine C2 - Level 4 - Class 0 - Page 1 of 3

	Rp 1	Rp 2	Rp 3	Rp 4	Rp 5	Rp 6	Rp 7	Rp 8	Rp 9
	004	022	040	103	121	202	220	301	400
Wt 1	1	0	0	0	0	0	0	0	0
	2	2	0	1	0	0	0	0	0
	9	12	3	6	3	2	0	0	0
	30	54	18	29	26	12	5	2	0
	102	211	87	114	128	62	34	16	3
	306	722	326	397	512	241	161	80	14
	890	2255	1090	1248	1755	840	606	317	64
	2406	6502	3280	3634	5432	2601	2005	1072	224
	6244	17618	9186	9906	15483	7486	5990	3268	719
	15422	45216	24144	25588	41406	20109	16614	9182	2066
						51290	43312	24205	5591
									14210
Wt 2	1	1	0	0	0	0	0	0	0
	4	6	2	3	2	1	0	0	0
	18	34	13	16	17	8	4	1	0
	64	140	62	74	90	42	25	12	2
	213	513	242	272	371	174	122	59	11
	638	1652	822	904	1310	620	462	242	50
	1803	4915	2524	2708	4139	1976	1554	827	177
	4772	13578	7168	7584	12012	5780	4691	2562	570
	12060	35481	19101	19942	32579	15788	13166	7267	1654
	29130	88184	48288	49954	83574	40755	34646	19368	4502
							86708	48830	11528
									28200
Wt 3	1	1	1	0	0	0	0	0	0
	4	8	4	3	4	1	1	0	0
	21	44	22	20	27	12	8	3	1
	76	184	92	93	132	58	44	20	4
	261	664	341	348	516	240	184	93	21
	784	2124	1108	1146	1768	827	664	348	76
	2222	6245	3324	3424	5446	2596	2124	1146	261
	5864	17100	9240	9504	15520	7450	6245	3424	784
	14776	44260	24262	24824	41446	20092	17100	9504	2222
			60532	61695	105004	51199	44260	24824	5864
								61695	14776
Wt 4	1	1	0	1	0	0	0	0	0
	5	8	2	5	3	1	0	0	0
	21	41	15	24	22	9	4	2	0
	76	169	72	96	110	48	29	15	2
	247	601	279	338	441	200	138	73	12
	737	1918	942	1077	1528	709	528	284	54
	2055	5627	2872	3168	4762	2247	1759	960	197
	5408	15426	8101	8715	13680	6535	5298	2925	631
	13555	39976	21462	22679	36801	17752	14776	8235	1840
			53972	56273	93774	45594	38737	21762	4992
								54576	12774

-continued next page-

Affine C2 - Level 4 - Class 0 - Page 2 of 3

	Rp 1	Rp 2	Rp 3	Rp 4	Rp 5	Rp 6	Rp 7	Rp 8	Rp 9
Wt 5	1	2	1	1	1	0	0	0	0
	7	15	7	8	9	3	2	1	0
	32	75	36	40	52	21	15	8	1
	119	299	149	162	227	99	75	40	7
	389	1031	528	565	837	380	299	162	32
	1156	3199	1676	1770	2735	1273	1031	565	119
	3191	9153	4889	5107	8156	3860	3199	1770	389
	8301	24537	13318	13791	22622	10840	9153	5107	1156
	20551	62332	34299	35254	59141	28610	24537	13791	3191
					147120	71719	62332	35254	8301
									20551
Wt 6	2	3	1	1	1	1	0	0	0
	9	19	8	9	11	5	3	1	0
	41	92	43	46	61	30	19	9	2
	142	354	173	186	266	124	92	46	9
	459	1202	612	645	967	461	354	186	41
	1330	3674	1918	2011	3135	1492	1202	645	142
	3643	10415	5557	5766	9267	4462	3674	2011	459
	9364	27684	15019	15490	25538	12351	10415	5766	1330
	23047	69883	38467	39399	66348	32332	27684	15490	3643
						80410	69883	39399	9364
									23047
Wt 7	2	4	2	1	2	1	1	0	0
	11	25	13	12	17	8	6	3	1
	50	122	62	59	90	42	34	16	4
	177	462	242	242	371	174	140	74	18
	570	1554	822	827	1310	620	513	272	64
	1654	4691	2524	2562	4139	1976	1652	904	213
	4502	13166	7168	7267	12012	5780	4915	2708	638
	11528	34646	19101	19368	32579	15788	13578	7584	1803
	28200	86708	48288	48830	83574	40755	35481	19942	4772
							88184	49954	12060
									29130
Wt 8	2	4	2	2	3	1	1	1	0
	12	29	15	15	22	9	8	5	1
	54	138	72	73	110	48	41	24	5
	197	528	279	284	441	200	169	96	21
	631	1759	942	960	1528	709	601	338	76
	1840	5298	2872	2925	4762	2247	1918	1077	247
	4992	14776	8101	8235	13680	6535	5627	3168	737
	12774	38737	21462	21762	36801	17752	15426	8715	2055
			53972	54576	93774	45594	39976	22679	5408
								56273	13555

continued next page

Affine C2 - Level 4 - Class 0 - Page 3 of 3

	Rp 1	Rp 2	Rp 3	Rp 4	Rp 5	Rp 6	Rp 7	Rp 8	Rp 9
Wt 9	3	5	3	2	3	2	2	1	1
	14	34	18	16	26	12	12	6	2
	64	161	87	80	128	62	54	29	9
	224	606	326	317	512	241	211	114	30
	719	2005	1090	1072	1755	840	722	397	102
	2066	5990	3280	3268	5432	2601	2255	1248	306
	5591	16614	9186	9182	15483	7486	6502	3634	890
	14210	43312	24144	24205	41406	20109	17618	9906	2406
						51290	45216	25588	6244
									15422

Affine C2 - Level 4 - Class 1

	Rp 1		Rp 2		Rp 3		Rp 4		Rp 5		Rp 6	
	013		031		112		130		211		310	
Wt 1	1	51199	0	61448	0	63220	0	38106	0	38694	0	11924
	5	121895	2	149628	2	155190	0	95246	0	97770	0	30868
	23	280023	15	350276	14	365930	4	227910	3	236128	0	76040
	90	623249	73	792244	70	833020	28	525264	24	548557	4	179574
	313	1348456	291	1738149	282	1838308	134	1171384	121	1231902	26	408934
	989	2844128	1004	3711190	988	3945852	512	2537312	482	2684923	120	902036
	2896	5861829	3123	7732398	3114	8261074	1712	5354708	1654	5697568	448	1934406
	7946	11829751	8963	15757418	9044	16909252	5174	11038290	5103	11803538	1464	4044894
	20648		24110		24578		14484		14516	23926346	4336	8267618
Wt 2	1	77897	1	96023	0	98788	0	61028	0	62409	0	19816
	7	182916	6	229292	4	237988	2	148840	1	153708	0	49670
	36	414791	34	527651	28	551866	14	348848	11	363282	2	119150
	145	912115	147	1175460	132	1237662	70	789712	61	828402	14	275212
	508	1951404	542	2544123	510	2694750	282	1733770	260	1830398	68	614974
	1593	4073126	1772	5365960	1718	5714046	982	3703748	937	3932558	262	1334488
	4605	8313768	5296	11056576	5240	11830722	3072	7719990	3006	8239064	880	2820914
	12452		14714	22303945	14794	23970560	8856	15737032	8823	16873359	2672	5823830
	31877		38538		39236		23892		24149		7508	11768362
Wt 3	2	97268	1	119904	1	123853	0	76411	0	78345	0	24876
	12	226039	9	283516	8	295201	3	184520	2	190964	0	61736
	55	508018	49	646844	45	678264	20	428727	17	447277	3	146802
	209	1108419	205	1430046	194	1508948	98	963072	88	1011838	20	336456
	700	2355048	735	3074181	711	3262107	383	2099809	359	2219895	93	746582
	2130	4885384	2349	6444389	2314	6873222	1304	4457812	1257	4739049	349	1609874
	6016	9916405	6885	13205299	6886	14149379	4004	9239347	3943	9871570	1151	3383540
	15974		18833		19072	28519586	11365	18737231	11373	20110624	3438	6948771
	40293		48672		49816		30255		30684		9533	13973926
Wt 4	2	119150	2	148840	1	153708	1	96023	0	98788	0	31877
	14	275212	14	348848	11	363282	6	229292	4	237988	1	77897
	68	614974	70	789712	61	828402	34	527651	28	551866	7	182916
	262	1334488	282	1733770	260	1830398	147	1175460	132	1237662	36	414791
	880	2820914	982	3703748	937	3932558	542	2544123	510	2694750	145	912115
	2672	5823830	3072	7719990	3006	8239064	1772	5365960	1718	5714046	508	1951404
	7508	11768362	8856	15737032	8823	16873359	5296	11056576	5240	11830722	1593	4073126
	19816		23892		24149		14714	22303945	14794	23970560	4605	8313768
	49670		61028		62409		38538		39236		12452	
Wt 5	3	146802	3	184520	2	190964	1	119904	1	123853	0	40293
	20	336456	20	428727	17	447277	9	283516	8	295201	2	97268
	93	746582	98	963072	88	1011838	49	646844	45	678264	12	226039
	349	1609874	383	2099809	359	2219895	205	1430046	194	1508948	55	508018
	1151	3383540	1304	4457812	1257	4739049	735	3074181	711	3262107	209	1108419
	3438	6948771	4004	9239347	3943	9871570	2349	6444389	2314	6873222	700	2355048
	9533	13973926	11365	18737231	11373	20110624	6885	13205299	6886	14149379	2130	4885384
	24876		30255		30684		18833		19072	28519586	6016	9916405
	61736		76411		78345		48672		49816		15974	
Wt 6	4	179574	4	227910	3	236128	2	149628	2	155190	1	51199
	26	408934	28	525264	24	548557	15	350276	14	365930	5	121895
	120	902036	134	1171384	121	1231902	73	792244	70	833020	23	280023
	448	1934406	512	2537312	482	2684923	291	1738149	282	1838308	90	623249
	1464	4044894	1712	5354708	1654	5697568	1004	3711190	988	3945852	313	1348456
	4336	8267618	5174	11038290	5103	11803538	3123	7732398	3114	8261074	989	2844128
	11924		14484		14516	23926346	8963	15757418	9044	16909252	2896	5861829
	30868		38106		38694		24110		24578		7946	11829751
	76040		95246		97770		61448		63220		20648	

Affine C2 - Level 5 - Class 0 - Page 1 of 3

	Rp 1	Rp 2	Rp 3	Rp 4	Rp 5	Rp 6	Rp 7	Rp 8	Rp 9	Rp10	Rp11	Rp12
	005	023	041	104	122	140	203	221	302	320	401	500
Wt 1	1	0	0	0	0	0	0	0	0	0	0	0
	2	2	0	1	0	0	0	0	0	0	0	0
	9	12	3	6	3	0	2	0	0	0	0	0
	30	54	22	29	26	5	12	5	2	0	0	0
	102	216	114	114	135	39	62	42	16	6	3	0
	312	758	468	406	569	199	252	222	90	46	22	3
	924	2452	1681	1309	2064	803	914	932	386	230	114	17
	2565	7340	5414	3938	6747	2816	2974	3351	1418	922	463	80
						8890		10819	4634	3198	1635	302
											5172	1009
Wt 2	1	1	0	0	0	0	0	0	0	0	0	0
	4	6	2	3	2	0	1	0	0	0	0	0
	18	34	16	16	17	4	8	4	1	0	0	0
	64	144	84	74	96	31	42	32	12	5	2	0
	218	544	359	280	420	157	184	174	68	37	18	2
	668	1824	1310	958	1577	642	686	737	303	188	92	14
	1945	5651	4307	2979	5274	2272	2307	2692	1128	756	381	66
	5315	16287	12988	8661	16172	7246	7091	8781	3750	2642	1350	253
									11373	8310	4307	847
												2577
Wt 3	1	1	1	0	0	0	0	0	0	0	0	0
	4	8	6	3	4	2	1	1	0	0	0	0
	21	47	35	20	31	14	12	12	3	2	1	0
	80	206	158	99	164	73	65	73	26	17	8	1
	284	785	619	388	687	311	286	340	131	91	45	7
	892	2634	2138	1344	2493	1149	1053	1317	537	385	193	35
	2630	8117	6754	4199	8112	3803	3485	4519	1891	1395	715	137
			19759	12182	24305	11572	10539	14090	6017	4535	2349	473
										13543	7094	1469
												4232
Wt 4	1	1	0	1	0	0	0	0	0	0	0	0
	5	8	2	5	3	0	1	0	0	0	0	0
	21	41	18	24	22	5	9	4	2	0	0	0
	76	173	96	96	117	35	48	36	15	6	2	0
	252	634	406	347	494	179	210	194	83	42	19	3
	768	2100	1476	1135	1817	721	778	825	350	212	99	16
	2203	6408	4815	3457	5986	2539	2595	2994	1287	841	415	76
			14438	9859	18159	8040	7924	9734	4207	2927	1472	281
										9150	4700	940
												2826
Wt 5	1	2	1	1	1	0	0	0	0	0	0	0
	7	15	9	8	9	3	3	2	1	0	0	0
	32	78	52	40	57	21	21	20	8	4	1	0
	123	324	234	169	266	109	107	114	48	27	12	2
	414	1171	891	611	1050	451	434	506	212	139	64	11
	1276	3801	3019	2000	3639	1631	1543	1898	810	558	275	53
	3652	11378	9337	6015	11483	5283	4932	6343	2727	1972	992	196
						15796	14583	19370	8414	6251	3217	661
										18348	9551	2000
												5674

continued next page

Affine C2 - Level 5 - Class 0 - Page 2 of 3

	Rp 1	Rp 2	Rp 3	Rp 4	Rp 5	Rp 6	Rp 7	Rp 8	Rp 9	Rp10	Rp11	Rp12
Wt 6	1	2	2	1	1	1	0	0	0	0	0	0
	7	17	14	8	12	6	3	4	1	1	0	0
	35	91	73	45	73	35	26	31	12	8	3	1
	137	385	311	193	340	158	131	164	65	47	20	4
	473	1395	1149	715	1317	619	537	687	286	206	99	21
	1469	4535	3803	2349	4519	2138	1891	2493	1053	785	388	80
	4232	13543	11572	7094	14090	6754	6017	8112	3485	2634	1344	284
						19759		24305	10539	8117	4199	892
											12182	2630
Wt 7	2	3	1	1	1	0	1	0	0	0	0	0
	9	19	10	9	11	3	5	3	1	0	0	0
	41	95	60	46	66	23	30	25	9	4	2	0
	146	380	265	193	307	121	133	135	54	31	15	2
	485	1349	1007	692	1192	504	519	584	239	155	78	12
	1455	4308	3380	2249	4100	1814	1781	2155	907	626	318	57
	4124	12768	10395	6713	12823	5861	5608	7122	3039	2192	1131	216
						17444		21566	9323	6928	3607	723
											10628	2194
Wt 8	2	4	3	1	2	1	1	1	0	0	0	0
	11	27	21	12	20	9	8	9	3	2	1	0
	53	139	109	64	114	52	48	57	21	15	8	1
	196	558	451	275	506	234	212	266	107	78	40	7
	661	1972	1631	992	1898	891	810	1050	434	324	169	32
	2000	6251	5283	3217	6343	3019	2727	3639	1543	1171	611	123
	5674	18348	15796	9551	19370	9337	8414	11483	4932	3801	2000	414
									14583	11378	6015	1276
												3652
Wt 9	2	4	3	2	3	1	1	1	1	0	0	0
	12	31	23	15	25	10	9	11	5	3	1	0
	57	155	121	78	135	60	54	66	30	19	9	2
	216	626	504	318	584	265	239	307	133	95	46	9
	723	2192	1814	1131	2155	1007	907	1192	519	380	193	41
	2194	6928	5861	3607	7122	3380	3039	4100	1781	1349	692	146
			17444	10628	21566	10395	9323	12823	5608	4308	2249	485
										12768	6713	1455
												4124
Wt10	2	5	4	2	4	2	1	2	1	1	0	0
	14	37	31	18	32	16	12	17	8	6	3	1
	66	188	157	92	174	84	68	96	42	34	16	4
	253	756	642	381	737	359	303	420	184	144	74	18
	847	2642	2272	1350	2692	1310	1128	1577	686	544	280	64
	2577	8310	7246	4307	8781	4307	3750	5274	2307	1824	958	218
						12988	11373	16172	7091	5651	2979	668
										16287	8661	1945
												5315

-continued next page-

Affine C2 - Level 5 - Class 0 - Page 3 of 3

	Rp 1	Rp 2	Rp 3	Rp 4	Rp 5	Rp 6	Rp 7	Rp 8	Rp 9	Rp10	Rp11	Rp12
Wt11	3	6	5	2	4	2	2	3	1	1	1	0
	16	42	35	19	36	18	15	22	9	8	5	1
	76	212	179	99	194	96	83	117	48	41	24	5
	281	841	721	415	825	406	350	494	210	173	96	21
	940	2927	2539	1472	2994	1476	1287	1817	778	634	347	76
	2826	9150	8040	4700	9734	4815	4207	5986	2595	2100	1135	252
						14438		18159	7924	6408	3457	768
											9859	2203
Wt12	3	6	5	3	5	3	2	3	2	2	1	1
	17	46	39	22	42	22	16	26	12	12	6	2
	80	230	199	114	222	114	90	135	62	54	29	9
	302	922	803	463	932	468	386	569	252	216	114	30
	1009	3198	2816	1635	3351	1681	1418	2064	914	758	406	102
			8890	5172	10819	5414	4634	6747	2974	2452	1309	312
										7340	3938	924
												2565

Affine C2 - Level 5 - Class 1 - Page 1 of 3

	Rp 1	Rp 2	Rp 3	Rp 4	Rp 5	Rp 6	Rp 7	Rp 8	Rp 9
	014	032	050	113	131	212	230	311	410
Wt 1	1	0	0	0	0	0	0	0	0
	5	2	0	2	0	0	0	0	0
	23	15	3	14	4	3	0	0	0
	90	77	21	70	34	24	6	4	0
	318	321	107	290	180	130	45	34	5
	1028	1166	430	1052	756	554	229	182	34
	3098	3818	1510	3452	2734	2033	923	764	165
	8780	11524	4779	10462	8868	6677	3232	2746	645
	23623	32550	13983	29692	26464	20149	10188	8838	2204
	60714	86979	38369	79738	73862	56774	29662	26146	6800
						151185	80974	72318	19433
									52168
Wt 2	1	1	0	0	0	0	0	0	0
	7	6	2	4	2	1	0	0	0
	36	37	13	28	18	11	4	2	0
	149	169	65	138	102	68	28	20	3
	539	662	268	560	450	315	145	112	22
	1755	2286	963	1982	1692	1222	595	486	110
	5275	7202	3119	6344	5668	4185	2126	1794	436
	14835	21031	9320	18772	17372	13045	6825	5900	1511
	39493	57765	26064	52094	49592	37739	20213	17780	4720
	100303	150607	68990	137020	133514	102713	56031	49982	13644
							147104	132694	37025
									95377

continued next page

Affine C2 - Level 5 - Class 1 - Page 2 of 3

	Rp 1	Rp 2	Rp 3	Rp 4	Rp 5	Rp 6	Rp 7	Rp 8	Rp 9
Wt 3	1	1	1	0	0	0	0	0	0
	7	8	4	4	4	1	1	0	0
	39	47	22	32	28	15	8	4	1
	167	213	95	164	144	90	47	32	7
	619	823	367	672	604	410	213	164	39
	2037	2814	1261	2384	2196	1559	823	672	167
	6158	8780	3974	7620	7184	5257	2814	2384	619
	17353	25432	11627	22468	21632	16158	8780	7620	2037
	46196	69332	32026	62084	60892	46205	25432	22468	6158
			83727	162548	162068	124486	69332	62084	17353
								162548	46196
Wt 4	2	1	0	1	0	0	0	0	0
	12	9	2	8	3	2	0	0	0
	55	52	16	45	25	17	5	3	0
	213	230	82	201	136	96	36	27	4
	734	872	337	767	585	422	184	146	28
	2308	2941	1202	2611	2155	1586	747	618	137
	6756	9082	3856	8132	7100	5307	2637	2242	538
	18619	26110	11410	23569	21468	16256	8378	7275	1844
	48767	70779	31625	64360	60585	46378	24578	21682	5713
			83025	167098	161511	124798	67563	60384	16392
								159022	44185
Wt 5	2	2	1	1	1	0	0	0	0
	14	16	7	11	9	4	2	1	0
	71	86	37	66	55	33	16	11	2
	287	368	159	299	259	172	86	66	14
	1010	1355	595	1141	1022	717	368	299	71
	3206	4466	1989	3855	3556	2579	1355	1141	287
	9404	13518	6105	11880	11241	8340	4466	3855	1010
	25864	38195	17467	34042	32917	24835	13518	11880	3206
	67454	101951	47151	91896	90542	69210	38195	34042	9404
					236305	182541	101951	91896	25864
									67454
Wt 6	3	3	1	2	1	1	0	0	0
	20	22	8	17	12	8	3	2	0
	96	115	46	93	73	51	22	17	3
	375	478	199	401	337	240	115	93	20
	1289	1725	742	1480	1307	952	478	401	96
	4014	5590	2462	4885	4475	3316	1725	1480	375
	11601	16691	7492	14792	13950	10493	5590	4885	1289
	31521	46618	21246	41798	40370	30732	16691	14792	4014
	81376	123220	56889	111554	109911	84542	46618	41798	11601
						220615	123220	111554	31521
									81376

continued next page

Affine C2 - Level 5 - Class 1 - Page 3 of 3

	Rp 1	Rp 2	Rp 3	Rp 4	Rp 5	Rp 6	Rp 7	Rp 8	Rp 9
Wt 7	3	4	2	2	2	1	1	0	0
	22	28	13	20	18	11	6	4	1
	110	145	65	112	102	68	37	28	7
	436	595	268	486	450	315	169	138	36
	1511	2126	963	1794	1692	1222	662	560	149
	4720	6825	3119	5900	5668	4185	2286	1982	539
	13644	20213	9320	17780	17372	13045	7202	6344	1755
	37025	56031	26064	49982	49592	37739	21031	18772	5275
	95377	147104	68990	132694	133514	102713	57765	52094	14835
							150607	137020	39493
									100303
Wt 8	4	5	2	3	3	2	1	1	0
	28	36	16	27	25	17	9	8	2
	137	184	82	146	136	96	52	45	12
	538	747	337	618	585	422	230	201	55
	1844	2637	1202	2242	2155	1586	872	767	213
	5713	8378	3856	7275	7100	5307	2941	2611	734
	16392	24578	11410	21682	21468	16256	9082	8132	2308
	44185	67563	31625	60384	60585	46378	26110	23569	6756
			83025	159022	161511	124798	70779	64360	18619
								167098	48767
Wt 9	5	6	3	4	4	3	2	2	1
	34	45	21	34	34	24	15	14	5
	165	229	107	182	180	130	77	70	23
	645	923	430	764	756	554	321	290	90
	2204	3232	1510	2746	2734	2033	1166	1052	318
	6800	10188	4779	8838	8868	6677	3818	3452	1028
	19433	29662	13983	26146	26464	20149	11524	10462	3098
	52168	80974	38369	72318	73862	56774	32550	29692	8780
						151185	86979	79738	23623
									60714

Weight Multiplicities for Affine C3

Affine C3 - Level 1 - Class 0

	Rp 1					Rp 2				
	0010					1000				
Wt 1	1	10733	2800049	236542618	1053970577*	0	4254	1247274	111242043	916533509*
	5	21774	4779579	370707777	887484834*	1	8805	2145646	175113880	248007396*
	19	42940	8060687	576864089	985105074*	5	17669	3644777	273650986	773033624*
	60	82593	13442049	891622437	268246606*	18	34506	6119019	424670481	877459665*
	168	155357	22181165	295528009*		55	65791	10160826	654677449	248886968*
	431	286421	36241846	1016141874*		150	122774	16699914	1002890823	
	1035	518535	58667418	1023514387*		376	224717	27183665	453303746*	
	2354	923304	94139831	489327647*		886	404140	43848484	164232668*	
	5122	1619202	149812387	736946476*		1983	715240	70125288	258986599*	
Wt 2	2	15286	3662556	296368926	185174148*	1	6211	1640621	139731471	996668620*
	9	30600	6213933	462808557	39739806*	3	12630	2803431	219141843	869722843*
	33	59598	10420641	717731802	645458161*	12	24921	4733097	341239161	31743719*
	98	113400	17285436	32010128*	57337690*	35	48028	7900590	527774066	193983675*
	267	211173	28381761	619097837*		99	90432	13049658	811005300	
	663	385851	46155861	428588923*		249	166998	21340716	164812706*	
	1557	692743	74386476	676353272*		601	302679	34575849	806593967*	
	3465	1224198	118864530	495787324*		1359	539712	55526328	691057087*	
	7413	2131776	188410935	187377805*		2967	947562	88432839	1040573160*	

* Multiplicity is given modulo 1073741527

Affine C3 - Level 2 - Class 0

	Rp 1		Rp 2		Rp 3		Rp 4		Rp 5		Rp 6	
	0002		0020		0101		0200		1010		2000	
Wt 1	1	91501	0	178479	0	205532	0	106770	0	122530	0	19358
	3	239275	3	474504	2	551228	0	291444	0	337420	0	54786
	18	600518	21	1208397	18	1414336	6	758706	4	885496	0	147228
	77	1454852	114	2963616	112	3490866	42	1896870	41	2228388	4	378066
	317	3413527	507	7030125	526	8327404	228	4575552	231	5407197	26	933908
	1133	7784948	1959	16187421	2112	19269458	966	10694034	1037	12702452	134	2228682
	3771	17300805	6783	36287337	7488	43388298	3600	24294480	3942	28991538	546	5159040
	11565	37558682	21558	79392864	24234	95308716	12012	53799660	13447	64467143	1972	11618654
	33447		63870		72756		37044		42000		6412	25526632
Wt 2	1	148233	1	292243	0	338238	0	177846	0	204938	0	32992
	6	379952	7	760318	6	886810	2	473362	1	550223	0	90770
	34	937016	49	1899401	44	2230912	18	1206382	16	1412450	2	237814
	149	2234195	238	4580286	244	5411198	106	2960140	104	3487758	12	598192
	587	5167917	1001	10701393	1058	12709672	488	7024230	511	8321891	68	1450530
	2046	11633732	3654	24307489	4004	29002820	1916	16177602	2071	19260581	286	3406736
	6580	25550417	12155	53819655	13548	64486544	6694	36271220	7414	43373220	1084	7773004
	19651		37269	116432134	42240	140037216	21380	79366784	24066	95284931	3644	17282288
	55380		107276		122920		63530		72463		11360	37527464
Wt 3	1	187701	1	372204	1	432655	0	227799	0	263874	0	42466
	8	476687	11	958385	11	1121781	3	599727	3	699799	0	115503
	45	1166101	70	2372942	73	2794472	27	1513809	28	1777773	2	299821
	203	2759660	339	5676092	364	6720639	153	3682953	163	4349799	17	747456
	785	6340608	1378	13166676	1511	15665373	684	8673027	745	10295548	91	1798534
	2715	14185469	4942	29711938	5510	35505363	2619	19838196	2903	23657416	390	4193740
	8606	30979398	16111	65396913	18205	78459364	8964	44202213	10077	52930967	1445	9507388
	25419		48680		55617	169429220	28140	96172800	32006	115600630	4814	21012956
	70810		138188		159356		82404		94653		14778	45382091
Wt 4	2	237814	2	473362	1	550223	1	292243	0	338238	0	55380
	12	598192	18	1206382	16	1412450	7	760318	6	886810	1	148233
	68	1450530	106	2960140	104	3487758	49	1899401	44	2230912	6	379952
	286	3406736	488	7024230	511	8321891	238	4580286	244	5411198	34	937016
	1084	7773004	1916	16177602	2071	19260581	1001	10701393	1058	12709672	149	2234195
	3644	17282288	6694	36271220	7414	43373220	3654	24307489	4004	29002820	587	5167917
	11360	37527464	21380	79366784	24066	95284931	12155	53819655	13548	64486544	2046	11633732
	32992		63530		72463		37269	116432134	42240	140037216	6580	25550417
	90770		177846		204938		107276		122920		19651	
Wt 5	2	299821	3	599727	3	699799	1	372204	1	432655	0	70810
	17	747456	27	1513809	28	1777773	11	958385	11	1121781	1	187701
	91	1798534	153	3682953	163	4349799	70	2372942	73	2794472	8	476687
	390	4193740	684	8673027	745	10295548	339	5676092	364	6720639	45	1166101
	1445	9507388	2619	19838196	2903	23657416	1378	13166676	1511	15665373	203	2759660
	4814	21012956	8964	44202213	10077	52930967	4942	29711938	5510	35505363	785	6340608
	14778	45382091	28140	96172800	32006	115600630	16111	65396913	18205	78459364	2715	14185469
	42466		82404		94653		48680		55617	169429220	8606	30979398
	115503		227799		263874		138188		159356		25419	
Wt 6	4	378066	6	758706	4	885496	3	474504	2	551228	1	91501
	26	933908	42	1896870	41	2228388	21	1208397	18	1414336	3	239275
	134	2228682	228	4575552	231	5407197	114	2963616	112	3490866	18	600518
	546	5159040	966	10694034	1037	12702452	507	7030125	526	8327404	77	1454852
	1972	11618654	3600	24294480	3942	28991538	1959	16187421	2112	19269458	317	3413527
	6412	25526632	12012	53799660	13447	64467143	6783	36287337	7488	43388298	1133	7784948
	19358		37044		42000		21558	79392864	24234	95308716	3771	17300805
	54786		106770		122530		63870		72756		11565	37558682
	147228		291444		337420		178479		205532		33447	

Affine C3 - Level 2 - Class 1

	Rp 1			Rp 2			Rp 3			Rp 4		
	0011			0110			1001			1100		
Wt 1	1	107428	116443829	0	120072	146580028	0	53105	69758299	0	36942	53791867
	8	292583	246613037	4	334684	313149460	1	150152	149828929	0	106618	116386607
	50	760832	512206456	36	887500	655619380	12	402995	315230866	5	291104	246522155
	244	1900435	1044841771	208	2254460	273552941*	78	1034384	650739929	41	758192	512063687
	1009	4581853	1022302570*	948	5515900	573870270*	377	2553850	246196378*	222	1895836	1044619861
	3681	10704318	918738474*	3688	13055208	40682069*	1525	6093209	486501777*	958	4573985	1021960921*
	12194	24311906	542873784*	12792	29998024	929389581*	5442	14101541	881938373*	3573	10691109	918217278*
	37371	53827448		40612	67116436		17654	31754566	367559802*	11973	24290063	542085324*
Wt 2	2	177846	169854358	1	201822	214811311	0	89928	102509752	0	63530	79366784
	18	473362	356186414	13	548158	454206821	4	247468	217864356	2	177846	169854358
	106	1206382	733053634	90	1421602	941952353	32	648996	453951116	18	473362	356186414
	488	2960140	408978889*	455	3541928	845038371*	176	1632720	928749956	106	1206382	733053634
	1916	7024230	803607332*	1901	8519173	622974574*	772	3960668	794519709*	488	2960140	408978889*
	6694	16177602	417300219*	6955	19859113	67296375*	2920	9302936	478241379*	1916	7024230	803607332*
	21380	36271220	448477936*	23019	45012751	786770423*	9904	21229816	776247566*	6694	16177602	417300219*
	63530	79366784		70395	99472320		30880	47204256		21380	36271220	448477936*
Wt 3	3	227799	204751422	2	260318	259528030	1	116890	124047129	0	82404	96172800
	27	599727	427309695	22	698456	546022312	9	317152	262277305	3	227799	204751422
	153	1513809	875532990	138	1792402	53414469*	56	821960	543900599	27	599727	427309695
	684	3682953	689878709*	660	4424774	138780588*	273	2047016	34166209*	153	1513809	875532990
	2619	8673027	275483634*	2658	10556086	267327848*	1121	4922105	72119635*	684	3682953	689878709*
	8964	19838196	388731105*	9460	24428498	376879062*	4065	11471633	83595868*	2619	8673027	275483634*
	28140	44202213		30646	55006952	194967596*	13391	25997975	997429157*	8964	19838196	388731105*
	82404	96172800		92114	120835348		40839	57446408		28140	44202213	
Wt 4	5	291104	246522155	4	334684	313149460	1	150152	149828929	1	107428	116443829
	41	758192	512063687	36	887500	655619380	12	402995	315230866	8	292583	246613037
	222	1895836	1044619861	208	2254460	273552941*	78	1034384	650739929	50	760832	512206456
	958	4573985	1021960921*	948	5515900	573870270*	377	2553850	246196378*	244	1900435	1044841771
	3573	10691109	918217278*	3688	13055208	40682069*	1525	6093209	486501777*	1009	4581853	1022302570*
	11973	24290063	542085324*	12792	29998024	929389581*	5442	14101541	881938373*	3681	10704318	918738474*
	36942	53791867		40612	67116436		17654	31754566	367559802*	12194	24311906	542873784*
	106618	116386607		120072	146580028		53105	69758299		37371	53827448	

* Multiplicity is given modulo 1073741527

Affine C3 - Level 3 - Class 0 - Page 1 of 2

	Rp 1	Rp 2	Rp 3	Rp 4	Rp 5	Rp 6	Rp 7	Rp 8	Rp 9	Rp10
	0012	0030	0111	0210	1002	1020	1101	1200	2010	3000
Wt 1	1	0	0	0	0	0	0	0	0	0
	8	3	4	0	1	0	0	0	0	0
	56	31	48	11	12	7	5	0	0	0
	321	216	364	126	96	87	68	13	9	0
	1590	1193	2124	888	581	632	527	141	106	5
	6954	5595	10352	4828	2917	3509	3041	947	740	50
	27475	23194	44148	22121	12730	16332	14534	4934	3956	319
	99690	87189	169644	89444	49848	66879	60677	21793	17802	1594
	336668	302577	599184	328268	178841	248040	228406	85371	70750	6799
						849403	791468	304823	255557	25852
										89954
Wt 2	1	1	0	0	0	0	0	0	0	0
	11	7	8	2	1	1	0	0	0	0
	84	59	88	31	19	19	13	3	2	0
	495	376	632	265	160	179	143	41	29	2
	2451	1972	3536	1643	951	1156	983	314	239	17
	10603	8915	16632	8276	4658	5983	5264	1803	1421	117
	41279	35904	68848	36010	19824	26524	23875	8587	6930	624
	147463	131725	257968	140149	75840	104714	95824	35741	29349	2813
	490504	447770	891616	499255	266511	377288	349737	134175	111674	11229
							1181871	463827	390245	40761
										136976
Wt 3	2	1	1	0	0	0	0	0	0	0
	21	13	19	5	4	3	2	0	0	0
	151	107	176	65	43	43	34	7	5	0
	847	653	1149	497	303	345	289	81	61	3
	4020	3281	6023	2861	1659	2045	1777	568	443	31
	16798	14290	27043	13683	7682	9986	8894	3066	2451	202
	63529	55775	108011	57226	31397	42427	38489	13946	11358	1030
	221488	199376	393249	215868	116426	162065	149135	56017	46291	4480
	721581	662932	1327384	749623	398940	568624	529325	204367	170892	17347
								689967	582647	61378
										201880
Wt 4	3	2	2	1	0	0	0	0	0	0
	33	24	36	15	7	8	6	2	1	0
	240	187	314	142	75	90	75	24	17	1
	1333	1091	1950	941	510	643	555	188	142	11
	6222	5286	9820	4988	2699	3541	3137	1109	871	75
	25528	22335	42666	22562	12125	16422	14822	5429	4367	404
	94832	84976	165798	90590	48271	67109	61468	23167	18969	1848
	325115	297182	589764	331077	174971	248609	230427	88930	73815	7493
				1120742	587787	850731	796359	313516	263137	27619
									872727	94210
Wt 5	3	1	2	0	1	0	0	0	0	0
	30	18	30	7	9	6	3	0	0	0
	207	144	252	90	72	65	48	11	6	0
	1122	863	1558	663	439	483	389	111	80	5
	5191	4230	7884	3720	2253	2718	2325	753	570	42
	21249	18087	34524	17415	10008	12888	11373	3942	3120	267
	79062	69463	135310	71631	39842	53541	48363	17598	14240	1312
	271949	245088	485352	266483	144891	201201	184698	69564	57345	5616
				914787	489273	696579	647754	250670	209319	21396
									706841	74873

continued next page

Affine C3 - Level 3 - Class 0 - Page 2 of 2

	Rp 1	Rp 2	Rp 3	Rp 4	Rp 5	Rp 6	Rp 7	Rp 8	Rp 9	Rp10
Wt 6	4	3	4	1	1	1	0	0	0	0
	47	34	56	21	15	14	9	3	1	0
	326	255	444	192	119	135	103	33	22	2
	1758	1437	2620	1242	725	882	739	252	182	15
	7994	6797	12764	6426	3613	4666	4064	1439	1114	101
	32166	28153	54164	28533	15668	21018	18826	6912	5506	518
	117589	105467	206752	112789	60861	84223	76803	28990	23646	2335
	397951	364101	725108	406973	216739	307164	284111	109824	90946	9296
				1362736		1038265	970862	382744	320991	33883
										114322
Wt 7	6	4	6	2	1	1	1	0	0	0
	63	47	78	32	18	20	17	4	3	0
	426	338	594	271	151	182	157	47	36	2
	2243	1862	3408	1666	918	1167	1026	341	267	20
	10037	8621	16254	8355	4545	6014	5385	1899	1518	130
	39826	35152	67824	36242	19492	26607	24188	8884	7223	670
	143947	129920	255360	140777	74947	104917	96600	36573	30157	2957
	482351	443588	885336	500841	264274	377771	351562	136342	113749	11631
						1262856	1186007	469130	395289	41793
										139490
Wt 8	7	5	8	4	2	2	1	1	0	0
	77	62	104	49	26	32	23	10	6	1
	527	435	776	382	206	258	213	83	56	6
	2771	2366	4368	2232	1199	1575	1357	510	384	40
	12336	10793	20496	10846	5799	7828	6964	2633	2057	206
	48607	43491	84344	45968	24407	33831	30671	11733	9454	964
	174430	158991	313800	175471	92545	130964	120548	46855	38434	3981
	580361	537781	1076816	615597	322514	464937	432857	170819	142220	15081
								578183	486686	52699
										172564
Wt 9	10	8	12	6	2	3	3	1	1	0
	104	85	144	71	33	44	39	14	11	1
	687	578	1032	523	262	349	310	112	88	8
	3532	3049	5652	2950	1524	2061	1848	680	540	51
	15451	13644	25980	13954	7280	10035	9110	3417	2758	268
	60051	54089	105192	57973	30309	42568	39085	14945	12235	1227
	213025	195238	386220	217791	113660	162431	150676	58650	48618	5017
			1310724	754289	392299	569529	533082	210904	176695	18720
								705400	596417	64698
Wt10	12	11	16	9	4	6	4	3	2	1
	130	111	192	103	48	66	54	27	18	3
	855	744	1344	714	354	488	417	176	132	18
	4377	3859	7216	3891	1986	2739	2427	975	751	87
	18994	17034	32640	17921	9246	12957	11694	4614	3681	405
	73245	66702	130368	72999	37827	53744	49275	19456	15816	1701
	257793	238277	473136	269943	139885	201747	187099	74367	61469	6645
							653685	262398	219495	23949
										80928

Weight Multiplicities for Affine C4

Affine C4 - Level 1 - Class 0

	Rp 1				Rp 2				Rp 3			
	00001				00100				10000			
Wt 1	1	15565	8195382	140432660*	0	22458	13000980	929780279*	0	5832	4069798	679631188
	4	37988	16125356	1070595287*	3	55947	25746546	329756172*	0	15080	8165520	139584610*
	20	89265	31192109	527915885*	18	133500	50094906	860610817*	2	37148	16078834	1069354313*
	74	202552	59388530	50278102*	84	307140	95895483	84681605*	13	87753	31119330	526102906*
	256	446068	111426244	405333606*	312	684060	180817506	374960449*	60	200018	59277220	47656616*
	786	955796	206214290	705615081*	1032	1480728	336186420	648191305*	221	441721	111255710	401550937*
	2264	1998925	376781657	857134991*	3084	3123912	616903548	101868106*	720	948712	205957704	700204721*
	6084	4088420	680203676		8583	6440283	44446127*		2124	1987220	376395556	849414754*
Wt 2	1	24102	11502236	541395657*	1	35669	18338138	523343733*	0	9604	5790338	909676305
	7	57905	22443615	690619457*	8	86874	35990026	414127060*	1	24102	11502236	541395657*
	33	133976	43076134	643438454*	41	203407	69446708	704154842*	7	57905	22443615	690619457*
	127	299934	81421691	997296039*	166	460166	131920764	255913865*	33	133976	43076134	643438454*
	426	652165	151728399	589043169*	578	1009933	246973268	737622745*	127	299934	81421691	997296039*
	1293	1381544	279015615	1028564453*	1806	2157402	456131356	589288098*	426	652165	151728399	589043169*
	3635	2858682	506748082		5195	4497961	831790886		1293	1381544	279015615	1028564453*
	9604	5790338	909676305		13988	9173642	425127853*		3635	2858682	506748082	
Wt 3	2	37148	16078834	1069354313*	3	55947	25746546	329756172*	1	15565	8195382	140432660*
	13	87753	31119330	526102906*	18	133500	50094906	860610817*	4	37988	16125356	1070595287*
	60	200018	59277220	47656616*	84	307140	95895483	84681605*	20	89265	31192109	527915885*
	221	441721	111255710	401550937*	312	684060	180817506	374960449*	74	202552	59388530	50278102*
	720	948712	205957704	700204721*	1032	1480728	336186420	648191305*	256	446068	111426244	405333606*
	2124	1987220	376395556	849414754*	3084	3123912	616903548	101868106*	786	955796	206214290	705615081*
	5832	4069798	679631188		8583	6440283	44446127*		2264	1998925	376781657	857134991*
	15080	8165520	139584610*		22458	13000980	929780279*		6084	4088420	680203676	

* Multiplicity is given modulo 1073741527

Affine C4 - Level 1 - Class 1

	Rp 1					Rp 2				
	00010					01000				
Wt 1	1	72779	58555543	224163923*	425136665*	0	46522	42196022	570205890*	35682901*
	7	170534	111295142	231569961*	992484525*	2	111310	80832850	741445978*	230364062*
	35	386101	208471357	710409568*	674390079*	14	256586	152517120	81799603*	471554136*
	140	848050	385218606	70627611*	478363090*	66	572488	283740898	481255306*	771040864*
	485	1812979	702814490	128136878*		252	1240974	520960194	306556597*	810911975*
	1512	3782669	193293732*	1037543891*		840	2621486	944776938	867143684*	
	4347	7720237	111218846*	221811289*		2534	5410360	619900541*	619991372*	
	11705	15443120	762867663*	12978776*		7084	10932740	855652426*	1476855*	
	29862	30327570	515037549*	1018014610*		18622	21669546	975568140*	223190601*	
Wt 2	2	111310	80832850	741445978*	230364062*	1	72779	58555543	224163923*	425136665*
	14	256586	152517120	81799603*	471554136*	7	170534	111295142	231569961*	992484525*
	66	572488	283740898	481255306*	771040864*	35	386101	208471357	710409568*	674390079*
	252	1240974	520960194	306556597*	810911975*	140	848050	385218606	70627611*	478363090*
	840	2621486	944776938	867143684*		485	1812979	702814490	128136878*	
	2534	5410360	619900541*	619991372*		1512	3782669	193293732*	1037543891*	
	7084	10932740	855652426*	1476855*		4347	7720237	111218846*	221811289*	
	18622	21669546	975568140*	223190601*		11705	15443120	762867663*	12978776*	
	46522	42196022	570205890*	35682901*		29862	30327570	515037549*	1018014610*	

* Multiplicity is given modulo 1073741527

Affine C4 - Level 2 - Class 0 - Page 1 of 3

	Rp 1 00002	Rp 2 00020	Rp 3 00101	Rp 4 00200	Rp 5 01010	Rp 6 02000	Rp 7 10001	Rp 8 10100	Rp 9 20000
Wt 1	1	0	0	0	0	0	0	0	0
	4	4	3	0	0	0	0	0	0
	30	38	36	12	11	0	2	0	0
	172	280	304	126	156	17	32	16	0
	930	1684	1948	952	1258	188	307	212	10
	4400	8664	10455	5536	7704	1360	2036	1612	96
	18992	39341	48772	27396	39172	7640	10976	9374	671
	74888	161364	204066	119299	174248	36407	50780	45624	3620
	274325	608414	780884	470996	698324	153508	209878	195550	16804
	941048	2136624	2774881	1714724	2573648	588598	792320	759208	69160
						2088936	2778250	2722361	260084
									907256
Wt 2	1	1	0	0	0	0	0	0	0
	8	10	9	3	2	0	0	0	0
	61	96	96	42	47	6	8	4	0
	366	656	745	364	464	71	104	76	4
	1929	3745	4442	2356	3204	572	813	651	40
	8904	18276	22462	12554	17714	3444	4828	4146	298
	37181	79401	99660	57978	83847	17416	23977	21512	1697
	142262	313192	399822	239739	352904	76950	104522	97084	8256
	506363	1141831	1476024	907008	1353444	306964	411897	392923	35380
	1692872	3893322	5085544	3188348	4809792	1126280	1496184	1459164	137754
						3857074		5045151	495281
									1667324
Wt 3	1	1	1	0	0	0	0	0	0
	11	16	17	5	6	0	1	0	0
	86	147	167	70	91	9	20	10	0
	531	998	1184	577	787	115	196	134	5
	2762	5529	6755	3586	5047	879	1351	1059	60
	12599	26376	32923	18482	26603	5143	7482	6344	438
	51761	112079	142217	83087	121661	25243	35498	31703	2469
	195182	434007	557959	336012	498747	108996	149767	138867	11758
	685051	1556546	2022671	1247769	1872840	426115	575664	549319	49513
			6861267	4317087	6540652	1537339	2049930	2001508	189513
								6811641	671427
Wt 4	2	2	1	1	0	0	0	0	0
	17	27	26	12	12	2	1	1	0
	134	234	255	132	156	27	32	26	2
	787	1520	1780	952	1263	234	302	255	17
	4010	8152	9902	5562	7704	1520	2036	1780	134
	17850	37866	47198	27396	39192	8152	10955	9902	787
	72024	157436	199765	119390	174248	37866	50780	47198	4010
	267054	598481	770046	470996	698390	157436	209808	199765	17850
	924152	2112780	2748490	1715005	2573648	598481	792320	770046	72024
				5836632	8846548	2112780	2778039	2748490	267054
									924152

continued next page

Affine C4 - Level 2 - Class 0 - Page 2 of 3

	Rp 1	Rp 2	Rp 3	Rp 4	Rp 5	Rp 6	Rp 7	Rp 8	Rp 9
Wt 5	2	3	3	1	1	0	0	0	0
	25	43	48	21	26	3	5	3	0
	193	363	426	211	281	43	66	48	2
	1148	2291	2777	1481	2053	363	531	426	25
	5731	11939	14809	8306	11828	2291	3250	2777	193
	25139	54130	68323	39780	57775	11939	16567	14809	1148
	99688	220390	282044	169079	249341	54130	73857	68323	5731
	364249	822954	1065139	653805	976102	220390	296704	282044	25139
	1243185	2860570	3737614	2339845	3529098	822954	1095791	1065139	99688
					11938265	2860570	3773873	3737614	364249
									1243185
Wt 6	4	6	4	3	2	1	0	0	0
	40	71	76	42	47	10	8	9	1
	298	572	651	364	464	96	104	96	8
	1697	3444	4146	2356	3204	656	813	745	61
	8256	17416	21512	12554	17714	3745	4828	4442	366
	35380	76950	97084	57978	83847	18276	23977	22462	1929
	137754	306964	392923	239739	352904	79401	104522	99660	8904
	495281	1126280	1459164	907008	1353444	313192	411897	399822	37181
	1667324	3857074	5045151	3188348	4809792	1141831	1496184	1476024	142262
						3893322		5085544	506363
									1692872
Wt 7	3	5	6	2	3	0	1	0	0
	38	70	84	37	52	5	14	6	0
	289	566	690	344	485	70	130	84	3
	1682	3444	4269	2290	3274	566	904	690	38
	8217	17420	21914	12358	17925	3444	5109	4269	289
	35342	77054	98179	57447	84424	17420	24764	21914	1682
	137717	307301	395888	238388	354385	77054	106575	98179	8217
	495486	1127506	1466446	903758	1357030	307301	416944	395888	35342
	1668159	3860389	5062732	3180866	4818098	1127506	1508012	1466446	137717
						3860389	5105548	5062732	495486
									1668159
Wt 8	5	9	10	5	6	1	1	1	0
	60	115	134	70	91	16	20	17	1
	438	879	1059	577	787	147	196	167	11
	2469	5143	6344	3586	5047	998	1351	1184	86
	11758	25243	31703	18482	26603	5529	7482	6755	531
	49513	108996	138867	83087	121661	26376	35498	32923	2762
	189513	426115	549319	336012	498747	112079	149767	142217	12599
	671427	1537339	2001508	1247769	1872840	434007	575664	557959	51761
			6811641	4317087	6540652	1556546	2049930	2022671	195182
								6861267	685051

--continued next page-

Affine C4 - Level 2 - Class 0 - Page 3 of 3

	Rp 1	Rp 2	Rp 3	Rp 4	Rp 5	Rp 6	Rp 7	Rp 8	Rp 9
Wt 9	10	17	16	12	11	4	2	3	1
	96	188	212	126	156	38	32	36	4
	671	1360	1612	952	1258	280	307	304	30
	3620	7640	9374	5536	7704	1684	2036	1948	172
	16804	36407	45624	27396	39172	8664	10976	10455	930
	69160	153508	195550	119299	174248	39341	50780	48772	4400
	260084	588598	759208	470996	698324	161364	209878	204066	18992
	907256	2088936	2722361	1714724	2573648	608414	792320	780884	74888
						2136624	2778250	2774881	274325
									941048

Affine C4 - Level 2 - Class 1 - Page 1 of 2

	Rp 1 00011	Rp 2 00110	Rp 3 01001	Rp 4 01100	Rp 5 10010	Rp 6 11000
Wt 1	1	0	0	0	0	0
	11	6	2	0	0	0
	94	78	34	15	6	0
	629	642	320	193	94	14
	3515	4040	2163	1502	794	166
	17025	21148	11861	8923	4960	1220
	73569	96518	55928	44369	25518	6922
	289535	395758	235015	193697	114280	33154
	1054107	1487908	900773	764385	460090	140286
	3592369	5204874	3200913	2780109	1700874	539168
	11566296	17126864	10672113	9448007	5859702	1916948
	35440117	53460512	33688587	30308841	19018192	6387254
	103954086	159352478	101401386	92500242	58632924	20137668
	293310164	456012886	292677944	270231433	172824772	60520420
	799263160	184577531*	813806396	759458916	489574066	174390350
	1036778508*	139108497*	40743900*	987919829*	264773029*	484075690
	47133836*	121403564*	339400796*	55661909*	323249205*	225689553*
					527949658*	162881943*
Wt 2	2	1	0	0	0	0
	26	21	8	4	1	0
	220	221	104	66	29	5
	1414	1606	836	590	297	65
	7540	9263	5102	3856	2082	520
	34936	45372	25954	20570	11602	3153
	145074	196444	115496	94868	55183	15946
	551280	771450	463278	391364	233008	70586
	1946148	2797388	1708764	1476870	895693	281806
	6454764	9488466	5878388	5178428	3188639	1035079
	20288306	30400136	19060714	17066040	10643697	3547569
	60846132	92699635	58726524	53325424	33625066	11464706
	175075428	270654624	173024694	159061578	101263841	35217286
	485481260	760334312	489990046	455403276	292388193	103479320
	228509753*	989689401*	265618147*	183330909*	813211120	292324691
	168426717*	59164892*	324929929*	136614715*	39547914*	797264813
		994083392*	531227368*	116513924*	337046559*	1032811379*
						39408779*
Wt 3	3	2	1	0	0	0
	41	38	19	8	4	0
	339	366	193	115	60	8
	2122	2512	1382	953	518	104
	11017	13898	7899	5915	3326	800
	49868	65944	38460	30394	17562	4696
	202919	278374	165789	136178	80448	23100
	757679	1070762	648861	548903	330286	99908
	2634275	3815832	2346426	2032096	1241868	391102
	8621116	12752346	7940743	7011141	4341974	1412582
	26779813	40335536	25392104	22787913	14275966	4771482
	79476843	121616906	77298655	70350185	44519904	15225216
	226544515	351552992	225355759	207624357	132572758	46246886
	622906303	978807750	632249418	588842794	379007936	134536932
	584349843*	487530236*	640069671*	537922511*	1044840374*	376670262
	1064509363*	438463448*	208436075*	1050844797*	640224182*	1019020614
			766563344*	262347731*	778184480*	525263962*
						373274036*

continued next page-

* Multiplicity is given modulo 1073741527

Affine C4 - Level 2 - Class 1 - Page 2 of 2

	Rp 1	Rp 2	Rp 3	Rp 4	Rp 5	Rp 6
Wt 4	5	4	1	1	0	0
	65	66	29	21	8	2
	520	590	297	221	104	26
	3153	3856	2082	1606	836	220
	15946	20570	11602	9263	5102	1414
	70586	94868	55183	45372	25954	7540
	281806	391364	233008	196444	115496	34936
	1035079	1476870	895693	771450	463278	145074
	3547569	5178428	3188639	2797388	1708764	551280
	11464706	17066040	10643697	9488466	5878388	1946148
	35217286	53325424	33625066	30400136	19060714	6454764
	103479320	159061578	101263841	92699635	58726524	20288306
	292324691	455403276	292388193	270654624	173024694	60846132
	797264813	183330909*	813211120	760334312	489990046	175075428
	1032811379*	136614715*	39547914*	989689401*	265618147*	485481260
	39408779*	116513924*	337046559*	59164892*	324929929*	228509753*
				994083392*	531227368*	168426717*
Wt 5	8	8	4	2	1	0
	104	115	60	38	19	3
	800	953	518	366	193	41
	4696	5915	3326	2512	1382	339
	23100	30394	17562	13898	7899	2122
	99908	136178	80448	65944	38460	11017
	391102	548903	330286	278374	165789	49868
	1412582	2032096	1241868	1070762	648861	202919
	4771482	7011141	4341974	3815832	2346426	757679
	15225216	22787913	14275966	12752346	7940743	2634275
	46246886	70350185	44519904	40335536	25392104	8621116
	134536932	207624357	132572758	121616906	77298655	26779813
	376670262	588842794	379007936	351552992	225355759	79476843
	1019020614	537922511*	1044840374*	978807750	632249418	226544515
	525263962*	1050844797*	640224182*	487530236*	640069671*	622906303
	373274036*	262347731*	778184480*	438463448*	208436075*	584349843*
					766563344*	1064509363*
Wt 6	14	15	6	6	2	1
	166	193	94	78	34	11
	1220	1502	794	642	320	94
	6922	8923	4960	4040	2163	629
	33154	44369	25518	21148	11861	3515
	140286	193697	114280	96518	55928	17025
	539168	764385	460090	395758	235015	73569
	1916948	2780109	1700874	1487908	900773	289535
	6387254	9448007	5859702	5204874	3200913	1054107
	20137668	30308841	19018192	17126864	10672113	3592369
	60520420	92500242	58632924	53460512	33688587	11566296
	174390350	270231433	172824772	159352478	101401386	35440117
	484075690	759458916	489574066	456012886	292677944	103954086
	225689553*	987919829*	264773029*	184577531*	813806396	293310164
	162881943*	55661909*	323249205*	139108497*	40743900*	799263160
			527949658*	121403564*	339400796*	1036778508*
						47133836*

* Multiplicity is given modulo 1073741527

Affine C5 - Level 1 - Class 0

	Rp 1 000010			
Wt 1	1	127893	165801000	898684305*
	9	355441	359463102	806996005*
	56	944188	764261878	21487343*
	273	2411027	522080208*	629607914*
	1133	5945625	55489172*	895152225*
	4173	14212763	181217034*	16225250*
	14013	33037234	309940289*	1051378197*
	43665	74871342	157027138*	
Wt 2	2	212178	244404713	980312982*
	19	577664	524773802	722115281*
	113	1506540	32008338*	1070971006*
	528	3784006	142139059*	353186972*
	2107	9193179	369713110*	545881437*
	7498	21679938	770587787*	1053019890*
	24430	49774337	265368765*	
	74145	111527958	719834772*	
Wt 3	5	348885	358555360	756055952*
	41	931065	762701450	1014366637*
	230	2385560	519432448*	502231435*
	1015	5897173	51045354*	695916975*
	3885	14122730	173841693*	780487887*
	13335	32872720	297817726*	
	42171	74576320	137295082*	
	124695	165279495	866843268*	

	Rp 2 001000			
Wt 1	0	99822	147189586	20917435*
	3	284590	322068143	849724859*
	27	772314	690568429	442974720*
	158	2008529	379483584*	926816337*
	729	5032346	858045501*	13055018*
	2884	12198955	747679778*	842487530*
	10194	28711093	449693747*	575694332*
	33037	65798112	611010110*	
Wt 2	1	170449	218425076	207772128*
	11	473084	472964586	179895753*
	72	1254811	1004394981	842253838*
	359	3199245	1021154544*	299339530*
	1504	7877171	1966325*	661464230*
	5562	18801662	87373028*	418935158*
	18698	43640660	89636693*	
	58251	98764770	616812854*	
Wt 3	4	286715	322385860	868247447*
	30	776575	691116955	472470443*
	170	2017001	380421188*	973451704*
	760	5048525	859625481*	86211453*
	2970	12229525	750317795*	956496269*
	10393	28767200	454045983*	752123143*
	33505	65899960	618129866*	
	100820	147370432	32445927*	

	Rp 3 100000			
Wt 1	0	20001	36883078	181105140*
	0	59619	81999186	697002728*
	2	167931	178402550	428628493*
	18	450827	380504397	157392747*
	103	1161101	796791220	729476738*
	465	2883643	566596003*	313592034*
	1804	6934576	102579776*	261930394*
	6267	16202209	193512709*	61982117*
Wt 2	0	35578	55294527	1073348352*
	1	102352	121517707	890804570*
	9	280031	261606861	696458343*
	53	733640	552592125	226480161*
	248	1850426	73127109*	728773675*
	994	4512959	194110707*	866401228*
	3557	10680717	413433261*	398970692*
	11657	24602441	743096152*	
Wt 3	1	62000	82388070	720925809*
	5	172780	179079580	466928824*
	30	460580	381670205	218252058*
	135	1179970	798770325	825413178*
	555	2919665	569922415*	463788020*
	2017	7001425	108104024*	495399531*
	6775	16324675	202604193*	
	21105	37102624	195913766*	

* Multiplicity is given modulo 1073741527

Affine C5 - Level 2 - Class 0 - Page 1 of 3

	Rp 1	Rp 2	Rp 3	Rp 4	Rp 5	Rp 6	Rp 7	Rp 8	Rp 9	Rp10	Rp11	Rp12
	000002	000020	000101	000200	001010	002000	010001	010100	020000	100010	101000	200000
Wt 1	1	0	0	0	0	0	0	0	0	0	0	0
	5	5	4	0	0	0	0	0	0	0	0	0
	45	60	60	20	21	0	5	0	0	0	0	0
	325	560	640	280	380	44	105	55	0	15	0	0
	2200	4285	5240	2740	4000	680	1310	980	55	314	60	0
	13124	27955	35760	20588	31650	6540	11370	9967	820	3550	1024	27
	70765	159710	210448	129520	205295	47680	78609	75235	7615	28590	9860	370
	346395	816330	1099180	708540	1148633	288760	459765	466305	53530	185475	71240	3305
								2501105	313749	1029060	424920	22520
												128805
Wt 2	1	1	0	0	0	0	0	0	0	0	0	0
	10	13	12	4	3	0	0	0	0	0	0	0
	96	159	168	76	93	12	21	11	0	2	0	0
	739	1415	1692	880	1235	212	370	287	17	79	16	0
	4963	10406	13092	7512	11288	2336	3885	3432	287	1145	340	10
	29051	64861	84672	51776	80967	18692	30161	28932	2922	10566	3676	139
	152088	355086	474648	304116	488297	121828	191807	194127	22111	75233	28892	1335
		1744505	2374320	1575256	2577746	680624	1051902	1107668	137478	447178	183856	9617
									739147	2323740	1005692	57866
												302800
Wt 3	1	1	1	0	0	0	0	0	0	0	0	0
	14	21	23	7	9	0	2	0	0	0	0	0
	141	254	301	134	189	21	55	29	0	8	0	0
	1123	2260	2821	1486	2222	378	744	559	29	183	35	0
	7487	16213	20959	12159	18877	3906	6872	6021	487	2161	616	14
	43304	98748	130891	80864	128874	29946	49633	47461	4749	18077	6188	223
	222964	528339	713587	460944	749009	188154	300520	304307	34674	120944	46158	2089
			3485981	2328499	3841138	1021197	1590377	1677660	209082	688830	282940	14768
									1095772	3466481	1502550	86703
												444271
Wt 4	2	2	1	1	0	0	0	0	0	0	0	0
	22	36	36	17	18	3	2	2	0	0	0	0
	224	416	477	258	336	63	90	76	6	16	4	0
	1724	3578	4433	2554	3736	794	1212	1116	100	338	111	4
	11302	24970	32227	19729	30320	7061	10944	10650	1108	3696	1335	54
	64058	148420	196966	125806	199846	49838	76814	78170	8956	29334	11439	540
	324176	777356	1051865	695090	1128358	296685	453298	477460	59114	188492	77794	4074
				3423576	5649374	1549537	2344640	2539836	334098	1040202	449516	25672
									1674766		2289155	140090
												683356
Wt 5	2	3	3	1	1	0	0	0	0	0	0	0
	33	59	68	31	41	5	10	6	0	1	0	0
	339	678	834	443	644	113	202	159	9	46	10	0
	2647	5697	7297	4239	6476	1353	2277	2031	169	687	205	5
	17066	38676	50920	31393	49495	11504	18635	17929	1805	6587	2293	84
	95148	224091	301033	193724	312292	78122	123242	124872	14179	48415	18554	838
	472693	1147150	1565123	1040708	1706037	450740	697275	734000	90763	295892	121315	6269
					8319165	2293816	3494903	3789760	499962	1574332	679175	38676
									2450349		3372019	206919
												990246

continued next page

Affine C5 - Level 2 - Class 0 - Page 2 of 3

	Rp 1	Rp 2	Rp 3	Rp 4	Rp 5	Rp 6	Rp 7	Rp 8	Rp 9	Rp10	Rp11	Rp12
Wt 6	4	6	4	3	2	1	0	0	0	0	0	0
	54	100	111	63	76	17	16	18	2	2	1	0
	540	1108	1335	794	1116	258	338	336	36	90	36	2
	4074	8956	11439	7061	10650	2554	3696	3736	416	1212	477	22
	25672	59114	77794	49838	78170	19729	29334	30320	3578	10944	4433	224
	140090	334098	449516	296685	477460	125806	188492	199846	24970	76814	32227	1724
	683356	1674766	2289155	1549537	2539836	695090	1040202	1128358	148420	453298	196966	11302
						3423576		5649374	777356	2341640	1051865	64058
												324176
Wt 7	3	5	6	2	3	0	1	0	0	0	0	0
	52	100	124	58	87	10	29	15	0	5	0	0
	533	1115	1426	772	1188	206	421	315	16	110	20	0
	4092	9065	11878	6996	11019	2304	4118	3603	289	1325	370	8
	25835	59705	79740	49676	79754	18626	31098	29707	2956	11417	3866	136
	141079	336900	456932	296458	483502	121678	195074	197291	22368	78603	29786	1327
	687521	1685640	2315376	1549744	2560755	680720	1062402	1119041	138778	459227	187604	9675
						3377838		5617802	744688	2362966	1019510	58244
												304816
Wt 8	5	9	10	5	6	1	1	1	0	0	0	0
	84	169	205	113	159	31	46	41	3	10	3	0
	838	1805	2293	1353	2031	443	687	644	59	202	68	2
	6269	14179	18554	11504	17929	4239	6587	6476	678	2277	834	33
	38676	90763	121315	78122	124872	31393	48415	49495	5697	18635	7297	339
	206919	499962	679175	450740	734000	193724	295892	312292	38676	123242	50920	2647
	990246	2450349	3372019	2293816	3789760	1040708	1574332	1706037	224091	697275	301033	17066
								8319165	1147150	3494903	1565123	95148
												472693
Wt 9	10	17	16	12	11	4	2	3	1	0	0	0
	139	287	340	212	287	76	79	93	13	21	12	1
	1335	2922	3676	2336	3432	880	1145	1235	159	370	168	10
	9617	22111	28892	18692	28932	7512	10566	11288	1415	3885	1692	96
	57866	137478	183856	121828	194127	51776	75233	80967	10406	30161	13092	739
	302800	739147	1005692	680624	1107668	304116	447178	488297	64861	191807	84672	4963
						1575256	2323740	2577746	355086	1051902	474648	29051
									1744505		2374320	152088
Wt10	8	16	20	10	15	2	5	3	0	1	0	0
	136	289	370	206	315	58	110	87	5	29	6	0
	1327	2956	3866	2304	3603	772	1325	1188	100	421	124	3
	9675	22368	29786	18626	29707	6996	11417	11019	1115	4118	1426	52
	58244	138778	187604	121678	197291	49676	78603	79754	9065	31098	11878	533
	304816	744688	1019510	680720	1119041	296458	459227	483502	59705	195074	79740	4092
				3377838	5617802	1549744	2362966	2560755	336900	1062402	456932	25835
									1685640		2315376	141079
												687521

continued next page

Affine C5 - Level 2 - Class 0 - Page 3 of 3

	Rp 1	Rp 2	Rp 3	Rp 4	Rp 5	Rp 6	Rp 7	Rp 8	Rp 9	Rp10	Rp11	Rp12
Wt11	14	29	35	21	29	7	8	9	1	2	1	0
	223	487	616	378	559	134	183	189	21	55	23	1
	2089	4749	6188	3906	6021	1486	2161	2222	254	744	301	14
	14768	34674	46158	29946	47461	12159	18077	18877	2260	6872	2821	141
	86703	209082	282940	188154	304307	80864	120944	128874	16213	49633	20959	1123
	444271	1095772	1502550	1021197	1677660	460944	688830	749009	98748	300520	130891	7487
						2328499	3466481	3841138	528339	1590377	713587	43304
											3485981	222964
Wt12	27	55	60	44	55	20	15	21	5	5	4	1
	370	820	1024	680	980	280	314	380	60	105	60	5
	3305	7615	9860	6540	9967	2740	3550	4000	560	1310	640	45
	22520	53530	71240	47680	75235	20588	28590	31650	4285	11370	5240	325
	128805	313749	424920	288760	466305	129520	185475	205295	27955	78609	35760	2200
					2501105	708540	1029060	1148633	159710	459765	210448	13124
									816330		1099180	70765
												346395

Affine C5 - Level 2 - Class 1 - Page 1 of 3

	Rp 1	Rp 2	Rp 3	Rp 4	Rp 5	Rp 6	Rp 7	Rp 8	Rp 9
	000011	000110	001001	001100	010010	011000	100001	100100	110000
Wt 1	1	0	0	0	0	0	0	0	0
	14	8	3	0	0	0	0	0	0
	152	136	66	30	16	0	2	0	0
	1304	1464	816	526	336	56	52	29	0
	9326	11968	7223	5468	3824	952	696	559	43
	57365	80376	51100	42408	31360	9408	6286	5878	681
	311550	464304	306144	269884	207296	69496	44384	45208	6410
	1524445	2378512	1611392	1482836	1171184	423248	263110	283414	45519
	6829132	11041024	7639409	7261232	5860288	2237200	1366348	1531823	268311
	28364901	47188208	33209388	32379762	26590640	10589440	6387380	7381659	1379444
						45831320	27391438	32426182	6375448
									27025554
Wt 2	2	1	0	0	0	0	0	0	0
	34	29	12	6	2	0	0	0	0
	378	413	216	144	82	15	10	6	0
	3182	4014	2350	1796	1194	307	198	166	14
	21944	30340	18932	15738	11308	3426	2156	2048	244
	129642	191077	124370	109408	82466	27653	17092	17502	2496
	676912	1045939	701672	643254	501196	180367	110000	118504	18962
	3193688	5118694	3513854	3324784	2655430	1007453	608144	680190	118198
	13841604	22846272	15972866	15499404	12621974	4991306	2988736	3441622	636948
	55804426	94371745	66977778	66353046	54897166	22452082	13356662	15746600	3064448
						93219067	55158218	66309186	13449248
									54675592

continued next page

Affine C5 - Level 2 - Class 1 - Page 2 of 3

	Rp 1	Rp 2	Rp 3	Rp 4	Rp 5	Rp 6	Rp 7	Rp 8	Rp 9
Wt 3	3 55 609 5028 33835 195199 997311 4615124 19662009	2 54 720 6646 48210 293654 1563812 7478684 32731460	1 29 420 4106 31026 194849 1062685 5180806 23038333 94837659	0 13 273 3127 25841 172043 978185 4921644 22439093 94276296	0 8 184 2240 19320 132784 773904 3972360 18412880 78448632	0 0 30 554 5774 44372 278782 1511540 7308748 32213444	0 1 30 417 3909 28503 173752 924170 4410418 19255414	0 0 17 334 3641 28933 186590 1033219 5083807 22739485 93966904	0 0 0 24 411 4042 29664 179528 943585 4444031 19149505
Wt 4	5 89 967 7793 51251 289457 1450860 6599869 27690088	4 96 1208 10688 74936 443776 2307856 10815232 46509764	1 46 681 6533 48103 294656 1571306 7509866 32816310	1 34 542 5515 42543 270223 1483664 7263109 32383910 133543351	0 16 336 3824 31360 207296 1171184 5860288 26590640 111250592	0 4 96 1208 10688 74936 443776 2307856 10815232 46509764	0 1 49 684 6254 44293 262893 1365824 6386229 27388922	0 1 46 681 6533 48103 294656 1571306 7509866 32816310	0 0 5 89 967 7793 51251 289457 1450860 6599869 27690088
Wt 5	8 148 1568 12276 78500 432460 2120976 9466888 39064520	8 176 2072 17424 117424 673728 3413992 15655376 66092056	4 100 1260 11084 77192 454256 2349084 10954088 46902776	2 68 978 9298 68246 417432 2224436 10627236 46427980	1 43 678 6805 51865 326341 1778239 8651341 38374106 157549102	0 8 176 2072 17424 117424 673728 3413992 15655376 66092056	0 6 118 1322 10822 71714 406726 2043504 9309030 39090904	0 4 100 1260 11084 77192 454256 2349084 10954088 46902776	0 0 8 148 1568 12276 78500 432460 2120976 9466888 39064520
Wt 6	14 244 2496 18962 118198 636948 3064448 13449248 54675592	15 307 3426 27653 180367 1007453 4991306 22452082 93219067	6 166 2048 17502 118504 680190 3441622 15746600 66309186	6 144 1796 15738 109408 643254 3324784 15499404 66353046	2 82 1194 11308 82466 501196 2655430 12621974 54897166	1 29 413 4014 30340 191077 1045939 5118694 22846272 94371745	0 10 198 2156 17092 110000 608144 2988736 13356662 55158218	0 12 216 2350 18932 124370 701672 3513854 15972866 66977778	0 2 34 378 3182 21944 129642 676912 3193688 13841604 55804426

continued next page

Affine C5 - Level 2 - Class 1 - Page 3 of 3

	Rp 1	Rp 2	Rp 3	Rp 4	Rp 5	Rp 6	Rp 7	Rp 8	Rp 9
Wt 7	14	16	10	5	4	0	1	0	0
	250	320	200	135	100	16	20	10	0
	2550	3520	2250	1744	1300	320	260	200	14
	19290	28160	18446	15500	11800	3520	2455	2250	250
	119760	182640	122320	108475	84420	28160	18315	18446	2550
	643336	1016384	693980	639985	508120	182640	114439	122320	19290
	3087750	5023040	3487410	3314280	2677900	1016384	622840	693980	119760
	13527140	22556160	15888390	15467927	12689900	5023040	3033925	3487410	643336
	54918090	93538640	66723718	66264020	55090780	22556160	13487380	15888390	3087750
						93538640	55517305	66723718	13527140
									54918090
Wt 8	24	30	17	13	8	2	1	1	0
	411	554	334	273	184	54	30	29	3
	4042	5774	3641	3127	2240	720	417	420	55
	29664	44372	28933	25841	19320	6646	3909	4106	609
	179528	278782	186590	172043	132784	48210	28503	31026	5028
	943585	1511540	1033219	978185	773904	293654	173752	194849	33835
	4444031	7308748	5083807	4921644	3972360	1563812	924170	1062685	195199
	19149505	32213444	22739485	22439093	18412880	7478684	4410418	5180806	997311
			93966904	94276296	78448632	32731460	19255414	23038333	4615124
								94837659	19662009
Wt 9	43	56	29	30	16	8	2	3	1
	681	952	559	526	336	136	52	66	14
	6410	9408	5878	5468	3824	1464	696	816	152
	45519	69496	45208	42408	31360	11968	6286	7223	1304
	268311	423248	283414	269884	207296	80376	44384	51100	9326
	1379444	2237200	1531823	1482836	1171184	464304	263110	306144	57365
	6375448	10589440	7381659	7261232	5860288	2378512	1366348	1611392	311550
	27025554	45831320	32426182	32379762	26590640	11041024	6387380	7639409	1524445
						47188208	27391438	33209388	6829132
									28364901

Weight Multiplicities for Affine C6

Affine C6 - Level 1 - Class 0

	Rp 1			Rp 2			Rp 3			Rp 4		
	0000001			0000100			0010000			1000000		
Wt 1	1	635990	176716503*	0	1139139	201181383*	0	623748	48412657*	0	102959	942827050
	6	1896741	223311146*	5	3453034	180290222*	0	1954200	389547858*	0	339423	71430241*
	42	5413338	118834988*	45	9991518	882541841*	9	5812455	256712346*	0	1053564	816327455*
	223	14861302	409278311*	300	27750522	216617240*	89	16528268	308656866*	5	3106728	805229756*
	1065	39401730	465609545*	1570	74322510	316334914*	588	45178470	498636425*	51	8763855	894467251*
	4479	101241615		7095	192665301		3027	119234067	775416030*	336	23778411	1008602943*
	17228	252834929		28500	484948542		13340	304933308		1717	62326243	417015477*
	61104	615232887		104605	114632214*		52287	757990455		7488	158383857	
	202923	388079702*		356145	694167973*		187338	762320444*		29040	391381710	
Wt 2	1	1087000	835885473*	1	1982098	406368957*	0	1121060	188949804*	0	193043	380453294*
	11	3181566	841447339*	14	5875179	287945208*	3	3411703	157399617*	0	612981	162297746*
	78	8926164	960648392*	112	16664706	365881690*	36	9899761	840312609*	2	1844676	196457639*
	432	24121161	54290527*	667	45467747	602158579*	260	27553330	139708793*	23	5299624	69486101*
	2023	63032343	630035129*	3281	119831876	960511957*	1438	73908831	177995680*	160	14618922	319759302*
	8367	159807303		14064	306142801		6681	191818962	5845420*	859	38898844	305121731*
	31404	394194000		54246	760387355		27351	483251614		3896	100219587	
	108936	948283729		192289	766986282*		101555	111297945*		15633	250802062	
	353862	81840643*		635770	57339970*		348584	687727853*		57068	611260297	
Wt 3	2	1844676	196457639*	3	3411703	157399617*	1	1982098	406368957*	0	353862	81840643*
	23	5299624	69486101*	36	9899761	840312609*	14	5875179	287945208*	1	1087000	835885473*
	160	14618922	319759302*	260	27553330	139708793*	112	16664706	365881690*	11	3181566	841447339*
	859	38898844	305121731*	1438	73908831	177995680*	667	45467747	602158579*	78	8926164	960648392*
	3896	100219587		6681	191818962	5845420*	3281	119831876	960511957*	432	24121161	54290527*
	15633	250802062		27351	483251614		14064	306142801		2023	63032343	630035129*
	57068	611260297		101555	111297945*		54246	760387355		8367	159807303	
	193043	380453294*		348584	687727853*		192289	766986282*		31404	394194000	
	612981	162297746*		1121060	188949804*		635770	57339970*		108936	948283729	
Wt 4	5	3106728	805229756*	9	5812455	256712346*	5	3453034	180290222*	1	635990	176716503*
	51	8763855	894467251*	89	16528268	308656866*	45	9991518	882541841*	6	1896741	223311146*
	336	23778411	1008602943*	588	45178470	498636425*	300	27750522	216617240*	42	5413338	118834988*
	1717	62326243	417015477*	3027	119234067	775416030*	1570	74322510	316334914*	223	14861302	409278311*
	7488	158383857		13340	304933308		7095	192665301		1065	39401730	465609545*
	29040	391381710		52287	757990455		28500	484948542		4479	101241615	
	102959	942827050		187338	762320444*		104605	114632214*		17228	252834929	
	339423	71430241*		623748	48412657*		356145	694167973*		61104	615232887	
	1053564	816327455*		1954200	389547858*		1139139	201181383*		202923	388079702*	

* Multiplicity is given modulo 1073741527

Affine C6 - Level 1 - Class 1

	Rp 1				Rp 2				Rp 3			
	0000010				0001000				0100000			
Wt 1	1	441863	226630583*	932443205*	0	393756	245777313*	735947749*	0	127102	532984871	1049771140*
	11	1380060	932693073*	945494621*	4	1262292	1006724998*	863602843*	0	426338	220440401*	164151555*
	82	4099170	696349422*	647602252*	44	3831740	929290642*	201161570*	5	1343780	920825763*	686791818*
	473	11652477	109089200*	970517000*	308	11095524	767478763*	863383783*	55	4017354	673954063*	505871811*
	2295	31862045	714995647*	875589682*	1672	30828032	303728795*	134427526*	375	11473683	67439150*	293677355*
	9779	84153602	808211120*	345423778*	7680	82571628	866769913*	461434604*	1980	31481994	638582569*	736447687*
	37648	215438210	1026592438*	781868841*	31196	214031180	784219356*	182205679*	8861	83365598	669789762*	59591168*
	133474	536159081	414646341*		115192	538616320	215799276*		35156	213840485	778809325*	714493231*
Wt 2	2	774096	928972361*	306032443*	1	710382	972380885*	151780093*	0	239688	835337064	631948840*
	24	2364792	398346212*	76174122*	15	2213511	540153002*	717713500*	2	774096	928972361*	306032443*
	174	6885732	32431078*	791631416*	123	6556902	443500057*	325996906*	24	2364792	398346212*	76174122*
	968	19225128	614733638*	485025079*	740	18585633	646268941*	282061493*	174	6885732	32431078*	791631416*
	4524	51718210	951680747*	1040547081*	3657	50673470	546561029*	221726079*	968	19225128	614733638*	485025079*
	18624	134582040	432029532*	25604710*	15708	133459911	871936028*	928949242*	4524	51718210	951680747*	1040547081*
	69506	339880272	919792203*		60631	340731489	1072693356*		18624	134582040	432029532*	25604710*
	239688	835337064	631948840*		214929	845753661	378003350*		69506	339880272	919792203*	
Wt 3	5	1343780	920825763*	686791818*	4	1262292	1006724998*	863602843*	1	441863	226630583*	932443205*
	55	4017354	673954063*	505871811*	44	3831740	929290642*	201161570*	11	1380060	932693073*	945494621*
	375	11473683	67439150*	293677355*	308	11095524	767478763*	863383783*	82	4099170	696349422*	647602252*
	1980	31481994	638582569*	736447687*	1672	30828032	303728795*	134427526*	473	11652477	109089200*	970517000*
	8861	83365598	669789762*	59591168*	7680	82571628	866769913*	461434604*	2295	31862045	714995647*	875589682*
	35156	213840485	778809325*	714493231*	31196	214031180	784219356*	182205679*	9779	84153602	808211120*	345423778*
	127102	532984871	1049771140*		115192	538616320	215799276*		37648	215438210	1026592438*	781868841*
	426338	220440401*	164151555*		393756	245777313*	735947749*		133474	536159081	414646341*	

* Multiplicity is given modulo 1073741527

Affine C7 - Level 1 - Class 0

	Rp 1			Rp 2			Rp 3			Rp 4		
	00000010			00001000			00100000			10000000		
Wt 1	1	4517929	294526550*	0	4512933	485566255*	0	1924650	647704160*	0	269059	462927631*
	13	14845451	407604197*	5	15159404	586031349*	0	6715494	639834661*	0	993330	915492618*
	113	46504016	443518599*	65	48381916	562716675*	9	22131486	323632094*	0	3432453	548780123*
	754	139640495	431410239*	530	147632615	326067381*	117	69419664	567653817*	5	11212110	137650488*
	4196	403751349	299430982*	3315	432891627	73406766*	927	208506519	809196265*	65	34884809	1054091343*
	20345	54576404*		17375	151106193*		5616	602613225	364997275*	509	104001976	209291007*
	88492	909848895*		79885	134609885*		28539	608843204*		3042	298529519	469131967*
	352235	538033710*		331566	339221313*		127530	258928916*		15252	828289324	
	1301931	275938199*		1266408	582814313*		515808	169495189*		67301	81204751*	
Wt 2	2	8111818	683863512*	1	8297099	842524554*	0	3671547	491132409*	0	537933	785486904*
	29	26101158	669945735*	19	27165863	776914276*	3	12396747	514594106*	0	1904029	1063948653*
	249	80218995	36879372*	188	84770013	445325563*	45	39731079	522044542*	2	6355239	165828702*
	1614	236727344	102485281*	1331	253541198	9662153*	390	121659267	252721642*	28	20162881	654184758*
	8679	673661220	787905943*	7610	730214618	452172234*	2529	357787563	393244911*	231	61186157	1019638796*
	40700	781549589*		37354	959131394*		13560	1014918927	844551666*	1445	178493604	729566178*
	171612	664923524*		163211	119396916*		63324	639349567*		7536	502639753	
	664037	19583377*		649946	448288618*		265779	987189996*		34419	297303700*	
	2392350	535530131*		2398077	354345599*		1023660	1026670603*		141839	411816906*	
Wt 3	5	14438479	996479159*	4	15057189	453160877*	1	6872158	867702947*	0	1047955	1012097924*
	69	45513832	551496094*	58	48129380	296574294*	17	22525372	783862304*	1	3574441	746464763*
	562	137306650	682975031*	494	147029530	874618118*	161	70375279	409204494*	13	11566721	535605026*
	3471	398404678	922693238*	3167	431494051	125129341*	1118	210752965	455957536*	108	35739939	769344176*
	17890	42637360*		16841	147953755*	922123795*	6334	607747125	613833474*	693	106001082	677711393*
	80840	883802291*		78147	127670221*		30959	620281195*		3724	303074968	222249462*
	329943	482404464*		326326	324279699*		135041	283829807*		17524	838368838	
	1240512	159424596*		1251584	551289974*		537602	222572861*		74279	103053050*	
	4356559	54849890*		4473125	420285603*		1984479	758668924*		289122	509310891*	
Wt 4	14	25474351	1060473806*	14	26995983	586534674*	6	12646032	839063414*	1	1996050	129675177*
	168	78714159	832861983*	162	84361697	67418018*	63	40346880	98508144*	7	6587938	448548817*
	1274	233223004	687722426*	1232	252587587	343862240*	483	123129006	462677433*	56	20729814	144712417*
	7426	665722743	9121277*	7252	728048538	97764386*	2898	361194078	744975661*	343	62525995	1055736246*
	36568	764004810*		36211	954327509*		14889	1022602665		1869	181569269	735438154*
	158998	627013787*		159747	108982415*		67599	656270089*		8939	509525287	
	628026	1013077290*		640164	426162711*		278649	1023630375*		38801	312358251*	
	2294824	368851851*		2371621	308227673*		1059903	29845572*		154527	444046791*	
	7859698	343656509*		8228873	748053350*		3768807	650472339*		572978	853136706*	

* Multiplicity is given modulo 1073741527

Weight Multiplcities for Affine C8

Affine C8 - Level 1 - Class 0 - Page 1 of 2

	Rp 1		Rp 2		Rp 3		Rp 4		Rp 5	
	000000001		000000100		000010000		001000000		100000000	
Wt 1	1	68881396	0	139357344	0	102488940	0	37666204	0	4916900
	8	226468144	7	463610644	0	349190640	0	133129460	0	18290100
	72	714877412	84	404647945*	20	62879333*	0	447226308	0	64192084
	500	27695072*	728	243395468*	265	331075219*	28	362954877*	0	214219873
	3064	1032369125*	4928	575410944*	2280	1040118737*	386	140441972*	14	683914284
	16436	17323354*	28392	34704028*	15110	65047437*	3312	327970716*	196	1025416369*
	79552	245050183*	143892	684146587*	84380	217337346*	21695	528260355*	1680	850622933*
	351788	792844163*	658770	801415133*	414120	203861158*	119164	427859874*	10948	666927572*
	1441772	350404413*	2770908	708513944*	1836880	816318983*	574600	1036461922*	59696	350736414*
	5530020		10850308		7499440	862084317*	2504012	574022084*	285580	775519417*
	20021406		39945668		28562020		10050632		1234577	968898944*
Wt 2	1	123601212	1	253049618	0	190376890	0	72232426	0	9803960
	15	399286532	20	825441064	5	633496260	0	247974802	0	35183707
	143	166262043*	219	437980813*	85	945542443*	9	812317848	0	119798756
	1045	495048389*	1740	291584800*	835	824821965*	145	405100006*	5	389517799
	6350	45915554*	11178	262120757*	6100	94987996*	1356	211512114*	79	141929919*
	33563	313923403*	61473	235760018*	36610	57741697*	9494	72096501*	724	436109509*
	158826	613399192*	299690	42933600*	190240	872332943*	54937	869058168*	4977	980466153*
	686474	454660484*	1325763	301432724*	883845	646796293*	276598	117675460*	28337	1066482232*
	2749764	991068734*	5410817	558100755*	3750860	1021629707*	1250115	227845926*	140650	961636493*
	10321056		20625353		14763025		5178025	5573615*	627702	995266930*
	36613531		74126726		54502265		19945950		2570891	717669651*
Wt 3	2	220220770	3	455378660	1	349286548	0	136224775	0	19129286
	33	699140482	54	383717702*	24	63082756*	3	455378660	0	66479160
	316	1062905436*	548	191747862*	284	331493539*	54	383717702*	2	220220770
	2265	940459073*	4092	451347381*	2340	1040965847*	548	191747862*	33	699140482
	13396	876961603*	24955	817678311*	15305	66722221*	4092	451347381*	316	1062905436*
	68805	830850858*	131290	17825503*	84912	220602343*	24955	817678311*	2265	940459073*
	316640	776751895*	616051	379582588*	415564	210104666*	131290	17825503*	13396	876961603*
	1333137	108904667*	2635916	636601319*	1840440	828107338*	616051	379582588*	68805	830850858*
	5211820		10446770	62255427*	7508090	883988937*	2635916	636601319*	316640	776751895*
	19129286		38797562		28581800		10446770	62255427*	1333137	108904667*
	66479160		136224775		102533418		38797562		5211820	
Wt 4	5	389517799	9	812317848	5	633496260	1	253049618	0	36613531
	79	141929919*	145	405100006*	85	945542443*	20	825441064	1	123601212
	724	436109509*	1356	211512114*	835	824821965*	219	437980813*	15	399286532
	4977	980466153*	9494	72096501*	6100	94987996*	1740	291584800*	143	166262043*
	28337	1066482232*	54937	869058168*	36610	57741697*	11178	262120757*	1045	495048389*
	140650	961636493*	276598	117675460*	190240	872332943*	61473	235760018*	6350	45915554*
	627702	995266930*	1250115	227845926*	883845	646796293*	299690	42933600*	33563	313923403*
	2570891	717669651*	5178025	5573615*	3750860	1021629707*	1325763	301432724*	158826	613399192*
	9803960		19945950		14763025		5410817	558100755*	686474	454660484*
	35183707		72232426		54502265		20625353		2749764	991068734*
	119798756		247974802		190376890		74126726		10321056	

* Multiplicity is given modulo 1073741527

Affine C8 - Level 1 - Class 0 - Page 2 of 2

	Rp 1		Rp 2		Rp 3		Rp 4		Rp 5	
Wt 5	14	683914284	28	362954877*	20	62879333*	7	463610644	1	68881396
	196	1025416369*	386	140441972*	265	331075219*	84	404647945*	8	226468144
	1680	850622933*	3312	327970716*	2280	1040118737*	728	243395468*	72	714877412
	10948	666927572*	21695	528260355*	15110	65047437*	4928	575410944*	500	27695072*
	59696	350736414*	119164	427859874*	84380	217337346*	28392	34704028*	3064	1032369125*
	285580	775519417*	574600	1036461922*	414120	203861158*	143892	684146587*	16436	17323354*
	1234577	968898944*	2504012	574022084*	1836880	816318983*	658770	801415133*	79552	245050183*
	4916900		10050632		7499440	862084317*	2770908	708513944*	351788	792844163*
	18290100		37666204		28562020		10850308		1441772	350404413*
	64192084		133129460		102488940		39945668		5530020	
	214219873		447226308		349190640		139357344		20021406	

* Multiplicity is given modulo 1073741527

Affine C8 - Level 1 - Class 1

	Rp 1			Rp 2			Rp 3			Rp 4		
	000000010			000001000			000100000			010000000		
Wt 1	1	12957069	703719615*	0	13777852	520195469*	0	6779906	382247762*	0	1519012	621332905*
	15	46693590	1020528026*	6	50771470	1025067271*	0	25961691	168060873*	0	6156290	954015543*
	149	159873675	460897114*	90	177138536	34954628*	14	93549617	230379168*	0	23260388	854181863*
	1130	523102510	286283141*	838	589028766	906542864*	210	319789028	93152862*	14	82791124	687793086*
	7103	569747422*	521760476*	5940	803072111*	905649211*	1897	1043694553*	518085177*	210	279827572	522818192*
	38668	683051183*		35030	386868061*		12985	47768333*	539009886*	1870	903797516	22176301*
	187765	627058914*		180068	944754567*		73969	204331520*		12580	656198994*	830082162*
	830077	658571143*		830882	631999703*		368102	893795110*		70408	872065925*	
	3391748	806201073*		3510574	181500057*		1648731	1072494384*		344522	666584648*	
Wt 2	2	24251514	608801106*	1	26421434	554062654*	0	13513814	616688540*	0	3185453	974563153*
	34	85518174	699494310*	23	94824392	189723270*	4	50019180	1022386897*	0	12353347	834279753*
	338	287046500	328572441*	267	323188334	463350541*	72	175085210	756385171*	5	45005626	905246467*
	2508	922263914	1001367119*	2180	1052450538*	920893783*	728	583631854	691169777*	83	155339309	646604519*
	15292	702011456*	477174350*	14194	69632808*	299792298*	5410	789353995*	629593278*	791	511345845	645156577*
	80662	982621267*		78571	257420630*		32830	353029755*	268493555*	5620	540215741*	653029905*
	379960	926760438*		384138	1019458389*		171906	863518487*		32913	610959706*	
	1632906	145977172*		1700701	289352487*		803096	441726943*		167400	455579368*	
	6500966	154032423*		6938504	1040027963*		3422392	819528172*		763215	260236376*	
Wt 3	5	45005626	905246467*	4	50019180	1022386897*	1	26421434	554062654*	0	6500966	154032423*
	83	155339309	646604519*	72	175085210	756385171*	23	94824392	189723270*	2	24251514	608801106*
	791	511345845	645156577*	728	583631854	691169777*	267	323188334	463350541*	34	85518174	699494310*
	5620	540215741*	653029905*	5410	789353995*	629593278*	2180	1052450538*	920893783*	338	287046500	328572441*
	32913	610959706*		32830	353029755*	268493555*	14194	69632808*	299792298*	2508	922263914	1001367119*
	167400	455579368*		171906	863518487*		78571	257420630*		15292	702011456*	477174350*
	763215	260236376*		803096	441726943*		384138	1019458389*		80662	982621267*	
	3185453	974563153*		3422392	819528172*		1700701	289352487*		379960	926760438*	
	12353347	834279753*		13513814	616688540*		6938504	1040027963*		1632906	145977172*	
Wt 4	14	82791124	687793086*	14	93549617	230379168*	6	50771470	1025067271*	1	12957069	703719615*
	210	279827572	522818192*	210	319789028	93152862*	90	177138536	34954628*	15	46693590	1020528026*
	1870	903797516	22176301*	1897	1043694553*	518085177*	838	589028766	906542864*	149	159873675	460897114*
	12580	656198994*	830082162*	12985	47768333*	539009886*	5940	803072111*	905649211*	1130	523102510	286283141*
	70408	872065925*		73969	204331520*		35030	386868061*		7103	569747422*	521760476*
	344522	666584648*		368102	893795110*		180068	944754567*		38668	683051183*	
	1519012	621332905*		1648731	1072494384*		830882	631999703*		187765	627058914*	
	6156290	954015543*		6779906	382247762*		3510574	181500057*		830077	658571143*	
	23260388	854181863*		25961691	168060873*		13777852	520195469*		3391748	806201073*	

-- * Multiplicity is given modulo 1073741527

Weight Multiplicities for Affine D4

Affine D4 - Level 1 - Class (0,0)

	Rp 1 1 0 0 0 0				
Wt 1	1	2580	353017	18468245	563251038
	4	5180	602348	29015700	839757660
	14	10108	1014580	45235414	171642403*
	40	19208	1688400	70005376	763819433*
	105	35693	2778517	107585845	550608368*
	252	64960	4524760	164245380	721790411*
	574	116090	7296752	249162620	367612195*
	1240	203984	11658920	375704920	792686671*

* Multiplicity is given modulo 1073741527

Affine D4 - Level 2 - Class (0,0)

	Rp 1 0 0 0 0 2		Rp 2 0 2 0 0 0		Rp 3 0 0 1 0 0		Rp 4 0 0 0 2 0		Rp 5 2 0 0 0 0	
Wt 1	1	763414	0	762718	0	1222212	0	762718	0	309684
	4	1815677	3	1814342	4	2931220	3	1814342	0	762718
	26	4187244	21	4185473	28	6811584	21	4185473	3	1814342
	115	9395758	109	9392558	152	15386688	109	9392558	21	4185473
	468	20564284	448	20560028	644	33882732	448	20560028	109	9392558
	1647	44004703	1622	43997367	2392	72903316	1622	43997367	448	20560028
	5351	92233413	5285	92223653	7952	153579700	5285	92223653	1622	43997367
	16031	189681045	15946	189664865	24404	317306652	15946	189664865	5285	92223653
	45278	383283131	45083	383261621	69976	643932012	45083	383261621	15946	189664865
	121153		120898		189948		120898		45083	383261621
	310211		309684		491692		309684		120898	
Wt 2	0	762718	1	763414	0	1222212	0	762718	0	309684
	3	1814342	4	1815677	4	2931220	3	1814342	0	762718
	21	4185473	26	4187244	28	6811584	21	4185473	3	1814342
	109	9392558	115	9395758	152	15386688	109	9392558	21	4185473
	448	20560028	468	20564284	644	33882732	448	20560028	109	9392558
	1622	43997367	1647	44004703	2392	72903316	1622	43997367	448	20560028
	5285	92223653	5351	92233413	7952	153579700	5285	92223653	1622	43997367
	15946	189664865	16031	189681045	24404	317306652	15946	189664865	5285	92223653
	45083	383261621	45278	383283131	69976	643932012	45083	383261621	15946	189664865
	120898		121153		189948		120898		45083	383261621
	309684		310211		491692		309684		120898	
Wt 3	1	1181397	1	1181397	1	1900702	1	1181397	0	488405
	9	2766115	9	2766115	11	4485071	9	2766115	1	1181397
	51	6291061	51	6291061	68	10271901	51	6291061	9	2766115
	228	13938604	228	13938604	319	22902307	228	13938604	51	6291061
	871	30159110	871	30159110	1261	49838180	871	30159110	228	13938604
	2970	63859485	2970	63859485	4411	106081537	2970	63859485	871	30159110
	9281	132561368	9281	132561368	14061	221267863	9281	132561368	2970	63859485
	27039	270188736	27039	270188736	41635	452997329	27039	270188736	9281	132561368
	74339		74339		116030	911550422	74339		27039	270188736
	194608		194608		307274		194608		74339	
	488405		488405		778910		488405		194608	
Wt 4	0	762718	0	762718	0	1222212	1	763414	0	309684
	3	1814342	3	1814342	4	2931220	4	1815677	0	762718
	21	4185473	21	4185473	28	6811584	26	4187244	3	1814342
	109	9392558	109	9392558	152	15386688	115	9395758	21	4185473
	448	20560028	448	20560028	644	33882732	468	20564284	109	9392558
	1622	43997367	1622	43997367	2392	72903316	1647	44004703	448	20560028
	5285	92223653	5285	92223653	7952	153579700	5351	92233413	1622	43997367
	15946	189664865	15946	189664865	24404	317306652	16031	189681045	5285	92223653
	45083	383261621	45083	383261621	69976	643932012	45278	383283131	15946	189664865
	120898		120898		189948		121153		45083	383261621
	309684		309684		491692		310211		120898	
Wt 5	3	1814342	3	1814342	4	2931220	3	1814342	1	763414
	21	4185473	21	4185473	28	6811584	21	4185473	4	1815677
	109	9392558	109	9392558	152	15386688	109	9392558	26	4187244
	448	20560028	448	20560028	644	33882732	448	20560028	115	9395758
	1622	43997367	1622	43997367	2392	72903316	1622	43997367	468	20564284
	5285	92223653	5285	92223653	7952	153579700	5285	92223653	1647	44004703
	15946	189664865	15946	189664865	24404	317306652	15946	189664865	5351	92233413
	45083	383261621	45083	383261621	69976	643932012	45083	383261621	16031	189681045
	120898		120898		189948		120898		45278	383283131
	309684		309684		491692		309684		121153	
	762718		762718		1222212		762718		310211	

Affine D4 - Level 2 - Class (0,1)

	Rp 1 0 1 0 1 0					Rp 2 1 0 0 0 1				
Wt 1	1	289575	433386532	690357086*	549484562*	0	178660	303398921	319143487*	718685933*
	10	735870	872994595	333091397*	84999378*	3	463842	616373760	455683311*	207082978*
	62	1799620	656710004*	80254379*	1017325612*	25	1155717	157729546*	109410208*	270196007*
	293	4254859	157837384*	815702927*	767400540*	136	2777795	275003233*	440435536*	403634654*
	1165	9761770	64032037*	1024622851*		590	6467115	401868797*	1029635515*	35188030*
	4097	21799465	553703184*	48255359*		2205	14633605	393812394*	1046673712*	
	13120	47506759	660967578*	449747105*		7391	32273399	859798920*	961640303*	
	38997	101252595	111557723*	326423091*		22780	69537814	519946876*	821553187*	
	109036	211451893	635222394*	897529387*		65613	146675875	427972469*	658778865*	
Wt 2	3	463842	616373760	455683311*	207082978*	1	289575	433386532	690357086*	549484562*
	25	1155717	157729546*	109410208*	270196007*	10	735870	872994595	333091397*	84999378*
	136	2777795	275003233*	440435536*	403634654*	62	1799620	656710004*	80254379*	1017325612*
	590	6467115	401868797*	1029635515*	35188030*	293	4254859	157837384*	815702927*	767400540*
	2205	14633605	393812394*	1046673712*		1165	9761770	64032037*	1024622851*	
	7391	32273399	859798920*	961640303*		4097	21799465	553703184*	48255359*	
	22780	69537814	519946876*	821553187*		13120	47506759	660967578*	449747105*	
	65613	146675875	427972469*	658778865*		38997	101252595	111557723*	326423091*	
	178660	303398921	319143487*	718685933*		109036	211451893	635222394*	897529387*	

* Multiplicity is given modulo 1073741527

Affine D4 - Level 3 - Class (0,0) - Page 1 of 2

	Rp 1 0 1 0 1 1	Rp 2 1 0 0 0 2	Rp 3 1 2 0 0 0	Rp 4 1 0 1 0 0	Rp 5 1 0 0 2 0	Rp 6 3 0 0 0 0
Wt 1	1	0	0	0	0	0
	19	3	3	2	3	0
	175	36	36	33	36	1
	1133	270	270	269	270	15
	5900	1536	1536	1622	1536	111
	26357	7317	7317	8006	7317	631
	104917	30564	30564	34373	30564	2971
	381220	115422	115422	132534	115422	12310
	1285730	401760	401760	469273	401760	46094
	4073863	1307313	1307313	1548681	1307313	159337
	12237695	4017624	4017624	4817126	4017624	515207
	35098599	11753874	11753874	14240198	11753874	1574751
	96652264	32938056	32938056	40271860	32938056	4584642
	256707575	88856490	88856490	109523851	88856490	12792387
	660080272	231697908	231697908	287668244	231697908	34376040
	574596410*	585962532	585962532	732287156	585962532	89324681
	786879135*	367611089*	367611089*	738287423*	367611089*	225184545
						552306978
Wt 2	3	1	0	0	0	0
	42	10	6	6	6	0
	336	83	69	69	69	3
	2004	520	480	504	480	30
	9861	2700	2595	2811	2595	204
	42252	12102	11850	13194	11850	1074
	162783	48466	47892	54471	47892	4836
	576066	177196	175956	203766	175956	19302
	1900626	601215	598635	703698	598635	70275
	5910786	1915400	1910220	2274582	1910220	237216
	17471562	5782450	5772342	6950160	5772342	752100
	49406964	16658924	16639716	20231544	16639716	2259978
	134362533	46060499	46024806	56444790	46024806	6483288
	352898046	122785750	122720790	151671060	122720790	17855526
	898320357	316778075	316661985	394091541	316661985	47428611
	75363706*	793464248	793260264	993465156	793260264	121961472
		861045492*	860692475*	289121899*	860692475*	304572201
						740617950
Wt 3	3	0	1	0	0	0
	42	6	10	6	6	0
	336	69	83	69	69	3
	2004	480	520	504	480	30
	9861	2595	2700	2811	2595	204
	42252	11850	12102	13194	11850	1074
	162783	47892	48466	54471	47892	4836
	576066	175956	177196	203766	175956	19302
	1900626	598635	601215	703698	598635	70275
	5910786	1910220	1915400	2274582	1910220	237216
	17471562	5772342	5782450	6950160	5772342	752100
	49406964	16639716	16658924	20231544	16639716	2259978
	134362533	46024806	46060499	56444790	46024806	6483288
	352898046	122720790	122785750	151671060	122720790	17855526
	898320357	316661985	316778075	394091541	316661985	47428611
	75363706*	793260264	793464248	993465156	793260264	121961472
		860692475*	861045492*	289121899*	860692475*	304572201
						740617950

continued next page-

* Multiplicity is given modulo 1073741527

Affine D4 - Level 3 - Class (0,0) - Page 2 of 2

	Rp 1	Rp 2	Rp 3	Rp 4	Rp 5	Rp 6
Wt 4	7	1	1	1	1	0
	84	17	17	16	17	1
	615	143	143	145	143	9
	3442	881	881	929	881	67
	16165	4420	4420	4826	4420	382
	66781	19217	19217	21521	19217	1855
	249860	74852	74852	85615	74852	7879
	863035	267257	267257	310945	267257	30249
	2789789	888228	888228	1048514	888228	106793
	8524911	2779515	2779515	3321763	2779515	351945
	24816186	8259875	8259875	9977717	8259875	1093660
	69237177	23467148	23467148	28615311	23467148	3231364
	186047100	64083365	64083365	78796738	64083365	9135104
	483416222	168935732	168935732	209278964	168935732	24838682
	144908174*	431468775	431468775	538118800	431468775	65229802
	841428877*	1070878645*	1070878645*	270024209*	1070878645*	166033201
						410817810
Wt 5	3	0	0	0	1	0
	42	6	6	6	10	0
	336	69	69	69	83	3
	2004	480	480	504	520	30
	9861	2595	2595	2811	2700	204
	42252	11850	11850	13194	12102	1074
	162783	47892	47892	54471	48466	4836
	576066	175956	175956	203766	177196	19302
	1900626	598635	598635	703698	601215	70275
	5910786	1910220	1910220	2274582	1915400	237216
	17471562	5772342	5772342	6950160	5782450	752100
	49406964	16639716	16639716	20231544	16658924	2259978
	134362533	46024806	46024806	56444790	46060499	6483288
	352898046	122720790	122720790	151671060	122785750	17855526
	898320357	316661985	316661985	394091541	316778075	47428611
	75363706*	793260264	793260264	993465156	793464248	121961472
		860692475*	860692475*	289121899*	861045492*	304572201
						740617950
Wt 6	14	3	3	4	3	1
	160	36	36	40	36	4
	1088	270	270	291	270	26
	5784	1536	1536	1684	1536	144
	26081	7317	7317	8162	7317	717
	104292	30564	30564	34736	30564	3180
	379870	115422	115422	133347	115422	12794
	1282944	401760	401760	470996	401760	47144
	4068275	1307313	1307313	1552206	1307313	161539
	12226828	4017624	4017624	4824120	4017624	519664
	35078006	11753874	11753874	14253681	11753874	1583490
	96614088	32938056	32938056	40297208	32938056	4601336
	256638216	88856490	88856490	109570538	88856490	12823570
	659956612	231697908	231697908	287752496	231697908	34433012
	574379449*	585962532	585962532	732436560	585962532	89426892
		367611089*	367611089*	738548173*	367611089*	225364808
						552619917

* Multiplicity is given modulo 1073741527

Affine D4 - Level 4 - Class (0,0) - Page 1 of 4

	Rp 1 0 0 0 0 4	Rp 2 0 2 0 0 2	Rp 3 0 4 0 0 0	Rp 4 0 0 1 0 2	Rp 5 0 2 1 0 0	Rp 6 0 0 2 0 0	Rp 7 0 0 0 2 2	Rp 8 0 2 0 2 0
Wt 1	1 3820 4 17354 26 144 779	0 12298 3 60784 36 318 2160	0 2948 0 14842 6 64 499	0 12036 4 59925 40 320 2122	0 10596 0 55623 15 204 1686	0 7310 0 39000 12 139 1156	0 12298 3 60784 36 318 2160	0 11193 0 57524 16 226 1820
Wt 2	0 7226 3 33052 30 231 1394	1 27374 10 128337 107 818 5149	0 7226 3 33052 30 231 1394	0 25647 6 123606 78 693 4620	0 25647 6 123606 78 693 4620	0 17156 2 85327 41 412 2959	0 26774 6 126661 87 754 4933	0 26774 6 126661 87 754 4933
Wt 3	0 2948 0 14842 6 64 499	0 12298 3 60784 36 318 2160	1 3820 4 17354 26 144 779	0 10596 0 55623 15 204 1686	0 12036 4 59925 40 320 2122	0 7310 0 39000 12 139 1156	0 11193 0 57524 16 226 1820	0 12298 3 60784 36 318 2160
Wt 4	1 11177 9 67 422 2306	1 40619 17 174 1303 7874	0 10217 3 40 322 1981	1 39278 16 161 1217 7480	0 37632 8 121 1059 6942	0 25966 5 125018 77 694 4678	1 40619 17 174 1303 7874	0 39360 10 141 1177 7455
Wt 5	0 10217 3 40 322 1981	1 40619 17 174 1303 7874	1 11177 9 67 422 2306	0 37632 8 121 1059 6942	1 39278 16 161 1217 7480	0 25966 5 125018 77 694 4678	0 39360 10 141 1177 7455	1 40619 17 174 1303 7874
Wt 6	1 15431 9 84 559 3166	1 59154 26 269 1990 11732	1 15431 9 84 559 3166	1 56694 21 233 1794 10956	1 56694 21 233 1794 10956	1 39000 12 146 1156 7343	1 59154 26 269 1990 11732	1 59154 26 269 1990 11732
Wt 7	0 7226 3 33052 30 231 1394	0 26774 6 126661 87 754 4933	0 6854 1 32044 22 194 1278	0 25647 6 123606 78 693 4620	0 25006 3 121834 66 633 4428	0 17156 2 85327 41 412 2959	1 27374 10 128337 107 818 5149	0 26774 6 126661 87 754 4933
Wt 8	0 6854 1 32044 22 194 1278	0 26774 6 126661 87 754 4933	0 7226 3 33052 30 231 1394	0 25006 3 121834 66 633 4428	0 25647 6 123606 78 693 4620	0 17156 2 85327 41 412 2959	0 26774 6 126661 87 754 4933	1 27374 10 128337 107 818 5149
Wt 9	0 10217 3 40 322 1981	0 39360 10 141 1177 7455	0 10217 3 40 322 1981	0 37632 8 121 1059 6942	0 37632 8 121 1059 6942	0 25966 5 125018 77 694 4678	1 40619 17 174 1303 7874	1 40619 17 174 1303 7874

continued next page-

Affine D4 - Level 4 - Class (0,0) - Page 2 of 4

	Rp 9	Rp10	Rp11	Rp12	Rp13	Rp14	Rp15	Rp16
	0 0	0 0	1 1	2 0	2 2	2 0	2 0	4 0
	1	0	0	0	0	1	0	0
	2 0	4 0	1 1	0 2	0 0	0 0	2 0	0 0
Wt 1	0 10596	0 2948	0 13372	0 2160	0 1820	0 1686	0 1820	0 64
	0 55623	0 14842	0 73688	0 12298	0 11193	0 10596	0 11193	0 499
	15	6	14	3 60784	0 57524	0 55623	0 57524	0 2948
	204	64	224	36	16	15	16	0 14842
	1686	499	1999	318	226	204	226	6
Wt 2	0 25006	0 6854	0 32095	0 4933	0 4933	0 4428	0 4822	0 194
	3 121834	1 32044	3 163886	0 26774	0 26774	0 25006	0 26426	0 1278
	66	22	66	6 126661	6 126661	3 121834	6 125559	0 6854
	633	194	710	87	87	66	81	1 32044
	4428	1278	5348	754	754	633	730	22
Wt 3	0 10596	0 2948	0 13372	0 1820	0 2160	0 1686	0 1820	0 64
	0 55623	0 14842	0 73688	0 11193	0 12298	0 10596	0 11193	0 499
	15	6	14	0 57524	3 60784	0 55623	0 57524	0 2948
	204	64	224	16	36	15	16	0 14842
	1686	499	1999	226	318	204	226	6
Wt 4	0 37632	0 10217	0 48873	0 7874	0 7455	0 6942	0 7455	0 322
	8	3	7 241353	1 40619	0 39360	0 37632	0 39360	0 1981
	121	40	124	17	10	8	10	0 10217
	1059	322	1205	174	141	121	141	3
	6942	1981	8515	1303	1177	1059	1177	40
Wt 5	0 37632	0 10217	0 48873	0 7455	0 7874	0 6942	0 7455	0 322
	8	3	7 241353	0 39360	1 40619	0 37632	0 39360	0 1981
	121	40	124	10	17	8	10	0 10217
	1059	322	1205	141	174	121	141	3
	6942	1981	8515	1177	1303	1059	1177	40
Wt 6	1 56694	1 15431	0 73710	0 11732	0 11732	0 10956	0 11732	0 559
	21	9	15	1 59154	1 59154	1 56694	1 59154	0 3166
	233	84	224	26	26	21	26	1 15431
	1794	559	2004	269	269	233	269	9
	10956	3166	13372	1990	1990	1794	1990	84
Wt 7	0 25647	0 7226	0 32095	0 4933	0 4822	0 4428	0 4933	0 194
	6 123606	3 33052	3 163886	0 26774	0 26426	0 25006	0 26774	0 1278
	78	30	66	6 126661	6 125559	3 121834	6 126661	0 6854
	693	231	710	87	81	66	87	1 32044
	4620	1394	5348	754	730	633	754	22
Wt 8	0 25647	0 7226	0 32095	0 4822	0 4933	0 4428	0 4933	0 194
	6 123606	3 33052	3 163886	0 26426	0 26774	0 25006	0 26774	0 1278
	78	30	66	6 125559	6 126661	3 121834	6 126661	0 6854
	693	231	710	81	87	66	87	1 32044
	4620	1394	5348	730	754	633	754	22
Wt 9	1 39278	1 11177	0 48873	0 7455	0 7455	0 6942	0 7874	0 322
	16	9	7 241353	0 39360	0 39360	0 37632	1 40619	0 1981
	161	67	124	10	10	8	17	0 10217
	1217	422	1205	141	141	121	174	3
	7480	2306	8515	1177	1177	1059	1303	40

continued next page

Affine D4 - Level 4 - Class (0,0) - Page 3 of 4

	Rp 1	Rp 2	Rp 3	Rp 4	Rp 5	Rp 6	Rp 7	Rp 8
Wt10	0 2948	0 11193	0 2948	0 10596	0 10596	0 7310	0 12298	0 12298
	0 14842	0 57524	0 14842	0 55623	0 55623	0 39000	3 60784	3 60784
	6	16	6	15	15	12	36	36
	64	226	64	204	204	139	318	318
	499	1820	499	1686	1686	1156	2160	2160
Wt11	1 22456	3 86902	1 22456	2 83692	2 83692	1 57873	3 86902	3 86902
	15	49	15	39	39	22	49	49
	131	458	131	395	395	243	458	458
	864	3159	864	2863	2863	1857	3159	3159
	4721	17825	4721	16725	16725	11253	17825	17825
Wt12	3 33052	6 126661	1 32044	6 123606	3 121834	2 85327	6 126661	6 125559
	30	87	22	78	66	41	87	81
	231	754	194	693	633	412	754	730
	1394	4933	1278	4620	4428	2959	4933	4822
	7226	26774	6854	25647	25006	17156	26774	26426
Wt13	1 32044	6 126661	3 33052	3 121834	6 123606	2 85327	6 125559	6 126661
	22	87	30	66	78	41	81	87
	194	754	231	633	693	412	730	754
	1278	4933	1394	4428	4620	2959	4822	4933
	6854	26774	7226	25006	25647	17156	26426	26774
Wt14	3	10	3	8	8	5 125018	10	10
	40	141	40	121	121	77	141	141
	322	1177	322	1059	1059	694	1177	1177
	1981	7455	1981	6942	6942	4678	7455	7455
	10217	39360	10217	37632	37632	25966	39360	39360
Wt15	1 32044	6 125559	1 32044	3 121834	3 121834	2 85327	6 126661	6 126661
	22	81	22	66	66	41	87	87
	194	730	194	633	633	412	754	754
	1278	4822	1278	4428	4428	2959	4933	4933
	6854	26426	6854	25006	25006	17156	26774	26774
Wt16	6	16	6	15	15	12	16	16
	64	226	64	204	204	139	226	226
	499	1820	499	1686	1686	1156	1820	1820
	2948	11193	2948	10596	10596	7310	11193	11193
	14842	57524	14842	55623	55623	39000	57524	57524

Affine D4 - Level 4 - Class (0,0) - Page 4 of 4

	Rp 9	Rp10	Rp11	Rp12	Rp13	Rp14	Rp15	Rp16
Wt10	0 12036 4 59925 40 320 2122	1 3820 4 17354 26 144 779	0 13372 0 73688 14 224 1999	0 1820 0 11193 0 57524 16 226	0 1820 0 11193 0 57524 16 226	0 1686 0 10596 0 55623 15 204	0 2160 0 12298 3 60784 36 318	0 64 0 499 0 2948 0 14842 6
Wt11	2 83692 39 395 2863 16725	1 22456 15 131 864 4721	1 110392 33 407 3305 20848	0 17825 3 86902 49 458 3159	0 17825 3 86902 49 458 3159	0 16725 2 83692 39 395 2863	0 17825 3 86902 49 458 3159	0 864 0 4721 1 22456 15 131
Wt12	3 121834 66 633 4428 25006	1 32044 22 194 1278 6854	3 163886 66 710 5348 32095	1 27374 10 128337 107 818 5149	0 26774 6 126661 87 754 4933	0 25647 6 123606 78 693 4620	0 26774 6 126661 87 754 4933	0 1394 0 7226 3 33052 30 231
Wt13	3 121834 66 633 4428 25006	1 32044 22 194 1278 6854	3 163886 66 710 5348 32095	0 26774 6 126661 87 754 4933	1 27374 10 128337 107 818 5149	0 25647 6 123606 78 693 4620	0 26774 6 126661 87 754 4933	0 1394 0 7226 3 33052 30 231
Wt14	8 121 1059 6942 37632	3 40 322 1981 10217	7 241353 124 1205 8515 48873	1 40619 17 174 1303 7874	1 40619 17 174 1303 7874	1 39278 16 161 1217 7480	1 40619 17 174 1303 7874	0 2306 1 11177 9 67 422
Wt15	6 123606 78 693 4620 25647	3 33052 30 231 1394 7226	3 163886 66 710 5348 32095	0 26774 6 126661 87 754 4933	0 26774 6 126661 87 754 4933	0 25647 6 123606 78 693 4620	1 27374 10 128337 107 818 5149	0 1394 0 7226 3 33052 30 231
Wt16	15 204 1686 10596 55623	6 64 499 2948 14842	14 224 1999 13372 73688	3 60784 36 318 2160 12298	3 60784 36 318 2160 12298	4 59925 40 320 2122 12036	3 60784 36 318 2160 12298	1 3820 4 17354 26 144 779

Affine D4 - Level 4 - Class (0,1) - Page 1 of 2

	Rp 1 0 1 0 1 2	Rp 2 0 3 0 1 0	Rp 3 0 1 1 1 0	Rp 4 0 1 0 3 0	Rp 5 1 0 0 0 3	Rp 6 1 2 0 0 1	Rp 7 1 0 1 0 1	Rp 8 1 0 0 2 1	Rp 9 2 1 0 1 0	Rp10 3 0 0 0 1
Wt 1	1 19 212 1673 10483 55529 258680 1087885 4207164	0 6 85 715 4565 24278 112915 473114 1821667	0 9 143 1301 8782 48715 234248 1008529 3972289	0 6 85 715 4565 24278 112915 473114 1821667	0 3 36 306 2030 11285 54733 238184 948402	0 3 57 573 4134 24067 120100 532631 2149663	0 2 44 472 3554 21354 109212 494040 2026926 7674968	0 3 57 573 4134 24067 120100 532631 2149663	0 0 12 177 1537 10021 54122 254718 1077434 4181005	0 0 1 25 257 1855 10725 53109 233781 937158
Wt 2	0 6 102 948 6484 36286 175628 760012 3005874	1 10 83 576 3416 17707 81948 344749 1337889	0 6 90 832 5770 32876 161896 711642 2853852	0 3 48 436 2935 16223 77730 333527 1309609	0 0 10 136 1106 6850 35490 161412 663454	0 3 42 393 2805 16488 83560 376738 1544645	0 0 17 245 2095 13520 72474 339117 1427770 5519215	0 0 21 295 2434 15263 79899 366602 1518259	0 0 6 102 948 6484 36286 175628 760012 3005874	0 0 0 10 136 1106 6850 35490 161412 663454
Wt 3	1 29 332 2576 15732 81305 370496 1528407	1 17 169 1230 7258 36703 164757 672126 2533793	1 23 274 2205 13889 73587 342233 1435949 5539616	1 17 169 1230 7258 36703 164757 672126 2533793	0 3 48 437 2939 16238 77776 333658 1309950	0 7 107 970 6561 36530 176328 761896 3010651	0 4 78 778 5574 32244 160050 706616 2840966	0 7 107 970 6561 36530 176328 761896 3010651	0 1 29 332 2576 15732 81305 370496 1528407	0 0 3 48 437 2939 16238 77776 333658 1309950
Wt 4	0 6 102 948 6484 36286 175628 760012 3005874	0 3 48 436 2935 16223 77730 333527 1309609	0 6 90 832 5770 32876 161896 711642 2853852	1 10 83 576 3416 17707 81948 344749 1337889	0 0 10 136 1106 6850 35490 161412 663454	0 0 21 295 2434 15263 79899 366602 1518259	0 0 17 245 2095 13520 72474 339117 1427770 5519215	0 3 42 393 2805 16488 83560 376738 1544645	0 0 6 102 948 6484 36286 175628 760012 3005874	0 0 0 10 136 1106 6850 35490 161412 663454
Wt 5	3 42 393 2805 16488 83560 376738 1544645	0 10 136 1106 6850 35490 161412 663454	0 17 245 2095 13520 72474 339117 1427770 5519215	0 10 136 1106 6850 35490 161412 663454	1 10 83 576 3416 17707 81948 344749 1337889	0 6 102 948 6484 36286 175628 760012 3005874	0 6 90 832 5770 32876 161896 711642 2853852	0 6 102 948 6484 36286 175628 760012 3005874	0 0 21 295 2434 15263 79899 366602 1518259	0 0 3 48 436 2935 16223 77730 333527 1309609

-continued next page-

Affine D4 - Level 4 - Class (0,1) - Page 2 of 2

	Rp 1	Rp 2	Rp 3	Rp 4	Rp 5	Rp 6	Rp 7	Rp 8	Rp 9	Rp10
Wt 6	3	3	2	1	0	1	0	0	0	0
	57	36	44	25	6	19	9	12	3	0
	573	306	472	257	85	212	143	177	57	6
	4134	2030	3554	1855	715	1673	1301	1537	573	85
	24067	11285	21354	10725	4565	10483	8782	10021	4134	715
	120100	54733	109212	53109	24278	55529	48715	54122	24067	4565
	532631	238184	494040	233781	112915	258680	234248	254718	120100	24278
	2149663	948402	2026926	937158	473114	1087885	1008529	1077434	532631	112915
			7674968		1821667	4207164	3972289	4181005	2149663	473114
										1821667
Wt 7	7	3	4	3	1	1	1	1	0	0
	107	48	78	48	17	29	23	29	7	1
	970	437	778	437	169	332	274	332	107	17
	6561	2939	5574	2939	1230	2576	2205	2576	970	169
	36530	16238	32244	16238	7258	15732	13889	15732	6561	1230
	176328	77776	160050	77776	36703	81305	73587	81305	36530	7258
	761896	333658	706616	333658	164757	370496	342233	370496	176328	36703
	3010651	1309950	2840966	1309950	672126	1528407	1435949	1528407	761896	164757
					2533793		5539616		3010651	672126
										2533793
Wt 8	3	1	2	3	0	0	0	1	0	0
	57	25	44	36	6	12	9	19	3	0
	573	257	472	306	85	177	143	212	57	6
	4134	1855	3554	2030	715	1537	1301	1673	573	85
	24067	10725	21354	11285	4565	10021	8782	10483	4134	715
	120100	53109	109212	54733	24278	54122	48715	55529	24067	4565
	532631	233781	494040	238184	112915	254718	234248	258680	120100	24278
	2149663	937158	2026926	948402	473114	1077434	1008529	1087885	532631	112915
			7674968		1821667	4181005	3972289	4207164	2149663	473114
										1821667
Wt 9	12	6	9	6	1	3	2	3	1	0
	177	85	143	85	25	57	44	57	19	3
	1537	715	1301	715	257	573	472	573	212	36
	10021	4565	8782	4565	1855	4134	3554	4134	1673	306
	54122	24278	48715	24278	10725	24067	21354	24067	10483	2030
	254718	112915	234248	112915	53109	120100	109212	120100	55529	11285
	1077434	473114	1008529	473114	233781	532631	494040	532631	258680	54733
	4181005	1821667	3972289	1821667	937158	2149663	2026926	2149663	1087885	238184
							7674968		4207164	948402
Wt10	21	10	17	10	3	6	6	6	3	1
	295	136	245	136	48	102	90	102	42	10
	2434	1106	2095	1106	436	948	832	948	393	83
	15263	6850	13520	6850	2935	6484	5770	6484	2805	576
	79899	35490	72474	35490	16223	36286	32876	36286	16488	3416
	366602	161412	339117	161412	77730	175628	161896	175628	83560	17707
	1518259	663454	1427770	663454	333527	760012	711642	760012	376738	81948
			5519215		1309609	3005874	2853852	3005874	1544645	344749
										1337889

Weight Multiplicities for Affine D5

Affine D5 - Level 1 - Class 0

	Rp 1 1 0 00 0 0				
Wt 1	1	6765	1633000	134992865	747688460*
	5	14725	2963840	223474667	961190039*
	20	31027	5302075	366772720	868131464*
	65	63505	9358470	597049255	337711778*
	190	126730	16313440	964375855	775031806*
	506	247170	28107365	472467168*	1043106197*
	1265	472295	47902010	314166807*	998827150*
	2990	885723	80803485	671549549*	873889867*

* Multiplicity is given modulo 1073741527

Affine D5 - Level 2 - Class 0

	Rp 1			Rp 2			Rp 3			Rp 4		
	0 1 00 0 1			0 0 10 0 0			0 0 00 2 0			2 0 00 0 0		
Wt 1	1	498507	97572835*	0	313965	832967971	0	197578	494044305	0	60483	208209123
	13	1512711	554668918*	4	975081	865389089*	3	606927	71171273*	0	197578	494044305
	102	4364970	733328996*	43	2869832	120760244*	31	1769373	448783562*	3	606927	71171273*
	595	12055731	625549187*	291	8063420	189015719*	202	4930075	401861218*	31	1769373	448783562*
	2859	32037888	379417728*	1526	21754026	98341171*	1029	13202483	778487401*	202	4930075	401861218*
	11949	82269522	491453114*	6769	56615850	243434486*	4462	34133040	152034560*	1029	13202483	778487401*
	44891	204854388	1025817784*	26571	142682211	526829415*	17196	85508968	49361155*	4462	34133040	152034560*
	154909	496091569		94889	349311387	746860487*	60483	208209123		17196	85508968	49361155*
Wt 2	3	872892	710095393*	1	557732	200889827*	1	349485	754316059	0	110599	321772211
	35	2581935	849513621*	15	1684053	786567698*	12	1044060	655015994*	1	349485	754316059
	244	7286107	510859492*	119	4838124	169835577*	87	2971998	659049424*	12	1044060	655015994*
	1305	19732348	635872775*	690	13310530	654008618*	480	8110875	1026852710*	87	2971998	659049424*
	5870	51529653	73937037*	3288	35249133	572385904*	2212	21326650	221128252*	480	8110875	1026852710*
	23288	130262092	1011249030*	13631	90231057	13730238*	8949	54245065	635555662*	2212	21326650	221128252*
	83868	319788831	332462603*	50835	224038391	357601761*	32751	133915083	657914269*	8949	54245065	635555662*
	279429	764493330		174294	541138775		110599	321772211		32751	133915083	657914269*
Wt 3	0	496530	97302748*	0	314505	833050130	1	198798	494202238	0	60917	208290357
	10	1508825	554208706*	5	976245	865532093*	5	609234	71438081*	0	198477	494185697
	90	4357270	732548805*	45	2872025	121000682*	40	1774019	449237667*	4	608821	71416178*
	565	12041280	624250591*	300	8067790	189422367*	221	4938574	402612441*	34	1773073	449198580*
	2775	32010730	377268156*	1545	21761985	99010775*	1086	13218631	779734610*	214	4937341	402560609*
	11760	82220445	487948029*	6825	56630835	244539854*	4578	34161715	154059315*	1059	13216035	779645504*
	44445	204765900	1020134521*	26685	142708695	528615923*	17481	85561062	52651730*	4545	34158305	153941120*
	153980	495936655		95165	349359120		61042	208299222		17382	85554348	52454919*
Wt 4	10	1508825	554208706*	5	976245	865532093*	4	608821	71416178*	1	198798	494202238
	90	4357270	732548805*	45	2872025	121000682*	34	1773073	449198580*	5	609234	71438081*
	565	12041280	624250591*	300	8067790	189422367*	214	4937341	402560609*	40	1774019	449237667*
	2775	32010730	377268156*	1545	21761985	99010775*	1059	13216035	779645504*	221	4938574	402612441*
	11760	82220445	487948029*	6825	56630835	244539854*	4545	34158305	153941120*	1086	13218631	779734610*
	44445	204765900	1020134521*	26685	142708695	528615923*	17382	85554348	52454919*	4578	34161715	154059315*
	153980	495936655		95165	349359120		60917	208290357		17481	85561062	52651730*
	496530	97302748*		314505	833050130		198477	494185697		61042	208299222	

* Multiplicity is given modulo 1073741527

Affine D5 - Level 2 - Class 1

	Rp 1 0 1 00 1 0					Rp 2 1 0 00 0 1				
Wt 1	1	1783683	659335143*	134745353*	635675897*	0	1031916	445399496*	668611039*	797777683*
	15	5136384	676875700*	649740521*	397476038*	4	3044316	923702861*	331898496*	466736404*
	121	14160091	834649136*	997785519*	321082465*	44	8571556	654286541*	446015816*	645845320*
	710	37566980	920058015*	131716565*	574857346*	300	23167896	570696554*	346391748*	603311661*
	3411	96320061	661026962*	103391723*		1584	60395940	604820782*	351797234*	854696778*
	14228	239505811	325607200*	687365415*		7064	152437508	633218257*	234392154*	
	53317	579263697	1024775416*	350250371*		27852	373703448	732742798*	770713636*	
	183513	292338671*	1041582946*	373731840*		99832	892243992	297740849*	40098436*	
	589122	1000555756*	908016070*	755286389*		331364	1005758469*	587059844*	897471796*	
Wt 2	4	3044316	923702861*	331898496*	466736404*	1	1783683	659335143*	134745353*	635675897*
	44	8571556	654286541*	446015816*	645845320*	15	5136384	676875700*	649740521*	397476038*
	300	23167896	570696554*	346391748*	603311661*	121	14160091	834649136*	997785519*	321082465*
	1584	60395940	604820782*	351797234*	854696778*	710	37566980	920058015*	131716565*	574857346*
	7064	152437508	633218257*	234392154*		3411	96320061	661026962*	103391723*	
	27852	373703448	732742798*	770713636*		14228	239505811	325607200*	687365415*	
	99832	892243992	297740849*	40098436*		53317	579263697	1024775416*	350250371*	
	331364	1005758469*	587059844*	897471796*		183513	292338671*	1041582946*	373731840*	
	1031916	445399496*	668611039*	797777683*		589122	1000555756*	908016070*	755286389*	

* Multiplicity is given modulo 1073741527

Affine D5 - Level 3 - Class 0 - Page 1 of 3

	Rp 1 0 0 00 1 2	Rp 2 0 2 00 1 0	Rp 3 0 0 01 1 0	Rp 4 1 1 00 0 1	Rp 5 1 0 10 0 0	Rp 6 1 0 00 2 0	Rp 7 3 0 00 0 0
Wt 1	1	0	0	0	0	0	0
	15	10	10	4	0	0	0
	175	155	166	80	15	15	0
	1480	1415	1620	870	235	219	6
	9955	9765	11670	6775	2150	1930	90
	56312	55806	68860	42345	14775	12865	780
	278850	277585	350840	225712	83950	71415	5145
	1241520	1238530	1595512	1064515	414990	346335	28255
	5066370	5059605	6619645	4551945	1840615	1512037	135854
	19217260	19202535	25450665	17953255	7477810	6061115	588665
	68481381	68450354	91727815	66141695	28237881	22626400	2344710
	231173940	231110435	312671900	229809611	100200445	79487880	8704260
	744145225	744018495	1015051091	758734390	336916595	264925780	30430225
	148940056*	148692886*	1008501536*	247240536*	6853708*	843063796	100984956
			851222479*	818576768*	102376959*	427389716*	320129695
							974367195

continued next page-

* Multiplicity is given modulo 1073741527

Affine D5 - Level 3 - Class 0 - Page 2 of 3

	Rp 1	Rp 2	Rp 3	Rp 4	Rp 5	Rp 6	Rp 7
Wt 2	0	1	0	0	0	0	0
	10	15	10	4	0	0	0
	155	175	166	80	15	15	0
	1415	1480	1620	870	235	219	6
	9765	9955	11670	6775	2150	1930	90
	55806	56312	68860	42345	14775	12865	780
	277585	278850	350840	225712	83950	71415	5145
	1238530	1241520	1595512	1064515	414990	346335	28255
	5059605	5066370	6619645	4551945	1840615	1512037	135854
	19202535	19217260	25450665	17953255	7477810	6061115	588665
	68450354	68481381	91727815	66141695	28237881	22626400	2344710
	231110435	231173940	312671900	229809611	100200445	79487880	8704260
	744018495	744145225	1015051091	758734390	336916595	264925780	30430225
	148692886*	148940056*	1008501536*	247240536*	6853708*	843063796	100984956
			851222479*	818576768*	102376959*	427389716*	320129695
							974367195
Wt 3	1	1	1	0	0	0	0
	28	28	29	10	2	3	0
	335	335	377	175	48	51	1
	2744	2744	3234	1730	544	523	21
	17676	17676	21565	12572	4319	3914	213
	96208	96208	120408	74581	27269	23822	1582
	460778	460778	588517	381541	146297	124418	9507
	1994185	1994185	2589322	1740558	693060	577498	49162
	7942293	7942293	10455725	7239932	2973517	2437281	226173
	29499413	29499413	39292868	27894820	11758632	9508077	947547
	103209617	103209617	138876346	100720286	43413121	34701847	3673032
	342812955	342812955	465400123	343867310	151110973	119595106	13332489
	13993995*	13993995*	414638696*	44135231*	499676696	392039117	45729030
	92471286*	92471286*	271307014*	258763044*	505471370*	155782828*	149271165
				758293276*	499074912*	485649545*	466391504
							327639899*
Wt 4	4	4	3	1	0	0	0
	70	70	71	31	6	6	0
	710	710	791	404	109	105	3
	5264	5264	6208	3487	1099	1005	45
	31759	31759	38836	23346	8072	7130	417
	164705	164705	206843	130830	48223	41441	2886
	759974	759974	974026	641354	247905	208537	16531
	3191563	3191563	4158109	2828999	1135089	938257	82228
	12398067	12398067	16373036	11448365	4735064	3857501	366707
	45084949	45084949	60227502	43104562	18286634	14714497	1496603
	154885065	154885065	208964405	152600000	66157343	52670175	5673877
	506297443	506297443	689019555	512137936	226245088	178452487	20202278
	510154637*	510154637*	24635160*	566240965*	736569362	576217177	68133355
	469581722*	469581722*	136504464*	742296692*	148420547*	709215789*	219107523
					441235878*	1015646297*	675521209
							931780940*

continued next page

* Multiplicity is given modulo 1073741527

Affine D5 - Level 3 - Class 0 - Page 3 of 3

	Rp 1	Rp 2	Rp 3	Rp 4	Rp 5	Rp 6	Rp 7
Wt 5	9	9	10	3	1	1	0
	147	147	167	73	22	23	1
	1375	1375	1626	835	270	260	12
	9605	9605	11698	6636	2289	2097	118
	55254	55254	68964	41863	15258	13452	885
	275869	275869	351183	224207	85461	73270	5496
	1233609	1233609	1596535	1060169	419365	351738	29318
	5046367	5046367	6622487	4540177	1852494	1526779	138855
	19168765	19168765	25458094	17923021	7508421	6099225	596640
	68367966	68367966	91746314	66067431	28313299	22720528	2364897
	230917026	230917026	312716055	229634087	100379252	79711418	8753232
	743579428	743579428	1015152766	758333327	337326425	265438761	30544794
	147725268*	147725268*	1008728322*	246351060*	7765442*	844205881	101244483
	378964573*	378964573*	851714370*	816656059*	104351738*	429864621*	320701104
							975593431
Wt 6	3	3	4	0	0	1	0
	63	63	76	24	8	13	0
	676	676	812	371	122	140	4
	5131	5131	6280	3359	1162	1145	52
	31304	31304	39060	22912	8316	7618	452
	163300	163300	207480	129495	49054	42974	3024
	755955	755955	975733	637546	250460	212989	17006
	3180768	3180768	4162441	2818778	1142376	950382	83698
	12370529	12370529	16383576	11422291	4754582	3888826	370916
	45017729	45017729	60252193	43040850	18336312	14791866	1507902
	154727075	154727075	209020428	152450030	66278354	52854015	5702668
	505938257	505938257	689143056	511796294	226528986	178874746	20272432
	509361665*	509361665*	24900607*	565484986*	737213710	577158400	68297902
	467876528*	467876528*	137061978*	740666854*	149840896*	711258237*	219480748
					444285834*	1019972544*	676343161
							933543950*
Wt 7	20	20	25	10	5	4	1
	300	300	365	170	65	60	5
	2611	2611	3200	1700	616	570	40
	17235	17235	21485	12438	4585	4110	285
	94880	94880	120246	74085	28130	24520	1820
	457055	457055	588230	379925	148850	126642	10228
	1984330	1984330	2588945	1735770	700145	584010	51205
	7917478	7917478	10455590	7226705	2992086	2455110	231600
	29439435	29439435	39294200	27860380	11805045	9554250	961310
	103069790	103069790	138882808	100634840	43524605	34815990	3706490
	342497150	342497150	465421335	343663875	151369455	119866213	13410909
	13300523*	13300523*	414697913*	43667853*	500257745	392661005	45907070
	90986412*	90986412*	271457712*	257722419*	506742091*	157166273*	149664150
					501785802*	488644224*	467236970
							329417663*

* Multiplicity is given modulo 1073741527

Weight Multiplicities for Affine D6

Affine D6 - Level 1 - Class (0,0)

	Rp 1 1 0 000 0 0				
Wt 1	1	15525	6190182	768291650	130256414*
	6	36280	11867310	258702803*	60002753*
	27	81816	22395359	141730441*	384755963*
	98	178794	41650050	676839645*	16416885*
	315	380051	76413078	139141995*	559125278*
	918	788004	138421358	285832610*	850407748*
	2492	1597725	247783113	67375610*	612112691*
	6372	3174210	438616728	162538712*	32649644*

* Multiplicity is given modulo 1073741527

Affine D6 - Level 2 - Class (0,0) - Page 1 of 2

	Rp 1	Rp 2	Rp 3	Rp 4	Rp 5	Rp 6
	0 0 000 0 2	0 2 000 0 0	0 0 001 0 0	0 0 100 0 0	0 0 000 2 0	2 0 000 0 0
Wt 1	1	0	0	0	0	0
	6	5	6	0	0	0
	57	50	66	15	10	0
	378	370	522	165	110	10
	2180	2145	3168	1221	800	110
	10713	10671	16260	6996	4510	800
	47072	46931	73194	34092	21650	4510
	187742	187567	297768	146883	92070	21650
	692740	692243	1114674	575400	356576	92070
	2391000	2390370	3894078	2085039	1279256	356576
	7796952	7795369	12824130	7080318	4305962	1279256
	24194045	24192008	40133166	22743534	13724128	4305962
	71873513	71868844	120100008	69621378	41719770	13724128
	205389687	205383618	345432942	204276120	121645864	41719770
	566883490	566870560	958903170	577209552	341786538	121645864
	442483332*	442466404*	430572106*	503034212*	928874614	341786538
	719982534*	719948542*	290301678*	956468868*	302045044*	928874614
				27834833*		302045044*
Wt 2	0	1	0	0	0	0
	5	6	6	0	0	0
	50	57	66	15	10	0
	370	378	522	165	110	10
	2145	2180	3168	1221	800	110
	10671	10713	16260	6996	4510	800
	46931	47072	73194	34092	21650	4510
	187567	187742	297768	146883	92070	21650
	692243	692740	1114674	575400	356576	92070
	2390370	2391000	3894078	2085039	1279256	356576
	7795369	7796952	12824130	7080318	4305962	1279256
	24192008	24194045	40133166	22743534	13724128	4305962
	71868844	71873513	120100008	69621378	41719770	13724128
	205383618	205389687	345432942	204276120	121645864	41719770
	566870560	566883490	958903170	577209552	341786538	121645864
	442466404*	442483332*	430572106*	503034212*	928874614	341786538
	719948542*	719982534*	290301678*	956468868*	302045044*	928874614
				27834833*		302045044*
Wt 3	1	1	1	0	0	0
	15	15	19	4	3	0
	133	133	185	58	41	3
	878	878	1286	486	330	41
	4760	4760	7211	3054	2015	330
	22355	22355	34692	15900	10264	2015
	93970	93970	148527	72214	45814	10264
	361339	361339	579515	295160	184612	45814
	1290774	1290774	2095092	1108274	684979	184612
	4332522	4332522	7103531	3878762	2373065	684979
	13783614	13783614	22795886	12789320	7756064	2373065
	41848510	41848510	69734935	40055878	24105061	7756064
	121917865	121917865	204517270	119934512	71682955	24105061
	342346304	342346304	577707043	345085830	204994947	71682955
	930003580	930003580	504041560*	958194706	566078750	204994947
	304274519*	304274519*	958462894*	429152106*	440883611*	566078750
	920489425*	920489425*	31717912*	287516110*	716849565*	440883611*
						716849565*

-continued next page-

* Multiplicity is given modulo 1073741527

Affine D6 - Level 2 - Class (0,0) - Page 2 of 2

	Rp 1	Rp 2	Rp 3	Rp 4	Rp 5	Rp 6
Wt 4	3	3	4	1	1	0
	41	41	58	19	15	1
	330	330	486	185	133	15
	2015	2015	3054	1286	878	133
	10264	10264	15900	7211	4760	878
	45814	45814	72214	34692	22355	4760
	184612	184612	295160	148527	93970	22355
	684979	684979	1108274	579515	361339	93970
	2373065	2373065	3878762	2095092	1290774	361339
	7756064	7756064	12789320	7103531	4332522	1290774
	24105061	24105061	40055878	22795886	13783614	4332522
	71682955	71682955	119934512	69734935	41848510	13783614
	204994947	204994947	345085830	204517270	121917865	41848510
	566078750	566078750	958194706	577707043	342346304	121917865
	440883611*	440883611*	429152106*	504041560*	930003580	342346304
	716849565*	716849565*	287516110*	958462894*	304274519*	930003580
				31717912*	920489425*	304274519*
						920489425*
Wt 5	0	0	0	0	1	0
	10	10	15	6	6	0
	110	110	165	66	57	5
	800	800	1221	522	378	50
	4510	4510	6996	3168	2180	370
	21650	21650	34092	16260	10713	2145
	92070	92070	146883	73194	47072	10671
	356576	356576	575400	297768	187742	46931
	1279256	1279256	2085039	1114674	692740	187567
	4305962	4305962	7080318	3894078	2391000	692243
	13724128	13724128	22743534	12824130	7796952	2390370
	41719770	41719770	69621378	40133166	24194045	7795369
	121645864	121645864	204276120	120100008	71873513	24192008
	341786538	341786538	577209552	345432942	205389687	71868844
	928874614	928874614	503034212*	958903170	566883490	205383618
	302045044*	302045044*	956468868*	430572106*	442483332*	566870560
			27834833*	290301678*	719982534*	442466404*
						719948542*
Wt 6	10	10	15	6	5	1
	110	110	165	66	50	6
	800	800	1221	522	370	57
	4510	4510	6996	3168	2145	378
	21650	21650	34092	16260	10671	2180
	92070	92070	146883	73194	46931	10713
	356576	356576	575400	297768	187567	47072
	1279256	1279256	2085039	1114674	692243	187742
	4305962	4305962	7080318	3894078	2390370	692740
	13724128	13724128	22743534	12824130	7795369	2391000
	41719770	41719770	69621378	40133166	24192008	7796952
	121645864	121645864	204276120	120100008	71868844	24194045
	341786538	341786538	577209552	345432942	205383618	71873513
	928874614	928874614	503034212*	958903170	566870560	205389687
	302045044*	302045044*	956468868*	430572106*	442466404*	566883490
			27834833*	290301678*	719948542*	442483332*
						719982534*

* Multiplicity is given modulo 1073741527

Affine D6 - Level 2 - Class (1,0)

	Rp 1 0 0 000 1 1					Rp 2 1 1 000 0 0				
Wt 1	1	8479580	342211257*	455771064*	899965980*	0	4635624	608476172*	406145255*	609185541*
	21	27247451	571499774*	972673141*	990509846*	5	15297926	524974256*	951171442*	921316439*
	212	83444383	357220108*	906688292*	549717152*	70	47951186	649740810*	946194679*	606459549*
	1498	244954437	253593410*	990014776*	682987377*	581	143681515	415449047*	953116921*	392152647*
	8475	692498450	187730000*	100540386*		3641	413697925	618919203*	160940849*	728213996*
	40999	818918583*	202564148*	461810739*		18947	75699479*	166726255*	364004426*	
	176183	722099433*	613293503*	399674844*		86088	945243867*	687534523*	568809402*	
	689264	49307955*	596398808*	574508572*		352135	566168204*	683442059*	570513876*	
	2496928	293636648*	161020709*	205915162*		1323360	166158632*	38458326*	658396834*	
Wt 2	5	15297926	524974256*	951171442*	921316439*	1	8479580	342211257*	455771064*	899965980*
	70	47951186	649740810*	946194679*	606459549*	21	27247451	571499774*	972673141*	990509846*
	581	143681515	415449047*	953116921*	392152647*	212	83444383	357220108*	906688292*	549717152*
	3641	413697925	618919203*	160940849*	728213996*	1498	244954437	253593410*	990014776*	682987377*
	18947	75699479*	166726255*	364004426*		8475	692498450	187730000*	100540386*	
	86088	945243867*	687534523*	568809402*		40999	818918583*	202564148*	461810739*	
	352135	566168204*	683442059*	570513876*		176183	722099433*	613293503*	399674844*	
	1323360	166158632*	38458326*	658396834*		689264	49307955*	596398808*	574508572*	
	4635624	608476172*	406145255*	609185541*		2496928	293636648*	161020709*	205915162*	

* Multiplicity is given modulo 1073741527

Affine D6 - Level 2 - Class (1,1)

	Rp 1 (0 1 / 000 / 0 1)				Rp 2 (0 0 / 010 / 0 0)				Rp 3 (1 0 / 000 / 1 0)			
Wt 1	1	1796353	993571642*	991434011*	0	1170858	558760258*	319375378*	0	486912	483014849*	20372145*
	16	6130340	522431182*	847625542*	5	4084989	607906874*	978034487*	0	1780828	987381460*	223142361*
	152	19787541	353886112*	901244954*	66	13434508	84179347*	76623754*	10	6094060	510563872*	588922739*
	1058	60848908	215103919*	437181219*	536	41984580	645380560*	865102060*	125	19705725	331490753*	759514513*
	5980	179298532	248220386*	885625176*	3312	125472196	569711518*	633142507*	960	60670114	173453869*	834083101*
	29034	508635500	464652745*	595264173*	17068	360422588	1021624510*	417178985*	5665	178918481	171807308*	746483181*
	125408	320803338*	561692503*	784320081*	76980	999312632	406016968*	98187554*	28116	507847496	326231387*	309431563*
	493284	486189059*	458988873*		313049	536188971*	1041807944*		122916	319205613*	313909390*	716944471*
Wt 2	3	3331458	221731042*	642482284*	1	2204525	457541668*	961305936*	0	944640	609782023*	533626970*
	45	11057931	392356396*	679021068*	20	7458946	174771400*	717123276*	3	3331458	221731042*	642482284*
	387	34835904	64264984*	370653045*	197	23891371	147256092*	433458615*	45	11057931	392356396*	679021068*
	2481	104846310	65275022*	1050336131*	1370	72961958	937543577*	398185687*	387	34835904	64264984*	370653045*
	13122	303073083	156497056*	552871851*	7666	213652333	899232853*	235708318*	2481	104846310	65275022*	1050336131*
	60363	845058564	347166932*	428550593*	36782	602667790	140350795*	131169525*	13122	303073083	156497056*	552871851*
	249345	133545220*	536496449*		157047	570113494*	155143340*		60363	845058564	347166932*	428550593*
	944640	609782023*	533626970*		611200	54664726*	333853018*		249345	133545220*	536496449*	
Wt 3	10	6094060	510563872*	588922739*	5	4084989	607906874*	978034487*	1	1796353	993571642*	991434011*
	125	19705725	331490753*	759514513*	66	13434508	84179347*	76623754*	16	6130340	522431182*	847625542*
	960	60670114	173453869*	834083101*	536	41984580	645380560*	865102060*	152	19787541	353886112*	901244954*
	5665	178918481	171807308*	746483181*	3312	125472196	569711518*	633142507*	1058	60848908	215103919*	437181219*
	28116	507847496	326231387*	309431563*	17068	360422588	1021624510*	417178985*	5980	179298532	248220386*	885625176*
	122916	319205613*	313909390*	716944471*	76980	999312632	406016968*	98187554*	29034	508635500	464652745*	595264173*
	486912	483014849*	20372145*		313049	536188971*	1041807944*		125408	320803338*	561692503*	784320081*
	1780828	987381460*	223142361*		1170858	558760258*	319375378*		493284	486189059*	458988873*	

* Multiplicity is given modulo 1073741527

Affine D6 - Level 3 - Class (0,0) - Page 1 of 2

	Rp 1	Rp 2	Rp 3	Rp 4	Rp 5	Rp 6	Rp 7	Rp 8
	0 1	0 0	1 0	1 2	1 0	1 0	1 0	3 0
	000	010	000	000	001	100	000	000
	1 1	1 0	0 2	0 0	0 0	0 0	2 0	0 0
Wt 1	1	0	0	0	0	0	0	0
	46	10	5	5	4	0	0	0
	802	250	120	120	129	15	15	0
	8858	3244	1580	1580	1878	340	310	6
	73745	29684	14740	14740	18557	4110	3586	126
	503574	216068	109320	109320	142780	35800	30256	1432
	2962058	1331907	685120	685120	918735	250985	207152	11933
	15487044	7221860	3768751	3768751	5156066	1502070	1216558	80810
	73549948	35319684	18663741	18663741	25944599	7950346	6340620	470425
	322280500	158587290	84721867	84721867	119325994	38107246	30000646	2433704
	244628075*	662335957	357254048	357254048	508759765	168195152	130958935	11443231
					958766001*	692131428	533759308	49683296
								201549900
Wt 2	3	1	0	0	0	0	0	0
	114	39	12	12	14	2	3	0
	1764	663	269	269	327	63	66	1
	17943	7239	3267	3267	4137	924	867	27
	140466	59874	28623	28623	37342	9271	8178	351
	914203	406992	201819	201819	269592	72432	61498	3271
	5172855	2385759	1213998	1213998	1652688	472611	390507	24255
	26191590	12439833	6454506	6454506	8925633	2685024	2173614	152316
	121059099	58945638	31059294	31059294	43520055	13655997	10876704	839998
	518246889	257797357	137562168	137562168	194937405	63398349	49827111	4171581
		1052863878*	567808167	567808167	812543822	272542444	211812804	18988809
						23046272*		80283635
Wt 3	5	0	1	0	0	0	0	0
	135	29	21	15	15	0	0	0
	1905	594	327	300	335	40	36	0
	18675	6873	3563	3465	4185	765	671	15
	143685	58227	29925	29610	37569	8430	7182	270
	926730	400455	206919	206001	270529	68690	57183	2850
	5217230	2362185	1232253	1229761	1656153	457920	373910	22416
	26337285	12361211	6515154	6508782	8937432	2632635	2115351	145206
	121508301	58699809	31248618	31233093	43557550	13483055	10686810	814992
	519559886	257069769	138122430	138086150	195049944	62862750	49245138	4090018
		1050809790*	569391087	569309271	812865345	270970935	210120264	18738900
								79557300
Wt 4	5	0	0	1	0	0	0	0
	135	29	15	21	15	0	0	0
	1905	594	300	327	335	40	36	0
	18675	6873	3465	3563	4185	765	671	15
	143685	58227	29610	29925	37569	8430	7182	270
	926730	400455	206001	206919	270529	68690	57183	2850
	5217230	2362185	1229761	1232253	1656153	457920	373910	22416
	26337285	12361211	6508782	6515154	8937432	2632635	2115351	145206
	121508301	58699809	31233093	31248618	43557550	13483055	10686810	814992
	519559886	257069769	138086150	138122430	195049944	62862750	49245138	4090018
		1050809790*	569309271	569391087	812865345	270970935	210120264	18738900
								79557300

-continued next page-

* Multiplicity is given modulo 1073741527

Affine D6 - Level 3 - Class (0,0) - Page 2 of 2

	Rp 1	Rp 2	Rp 3	Rp 4	Rp 5	Rp 6	Rp 7	Rp 8
Wt 5	13	3	1	1	1	0	0	0
	314	100	42	42	46	6	6	0
	4038	1484	676	676	810	149	141	3
	36884	14808	7080	7080	8918	1974	1762	60
	268618	114671	56657	56657	73889	18383	15794	721
	1658066	741202	375262	375262	501974	135822	113412	6304
	9004475	4174010	2153899	2153899	2938719	847624	692535	44585
	44107426	21060026	11036690	11036690	15299830	4643786	3728578	269166
	198341747	97076079	51531957	51531957	72387555	22911436	18134757	1436537
	829521860	414668350	222542658	222542658	316128606	103653578	81061808	6937092
			898942001	898942001	215624011*	435761088	337290582	30824129
								127603364
Wt 6	34	13	3	3	4	1	1	0
	717	279	100	100	127	28	29	1
	8398	3429	1450	1450	1861	435	413	15
	71660	30604	14083	14083	18456	4634	4156	184
	495253	219981	106490	106490	142307	38207	32867	1726
	2931961	1346715	674294	674294	916816	260686	217636	13215
	15386431	7273104	3730943	3730943	5149095	1537430	1254593	85782
	73234849	35484581	18541106	18541106	25921305	8069287	6467955	488043
	321346635	159086905	84347700	84347700	119253322	38481663	30399590	2491657
	241988816*	663774046	356170355	356170355	508545535	169309636	132141048	11622433
						695292901	537098117	50208861
								203022327
Wt 7	10	5	0	0	0	0	1	0
	280	120	30	30	40	10	16	0
	3815	1610	595	595	779	190	212	5
	35750	15440	6656	6656	8784	2240	2148	80
	263740	117378	54786	54786	73363	19730	17570	850
	1639450	751548	367952	367952	500082	141620	120600	6950
	8939696	4210161	2127918	2127918	2932350	869790	718904	47340
	43898076	21177524	10951090	10951090	15279640	4721110	3817856	279600
	197705417	97435500	51267170	51267170	72326731	23161965	18417651	1472625
	827685538	415712520	221766194	221766194	315953564	104417020	81909110	7053072
					215140247*	437969590	339708272	31174922
								128611074
Wt 8	85	39	10	10	15	6	5	1
	1590	694	240	240	330	96	90	6
	17091	7473	3075	3075	4135	1122	1035	57
	136866	61122	27650	27650	37278	10248	9070	498
	900582	412420	197661	197661	269163	76590	65475	3900
	5125548	2406420	1198266	1198266	1650666	488468	406110	26700
	26038465	12511221	6400247	6400247	8917674	2740497	2229162	161079
	120591990	59174294	30885378	30885378	43492200	13836642	11059632	869226
	516892404	258485619	137037630	137037630	194847961	63952335	50392074	4263456
					812275488	274156230	213465546	19262890
								81064590

* Multiplicity is given modulo 1073741527

Weight Multiplicities for Affine D7

Affine D7 - Level 1 - Class 0

Rp 1

```
 1   0
  0000
 0   0
```

Wt 1	1	32305	20247710	409083254*	52130125*
	7	80465	40786690	124757309*	1000298023*
	35	192899	80782800	1023243108*	594048580*
	140	447146	157510780	459326577*	696348890*
	490	1006145	302666903	158673252*	871120898*
	1547	2204475	573720808	540260062*	512278697*
	4522	4715510	1073720305*	894147031*	419189929*
	12405	9869132	911765248*	342295921*	69617194*

* Multiplicity is given modulo 1073741527

Affine D7 - Level 2 - Class 0

	Rp 1 0 1 0000 0 1		Rp 2 0 0 0010 0 0		Rp 3 0 0 1000 0 0		Rp 4 0 0 0000 2 0		Rp 5 2 0 0000 0 0	
Wt 1	1	250119164	0	176855497	0	87458097	0	52081915	0	14762775
	19	809004259	6	579783283	0	294870540	0	174468363	0	52081915
	212	363505381*	94	747283854*	15	949309591	10	558438720	0	174468363
	1717	1067872289*	892	135935397*	221	785900410*	145	642822292*	10	558438720
	11181	248587406*	6362	1042156934*	1964	146945966*	1265	793386508*	145	642822292*
	61965	818228168*	37446	393579076*	13237	475407769*	8386	637469103*	1265	793386508*
	303015	933626590*	191195	546495414*	74277	566925190*	46389	908170905*	8386	637469103*
	1339863	893521828*	873685	82796245*	364231	177876760*	224697	559224191*	46389	908170905*
	5451580	496036022*	3648878	917495638*	1607951	256218766*	981457	870264981*	224697	559224191*
	20676424		14138925		6518559	621227401*	3942031		981457	870264981*
	73828546		51402949		24611735		14762775		3942031	
Wt 2	3	451178138	1	321824699	0	161937482	0	96229122	0	28052467
	55	355880396*	25	1032221317*	4	532825556	3	314598905	0	96229122
	565	60788009*	296	1031840574*	73	605138330*	51	985608226	3	314598905
	4244	999524102*	2417	852796898*	737	794284380*	493	824868229*	51	985608226
	25914	1056794304*	15616	293592985*	5438	940196270*	3532	71107772*	493	824868229*
	136055	15629083*	85460	522964388*	32683	382902079*	20771	839303456*	3532	71107772*
	635599	220461678*	412280	280342993*	169262	487054371*	105764	573275653*	20771	839303456*
	2702950	324356378*	1799263	14819951*	781452	48210177*	481657	228142841*	105764	573275653*
	10633568	590883760*	7232049	861794330*	3289379	195120909*	2004545	468547218*	481657	228142841*
	39163427		27124884		12825500		7740926		2004545	468547218*
	136270072		95875211		46866154		28052467		7740926	
Wt 3	10	805587713	5	579125374	1	295992987	1	175488053	0	52498863
	160	355575676*	87	745743528*	23	951980155	18	560854966	1	175488053
	1481	1049943455*	856	132425736*	266	792074789*	189	648388156*	18	560854966
	10263	209002832*	6213	1034353902*	2163	160857292*	1457	805884723*	189	648388156*
	58776	732706220*	36904	376614090*	13999	506024215*	9112	664890918*	1457	805884723*
	292855	752501280*	189414	510358051*	76889	632878209*	48853	967074673*	9112	664890918*
	1309675	516903418*	868261	7256711*	372469	317180770*	232401	683312739*	48853	967074673*
	5366907	799882810*	3633372	762320006*	1632217	545154338*	1003984	53273276*	232401	683312739*
	20450253		14096844		6586135	136766686*	4004360	391975302*	1003984	53273276*
	73249311		51293748		24791111		14927291		4004360	391975302*
	248689321		176582853		87915002		52498863		14927291	
Wt 4	0	448963984	0	321394675	0	162655976	1	96887566	0	28315604
	35	350664547*	21	1031201171*	7	534561391	7	316182637	0	96884376
	455	48842018*	273	1029491950*	91	609210366*	77	989299852	6	316174733
	3787	972831000*	2331	847517281*	833	803574346*	596	833259306*	69	989289684
	24206	998503838*	15288	282006684*	5824	960883852*	3949	89709052*	587	833236008*
	130389	964681967*	84399	498050721*	34104	427934508*	22178	879669100*	3905	89678819*
	618072	1032573213*	408954	227830313*	173901	583108252*	110336	659077941*	22126	879604107*
	2652405	858712216*	1789704	979804324*	795662	249219321*	495184	407225631*	110143	658993068*
	10494939	570797267*	7205562	640338758*	3329963	608519910*	2042952	835832087*	494948	407052641*
	38801021		27055427		12936035		7843762		2042218	835605146*
	135358132		95698967		47153197		28318113		7842844	
Wt 5	35	350664547*	21	1031201171*	7	534561391	6	316174733	1	96887566
	455	48842018*	273	1029491950*	91	609210366*	69	989289684	7	316182637
	3787	972831000*	2331	847517281*	833	803574346*	587	833236008*	77	989299852
	24206	998503838*	15288	282006684*	5824	960883852*	3905	89678819*	596	833259306*
	130389	964681967*	84399	498050721*	34104	427934508*	22126	879604107*	3949	89709052*
	618072	1032573213*	408954	227830313*	173901	583108252*	110143	658993068*	22178	879669100*
	2652405	858712216*	1789704	979804324*	795662	249219321*	494948	407052641*	110336	659077941*
	10494939	570797267*	7205562	640338758*	3329963	608519910*	2042218	835605146*	495184	407225631*
	38801021		27055427		12936035		7842844		2042952	835832087*
	135358132		95698967		47153197		28315604		7843762	
	448963984		321394675		162655976		96884376		28318113	

* Multiplicity is given modulo 1073741527

Affine D7 - Level 2 - Class 3

	Rp 1 0 0 0000 1 1					Rp 2 1 1 0000 0 0				
Wt 1	1	33471660	824986083*	466938557*	1024636676*	0	17391614	864820335*	169370870*	152291977*
	28	118670880	162747829*	462533591*	598350998*	6	63463920	647589457*	136846639*	896379342*
	344	399404752	314711768*	102290336*	714849323*	104	219038284	669085616*	1051540565*	831461650*
	2864	210319384*	962516061*	546478803*	240174509*	1028	720010304	946231884*	612057042*	885451355*
	18733	742060019*	754801803*	725401727*		7504	119310670*	249400864*	222901262*	467380523*
	103412	1055847122*	763486627*	906542231*		44812	423963766*	649344456*	1015659734*	
	502168	663980791*	560877819*	636793688*		231104	761176058*	619862452*	467280825*	
	2202960	343485149*	740567493*	574348721*		1063816	37514621*	716063020*	582217156*	
	8892418	1063335395*	879785913*	160343198*		4468096	75264124*	599508901*	475289130*	
Wt 2	6	63463920	647589457*	136846639*	896379342*	1	33471660	824986083*	466938557*	1024636676*
	104	219038284	669085616*	1051540565*	831461650*	28	118670880	162747829*	462533591*	598350998*
	1028	720010304	946231884*	612057042*	885451355*	344	399404752	314711768*	102290336*	714849323*
	7504	119310670*	249400864*	222901262*	467380523*	2864	210319384*	962516061*	546478803*	240174509*
	44812	423963766*	649344456*	1015659734*		18733	742060019*	754801803*	725401727*	
	231104	761176058*	619862452*	467280825*		103412	1055847122*	763486627*	906542231*	
	1063816	37514621*	716063020*	582217156*		502168	663980791*	560877819*	636793688*	
	4468096	75264124*	599508901*	475289130*		2202960	343485149*	740567493*	574348721*	
	17391614	864820335*	169370870*	152291977*		8892418	1063335395*	879785913*	160343198*	

* Multiplicity is given modulo 1073741527

Affine D7 - Level 3 - Class 0 - Page 1 of 3

	Rp 1 0 0 0000 1 2	Rp 2 0 2 0000 1 0	Rp 3 0 0 0001 1 0	Rp 4 0 0 0100 1 0	Rp 5 1 1 0000 0 1	Rp 6 1 0 0010 0 0	Rp 7 1 0 1000 0 0	Rp 8 1 0 0000 2 0	Rp 9 3 0 0000 0 0
Wt 1	1	0	0	0	0	0	0	0	0
	28	21	21	0	6	0	0	0	0
	560	525	603	84	210	49	0	0	0
	7602	7462	9345	2016	3717	1267	105	91	0
	77609	77119	101850	27048	44625	18011	2415	2029	36
	641032	639485	875448	263382	412839	183707	31185	25592	798
	4499124	4494602	6313839	2068290	3155796	1497979	294330	236838	9954
	27736102	27723697	39714318	13851369	20813268	10350347	2252040	1782704	91287
	153745704	153713399	223620579	81888261	121916883	62824237	14756175	11519662	681723
	779567712	779487247	74612564*	437401377	647449362	343229334	85627710	66054618	4375749
						643690077*	450081555	343626514	24945480
									129115536

-continued next page-

* Multiplicity is given modulo 1073741527

Affine D7 - Level 3 - Class 0 - Page 2 of 3

	Rp 1	Rp 2	Rp 3	Rp 4	Rp 5	Rp 6	Rp 7	Rp 8	Rp 9
Wt 2	0 21 525 7462 77119 639485 4494602 27723697 153713399 779487247	1 28 560 7602 77609 641032 4499124 27736102 153745704 779567712	0 21 603 9345 101850 875448 6313839 39714318 223620579 74612564*	0 0 84 2016 27048 263382 2068290 13851369 81888261 437401377	0 6 210 3717 44625 412839 3155796 20813268 121916883 647449362	0 0 49 1267 18011 183707 1497979 10350347 62824237 343229334 643690077*	0 0 0 105 2415 31185 294330 2252040 14756175 85627710 450081555	0 0 0 91 2029 25592 236838 1782704 11519662 66054618 343626514	0 0 0 0 36 798 9954 91287 681723 4375749 24945480 129115536
Wt 3	1 59 1227 15964 154920 1222801 8253898 49214271 265080985 237237790*	1 59 1227 15964 154920 1222801 8253898 49214271 265080985 237237790*	1 67 1532 21007 211158 1709478 11765606 71251091 388719305 869552963*	0 10 334 5540 62650 550974 4035975 25675291 145828336 754054437	0 16 511 8291 92822 811981 5930240 37659164 213659476 30281192*	0 4 178 3354 41079 382087 2920163 19212226 112168094 593573429	0 0 15 445 6820 73175 619180 4402710 27347520 152293465 774478015	0 0 15 401 5845 60620 500476 3491319 21355484 117417895 590704957	0 0 0 6 162 2316 23685 193311 1336366 8113513 44333654 221854932
Wt 4	3 147 2772 33477 306573 2309157 14996106 86553585 453411144	3 147 2772 33477 306573 2309157 14996106 86553585 453411144	4 193 3715 45908 428835 3285041 21641640 126454435 669517150	1 49 1033 13820 137984 1116955 7706620 46831406 256444598 212995208*	0 44 1233 18220 190234 1575856 11014309 67448256 371111563 794098082*	0 18 528 8229 89919 773856 5584989 35143830 197934012 1016605176*	0 2 78 1409 17228 162045 1257432 8404141 49809587 267269805 244793127*	0 3 81 1302 14931 134964 1018500 6667098 38874663 205847037	0 0 1 33 525 5930 52790 393630 2551987 14764945 77681444
Wt 5	6 189 3129 35769 318816 2366274 15236037 87479460 456740733	6 189 3129 35769 318816 2366274 15236037 87479460 456740733	5 211 3879 47009 434877 3313793 21764213 126933039 671254873	0 29 832 12375 129614 1075380 7522990 46092772 253693668 203393125*	1 64 1434 19665 198604 1617431 11197939 68186890 373862493 803700165*	0 15 489 7911 87942 763551 5537862 34949091 197192925 1013971797*	0 0 40 1045 14700 147795 1188170 8103655 48618955 262892165 229682453*	0 0 36 903 12285 120471 949494 6372297 37720704 201646089	0 0 0 15 361 4829 46748 364878 2429414 14286341 75943721

continued next page-

* Multiplicity is given modulo 1073741527

Affine D7 - Level 3 - Class 0 - Page 3 of 3

	Rp 1	Rp 2	Rp 3	Rp 4	Rp 5	Rp 6	Rp 7	Rp 8	Rp 9
Wt 6	15	15	18	3	3	1	0	0	0
	445	445	564	129	165	61	6	6	0
	6820	6820	9027	2415	3318	1370	189	177	3
	73175	73175	99873	29694	41979	18691	3129	2751	75
	619180	619180	865143	277875	398346	187473	35769	30172	1116
	4402710	4402710	6266712	2137296	3086790	1516167	318816	261007	11931
	27347520	27347520	39519579	14146170	20518467	10429240	2366274	1894265	101592
	152293465	152293465	222879492	83042220	120762924	63137578	15236037	11983873	728850
	774478015	774478015	71979185*	441602325	643248414	344385651	87479460	67831140	4570488
				13133575*		647698031*	456740733	349968282	25686567
									131748915
Wt 7	40	40	55	16	10	5	1	1	0
	1045	1045	1429	420	425	196	34	35	1
	14700	14700	20327	6262	7569	3518	640	602	18
	147795	147795	207392	67230	88242	42180	8244	7290	265
	1188170	1188170	1691290	575143	787812	388129	81553	68990	2996
	8103655	8103655	11686713	4147536	5818679	2948915	661337	542051	27451
	48618955	48618955	70937750	26139502	37194953	19334799	4591016	3674949	211499
	262892165	262892165	387562988	147604858	211882954	112646698	28112279	22094118	1415259
				760396205	23939424*	595311152	155165638	120168825	8426854
							784576944	600307040	45489971
									225862886
Wt 8	10	10	15	6	0	0	0	1	0
	370	370	525	174	120	60	12	19	0
	6200	6200	8709	2814	2919	1411	273	299	6
	69310	69310	97896	32340	39333	19129	3843	3620	114
	599130	599130	854838	292368	383853	190409	40353	35275	1434
	4311570	4311570	6219585	2206302	3017784	1531776	343302	286850	13908
	26973425	26973425	39324840	14440971	20223666	10500689	2480508	2010777	111897
	150878975	150878975	222138405	84196179	119608965	63430714	15715899	12461801	775977
	769482395	769482395	69345806*	445803273	639047466	345489827	89331210	69643694	4765227
							463399911	356400481	26427654
									134382294
Wt 9	105	105	154	56	35	21	7	6	1
	2415	2415	3458	1176	1085	567	133	126	7
	31185	31185	44450	15036	16989	8547	1849	1701	77
	294330	294330	421596	145572	182616	91896	20048	17577	798
	2252040	2252040	3252648	1156105	1536645	784161	177275	149457	7420
	14756175	14756175	21508734	7883239	10837554	5632116	1329692	1087506	60095
	85627710	85627710	125948179	47550979	66728431	35338569	8713201	6961899	426174
	450081555	450081555	667709035	259146300	368409391	198675099	51023287	40028622	2685175
							271706540	210047985	15271732
									79490600

Weight Multiplicities for Affine D8

Affine D8 - Level 1 - Class (0,0)

	Rp 1 1 0 00000 0 0				
Wt 1	1	62337	59068372	882071914*	754353614*
	8	164560	124505880	1063503217*	1005972877*
	44	417140	257770964	377995504*	1006623135*
	192	1020416	524871424	45655873*	700664219*
	726	2418710	1052316364	427920852*	853810195*
	2464	5573568	1005750217*	509756845*	266184294*
	7704	12520744	832753459*	766484293*	194507021*
	22528	27484160	287029279*	942242413*	640722352*

* Multiplicity is given modulo 1073741527

Affine D8 - Level 2 - Class (0,0) - Page 1 of 3

	Rp 1	Rp 2	Rp 3	Rp 4	Rp 5	Rp 6	Rp 7
	0 0 00000 0 2	0 2 00000 0 0	0 0 00001 0 0	0 0 00100 0 0	0 0 10000 0 0	0 0 00000 2 0	2 0 00000 0 0
Wt 1	1	0	0	0	0	0	0
	8	7	8	0	0	0	0
	100	91	120	28	0	0	0
	885	875	1248	420	56	35	0
	6627	6573	9880	4068	840	525	35
	41888	41825	65056	30080	7896	4893	525
	232936	232680	370944	185796	56560	34720	4893
	1162611	1162302	1887120	1001304	338688	205968	34720
	5306940	5305896	8738264	4847036	1774248	1069635	205968
	22445595	22444305	37388256	21484764	8371224	5006736	1069635
	88907486	88903678	149502968	88444332	36261408	21531699	5006736
	332551378	332546590	563692688	341729048	146216632	86256870	21531699
	108852046*	108839299*	944602433*	175790333*	554513984	325190474	86256870
	799121091*	799104855*	460278334*	57260840*	919943297*	89163139*	325190474
	241387799*	241348001*	107895701*	558806833*	396245750*	748213227*	89163139*
				408703895*	1020415020*	113669161*	748213227*
							113669161*
Wt 2	0	1	0	0	0	0	0
	7	8	8	0	0	0	0
	91	100	120	28	0	0	0
	875	885	1248	420	56	35	0
	6573	6627	9880	4068	840	525	35
	41825	41888	65056	30080	7896	4893	525
	232680	232936	370944	185796	56560	34720	4893
	1162302	1162611	1887120	1001304	338688	205968	34720
	5305896	5306940	8738264	4847036	1774248	1069635	205968
	22444305	22445595	37388256	21484764	8371224	5006736	1069635
	88903678	88907486	149502968	88444332	36261408	21531699	5006736
	332546590	332551378	563692688	341729048	146216632	86256870	21531699
	108839299*	108852046*	944602433*	175790333*	554513984	325190474	86256870
	799104855*	799121091*	460278334*	57260840*	919943297*	89163139*	325190474
	241348001*	241387799*	107895701*	558806833*	396245750*	748213227*	89163139*
				408703895*	1020415020*	113669161*	748213227*
							113669161*
Wt 3	1	1	1	0	0	0	0
	21	21	27	6	0	0	0
	255	255	362	122	15	10	0
	2252	2252	3377	1382	277	180	10
	15991	15991	24831	11412	2917	1856	180
	96558	96558	153635	76376	22740	14222	1856
	513685	513685	832288	438210	145255	89562	14222
	2466535	2466535	4053258	2230788	801804	488502	89562
	10876446	10876446	18078519	10309930	3950658	2382506	488502
	44618395	44618395	74869588	43969646	17754702	10613160	2382506
	171985286	171985286	290919738	175146618	73901716	43836328	10613160
	627811605	627811605	1069358533*	657706686	288159342	169766848	43836328
	36624222*	36624222*	521644432*	197939434*	1061801993*	621767544	169766848
	836599213*	836599213*	730091269*	473446793*	501695089*	20735384*	621767544
			678594231*	354759666*	679103197*	796143690*	20735384*
					552022036*		796143690*

-continued next page-

* Multiplicity is given modulo 1073741527

Affine D8 - Level 2 - Class (0,0) - Page 2 of 3

	Rp 1	Rp 2	Rp 3	Rp 4	Rp 5	Rp 6	Rp 7
Wt 4	3	3	4	1	0	0	0
	61	61	88	31	4	3	0
	691	691	1044	435	88	61	3
	5706	5706	8888	4108	1044	691	61
	38188	38188	60808	30213	8888	5706	691
	219105	219105	354816	186120	60808	38188	5706
	1115394	1115394	1830684	1002187	354816	219105	38188
	5154965	5154965	8554744	4849041	1830684	1115394	219105
	21984823	21984823	36824832	21489606	8554744	5154965	1115394
	87573309	87573309	147859800	88454723	36824832	21984823	5154965
	328853343	328853343	559103336	341752134	147859800	87573309	21984823
	98969717*	98969717*	932272865*	175837622*	559103336	328853343	87573309
	773594160*	773594160*	428262042*	57359606*	932272865*	98969717*	328853343
	177379833*	177379833*	27284597*	559001376*	428262042*	773594160*	98969717*
				409091072*	27284597*	177379833*	773594160*
							177379833*
Wt 5	10	10	15	6	1	1	0
	180	180	277	122	27	21	1
	1856	1856	2917	1382	362	255	21
	14222	14222	22740	11412	3377	2252	255
	89562	89562	145255	76376	24831	15991	2252
	488502	488502	801804	438210	153635	96558	15991
	2382506	2382506	3950658	2230788	832288	513685	96558
	10613160	10613160	17754702	10309930	4053258	2466535	513685
	43836328	43836328	73901716	43969646	18078519	10876446	2466535
	169766848	169766848	288159342	175146618	74869588	44618395	10876446
	621767544	621767544	1061801993*	657706686	290919738	171985286	44618395
	20735384*	20735384*	501695089*	197939434*	1069358533*	627811605	171985286
	796143690*	796143690*	679103197*	473446793*	521644432*	36624222*	627811605
			552022036*	354759666*	730091269*	836599213*	36624222*
					678594231*		836599213*
Wt 6	0	0	0	0	0	1	0
	35	35	56	28	8	8	0
	525	525	840	420	120	100	7
	4893	4893	7896	4068	1248	885	91
	34720	34720	56560	30080	9880	6627	875
	205968	205968	338688	185796	65056	41888	6573
	1069635	1069635	1774248	1001304	370944	232936	41825
	5006736	5006736	8371224	4847036	1887120	1162611	232680
	21531699	21531699	36261408	21484764	8738264	5306940	1162302
	86256870	86256870	146216632	88444332	37388256	22445595	5305896
	325190474	325190474	554513984	341729048	149502968	88907486	22444305
	89163139*	89163139*	919943297*	175790333*	563692688	332551378	88903678
	748213227*	748213227*	396245750*	57260840*	944602433*	108852046*	332546590
	113669161*	113669161*	1020415020*	558806833*	460278334*	799121091*	108839299*
				408703895*	107895701*	241387799*	799104855*
							241348001*

continued next page

* Multiplicity is given modulo 1073741527

Affine D8 - Level 2 - Class (0,0) - Page 3 of 3

	Rp 1	Rp 2	Rp 3	Rp 4	Rp 5	Rp 6	Rp 7
Wt 7	35	35	56	28	8	7	1
	525	525	840	420	120	91	8
	4893	4893	7896	4068	1248	875	100
	34720	34720	56560	30080	9880	6573	885
	205968	205968	338688	185796	65056	41825	6627
	1069635	1069635	1774248	1001304	370944	232680	41888
	5006736	5006736	8371224	4847036	1887120	1162302	232936
	21531699	21531699	36261408	21484764	8738264	5305896	1162611
	86256870	86256870	146216632	88444332	37388256	22444305	5306940
	325190474	325190474	554513984	341729048	149502968	88903678	22445595
	89163139*	89163139*	919943297*	175790333*	563692688	332546590	88907486
	748213227*	748213227*	396245750*	57260840*	944602433*	108839299*	332551378
	113669161*	113669161*	1020415020*	558806833*	460278334*	799104855*	108852046*
				408703895*	107895701*	241348001*	799121091*
							241387799*

* Multiplicity is given modulo 1073741527

Affine D8 - Level 2 - Class (0,1)

	Rp 1					Rp 2				
	0 1 / 00000 / 1 0					1 0 / 00000 / 0 1				
Wt 1	1	114728487	162863353*	334535022*	312336513*	0	56914641	1013750658*	872049140*	584630865*
	36	444907728	1057409910*	33364849*	275720260*	7	227611094	1019351813*	360415362*	701489831*
	527	558297150*	959782216*	59580591*	310571192*	147	857674479	198699270*	617650788*	174454366*
	5085	332079898*	228124177*	55129233*	565292667*	1701	920354317*	558454138*	396323426*	354833997*
	37917	810845416*	545018884*	646078990*		14253	815202981*	599341114*	86316405*	669712051*
	235900	106898271*	493249378*	760921174*		96419	1061369296*	366451004*	541206471*	
	1279926	265493226*	624221750*	330922976*		557811	28456099*	459099283*	457388957*	
	6230808	825275465*	875478654*	356326483*		2858558	517546456*	978262491*	520421203*	
	27753323	908066072*	704417633*	476899628*		13283793	398431733*	961196460*	659295823*	
Wt 2	7	227611094	1019351813*	360415362*	701489831*	1	114728487	162863353*	334535022*	312336513*
	147	857674479	198699270*	617650788*	174454366*	36	444907728	1057409910*	33364849*	275720260*
	1701	920354317*	558454138*	396323426*	354833997*	527	558297150*	959782216*	59580591*	310571192*
	14253	815202981*	599341114*	86316405*	669712051*	5085	332079898*	228124177*	55129233*	565292667*
	96419	1061369296*	366451004*	541206471*		37917	810845416*	545018884*	646078990*	
	557811	28456099*	459099283*	457388957*		235900	106898271*	493249378*	760921174*	
	2858558	517546456*	978262491*	520421203*		1279926	265493226*	624221750*	330922976*	
	13283793	398431733*	961196460*	659295823*		6230808	825275465*	875478654*	356326483*	
	56914641	1013750658*	872049140*	584630865*		27753323	908066072*	704417633*	476899628*	

* Multiplicity is given modulo 1073741527

Affine D8 - Level 2 - Class (1,1)

	Rp 1			Rp 2			Rp 3			Rp 4		
	0 1 / 00000 / 0 1			0 0 / 00010 / 0 0			0 0 / 01000 / 0 0			1 0 / 00000 / 1 0		
Wt 1	1	60624038	765447185*	0	42318993	623384852*	0	20388980	721097456*	0	6612809	790637323*
	22	237366037	406695369*	7	168632782	642220415*	0	84275336	389911980*	0	28994665	550937501*
	282	878615706	968344063*	127	633622394	864973481*	21	326660859	302440659*	0	118127604	539980737*
	2607	947487944*	99973105*	1380	113668768*	824522013*	357	123757605*	705687135*	35	452008264	96445170*
	19241	767655603*	1033563298*	11174	192316729*	772338750*	3612	958591433*	478873669*	581	564174866*	840431330*
	120015	504731942*		74077	529117222*		27447	8506268*	901566353*	5712	288817114*	765150145*
	656268	374487779*		423036	96576100*		172236	858806701*		42259	472462622*	659060612*
	3226128	507318440*		2148895	1020873928*		938211	701944724*		258944	598674198*	
	14518059	244133022*		9923435	291320253*		4575090	450612964*		1381296	1031606524*	
Wt 2	3	119945208	584823633*	1	84857938	443634093*	0	41983692	230744058*	0	14161839	477416840*
	65	457119530	443553228*	30	328313259	882511337*	5	167657121	584722537*	0	59551833	718793204*
	778	577992709*	710036598*	416	128259899*	577836685*	108	630904839	108571659*	10	234288980	410058469*
	6720	324889528*	330842276*	3915	970430158*	645763604*	1260	106384720*	316532174*	195	870145224	254474801*
	46661	563730430*	26376522*	28785	38659673*	479101085*	10578	173447496*	1038261313*	2128	925013060*	880058073*
	276029	823161524*		177488	933431074*		71535	481709261*	884604802*	17031	709949917*	526908999*
	1441860	496027251*		957046	881862964*		413361	1054454723*		111048	360886502*	
	6812264	977084721*		4637793	874153630*		2115157	744744297*		623301	25405720*	
	29612626	217603090*		20585226	622749166*		9813880	721957216*		3114156	754389445*	
Wt 3	10	234288980	410058469*	5	167657121	584722537*	1	84857938	443634093*	0	29612626	217603090*
	195	870145224	254474801*	108	630904839	108571659*	30	328313259	882511337*	3	119945208	584823633*
	2128	925013060*	880058073*	1260	106384720*	316532174*	416	128259899*	577836685*	65	457119530	443553228*
	17031	709949917*	526908999*	10578	173447496*	1038261313*	3915	970430158*	645763604*	778	577992709*	710036598*
	111048	360886502*		71535	481709261*	884604802*	28785	38659673*	479101085*	6720	324889528*	330842276*
	623301	25405720*		413361	1054454723*		177488	933431074*		46661	563730430*	26376522*
	3114156	754389445*		2115157	744744297*		957046	881862964*		276029	823161524*	
	14161839	477416840*		9813880	721957216*		4637793	874153630*		1441860	496027251*	
	59551833	718793204*		41983692	230744058*		20585226	622749166*		6812264	977084721*	
Wt 4	35	452008264	96445170*	21	326660859	302440659*	7	168632782	642220415*	1	60624038	765447185*
	581	564174866*	840431330*	357	123757605*	705687135*	127	633622394	864973481*	22	237366037	406695369*
	5712	288817114*	765150145*	3612	958591433*	478873669*	1380	113668768*	824522013*	282	878615706	968344063*
	42259	472462622*	659060612*	27447	8506268*	901566353*	11174	192316729*	772338750*	2607	947487944*	99973105*
	258944	598674198*		172236	858806701*		74077	529117222*		19241	767655603*	1033563298*
	1381296	1031606524*		938211	701944724*		423036	96576100*		120015	504731942*	
	6612809	790637323*		4575090	450612964*		2148895	1020873928*		656268	374487779*	
	28994665	550937501*		20388980	721097456*		9923435	291320253*		3226128	507318440*	
	118127604	539980737*		84275336	389911980*		42318993	623384852*		14518059	244133022*	

* Multiplicity is given modulo 1073741527

Affine D8 - Level 3 - Class (0,0) - Page 1 of 2

	Rp 1 0 1 00000 1 1	Rp 2 0 0 00010 1 0	Rp 3 0 0 01000 1 0	Rp 4 1 0 00000 0 2	Rp 5 1 2 00000 0 0	Rp 6 1 0 00001 0 0	Rp 7 1 0 00100 0 0	Rp 8 1 0 10000 0 0	Rp 9 1 0 00000 2 0	R10 3 0 00000 0 0
Wt 1	1	0	0	0	0	0	0	0	0	0
	85	21	0	7	7	6	0	0	0	0
	2375	861	84	280	280	321	49	0	0	0
	39504	16947	2688	5684	5684	7257	1735	105	91	0
	474710	224427	45108	77973	77973	105627	31079	3165	2632	36
	4533907	2285843	531867	820463	820463	1154022	384669	50835	41132	1030
	36394313	19218926	4951450	7098427	7098427	10251066	3719678	579735	459249	15892
	254574303	139274953	38724322	52736579	52736579	77675118	30001771	5255379	4093166	175682
	517327299*	895318170	264308618	346444014	346444014	518221106	210100121	40213290	30884239	1553004
			541447013*				238492007*	269481639	204530957	11638256
										76629256
Wt 2	3	1	0	0	0	0	0	0	0	0
	226	88	10	18	18	22	4	0	0	0
	5727	2520	418	685	685	893	227	15	15	0
	88144	41982	8508	12938	12938	17660	5298	550	492	6
	995470	501567	115881	166748	166748	234927	78466	10280	8707	198
	9039148	4752108	1207712	1665030	1665030	2401752	866328	132400	108170	3432
	69578981	37828678	10348249	13786247	13786247	20254393	7748641	1325639	1055561	41986
	469773762	262503934	76189054	98679800	98679800	147141562	58996744	11018892	8603786	404800
	700887632*	554816674*	496416679	627900323	627900323	947837631	395003961	79200418	60887176	3268644
			775362488*				231576674*	506008906	384078836	22956938
										143911581
Wt 3	10	5	1	0	0	0	0	0	0	0
	605	293	59	50	50	70	22	2	3	0
	13669	6781	1487	1685	1685	2371	778	93	96	1
	194043	99433	23649	29223	29223	41724	14474	1999	1828	39
	2060173	1088858	278365	352653	352653	511380	186965	28940	24776	735
	17805843	9675153	2632756	3341049	3341049	4914374	1882771	320158	263052	9791
	131603518	73272968	21040998	26493818	26493818	39466986	15750517	2898691	2314394	102272
	858685507	488428549	146943027	182886543	182886543	275498899	113927781	22424798	17528128	888461
			918259545	54507844*	54507844*	642657778*	732212213	152693198	117393453	6661847
								934944740	709285686	44263127
										265700434
Wt 4	7	0	0	1	0	0	0	0	0	0
	308	76	0	36	28	28	0	0	0	0
	6664	2414	238	898	854	1001	154	0	0	0
	95900	41228	6608	14808	14616	18788	4508	280	232	0
	1047312	496927	101024	180003	179277	243642	72128	7490	6077	91
	9335424	4726780	1112608	1744820	1742356	2456916	824788	111020	88356	2352
	71078546	37703512	9823128	14208646	14200942	20555920	7514408	1191275	932718	33880
	476655256	261936076	73605224	100694660	100672132	148611988	57822672	10288804	7943836	354788
	730000779*	552423905*	484837437	636715124	636652787	954375212*	389655809	75658088	57715812	3001180
							209081970*	490327848	370156028	21677040
										138313742
Wt 5	7	0	0	0	1	0	0	0	0	0
	308	76	0	28	36	28	0	0	0	0
	6664	2414	238	854	898	1001	154	0	0	0
	95900	41228	6608	14616	14808	18788	4508	280	232	0
	1047312	496927	101024	179277	180003	243642	72128	7490	6077	91
	9335424	4726780	1112608	1742356	1744820	2456916	824788	111020	88356	2352
	71078546	37703512	9823128	14200942	14208646	20555920	7514408	1191275	932718	33880
	476655256	261936076	73605224	100672132	100694660	148611988	57822672	10288804	7943836	354788
	730000779*	552423905*	484837437	636652787	636715124	954375212*	389655809	75658088	57715812	3001180
							209081970*	490327848	370156028	21677040
										138313742

* Multiplicity is given modulo 1073741527

continued next page

Affine D8 - Level 3 - Class (0,0) - Page 2 of 2

	Rp 1	Rp 2	Rp 3	Rp 4	Rp 5	Rp 6	Rp 7	Rp 8	Rp 9	Rp10
Wt 6	19 776 15496 208390 2152048 18312657 134093408 869824441	5 287 6690 98587 1082897 9640139 73093210 487597557 779818944*	0 29 1070 19767 250246 2462303 20138032 142644896 899517542	1 79 2059 32534 375724 3476737 27195149 186157343 68524793*	1 79 2059 32534 375724 3476737 27195149 186157343 68524793*	1 92 2646 44147 528357 5014971 39991070 277961453 653284375*	0 15 643 13035 175716 1811095 15356988 112002735 723633356	0 0 40 1325 22995 278700 2654320 21158178 146768860 909483696	0 0 36 1135 18973 223844 2088274 16374782 112065951 686619572	0 0 0 15 452 7411 86241 796401 6194419 42113122 256588369
Wt 7	54 1954 35638 446990 4367087 35515498 250409366 499230509*	23 890 17224 226458 2298313 19285766 139596730 896736102	3 158 3584 52756 583895 5252226 40258809 271391662 571547325*	3 204 4857 71370 777540 6858566 51544853 341060140	3 204 4857 71370 777540 6858566 51544853 341060140	4 274 6707 101042 1123390 10076492 76794833 514197322 948237556*	1 76 2084 34204 406725 3850974 30687860 213331172 252459241*	0 6 229 4564 62255 655056 5680703 42343168 279163679 588053663*	0 6 213 3972 51849 528768 4480485 32802234 213167955	0 0 3 90 1602 20370 204359 1711316 12416869 80116436
Wt 8	155 4864 80770 945294 8749299 68101988 462965781	81 2506 42142 503688 4767917 37920922 262965838 556888211*	19 590 10363 130536 1302026 10870589 78763843 507968387	10 535 11379 154803 1590158 13380742 96718970 619248368	10 535 11379 154803 1590158 13380742 96718970 619248368	15 776 16476 225938 2345434 19948530 145656209 941252214	6 280 5960 84164 905139 7972522 60134931 400236456	1 40 885 13399 155201 1465457 11768099 82805014 521880748	1 41 827 11731 129731 1185250 9288128 64139040 398250992	0 1 21 361 4786 51246 459150 3551056 24284806 149654543
Wt 9	35 1610 32158 420910 4206461 34658050 246312998 481340933*	21 882 17276 227402 2306269 19335846 139861426 897972806	7 238 4508 60508 636258 5553954 41795733 278480286	0 140 4109 65170 736352 6625326 50376153 335750628	0 140 4109 65170 736352 6625326 50376153 335750628	0 210 6062 96026 1091062 9895900 75894833 510113002	0 84 2316 36756 426375 3973246 31342936 216462932 266126803*	0 14 371 6076 74235 732760 6115067 44504432 288947071	1 22 401 5652 64055 604072 4888293 34787554 222012837	0 0 7 154 2240 25186 234507 1875300 13215412 83668564
Wt10	434 11872 179928 1969536 17304408 129134016 847616664	252 6608 99056 1089920 9692534 73394192 489077640	76 1840 27256 305456 2799715 21932904 151208284	35 1400 26348 331408 3212146 25815552 179688824	35 1400 26348 331408 3212146 25815552 179688824	56 2128 39410 494704 4814424 38943792 273035656	28 896 15792 197568 1951572 16132816 115813080 740660912	8 176 2837 35640 364628 3153920 23727610 158730464	7 168 2604 31024 304262 2548448 18709992 122815728	1 8 100 1200 12934 121184 990504 7161424 46510769

* Multiplicity is given modulo 1073741527

Weight Multiplicities for Affine E6

Affine E6 - Level 1 - Class 0

	Rp 1 1 0 00000				
Wt 1	1	15525	6190182	768291650	130256414*
	6	36280	11867310	258702803*	60002753*
	27	81816	22395359	141730441*	384755963*
	98	178794	41650050	676839645*	16416885*
	315	380051	76413078	139141995*	559125278*
	918	788004	138421358	285832610*	850407748*
	2492	1597725	247783113	67375610*	612112691*
	6372	3174210	438616728	162538712*	32649644*

* Multiplicity is given modulo 1073741527

Affine E6 - Level 2 - Class 0

	Rp 1 0 1 00000			
Wt 1	1	2621211	341605215*	86315890*
	21	8964862	714926515*	30021612*
	213	29005994	126869814*	701616633*
	1515	89427944	844898049*	76000155*
	8635	264241415	343060761*	689234248*
	42092	751801049	211621224*	975008243*
	182254	993819804*	319444565*	
	718336	145386553*	218664582*	
Wt 2	0	1384156	319678802*	234178180*
	5	4883574	257364700*	101330545*
	70	16229306	51855771*	509057885*
	585	51218023	563954586*	103514136*
	3694	154490784	735194497*	212159260*
	19373	447702648	122522616*	393424306*
	88706	178039423*	820590179*	28597826*
	365602	167633360*	918824044*	
Wt 3	6	4884669	257491482*	107576320*
	72	16231286	52064713*	518735265*
	594	51222060	564311430*	118695700*
	3712	154497888	735772299*	235393978*
	19428	447716275	123483981*	429325692*
	88812	178062827*	822122067*	82921870*
	365867	167676180*	921316497*	
	1384650	319750764*	238092134*	

	Rp 2 0 0 10001			
Wt 1	5	5576968	844509590*	217645731*
	74	18607476	12277351*	692061893*
	634	58926729	694211261*	246465458*
	4067	178284960	108117338*	1008927828*
	21569	518058266	545317838*	738753126*
	99603	378282018*	262135754*	706130019*
	413288	718393763*	408565170*	
	1573363	714702801*	608088489*	
Wt 2	1	2986831	567166627*	357933692*
	22	10259072	345269273*	152070964*
	229	33316280	256935023*	283588316*
	1658	103049854	270893939*	12606227*
	9569	305365452	738580257*	860618459*
	47086	871033878	252346833*	682419962*
	205394	253526244*	557366935*	595715962*
	814432	1048119157*	151178089*	
Wt 3	20	10256598	344988843*	138542634*
	219	33311149	256452428*	262270424*
	1638	103040772	270109051*	1053614708*
	9505	305347797	737265766*	809837319*
	46962	871003388	250243649*	605346212*
	205074	253469851*	553925394*	
	813832	1048023895*	145753197*	
	2985471	566997280*	349233217*	

	Rp 3 2 0 00000			
Wt 1	0	481796	290964568*	95847447*
	1	1740615	484064332*	265349182*
	17	5902905	308990370*	598037935*
	160	18959778	417577234*	667622050*
	1093	58081372	191066442*	808933945*
	6066	170647634	935598353*	586681309*
	29003	483057007	197250735*	893026552*
	123719	248669955*	544572811*	
Wt 2	0	246504	15407306*	706834240*
	0	923309	295219004*	718031433*
	4	3228226	486728505*	954669993*
	53	10644874	524778050*	986458759*
	426	33366474	552938111*	83266026*
	2617	100042382	361241064*	263061427*
	13446	288368946	471443966*	83671358*
	60588	802416006	705085081*	256760311*
Wt 3	1	924723	295400884*	727567831*
	6	3230804	487030361*	969523703*
	63	10650261	525299438*	1009917264*
	446	33376032	553787877*	119344060*
	2682	100061077	362668774*	319144002*
	13572	288401310	473732812*	168919910*
	60917	802476171	708840964*	
	247122	15509178*	712765568*	

* Multiplicity is given modulo 1073741527

Affine E6 - Level 3 - Class 0 - Page 1 of 2

	Rp 1	Rp 2	Rp 3	Rp 4	Rp 5	Rp 6	Rp 7	Rp 8
	0	0	0	0	0	1	1	3
	0	0	0	0	0	1	0	0
	00003	00011	00100	11000	30000	00000	10001	00000
Wt 1	1	0	0	0	0	0	0	0
	6	6	0	0	0	0	0	0
	63	108	45	64	24	0	20	0
	626	1422	870	1212	448	64	480	0
	5463	14196	9955	13344	4776	1212	6375	24
	40470	112885	84930	109886	38112	13344	60050	448
	258576	753270	591735	743724	251188	109886	449745	4776
	1458060	4378029	3544790	4349850	1436592	743724	2847870	38112
	7408836	22757614	18854100	22679484	7349985	4349850	15834715	251188
	34483714	107859135	91015980	107653614	34330358	22679484	79286550	1436592
	148923237	472837386	405129975	472320036	148540431	107653614	363968730	7349985
			608909333*			472320036	478490873*	34330358
								148540431
Wt 2	1	1	0	0	0	0	0	0
	17	32	15	24	10	0	5	0
	240	566	355	523	204	24	180	0
	2489	6542	4610	6360	2342	523	2815	10
	20394	56915	42690	56249	19868	6360	29275	204
	139111	404039	315790	401854	137417	56249	234720	2342
	822587	2459435	1980580	2452833	817540	401854	1564070	19868
	4335711	13259202	10928695	13240546	4321618	2452833	9052190	137417
	20787424	64744886	54379215	64694988	20750099	13240546	46830100	817540
	92023602	291039154	248317520	290911794	91929143	64694988	220924895	4321618
	380471522	145402829*	1054237015*	145090558*	380241620	290911794	964383565	20750099
						145090558*		91929143
								380241620
Wt 3	1	2	1	2	1	0	0	0
	34	79	48	79	34	2	19	0
	498	1283	879	1283	498	79	477	1
	4982	13631	9982	13631	4982	1283	6366	34
	38844	110894	84996	110894	38844	13631	60024	498
	253525	746926	591888	746926	253525	110894	449682	4982
	1443479	4359288	3545140	4359288	1443479	746926	2847726	38844
	7369044	22705627	18854841	22705627	7369044	4359288	15834391	253525
	34380340	107722327	91017501	107722327	34380340	22705627	79285872	1443479
	148665810	472492900	405133036	472492900	148665810	107722327	363967353	7369044
		863961349*	608915264*	863961349*		472492900	478488135*	34380340
						863961349*		148665810
Wt 4	0	0	0	1	1	0	0	0
	10	24	15	32	17	0	5	0
	204	523	355	566	240	24	180	0
	2342	6360	4610	6542	2489	523	2815	10
	19868	56249	42690	56915	20394	6360	29275	204
	137417	401854	315790	404039	139111	56249	234720	2342
	817540	2452833	1980580	2459435	822587	401854	1564070	19868
	4321618	13240546	10928695	13259202	4335711	2452833	9052190	137417
	20750099	64694988	54379215	64744886	20787424	13240546	46830100	817540
	91929143	290911794	248317520	291039154	92023602	64694988	220924895	4321618
	380241620	145090558*	1054237015*	145402829*	380471522	290911794	964383565	20750099
						145090558*		91929143
								380241620

continued next page

* Multiplicity is given modulo 1073741527

Affine E6 - Level 3 - Class 0 - Page 2 of 2

	Rp 1	Rp 2	Rp 3	Rp 4	Rp 5	Rp 6	Rp 7	Rp 8
Wt 5	0	0	0	0	1	0	0	0
	0	0	0	6	6	0	0	0
	24	64	45	108	63	0	20	0
	448	1212	870	1422	626	64	480	0
	4776	13344	9955	14196	5463	1212	6375	24
	38112	109886	84930	112885	40470	13344	60050	448
	251188	743724	591735	753270	258576	109886	449745	4776
	1436592	4349850	3544790	4378029	1458060	743724	2847870	38112
	7349985	22679484	18854100	22757614	7408836	4349850	15834715	251188
	34330358	107653614	91015980	107859135	34483714	22679484	79286550	1436592
	148540431	472320036	405129975	472837386	148923237	107653614	363968730	7349985
			608909333*			472320036	478490873*	34330358
								148540431
Wt 6	10	24	15	24	10	1	5	0
	204	523	355	523	204	32	180	1
	2342	6360	4610	6360	2342	566	2815	17
	19868	56249	42690	56249	19868	6542	29275	240
	137417	401854	315790	401854	137417	56915	234720	2489
	817540	2452833	1980580	2452833	817540	404039	1564070	20394
	4321618	13240546	10928695	13240546	4321618	2459435	9052190	139111
	20750099	64694988	54379215	64694988	20750099	13259202	46830100	822587
	91929143	290911794	248317520	290911794	91929143	64744886	220924895	4335711
	380241620	145090558*	1054237015*	145090558*	380241620	291039154	964383565	20787424
						145402829*		92023602
								380471522
Wt 7	4	8	4	8	4	0	1	0
	88	213	136	213	88	8	62	0
	1112	2926	2054	2926	1112	213	1191	4
	10112	28103	20932	28103	10112	2926	13874	88
	73873	213290	165462	213290	73873	28103	120047	1112
	458910	1364055	1091036	1364055	458910	213290	845766	10112
	2513159	7643846	6262571	7643846	2513159	1364055	5111160	73873
	12427538	38518572	32182330	38518572	12427538	7643846	27379188	458910
	56453609	177768412	150977623	177768412	56453609	38518572	132951335	2513159
	238604442	761605728	655925898	761605728	238604442	177768412	594756188	12427538
			520360523*			761605728	333404880*	56453609
								238604442
Wt 8	24	64	45	64	24	6	20	1
	448	1212	870	1212	448	108	480	6
	4776	13344	9955	13344	4776	1422	6375	63
	38112	109886	84930	109886	38112	14196	60050	626
	251188	743724	591735	743724	251188	112885	449745	5463
	1436592	4349850	3544790	4349850	1436592	753270	2847870	40470
	7349985	22679484	18854100	22679484	7349985	4378029	15834715	258576
	34330358	107653614	91015980	107653614	34330358	22757614	79286550	1458060
	148540431	472320036	405129975	472320036	148540431	107859135	363968730	7408836
			608909333*			472837386	478490873*	34483714
								148923237

* Multiplicity is given modulo 1073741527

Affine E6 - Level 3 - Class 1 - Page 1 of 2

	Rp 1 0 0 01001	Rp 2 0 1 10000	Rp 3 0 0 20001	Rp 4 1 0 00002	Rp 5 1 0 00010	Rp 6 2 0 10000
Wt 1	1	0	1	0	0	0
	51	15	48	5	4	0
	948	376	826	135	143	14
	10906	4991	9099	1924	2216	332
	93728	46842	75847	18978	22901	4203
	657420	349619	520228	147026	182782	38204
	3960141	2207414	3079458	955420	1213784	278375
	21160319	12243896	16224502	5423349	7005749	1725275
	102543934	61182098	77711149	27620742	36162630	9427672
	457948420	280373169	343628921	128618494	170283352	46532354
	833704951*	120192773*	345375502*	555272364	742145860	211031691
	1036625914*	477474261*	154276718*	98704298*	879319166*	890660348
	966413608*	869626061*	388274*	1068566194*	914064368*	311363819*
	939537007*	577477663*	4450333*	67715517*	748144490*	384136883*
	474849875*	737854070*	707206668*	30594594*	837771918*	250549294*
	876291730*	652806980*	469224580*	384694825*	863107047*	696264526*
	372773167*	661917377*		262683272*	886351841*	96945461*
						1050802871*
Wt 2	4	1	5	0	0	0
	143	51	135	14	15	1
	2216	948	1924	332	376	48
	22901	10906	18978	4203	4991	826
	182782	93728	147026	38204	46842	9099
	1213784	657420	955420	278375	349619	75847
	7005749	3960141	5423349	1725275	2207414	520228
	36162630	21160319	27620742	9427672	12243896	3079458
	170283352	102543934	128618494	46532354	61182098	16224502
	742145860	457948420	555272364	211031691	280373169	77711149
	879319166*	833704951*	98704298*	890660348	120192773*	343628921
	914064368*	1036625914*	1068566194*	311363819*	477474261*	345375502*
	748144490*	966413608*	67715517*	384136883*	869626061*	154276718*
	837771918*	939537007*	30594594*	250549294*	577477663*	388274*
	863107047*	474849875*	384694825*	696264526*	737854070*	4450333*
	886351841*	876291730*	262683272*	96945461*	652806980*	707206668*
		372773167*		1050802871*	661917377*	469224580*
Wt 3	0	0	1	0	0	0
	16	5	22	0	0	0
	382	150	375	40	44	4
	5020	2251	4386	784	912	128
	46954	23042	38877	8924	10758	1889
	350004	183279	280594	75216	93196	18838
	2208611	1215366	1732010	518183	655698	146535
	12247362	7010412	9446784	3073346	3954988	953866
	61191535	36175532	46583666	16207393	21145870	5418791
	280397626	170317240	211163118	77665774	102505524	27608190
	120253478*	742230994	890983617	343513973	457850880	128585664
	477619490*	879524995*	312131259*	345095571*	833466809*	555190212
	869962320*	914545506*	385902586*	153618157*	1036064269*	98506387*
	578234084*	749236113*	254499578*	1072627102*	965128911*	1068105095*
	739511958*	840183252*	704882188*	1113900*	936677898*	66672537*
	656356501*	868306572*	115319926*	699978934*	468642174*	28297207*
	669356086*	897320164*	15429622*	453913176*	863112891*	379753936*
						252285898*

continued next page

* Multiplicity is given modulo 1073741527

Affine E6 - Level 3 - Class 1 - Page 2 of 2

	Rp 1	Rp 2	Rp 3	Rp 4	Rp 5	Rp 6
Wt 4	5	0	4	1	0	0
	150	44	128	22	16	0
	2251	912	1889	375	382	40
	23042	10758	18838	4386	5020	784
	183279	93196	146535	38877	46954	8924
	1215366	655698	953866	280594	350004	75216
	7010412	3954988	5418791	1732010	2208611	518183
	36175532	21145870	27608190	9446784	12247362	3073346
	170317240	102505524	128585664	46583666	61191535	16207393
	742230994	457850880	555190212	211163118	280397626	77665774
	879524995*	833466809*	98506387*	890983617	120253478*	343513973
	914545506*	1036064269*	1068105095*	312131259*	477619490*	345095571*
	749236113*	965128911*	66672537*	385902586*	869962320*	153618157*
	840183252*	936677898*	28297207*	254499578*	578234084*	1072627102*
	868306572*	468642174*	379753936*	704882188*	739511958*	1113900*
	897320164*	863112891*	252285898*	115319926*	656356501*	699978934*
				15429622*	669356086*	453913176*
Wt 5	15	4	14	1	1	0
	376	143	332	48	51	5
	4991	2216	4203	826	948	135
	46842	22901	38204	9099	10906	1924
	349619	182782	278375	75847	93728	18978
	2207414	1213784	1725275	520228	657420	147026
	12243896	7005749	9427672	3079458	3960141	955420
	61182098	36162630	46532354	16224502	21160319	5423349
	280373169	170283352	211031691	77711149	102543934	27620742
	120192773*	742145860	890660348	343628921	457948420	128618494
	477474261*	879319166*	311363819*	345375502*	833704951*	555272364
	869626061*	914064368*	384136883*	154276718*	1036625914*	98704298*
	577477663*	748144490*	250549294*	388274*	966413608*	1068566194*
	737854070*	837771918*	696264526*	4450333*	939537007*	67715517*
	652806980*	863107047*	96945461*	707206668*	474849875*	30594594*
	661917377*	886351841*	1050802871*	469224580*	876291730*	384694825*
					372773167*	262683272*
Wt 6	44	16	40	4	5	1
	912	382	784	128	150	22
	10758	5020	8924	1889	2251	375
	93196	46954	75216	18838	23042	4386
	655698	350004	518183	146535	183279	38877
	3954988	2208611	3073346	953866	1215366	280594
	21145870	12247362	16207393	5418791	7010412	1732010
	102505524	61191535	77665774	27608190	36175532	9446784
	457850880	280397626	343513973	128585664	170317240	46583666
	833466809*	120253478*	345095571*	555190212	742230994	211163118
	1036064269*	477619490*	153618157*	98506387*	879524995*	890983617
	965128911*	869962320*	1072627102*	1068105095*	914545506*	312131259*
	936677898*	578234084*	1113900*	66672537*	749236113*	385902586*
	468642174*	739511958*	699978934*	28297207*	840183252*	254499578*
	863112891*	656356501*	453913176*	379753936*	868306572*	704882188*
		669356086*		252285898*	897320164*	115319926*
						15429622*

* Multiplicity is given modulo 1073741527

Affine E6 - Level 4 - Class 0 - Page 1 of 3

	Rp 1 0 0 01002	Rp 2 0 0 01010	Rp 3 0 2 00000	Rp 4 0 1 10001	Rp 5 0 0 20002	Rp 6 0 0 20010	Rp 7 1 0 00003
Wt 1	1 2344505 51 18384556 1303 21181 250348	0 2507292 30 20156988 1110 20520 258064	0 254880 0 2303046 46 1382 22316	0 2131314 15 17887257 720 15177 207156	1 2035752 48 1212 19231 221922	0 2312173 31 18250678 1112 19896 243391	0 330026 5 2793163 135 2440 32129
Wt 2	1 3947815 92 2472 38799 438734	1 4285314 83 33110958 2407 39703 464516	0 452552 2 3888488 128 2966 42574	0 3686719 43 29657966 1614 30045 378953	1 3427137 95 2378 35632 390099	1 3947815 92 2472 38799 438734	0 558506 5 4587318 215 4260 55856
Wt 3	10 11054378 500 10120 133910 1343606	9 12133359 474 10192 140009 1444107	1 1362846 26 10724978 719 12029 144515	5 10671617 301 7408 110822 1214097	15 9499400 506 9547 120863 1179291	10 11054378 500 10120 133910 1343606	0 1629422 30 974 15772 180376
Wt 4	4 6645742 226 5122 73030 774310	3 7250971 205 5042 75384 825233	0 788674 7 6482126 301 5992 78772	1 6308292 119 48748682 3524 58363 683553	5 5741117 233 4884 66550 684258	4 6645742 226 5122 73030 774310	0 960990 14 7573508 476 8358 101490
Wt 5	0 1354160 16 11098608 552 10520 136080	0 1448538 10 12153034 490 10330 140874	0 142452 0 1353776 20 652 11660	0 1216342 5 10681959 310 7472 111274	1 1192147 22 591 10047 123713	0 1354160 16 11098608 552 10520 136080	0 182536 0 1639216 40 1040 16224
Wt 6	0 2312173 31 18250678 1112 19896 243391	0 2507292 30 20156988 1110 20520 258064	0 254880 0 2303046 46 1382 22316	0 2131314 15 17887257 720 15177 207156	1 2035752 48 1212 19231 221922	1 2344505 51 18384556 1303 21181 250348	0 314785 0 2726538 70 1929 29074
Wt 7	5 4017009 150 2955 41805 454190	0 4292640 80 2416 39888 465872	0 447848 0 3868520 104 2792 41592	0 3691180 44 1632 30192 379848	4 3455844 128 2620 37012 396792	0 3923774 75 2319 37803 433465	1 586555 22 4704933 375 5346 61812
Wt 8	15 11099283 550 10505 136075 1354179	10 12144262 486 10282 140530 1446646	0 1357010 19 10701976 663 11745 143103	4 10674692 298 7400 110878 1214590	14 9521292 522 9676 121816 1184080	10 11051965 495 10099 133767 1343053	1 1647159 48 12442120 1107 16590 184395
Wt 9	31 18254936 1115 19925 243579 2313138	28 20128960 1082 20298 256746 2500794	1 2316173 58 17542170 1478 22899 257838	14 17872598 706 15065 206483 2127955	33 15613117 1085 18432 217778 2017062	31 18254936 1115 19925 243579 2313138	1 2745978 86 2065 29925 319150
Wt10	10 11051965 495 10099 133767 1343053	10 12144262 486 10282 140530 1446646	0 1357010 19 10701976 663 11745 143103	4 10674692 298 7400 110878 1214590	14 9521292 522 9676 121816 1184080	15 11099283 550 10505 136075 1354179	0 1623898 30 12348118 951 15597 179235

continued next page

Affine E6 - Level 4 - Class 0 - Page 2 of 3

	Rp 8 1 0 00011	Rp 9 1 0 00100	Rp10 1 0 11000	Rp11 1 0 30000	Rp12 2 1 00000	Rp13 2 0 10001	Rp14 4 0 00000
Wt 1	0 782160 4 6921768 204 4776 71076	0 488460 0 4521150 75 2445 41400	0 766032 0 6849696 144 4272 67936	0 314785 0 2726538 70 1929 29074	0 30222 0 336666 0 2980412 66 1928	0 188690 0 1896850 14 15618606 692 14208	0 531 0 7773 0 82197 0 698376 20
Wt 2	0 1361920 8 11563080 424 9320 130288	0 872193 3 7668387 213 5302 79560	0 1361920 8 11563080 424 9320 130288	0 558506 5 4587318 215 4260 55856	0 57779 0 597197 3 5021865 190 4180	0 345510 0 3276216 46 1602 28362	0 1151 0 14803 0 145021 1 1170066 58
Wt 3	0 4056768 64 2112 36288 433664	0 2654271 35 21333688 1186 21791 273323	0 4056768 64 2112 36288 433664	0 1629422 30 974 15772 180376	0 194596 1 1784778 43 1067 16875	0 1097011 5 9407901 325 7289 103659	0 4606 0 49149 1 426869 17 336
Wt 4	0 2364792 24 19225128 968 18624 239688	0 1529253 10 12844516 507 10824 148379	0 2364792 24 19225128 968 18624 239688	0 960990 14 7573508 476 8358 101490	0 106113 0 1034763 11 8333106 442 8365	0 618770 1 5575435 124 3444 54543	0 2268 0 26829 0 248610 4 1924788 133
Wt 5	0 433664 0 4056768 64 2112 36288	0 270585 0 2641155 25 21280150 1115 21255	0 433664 0 4056768 64 2112 36288	0 182536 0 1639216 40 1040 16224	0 15856 0 189162 0 1759808 26 912	0 101788 0 1087664 4 9365920 288 7004	0 261 0 4177 0 46826 0 416687 10
Wt 6	0 766032 0 6849696 144 4272 67936	0 488460 0 4521150 75 2445 41400	0 782160 4 6921768 204 4776 71076	0 330026 5 2793163 135 2440 32129	0 30222 0 336666 0 2980412 66 1928	0 188690 0 1896850 14 15618606 692 14208	0 531 0 7773 0 82197 0 698376 20
Wt 7	0 1385600 16 11665936 528 10128 135072	0 865500 0 7639680 180 5060 78180	0 1338240 0 11460224 320 8512 125504	0 541914 0 4515309 145 3701 52513	0 55520 0 585872 0 4972272 144 3808	0 341736 0 3257768 40 1520 27720	0 1004 0 13908 0 140560 0 1150668 40
Wt 8	1 4070808 77 31723615 2215 36912 436825	0 2647660 30 21306740 1150 21520 271940	0 4045716 60 31620424 2060 35892 431376	0 1623898 30 12348118 951 15597 179235	0 191446 0 1770816 30 13698250 964 16252	0 1092622 5 9387850 310 7165 102801	0 4314 0 47702 0 420832 10 3136968 280
Wt 9	2 6885732 174 4524 69506 774096	1 4539840 90 35060353 2572 42199 492604	2 6885732 174 4524 69506 774096	1 2745978 86 2065 29925 319150	0 344704 2 3016254 96 2180 31790	0 1910724 19 15678300 755 14692 191522	0 8446 0 85556 1 713035 34 643
Wt10	0 4045716 60 31620424 2060 35892 431376	0 2647660 30 21306740 1150 21520 271940	1 4070808 77 31723615 2215 36912 436825	1 1647159 48 12442120 1107 16590 184395	0 191446 0 1770816 30 13698250 964 16252	0 1092622 5 9387850 310 7165 102801	0 4314 0 47702 0 420832 10 3136968 280

-continued next page-

Affine E6 - Level 4 - Class 0 - Page 3a of 3

	Rp 1	Rp 2	Rp 3	Rp 4	Rp 5	Rp 6	Rp 7
Wt11	0 3923774 75 2319 37803 433465	0 4292640 80 2416 39888 465872	0 447848 0 3868520 104 2792 41592	0 3691180 44 1632 30192 379848	4 3455844 128 2620 37012 396792	5 4017009 150 2955 41805 454190	0 541914 0 4515309 145 3701 52513
Wt12	170 4670 70290 760526 6585158	169 4773 73825 817605 7218035	11 6531566 353 6392 81116 800162	106 3417 57696 680107 6292778	165 4342 63297 668031 5670318	170 4670 70290 760526 6585158	10 7526308 424 7962 99198 949894
Wt13	76 2332 37904 434056 3926644	72 33064632 2320 39184 461896 4273728	3 3905847 143 3091 43345 456473	40 29635740 1584 29836 377812 3681376	76 2212 34572 384596 3402384	76 2332 37904 434056 3926644	4 4570700 200 4136 55100 554708
Wt14	360 9080 127776 1313856	374 9498 136224 1426444	36 828 12924 149508 1387184	261 7113 109197 1206090	351 8307 113828 1144686	360 9080 127776 1313856	24 864 14976 175656 1607040

Affine E6 - Level 4 - Class 0 - Page 3b of 3

	Rp 8	Rp 9	Rp10	Rp11	Rp12	Rp13	Rp14
Wt11	0 1338240 0 11460224 320 8512 125504	0 865500 0 7639680 180 5060 78180	0 1385600 16 11665936 528 10128 135072	1 586555 22 4704933 375 5346 61812	0 55520 0 585872 0 4972272 144 3808	0 341736 0 3257768 40 1520 27720	0 1004 0 13908 0 140560 0 1150668 40
Wt12	24 19225128 968 18624 239688 2364792	15 12915315 575 11366 151632 1545480	24 19225128 968 18624 239688 2364792	10 7526308 424 7962 99198 949894	1 1070539 32 8482473 646 9757 113737	5 5638703 180 3900 57339 633002	0 30275 1 264124 17 1987568 240 2935
Wt13	8 11563080 424 9320 130288 1361920	4 7693140 232 5468 80620 877696	8 11563080 424 9320 130288 1361920	4 4570700 200 4136 55100 554708	0 609464 8 5074520 248 4616 60296	1 3298266 62 1743 29273 350327	0 15944 0 150364 4 1192292 88 1360
Wt14	64 2112 36288 433664 4056768	45 1350 22986 280447 2688576	64 2112 36288 433664 4056768	24 864 14976 175656 1607040	6 1863324 108 1602 20226 211986	20 480 8460 110284 1129392	1 57105 6 461387 63 626 6273

Weight Multiplicities for Affine E7

Affine E7 - Level 1 - Class 0

	Rp 1 0 1000000				
Wt 1	1	32305	20247710	409083254*	52130125*
	7	80465	40786690	124757309*	1000298023*
	35	192899	80782800	1023243108*	594048580*
	140	447146	157510780	459326577*	696348890*
	490	1006145	302666903	158673252*	871120898*
	1547	2204475	573720808	540260062*	512278697*
	4522	4715510	1073720305	894147031*	419189929*
	12405	9869132	911765248*	342295921*	69617194*

* Multiplicity is given modulo 1073741527

Affine E7 - Level 2 - Class 0

	Rp 1			Rp 2			Rp 3			Rp 4		
	0 0000002			0 0000010			0 0100000			0 2000000		
Wt 1	1	15185366	860746153*	0	21487060	5489886*	0	10887281	442129889*	0	1890966	484751538*
	7	56010206	438779003*	7	79736895	6825976*	0	41699350	635752235*	0	7717248	50464389*
	98	195398924	1005253341*	112	279649272	992344255*	27	150297182	1069901860*	0	29369361	864462500*
	875	649221356	1004434232*	1127	933461926	829109605*	357	513863497	436676431*	21	105269769	272183012*
	6328	992006864*	102916940*	8393	834962363*	28069590*	3129	603361649*	1034426821*	273	358163253	283967645*
	37730	954810585*		51253	573512081*		21063	956953446*	583209155*	2346	90147962*	837799612*
	195979	439567071*		269906	335555151*		119245	815118976*		15540	409434913*	880336127*
	909960	922343527*		1267378	766417110*		591682	70311827*		86835	179709984*	
	3860731	769025281*		5423439	930482992*		2647883	943767129*		426300	615311937*	
Wt 2	1	29376311	866682427*	1	41703574	637091639*	0	21484233	4523035*	0	3858105	768070971*
	24	105284542	276032603*	29	150306456	1072197406*	6	79731159	5109986*	0	15180260	859015094*
	287	358192150	290658078*	368	513880938	440697712*	109	279637227	989381348*	5	55998828	435801064*
	2382	90205830*	849159411*	3152	603397407*	1041193535*	1114	933438538	823984507*	86	195377421	1000009207*
	15653	409543944*	899632922*	21143	957018937*	594775892*	8360	834915972*	19401073*	850	649176766	995579280*
	87098	179918407*		119409	815246372*		51153	573425227*	995556458*	6237	991924812*	87711835*
	426981	615692363*		592136	70539571*		269676	335390404*		37543	954649524*	
	1892440	485451952*		2648788	944192007*		1266796	766118062*		195446	439277875*	
	7720644	51708338*		10889464	442872854*		5422192	929935947*		908893	921800006*	
Wt 3	5	55998828	435801064*	6	79731159	5109986*	1	41703574	637091639*	0	7720644	51708338*
	86	195377421	1000009207*	109	279637227	989381348*	29	150306456	1072197406*	1	29376311	866682427*
	850	649176766	995579280*	1114	933438538	823984507*	368	513880938	440697712*	24	105284542	276032603*
	6237	991924812*	87711835*	8360	834915972*	19401073*	3152	603397407*	1041193535*	287	358192150	290658078*
	37543	954649524*		51153	573425227*	995556458*	21143	957018937*	594775892*	2382	90205830*	849159411*
	195446	439277875*		269676	335390404*		119409	815246372*		15653	409543944*	899632922*
	908893	921800006*		1266796	766118062*		592136	70539571*		87098	179918407*	
	3858105	768070971*		5422192	929935947*		2648788	944192007*		426981	615692363*	
	15180260	859015094*		21484233	4523035*		10889464	442872854*		1892440	485451952*	
Wt 4	21	105269769	272183012*	27	150297182	1069901860*	7	79736895	6825976*	1	15185366	860746153*
	273	358163253	283967645*	357	513863497	436676431*	112	279649272	992344255*	7	56010206	438779003*
	2346	90147962*	837799612*	3129	603361649*	1034426821*	1127	933461926	829109605*	98	195398924	1005253341*
	15540	409434913*	880336127*	21063	956953446*	583209155*	8393	834962363*	28069590*	875	649221356	1004434232*
	86835	179709984*		119245	815118976*		51253	573512081*		6328	992006864*	102916940*
	426300	615311937*		591682	70311827*		269906	335555151*		37730	954810585*	
	1890966	484751538*		2647883	943767129*		1267378	766417110*		195979	439567071*	
	7717248	50464389*		10887281	442129889*		5423439	930482992*		909960	922343527*	
	29369361	864462500*		41699350	635752235*		21487060	5489886*		3860731	769025281*	

* Multiplicity is given modulo 1073741527

Affine E7 - Level 2 - Class 1

	Rp 1					Rp 2				
	1					0				
	0000000					1000001				
Wt 1	1	38284904	559126805*	806786400*	693174487*	0	27935059	653456644*	55722454*	1027832698*
	28	137694781	1056432345*	659085371*	408085434*	7	104065696	108059360*	454501*	18788469*
	350	469895244	332140582*	957780602*	459716524*	132	366192582	577824305*	1048686945*	1000522957*
	2968	457337540*	595010693*	953156416*	375837778*	1379	152183789*	1031476250*	276590446*	1063736520*
	19768	492625624*	541456924*	944919288*		10500	705843795*	1023411479*	552303235*	446210568*
	111048	468673007*	953889510*	704122014*		64925	282750847*	617684531*	219233328*	
	548359	170857779*	379797601*	151132224*		345044	513077673*	555097769*	31026827*	
	2444564	754173822*	769486725*	857192040*		1631259	380392444*	886559457*	90402271*	
	10021165	373426668*	695680881*	905046830*		7019068	315377796*	486564551*	225491773*	
Wt 2	6	73116745	1052576668*	953481507*	514657450*	1	54330972	1070464310*	963255475*	382943555*
	105	255932796	258728258*	369118114*	53420666*	34	196513512	419074333*	743088091*	234367132*
	1056	852828343	373776397*	226539375*	957923756*	449	673945162	1057988615*	396024131*	846227220*
	7849	573149812*	606404929*	498772947*	510771322*	3926	58098623*	325991450*	492407941*	845195466*
	47710	831298934*	766470837*	121920067*		26686	481321054*	581551666*	428225205*	
	250286	31902173*	274923293*	471469486*		152156	537284385*	223652083*	660418365*	
	1171154	86618761*	840700188*	968236692*		760063	14924775*	659116652*	682575295*	
	4996951	180340914*	843378599*	246828076*		3419988	697352133*	1037458564*	545445311*	
	19746758	790750326*	983621678*	504716704*		14128426	652828105*	767448507*	616279891*	

* Multiplicity is given modulo 1073741527

Affine E7 - Level 3 - Class 0 - Page 1 of 2

	Rp 1 0 0010000	Rp 2 1 0000001	Rp 3 0 1000002	Rp 4 0 1000010	Rp 5 0 1100000	Rp 6 0 3000000
Wt 1	1	5	0	0	0	0
	78	265	25	27	2	0
	1954	5539	826	945	130	1
	28776	73475	13836	16497	2946	56
	307293	730888	160179	196236	41062	1160
	2624764	5925417	1448931	1810458	422623	15361
	18966351	41113495	10937534	13873977	3514414	152651
	120138185	252028649	71737648	92096568	24874164	1237134
	683527203	321687419*	420025671	544570749	154957263	8582324
	333699636*	649381463*	90452832*	778200577*	869552640	52606129
	1018614861*	203086781*	269207487*	533035454*	174940651*	291256824
	954504401*	169494849*	52085797*	353680705*	915708591*	406449315*
	721184206*	495225738*	286098767*	809226018*	755538768*	547234038*
	92696341*	513978771*	420493475*	142833188*	265422853*	897091826*
	379478319*	555998070*	423949177*	155175486*	533633742*	786550654*
	116605357*	197727177*	174942421*	304435289*	135787193*	822025864*
	579678906*		60584202*	108236182*	1038816384*	107319910*
						244762071*
Wt 2	0	1	0	0	0	0
	21	84	7	6	0	0
	714	2157	294	321	35	0
	12222	32306	5733	6699	1092	15
	143451	349077	73304	88836	17640	435
	1310582	3008160	712152	883539	199122	6665
	9968070	21887540	5675607	7162551	1767801	72537
	65772630	139419588	38864007	49695819	13152153	626634
	387000103	796920723	235719267	304634778	85261687	4564128
	996778421*	939785251*	219913220*	612944243*	494417532	29085639
	555701472*	785103888*	83032647*	1055830778*	465460216*	166288752
	878472219*	732018261*	550902443*	750320819*	950280052*	868344890
	884994768*	11178557*	632266066*	1066586675*	234342951*	976196046*
	238795267*	981026134*	211749373*	208034865*	1060274378*	723058231*
	464578106*	456948811*	64656722*	262263108*	588727140*	370958546*
	134612995*	768077147*	251251410*	663317980*	907418031*	972917588*
	135298914*	201624138*	73141658*	442489248*	163431432*	679172618*
					184177552*	803799384*
						342212342*
Wt 3	0	6	1	0	0	0
	71	276	34	27	0	0
	1916	5609	881	945	111	0
	28609	73820	14100	16497	2829	45
	306645	732325	161270	196236	40497	1090
	2622490	5930730	1452955	1810458	420291	15016
	18958955	41131414	10951142	13873977	3505839	151214
	120115624	252084818	71780508	92096568	24845314	1231821
	683461953	321853134*	420152941	544570749	154866928	8564405
	333519449*	649846002*	90812190*	778200577*	869286131	52549960
	1018136885*	204332875*	270178945*	533035454*	174193151*	291091109
	953280924*	172711316*	54613348*	353680705*	913701520*	405984776*
	718150542*	503250612*	292455495*	809226018*	750351735*	545987944*
	85387125*	533401629*	436001285*	142833188*	252463611*	893875359*
	362320354*	601739831*	460756580*	155175486*	502221705*	778525780*
	77275056*	302812329*	260147899*	304435289*	61696832*	802603006*
	491471930*		253369916*	108236182*	868339614*	61578149*
						139676919*

continued next page

* Multiplicity is given modulo 1073741527

Affine E7 - Level 3 - Class 0 - Page 2 of 2

	Rp 1	Rp 2	Rp 3	Rp 4	Rp 5	Rp 6
Wt 4	5	22	1	1	0	0
	247	780	91	99	10	0
	5005	13637	2232	2592	396	5
	65262	162305	32371	38971	7396	165
	641740	1497652	341697	421614	91869	2860
	5158907	11477675	2893327	3634794	874071	33931
	35560570	76191852	20764188	26451668	6856324	312806
	216833837	450531692	130798820	168505227	46362041	2396302
	121527145*	272995241*	740758579	963209475	278319404	15904716
	683098749*	178145341*	616324541*	734388614*	440608379*	94039057
	567960735*	608441211*	170400606*	687558407*	66569466*	505205336
	194275142*	1558102*	124319683*	381334965*	990320676*	354890113*
	425817501*	806792571*	46727731*	382137103*	215127085*	818981021*
	458554328*	946350862*	328407701*	880202195*	810176596*	807667916*
	754925190*	87091388*	521430528*	62379902*	105979800*	409782144*
	482513311*	412286650*	927371953*	498135004*	23879733*	282566151*
				543836480*	81089540*	217343901*
						769691059*
Wt 5	22	75	5	6	1	0
	723	2102	276	321	46	1
	12276	32042	5624	6699	1164	24
	143704	347988	72787	88836	18003	490
	1311601	3004153	710046	883539	200667	6929
	9971733	21874032	5667954	7162551	1773624	73626
	65784711	139377189	38838516	49695819	13172140	630641
	387037250	796795284	235640166	304634778	85325342	4577636
	996886232*	939432389*	219681714*	612944243*	494608089	29128038
	555999351*	784153587*	82388025*	1055830778*	466001573*	166414191
	879261115*	729554975*	549182984*	750320819*	951750202*	868697752
	887007861*	5006140*	627849146*	1066586675*	238181226*	977146347*
	243766078*	966021694*	200775432*	208034865*	1069952444*	725521517*
	476496187*	421459489*	38193572*	262263108*	612383086*	377130963*
	162439483*	686198064*	189131583*	663317980*	963643785*	987922028*
	198720216*	16963641*	1004585572*	442489248*	293719627*	714661940*
					479170925*	885678467*
Wt 6	77	225	21	27	7	1
	1967	5329	777	945	175	7
	28882	72535	13538	16497	3206	98
	307867	727232	158753	196236	42287	1379
	2627219	5912566	1443127	1810458	427588	16312
	18975439	41071683	10916535	13873977	3532424	156326
	120168671	251901135	71668156	92096568	24934189	1250041
	683621841	321319243*	419811392	544570749	155143884	8624208
	333975515*	648366617*	89830003*	778200577*	870100126	52733761
	1019378649*	200401123*	267484804*	533035454*	176469588*	291625275
	956527504*	162638513*	47520574*	353680705*	919799625*	407464526*
	726341400*	478273132*	274445031*	809226018*	766081709*	549920269*
	105408013*	473253139*	391713320*	142833188*	291702683*	903949325*
	409886560*	460684349*	354943717*	155175486*	597212634*	803504823*
	187416177*	1053644856*	13842291*	304435289*	285510057*	862753877*
					309118079*	202638020*
						462591853*

* Multiplicity is given modulo 1073741527

Affine E7 - Level 4 - Class 0 - Page 1 of 6 (See page guide at right.)

	Rp 1 0 0000004	Rp 2 0 0000012	Rp 3 0 0000020	Rp 4 0 0000101	Rp 5 0 0001000	Rp 6 2 0000000	Rp 7 0 0100002	Rp 8 0 0100010
Wt 1	1	0	0	0	0	0	0	0
	7	7	0	0	0	0	0	0
	98	175	63	77	0	0	27	0
	1379	3710	2079	3311	980	603	945	588
	20062	67643	45738	79618	33320	19845	24570	22755
	259112	987238	722799	1319563	629308	364581	449253	473235
	2850379	11628596	8867214	16649906	8505595	4822335	6166818	6867042
					91152425	50810585	67677204	77694708
Wt 2	1	1	0	0	0	0	0	0
	24	46	17	22	0	0	6	0
	490	1367	796	1267	390	248	321	212
	8828	30306	20789	36150	15240	9217	10482	9951
	129588	495753	364550	663394	315692	184502	218016	231280
	1541658	6281642	4790987	8966946	4558496	2599530	3244018	3616149
	15286681	64941494	50677224	96731801	51355455	28748408	37679470	43208862
					480283924			423007152
Wt 3	1	1	1	0	0	0	0	0
	24	63	39	54	15	10	6	6
	763	2510	1734	2863	1150	723	663	644
	15616	58154	42404	75372	34541	20715	21915	23149
	232992	927094	700012	1288819	636104	369421	430949	475435
	2714437	11302018	8735386	16469468	8537644	4845055	6059862	6877586
	26158386	112604048	88550114	169866651	91286125	50904890	67136862	77739428
Wt 4	1	2	1	1	0	0	0	0
	56	162	101	156	50	34	27	21
	1648	5704	3980	6817	2855	1781	1707	1681
	31434	119546	87991	158323	74186	44150	48084	51280
	438862	1769989	1344879	2494445	1248456	720702	857366	952857
	4858553	20419192	15856290	30043351	15730425	8885413	11277057	12867020
	44950022	194887709	153813072	296170939	160379057	89104038	118863094	138198277
Wt 5	1	3	2	3	1	1	0	0
	109	362	249	410	160	108	75	73
	3321	12158	8789	15400	6849	4229	4035	4212
	60673	237229	177488	323032	155633	91906	101297	110555
	806759	3306929	2535701	4736486	2409228	1382798	1666524	1874048
	8556804	36350471	28393244	54063005	28615093	16092047	20659110	23740000
	76358324	333654612	264429077	511056974	278948777	154435979	208030998	243044680

1 2
3 4
5 6

page guide

continued next page

Affine E7 - Level 4 - Class 0 - Page 2 of 6

	Rp 9	Rp10	Rp11	Rp12	Rp13	Rp14	Rp15
	0	0	1	0	0	0	0
	0200000	1010000	1000001	2000002	2000010	2100000	4000000
Wt 1	0	0	0	0	0	0	0
	0	0	0	0	0	0	0
	0	0	0	0	0	0	0
	0	0	225	21	0	0	0
	1197	2079	11025	777	504	0	0
	37674	68355	260127	20958	20048	1575	0
	667926	1240890	4100670	384657	415772	49575	420
	8553404	16196271	49267449	5232423	5962243	867405	12516
					66604909	10957695	211575
							2591715
Wt 2	0	0	0	0	0	0	0
	0	0	0	0	0	0	0
	0	0	75	5	0	0	0
	510	879	4630	276	190	0	0
	18022	32290	123932	9199	9004	695	0
	344628	635067	2121106	189618	206216	24120	190
	4673066	8796882	27033875	2778866	3168702	452380	6256
	50072508	95850013	277986742	31801870	37270159	6029260	112315
						63663055	1444108
							14803513
Wt 3	0	0	0	0	0	0	0
	0	0	0	0	0	0	0
	30	39	235	5	5	0	0
	1585	2687	11130	601	600	40	0
	41252	73704	260843	19684	21319	2205	15
	691928	1277580	4104465	377512	426166	55600	626
	8690556	16404138	49284621	5198459	6031111	909975	14603
	88325338	169540364	484355805	56620914	66983677	11206410	226201
						112132945	2678025
							25981002
Wt 4	0	0	0	0	0	0	0
	0	0	5	0	0	0	0
	80	129	695	25	20	0	0
	3740	6491	25791	1576	1595	115	0
	86057	155612	536195	43045	47019	5175	35
	1332440	2476672	7812033	747615	849785	115345	1426
	15788287	29944837	88742820	9632713	11232528	1743335	29744
	153484727	295690267	835757698	99885972	118636068	20275080	428463
						194198260	4803638
							44692204
Wt 5	0	0	0	0	0	0	0
	2	3	23	0	0	0	0
	249	410	1863	76	75	3	0
	8789	15400	57523	3779	4035	362	1
	177488	323032	1077719	90746	101297	12158	109
	2535701	4736486	14639358	1449767	1666524	237229	3321
	28393244	54063005	157937548	17595276	20659110	3306929	60673
	264429077	511056974		174327459	208030998	36350471	806759
						333654612	8556804
							76358324

-continued next page-

Affine E7 - Level 4 - Class 0 - Page 3 of 6

	Rp 1	Rp 2	Rp 3	Rp 4	Rp 5	Rp 6	Rp 7	Rp 8
Wt 6	0	0	0	0	0	1	0	0
	35	120	84	140	56	42	21	20
	1470	5320	3815	6636	2892	1848	1638	1680
	30184	116725	86778	156912	74473	44492	47432	51268
	431860	1753430	1337560	2485336	1249794	722428	853020	952896
	4823448	20333376	15817900	29993663	15736840	8892408	11251912	12867232
	44793140	194492368	153633193	295929690	160403796	89131504	118738220	138200160
Wt 7	5	10	4	5	0	0	1	0
	165	475	287	455	145	96	99	70
	3815	13240	9190	15911	6726	4135	4311	4181
	64014	244980	180551	327112	154881	91350	103519	110341
	825680	3352726	2554644	4762476	2405434	1380043	1680778	1872895
	8650676	36585279	28493732	54203869	28598405	16080092	20736913	23734693
			264902711	511731865	278882727	154389119	208406441	243022985
Wt 8	5	14	9	13	4	3	1	1
	281	948	658	1097	441	286	228	226
	7172	26665	19440	34396	15630	9503	9524	10065
	119041	471140	354778	650226	317963	186146	210803	231976
	1484316	6142004	4732910	8886659	4569562	2607458	3199818	3619665
	15010627	64247148	50378533	96310865	51402824	28782056	37446308	43224350
				874431052	480467374	265101358	360838346	423068552
Wt 9	15	40	30	39	15	10	5	6
	626	2205	1585	2687	1150	723	600	644
	14603	55600	41252	73704	34541	20715	21319	23149
	226201	909975	691928	1277580	636104	369421	426166	475435
	2678025	11206410	8690556	16404138	8537644	4845055	6031111	6877586
	25981002	112132945	88325338	169540364	91286125	50904890	66983677	77739428
Wt10	35	115	80	129	50	34	20	21
	1426	5175	3740	6491	2855	1781	1595	1681
	29744	115345	86057	155612	74186	44150	47019	51280
	428463	1743335	1332440	2476672	1248456	720702	849785	952857
	4803638	20275080	15788287	29944837	15730425	8885413	11232528	12867020
	44692204	194198260	153484727	295690267	160379057	89104038	118636068	138198277

continued next page

Affine E7 - Level 4 - Class 0 - Page 4 of 6

	Rp 9	Rp10	Rp11	Rp12	Rp13	Rp14	Rp15
Wt 6	0	0	0	0	0	0	0
	0	0	7	0	0	0	0
	84	140	714	28	21	0	0
	3815	6636	25928	1575	1638	120	0
	86778	156912	536900	43050	47432	5320	35
	1337560	2485336	7815276	747312	853020	116725	1470
	15817900	29993663	88755604	9631216	11251912	1753430	30184
	153633193	295929690		99876861	118738220	20333376	431860
						194492368	4823448
							44793140
Wt 7	0	0	0	0	0	0	0
	0	0	22	1	0	0	0
	209	351	1852	91	67	0	0
	8370	14778	57443	3905	3896	295	0
	174311	318292	1077271	91527	99939	11415	85
	2516184	4707236	14637231	1453754	1656733	231400	3054
	28290251	53908049	157928631	17612972	20601309	3270335	58582
	263945647	510327104		174398031	207735945	36154527	793695
							8487236
Wt 8	0	0	0	0	0	0	0
	9	13	78	1	1	0	0
	658	1097	4664	224	228	14	0
	19440	34396	124171	8796	9524	948	5
	354778	650226	2122412	187244	210803	26665	281
	4732910	8886659	27039928	2767187	3199818	471140	7172
	50378533	96310865	278011616	31751476	37446308	6142004	119041
		874431052			360838346	64247148	1484316
							15010627
Wt 9	1	0	0	0	0	0	0
	39	54	235	5	6	1	0
	1734	2863	11130	601	663	63	1
	42404	75372	260843	19684	21915	2510	24
	700012	1288819	4104465	377512	430949	58154	763
	8735386	16469468	49284621	5198459	6059862	927094	15616
	88550114	169866651	484355805	56620914	67136862	11302018	232992
						112604048	2714437
							26158386
Wt10	1	1	5	0	0	0	0
	101	156	695	25	27	2	0
	3980	6817	25791	1576	1707	162	1
	87991	158323	536195	43045	48084	5704	56
	1344879	2494445	7812033	747615	857366	119546	1648
	15856290	30043351	88742820	9632713	11277057	1769989	31434
	153813072	296170939	835757698	99885972	118863094	20419192	438862
						194887709	4858553
							44950022

continued next page

Affine E7 - Level 4 - Class 0 - Page 5 of 6

	Rp 1	Rp 2	Rp 3	Rp 4	Rp 5	Rp 6	Rp 7	Rp 8
Wt11	15	45	29	45	15	11	6	5
	665	2310	1621	2761	1151	734	636	635
	14959	56615	41621	74450	34549	20788	21645	23092
	228936	917330	695030	1282765	636150	369796	428695	475151
	2693656	11249351	8709204	16434961	8537855	4846685	6046070	6876366
	26060520	112350726	88424014	169696543	91286969	50911171	67062483	77734753
					820071622	451039187	619982709	729341570
Wt12	45	135	84	135	45	36	27	20
	1530	5410	3810	6606	2826	1779	1674	1661
	30609	117385	86748	156681	73944	44142	47736	51154
	433770	1756430	1337370	2484270	1247324	720444	854460	952114
	4832133	20347292	15817080	29988594	15724835	8884164	11259091	12863756
		194544964	153629709	295910280	160357306	89097117	118766374	138184645
Wt13	85	295	209	351	145	96	67	70
	3054	11415	8370	14778	6726	4135	3896	4181
	58582	231400	174311	318292	154881	91350	99939	110341
	793695	3270335	2516184	4707236	2405434	1380043	1656733	1872895
	8487236	36154527	28290251	53908049	28598405	16080092	20601309	23734693
			263945647	510327104	278882727	154389119	207735945	243022985
Wt14	190	695	510	879	390	248	190	212
	6256	24120	18022	32290	15240	9217	9004	9951
	112315	452380	344628	635067	315692	184502	206216	231280
	1444108	6029260	4673066	8796882	4558496	2599530	3168702	3616149
	14803513	63663055	50072508	95850013	51355455	28748408	37270159	43208862
					480283924			423007152
Wt15	420	1575	1197	2079	980	603	504	588
	12516	49575	37674	68355	33320	19845	20048	22755
	211575	867405	667926	1240890	629308	364581	415772	473235
	2591715	10957695	8553404	16196271	8505595	4822335	5962243	6867042
					91152425	50810585	66604909	77694708

continued next page

Affine E7 - Level 4 - Class 0 - Page 6 of 6

	Rp 9	Rp10	Rp11	Rp12	Rp13	Rp14	Rp15
Wt11	0 29 1621 41621 695030 8709204 88424014	0 45 2761 74450 1282765 16434961 169696543	1 246 11206 261247 4106280 49291825 484381801	0 7 621 19815 378179 5201355 56632081	0 6 636 21645 428695 6046070 67062483 619982709	0 0 45 2310 56615 917330 11249351 112350726	0 0 0 15 665 14959 228936 2693656 26060520
Wt12	0 84 3810 86748 1337370 15817080 153629709	0 135 6606 156681 2484270 29988594 295910280	6 708 25869 536610 7813734 88749372	1 34 1664 43520 750136 9643220 99928228	0 27 1674 47736 854460 11259091 118766374	0 0 135 5410 117385 1756430 20347292 194544964	0 0 0 45 1530 30609 433770 4832133
Wt13	4 287 9190 180551 2554644 28493732 264902711	5 455 15911 327112 4762476 54203869 511731865	22 1852 57443 1077271 14637231 157928631	1 91 3905 91527 1453754 17612972 174398031	1 99 4311 103519 1680778 20736913 208406441	0 10 475 13240 244980 3352726 36585279	0 0 5 165 3815 64014 825680 8650676
Wt14	17 796 20789 364550 4790987 50677224	22 1267 36150 663394 8966946 96731801	75 4630 123932 2121106 27033875 277986742	5 276 9199 189618 2778866 31801870	6 321 10482 218016 3244018 37679470	1 46 1367 30306 495753 6281642 64941494	0 1 24 490 8828 129588 1541658 15286681
Wt15	63 2079 45738 722799 8867214	77 3311 79618 1319563 16649906	225 11025 260127 4100670 49267449	21 777 20958 384657 5232423	27 945 24570 449253 6166818 67677204	7 175 3710 67643 987238 11628596	1 7 98 1379 20062 259112 2850379

Affine E7 - Level 4 - Class 1 - Page 1 of 2

	Rp 1 0 0010001	Rp 2 1 0000002	Rp 3 1 0000010	Rp 4 1 0100000	Rp 5 0 1000003	Rp 6 0 1000011	Rp 7 0 1000100	Rp 8 0 1100001	Rp 9 1 2000000	Rp10 0 3000001
Wt 1	1 217 9583 220053 3426305 40856790 399474504 122036333* 1037871753*	5 485 16388 325155 4599317 51157562 474088488 579459005*	4 468 17007 351394 5105288 57900153 544681454 124605362* 637383620*	0 51 3052 81182 1386050 17640153 181205231 503418660* 202243239*	0 25 1365 34976 585249 7352913 74838837 646800474 604008841*	0 45 2911 80978 1424591 18529678 193550814 633496495* 262687623*	0 20 1598 48635 902140 12166002 130532672 102050630* 620539330*	0 2 405 16330 353883 5293689 61279393 585496767 514921049*	0 0 53 2980 76020 1260892 15709607 158749069 289953635* 612134594*	0 0 1 193 7191 147778 2128724 23952952 223823872 732111959*
Wt 2	0 21 1624 48944 904645 12182016 130619328 102464104* 622325167*	1 84 3374 79380 1283296 15835337 159368587 292707562* 623386943*	0 56 3151 82215 1393805 17687229 181450521 504555459* 207040149*	0 0 399 16331 346486 5076029 57749106 543984077 121663452* 625870821*	0 7 294 8176 154924 2171332 24172988 224840056 736399428*	0 6 483 17150 360129 5332250 61483849 586460798 519057458*	0 0 196 9359 218343 3415755 40800914 399211267 120907331* 1033391480*	0 0 35 2737 79254 1411996 18454429 193162592 631708667* 255174657*	0 0 0 385 15323 316990 4548761 50889286 472824149 574041180*	0 0 0 15 1190 33250 572712 7278446 74456781 645050182 596689155*
Wt 3	0 71 4021 104980 1774887 22447384 229547395 917236171* 90656926*	1 176 7220 159827 2432554 28547319 275772622 138525445* 623853862*	1 168 7412 171401 2685065 32171914 315761929 503395745* 276331279*	0 15 1219 37620 704028 9552859 102964642 930783190* 869860782*	0 7 546 16269 298207 3991639 42606846 382311217 838252754*	0 12 1137 37114 718547 9984122 109554489 1004424640* 465945016*	0 5 606 21926 450108 6504015 73441843 688444959 201084093* 219251431*	0 0 135 6967 170826 2766019 33891302 338211654 729224863* 27218375*	0 0 15 1199 35496 644290 8546972 90538264 807231010 901691783*	0 0 0 65 3111 72097 1121311 13331251 129947729 10817332* 464360671*
Wt 4	5 606 21926 450108 6504015 73441843 688444959 201084093* 219251431*	15 1199 35496 644290 8546972 90538264 807231010 901691783*	15 1219 37620 704028 9552859 102964642 930783190* 869860782*	1 168 7412 171401 2685065 32171914 315761929 503395745* 276331279*	0 65 3111 72097 1121311 13331251 129947729 10817332* 464360671*	0 135 6967 170826 2766019 33891302 338211654 729224863* 27218375*	0 71 4021 104980 1774887 22447384 229547395 917236171* 90656926*	0 12 1137 37114 718547 9984122 109554489 1004424640* 465945016*	0 1 176 7220 159827 2432554 28547319 275772622 138525445* 623853862*	0 0 7 546 16269 298207 3991639 42606846 382311217 838252754*
Wt 5	0 71 4061 105535 1779637 22478360 229716352 918043861*	6 276 8310 168357 2486310 28836876 277155087 144517435*	0 171 7577 173351 2700235 32265046 316248185 505644950*	0 0 999 35676 690930 9479079 102600432 929161480* 863221486*	1 34 881 19109 317107 4097747 43130545 384645132 847818869*	0 27 1377 39379 734581 10077980 110032169 1006604785* 475053556*	0 0 531 21240 445356 6476625 73304073 687821769 198498280*	0 0 111 6564 166956 2738454 33729987 337393752 725513942*	0 0 0 934 32881 625205 8432987 89949949 804519644 890286371*	0 0 0 45 2745 68490 1095271 13177921 129168119 7277323*

continued next page

* Multiplicity is given modulo 1073741527

Affine E7 - Level 4 - Class 1 - Page 2 of 2

	Rp 1	Rp 2	Rp 3	Rp 4	Rp 5	Rp 6	Rp 7	Rp 8	Rp 9	Rp10
Wt 6	5	22	16	0	1	1	0	0	0	0
	612	1306	1245	149	91	156	66	10	0	0
	22011	36500	37920	7211	3395	7207	3959	1091	150	5
	450865	651369	706432	169863	74309	172761	104459	36611	6913	500
	6509124	8588321	9568117	2675527	1135111	2778484	1771506	714630	157312	15766
	73470586	90748232	103046865	32121141	13404974	33959587	22428790	9959679	2416257	294308
	688586164	808187112	931174450*	315521628	130297236	338542164	229457333	109424156	28457821	3967434
	201708368*	905676398*	871547726*	502361009*	12323789*	730673825*	916841122*	1003808674*	275339322	42478468
				272211028*		33073404*	89060033*	463301065*	136627490*	381707582
									616196831*	835674221*
Wt 7	20	53	51	4	1	2	1	0	0	0
	1598	2980	3052	468	193	405	217	45	5	0
	48635	76020	81182	17007	7191	16330	9583	2911	485	25
	902140	1260892	1386050	351394	147778	353883	220053	80978	16388	1365
	12166002	15709607	17640153	5105288	2128724	5293689	3426305	1424591	325155	34976
	130532672	158749069	181205231	57900153	23952952	61279393	40856790	18529678	4599317	585249
	102050630*	289953635*	503418660*	544681454	223823872	585496767	399474504	193550814	51157562	7352913
	620539330*	612134594*	202243239*	124605362*	732111959*	514921049*	122036333*	633496495*	474088488	74838837
				637383620*			1037871753*	262687623*	579459005*	646800474
										604008841*
Wt 8	66	150	149	16	5	10	5	1	0	0
	3959	6913	7211	1245	500	1091	612	156	22	1
	104459	157312	169863	37920	15766	36611	22011	7207	1306	91
	1771506	2416257	2675527	706432	294308	714630	450865	172761	36500	3395
	22428790	28457821	32121141	9568117	3967434	9959679	6509124	2778484	651369	74309
	229457333	275339322	315521628	103046865	42478468	109424156	73470586	33959587	8588321	1135111
	916841122*	136627490*	502361009*	931174450*	381707582	1003808674*	688586164	338542164	90748232	13404974
	89060033*	616196831*	272211028*	871547726*	835674221*	463301065*	201708368*	730673825*	808187112	130297236
								33073404*	905676398*	12323789*
Wt 9	196	385	399	56	15	35	21	6	1	0
	9359	15323	16331	3151	1190	2737	1624	483	84	7
	218343	316990	346486	82215	33250	79254	48944	17150	3374	294
	3415755	4548761	5076029	1393805	572712	1411996	904645	360129	79380	8176
	40800914	50889286	57749106	17687229	7278446	18454429	12182016	5332250	1283296	154924
	399211267	472824149	543984077	181450521	74456781	193162592	130619328	61483849	15835337	2171332
	120907331*	574041180*	121663452*	504555459*	645050182	631708667*	102464104*	586460798	159368587	24172988
	1033391480*		625870821*	207040149*	596689155*	255174657*	622325167*	519057458*	292707562*	224840056
									623386943*	736399428*
Wt10	531	934	999	171	45	111	71	27	6	1
	21240	32881	35676	7577	2745	6564	4061	1377	276	34
	445356	625205	690930	173351	68490	166956	105535	39379	8310	881
	6476625	8432987	9479079	2700235	1095271	2738454	1779637	734581	168357	19109
	73304073	89949949	102600432	32265046	13177921	33729987	22478360	10077980	2486310	317107
	687821769	804519644	929161480*	316248185	129168119	337393752	229716352	110032169	28836876	4097747
	198498280*	890286371*	863221486*	505644950*	7277323*	725513942*	918043861*	1006604785*	277155087	43130545
								475053556*	144517435*	384645132
										847818869*

* Multiplicity is given modulo 1073741527

Weight Multiplicities for Affine E8

Affine E8 - Level 1

	Rp 1 0 10000000				
Wt 1	1	62337	59068372	882071914*	754353614*
	8	164560	124505880	1063503217*	1005972877*
	44	417140	257770964	377995504*	1006623135*
	192	1020416	524871424	45655873*	700664219*
	726	2418710	1052316364	427920852*	853810195*
	2464	5573568	1005750217*	509756845*	266184294*
	7704	12520744	832753459*	766484293*	194507021*
	22528	27484160	287029279*	942242413*	640722352*

* Multiplicity is given modulo 1073741527

Affine E8 - Level 2

	Rp 1 0 01000000				Rp 2 0 00000001				Rp 3 0 20000000			
Wt 1	1	46185805	551001974*	151700007*	7	74415792	701020338*	596919661*	0	7834368	651953488*	762393186*
	43	196645749	593144965*	40737251*	154	304997978	417246826*	343998386*	1	35829240	228161043*	336801103*
	682	784082225	868296946*	645610761*	1855	102885215*	846373080*	598862115*	37	151881220	617028321*	898012847*
	6977	805581129*	557935949*	593609614*	16115	9943397*	162061137*	618447837*	564	603299259	272443177*	321951579*
	54590	913948849*	733743163*	330848928*	112681	1069586276*	877330912*	435948798*	5650	117168161*	866756525*	165512120*
	354120	800618597*	1012924557*	63119583*	672200	914987008*	949119156*	916617581*	43603	571772396*	151815398*	475829967*
	1994549	62379372*	991831121*		3545165	18324994*	823628912*		280031	789371139*	313397785*	954027449*
	10046718	530307477*	308030948*		16927232	366149445*	186011447*		1565079	531377074*	663313063*	
Wt 2	0	21766280	169962116*	181836535*	1	35829240	228161043*	336801103*	0	3545165	18324994*	823628912*
	8	96145624	405268299*	494056841*	37	151881220	617028321*	898012847*	0	16927232	366149445*	186011447*
	184	395668264	443027757*	330098545*	564	603299259	272443177*	321951579*	7	74415792	701020338*	596919661*
	2272	458094377*	1033728022*	900840535*	5650	117168161*	866756525*	165512120*	154	304997978	417246826*	343998386*
	20056	253321653*	362211687*	53451814*	43603	571772396*	151815398*	475829967*	1855	102885215*	846373080*	598862115*
	141848	353466018*	973218186*	59961679*	280031	789371139*	313397785*	954027449*	16115	9943397*	162061137*	618447837*
	853576	547914021*	118709923*	943405423*	1565079	531377074*	663313063*	685358183*	112681	1069586276*	877330912*	435948798*
	4532904	764369414*	268996636*		7834368	651953488*	762393186*		672200	914987008*	949119156*	916617581*
Wt 3	8	96145624	405268299*	494056841*	35	151878010	616561106*	869889598*	1	16928758	366383076*	200843507*
	184	395668264	443027757*	330098545*	553	603291909	271581030*	275269045*	8	74417948	701364659*	618707187*
	2272	458094377*	1033728022*	900840535*	5629	117155575*	865371330*	93833032*	164	305004026	417949080*	382270552*
	20056	253321653*	362211687*	53451814*	43527	571745614*	149341066*	359069948*	1867	102893928*	847409378*	654906002*
	141848	353466018*	973218186*	59961679*	279891	789326115*	309472255*	776440633*	16181	9965307*	164078376*	714399893*
	853576	547914021*	118709923*	943405423*	1564673	531286622*	656493785*		112767	1069618221*	880306446*	575954731*
	4532904	764369414*	268996636*		7833641	651803922*	751700181*		672543	915060776*	954686991*	
	21766280	169962116*	181836535*		35827404	227874214*	318670003*		3545635	18433314*	831828334*	

* Multiplicity is given modulo 1073741527

Affine E8 - Level 3

	Rp 1 0 00100000	Rp 2 1 00000000	Rp 3 0 11000000	Rp 4 0 10000001	Rp 5 0 30000000
Wt 1	1 684609763* 147 977189383* 5087 599587358* 96673 290036965* 1286013 26926273* 13374878 577056935* 115745531 401047756* 866932282 813971545* 404170568* 907166022* 483197851* 117393148*	6 517623830* 421 972002921* 11185 4857186* 183363 736684923* 2209375 822928159* 21370477 891719935* 174817916 768156722* 177525736* 115539472* 507864837* 912708999* 736216567* 942973502*	0 411971949* 2 769430645* 253 1041850382* 8154 279191467* 149063 741549680* 1932331 286355731* 19722537 909739201* 168218836 868997510* 171612310* 161059477* 696990371* 850512650*	0 643983768* 41 436894284* 2010 1003865065* 45606 161737105* 679180 1049898215* 7656196 221973503* 70451136 452940123* 554124684 124989398* 620058256* 910315074* 360529640* 670187118*	0 662550144* 0 1022356827* 1 816163735* 111 799122673* 3348 776383657* 58980 242398143* 745677 951510534* 7472734 805841202* 62838097 48342366* 459907818 615090036* 856985618*
Wt 2	0 1036112533* 36 644291668* 1702 347927302* 37992 560750386* 560097 517661723* 6269296 529098783* 57381954 766284932* 449428752 735712422* 957283850* 257568803* 989346* 644900496*	1 239930014* 120 135833330* 4014 246994212* 75172 99390148* 990960 334745341* 10240272 1013416058* 88185092 682485715* 657919500 62040663* 71942819* 328379124* 515247936* 593613597*	0 100281312* 0 300083454* 63 856941168* 2772 693505967* 59192 88038464* 847896 560346187* 9296883 961789159* 83767560 373395320* 647948670 490697649* 135623660* 589161446* 502186438*	0 754331214* 8 371401132* 618 533161929* 17064 838487844* 286205 775369405* 3502776 580380970* 34275672 448204791* 282936352 540816307* 966307493* 186628237* 277653480* 522723780* 37642524*	0 284799933* 0 44936536* 0 450680975* 28 529502496* 1155 68037820* 23652 355970009* 329566 1029538235* 3541884 406428200* 31427739 783213774* 240138824 550536788* 551847677* 217491977*
Wt 3	35 642969334* 1694 344709265* 37947 553138626* 559896 500113829* 6268516 489578883* 57379236 679164164* 449420022 547397718* 957257624* 931594810* 914790* 644698316* 1035586137*	111 130767534* 3960 234845517* 74918 71023606* 989934 270099517* 10236573 869323808* 88172864 367792186* 657881826 461313617* 71833308*1055311034* 514944983* 592810518* 237879349*	1 302231403* 72 862272252* 2826 706348283* 59447 118152964* 848931 629255709* 9300635 42241410* 83780033 711401042* 647987316 143956771* 135736616*1051797499* 502500593* 101118366*	7 369552446* 609 528621750* 17011 827658774* 285959 750212224* 3501795 523320933* 34272174 321586896* 282924904 265444144* 966272537* 672507544* 277552698* 37365789* 753602646*	0 45765640* 1 452805416* 37 534767754* 1209 80704182* 23907 385632668* 330600 23586067* 3545628 557930284* 31440168 41188298* 240177277 189465081* 551959898* 285111571*
Wt 4	7 407337245* 526 1056326933* 14268 321852671* 236668 338671814* 2874146 57602244* 27962190 51489011* 229776738 175029304* 576931641* 90960408* 954015028* 625618043* 1027018909* 682800351*	29 363646903* 1346 914349711* 29622 298566673* 432574 582944161* 4808675 155286045* 43782446 755810122* 341483550 39253574* 203493680* 375771915* 634953001* 296650483* 343987794*	0 596238144* 14 55188904* 882 834994517* 22554 34380236* 361501 62636495* 4289426 273627928* 41019183 667460563* 332566280 921149457* 215664279* 911184095* 30425956* 628590291*	1 670978645* 170 951969871* 6045 958192986* 116488 190614195* 1563479 611324963* 16365208 205867848* 142321953 582057612* 1070195962* 867533124* 707164195* 987528661* 321403896* 219953203*	0 573567367* 0 762939051* 7 1010096226* 378 412980325* 9141 177959269* 141748 829217829* 1644423 582719152* 15463246 691099708* 123726929 754552812* 869793382 767072542* 125913000*
Wt 5	140 971833885* 5040 586683033* 96440 259784080* 1285040 1031484349* 13371270 422390712* 115733344 62384761* 866894152 86484088* 404058301* 482884104* 116554644* 682454065*	370 951597721* 10912 1030110692* 182136 624409047* 2204608 568995251* 21353824 329651728* 174764192 622231776* 177363497* 738236213* 507401407* 734953284* 939667771* 509278920*	8 778185182* 304 1063356535* 8427 330521703* 150296 860907901* 1937148 557391907* 19739456 438223089* 168273734 33099604* 171779033* 817695893* 697469215* 851824869* 415423199*	35 429386905* 1960 985611961* 45340 118592446* 678000 950488807* 7651662 1071923961* 70435456 1033487864* 554074568 132908558* 619908203* 360104333* 669036055* 640991551*	1 1025774634* 8 824820481* 164 820356501* 3632 826992853* 60226 359924329* 750512 144302452* 7489688 323598233* 62892880 261326474* 460073803 141757495* 857461554* 663851812*

* Multiplicity is given modulo 1073741527

Affine E8 - Level 4 - Page 1 of 3

	Rp 1 0 02000000	Rp 2 0 00010000	Rp 3 0 00000010	Rp 4 0 01000001	Rp 5 0 00000002	Rp 6 0 10100000	Rp 7 1 10000000	Rp 8 0 21000000	Rp 9 0 20000001	Rp10 0 40000000
Wt 1	1 64 5323 197889 4539345 75144514 980446559* 979001091* 859157748*	27 3408 146255 3619904 63070911 854624671* 965746271* 377629115*	77 7295 268151 6041502 98555112 197244871* 782383422* 270858313*	7 1582 84918 2398938 45726654 662156883 280888475* 669171513*	63 5035 178366 3927323 63064782 803645544 1044373435* 533340820*	0 91 8673 339093 7972906 134217866 699311291* 107393373* 872962430*	0 447 30832 984850 20325959 311714121 614812017* 16346371*	0 1 101 7595 270775 6059636 98660808 197785567* 784875838*	0 7 1568 79100 2167872 40521863 578637456 298362613* 264506415*	0 0 1 37 2274 73980 1578594 24881502 313051172 81156170*
Wt 2	0 2 621 34570 991981 19094252 278445564 75285439* 899283397*	1 382 24565 768438 15678868 238599204 771899979* 45130988* 805509490*	5 934 48224 1339042 25262168 363195670 1032267578* 478818481* 1063812658*	0 143 12940 479849 10906688 179217012 177107948* 418023314* 142965576*	4 662 32558 879321 16283606 230940932 528685068* 652221299* 83343261*	0 3 1002 58473 1725826 33859362 500717326 623848014* 589273388*	0 34 4314 187041 4678108 82161185 46493253* 776462910* 955063164*	0 0 3 894 47788 1335602 25240210 363075502 1031683614* 476239297*	0 0 142 12199 437920 9740071 157564100 945764419* 137374730* 283326264*	0 0 0 1 270 13325 353800 6450804 90347326 1036415787* 484323999*
Wt 3	0 0 189 13664 447693 9387836 145604090 733547101* 599947352* 125849087*	0 112 9464 340844 7613011 123591349 514848944* 933913120* 150834413*	1 303 19381 608991 12478735 190550722 190371182* 543346651* 453805145*	0 36 4693 205576 5173673 91255964 174497431* 102690185* 391474765*	1 220 13217 402300 8076754 121537016 400738796* 72673106* 638052174*	0 0 300 22908 774403 16578792 261018035 56054136* 143692370* 432101170*	0 7 1476 77663 2174216 41223786 594773260 541931799* 978863676* 83319522*	0 0 0 273 19012 605857 12457739 190431622 189776089* 540658401* 442628112*	0 0 35 4452 188711 4640812 80503226 13580873* 315773632* 104718116*	0 0 0 0 84 5377 162144 3208353 47668512 572289746 443625576* 1052438622*
Wt 4	0 13 1866 84025 2144659 38175171 525642625 583034785* 760291618*	6 1188 61117 1688113 31713599 454455462 1013460713* 134124045* 688028657*	22 2701 115752 2876285 50295143 683540663 145112279* 610225217* 572819002*	1 501 33910 1087488 22539616 346857692 1059050170* 423221564* 754807770*	17 1886 77513 1878836 32297079 433383016 514230852* 745221201*	0 19 3045 143215 3750838 67954308 948018480* 107300430* 727618373*	0 133 11852 435764 9849869 161212162 1011036480* 994765029* 841195591*	0 0 20 2665 115386 2873537 50278249 683450915 144686583* 608382713*	0 1 493 31712 986773 20044156 303973011 487861791* 588963416*	0 0 0 7 792 31758 753976 12758051 169119120 784773100* 395154021*

continued next page

* Multiplicity is given modulo 1073741527

Affine E8 - Level 4 - Page 2 of 3

	Rp 1	Rp 2	Rp 3	Rp 4	Rp 5	Rp 6	Rp 7	Rp 8	Rp 9	Rp10
Wt 5	0	0	0	0	1	0	0	0	0	0
	0	28	93	8	72	0	0	0	0	0
	56	3472	7480	1605	5200	76	462	0	7	0
	5166	146860	270000	85224	179648	8552	31008	77	1512	0
	196736	3624936	6054208	2401516	3936816	337835	986548	7224	78636	28
	4530724	63101619	98631798	45744144	63118624	7965064	20337496	267568	2163808	2080
	75096496	854792904*	197637929*	662255712	803926240	134170332	311782416	6036800	40497928	72494
	980195512*	966545664*	784219584*	281382527*	1045655583*	699075961*	615157931*	98528566	578504544	1567600
	977861153*	381134605*	278646846*	671393776*	538798146*	106295688*	17940901*	197104937*	297737746*	24820124
	854297556*			641657747*		868373834*	599323557*	781749696*	261749142*	312733808
								268163134*		79722350*
										135477284*
Wt 6	1	105	275	34	198	1	6	0	0	0
	210	9351	19032	4646	12955	321	1449	2	41	0
	13906	339746	605990	205045	400108	23145	77335	327	4545	1
	449683	7605021	12458420	5169418	8062144	776332	2171457	19658	189595	111
	9400970	123543340	190434565	91228663	121454404	16591425	41205447	611258	4647161	5693
	145677943	514598565*	189787185*	174347898*	400326836*	261088721	594670025	12493580	80541101	164739
	733914258*	932743672*	540695721*	101964679*	70815170*	56404339*	541419750*	190633613	13777692*	3225325
	601600785*	145839690*	442740843*	388279570*	630338201*	145265854*	976566912*	190780699*	316691812*	47763034
						438621229*		545180997*	108641120*	572755230
										445703619*
Wt 7	0	28	84	7	62	0	1	0	0	0
	55	3444	7378	1588	5080	84	464	0	8	0
	5212	146580	269016	85034	178810	8617	30980	84	1550	0
	197148	3622521	6047563	2400036	3931196	338492	985996	7378	78932	28
	4534324	63086660	98592228	45734566	63088280	7969122	20332987	269016	2166160	2149
	75117778	854710178*	197436805*	662203084	803773696	134194676	311753028	6047563	40511260	73100
	980311476*	966150488*	783285752*	281124215*	1044974259*	699195794*	615001435*	98592228	578576176	1572520
	978399989*	379395627*	274702325*	670245888*	535944809*	106851994*	17196613*	197436805*	298068122*	24848644
	856632260*	450576757*		636938653*		870692050*	596136740*	783285752*	263185607*	312885016
								274702325*		80415275*
										138434458*
Wt 8	29	1056	2296	455	1584	35	111	1	7	0
	2144	59603	111568	33237	74456	3318	11464	72	609	1
	86798	1675781	2844496	1081250	1855936	145907	431866	3471	32970	37
	2165243	31633179	50097967	22495437	32156528	3770667	9820880	123184	996737	1209
	38300816	454008661	682486048	346597496	432637408	68074700	161035472	2930737	20108176	35612
	526307059	1011265381*	140083545*	1057714274*	510698932*	948652783*	1010105343*	50625343	304325939	781515
	586177092*	124347360*	588369028*	417062870*	729953809*	110292608*	990381809*	685279902	489589557*	12922241
	773866838*					740522793*	822359864*	153314769*	596649237*	169973092
										788760494*

--continued next page-

* Multiplicity is given modulo 1073741527

Affine E8 - Level 4 - Page 3 of 3

	Rp 1	Rp 2	Rp 3	Rp 4	Rp 5	Rp 6	Rp 7	Rp 8	Rp 9	Rp10
Wt 9	6	350	826	133	580	7	29	0	1	0
	698	24143	47004	12760	31658	1078	4210	14	170	0
	35394	764722	1329214	478013	872200	59276	185899	1106	12549	7
	998413	15653358	25198581	10892896	16238106	1732038	4669108	50056	440920	378
	19135036	238452088	362844394	179132349	230691908	33898508	82103993	1353294	9760369	14441
	278667810	771156033*	1030548802*	176659432*	527475448*	500929738	46182085*	25352287	157680256	362312
	76362495*	41740356*	471187957*	415903468*	646882689*	624874402*	774960366*	363685570	946349623*	6503858
	904032131*	791316343*	1032551968*	133827244*	61578561*	593789768*	948469804*	1034636550*	140038654*	90632514
								489242245*	294481993*	1037782288*
										490290715*
Wt10	120	2960	6000	1407	4104	140	370	8	35	1
	6155	141580	255880	82776	169392	9520	29600	304	1960	8
	205960	3584680	5952560	2380668	3863896	346765	973460	10192	83020	164
	4595980	62850910	98025674	45604224	62687888	8027800	20245200	293896	2196320	3632
	75480640	853447040	194493465	661462812	801708224	134537900	311241270	6219488	40697500	85546
	982169480*	960108824*	769589312	277430111*	1035398399*	700958601*	612390619*	99593866	579567840	1655552
	986980657*					114973828*	5225301*	202549977*	302805618*	25323548
								806817856*		315272720

* Multiplicity is given modulo 1073741527

Affine E8 - Level 5 - Page 1 of 6

	Rp 1	Rp 2	Rp 3	Rp 4	Rp 5	Rp 6	Rp 7	Rp 8
	0	0	1	0	1	0	0	0
	01100000	00001000	01000000	00100001	00000001	12000000	10010000	10000010
Wt 1	1	71	6	27	171	0	0	0
	425	14641	3254	8531	28424	2	321	1081
	56199	934855	303297	656503	1635750	465	50049	128007
	3008464	32755425	13225978	25708742	53751748	51423	2855060	6297054
	94994326	779379234	364937662	660185911	149157111	2539708	93623148	187783967
	1030127497*	83796879*	891486711*	776755001*	902738390	76367399	1054902234*	769151247*
	493354828*		296076476*			560850477*	566838474*	521171735*
Wt 2	0	1	0	0	4	0	0	0
	5	792	111	370	1733	0	4	30
	3320	86110	22826	54620	159706	5	2853	8705
	296780	4121948	1474437	3040590	7014428	3130	274366	660200
	12568160	121206179	52210234	98337475	195104838	257775	12156783	25758000
	339769645	404914818*	178935010*	67794906*	726546425*	10332220	338983538	660677290
	283128638*	76011553*	170017403*	740994685*	656603958*	268634700	422430964*	780750198*
	143213014*					869162152*	1024999167*	489706214*
Wt 3	0	21	1	6	56	0	0	0
	118	5852	1149	3204	11753	0	91	364
	22831	434056	132597	295606	773065	125	20188	54405
	1422274	16700460	6496207	12858333	27717801	20955	1340220	3036883
	49143713	424231080	193514225	354516445	671041758	1208975	48167617	98293915
	84657320*	511190078*	889448687*	680060686*	442316189*	39756956	93272477*	67396836*
	326173286*	136222935*	354697829*	153071816*	1020128688*	904735494*	914966922*	737964449*
						730330228*		
Wt 4	0	5	0	1	16	0	0	0
	29	2213	374	1129	4622	0	22	110
	8936	196044	56031	129136	355937	31	7789	22219
	658015	8370210	3127804	6313648	14060080	8313	614415	1432465
	25054595	228166885	101234922	187963198	363984512	565133	24401163	50703856
	631001844	252696685*	133638333*	770704439*	544074655*	20426279	632704788	142654188*
	47852529*	781936446*	810001888*	532563060*	613392104*	495756758	340176161*	584908345*
						471000227*		
Wt 5	0	0	0	0	1	0	0	0
	0	266	28	112	624	0	0	7
	1154	36708	8910	22316	69822	0	972	3256
	130008	1992459	679172	1434329	3439556	1106	118886	296746
	6193761	63554036	26519898	50727168	103296483	114240	5947412	12874112
	180704748	344889141*	680356380	142876553*	62404554*	5136125	179336657	354679245
	557756077*	753711758*	82901692*	586656260*	263999102*	143813152	620507843*	681422482*
	552758753*		845749919*			769076162*	605026599*	162782749*

continued next page-

* Multiplicity is given modulo 1073741527

Affine E8 - Level 5 - Page 2 of 6

	Rp 9	Rp10	Rp11	Rp12	Rp13	Rp14	Rp15
	0	0	0	1	0	0	0
	11000001	10000002	20100000	20000000	31000000	30000001	50000000
Wt 1	0	0	0	0	0	0	0
	68	828	1	6	0	0	0
	17187	91782	540	3411	2	41	0
	1217405	4361369	66102	298690	401	8735	1
	45561349	127164492	3397486	12593101	39692	572774	111
	61693262*	511338905*	104638544	340013951	1851506	20518664	9098
	779683405*	497883648*	130217243*	285088166*	53730569	496617836	387594
			812075514*		47973216*	477658116*	10629383
							213177701
Wt 2	0	0	0	0	0	0	0
	0	26	0	0	0	0	0
	749	6491	6	115	0	0	0
	103480	466638	3932	22720	4	380	0
	5488733	17672114	339068	1420390	2444	49550	1
	171605668	444352178	13987426	49121950	191132	2513430	576
	549513039*	830135523*	370950661	84461248*	7378406	76092315	41242
	640208091*		794164864*	324696727*	186606652	558518653*	1496604
					291105162*	343920715*	36184807
							658084680
Wt 3	0	0	0	0	0	0	0
	14	286	0	1	0	0	0
	6419	39456	147	1193	0	8	0
	550480	2116613	26817	130711	105	3237	0
	22899170	66837747	1609888	6202556	16149	259708	28
	612312603	406449360*	54280639	180790303	884372	10356055	3703
	215577440*	579555411*	183443983*	558446778*	28082583	268862196	186388
			721476159*	557574806*	623068832	870961169*	5593586
					983210120*		119117946
							898635382*
Wt 4	0	0	0	0	0	0	0
	2	91	0	0	0	0	0
	2275	16326	36	387	0	1	0
	242435	1005130	10534	55481	26	1151	0
	11316834	34626696	748104	2999052	6448	115175	7
	326263788	815435109*	27774041	94899170	416071	5148570	1501
	324451005*	626875043*	686790661	1029334762*	14504127	143938130	88689
	376565357*		917803951*	487682843*	342845429	770101343*	2914429
					772140185*		65997185
							70172172*
Wt 5	0	0	0	0	0	0	0
	0	8	0	0	0	0	0
	224	2488	0	28	0	0	0
	42581	211638	1365	8883	0	112	0
	2607332	8878220	149156	657040	868	20504	0
	89008780	239388384	6920046	25042742	85275	1201502	210
	1005998793*	439412272*	197910188	630891136	3688800	39671588	18648
	457376967*	60168908*	855160396*	46994924*	100357328	903971427*	755636
					917213997*	724602606*	19608464
							375286720

continued next page

* Multiplicity is given modulo 1073741527

Affine E8 - Level 5 - Page 3 of 6

	Rp 1	Rp 2	Rp 3	Rp 4	Rp 5	Rp 6	Rp 7	Rp 8
Wt 6	56 9660 669500 25185542 632189349 56909626*	1897 189119 8275022 227193743 244622260 724757410	336 54761 3105863 100975400 131250352 791719105	1008 125629 6258715 187354437 765365604 493166325*	3880 341085 13867164 362085913 528760170	1 93 10092 594150 20762891 498834438 494562470*	27 7908 615917 24413978 632788680 340616360*	77 21084 1412244 50457465 140329953 566735240
Wt 7	2 1231 131434 6211859 180882259 559195693* 562823470*	243 35968 1980361 63416000 343651176 744408264*	28 8814 676773 26486053 680009551 80037965* 825615395*	103 21946 1427446 50642015 142068023 580314570	550 68046 3413600 103018210 60006781 246485613	0 3 1262 117555 5180695 144262728 772772691*	1 999 119219 5950294 179356004 620614062* 605510548*	5 3149 294377 12841164 354338523 678568930* 142420460*
Wt 8	0 391 55605 3000985 94920330 1029518192* 489024004*	77 14889 939295 32809119 779880892 87686653* 483538946*	6 3288 304181 13239064 365076984 892673435* 304635436*	28 8632 658820 25740375 660504723 779359990* 480074365*	190 29003 1645174 53859496 150127440 910061758 400725379*	0 0 399 50099 2521820 76183269 559302705* 349630147*	0 315 49958 2854149 93616452 1054862934* 566650459*	1 1121 128894 6309819 187920639 770330741* 529803974*
Wt 9	12 3546 300659 12614860 340210158 286593087*	705 83901 4089159 120853275 401875550 53859899*	104 22467 1467321 52119431 178057468 163056079	338 53522 3021879 98118630 65799288 725820029	1508 154779 6946521 194407266 720731444 615245661*	0 19 3649 267301 10450493 269767344 878134627*	6 2905 275019 12162441 339021248 422634281* 1025895463*	22 8367 653524 25671291 659821939 773833821*
Wt10	0 111 22664 1419756 49116252 84416674* 324388639*	21 5922 435589 16720788 424431343 512801261* 147343877*	0 1147 132794 6500403 193565316 889916420* 358228355*	7 3248 296549 12871196 354648243 681158904* 160886195*	64 11986 776858 27761830 671446176 445429536* 1040979469*	0 0 119 20636 1203300 39690302 904131719* 725776426*	0 84 20096 1339398 48162030 93240973* 914816461*	0 372 54672 3041372 98346075 67870678* 741559843*

continued next page-

* Multiplicity is given modulo 1073741527

Affine E8 - Level 5 - Page 4 of 6

	Rp 9	Rp10	Rp11	Rp12	Rp13	Rp14	Rp15
Wt 6	7	63	0	0	0	0	0
	2541	15318	91	447	1	7	0
	247863	987307	12397	57271	101	1568	1
	11388209	34412358	780444	3026806	8660	124355	37
	326972863	813438036	28160419	95202004	452168	5273885	2274
	330222527*	611424303*	690385269	1031956655*	14920264	145210492	100469
			945640851*		346621550	780599331*	3044948
							67148848
Wt 7	0	4	0	0	0	0	0
	244	2360	3	34	0	0	0
	43173	209260	1539	9100	3	142	0
	2616530	8847615	152950	660770	1053	21479	1
	89110311	239084682	6971868	25086848	89331	1217402	270
	1006889444*	436931832*	198438580	631298011	3743560	39852016	19936
	463970283*	42764472*	859540784*	50126520*	100907856	905581415*	772619
					921715775*	736652437*	19775722
							376630392
Wt 8	0	1	0	0	0	0	0
	63	864	0	7	0	0	0
	16989	92567	483	3374	0	35	0
	1214017	4372515	64750	297632	329	8435	0
	45521707	127282401	3377836	12577996	38144	567155	84
	61331258*	512345717*	104428793	339860425	1830041	20449816	8575
	776920581*	505184991*	128416495*	283834133*	53507902	495972246	380731
			798959504*	147991312*	46101061*	472650733*	10560539
					52389815*		212612749
							151563620*
Wt 9	1	17	0	0	0	0	0
	824	6160	19	133	0	1	0
	105233	460486	4487	23313	20	493	0
	5513508	17594901	349772	1430112	3079	52467	7
	171864084	443607208	14123716	49232609	202919	2556574	792
	551692687*	824205403*	372275897	85451087*	7524386	76553830	45057
	655871684*		804773010*	332122420*	187996137	562471789*	1542262
					302037889*		36608450
							661351711
Wt10	0	1	0	0	0	0	0
	14	310	0	0	0	0	0
	6363	39856	133	1155	0	7	0
	549304	2121684	26376	130080	98	3136	0
	22884099	66889351	1602713	6195012	15758	257629	28
	612167626	406886549*	54198984	180718898	877382	10329118	3598
	214430364*	582741447*	182712294*	557879123*	28000755	268599820	184359
			715975580*	553652817*	622331732	868865334*	5569446
					977684331*		118901080
							897021949*

continued next page

* Multiplicity is given modulo 1073741527

Affine E8 - Level 5 - Page 5 of 6

	Rp 1	Rp 2	Rp 3	Rp 4	Rp 5	Rp 6	Rp 7	Rp 8
Wt11	174 24492 1448593 49438701 87275938 345737087	4938 416176 16471130 421993724 493262563	1028 129141 6441613 192903892 884056353 314671868*	2790 285936 12721229 353086820 668038603 67098711*	9658 735260 27256972 666703613 408473109	1 284 25410 1278710 40535049 911624166* 781677302*	105 20493 1343735 48195981 93449688 915858876	275 51469 2987421 97718110 62157700 698171422
Wt12	48 9539 668060 25171754 632075804 56090888*	1925 189700 8282988 227274656 245292146 729494478	344 54910 3108012 100998752 131457254 793272268	1007 125808 6262438 187400044 765783568 496320729*	3920 342131 13881916 362236402 529995742 515937100	0 76 9818 590848 20729718 498553156 492494009*	28 7924 616007 24414328 632789250 340615183*	84 21224 1414253 50479804 140531893 568281160
Wt13	1624 142288 6370369 182539213 573074093	28441 1857905 62021540 331152901 650509309	7504 649773 26127235 676419135 50724886	18081 1357035 49777214 133882138 516203715	52040 3165146 100286871 36187676 71409643	29 2789 150535 5627260 148781070 809984159*	1056 120551 5965280 179471868 621299344	2296 272497 12522113 350973992 650125503
Wt14	552 60120 3069268 95657424 1035864223 535250447*	12103 889868 32213251 774316494 44549258	2864 293981 13091644 363519663 879456546	7279 631194 25377176 656903012 750161398	22840 1542674 52676494 139426361 829110125	6 922 62896 2707734 78156309 576137221*	350 50680 2862154 93678922 1055240682 568482130*	826 120400 6177221 186454454 757487167
Wt15	4480 326045 12973160 343840100	64470 3796128 117685344 374551314	18688 1398764 51273872 170010889	43065 2849080 96122900 47707026	114816 6363558 188280464 669107491	120 7920 349560 11492400 279840240	2960 277120 12187915 339220120	6000 599040 24926310 652310880

-continued next page-

* Multiplicity is given modulo 1073741527

Affine E8 - Level 5 - Page 6 of 6

	Rp 9	Rp10	Rp11	Rp12	Rp13	Rp14	Rp15
Wt11	34	198	1	6	0	0	0
	7181	36720	321	1449	2	41	0
	564380	2072149	31865	136090	327	4545	1
	23070134	66332357	1690439	6274630	22003	284215	111
	613936189	401928822	55190978	181516529	973323	10662951	5693
	228355319*	545658891*	191579182*	564420277*	29056525	271812904	215069
			782557431*		631575298	894383749*	5896766
							121698864
Wt12	7	62	0	1	0	0	0
	2520	15356	84	464	0	8	0
	247422	988396	12222	57350	84	1550	0
	11382308	34427636	777644	3026128	8372	123760	28
	326914189	813590213	28127841	95186346	448496	5264625	2149
	329744714*	612642045*	690085340	1031786301*	14881984	145111544	99120
	413567386*		943328401*	505460505*	346290667	779761554*	3032084
					798382597*		67043820
							77990495*
Wt13	455	1584	35	111	1	7	0
	49196	189711	3318	11464	72	609	1
	2709644	8567654	191387	699805	3471	32970	37
	90136081	236175959	7495607	25540325	134889	1389822	1209
	1015881687	412636694	203773941	635447186	4326036	41742009	35612
			903754984*		106623188	922172210*	960841
					967892015*		21551981
Wt14	133	580	7	29	0	1	0
	19264	84774	1078	4210	14	170	0
	1252013	4254433	79492	313006	1106	12549	7
	45962910	126003370	3594458	12768886	55356	635744	378
	65351519	501310406	106749104	341689003	2069570	21251715	14441
	807577235		148363926*	298333182*	55988762	503331182	457898
					66930481*	529092469*	11331326
							218918648
Wt15	1407	4104	140	370	8	35	1
	120456	412992	9520	29600	304	1960	8
	5733868	16950576	446320	1525160	10192	83020	164
	174179264	437172480	15351280	50274640	319216	2974720	3632
			384206690	94596613	8898608	80876090	85546
					200808016		1988192
							40594328

* Multiplicity is given modulo 1073741527

Weight Multiplicities for Affine F4

Affine F4 - Level 1

	Rp 1					Rp 2				
	00001					10000				
Wt 1	1	28084	13204382	721487893*	385997708*	0	10567	5674332	819593385	191900862*
	6	61363	23864370	805516426*	328010539*	1	23628	10344854	281084921*	297818754*
	26	129978	42560772	524609216*	474271877*	6	51056	18598828	74111144*	414739426*
	92	267869	74969354	291398513*	591210491*	25	107066	33007966	393798965*	717761702*
	286	538682	130532323	747299247*		86	218650	57875764	470836228*	301424785*
	808	1059656	224813130	736021619*		261	436066	100334951	777319546*	
	2123	2043198	383242038	706917481*		722	851288	172101094	967389758*	
	5258	3868388	647029798	840286057*		1863	1629958	292253622	9604926*	
	12405	7202390	8698319*	380300685*		4542	3066059	491614220	679801959*	
Wt 2	2	41315	17758170	156562977*	514614368*	1	16298	7705688	1055792254	683621638*
	11	89020	31884949	553917081*	336982040*	4	35610	13939718	664205291*	988014206*
	46	186167	56517918	769709261*	962339674*	17	75412	24884407	690986648*	955925596*
	155	379306	98983458	245005723*	741687080*	56	155513	43873056	306373436*	552813518*
	468	754815	171414910	851408512*		172	312902	76457960	963562450*	
	1284	1470704	293722005	597161415*		476	616070	131795534	28060144*	
	3298	2810855	498304628	292591268*		1245	1188884	224862471	549395001*	
	7999	5278754	837458755	662169778*		3062	2253132	379942694	685709808*	
	18544	9754389	321220785*	692659842*		7214	4198948	636116539	387708457*	

* Multiplicity is given modulo 1073741527

Affine F4 - Level 2

	Rp 1		Rp 2		Rp 3		Rp 4		Rp 5	
	00002		00010		01000		10001		20000	
Wt 1	1	29380880	0	32104849	0	14350336	0	19551284	0	3096293
	12	81992635	10	89948152	4	40616924	1	56036364	0	9073980
	120	220492440	108	242726400	42	110619752	32	154286780	4	25486226
	796	573606004	785	633379696	308	291082324	290	409913088	33	68944948
	4300	374472213*	4364	529775749*	1744	742583320	1876	1054740088*	232	180327995
	19724	337071209*	20508	728396707*	8368	768181498*	9660	489004826*	1240	457550072
	80490	1011959715*	84844	897412424*	35300	158151576*	42856	1050184871*	5784	55605253*
	298676	650496498*	318380	795740676*	134836	852126327*	169508	222786214*	23787	570655130*
	1026785	726010394*	1103364	712941553*	474753	677619038*	613240	25989579*	89276	1023637355*
	3309400		3580172		1562416		2061425		309640	756802245*
	10098156		10983436		4854170		6517548		1006940	
Wt 2	1	38071751	1	41714277	0	18623430	0	25588643	0	4042935
	20	105255490	20	115716091	5	52230590	5	72537591	0	11737185
	185	280681885	189	309515235	60	141069209	65	197802967	4	32685332
	1200	724651258	1245	801290415	441	368399243	505	521041335	45	87722809
	6258	743204304*	6570	940396140*	2484	933318296	2965	256678848*	316	227784990
	28017	141114656*	29697	633829625*	11734	152813391*	14470	81339954*	1725	574107370
	111815	906517552*	119459	1031557377*	48760	159949890*	61734	473664213*	7960	334577453*
	407525	1006322161*	438278	631645050*	183475	99805251*	237415	625418355*	32473	149011554*
	1379065	375119435*	1491545	542493304*	637289	862668866*	839971	699715010*	120376	367749374*
	4385330		4766275		2071372		2773605		412942	879760788*
	13222935		14433275		6363793		8639005		1328040	
Wt 3	3	49322776	2	54005708	1	24266156	0	33276630	0	5330610
	40	135045849	35	148414628	14	67335134	8	93421864	1	15283198
	317	357034634	304	393655734	119	180174398	102	252546743	12	42105056
	1892	914679420	1897	1011418376	752	466637385	765	660037280	92	111947896
	9358	129955022*	9634	377375201*	3893	99704393*	4352	599593092*	548	288284120
	40366	156391705*	42286	771075099*	17430	725275040*	20684	905540771*	2729	721211346
	156670	203652956*	166164	617328591*	69774	419975284*	86302	260475842*	11884	683612860*
	558628	268393194*	598030	560982707*	255356	994715721*	325685	771607657*	46591	958730635*
	1856984		2002602		867987	151515214*	1133752	1028767494*	167740	60032682*
	5817742		6311509		2772360		3691467		562622	
	17320480		18884452		8394740		11357338		1777164	
Wt 4	5	63605587	5	69812205	1	31344390	1	43305175	0	6926933
	64	172615582	65	190066493	19	86219857	20	120332390	1	19682640
	486	452728948	501	499939115	170	228879520	190	322352486	15	53780590
	2800	77713833*	2924	201121708*	1075	588491835	1259	835673856	125	141912223
	13455	700584221*	14199	1013376951*	5500	396310842*	6680	1029458517*	759	362910248
	56675	425182830*	60325	122429211*	24257	355682073*	30331	857686242*	3775	902091566
	215637	72024430*	231180	850933455*	95638	980068840*	122475	505340643*	16301	37595376*
	756003	792593559*	815453	846919683*	345059	411702969*	450805	864189870*	63205	873797517*
	2476821		2685643		1157720		1538468	367539741*	224940	151351298*
	7661985		8346018		3654415		4928175		746077	
	22558310		24671250		10948366		14955422		2331976	
Wt 5	12	81991390	9	89948152	4	40617240	2	56036364	1	9073980
	115	220492440	108	242725325	43	110619752	32	154287980	4	25487485
	796	573603080	780	633379696	308	291083083	295	409913088	38	68944948
	4280	374472213*	4364	529773265*	1748	742583320	1876	1054742888*	232	180330960
	19724	337064639*	20489	728396707*	8368	768183223*	9680	489004826*	1260	457550072
	80425	1011959715*	84844	897406924*	35315	158151576*	42856	1050191129*	5784	55611933*
	298676	650482299*	318320	795740676*	134836	852130102*	169572	222786214*	23852	570655130*
	1026596		1103364	712929819*	474798	677619038*	613240	26003034*	89276	1023651825*
	3309400		3580002		1562416		2061610		309830	756802245*
	10097655		10983436		4854295		6517548		1006940	
	29380880		32104408		14350336		19551770		3096798	

-* Multiplicity is given modulo 1073741527

Affine F4 - Level 3 - Page 1 of 2

	Rp 1 00003	Rp 2 00011	Rp 3 00100	Rp 4 01001	Rp 5 10002	Rp 6 10010	Rp 7 11000	Rp 8 20001	Rp 9 30000
Wt 1	1	0	0	0	0	0	0	0	0
	12	16	4	4	1	0	0	0	0
	190	320	108	116	52	44	7	4	0
	2016	3680	1360	1536	822	748	165	108	4
	15994	30432	11748	13796	8095	7572	1918	1368	68
	103255	201616	80032	96740	60206	57312	15646	11804	732
	571288	1136384	460096	568828	369146	355884	102134	80304	5664
	2801657	5653296	2323548	2925364	1959876	1908248	567914	461160	35524
	12463059	25439776	10582144	13524796	9293614	9122024	2792108	2327196	191464
	51127880	105366176	44264852	57296156	40199927	39724408	12437332	10593456	917784
							51061545	44297404	4002888
									16142480
Wt 2	1	1	0	0	0	0	0	0	0
	29	46	14	14	5	4	0	0	0
	426	753	266	290	141	126	23	14	0
	4108	7654	2882	3302	1834	1694	396	270	10
	30320	58324	22768	27020	16215	15288	3987	2910	158
	185278	364404	145744	177610	112259	107406	29886	22908	1492
	982067	1963965	799578	994992	653205	632092	183868	146304	10666
	4652847	9428058	3891838	4926060	3330300	3252140	977809	801542	63264
	20116241	41202113	17199316	22080536	15285066	15039130	4640767	3898054	326776
	80561851	166502204	70155102	91155378	64351835	63720056	20083686	17217548	1514334
								70205310	6423900
Wt 3	1	2	1	0	0	0	0	0	0
	43	76	29	21	12	7	2	0	0
	612	1134	439	423	240	196	53	20	1
	5728	10908	4281	4650	2784	2462	672	387	25
	41107	80082	31985	36957	23076	21321	6005	4046	275
	245647	486892	197359	237213	153540	145305	42188	31078	2280
	1278137	2569412	1055065	1303197	868560	835598	249389	194238	15141
	5963177	12127576	5034936	6347895	4335840	4220263	1289959	1045232	86015
	25447244	52263476	21905160	28065573	19570524	19221564	5997904	5006075	431239
			88205290	114507624	81277956	80404962	25543330	21825268	1956102
									8161464
Wt 4	3	4	1	1	0	0	0	0	0
	74	124	41	43	17	15	2	1	0
	935	1692	616	686	350	319	67	43	1
	8182	15444	5903	6857	3919	3655	899	630	29
	56400	109424	43122	51685	31611	29984	8033	5975	353
	327489	648108	260942	320429	205330	197311	55859	43416	2964
	1667838	3351380	1371335	1716929	1139075	1105974	325716	261996	19701
	7650176	15561344	6449468	8204515	5593536	5477140	1662462	1374737	110982
	32196523	66156048	27707510	35722002	24899972	24554840	7634890	6459634	551080
					102188643	101380227	32155312	27735988	2474304
									10224849

continued next page

Affine F4 - Level 3 - Page 2 of 2

	Rp 1	Rp 2	Rp 3	Rp 4	Rp 5	Rp 6	Rp 7	Rp 8	Rp 9
Wt 5	5	8	3	1	1	0	0	0	0
	111	200	77	63	35	25	7	1	0
	1342	2504	970	985	573	480	129	61	4
	11319	21736	8585	9550	5800	5227	1442	884	57
	75948	148840	59739	70080	44341	41325	11757	8215	577
	431520	859632	350047	424975	277926	264675	77592	58363	4368
	2159412	4358720	1796647	2235825	1503185	1452231	437040	345449	27462
	9760773	19919560	8297475	10521600	7239015	7069153	2176428	1782501	149034
	40572375	83572008	35128049	45217763	31727790	31241664	9809751	8256771	721258
						127435300	40705581	35017259	3176091
									12927676
Wt 6	9	14	4	4	1	1	0	0	0
	179	312	108	116	52	47	8	5	0
	1981	3652	1357	1536	822	758	171	115	4
	15897	30340	11738	13796	8095	7604	1938	1394	72
	103002	201358	79999	96740	60206	57399	15714	11892	747
	570652	1135712	459998	568828	369146	356106	102328	80563	5714
	2800163	5651626	2323291	2925364	1959876	1908803	568423	461853	35677
	12459661	25435868	10581519	13524796	9293614	9123316	2793373	2328936	191874
	51120416	105357404	44263381	57296156	40199927	39727316	12440311	10597599	918801
						160057697	51068227	44306799	4005314
									16147945
Wt 7	13	24	10	6	3	2	1	0	0
	261	480	186	171	96	75	22	8	1
	2786	5280	2070	2187	1284	1124	312	168	12
	21734	42096	16738	19041	11733	10736	3014	1968	139
	137469	271200	109488	130164	83295	78387	22607	16286	1186
	746903	1495296	611658	749583	495036	474140	140457	107778	8307
	3605778	7307736	3023838	3789024	2568096	2491467	756282	605886	49165
	15825954	32405280	13542060	17265462	11961081	11716488	3633237	3005982	255642
	64173878	132563904	55876848	72243498	50990574	50333428	15901836	13490544	1194888
								55736526	5115665
Wt 8	24	40	13	14	5	5	1	1	0
	405	728	262	290	141	131	28	20	1
	4038	7568	2867	3302	1834	1714	417	296	15
	30110	58064	22722	27020	16215	15353	4057	3000	178
	184707	363688	145614	177610	112259	107597	30096	23184	1557
	980611	1962128	799242	994992	653205	632603	184439	147070	10857
	4649346	9423608	3891014	4926060	3330300	3253425	979265	803523	63775
	20108191	41191832	17197402	22080536	15285066	15042185	4644268	3902880	328061
				91155378	64351835	63727011	20091736	17228759	1517389
									6430855
Wt 9	36	64	26	21	12	9	4	2	1
	580	1088	428	423	240	204	64	32	4
	5616	10752	4246	4650	2784	2495	712	436	38
	40768	79616	31888	36957	23076	21431	6136	4212	320
	244708	485632	197102	237213	153540	145623	42572	31578	2413
	1275732	2566208	1054420	1303197	868560	836454	250412	195604	15511
	5957376	12119872	5033414	6347895	4335840	4222419	1292528	1048718	86979
	25433880	52245824	21901692	28065573	19570524	19226674	6004028	5014492	433585
							25557260	21844648	1961584
									8173764

Affine F4 - Level 4 - Page 1 of 6 (See page guide at right.)

	Rp 1 00004	Rp 2 00012	Rp 3 00020	Rp 4 00101	Rp 5 01002	Rp 6 01010	Rp 7 02000	Rp 8 10003
Wt 1	1	0	0	0	0	0	0	0
	12	16	6	4	4	0	0	1
	218	436	240	226	160	100	15	52
	2972	7020	4127	4336	3092	2347	352	1152
	30784	78046	47440	51932	38262	31088	5000	15776
	253884	672028	416190	467164	354148	297282	50144	155748
	1759172	4792585	3008284	3431690	2663568	2280636	399809	1227088
Wt 2	1	1	0	0	0	0	0	0
	29	55	26	26	14	9	0	5
	557	1279	718	763	486	377	41	181
	7164	17830	10581	11605	8068	6592	927	3268
	68660	179038	109297	122403	89634	75239	11825	38695
	531163	1429557	888487	1010154	766224	654535	109822	346969
		9620127	6056106	6965704	5421934	4691229	823137	2547589
Wt 3	1	1	1	0	0	0	0	0
	39	78	44	39	26	15	2	5
	774	1778	1052	1063	740	548	83	246
	9616	23961	14487	15618	11295	9033	1426	4391
	89657	233881	144112	160074	119668	99457	16506	50894
	678005	1826568	1140946	1292482	991988	843257	145568	446547
				8756763	6866172	5926744	1057360	3220183
Wt 4	1	2	1	1	0	0	0	0
	55	120	65	67	36	27	2	12
	1050	2519	1461	1572	1021	819	102	383
	12740	32338	19439	21470	15135	12528	1845	6212
	115924	305698	187882	211388	156309	132107	21250	68263
	860091	2332214	1455668	1660597	1268615	1088408	185258	579898
					8630920	7490828	1327585	4086576
Wt 5	3	4	2	1	1	0	0	0
	90	184	104	96	66	41	6	17
	1508	3547	2100	2175	1532	1158	176	536
	17232	43454	26396	28706	20962	16951	2714	8387
	151313	398236	246212	275090	207147	173248	29048	89564
	1095552	2970256	1860707	2115768	1633358	1394382	242852	743851
						9422918	1693252	5148114
Wt 6	3	6	3	3	1	1	0	0
	118	267	149	156	91	72	7	29
	2003	4902	2883	3128	2090	1703	227	795
	22524	57885	35064	38944	27861	23248	3531	11601
	194105	516221	318827	360153	268853	228367	37402	118784
	1381890	3770633	2361619	2702294	2078512	1789577	308317	958897
							2120262	

| 1 2 |
| 3 4 |
| 5 6 |
page guide

--continued next page-

Affine F4 - Level 4 - Page 2 of 6

	Rp 9 10011	Rp10 10100	Rp11 11001	Rp12 20002	Rp13 20010	Rp14 21000	Rp15 30001	Rp16 40000
Wt 1	0	0	0	0	0	0	0	0
	0	0	0	0	0	0	0	0
	64	18	7	4	0	0	0	0
	1792	588	432	160	108	22	4	0
	26240	8956	7846	3092	2472	524	217	10
	268800	94428	89564	37061	31788	7136	3684	164
	2168768	777620	772929	333964	295769	69360	40162	2017
				2454416	2217932	539078	333504	18136
								133658
Wt 2	0	0	0	0	0	0	0	0
	4	0	0	0	0	0	0	0
	252	67	55	14	10	0	0	0
	5180	1635	1477	497	412	62	30	0
	64960	21911	20931	7993	6928	1348	735	20
	602348	210788	209132	86046	76432	16624	9767	390
	4524760	1622452	1655291	716341	646827	150134	92938	4465
					4533513	1099847	707973	38095
								266757
Wt 3	0	0	0	0	0	0	0	0
	8	3	0	0	0	0	0	0
	384	127	79	30	16	3	0	0
	7264	2476	2039	773	602	128	40	1
	87040	30342	27914	11282	9484	2098	1001	44
	782992	278729	271750	115075	100884	23320	12907	620
	5752832	2082767	2107180	927374	831713	199364	120059	6349
					5716872	1413189	897098	50836
								343390
Wt 4	0	0	0	0	0	0	0	0
	14	3	2	0	0	0	0	0
	574	169	144	39	31	3	1	0
	10108	3289	3031	1057	899	155	77	1
	116090	39711	38407	14988	13136	2679	1495	49
	1014580	357971	358207	149537	133660	29769	17780	765
	7296752	2630896	2701221	1181310	1071057	252246	158017	7908
						1767027	1147725	63390
								424988
Wt 5	0	0	0	0	0	0	0	0
	24	8	2	1	0	0	0	0
	848	283	196	72	47	10	1	0
	13976	4784	4090	1584	1260	271	102	4
	154192	54052	50611	20712	17656	3958	1980	89
	1310720	468648	461870	197705	174442	40721	23146	1150
	9233760	3355598	3419800	1517652	1367229	330478	202074	10889
						2251377	1444520	83216
								541389
Wt 6	0	0	0	0	0	0	0	0
	40	10	8	1	1	0	0	0
	1240	381	336	100	84	11	5	0
	19208	6366	5968	2159	1866	344	181	4
	203984	70495	68921	27454	24254	5104	2924	107
	1688400	599700	604701	255879	229840	52129	31636	1438
	11658920	4223864	4361988	1926406	1752663	417814	264663	13684
							1840694	103940

continued next page

Affine F4 - Level 4 - Page 3 of 6

	Rp 1	Rp 2	Rp 3	Rp 4	Rp 5	Rp 6	Rp 7	Rp 8
Wt 7	6 174 2779 29928 250835 1748528	9 386 6732 76762 666900 4774444	6 227 4056 47002 414428 3001272	4 222 4272 51606 465553 3425654	3 152 3039 38008 353042 2659474	2 108 2386 31201 297634 2281435	1 18 383 5112 50663 401504	0 39 1092 15471 154644 1222971
Wt 8	5 131 2071 22815 195209 1385686	8 284 5000 58339 518040 3777168	3 152 2908 35202 319445 2364012	3 159 3150 39069 360715 2704497	1 92 2102 27932 269190 2079851	0 67 1682 23180 228170 1789075	0 6 219 3491 37233 307695 2118168	1 35 826 11729 119265 960537
Wt 9	9 254 3791 39447 322192 2203553	18 586 9397 102310 862795 6045176	9 331 5569 62296 534886 3797127	9 350 6081 69513 606342 4356866	4 214 4151 50353 456468 3372075	3 169 3398 42203 389105 2911563	0 18 470 6548 64632 506757	1 74 1629 21345 204177 1570096
Wt10	14 366 5179 51990 413855 2775784	27 833 12777 134679 1108516 7621331	16 493 7719 82790 690904 4802985	13 492 8243 91487 779690 5499527	9 340 5916 67921 595017 4291708	5 248 4698 56174 504033 3694383	1 39 747 9281 86493 655093	1 102 2217 28291 264322 1993563
Wt11	19 497 6873 67458 526723	42 1189 17358 176975 1421632 9592558	23 690 10413 108505 885292 6044613	24 740 11461 121698 1007253 6955103	12 470 7972 89184 764395 5415314	10 382 6611 75289 654637 4691324	1 48 966 12000 110514 825602	3 166 3182 38307 345432 2542129
Wt12	32 714 9343 88404 673457	64 1681 23436 231595 1817701	39 1003 14267 143045 1137036	33 1030 15399 159165 1288708	23 720 11172 119127 989785	15 545 9021 99383 843001 5925728	3 87 1465 16655 146280 1059760	5 233 4330 50524 445147 3214666

--continued next page-

Affine F4 - Level 4 - Page 4 of 6

	Rp 9	Rp10	Rp11	Rp12	Rp13	Rp14	Rp15	Rp16
Wt 7	0	0	0	0	0	0	0	0
	64	24	12	3	2	1	0	0
	1792	613	471	174	127	30	8	1
	26240	9114	8076	3165	2620	590	252	12
	268800	95005	90596	37580	32512	7456	3934	197
	2168768	779960	777072	336072	299012	70843	41330	2146
			5504044	2463148	2229963	544634	338464	18797
								135970
Wt 8	0	0	0	0	0	0	0	0
	40	9	7	1	0	0	0	0
	1240	372	324	97	76	10	4	0
	19208	6319	5895	2132	1813	329	169	4
	203984	70287	68570	27294	23987	5011	2844	102
	1688400	598906	603272	255137	228689	51675	31232	1403
	11658920	4221114	4356796	1923458	1748299	415971	262965	13504
							1834396	103195
Wt 9	1	0	0	0	0	0	0	0
	105	28	23	5	4	0	0	0
	2580	810	727	234	197	29	14	0
	35693	11970	11356	4241	3683	707	383	10
	353017	122933	121223	49171	43626	9384	5470	214
	2778517	992271	1007096	431376	388794	89431	54940	2594
		6718450	6974089	3107811	2835218	682729	436486	23138
							2920157	167925
Wt10	2	1	0	0	0	0	0	0
	164	57	33	12	7	2	0	0
	3670	1258	1009	381	292	66	20	1
	48268	16818	15240	6093	5099	1144	527	25
	461906	163985	158363	66473	57996	13398	7287	360
	3550204	1281631	1287777	562307	502681	120001	71349	3768
				3952020	3590570	883122	555497	31277
								217787
Wt11	4	1	1	0	0	0	0	0
	252	75	66	17	15	2	1	0
	5180	1682	1553	529	458	81	43	1
	64960	22132	21324	8194	7209	1470	830	29
	602348	211693	210842	87023	77770	17220	10273	447
	4524760	1625763	1661828	720356	652280	152622	95165	4737
					4553287	1109035	716526	39209
								270800
Wt12	8	3	1	1	0	0	0	0
	384	135	92	35	25	7	1	0
	7264	2512	2124	833	661	153	61	4
	87040	30575	28367	11552	9857	2264	1126	57
	782992	279604	273687	116437	102511	24049	13583	711
	5752832	2086331	2114634	932377	838376	202433	122829	6697
						1424043	907680	52317
								348299

continued next page

Affine F4 - Level 4 - Page 5 of 6

	Rp 1	Rp 2	Rp 3	Rp 4	Rp 5	Rp 6	Rp 7	Rp 8
Wt13	41	95	54	57	32	27	3	9
	957	2347	1385	1501	988	816	111	360
	12265	31446	19042	21100	14941	12509	1901	6076
	113905	301840	186152	209762	155418	132008	21519	67634
	852512	2317544	1449046	1654347	1265067	1088000	186367	577361
						7489325	1331640	4077448
Wt14	62	138	84	80	56	40	8	13
	1336	3244	1966	2060	1463	1150	193	500
	16398	41924	25716	28096	20602	16907	2814	8171
	147863	391740	243300	272421	205569	173052	29515	88548
	1082862	2945900	1849716	2105544	1627286	1393621	244732	739754
Wt15	84	201	117	126	77	66	9	24
	1783	4472	2678	2935	1993	1672	247	749
	21428	55734	34049	37988	27350	23119	3656	11319
	189519	507159	314575	356129	266628	227882	38003	117446
	1364912	3736882	2345833	2687301	2069997	1787924	310776	953464
Wt16	126	285	176	174	125	96	21	36
	2448	6096	3725	3984	2876	2321	404	1032
	28365	73600	45472	50151	37198	30956	5294	15120
	244220	653776	408022	459636	349664	296722	51472	152814
							405011	

Affine F4 - Level 4 - Page 6 of 6

	Rp 9	Rp10	Rp11	Rp12	Rp13	Rp14	Rp15	Rp16
Wt13	14	4	4	1	1	0	0	0
	574	180	166	52	47	8	5	0
	10108	3362	3180	1151	1016	198	115	4
	116090	40074	39166	15489	13773	2939	1732	72
	1014580	359501	361459	151751	136500	30990	18930	896
	7296752	2636584	2713485	1189862	1082124	257188	162776	8496
						1784841	1165215	65692
								433065
Wt14	24	10	6	3	2	1	0	0
	848	304	237	96	75	22	8	1
	13976	4914	4354	1749	1462	359	168	12
	154192	54681	51919	21570	18733	4436	2384	139
	1310720	471238	467360	201427	179186	42870	25080	1398
		3365079	3440201	1531840	1385506	338908	209980	11942
							1473292	87180
Wt15	40	13	14	5	5	1	1	0
	1240	409	396	141	131	28	20	1
	19208	6541	6347	2423	2181	470	296	15
	203984	71343	70776	28767	25869	5791	3567	178
	1688400	603199	612410	261409	236769	55190	34571	1795
				1947030	1778873	429725	276301	15190
								109571
Wt16	64	26	21	12	9	4	2	1
	1792	656	564	240	204	64	32	4
	26240	9349	8642	3600	3102	800	436	38
	268800	96244	93344	39570	34960	8552	4912	320
	2168768	784856	788296	344472	309220	75484	45758	2749
							355600	21116

Weight Multiplicities for Affine G2

Affine G2 - Level 1

	Rp 1				
	001				
Wt 1	1	1456	133422	4926146	109851276
	3	2586	206061	7100979	151613745
	9	4482	315207	10177137	208438650
	22	7625	477747	14506500	285484972
	51	12729	718029	20569581	
	108	20925	1070496	29021841	
	220	33881	1584112	40751719	
	426	54153	2327472	56961591	
	801	85473	3396906	79269921	
Wt 2	1	1752	154136	5566162	122345236
	4	3088	237358	8008214	168638266
	11	5330	361990	11456838	231552098
	28	9014	547214	16302157	316758508
	63	14990	820261	23077770	
	134	24522	1220006	32508747	
	269	39564	1801110	45578860	
	520	62981	2640604	63615341	
	967	99084	3845771	88405622	

	Rp 2				
	100				
Wt 1	0	702	68931	2624481	59604768
	1	1258	106967	3792288	82399071
	3	2211	164259	5447949	113458503
	9	3789	249957	7782489	155631240
	21	6388	376949	11058736	212693727
	48	10566	563913	15633978	
	99	17235	836982	21995343	
	199	27691	1233420	30800605	
	378	43962	1804986	42939264	
Wt 2	1	878	80280	2972356	66451984
	2	1559	124004	4286028	91737318
	6	2696	189764	6144536	126150746
	14	4589	287690	8761152	172819936
	32	7652	432551	12426414	
	66	12584	645064	17537655	
	135	20370	954922	24632534	
	258	32570	1403442	34440023	
	485	51406	2049022	47940250	

Affine G2 - Level 2

	Rp 1			Rp 2			Rp 3			Rp 4		
	002			010			101			200		
Wt 1	1	13559	3971714	0	9132	2750748	0	10605	3425433	0	2742	963606
	5	30579	7312216	3	20736	5074275	3	24390	6353691	0	6393	1797798
	26	66511	13252283	15	45270	9212070	15	53832	11593380	3	14355	3298743
	94	139904	23669034	60	95643	16480023	63	114861	20837970	12	30960	5958561
	305	286126	41707031	195	196188	29081976	210	237672	36933630	48	64815	10610817
	870	570243	72570229	573	392235	50673426	633	478953	64618167	147	131739	18643731
	2315	1111105	124797377	1533	766119	87253449	1725	942189	111691194	426	261411	32355798
	5741	2120387		3849	1465587		4407	1814217	190880742	1107	506889	55502529
Wt 2	1	15565	4402001	1	10598	3052840	0	12146	3798913	0	3231	1072717
	7	34900	8082666	5	23830	5614820	3	27766	7027408	1	7454	1994891
	32	75436	14611527	22	51644	10166154	18	61015	12791075	5	16527	3648976
	116	157935	26036581	79	108398	18141828	73	129578	22937213	19	35391	6573595
	364	321498	45779242	246	221186	31942026	247	267086	40566959	64	73511	11676263
	1031	638294	79496235	697	440109	55539411	736	536173	70830884	191	148634	20469456
	2701	1239129	136449225	1830	856159	95445114	2000	1051253	122198050	522	293357	35448223
	6651	2357200		4516	1631818		5070	2017774		1336	566532	60689011
Wt 3	2	20462	5400128	1	13834	3743823	1	16172	4675178	0	4210	1319083
	12	45296	9862636	7	30775	6850086	7	36402	8599428	1	9630	2440204
	50	96826	17741746	31	66059	12343167	32	78954	15571152	6	21169	4441733
	172	200752	31470456	110	137451	21928285	117	165810	27788756	25	44968	7965205
	520	405140	55098994	339	278254	38447484	370	338464	48929312	85	92706	14087609
	1432	798152	95301532	946	549683	66589209	1055	673663	85079029	254	186131	24597690
	3670	1538676		2448	1062272	114013819	2780	1310775	146213226	692	364975	42436666
	8876	2908508		5965	2012401		6880	2498694		1756	700586	
Wt 4	3	23430	5977083	2	15966	4148607	1	18458	5177703	1	4944	1466111
	15	51516	10887591	11	35233	7569478	8	41358	9499421	2	11143	2703266
	63	109575	19538694	42	75095	13604226	37	89261	17159317	11	24300	4906795
	207	226056	34579845	141	155362	24111671	138	186701	30554930	35	51171	8775706
	621	454371	60417438	418	312917	42183026	431	379588	53687048	114	104897	15484609
	1680	891669	104296191	1139	615505	72911020	1227	752896	93169563	317	209375	26976623
	4272	1713246		2892	1184842		3208	1460083	159821921	847	408786	46447041
	10227	3228282		6960	2236953		7901	2775068		2086	781348	

Affine G2 - Level 3

	Rp 1		Rp 2		Rp 3		Rp 4		Rp 5		Rp 6	
	003		011		102		110		201		300	
Wt 1	1	146744	0	162228	0	143847	0	86352	0	65347	0	13224
	6	360015	5	400617	3	360222	2	217002	0	166728	0	34062
	39	849789	36	950994	24	865758	15	523044	6	407004	1	84003
	179	1939427	177	2180919	132	2007030	78	1216008	43	956549	9	199116
	696	4295334	716	4850998	558	4507728	332	2737941	207	2174088	39	456306
	2373	9261636	2505	10499529	2034	9841500	1203	5991789	807	4797432	157	1014230
	7371	19492064	7902	22173117	6600	20947803	3924	12780783	2757	10308936	537	2194386
	21177	40131570	23005	45792773	19683	43572660	11730	26637883	8505	21629040	1683	4632429
	57228		62763		54732		32754		24327	44403189	4859	9565057
Wt 2	1	179592	1	199407	0	177087	0	105887	0	81109	0	16193
	9	436781	9	487636	5	439190	2	263879	1	204572	0	41433
	54	1023119	55	1147871	36	1046705	19	631281	12	494565	1	101492
	241	2319121	252	2613167	185	2408658	101	1457710	67	1152704	10	239059
	915	5104983	970	5774801	752	5374390	428	3262017	296	2601020	48	544622
	3059	10946806	3288	12426697	2665	11664896	1538	7098835	1095	5702927	195	1203929
	9343	22923463	10142	26106101	8470	24697480	4960	15064927	3617	12184965	668	2591603
	26485		29009	53659966	24842	51124970	14664	31251554	10900	25434064	2088	5445349
	70741		78042		68150		40536		30625		5993	11194609
Wt 3	2	242596	2	269871	1	240840	0	144281	0	110999	0	22265
	17	582300	17	651236	10	589128	5	354410	3	275781	0	56144
	90	1348623	93	1515318	65	1386930	34	837544	22	658318	3	135578
	382	3026025	401	3414370	300	3157789	168	1912955	114	1517097	17	315747
	1381	6601313	1473	7476483	1165	6978730	664	4239792	465	3389492	80	711700
	4479	14040163	4830	15955958	3956	15018300	2300	9146818	1654	7365478	297	1559056
	13312	29184789	14501	33270063	12230	31551450	7177	19260064	5272	15611433	992	3327872
	36972		40595		35002	64854575	20730	39670048	15504	32348483	2989	6940418
	96984		107247		94235		56143		42630		8415	14169816
Wt 4	3	268075	2	297728	1	266578	1	160405	0	122705	0	25019
	21	640698	19	715732	13	649051	8	391807	3	303739	1	62494
	109	1478060	106	1659624	77	1522152	47	921380	25	722403	5	150026
	446	3305395	456	3727909	353	3453426	209	2095990	128	1659528	26	347406
	1592	7188799	1668	8139875	1339	7608758	796	4628881	527	3696620	102	779819
	5088	15248572	5435	17326946	4508	16328091	2675	9955556	1860	8011472	366	1701625
	14997	31618468	16229	36043135	13786	34217107	8217	20904900	5913	16938335	1165	3620857
	41314		45189		39195		23422		17287	35018550	3454	7529525
	107748		118828		104824		62885		47342		9543	
Wt 5	4	326094	4	363480	2	326015	1	195597	1	151151	0	30522
	29	773193	30	866158	20	786814	10	474051	7	370291	1	75710
	147	1771194	153	1993126	110	1831065	61	1106975	42	872966	6	180547
	589	3935993	624	4446948	481	4126070	271	2502201	189	1990185	32	415532
	2060	8511698	2207	9651724	1770	9035539	1022	5494126	729	4403621	128	927442
	6477	17961632	7017	20434448	5822	19283830	3396	11754321	2471	9487184	458	2013172
	18823	37069035	20566	42299744	17490	40210481	10313	24563103	7643	19951759	1453	4263032
	51239		56420		49010		29088		21900	41050099	4276	8825226
	132253		146565		129512		77348		59051		11730	
Wt 6	5	359812	4	400461	3	360222	2	217056	1	166932	1	34166
	36	849477	34	950760	24	865758	16	523133	8	407317	2	84159
	173	1938950	174	2180572	132	2007030	79	1216140	47	957022	11	199350
	686	4294641	708	4850466	558	4507728	334	2738133	216	2174810	44	456673
	2355	9260608	2492	10498752	2034	9841500	1208	5992088	823	4798496	165	1014762
	7338	19490566	7881	22172002	6600	20947803	3932	12781209	2784	10310490	550	2195163
	21127		22966		19683	43572660	11742	26638494	8554	21631312	1709	4633580
	57144		62701		54732		32778		24406		4898	9566695
	146610		162134		143847		86387		65473		13286	

Affine G2 - Level 4 - Page 1 of 3

	Rp 1 004	Rp 2 012	Rp 3 020	Rp 4 103	Rp 5 111	Rp 6 202	Rp 7 210	Rp 8 301	Rp 9 400
Wt 1	1	0	0	0	0	0	0	0	0
	6	6	2	3	2	0	0	0	0
	45	51	21	28	24	9	3	1	0
	236	291	123	183	162	72	36	18	1
	1047	1332	584	897	820	414	212	114	15
	4002	5241	2328	3721	3456	1845	990	562	78
	13831	18420	8289	13557	12780	7137	3876	2277	352
	43854	59274	26877	44928	42786	24606	13566	8181	1308
	129966	177594	81087	137744	132360	78003	43323	26655	4428
	363444	501417	230085	396540	383760	230445	128940	80655	13670
						642978	361523	229266	39633
									108450
Wt 2	1	1	0	0	0	0	0	0	0
	10	11	4	5	4	1	0	0	0
	68	81	33	46	40	16	7	3	0
	346	432	185	276	248	117	58	30	3
	1473	1897	833	1297	1192	613	321	177	23
	5497	7236	3228	5187	4844	2629	1413	813	119
	18561	24839	11198	18435	17432	9823	5367	3183	496
	57845	78427	35637	59819	57128	33110	18288	11103	1803
	168814	231328	105759	180385	173688	102917	57305	35445	5921
	466176	644525	296132	511950	496352	299459	167774	105387	17985
								295434	51267
Wt 3	1	1	1	0	0	0	0	0	0
	11	13	6	7	4	1	1	0	0
	77	93	43	55	44	20	11	3	1
	386	490	220	322	276	135	77	33	5
	1641	2126	962	1473	1324	701	386	195	34
	6074	8039	3639	5817	5360	2948	1641	901	147
	20413	27396	12477	20447	19196	10927	6074	3506	591
	63271	85993	39306	65872	62612	36505	20413	12196	2061
	183897	252386	115873	197390	189544	112828	63271	38747	6664
			322603	557514	539556	326531	183897	114791	19926
								320587	56346
Wt 4	2	2	1	1	0	0	0	0	0
	19	22	10	12	8	3	2	0	0
	117	144	65	89	72	33	19	6	1
	564	718	323	479	416	210	117	53	9
	2299	2998	1353	2105	1904	1017	564	292	48
	8297	11018	4997	8039	7448	4140	2299	1280	212
	27265	36722	16732	27609	26000	14892	8297	4837	812
	83091	113210	51809	87198	83120	48744	27265	16400	2788
	237945	327334	150403	257240	247520	147975	83091	51150	8807
				716803	694952	422217	237945	149141	25986
									72409

continued next page

Affine G2 - Level 4 - Page 2 of 3

	Rp 1	Rp 2	Rp 3	Rp 4	Rp 5	Rp 6	Rp 7	Rp 8	Rp 9
Wt 5	3	3	1	1	1	0	0	0	0
	26	30	12	16	14	5	2	1	0
	154	189	80	116	104	47	23	12	1
	719	915	399	610	558	280	145	80	10
	2867	3741	1660	2631	2446	1305	697	399	57
	10173	13521	6072	9889	9312	5179	2816	1660	255
	32981	44466	20142	33513	31888	18287	10065	6072	975
	99401	135573	61824	104675	100460	59004	32759	20142	3336
	282018	388350	178048	305878	295704	177068	98969	61824	10475
						500424	281199	178048	30710
									85046
Wt 6	5	5	2	2	2	1	0	0	0
	39	47	19	27	24	10	5	3	0
	226	280	121	177	162	80	39	22	3
	1013	1305	572	889	820	425	226	130	17
	3946	5179	2312	3690	3456	1888	1013	594	91
	13684	18287	8235	13513	12780	7202	3946	2363	370
	43608	59004	26799	44802	42786	24791	13684	8343	1369
	129415	177068	80881	137555	132360	78292	43608	27021	4518
	362534	500424	229779	396089	383760	231129	129415	81313	13907
						644056	362534	230598	39993
									109260
Wt 7	5	6	3	3	2	1	1	0	0
	44	54	25	33	26	13	8	3	1
	251	318	144	208	180	93	54	25	5
	1128	1464	663	1013	912	490	276	144	26
	4359	5760	2613	4150	3828	2126	1189	663	115
	15058	20190	9193	15016	14090	8039	4492	2613	448
	47723	64758	29604	49411	46948	27396	15339	9193	1578
	141072	193356	88732	150710	144592	85993	48282	29604	5119
	393598	544110	250584	431799	417554	252386	142154	88732	15472
								250584	44091
Wt 8	7	8	3	4	4	2	1	1	0
	58	71	30	43	40	20	10	7	1
	321	407	177	267	248	129	68	42	6
	1413	1836	813	1275	1192	644	346	209	32
	5367	7103	3183	5136	4844	2702	1473	890	141
	18288	24558	11103	18327	17432	9983	5497	3354	547
	57305	77868	35445	59597	57128	33442	18561	11463	1911
	167774	230246	105387	179953	173688	103576	57845	36168	6143
				511131	496352	300721	168814	106785	18417
									52086

continued next page-

Affine G2 - Level 4 - Page 3 of 3

	Rp 1	Rp 2	Rp 3	Rp 4	Rp 5	Rp 6	Rp 7	Rp 8	Rp 9
Wt 9	8	9	5	5	4	3	2	1	1
	64	81	37	52	44	24	16	8	2
	358	459	211	308	276	153	88	47	11
	1564	2052	931	1449	1324	735	420	236	44
	5924	7875	3588	5746	5360	3048	1710	992	181
	20068	27054	12344	20325	19196	11109	6244	3721	649
	62638	85311	39085	65579	62612	36939	20732	12625	2211
	182600	251073	115381	196892	189544	113592	63956	39644	6920
						328137	185124	116463	20495
									57301

Affine G2 - Level 5 - Page 1 of 3

	Rp 1	Rp 2	Rp 3	Rp 4	Rp 5	Rp 6	Rp 7	Rp 8	Rp 9	Rp10	Rp11	Rp12
	005	013	021	104	112	120	203	211	302	310	401	500
Wt 1	1	0	0	0	0	0	0	0	0	0	0	0
	6	6	3	3	2	0	0	0	0	0	0	0
	47	57	36	30	28	9	9	6	1	0	0	0
	268	360	240	213	219	84	87	69	23	9	3	0
	1286	1812	1245	1143	1247	504	561	465	182	87	39	3
	5323	7755	5427	5154	5807	2418	2799	2391	1034	519	252	30
	19779	29475	20913	20337	23463	9921	11874	10329	4721	2448	1254	165
	67232	101985	73113	72555	85101	36390	44589	39336	18719	9858	5229	744
								136017	66603	35502	19332	2862
												9942
Wt 2	1	1	0	0	0	0	0	0	0	0	0	0
	10	12	6	5	4	2	1	0	0	0	0	0
	74	97	59	50	50	21	19	11	3	2	0	0
	408	567	371	333	355	151	152	114	43	23	6	1
	1887	2710	1845	1714	1907	808	892	721	301	161	64	9
	7593	11191	7799	7470	8519	3637	4206	3550	1590	843	389	58
	27549	41382	29303	28695	33372	14319	17156	14839	6935	3705	1848	274
	91828	140088	100367	100139	118128	50976	62594	55068	26596	14256	7463	1137
									92266	49728	26896	4153
												13923
Wt 3	1	1	1	0	0	0	0	0	0	0	0	0
	12	15	10	7	6	2	1	1	0	0	0	0
	91	120	81	65	66	25	24	19	5	2	1	0
	496	691	474	419	449	181	191	159	59	27	12	1
	2266	3261	2274	2102	2345	969	1097	932	388	194	94	10
	9013	13307	9399	8992	10277	4328	5086	4406	1974	1013	514	67
	32373	48700	34772	34058	39675	16894	20454	17960	8409	4411	2321	322
			117692	117523	138832	59642	73764	65502	31700	16820	9087	1332
										58156	32086	4847
												16154

continued next page

Affine G2 - Level 5 - Page 2 of 3

	Rp 1	Rp 2	Rp 3	Rp 4	Rp 5	Rp 6	Rp 7	Rp 8	Rp 9	Rp10	Rp11	Rp12
Wt 4	2	2	1	1	0	0	0	0	0	0	0	0
	20	24	15	12	10	3	3	2	0	0	0	0
	132	175	116	100	99	38	38	30	9	3	1	0
	689	961	657	592	636	255	279	230	86	41	18	1
	3029	4377	3050	2859	3189	1323	1511	1285	544	270	130	15
	11769	17422	12317	11861	13594	5727	6790	5884	2651	1369	696	89
	41440	62516	44682	44002	51346	21900	26629	23417	11032	5787	3051	429
				149229	176639	75945	94309	83835	40726	21655	11728	1719
										73631	40713	6181
Wt 5	3	3	2	1	1	0	0	0	0	0	0	0
	28	35	23	18	17	6	5	4	1	0	0	0
	182	245	166	141	148	57	59	48	16	7	3	0
	921	1301	898	814	890	363	399	336	135	64	30	3
	3973	5781	4054	3810	4305	1793	2075	1781	774	393	196	23
	15147	22551	16006	15483	17860	7562	9023	7874	3619	1876	973	132
	52569	79619	57074	56383	66130	28272	34601	30556	14561	7687	4107	586
						96452	120295	107293	52589	28033	15322	2286
											52116	7975
Wt 6	3	4	2	1	1	1	0	0	0	0	0	0
	31	41	25	19	20	9	7	4	1	1	0	0
	200	277	181	156	167	73	70	52	20	11	3	1
	1012	1447	981	892	993	424	458	369	154	85	33	5
	4333	6355	4414	4170	4745	2033	2320	1952	874	467	214	34
	16453	24601	17369	16844	19538	8387	9965	8594	4006	2150	1068	162
	56820	86327	61687	61092	71869	31009	37823	33198	15984	8581	4475	690
								116064	57236	30868	16639	2581
												8862
Wt 7	5	6	3	2	2	1	1	0	0	0	0	0
	44	59	36	30	31	14	12	8	3	2	0	0
	276	384	253	222	242	104	108	82	33	19	6	1
	1349	1943	1326	1224	1370	587	649	532	233	126	53	9
	5662	8344	5821	5537	6347	2719	3160	2682	1225	661	314	48
	21099	31688	22448	21897	25508	10977	13158	11426	5413	2906	1475	229
	71839	109503	78456	77998	92112	39789	48897	43096	20956	11290	5974	925
								147927	73527	39710	21615	3388
												11347
Wt 8	6	7	5	3	3	1	1	1	0	0	0	0
	54	72	49	40	41	16	16	14	5	2	1	0
	335	466	322	281	307	124	137	117	47	23	12	1
	1619	2334	1633	1505	1689	703	805	697	306	154	80	10
	6718	9915	7019	6679	7673	3237	3841	3357	1541	802	425	57
	24794	37296	26656	26028	30382	12965	15748	13910	6620	3487	1881	272
	83702	127776	92080	91654	108429	46615	57788	51478	25133	13396	7352	1095
									86903	46642	25985	3984
												13243

-continued next page-

Affine G2 - Level 5 - Page 3 of 3

	Rp 1	Rp 2	Rp 3	Rp 4	Rp 5	Rp 6	Rp 7	Rp 8	Rp 9	Rp10	Rp11	Rp12
Wt 9	8	10	7	5	5	2	2	2	1	0	0	0
	74	100	69	57	62	24	27	24	10	5	3	0
	445	629	438	390	433	179	203	178	80	39	22	3
	2120	3083	2172	2016	2298	962	1131	992	456	237	130	17
	8644	12851	9140	8760	10147	4311	5180	4575	2171	1138	624	91
	31483	47595	34143	33484	39352	16850	20687	18392	8924	4747	2621	389
				116180	138068	59544	74319	66541	32959	17643	9838	1507
										60073	33853	5263
Wt10	9	12	7	5	6	3	3	2	1	1	0	0
	81	113	74	63	70	32	33	26	13	8	3	1
	489	700	476	427	485	211	237	196	93	54	25	5
	2311	3392	2359	2208	2539	1100	1275	1092	523	289	144	26
	9391	14031	9914	9536	11125	4805	5755	5012	2426	1327	693	115
	34038	51648	36911	36318	42844	18549	22704	20044	9866	5356	2880	470
						64899	80947	72167	36049	19577	10753	1727
											36759	5915
Wt11	11	14	10	7	8	3	4	4	2	1	1	0
	98	136	95	81	90	37	43	40	20	10	7	1
	586	839	591	530	604	252	299	268	129	68	42	6
	2744	4034	2862	2682	3095	1312	1572	1403	681	359	209	32
	11053	16547	11833	11399	13341	5698	6960	6211	3039	1622	925	141
	39741	60418	43512	42879	50722	21825	27058	24247	12033	6447	3654	569
								85811	43141	23252	13205	2069
												7000
Wt12	12	16	10	8	9	5	5	4	3	2	1	1
	108	153	102	88	102	46	52	44	24	16	8	2
	639	928	640	582	672	297	343	296	153	88	47	11
	2988	4424	3106	2926	3414	1484	1770	1545	774	436	236	44
	11973	18020	12811	12400	14592	6339	7703	6804	3405	1873	1027	181
				46422	55152	23950	29664	26417	13278	7268	4032	674
										25738	14441	2382

Affine G2 - Level 6 - Page 1 of 3

	Rp 1	Rp 2	Rp 3	Rp 4	Rp 5	Rp 6	Rp 7	Rp 8	Rp 9	Rp10	Rp11	Rp12	Rp13	Rp14	Rp15	Rp16
	006	014	022	030	105	113	121	204	212	220	303	311	402	410	501	600
Wt 1	1	0	0	0	0	0	0	0	0	0	0	0	0	0	0	0
	6	6	3	1	3	2	0	0	0	0	0	0	0	0	0	0
	48	59	42	15	30	30	13	9	6	3	1	0	0	0	0	0
	284	396	309	117	225	252	141	93	84	33	27	14	3	2	0	0
	1431	2118	1743	665	1278	1540	939	649	633	258	228	162	54	27	6	1
	6225	9614	8145	3150	6117	7662	4888	3474	3573	1455	1418	1092	405	210	76	9
	24335	38622	33426	13023	25542	32928	21546	15765	16674	6875	7008	5656	2277	1169	498	69
									68037	28227	29847	24726	10422	5421	2511	343
														21816	10636	1512
																5778
Wt 2	1	1	0	0	0	0	0	0	0	0	0	0	0	0	0	0
	10	12	7	2	5	4	2	1	0	0	0	0	0	0	0	0
	76	103	75	26	52	54	31	19	14	4	3	2	0	0	0	0
	441	641	509	186	364	418	258	168	153	56	49	36	9	3	1	0
	2146	3259	2704	1020	1985	2418	1545	1062	1052	408	388	304	101	46	18	1
	9102	14261	12164	4668	9147	11558	7553	5390	5579	2236	2244	1818	687	333	147	15
	34725	55692	48415	18800	37125	48140	31956	23460	24977	10192	10622	8802	3573	1795	842	101
										40779	43694	36786	15652	8017	3921	513
															15838	2180
Wt 3	1	1	1	0	0	0	0	0	0	0	0	0	0	0	0	0
	12	16	12	4	7	6	4	1	1	0	0	0	0	0	0	0
	97	136	108	38	69	76	47	27	23	8	5	4	1	0	0	0
	559	831	681	254	477	562	358	227	221	82	73	56	16	7	3	0
	2700	4147	3500	1329	2544	3150	2043	1402	1421	562	535	430	154	70	30	3
	11305	17860	15371	5933	11535	14714	9700	6923	7274	2933	2974	2438	948	470	216	23
	42674	68807	60170	23434	46114	60200	40174	29576	31739	13025	13653	11406	4726	2386	1145	146
												46528	20039	10330	5127	683
															20110	2819
Wt 4	1	1	1	1	0	0	0	0	0	0	0	0	0	0	0	0
	13	17	14	6	7	8	4	1	1	1	0	0	0	0	0	0
	105	147	120	48	75	88	51	28	27	12	7	4	1	1	0	0
	607	897	749	293	517	628	388	248	247	104	86	60	20	11	3	1
	2911	4469	3803	1483	2756	3460	2212	1519	1569	649	607	468	174	93	33	5
	12153	19184	16601	6489	12430	15988	10460	7492	7933	3287	3289	2648	1060	553	234	34
			64616	25371	49509	64936	43160	31835	34360	14288	14917	12332	5181	2706	1247	178
												50080	21785	11413	5540	796
																3151
Wt 5	2	2	1	0	1	0	0	0	0	0	0	0	0	0	0	0
	20	25	17	5	12	10	5	3	2	0	0	0	0	0	0	0
	138	192	147	51	104	110	65	41	35	11	9	6	1	0	0	0
	756	1117	906	334	655	766	479	320	306	113	103	78	23	9	3	0
	3506	5387	4525	1715	3354	4136	2666	1862	1885	739	720	572	203	94	39	3
	14338	22660	19480	7505	14755	18818	12372	8932	9365	3775	3851	3150	1231	605	274	30
			74940	29183	57811	75490	50339	37274	40017	16403	17293	14424	5979	3025	1445	180
												57864	24973	12858	6368	851
																3451

-continued next page-

Affine G2 - Level 6 - Page 2 of 3

	Rp 1	Rp 2	Rp 3	Rp 4	Rp 5	Rp 6	Rp 7	Rp 8	Rp 9	Rp10	Rp11	Rp12	Rp13	Rp14	Rp15	Rp16
Wt 6	3	3	2	1	1	1	0	0	0	0	0	0	0	0	0	0
	29	37	28	11	18	19	9	5	4	2	1	0	0	0	0	0
	197	277	222	86	152	176	101	64	60	25	19	11	3	2	0	0
	1051	1565	1301	503	933	1136	702	474	473	195	174	125	43	23	6	1
	4770	7373	6281	2441	4648	5860	3759	2652	2750	1132	1099	856	327	171	64	9
	19136	30389	26361	10300	19988	25835	16957	12341	13128	5433	5550	4502	1834	954	414	58
							67471	50284	54501	22678	23983	19938	8482	4428	2074	293
													34242	17954	8817	1267
																4860
Wt 7	3	4	3	1	1	1	1	0	0	0	0	0	0	0	0	0
	33	46	36	13	21	23	15	7	6	2	1	1	0	0	0	0
	228	334	272	101	181	212	136	82	78	29	24	19	5	2	1	0
	1219	1854	1553	589	1102	1351	877	583	587	230	215	173	59	27	12	1
	5506	8616	7374	2837	5437	6885	4529	3179	3314	1333	1329	1090	417	204	94	10
	21959	35142	30577	11883	23160	30032	19991	14534	15519	6345	6590	5493	2246	1132	543	67
								58440	63517	26259	28073	23706	10126	5198	2573	341
														20863	10585	1469
																5602
Wt 8	5	6	4	1	2	2	1	1	0	0	0	0	0	0	0	0
	46	64	48	17	32	34	21	12	10	3	3	2	0	0	0	0
	305	445	360	132	248	292	184	121	113	43	38	30	9	3	1	0
	1571	2398	2005	761	1453	1786	1156	791	801	313	307	246	86	41	18	1
	6956	10913	9352	3596	6960	8852	5823	4160	4347	1757	1782	1468	577	281	130	15
	27263	43769	38139	14839	29079	37830	25215	18498	19835	8121	8547	7144	2960	1504	728	89
									79582	32973	35509	30078	12991	6680	3332	450
														26254	13408	1870
																7020
Wt 9	6	8	6	2	3	3	2	1	1	0	0	0	0	0	0	0
	59	84	67	24	43	48	31	18	17	6	5	4	1	0	0	0
	386	575	475	178	327	393	254	166	166	63	59	48	16	7	3	0
	1975	3045	2581	985	1865	2329	1526	1055	1092	435	432	355	135	64	30	3
	8630	13649	11786	4558	8786	11280	7480	5373	5705	2318	2395	1997	812	406	196	23
	33481	54038	47348	18474	36141	47352	31725	23415	25323	10431	11087	9353	3976	2031	1009	132
										41408	44990	38332	16815	8714	4425	609
															17223	2455
Wt10	7	8	7	3	3	4	2	1	1	1	0	0	0	0	0	0
	64	90	74	30	47	56	33	19	20	9	7	4	1	1	0	0
	420	619	521	204	355	440	274	182	186	81	70	52	20	11	3	1
	2130	3277	2800	1097	2022	2560	1649	1146	1209	505	494	388	154	85	33	5
	9278	14649	12720	4983	9471	12264	8060	5824	6234	2608	2661	2176	914	482	214	34
	35832	57827	50828	19996	38812	51100	34078	25231	27455	11470	12153	10140	4380	2318	1104	162
												41344	18347	9673	4802	716
															18606	2761

--continued next page-

Affine G2 - Level 6 - Page 3 of 3

	Rp 1	Rp 2	Rp 3	Rp 4	Rp 5	Rp 6	Rp 7	Rp 8	Rp 9	Rp10	Rp11	Rp12	Rp13	Rp14	Rp15	Rp16
Wt11	9	11	9	4	5	6	3	2	2	1	1	0	0	0	0	0
	84	119	98	39	64	78	46	30	31	14	12	8	3	2	0	0
	534	798	673	264	469	586	369	253	265	113	108	82	33	19	6	1
	2678	4146	3562	1396	2592	3314	2150	1529	1628	684	691	554	233	126	53	9
	11490	18258	15914	6248	11925	15544	10273	7509	8119	3400	3557	2942	1270	678	314	48
			62779	24729	48107	63694	42652	31853	34878	14623	15701	13218	5839	3096	1515	229
												52516	23646	12530	6340	953
																3587
Wt12	10	14	11	4	6	7	5	3	3	1	1	1	0	0	0	0
	97	143	118	44	77	93	62	40	41	16	16	14	5	2	1	0
	618	943	797	304	556	696	461	317	333	133	137	117	47	23	12	1
	3085	4839	4167	1609	3037	3894	2595	1851	1980	808	850	723	306	154	80	10
	13169	21098	18431	7183	13832	18082	12137	8901	9663	3987	4277	3650	1590	819	425	57
							49649	37190	40860	17000	18547	15906	7085	3688	1926	272
													28069	14728	7752	1123
																4192
Wt13	13	18	15	5	8	10	7	5	5	2	2	2	1	0	0	0
	122	183	153	58	102	126	85	57	62	24	27	24	10	5	3	0
	774	1192	1020	390	715	912	610	433	464	190	203	178	80	39	22	3
	3817	6040	5240	2035	3842	4984	3349	2426	2644	1087	1183	1022	456	237	130	17
	16153	26035	22883	8944	17226	22726	15345	11390	12501	5201	5685	4914	2227	1157	624	91
									51690	21602	23926	20696	9452	4975	2671	389
														19152	10285	1538
																5493
Wt14	14	19	16	7	9	12	7	5	6	3	3	2	1	1	0	0
	133	196	167	67	111	142	90	63	70	32	33	26	13	8	3	1
	834	1279	1103	437	777	1006	656	471	518	224	237	196	93	54	25	5
	4103	6474	5651	2230	4145	5432	3603	2639	2905	1238	1331	1122	523	289	144	26
	17283	27845	24559	9700	18516	24570	16485	12303	13613	5762	6289	5362	2484	1349	693	115
										23616	26115	22442	10417	5600	2930	470
															11212	1761
Wt15	16	23	19	7	11	14	10	7	8	3	4	4	2	1	1	0
	153	232	197	75	132	168	115	81	90	37	43	40	20	10	7	1
	958	1494	1290	499	913	1184	802	580	641	265	299	268	129	68	42	6
	4697	7491	6553	2556	4821	6338	4295	3166	3504	1461	1632	1438	681	359	209	32
			28258	11094	21352	28426	19322	14493	16114	6747	7543	6602	3102	1644	925	141
												26900	12632	6705	3710	569
																2103
Wt16	18	24	21	9	12	16	10	8	9	5	5	4	3	2	1	1
	165	248	213	86	144	188	122	88	102	46	52	44	24	16	8	2
	1033	1600	1395	553	987	1300	861	634	711	313	343	296	153	88	47	11
	5033	8015	7047	2794	5196	6888	4618	3433	3846	1647	1834	1580	774	436	236	44
									17517	7467	8318	7208	3471	1898	1027	181
														7544	4088	674
																2420

32. TABLES OF BRANCHING RULES

These are the tables of branching rules for affine algebras of rank ≤ 8 to representations of maximal finite dimensional semisimple subalgebras. Branching rules are given for a subset of the representations of the previous section. The tables for each algebra are ordered by subalgebra, level and congruence class. The subalgebras considered are the maximal reductive finite dimensional subalgebras, i.e., those obtained by the elimination of a single node of the affine Coxeter-Dynkin diagram.

The highest weight Λ of the representation is given in bold print with coordinates corresponding to the nodes of the Coxeter-Dynkin diagram in the upper right hand corner of the page. The algebra, level and class are named before each branching rule. The representations of the subalgebra are given by their highest weights in the same configuration of the Coxeter-Dynkin diagram with the missing node denoted by $\bullet$. If a representation appears more than once, its multiplicity precedes the diagram of its highest weight.

When several branching rules are equivalent up to a diagram symmetry, only one is included. The congruence class is not named when a branching rule can be interpreted to be in various congruence classes by diagram automorphisms.

Affine A1 Branching Rules

Affine A1 ⊃ A1 - Level 1 - Class 0

Sl 0: $(\dot{0})$ **Highest Weight** $(\overset{1}{0})$

Sl 1: $(\dot{2})$

Sl 2: $(\dot{2}) + (\dot{0})$

Sl 3: $2(\dot{2}) + (\dot{0})$

Sl 4: $(\dot{4}) + 2(\dot{2}) + 2(\dot{0})$

Sl 5: $(\dot{4}) + 4(\dot{2}) + 2(\dot{0})$

Sl 6: $2(\dot{4}) + 5(\dot{2}) + 4(\dot{0})$

Sl 7: $3(\dot{4}) + 8(\dot{2}) + 4(\dot{0})$

Sl 8: $5(\dot{4}) + 10(\dot{2}) + 7(\dot{0})$

Sl 9: $(\dot{6}) + 6(\dot{4}) + 15(\dot{2}) + 8(\dot{0})$

Sl10: $(\dot{6}) + 10(\dot{4}) + 19(\dot{2}) + 12(\dot{0})$

Sl11: $2(\dot{6}) + 13(\dot{4}) + 27(\dot{2}) + 14(\dot{0})$

Affine A1 ⊃ A1 - Level 1 - Class 1

Sl 0: $(\dot{1})$ **Highest Weight** $(\overset{0}{1})$

Sl 1: $(\dot{1})$

Sl 2: $(\dot{3}) + (\dot{1})$

Sl 3: $(\dot{3}) + 2(\dot{1})$

Sl 4: $2(\dot{3}) + 3(\dot{1})$

Sl 5: $3(\dot{3}) + 4(\dot{1})$

Sl 6: $(\dot{5}) + 4(\dot{3}) + 6(\dot{1})$

Sl 7: $(\dot{5}) + 6(\dot{3}) + 8(\dot{1})$

Sl 8: $2(\dot{5}) + 9(\dot{3}) + 11(\dot{1})$

Sl 9: $3(\dot{5}) + 12(\dot{3}) + 15(\dot{1})$

Sl10: $5(\dot{5}) + 17(\dot{3}) + 20(\dot{1})$

Sl11: $7(\dot{5}) + 23(\dot{3}) + 26(\dot{1})$

Affine A1 ⊃ A1 - Level 2 - Class 0

Sl 0: (2) **Highest Weight** $\overset{0}{(2)}$

Sl 1: (2) + (0)

Sl 2: (4) + 2(2) + (0)

Sl 3: 2(4) + 3(2) + 2(0)

Sl 4: (6) + 3(4) + 6(2) + 3(0)

Sl 5: (6) + 6(4) + 9(2) + 5(0)

Sl 6: 3(6) + 10(4) + 15(2) + 7(0)

Sl 7: 5(6) + 16(4) + 22(2) + 12(0)

Sl 8: (8) + 9(6) + 25(4) + 35(2) + 16(0)

Sl 9: 2(8) + 14(6) + 39(4) + 50(2) + 25(0)

Affine A1 ⊃ A1 - Level 2 - Class 0

Sl 0: (0) **Highest Weight** $\overset{2}{(0)}$

Sl 1: (2)

Sl 2: (4) + (2) + (0)

Sl 3: (4) + 3(2) + (0)

Sl 4: 3(4) + 4(2) + 3(0)

Sl 5: (6) + 4(4) + 8(2) + 3(0)

Sl 6: 2(6) + 8(4) + 11(2) + 7(0)

Sl 7: 4(6) + 12(4) + 19(2) + 8(0)

Sl 8: (8) + 6(6) + 21(4) + 27(2) + 15(0)

Sl 9: (8) + 12(6) + 30(4) + 43(2) + 19(0)

Affine A1 ⊃ A1 - Level 2 - Class 1

```
          •                      1
Sl 0:    (1)  Highest Weight    (1)

          •         •
Sl 1:    (3)  +    (1)

          •         •
Sl 2:   2(3)  +   2(1)

          •         •         •
Sl 3:    (5)  +   3(3)  +   4(1)

          •         •         •
Sl 4:   2(5)  +   6(3)  +   6(1)

          •         •         •
Sl 5:   4(5)  +  10(3)  +  10(1)

          •         •         •         •
Sl 6:    (7)  +   7(5)  +  16(3)  +  16(1)

          •         •         •         •
Sl 7:   2(7)  +  12(5)  +  26(3)  +  24(1)

          •         •         •         •
Sl 8:   4(7)  +  20(5)  +  40(3)  +  36(1)

          •         •         •         •
Sl 9:   8(7)  +  32(5)  +  60(3)  +  54(1)
```

Affine A1 ⊃ A1 - Level 3 - Class 0

```
          •                      1
Sl 0:    (2)  Highest Weight    (2)

          •         •         •
Sl 1:    (4)  +    (2)  +    (0)

          •         •         •
Sl 2:   2(4)  +   3(2)  +    (0)

          •         •         •         •
Sl 3:    (6)  +   4(4)  +   5(2)  +   3(0)

          •         •         •         •
Sl 4:   3(6)  +   7(4)  +  10(2)  +   4(0)

          •         •         •         •         •
Sl 5:    (8)  +   5(6)  +  14(4)  +  16(2)  +   8(0)

          •         •         •         •         •
Sl 6:   2(8)  +  11(6)  +  23(4)  +  28(2)  +  12(0)

          •         •         •         •         •
Sl 7:   5(8)  +  19(6)  +  40(4)  +  44(2)  +  21(0)

          •         •         •         •         •         •
Sl 8:   (10)  +   9(8)  +  34(6)  +  64(4)  +  72(2)  +  30(0)
```

Affine A1 ⊃ A1 - Level 3 - Class 0

Sl 0: $\dot{(0)}$ **Highest Weight** $\overset{3}{(0)}$

Sl 1: $\dot{(2)}$

Sl 2: $\dot{(4)} + \dot{(2)} + \dot{(0)}$

Sl 3: $\dot{(6)} + \dot{(4)} + 3\dot{(2)} + \dot{(0)}$

Sl 4: $\dot{(6)} + 4\dot{(4)} + 4\dot{(2)} + 3\dot{(0)}$

Sl 5: $3\dot{(6)} + 6\dot{(4)} + 9\dot{(2)} + 3\dot{(0)}$

Sl 6: $\dot{(8)} + 5\dot{(6)} + 12\dot{(4)} + 13\dot{(2)} + 8\dot{(0)}$

Sl 7: $2\dot{(8)} + 10\dot{(6)} + 19\dot{(4)} + 24\dot{(2)} + 9\dot{(0)}$

Sl 8: $5\dot{(8)} + 16\dot{(6)} + 34\dot{(4)} + 35\dot{(2)} + 18\dot{(0)}$

Affine A1 ⊃ A1 - Level 3 - Class 1

Sl 0: $\dot{(3)}$ **Highest Weight** $\overset{0}{(3)}$

Sl 1: $\dot{(3)} + \dot{(1)}$

Sl 2: $\dot{(5)} + 2\dot{(3)} + 2\dot{(1)}$

Sl 3: $2\dot{(5)} + 4\dot{(3)} + 3\dot{(1)}$

Sl 4: $\dot{(7)} + 4\dot{(5)} + 7\dot{(3)} + 6\dot{(1)}$

Sl 5: $2\dot{(7)} + 7\dot{(5)} + 12\dot{(3)} + 10\dot{(1)}$

Sl 6: $\dot{(9)} + 4\dot{(7)} + 13\dot{(5)} + 21\dot{(3)} + 16\dot{(1)}$

Sl 7: $\dot{(9)} + 8\dot{(7)} + 22\dot{(5)} + 33\dot{(3)} + 26\dot{(1)}$

Sl 8: $3\dot{(9)} + 15\dot{(7)} + 37\dot{(5)} + 53\dot{(3)} + 41\dot{(1)}$

Affine A1 ⊃ A1 - Level 3 - Class 1

Sl 0: $\dot{(1)}$ **Highest Weight** $\overset{2}{(1)}$

Sl 1: $\dot{(3)} + \dot{(1)}$

Sl 2: $\dot{(5)} + 2\dot{(3)} + 2\dot{(1)}$

Sl 3: $2\dot{(5)} + 4\dot{(3)} + 4\dot{(1)}$

Sl 4: $\dot{(7)} + 4\dot{(5)} + 8\dot{(3)} + 7\dot{(1)}$

Sl 5: $2\dot{(7)} + 8\dot{(5)} + 14\dot{(3)} + 12\dot{(1)}$

Sl 6: $5\dot{(7)} + 15\dot{(5)} + 24\dot{(3)} + 20\dot{(1)}$

Sl 7: $\dot{(9)} + 9\dot{(7)} + 26\dot{(5)} + 40\dot{(3)} + 32\dot{(1)}$

Sl 8: $3\dot{(9)} + 17\dot{(7)} + 44\dot{(5)} + 65\dot{(3)} + 51\dot{(1)}$

Affine A1 ⊃ A1 - Level 4 - Class 0

Sl 0: $\dot{(4)}$ **Highest Weight** $\overset{0}{(4)}$

Sl 1: $\dot{(4)} + \dot{(2)}$

Sl 2: $\dot{(6)} + 2\dot{(4)} + 2\dot{(2)} + \dot{(0)}$

Sl 3: $2\dot{(6)} + 4\dot{(4)} + 4\dot{(2)} + \dot{(0)}$

Sl 4: $\dot{(8)} + 4\dot{(6)} + 8\dot{(4)} + 7\dot{(2)} + 3\dot{(0)}$

Sl 5: $2\dot{(8)} + 8\dot{(6)} + 13\dot{(4)} + 13\dot{(2)} + 5\dot{(0)}$

Sl 6: $\dot{(10)} + 5\dot{(8)} + 14\dot{(6)} + 24\dot{(4)} + 22\dot{(2)} + 9\dot{(0)}$

Affine A1 ⊃ A1 - Level 4 - Class 0

Sl 0: $\dot{(2)}$ **Highest Weight** $\overset{2}{(2)}$

Sl 1: $\dot{(4)} + \dot{(2)} + \dot{(0)}$

Sl 2: $\dot{(6)} + 2\dot{(4)} + 3\dot{(2)} + \dot{(0)}$

Sl 3: $2\dot{(6)} + 5\dot{(4)} + 5\dot{(2)} + 3\dot{(0)}$

Sl 4: $\dot{(8)} + 5\dot{(6)} + 9\dot{(4)} + 11\dot{(2)} + 4\dot{(0)}$

Sl 5: $3\dot{(8)} + 9\dot{(6)} + 18\dot{(4)} + 18\dot{(2)} + 9\dot{(0)}$

Sl 6: $\dot{(10)} + 6\dot{(8)} + 19\dot{(6)} + 31\dot{(4)} + 33\dot{(2)} + 13\dot{(0)}$

Affine A1 ⊃ A1 - Level 4 - Class 0

Sl 0: $(\dot{0})$ **Highest Weight** $\overset{4}{(0)}$

Sl 1: $(\dot{2})$

Sl 2: $(\dot{4}) + (\dot{2}) + (\dot{0})$

Sl 3: $(\dot{6}) + (\dot{4}) + 3(\dot{2}) + (\dot{0})$

Sl 4: $(\dot{8}) + (\dot{6}) + 4(\dot{4}) + 4(\dot{2}) + 3(\dot{0})$

Sl 5: $(\dot{8}) + 4(\dot{6}) + 6(\dot{4}) + 9(\dot{2}) + 3(\dot{0})$

Sl 6: $3(\dot{8}) + 7(\dot{6}) + 13(\dot{4}) + 13(\dot{2}) + 8(\dot{0})$

Sl 7: $(\dot{10}) + 5(\dot{8}) + 14(\dot{6}) + 21(\dot{4}) + 25(\dot{2}) + 9(\dot{0})$

Affine A1 ⊃ A1 - Level 4 - Class 1

Sl 0: $(\dot{3})$ **Highest Weight** $\overset{1}{(3)}$

Sl 1: $(\dot{5}) + (\dot{3}) + (\dot{1})$

Sl 2: $2(\dot{5}) + 3(\dot{3}) + 2(\dot{1})$

Sl 3: $(\dot{7}) + 4(\dot{5}) + 6(\dot{3}) + 4(\dot{1})$

Sl 4: $3(\dot{7}) + 8(\dot{5}) + 11(\dot{3}) + 8(\dot{1})$

Sl 5: $(\dot{9}) + 6(\dot{7}) + 15(\dot{5}) + 20(\dot{3}) + 14(\dot{1})$

Sl 6: $3(\dot{9}) + 12(\dot{7}) + 27(\dot{5}) + 35(\dot{3}) + 24(\dot{1})$

Sl 7: $(\dot{11}) + 6(\dot{9}) + 23(\dot{7}) + 47(\dot{5}) + 58(\dot{3}) + 41(\dot{1})$

Affine A1 ⊃ A1 - Level 4 - Class 1

Sl 0: $(\dot{1})$ **Highest Weight** $\overset{3}{(1)}$

Sl 1: $(\dot{3}) + (\dot{1})$

Sl 2: $(\dot{5}) + 2(\dot{3}) + 2(\dot{1})$

Sl 3: $(\dot{7}) + 2(\dot{5}) + 4(\dot{3}) + 4(\dot{1})$

Sl 4: $2(\dot{7}) + 5(\dot{5}) + 8(\dot{3}) + 7(\dot{1})$

Sl 5: $(\dot{9}) + 4(\dot{7}) + 10(\dot{5}) + 15(\dot{3}) + 12(\dot{1})$

Sl 6: $2(\dot{9}) + 9(\dot{7}) + 19(\dot{5}) + 26(\dot{3}) + 21(\dot{1})$

Sl 7: $5(\dot{9}) + 17(\dot{7}) + 34(\dot{5}) + 45(\dot{3}) + 34(\dot{1})$

Affine A1 ⊃ A1 - Level 5 - Class 0

```
         .                      1
Sl 0:   (4)  Highest Weight  (4)
         .        .        .
Sl 1:   (6) +    (4) +    (2)
         .        .        .         .
Sl 2:  2(6) +   3(4) +   2(2) +     (0)
         .        .        .         .         .
Sl 3:   (8) +   4(6) +   6(4) +    5(2) +     (0)
         .        .        .         .         .
Sl 4:  3(8) +   8(6) + 12(4) +     9(2) +    4(0)
         .        .        .         .         .         .
Sl 5:  (10) +   6(8) + 16(6) +   21(4) +   18(2) +    6(0)
         .        .        .         .         .         .
Sl 6: 3(10) + 13(8) + 28(6) +   39(4) +   31(2) +  12(0)
```

Affine A1 ⊃ A1 - Level 5 - Class 0

```
         .                      3
Sl 0:   (2)  Highest Weight  (2)
         .        .        .
Sl 1:   (4) +    (2) +    (0)
         .        .        .         .
Sl 2:   (6) +   2(4) +   3(2) +     (0)
         .        .        .         .         .
Sl 3:   (8) +   2(6) +   5(4) +    5(2) +    3(0)
         .        .        .         .         .
Sl 4:  2(8) +   6(6) +   9(4) +   11(2) +    4(0)
         .        .        .         .         .         .
Sl 5:  (10) +   5(8) + 11(6) +   19(4) +   18(2) +    9(0)
         .        .        .         .         .         .
Sl 6: 3(10) + 10(8) + 23(6) +   33(4) +   34(2) +  13(0)
```

Affine A1 ⊃ A1 - Level 5 - Class 0

```
         .                      5
Sl 0:   (0)  Highest Weight  (0)
         .
Sl 1:   (2)
         .        .        .
Sl 2:   (4) +    (2) +    (0)
         .        .        .         .
Sl 3:   (6) +    (4) +   3(2) +     (0)
         .        .        .         .         .
Sl 4:   (8) +    (6) +   4(4) +    4(2) +    3(0)
         .        .        .         .         .         .
Sl 5:  (10) +    (8) +   4(6) +    6(4) +    9(2) +    3(0)
         .        .        .         .         .         .
Sl 6:  (10) +   4(8) +   7(6) +   13(4) +   13(2) +    8(0)
```

Affine A1 $\supset$ A1 - Level 5 - Class 1

Sl 0: $\overset{\bullet}{(5)}$ **Highest Weight** $\overset{0}{(5)}$

Sl 1: $\overset{\bullet}{(5)} + \overset{\bullet}{(3)}$

Sl 2: $\overset{\bullet}{(7)} + 2\overset{\bullet}{(5)} + 2\overset{\bullet}{(3)} + \overset{\bullet}{(1)}$

Sl 3: $2\overset{\bullet}{(7)} + 4\overset{\bullet}{(5)} + 4\overset{\bullet}{(3)} + 2\overset{\bullet}{(1)}$

Sl 4: $\overset{\bullet}{(9)} + 4\overset{\bullet}{(7)} + 8\overset{\bullet}{(5)} + 8\overset{\bullet}{(3)} + 4\overset{\bullet}{(1)}$

Sl 5: $2\overset{\bullet}{(9)} + 8\overset{\bullet}{(7)} + 14\overset{\bullet}{(5)} + 14\overset{\bullet}{(3)} + 8\overset{\bullet}{(1)}$

Sl 6: $\overset{\bullet}{(11)} + 5\overset{\bullet}{(9)} + 15\overset{\bullet}{(7)} + 25\overset{\bullet}{(5)} + 25\overset{\bullet}{(3)} + 15\overset{\bullet}{(1)}$

Affine A1 $\supset$ A1 - Level 5 - Class 1

Sl 0: $\overset{\bullet}{(3)}$ **Highest Weight** $\overset{2}{(3)}$

Sl 1: $\overset{\bullet}{(5)} + \overset{\bullet}{(3)} + \overset{\bullet}{(1)}$

Sl 2: $\overset{\bullet}{(7)} + 2\overset{\bullet}{(5)} + 3\overset{\bullet}{(3)} + 2\overset{\bullet}{(1)}$

Sl 3: $2\overset{\bullet}{(7)} + 5\overset{\bullet}{(5)} + 6\overset{\bullet}{(3)} + 4\overset{\bullet}{(1)}$

Sl 4: $\overset{\bullet}{(9)} + 5\overset{\bullet}{(7)} + 10\overset{\bullet}{(5)} + 12\overset{\bullet}{(3)} + 8\overset{\bullet}{(1)}$

Sl 5: $3\overset{\bullet}{(9)} + 10\overset{\bullet}{(7)} + 19\overset{\bullet}{(5)} + 22\overset{\bullet}{(3)} + 15\overset{\bullet}{(1)}$

Sl 6: $\overset{\bullet}{(11)} + 7\overset{\bullet}{(9)} + 20\overset{\bullet}{(7)} + 35\overset{\bullet}{(5)} + 40\overset{\bullet}{(3)} + 26\overset{\bullet}{(1)}$

Affine A1 $\supset$ A1 - Level 5 - Class 1

Sl 0: $\overset{\bullet}{(1)}$ **Highest Weight** $\overset{4}{(1)}$

Sl 1: $\overset{\bullet}{(3)} + \overset{\bullet}{(1)}$

Sl 2: $\overset{\bullet}{(5)} + 2\overset{\bullet}{(3)} + 2\overset{\bullet}{(1)}$

Sl 3: $\overset{\bullet}{(7)} + 2\overset{\bullet}{(5)} + 4\overset{\bullet}{(3)} + 4\overset{\bullet}{(1)}$

Sl 4: $\overset{\bullet}{(9)} + 2\overset{\bullet}{(7)} + 5\overset{\bullet}{(5)} + 8\overset{\bullet}{(3)} + 7\overset{\bullet}{(1)}$

Sl 5: $2\overset{\bullet}{(9)} + 5\overset{\bullet}{(7)} + 10\overset{\bullet}{(5)} + 15\overset{\bullet}{(3)} + 12\overset{\bullet}{(1)}$

Sl 6: $\overset{\bullet}{(11)} + 4\overset{\bullet}{(9)} + 11\overset{\bullet}{(7)} + 20\overset{\bullet}{(5)} + 26\overset{\bullet}{(3)} + 21\overset{\bullet}{(1)}$

Affine A2 Branching Rules

Affine A2 ⊃ A2 - Level 1 - Class 0

Sl 0: $\dot{(00)}$ **Highest Weight** $\overset{1}{(00)}$

Sl 1: $\dot{(11)}$

Sl 2: $2\dot{(11)} + \dot{(00)}$

Sl 3: $\dot{(30)} + \dot{(03)} + 3\dot{(11)} + 2\dot{(00)}$

Sl 4: $\dot{(22)} + \dot{(30)} + \dot{(03)} + 6\dot{(11)} + 3\dot{(00)}$

Sl 5: $2\dot{(22)} + 3\dot{(30)} + 3\dot{(03)} + 10\dot{(11)} + 4\dot{(00)}$

Sl 6: $5\dot{(22)} + 5\dot{(30)} + 5\dot{(03)} + 16\dot{(11)} + 8\dot{(00)}$

Sl 7: $\dot{(41)} + \dot{(14)} + 8\dot{(22)} + 9\dot{(30)} + 9\dot{(03)} + 27\dot{(11)} + 10\dot{(00)}$

Sl 8: $2\dot{(41)} + 2\dot{(14)} + 16\dot{(22)} + 14\dot{(30)} + 14\dot{(03)} + 42\dot{(11)} + 17\dot{(00)}$

Sl 9: $\dot{(33)} + 4\dot{(41)} + 4\dot{(14)} + 26\dot{(22)} + 25\dot{(30)} + 25\dot{(03)} + 64\dot{(11)} + 24\dot{(00)}$

Affine A2 ⊃ A2 - Level 1 - Class 1

Sl 0: $\dot{(10)}$ **Highest Weight** $\overset{0}{(10)}$

Sl 1: $\dot{(02)} + \dot{(10)}$

Sl 2: $\dot{(21)} + \dot{(02)} + 2\dot{(10)}$

Sl 3: $2\dot{(21)} + 3\dot{(02)} + 3\dot{(10)}$

Sl 4: $\dot{(13)} + 4\dot{(21)} + 4\dot{(02)} + 6\dot{(10)}$

Sl 5: $\dot{(40)} + 2\dot{(13)} + 7\dot{(21)} + 8\dot{(02)} + 9\dot{(10)}$

Sl 6: $\dot{(32)} + \dot{(40)} + 4\dot{(13)} + 13\dot{(21)} + 12\dot{(02)} + 15\dot{(10)}$

Sl 7: $2\dot{(32)} + 3\dot{(40)} + 8\dot{(13)} + 21\dot{(21)} + 21\dot{(02)} + 22\dot{(10)}$

Sl 8: $5\dot{(32)} + \dot{(05)} + 5\dot{(40)} + 14\dot{(13)} + 35\dot{(21)} + 31\dot{(02)} + 35\dot{(10)}$

Sl 9: $\dot{(24)} + 9\dot{(32)} + \dot{(05)} + 10\dot{(40)} + 24\dot{(13)} + 55\dot{(21)} + 50\dot{(02)} + 51\dot{(10)}$

Affine A2 ⊃ A2 - Level 2 - Class 0

Sl 0: (11) **Highest Weight** $\overset{0}{(11)}$

Sl 1: (30) + (03) + 2(11) + (00)

Sl 2: 2(22) + 2(30) + 2(03) + 5(11) + 2(00)

Sl 3: (41) + (14) + 5(22) + 5(30) + 5(03) + 12(11) + 3(00)

Sl 4: (33) + 3(41) + 3(14) + 14(22) + 11(30) + 11(03) + 24(11) + 8(00)

Sl 5: 4(33) + 9(41) + 9(14) + 30(22) + 24(30) + 24(03) + 48(11) + 14(00)

Sl 6: (52) + (25) + (60) + 11(33) + (06) + 21(41) + 21(14) + 66(22) + 46(30) + 46(03) + 92(11) + 26(00)

Sl 7: 4(52) + 4(25) + 3(60) + 28(33) + 3(06) + 46(41) + 46(14) +130(22) + 91(30) + 91(03) +168(11) + 46(00)

Affine A2 ⊃ A2 - Level 2 - Class 0

Sl 0: (00) **Highest Weight** $\overset{2}{(00)}$

Sl 1: (11)

Sl 2: (22) + 2(11) + (00)

Sl 3: 2(22) + 2(30) + 2(03) + 4(11) + 2(00)

Sl 4: (41) + (14) + 6(22) + 3(30) + 3(03) + 10(11) + 4(00)

Sl 5: (33) + 3(41) + 3(14) + 12(22) + 9(30) + 9(03) + 20(11) + 6(00)

Sl 6: (60) + 4(33) + (06) + 7(41) + 7(14) + 28(22) + 18(30) + 18(03) + 38(11) + 14(00)

Sl 7: (52) + (25) + (60) + 10(33) + (06) + 18(41) + 18(14) + 54(22) + 37(30) + 37(03) + 73(11) + 20(00)

Affine A2 ⊃ A2 - Level 2 - Class 1

Sl 0: (02) **Highest Weight** $\overset{0}{(02)}$

Sl 1: (21) + (02) + (10)

Sl 2: (40) + (13) + 2(21) + 3(02) + 2(10)

Sl 3: (32) + (40) + 3(13) + 6(21) + 5(02) + 5(10)

Sl 4: (24) + 3(32) + 4(40) + 7(13) + 13(21) + 12(02) + 9(10)

Sl 5: (51) + 2(24) + 9(32) + 2(05) + 7(40) + 16(13) + 28(21) + 21(02) + 19(10)

Sl 6: (43) + 3(51) + 7(24) + 20(32) + 4(05) + 17(40) + 34(13) + 54(21) + 43(02) + 34(10)

Sl 7: 4(43) + (16) + 8(51) + 15(24) + 45(32) + 11(05) + 31(40) + 68(13) +105(21) + 74(02) + 64(10)

Affine A2 ⊃ A2 - Level 2 - Class 1

Sl 0: (10) **Highest Weight** $\overset{1}{(10)}$

Sl 1: (21) + (02) + (10)

Sl 2: (13) + 3(21) + 2(02) + 3(10)

Sl 3: (32) + (40) + 3(13) + 7(21) + 6(02) + 6(10)

Sl 4: 4(32) + (05) + 3(40) + 8(13) + 16(21) + 12(02) + 13(10)

Sl 5: (51) + 2(24) + 10(32) + 2(05) + 8(40) + 19(13) + 34(21) + 26(02) + 24(10)

Sl 6: (43) + 3(51) + 6(24) + 25(32) + 6(05) + 17(40) + 41(13) + 69(21) + 49(02) + 46(10)

Sl 7: 4(43) + (16) + 9(51) + 17(24) + 54(32) + 13(05) + 38(40) + 84(13) +132(21) + 94(02) + 82(10)

Affine A2 ⊃ A2 - Level 3 - Class 0

Sl 0: (03) **Highest Weight** $\overset{0}{(03)}$

Sl 1: (22) + (03) + (11)

Sl 2: (41) + (14) + 2(22) + (30) + 3(03) + 3(11)

Sl 3: (60) + (33) + 2(41) + 3(14) + 7(22) + 3(30) + 6(03) + 7(11) + 2(00)

Sl 4: (52) + (25) + (60) + 4(33) + 7(41) + 8(14) + 16(22) + 9(30) + 14(03) + 17(11) + 3(00)

Sl 5: (44) + 3(52) + 3(25) + 4(60) + 12(33) + 2(06) + 17(41) + 19(14) + 40(22) + 20(30) + 27(03)
+ 37(11) + 8(00)

Affine A2 ⊃ A2 - Level 3 - Class 0

sl 0: (11) **Highest Weight** $\overset{1}{(11)}$

sl 1: (22) + (30) + (03) + 2(11) + (00)

sl 2: (41) + (14) + 4(22) + 3(30) + 3(03) + 6(11) + 2(00)

sl 3: 2(33) + 4(41) + 4(14) + 12(22) + 8(30) + 8(03) + 16(11) + 4(00)

sl 4: (52) + (25) + (60) + 8(33) + (06) + 12(41) + 12(14) + 32(22) + 21(30) + 21(03) + 36(11) + 10(00)

sl 5: (44) + 5(52) + 5(25) + 4(60) + 24(33) + 4(06) + 33(41) + 33(14) + 77(22) + 48(30) + 48(03) + 80(11) + 20(00)

Affine A2 ⊃ A2 - Level 3 - Class 0

sl 0: (00) **Highest Weight** $\overset{3}{(00)}$

sl 1: (11)

sl 2: (22) + 2(11) + (00)

sl 3: (33) + 2(22) + 2(30) + 2(03) + 4(11) + 2(00)

sl 4: 2(33) + 2(41) + 2(14) + 7(22) + 3(30) + 3(03) + 10(11) + 4(00)

sl 5: (52) + (25) + 6(33) + 6(41) + 6(14) + 16(22) + 10(30) + 10(03) + 21(11) + 6(00)

sl 6: (44) + 3(52) + 3(25) + 2(60) + 16(33) + 2(06) + 15(41) + 15(14) + 40(22) + 22(30) + 22(03) + 42(11) + 15(00)

Affine A2 ⊃ A2 - Level 3 - Class 1

sl 0: (21) **Highest Weight** $\overset{0}{(21)}$

sl 1: (40) + (13) + 2(21) + (02) + (10)

sl 2: 2(32) + (05) + 2(40) + 3(13) + 6(21) + 3(02) + 3(10)

sl 3: (51) + 2(24) + 6(32) + 2(05) + 6(40) + 9(13) + 15(21) + 9(02) + 7(10)

sl 4: 2(43) + (16) + 4(51) + 6(24) + 18(32) + 6(05) + 13(40) + 24(13) + 36(21) + 20(02) + 17(10)

sl 5: (62) + (35) + 7(43) + 3(16) + 12(51) + 20(24) + 45(32) + 14(05) + 32(40) + 57(13) + 78(21) + 47(02) + 36(10) (continued next page)

Sl 6: (54) + 3(62) + 5(35) + 2(70) + 23(43) + 10(16) + 32(51) + 50(24) +107(32) + 35(05) + 67(40)

+127(13) +166(21) + 96(02) + 76(10)

Affine A2 ⊃ A2 - Level 3 - Class 1

Sl 0: (02) **Highest Weight** $\overset{1}{\mathbf{(02)}}$

Sl 1: (13) + (21) + (02) + (10)

Sl 2: (32) + (40) + 3(13) + 3(21) + 4(02) + 2(10)

Sl 3: (51) + (24) + 4(32) + (05) + 2(40) + 8(13) + 10(21) + 8(02) + 6(10)

Sl 4: (43) + 3(51) + 5(24) + 12(32) + 3(05) + 8(40) + 20(13) + 24(21) + 20(02) + 13(10)

Sl 5: (35) + (70) + 5(43) + (16) + 9(51) + 14(24) + 33(32) + 10(05) + 19(40) + 47(13) + 57(21)

+ 40(02) + 30(10)

Sl 6: 2(62) + 4(35) + 2(70) + 16(43) + 5(16) + 24(51) + 39(24) + 79(32) + 23(05) + 47(40) +104(13)

+123(21) + 86(02) + 60(10)

Affine A2 ⊃ A2 - Level 3 - Class 1

Sl 0: (10) **Highest Weight** $\overset{2}{\mathbf{(10)}}$

Sl 1: (21) + (02) + (10)

Sl 2: (32) + (13) + 3(21) + 2(02) + 3(10)

Sl 3: (24) + 3(32) + 2(40) + 4(13) + 8(21) + 6(02) + 6(10)

Sl 4: (43) + (51) + 3(24) + 10(32) + 2(05) + 5(40) + 12(13) + 20(21) + 13(02) + 14(10)

Sl 5: 4(43) + (16) + 5(51) + 11(24) + 25(32) + 5(05) + 15(40) + 31(13) + 46(21) + 31(02) + 27(10)

Sl 6: (62) + 2(35) + (70) + 13(43) + 4(16) + 15(51) + 28(24) + 63(32) + 16(05) + 34(40) + 73(13)

+101(21) + 63(02) + 56(10)

Affine A2 ⊃ A2 - Level 4 - Class 0

```
         .                            0
Sl 0:   (22)   Highest Weight (22)

         .         .          .          .          .          .
Sl 1:   (41) +   (14) +   2(22) +    (30) +     (03) +     (11)

         .         .          .          .          .          .          .          .          .          .
Sl 2:   (60) +  2(33) +    (06) +   3(41) +   3(14) +   7(22) +   3(30) +   3(03) +   4(11) +    (00)

         .         .          .          .          .          .          .          .          .          .          .
Sl 3:  2(52) +  2(25) +   2(60) +   7(33) +   2(06) +  10(41) +  10(14) +  18(22) +  10(30) +  10(03) +  12(11)

         .
    +  2(00)
```

Affine A2 ⊃ A2 - Level 4 - Class 0

```
         .                            1
Sl 0:   (03)   Highest Weight (03)

         .         .          .          .
Sl 1:   (14) +   (22) +    (03) +     (11)

         .         .          .          .          .          .          .
Sl 2:   (33) +   (41) +   3(14) +   3(22) +    (30) +   4(03) +   3(11)

         .         .          .          .          .          .          .          .          .          .          .
Sl 3:   (52) +   (25) +    (60) +   4(33) +    (06) +   3(41) +   8(14) +  11(22) +   4(30) +   9(03) +   8(11)

         .
    +  2(00)

         .         .          .          .          .          .          .          .          .          .          .
Sl 4:   (71) +   (44) +   4(52) +   5(25) +   2(60) +  13(33) +   3(06) +  12(41) +  21(14) +  28(22) +  12(30)

         .         .          .
    + 22(03) +  22(11) +   3(00)
```

Affine A2 ⊃ A2 - Level 4 - Class 0

```
         .                            2
Sl 0:   (11)   Highest Weight (11)

         .         .          .          .          .
Sl 1:   (22) +   (30) +    (03) +   2(11) +    (00)

         .         .          .          .          .          .          .          .
Sl 2:   (33) +   (41) +    (14) +   4(22) +   3(30) +   3(03) +   6(11) +   2(00)

         .         .          .          .          .          .          .          .          .          .
Sl 3:   (52) +   (25) +   4(33) +   5(41) +   5(14) +  13(22) +   8(30) +   8(03) +  16(11) +   4(00)

         .         .          .          .          .          .          .          .          .          .          .
Sl 4:  2(44) +  4(52) +   4(25) +   2(60) +  15(33) +   2(06) +  16(41) +  16(14) +  36(22) +  22(30) +  22(03)

         .         .
    + 37(11) +  10(00)
```

Affine A2 ⊃ A2 - Level 4 - Class 0

Sl 0: (00) **Highest Weight** $\overset{4}{(00)}$

Sl 1: (11)

Sl 2: (22) + 2(11) + (00)

Sl 3: (33) + 2(22) + 2(30) + 2(03) + 4(11) + 2(00)

Sl 4: (44) + 2(33) + 2(41) + 2(14) + 7(22) + 3(30) + 3(03) + 10(11) + 4(00)

Sl 5: 2(44) + 2(52) + 2(25) + 7(33) + 6(41) + 6(14) + 16(22) + 10(30) + 10(03) + 21(11) + 6(00)

Affine A2 ⊃ A2 - Level 4 - Class 1

Sl 0: (13) **Highest Weight** $\overset{0}{(13)}$

Sl 1: (32) + (05) + 2(13) + (21) + (02)

Sl 2: (51) + 2(24) + 3(32) + 2(05) + (40) + 6(13) + 4(21) + 3(02) + (10)

Sl 3: (70) + 2(43) + (16) + 3(51) + 6(24) + 10(32) + 6(05) + 4(40) + 16(13) + 12(21) + 8(02) + 4(10)

Sl 4: 2(62) + 2(35) + 2(70) + 7(43) + 4(16) + 10(51) + 19(24) + 28(32) + 14(05) + 13(40) + 39(13) + 32(21) + 21(02) + 11(10)

Affine A2 ⊃ A2 - Level 4 - Class 1

Sl 0: (40) **Highest Weight** $\overset{0}{(40)}$

Sl 1: (32) + (40) + (21)

Sl 2: (51) + (24) + 2(32) + 3(40) + (13) + 3(21) + (02)

Sl 3: (43) + (16) + 3(51) + 2(24) + 7(32) + (05) + 6(40) + 4(13) + 8(21) + 2(02) + 2(10)

Sl 4: (62) + (35) + (08) + 4(43) + 2(16) + 8(51) + 8(24) + 17(32) + 3(05) + 15(40) + 12(13) + 19(21) + 8(02) + 5(10)

Affine A2 ⊃ A2 - Level 4 - Class 1

Sl 0: (21) **Highest Weight** $\overset{1}{(21)}$

Sl 1: (32) + (40) + (13) + 2(21) + (02) + (10)

Sl 2: (51) + (24) + 4(32) + (05) + 3(40) + 4(13) + 7(21) + 3(02) + 3(10)

Sl 3: 2(43) + (16) + 4(51) + 5(24) + 13(32) + 3(05) + 9(40) + 13(13) + 19(21) + 10(02) + 8(10)

Sl 4: (62) + 2(35) + (70) + 9(43) + 4(16) + 13(51) + 17(24) + 37(32) + 10(05) + 23(40) + 37(13) + 49(21) + 25(02) + 20(10)

Affine A2 ⊃ A2 - Level 4 - Class 1

Sl 0: (02) **Highest Weight** $\overset{2}{(02)}$

Sl 1: (13) + (21) + (02) + (10)

Sl 2: (24) + (32) + (40) + 3(13) + 3(21) + 4(02) + 2(10)

Sl 3: (43) + (51) + 3(24) + 5(32) + 2(05) + 2(40) + 9(13) + 10(21) + 8(02) + 6(10)

Sl 4: (62) + (35) + 4(43) + (16) + 4(51) + 11(24) + 16(32) + 5(05) + 9(40) + 24(13) + 25(21) + 21(02) + 13(10)

Affine A2 ⊃ A2 - Level 4 - Class 1

Sl 0: (10) **Highest Weight** $\overset{3}{(10)}$

Sl 1: (21) + (02) + (10)

Sl 2: (32) + (13) + 3(21) + 2(02) + 3(10)

Sl 3: (43) + (24) + 3(32) + 2(40) + 4(13) + 8(21) + 6(02) + 6(10)

Sl 4: (35) + 3(43) + 2(51) + 4(24) + 11(32) + 2(05) + 5(40) + 12(13) + 20(21) + 13(02) + 14(10)

Affine A3 Branching Rules

Affine A3 ⊃ A3 - Level 1 - Class 0

Sl 0: (000) **Highest Weight** $\overset{1}{(000)}$

Sl 1: (101)

Sl 2: (020) + 2(101) + (000)

Sl 3: (210) + (012) + (020) + 4(101) + 2(000)

Sl 4: (202) + 2(210) + 2(012) + 4(020) + 7(101) + 4(000)

Sl 5: (121) + 2(202) + 5(210) + 5(012) + 6(020) + 14(101) + 5(000)

Sl 6: 3(121) + (400) + 6(202) + (004) + 9(210) + 9(012) + 13(020) + 24(101) + 10(000)

Sl 7: (311) + (113) + 7(121) + (400) + 11(202) + (004) + 19(210) + 19(012) + 21(020) + 42(101) + 14(000)

Affine A3 ⊃ A3 - Level 1 - Class 1

Sl 0: (100) **Highest Weight** $\overset{0}{(100)}$

Sl 1: (011) + (100)

Sl 2: (201) + 2(011) + 2(100)

Sl 3: (120) + 2(201) + (003) + 4(011) + 4(100)

Sl 4: (112) + 2(120) + 5(201) + (003) + 8(011) + 7(100)

Sl 5: (310) + 3(112) + 5(120) + 9(201) + 3(003) + 15(011) + 12(100)

Sl 6: (302) + (031) + 2(310) + 7(112) + 10(120) + 18(201) + 6(003) + 26(011) + 21(100)

Sl 7: (221) + 2(302) + (104) + 2(031) + 5(310) + 15(112) + 20(120) + 32(201) + 11(003) + 46(011) + 33(100)

Affine A3 ⊃ A3 - Level 1 - Class 2

- Sl 0: (010) **Highest Weight (010)**
- Sl 1: (200) + (002) + (010)
- Sl 2: (111) + (200) + (002) + 3(010)
- Sl 3: 3(111) + 3(200) + 3(002) + 4(010)
- Sl 4: (301) + (103) + (030) + 6(111) + 5(200) + 5(002) + 9(010)
- Sl 5: (220) + (022) + 2(301) + 2(103) + (030) + 13(111) + 10(200) + 10(002) + 14(010)
- Sl 6: (212) + 2(220) + 2(022) + 5(301) + 5(103) + 4(030) + 25(111) + 16(200) + 16(002) + 26(010)
- Sl 7: 3(212) + 6(220) + 6(022) + 10(301) + 10(103) + 7(030) + 45(111) + 30(200) + 30(002) + 40(010)

Affine A3 ⊃ A3 - Level 2 - Class 0

- Sl 0: (020) **Highest Weight (020)**
- Sl 1: (210) + (012) + (020) + (101)
- Sl 2: (121) + (400) + (202) + (004) + 2(210) + 2(012) + 4(020) + 3(101) + (000)
- Sl 3: (311) + (113) + 4(121) + (400) + 3(202) + (004) + 7(210) + 7(012) + 7(020) + 9(101) + (000)
- Sl 4: (222) + 4(311) + 4(113) + (040) + 12(121) + 4(400) + 11(202) + 4(004) + 16(210) + 16(012) + 19(020) + 20(101) + 5(000)
- Sl 5: 3(222) + (501) + 3(303) + 3(230) + (105) + 3(032) + 12(311) + 12(113) + 2(040) + 32(121) + 8(400) + 28(202) + 8(004) + 40(210) + 40(012) + 36(020) + 47(101) + 9(000)

Affine A3 ⊃ A3 - Level 2 - Class 0

- Sl 0: (101) **Highest Weight (101)**
- Sl 1: (210) + (012) + (020) + 2(101) + (000)
- Sl 2: (121) + 2(202) + 3(210) + 3(012) + 3(020) + 6(101) + 2(000)
- Sl 3: (311) + (113) + 5(121) + (400) + 6(202) + (004) + 9(210) + 9(012) + 9(020) + 15(101) + 4(000) (continued next page)

Sl 4: $(3\dot{0}3) + (2\dot{3}0) + (0\dot{3}2) + 5(3\dot{1}1) + 5(1\dot{1}3) + (0\dot{4}0) + 16(1\dot{2}1) + 3(4\dot{0}0) + 19(2\dot{0}2) + 3(0\dot{0}4)$

$+ 23(2\dot{1}0) + 23(0\dot{1}2) + 22(0\dot{2}0) + 36(1\dot{0}1) + 9(0\dot{0}0)$

Sl 5: $3(2\dot{2}2) + (5\dot{0}1) + 4(3\dot{0}3) + 4(2\dot{3}0) + (1\dot{0}5) + 4(0\dot{3}2) + 17(3\dot{1}1) + 17(1\dot{1}3) + 4(0\dot{4}0) + 45(1\dot{2}1)$

$+ 9(4\dot{0}0) + 48(2\dot{0}2) + 9(0\dot{0}4) + 57(2\dot{1}0) + 57(0\dot{1}2) + 52(0\dot{2}0) + 78(1\dot{0}1) + 19(0\dot{0}0)$

Affine A3 ⊃ A3 - Level 2 - Class 0

Sl 0: $(0\dot{0}0)$ **Highest Weight** $\mathbf{\overset{2}{(000)}}$

Sl 1: $(1\dot{0}1)$

Sl 2: $(2\dot{0}2) + (0\dot{2}0) + 2(1\dot{0}1) + (0\dot{0}0)$

Sl 3: $(1\dot{2}1) + 2(2\dot{0}2) + 2(2\dot{1}0) + 2(0\dot{1}2) + (0\dot{2}0) + 5(1\dot{0}1) + 2(0\dot{0}0)$

Sl 4: $(3\dot{1}1) + (1\dot{1}3) + (0\dot{4}0) + 3(1\dot{2}1) + 7(2\dot{0}2) + 5(2\dot{1}0) + 5(0\dot{1}2) + 6(0\dot{2}0) + 11(1\dot{0}1) + 5(0\dot{0}0)$

Sl 5: $(3\dot{0}3) + (2\dot{3}0) + (0\dot{3}2) + 4(3\dot{1}1) + 4(1\dot{1}3) + (0\dot{4}0) + 11(1\dot{2}1) + (4\dot{0}0) + 15(2\dot{0}2) + (0\dot{0}4)$

$+ 15(2\dot{1}0) + 15(0\dot{1}2) + 12(0\dot{2}0) + 26(1\dot{0}1) + 7(0\dot{0}0)$

Affine A3 ⊃ A3 - Level 2 - Class 1

Sl 0: $(0\dot{1}1)$ **Highest Weight** $\mathbf{\overset{0}{(011)}}$

Sl 1: $(1\dot{2}0) + (2\dot{0}1) + (0\dot{0}3) + 2(0\dot{1}1) + (1\dot{0}0)$

Sl 2: $(3\dot{1}0) + 2(1\dot{1}2) + 3(1\dot{2}0) + 3(2\dot{0}1) + 2(0\dot{0}3) + 6(0\dot{1}1) + 3(1\dot{0}0)$

Sl 3: $(2\dot{2}1) + (3\dot{0}2) + (1\dot{0}4) + (0\dot{3}1) + 3(3\dot{1}0) + 7(1\dot{1}2) + 9(1\dot{2}0) + 10(2\dot{0}1) + 5(0\dot{0}3) + 15(0\dot{1}1)$

$+ 7(1\dot{0}0)$

Sl 4: $(2\dot{1}3) + 5(2\dot{2}1) + (0\dot{2}3) + (5\dot{0}0) + 5(3\dot{0}2) + 3(1\dot{0}4) + 5(0\dot{3}1) + 9(3\dot{1}0) + 21(1\dot{1}2) + 23(1\dot{2}0)$

$+ 26(2\dot{0}1) + 13(0\dot{0}3) + 35(0\dot{1}1) + 17(1\dot{0}0)$

Sl 5: $(1\dot{3}2) + 2(4\dot{1}1) + 5(2\dot{1}3) + (1\dot{4}0) + 17(2\dot{2}1) + 5(0\dot{2}3) + 2(5\dot{0}0) + 16(3\dot{0}2) + 10(1\dot{0}4) + 15(0\dot{3}1)$

$+ 24(3\dot{1}0) + 56(1\dot{1}2) + 56(1\dot{2}0) + 63(2\dot{0}1) + 29(0\dot{0}3) + 77(0\dot{1}1) + 37(1\dot{0}0)$

Affine A3 ⊃ A3 - Level 2 - Class 1

Sl 0: (100) **Highest Weight** $(\overset{1}{1}00)$

Sl 1: (201) + (011) + (100)

Sl 2: (112) + (120) + 3(201) + 3(011) + 3(100)

Sl 3: (302) + (031) + (310) + 4(112) + 4(120) + 8(201) + 2(003) + 8(011) + 7(100)

Sl 4: 2(221) + (023) + 4(302) + (104) + 3(031) + 4(310) + 13(112) + 12(120) + 20(201) + 5(003) + 21(011) + 15(100)

Sl 5: (411) + 2(213) + (140) + 8(221) + 4(023) + 12(302) + 4(104) + 10(031) + 13(310) + 36(112) + 32(120) + 47(201) + 14(003) + 49(011) + 31(100)

Affine A3 ⊃ A3 - Level 2 - Class 2

Sl 0: (200) **Highest Weight** $(\overset{0}{2}00)$

Sl 1: (111) + (200) + (010)

Sl 2: (022) + (301) + 3(111) + 3(200) + (002) + 2(010)

Sl 3: (212) + (220) + 2(022) + 3(301) + 2(103) + 2(030) + 8(111) + 6(200) + 3(002) + 6(010)

Sl 4: (402) + (131) + 4(212) + (014) + 4(220) + 7(022) + 8(301) + 5(103) + 4(030) + 22(111) + 14(200) + 8(002) + 12(010)

Sl 5: (321) + (123) + 2(402) + (204) + 4(131) + 2(410) + 14(212) + 3(014) + 12(220) + 16(022) + 20(301) + 16(103) + 13(030) + 52(111) + 27(200) + 19(002) + 28(010)

Affine A3 ⊃ A3 - Level 2 - Class 2

Sl 0: (010) **Highest Weight** $(\overset{1}{0}10)$

Sl 1: (111) + (200) + (002) + (010)

Sl 2: (301) + (103) + (030) + 4(111) + 2(200) + 2(002) + 4(010)

Sl 3: (212) + 2(220) + 2(022) + 3(301) + 3(103) + 2(030) + 12(111) + 7(200) + 7(002) + 8(010)

Sl 4: (131) + (410) + 6(212) + (014) + 6(220) + 6(022) + 10(301) + 10(103) + 8(030) + 32(111) + 15(200) + 15(002) + 21(010) (continued next page)

Sl 5: (321) + (123) + 2(402) + 2(204) + 5(131) + 3(410) + 19(212) + 3(014) + 21(220) + 21(022)
+ 26(301) + 26(103) + 18(030) + 78(111) + 37(200) + 37(002) + 42(010)

Sl 6: (313) + 6(321) + 6(123) + (050) + 7(402) + 7(204) + 18(131) + 11(410) + 56(212) + 11(014)
+ 52(220) + 52(022) + 66(301) + 66(103) + 49(030) +179(111) + 76(200) + 76(002) + 92(010)

Affine A3 ⊃ A3 - Level 3 - Class 0

Sl 0: (210) **Highest Weight** $\overset{0}{(210)}$

Sl 1: (121) + (400) + (202) + 2(210) + (020) + (101)

Sl 2: (032) + 2(311) + (113) + 4(121) + 2(400) + 3(202) + 7(210) + 2(012) + 3(020) + 4(101)

Sl 3: 2(222) + (024) + (501) + (303) + (230) + 3(032) + 8(311) + 5(113) + 2(040) + 14(121)
+ 6(400) + 12(202) + (004) + 18(210) + 8(012) + 11(020) + 12(101) + 2(000)

Sl 4: (133) + 2(412) + (214) + (141) + (420) + 9(222) + 3(024) + 4(501) + 7(303) + 7(230)
+ 2(105) + 12(032) + 26(311) + 19(113) + 6(040) + 44(121) + 15(400) + 35(202) + 4(004) + 49(210)
+ 28(012) + 28(020) + 34(101) + 5(000)

Affine A3 ⊃ A3 - Level 3 - Class 0

Sl 0: (020) **Highest Weight** $\overset{1}{(020)}$

Sl 1: (121) + (210) + (012) + (020) + (101)

Sl 2: (311) + (113) + (040) + 4(121) + (400) + 2(202) + (004) + 3(210) + 3(012) + 5(020)
+ 3(101) + (000)

Sl 3: (222) + (501) + (303) + 2(230) + (105) + 2(032) + 5(311) + 5(113) + 2(040) + 14(121)
+ 2(400) + 8(202) + 2(004) + 12(210) + 12(012) + 11(020) + 11(101) + (000)

Sl 4: (412) + (214) + (141) + 2(420) + 8(222) + 2(024) + 3(501) + 5(303) + 7(230) + 3(105)
+ 7(032) + 19(311) + 19(113) + 9(040) + 43(121) + 9(400) + 30(202) + 9(004) + 33(210) + 33(012)
+ 33(020) + 30(101) + 6(000)

Affine A3 ⊃ A3 - Level 3 - Class 0

Sl 0: (101) **Highest Weight** $\overset{1}{(101)}$

Sl 1: (202) + (210) + (012) + (020) + 2(101) + (000)

Sl 2: (311) + (113) + 2(121) + 4(202) + 4(210) + 4(012) + 3(020) + 7(101) + 2(000)

Sl 3: (222) + 2(303) + (230) + (032) + 5(311) + 5(113) + (040) + 10(121) + 2(400) + 14(202)
+ 2(004) + 13(210) + 13(012) + 11(020) + 19(101) + 5(000)

Sl 4: (412) + (214) + (141) + (420) + 7(222) + (024) + (501) + 9(303) + 6(230) + (105)
+ 6(032) + 20(311) + 20(113) + 4(040) + 37(121) + 7(400) + 42(202) + 7(004) + 39(210) + 39(012)
+ 30(020) + 50(101) + 11(000)

Affine A3 ⊃ A3 - Level 3 - Class 0

Sl 0: (000) **Highest Weight** $\overset{3}{(000)}$

Sl 1: (101)

Sl 2: (202) + (020) + 2(101) + (000)

Sl 3: (303) + (121) + 2(202) + 2(210) + 2(012) + (020) + 5(101) + 2(000)

Sl 4: (222) + 2(303) + 2(311) + 2(113) + (040) + 3(121) + 8(202) + 5(210) + 5(012) + 6(020)
+ 11(101) + 5(000)

Affine A3 ⊃ A3 - Level 3 - Class 1

Sl 0: (003) **Highest Weight** $\overset{0}{(003)}$

Sl 1: (112) + (003) + (011)

Sl 2: (221) + (104) + 3(112) + (120) + (201) + 3(003) + 3(011)

Sl 3: (330) + (213) + 3(221) + (023) + 2(302) + 3(104) + 2(031) + (310) + 9(112) + 4(120)
+ 4(201) + 7(003) + 8(011) + 2(100)

Sl 4: (322) + 2(330) + (205) + (132) + 2(411) + 5(213) + 2(140) + 11(221) + 4(023) + 6(302)
+ 9(104) + 6(031) + 5(310) + 26(112) + 14(120) + 13(201) + 16(003) + 21(011) + 5(100)

Affine A3 ⊃ A3 - Level 3 - Class 1

Sl 0: (120) **Highest Weight** $\overset{0}{(120)}$

Sl 1: (031) + (310) + (112) + 2(120) + (201) + (011)

Sl 2: 2(221) + (023) + (500) + (302) + (104) + 3(031) + 3(310) + 4(112) + 7(120) + 4(201) + (003) + 4(011) + (100)

Sl 3: (132) + 2(411) + 2(213) + (140) + (015) + 8(221) + 4(023) + 2(500) + 5(302) + 3(104) + 10(031) + 10(310) + 16(112) + 19(120) + 14(201) + 4(003) + 13(011) + 4(100)

Sl 4: 2(322) + (124) + (601) + (403) + (330) + (205) + 6(132) + 8(411) + 10(213) + 5(140) + 3(015) + 29(221) + 14(023) + 6(500) + 19(302) + 12(104) + 28(031) + 29(310) + 50(112) + 52(120) + 40(201) + 14(003) + 37(011) + 13(100)

Affine A3 ⊃ A3 - Level 3 - Class 1

Sl 0: (201) **Highest Weight** $\overset{0}{(201)}$

Sl 1: (310) + (112) + (120) + 2(201) + (011) + (100)

Sl 2: (221) + (023) + 2(302) + (031) + 3(310) + 4(112) + 4(120) + 7(201) + (003) + 4(011) + 3(100)

Sl 3: (132) + (411) + 2(213) + 6(221) + 3(023) + (500) + 7(302) + 2(104) + 5(031) + 10(310) + 15(112) + 14(120) + 19(201) + 4(003) + 14(011) + 8(100)

Sl 4: (322) + (124) + 2(403) + (330) + 5(132) + 6(411) + 10(213) + 3(140) + (015) + 23(221) + 12(023) + 3(500) + 24(302) + 8(104) + 18(031) + 28(310) + 48(112) + 41(120) + 52(201) + 13(003) + 39(011) + 21(100)

Affine A3 ⊃ A3 - Level 3 - Class 1

Sl 0: (011) **Highest Weight** $\overset{1}{(011)}$

Sl 1: (112) + (120) + (201) + (003) + 2(011) + (100)

Sl 2: (221) + (302) + (104) + (031) + (310) + 5(112) + 4(120) + 4(201) + 3(003) + 7(011) + 3(100) (continued next page)

Sl 3: $(\dot{4}11) + 2(\dot{2}13) + (\dot{1}40) + 6(\dot{2}21) + 2(\dot{0}23) + 5(\dot{3}02) + 4(\dot{1}04) + 5(\dot{0}31) + 5(\dot{3}10) + 18(\dot{1}12)$

$+ 14(\dot{1}20) + 15(\dot{2}01) + 9(\dot{0}03) + 20(\dot{0}11) + 8(\dot{1}00)$

Sl 4: $(\dot{3}22) + (\dot{4}03) + 2(\dot{3}30) + (\dot{2}05) + 3(\dot{1}32) + 5(\dot{4}11) + 11(\dot{2}13) + 4(\dot{1}40) + (\dot{0}15) + 24(\dot{2}21)$

$+ 9(\dot{0}23) + 2(\dot{5}00) + 20(\dot{3}02) + 14(\dot{1}04) + 19(\dot{0}31) + 19(\dot{3}10) + 57(\dot{1}12) + 43(\dot{1}20) + 44(\dot{2}01) + 25(\dot{0}03)$

$+ 53(\dot{0}11) + 22(\dot{1}00)$

Affine A3 ⊃ A3 - Level 3 - Class 1

Sl 0: $(\dot{1}00)$ **Highest Weight** $\overset{2}{\mathbf{(100)}}$

Sl 1: $(\dot{2}01) + (\dot{0}11) + (\dot{1}00)$

Sl 2: $(\dot{3}02) + (\dot{1}12) + (\dot{1}20) + 3(\dot{2}01) + 3(\dot{0}11) + 3(\dot{1}00)$

Sl 3: $(\dot{2}13) + (\dot{2}21) + 3(\dot{3}02) + (\dot{0}31) + 2(\dot{3}10) + 5(\dot{1}12) + 4(\dot{1}20) + 9(\dot{2}01) + 2(\dot{0}03) + 8(\dot{0}11)$

$+ 7(\dot{1}00)$

Sl 4: $(\dot{4}03) + (\dot{1}32) + (\dot{4}11) + 4(\dot{2}13) + (\dot{1}40) + 6(\dot{2}21) + 2(\dot{0}23) + 11(\dot{3}02) + 2(\dot{1}04) + 4(\dot{0}31)$

$+ 7(\dot{3}10) + 18(\dot{1}12) + 14(\dot{1}20) + 24(\dot{2}01) + 5(\dot{0}03) + 22(\dot{0}11) + 16(\dot{1}00)$

Affine A3 ⊃ A3 - Level 3 - Class 2

Sl 0: $(\dot{0}30)$ **Highest Weight** $\overset{0}{\mathbf{(030)}}$

Sl 1: $(\dot{2}20) + (\dot{0}22) + (\dot{0}30) + (\dot{1}11)$

Sl 2: $(\dot{1}31) + (\dot{4}10) + (\dot{2}12) + (\dot{0}14) + 2(\dot{2}20) + 2(\dot{0}22) + (\dot{3}01) + (\dot{1}03) + 4(\dot{0}30) + 4(\dot{1}11)$

$+ (\dot{0}10)$

Sl 3: $(\dot{3}21) + (\dot{1}23) + (\dot{6}00) + (\dot{4}02) + (\dot{2}04) + 4(\dot{1}31) + (\dot{0}06) + 2(\dot{4}10) + 4(\dot{2}12) + 2(\dot{0}14)$

$+ 8(\dot{2}20) + 8(\dot{0}22) + 4(\dot{3}01) + 4(\dot{1}03) + 8(\dot{0}30) + 12(\dot{1}11) + 3(\dot{2}00) + 3(\dot{0}02) + 2(\dot{0}10)$

Affine A3 ⊃ A3 - Level 3 - Class 2

Sl 0: $(\dot{1}11)$ **Highest Weight** $\overset{0}{\mathbf{(111)}}$

Sl 1: $(\dot{2}20) + (\dot{0}22) + (\dot{3}01) + (\dot{1}03) + (\dot{0}30) + 3(\dot{1}11) + (\dot{2}00) + (\dot{0}02) + (\dot{0}10)$

Sl 2: $(\dot{1}31) + (\dot{4}10) + 3(\dot{2}12) + (\dot{0}14) + 4(\dot{2}20) + 4(\dot{0}22) + 4(\dot{3}01) + 4(\dot{1}03) + 4(\dot{0}30) + 11(\dot{1}11)$

$+ 4(\dot{2}00) + 4(\dot{0}02) + 4(\dot{0}10)$ (continued next page)

Sl 3: 2(321) + 2(123) + 2(402) + 2(204) + 6(131) + 4(410) + 13(212) + 4(014) + 15(220) + 15(022)
+ 14(301) + 14(103) + 12(030) + 36(111) + 12(200) + 12(002) + 12(010)

Sl 4: (232) + (511) + 3(313) + (240) + (115) + (042) + 11(321) + 11(123) + (050) + (600)
+ 9(402) + 9(204) + 24(131) + (006) + 14(410) + 49(212) + 14(014) + 46(220) + 46(022) + 44(301)
+ 44(103) + 37(030) +101(111) + 34(200) + 34(002) + 34(010)

Affine A3 ⊃ A3 - Level 3 - Class 2

Sl 0: (200) **Highest Weight** $\overset{1}{(200)}$

Sl 1: (301) + (111) + (200) + (010)

Sl 2: (212) + (220) + (022) + 3(301) + 4(111) + 4(200) + (002) + 2(010)

Sl 3: (123) + (402) + (131) + (410) + 5(212) + 4(220) + 3(022) + 9(301) + 3(103) + 3(030)
+ 13(111) + 9(200) + 3(002) + 7(010)

Sl 4: (313) + (042) + 2(321) + 4(123) + 5(402) + 2(204) + 6(131) + 4(410) + 18(212) + 2(014)
+ 15(220) + 13(022) + 24(301) + 10(103) + 8(030) + 39(111) + 23(200) + 11(002) + 16(010)

Affine A3 ⊃ A3 - Level 3 - Class 2

Sl 0: (010) **Highest Weight** $\overset{2}{(010)}$

Sl 1: (111) + (200) + (002) + (010)

Sl 2: (212) + (301) + (103) + (030) + 4(111) + 2(200) + 2(002) + 4(010)

Sl 3: (402) + (204) + (131) + 4(212) + 3(220) + 3(022) + 4(301) + 4(103) + 2(030) + 13(111)
+ 7(200) + 7(002) + 8(010)

Sl 4: (313) + 2(321) + 2(123) + (050) + 3(402) + 3(204) + 5(131) + 2(410) + 17(212) + 2(014)
+ 10(220) + 10(022) + 15(301) + 15(103) + 10(030) + 38(111) + 16(200) + 16(002) + 22(010)

Affine A4 Branching Rules

Affine A4 $\supset$ A4 - Level 1 - Class 0

sl 0: (0000) **Highest Weight** $\overset{1}{(0000)}$

sl 1: (1001)

sl 2: (0110) + 2(1001) + (0000)

sl 3: (2010) + (0102) + 2(0110) + 4(1001) + 2(0000)

sl 4: (2002) + (1200) + (0021) + 2(2010) + 2(0102) + 5(0110) + 8(1001) + 4(0000)

sl 5: (1111) + 2(2002) + 2(1200) + 2(0021) + 6(2010) + 6(0102) + 10(0110) + 15(1001) + 6(0000)

sl 6: 4(1111) + (3100) + (0013) + 6(2002) + 5(1200) + 5(0021) + 11(2010) + 11(0102) + 21(0110) + 28(1001) + 11(0000)

sl 7: (3011) + (1103) + (1030) + (0301) + 10(1111) + 2(3100) + 2(0013) + 12(2002) + 11(1200) + 11(0021) + 24(2010) + 24(0102) + 38(0110) + 50(1001) + 16(0000)

Affine A4 $\supset$ A4 - Level 1 - Class 1

sl 0: (1000) **Highest Weight** $\overset{0}{(1000)}$

sl 1: (0101) + (1000)

sl 2: (2001) + (0020) + 2(0101) + 2(1000)

sl 3: (1110) + (0012) + 2(2001) + (0020) + 5(0101) + 4(1000)

sl 4: (1102) + 3(1110) + 2(0012) + 5(2001) + 4(0020) + 9(0101) + 8(1000)

sl 5: (1021) + (3010) + 3(1102) + (0300) + 7(1110) + 5(0012) + 10(2001) + 7(0020) + 19(0101) + 13(1000)

sl 6: (0211) + (3002) + (2200) + 3(1021) + 2(3010) + 8(1102) + (0300) + 16(1110) + (0004) + 10(0012) + 20(2001) + 15(0020) + 34(0101) + 24(1000)

sl 7: (2111) + (1013) + 3(0211) + 2(3002) + 2(2200) + 8(1021) + 6(3010) + 18(1102) + 4(0300) + 33(1110) + (0004) + 21(0012) + 37(2001) + 26(0020) + 63(0101) + 39(1000)

Affine A4 ⊃ A4 - Level 1 - Class 2

Sl 0: (0100) **Highest Weight** $(0100)^{0}$

Sl 1: (0011) + (2000) + (0100)

Sl 2: (1101) + 2(0011) + (2000) + 3(0100)

Sl 3: (1020) + 3(1101) + (0003) + 4(0011) + 3(2000) + 5(0100)

Sl 4: (1012) + (0210) + (3001) + 2(1020) + 7(1101) + (0003) + 9(0011) + 5(2000) + 10(0100)

Sl 5: (2110) + (0202) + 3(1012) + 2(0210) + 2(3001) + 6(1020) + 15(1101) + 3(0003) + 17(0011) + 11(2000) + 17(0100)

Sl 6: (2102) + (0121) + 3(2110) + 2(0202) + 8(1012) + 6(0210) + 5(3001) + 12(1020) + 31(1101) + 6(0003) + 31(0011) + 18(2000) + 32(0100)

Sl 7: (2021) + 3(2102) + (1300) + 3(0121) + 8(2110) + (1004) + 7(0202) + 17(1012) + 12(0210) + 11(3001) + 26(1020) + 59(1101) + 12(0003) + 57(0011) + 34(2000) + 52(0100)

Affine A4 ⊃ A4 - Level 2 - Class 0

Sl 0: (0110) **Highest Weight** $(0110)^{0}$

Sl 1: (1200) + (0021) + (2010) + (0102) + 2(0110) + (1001)

Sl 2: 2(1111) + (3100) + (0013) + (2002) + 3(1200) + 3(0021) + 3(2010) + 3(0102) + 7(0110) + 4(1001) + (0000)

Sl 3: (2201) + (1022) + (3011) + (1103) + (1030) + (0301) + 9(1111) + 3(3100) + 3(0013) + 4(2002) + 9(1200) + 9(0021) + 11(2010) + 11(0102) + 18(0110) + 12(1001) + 2(0000)

Sl 4: (2112) + (2120) + (0212) + 5(2201) + 5(1022) + 2(0220) + 6(3011) + 6(1103) + 5(1030) + 5(0301) + (5000) + 30(1111) + (0005) + 9(3100) + 9(0013) + 16(2002) + 25(1200) + 25(0021) + 30(2010) + 30(0102) + 47(0110) + 32(1001) + 6(0000)

Affine A4 ⊃ A4 - Level 2 - Class 0

sl 0: (1001) **Highest Weight** $\overset{0}{(1001)}$

sl 1: (2010) + (0102) + (0110) + 2(1001) + (0000)

sl 2: (1111) + 2(2002) + (1200) + (0021) + 3(2010) + 3(0102) + 4(0110) + 6(1001) + 2(0000)

sl 3: (3011) + (1103) + (1030) + (0301) + 6(1111) + (3100) + (0013) + 6(2002) + 4(1200) + 4(0021) + 10(2010) + 10(0102) + 12(0110) + 16(1001) + 4(0000)

sl 4: (2120) + (0212) + (3003) + 2(2201) + 2(1022) + 2(0220) + 5(3011) + 5(1103) + 3(1030) + 3(0301) + 23(1111) + 4(3100) + 4(0013) + 20(2002) + 14(1200) + 14(0021) + 27(2010) + 27(0102) + 34(0110) + 39(1001) + 10(0000)

Affine A4 ⊃ A4 - Level 2 - Class 0

sl 0: (0000) **Highest Weight** $\overset{2}{(0000)}$

sl 1: (1001)

sl 2: (2002) + (0110) + 2(1001) + (0000)

sl 3: (1111) + 2(2002) + 2(2010) + 2(0102) + 2(0110) + 5(1001) + 2(0000)

sl 4: (0220) + (3011) + (1103) + 4(1111) + 7(2002) + 2(1200) + 2(0021) + 5(2010) + 5(0102) + 7(0110) + 12(1001) + 5(0000)

Affine A4 ⊃ A4 - Level 2 - Class 1

sl 0: (0020) **Highest Weight** $\overset{0}{(0020)}$

sl 1: (1110) + (0012) + (0020) + (0101)

sl 2: (2200) + (1021) + (1102) + 3(1110) + (0004) + 2(0012) + (2001) + 4(0020) + 3(0101) + (1000)

sl 3: (2111) + (1013) + (0211) + 2(2200) + 4(1021) + 2(3010) + 4(1102) + 2(0300) + 9(1110) + (0004) + 7(0012) + 4(2001) + 8(0020) + 10(0101) + 2(1000)

sl 4: (2022) + (1301) + 5(2111) + (0203) + (0130) + (4100) + 4(1013) + 5(0211) + 2(3002) + 7(2200) + 13(1021) + 5(3010) + 14(1102) + 4(0300) + 26(1110) + 4(0004) + 17(0012) + 13(2001) + 21(0020) + 24(0101) + 8(1000)

Affine A4 ⊃ A4 - Level 2 - Class 1

Sl 0: (0101) **Highest Weight** $\overset{0}{(0101)}$

Sl 1: (1110) + (0012) + (2001) + (0020) + 2(0101) + (1000)

Sl 2: (1021) + (3010) + 2(1102) + (0300) + 4(1110) + 3(0012) + 3(2001) + 3(0020) + 7(0101) + 3(1000)

Sl 3: (2111) + (1013) + 2(0211) + (3002) + 2(2200) + 5(1021) + 3(3010) + 8(1102) + 2(0300) + 14(1110) + (0004) + 9(0012) + 11(2001) + 10(0020) + 18(0101) + 8(1000)

Sl 4: (2103) + (2030) + (1301) + (0122) + 7(2111) + (0203) + (0130) + (4100) + 5(1013) + 9(0211) + 5(3002) + 6(2200) + 18(1021) + 11(3010) + 27(1102) + 8(0300) + 41(1110) + 3(0004) + 25(0012) + 30(2001) + 25(0020) + 48(0101) + 19(1000)

Affine A4 ⊃ A4 - Level 2 - Class 1

Sl 0: (1000) **Highest Weight** $\overset{1}{(1000)}$

Sl 1: (2001) + (0101) + (1000)

Sl 2: (1102) + (1110) + 3(2001) + (0020) + 3(0101) + 3(1000)

Sl 3: (0211) + (3002) + (1021) + (3010) + 4(1102) + 5(1110) + 2(0012) + 8(2001) + 2(0020) + 9(0101) + 7(1000)

Sl 4: 2(2111) + (0203) + (0130) + (1013) + 4(0211) + 4(3002) + (2200) + 5(1021) + 4(3010) + 14(1102) + 2(0300) + 16(1110) + 7(0012) + 21(2001) + 9(0020) + 23(0101) + 16(1000)

Affine A4 ⊃ A4 - Level 2 - Class 2

Sl 0: (0011) **Highest Weight** $\overset{0}{(0011)}$

Sl 1: (1020) + (1101) + (0003) + 2(0011) + (0100)

Sl 2: (2110) + 2(1012) + (0210) + 3(1020) + 4(1101) + 2(0003) + 6(0011) + (2000) + 3(0100)

Sl 3: (2021) + (2102) + (1300) + (0121) + 4(2110) + (1004) + 2(0202) + 7(1012) + 4(0210) + 2(3001) + 10(1020) + 13(1101) + 5(0003) + 16(0011) + 4(2000) + 8(0100)

Sl 4: (2013) + 2(1211) + (3200) + 5(2021) + (0113) + (4010) + 6(2102) + 3(1300) + 6(0121) + 14(2110) + 3(1004) + 7(0202) + 23(1012) + 14(0210) + 7(3001) + 27(1020) + 39(1101) + 13(0003) + 38(0011) + 11(2000) + 21(0100)

Affine A4 ⊃ A4 - Level 2 - Class 2

sl 0: (2000) **Highest Weight** $\overset{0}{(2000)}$

sl 1: (1101) + (2000) + (0100)

sl 2: (0202) + (3001) + (1020) + 3(1101) + (0011) + 3(2000) + 2(0100)

sl 3: (2102) + (0121) + (2110) + 2(0202) + 2(1012) + 2(0210) + 3(3001) + 2(1020) + 9(1101) + 4(0011) + 6(2000) + 6(0100)

sl 4: (1211) + (4002) + (2021) + (0113) + (0040) + 4(2102) + 3(0121) + 5(2110) + 8(0202) + 7(1012) + 6(0210) + 8(3001) + 9(1020) + 24(1101) + 2(0003) + 11(0011) + 15(2000) + 13(0100)

Affine A4 ⊃ A4 - Level 2 - Class 2

sl 0: (0100) **Highest Weight** $\overset{1}{(0100)}$

sl 1: (1101) + (0011) + (2000) + (0100)

sl 2: (1012) + (0210) + (3001) + (1020) + 4(1101) + 3(0011) + 2(2000) + 4(0100)

sl 3: (2102) + (0121) + 2(2110) + 2(0202) + 4(1012) + 3(0210) + 3(3001) + 5(1020) + 13(1101) + 2(0003) + 9(0011) + 7(2000) + 9(0100)

sl 4: (1211) + 2(2021) + (0113) + (4010) + 6(2102) + (1300) + 5(0121) + 8(2110) + (1004) + 7(0202) + 15(1012) + 12(0210) + 10(3001) + 16(1020) + 37(1101) + 5(0003) + 25(0011) + 16(2000) + 23(0100)

Affine A4 ⊃ A4 - Level 3 - Class 0

sl 0: (0021) **Highest Weight** $\overset{0}{(0021)}$

sl 1: (1030) + (1111) + (0013) + 2(0021) + (0102) + (0110)

sl 2: (2120) + (2201) + 2(1022) + (0220) + (1103) + 3(1030) + 5(1111) + (0005) + 3(0013) + (2002) + (1200) + 7(0021) + (2010) + 4(0102) + 4(0110) + (1001)

sl 3: (3210) + (2031) + 2(2112) + (1310) + (0131) + 5(2120) + 2(1014) + 2(0212) + 4(2201) + 8(1022) + 4(0220) + 3(3011) + 6(1103) + 11(1030) + 3(0301) + 20(1111) + 2(0005) + (3100) + 10(0013) + 5(2002) + 5(1200) + 20(0021) + 6(2010) + 15(0102) + 14(0110) + 5(1001)

Affine A4 ⊃ A4 - Level 3 - Class 0

Sl 0: (0102) **Highest Weight** ^{0}**(0102)**

Sl 1: (1111) + (0013) + (2002) + (0021) + 2(0102) + (0110) + (1001)

Sl 2: (2120) + (1022) + (3011) + 2(1103) + (1030) + (0301) + 5(1111) + 3(0013) + 3(2002)
+ (1200) + 4(0021) + 2(2010) + 8(0102) + 4(0110) + 4(1001)

Sl 3: (2031) + (4020) + 2(2112) + (1310) + 4(2120) + (1014) + 2(0212) + (3003) + 3(2201)
+ 6(1022) + 3(0220) + 5(3011) + 9(1103) + 5(1030) + 3(0301) + 21(1111) + (0005) + (3100)
+ 10(0013) + 13(2002) + 6(1200) + 15(0021) + 9(2010) + 22(0102) + 16(0110) + 13(1001) + 2(0000)

Affine A4 ⊃ A4 - Level 3 - Class 0

Sl 0: (1200) **Highest Weight** ^{0}**(1200)**

Sl 1: (0301) + (1111) + (3100) + 2(1200) + (2010) + (0110)

Sl 2: (0212) + 2(2201) + (1022) + (0220) + (3011) + 3(0301) + (5000) + 5(1111) + 3(3100)
+ (2002) + 7(1200) + (0021) + 4(2010) + (0102) + 4(0110) + (1001)

Sl 3: (1302) + (0123) + 2(2112) + (1310) + (0131) + 2(4101) + 2(2120) + 5(0212) + 8(2201)
+ 4(1022) + 4(0220) + 6(3011) + 3(1103) + 3(1030) + 11(0301) + 2(5000) + 20(1111) + 10(3100)
+ (0013) + 5(2002) + 20(1200) + 5(0021) + 15(2010) + 6(0102) + 14(0110) + 5(1001)

Affine A4 ⊃ A4 - Level 3 - Class 0

Sl 0: (2010) **Highest Weight** ^{0}**(2010)**

Sl 1: (1111) + (3100) + (2002) + (1200) + 2(2010) + (0110) + (1001)

Sl 2: (0212) + (2201) + 2(3011) + (1103) + (1030) + (0301) + 5(1111) + 3(3100) + 3(2002)
+ 4(1200) + (0021) + 8(2010) + 2(0102) + 4(0110) + 4(1001)

Sl 3: (1302) + 2(2112) + (0204) + (0131) + (4101) + 2(2120) + 4(0212) + (3003) + 6(2201)
+ 3(1022) + 3(0220) + 9(3011) + 5(1103) + 3(1030) + 5(0301) + (5000) + 21(1111) + 10(3100)
+ (0013) + 13(2002) + 15(1200) + 6(0021) + 22(2010) + 9(0102) + 16(0110) + 13(1001) + 2(0000)

Affine A4 ⊃ A4 - Level 3 - Class 0

Sl 0: (0110) **Highest Weight** $\overset{1}{(0110)}$

Sl 1: (1111) + (1200) + (0021) + (2010) + (0102) + 2(0110) + (1001)

Sl 2: (2201) + (1022) + (0220) + (3011) + (1103) + (1030) + (0301) + 6(1111) + (3100)
+ (0013) + 2(2002) + 4(1200) + 4(0021) + 4(2010) + 4(0102) + 8(0110) + 4(1001) + (0000)

Sl 3: 2(2112) + (1310) + (0131) + (4101) + 3(2120) + (1014) + 3(0212) + (3003) + 6(2201)
+ 6(1022) + 4(0220) + 6(3011) + 6(1103) + 6(1030) + 6(0301) + 26(1111) + 5(3100) + 5(0013)
+ 10(2002) + 15(1200) + 15(0021) + 17(2010) + 17(0102) + 24(0110) + 14(1001) + 2(0000)

Affine A4 ⊃ A4 - Level 3 - Class 0

Sl 0: (1001) **Highest Weight** $\overset{1}{(1001)}$

Sl 1: (2002) + (2010) + (0102) + (0110) + 2(1001) + (0000)

Sl 2: (3011) + (1103) + 2(1111) + 4(2002) + (1200) + (0021) + 4(2010) + 4(0102) + 4(0110)
+ 7(1001) + 2(0000)

Sl 3: (2112) + (2120) + (0212) + 2(3003) + (2201) + (1022) + (0220) + 5(3011) + 5(1103)
+ (1030) + (0301) + 12(1111) + 2(3100) + 2(0013) + 14(2002) + 5(1200) + 5(0021) + 14(2010)
+ 14(0102) + 14(0110) + 20(1001) + 5(0000)

Affine A4 ⊃ A4 - Level 3 - Class 0

Sl 0: (0000) **Highest Weight** $\overset{3}{(0000)}$

Sl 1: (1001)

Sl 2: (2002) + (0110) + 2(1001) + (0000)

Sl 3: (3003) + (1111) + 2(2002) + 2(2010) + 2(0102) + 2(0110) + 5(1001) + 2(0000)

Sl 4: (2112) + 2(3003) + (0220) + 2(3011) + 2(1103) + 4(1111) + 8(2002) + 2(1200) + 2(0021)
+ 5(2010) + 5(0102) + 7(0110) + 12(1001) + 5(0000)

Affine A4 ⊃ A4 - Level 3 - Class 1

Sl 0: (0012) **Highest Weight (0012)**

Sl 1: (1021) + (1102) + (0004) + 2(0012) + (0020) + (0101)

Sl 2: (2030) + (2111) + 2(1013) + (0211) + 4(1021) + 4(1102) + 2(1110) + 2(0004) + 7(0012) + (2001) + 3(0020) + 4(0101)

Sl 3: (3120) + 2(2022) + (1220) + (2103) + 3(2030) + (1301) + (0122) + 6(2111) + (1005) + 2(0203) + 2(0130) + 8(1013) + 5(0211) + 2(3002) + (2200) + 15(1021) + (3010) + 15(1102) + (0300) + 10(1110) + 6(0004) + 19(0012) + 5(2001) + 11(0020) + 13(0101) + 2(1000)

Affine A4 ⊃ A4 - Level 3 - Class 1

Sl 0: (0300) **Highest Weight (0300)**

Sl 1: (0211) + (2200) + (0300) + (1110)

Sl 2: (1301) + (0122) + (2111) + (4100) + 3(0211) + 2(2200) + (1021) + (3010) + (1102) + 4(0300) + 4(1110) + (0101)

Affine A4 ⊃ A4 - Level 3 - Class 1

Sl 0: (1110) **Highest Weight (1110)**

Sl 1: (0211) + (2200) + (1021) + (3010) + (1102) + (0300) + 3(1110) + (2001) + (0020) + (0101)

Sl 2: (1301) + (0122) + 3(2111) + (0203) + (0130) + (4100) + (1013) + 5(0211) + (3002) + 4(2200) + 5(1021) + 4(3010) + 5(1102) + 4(0300) + 12(1110) + 2(0012) + 5(2001) + 4(0020) + 5(0101) + (1000)

Sl 3: 2(1212) + 2(3201) + 2(2022) + 2(1220) + (0114) + (0041) + 2(4011) + 2(2103) + 2(2030) + 6(1301) + 6(0122) + 16(2111) + 5(0203) + 5(0130) + 4(4100) + 6(1013) + 21(0211) + 6(3002) + 15(2200) + 21(1021) + 15(3010) + 23(1102) + 13(0300) + 41(1110) + (0004) + 10(0012) + 18(2001) + 14(0020) + 19(0101) + 5(1000)

Affine A4 ⊃ A4 - Level 3 - Class 1

Sl 0: (2001) **Highest Weight** $\overset{0}{(2001)}$

Sl 1: (3010) + (1102) + (1110) + 2(2001) + (0101) + (1000)

Sl 2: (2111) + (0203) + (0211) + 2(3002) + (2200) + (1021) + 3(3010) + 4(1102) + 5(1110)
+ (0012) + 7(2001) + (0020) + 4(0101) + 3(1000)

Sl 3: (1212) + (4011) + 2(2103) + (2030) + (1301) + (0122) + 7(2111) + 3(0203) + (0130)
+ (4100) + 2(1013) + 6(0211) + 7(3002) + 4(2200) + 6(1021) + 11(3010) + 16(1102) + 3(0300)
+ 18(1110) + 5(0012) + 20(2001) + 5(0020) + 15(0101) + 8(1000)

Affine A4 ⊃ A4 - Level 3 - Class 1

Sl 0: (0020) **Highest Weight** $\overset{1}{(0020)}$

Sl 1: (1021) + (1110) + (0012) + (0020) + (0101)

Sl 2: (2111) + (0130) + (1013) + (0211) + (2200) + 4(1021) + 2(1102) + 4(1110) + (0004)
+ 3(0012) + (2001) + 5(0020) + 3(0101) + (1000)

Sl 3: (3201) + (2022) + (1220) + (2103) + 2(2030) + (1301) + 2(0122) + 6(2111) + (1005)
+ (0203) + 3(0130) + 5(1013) + 6(0211) + (3002) + 3(2200) + 15(1021) + 3(3010) + 10(1102)
+ 3(0300) + 15(1110) + 2(0004) + 12(0012) + 5(2001) + 12(0020) + 12(0101) + 2(1000)

Affine A4 ⊃ A4 - Level 3 - Class 1

Sl 0: (0101) **Highest Weight** $\overset{1}{(0101)}$

Sl 1: (1102) + (1110) + (0012) + (2001) + (0020) + 2(0101) + (1000)

Sl 2: (2111) + (1013) + (0211) + (3002) + 2(1021) + (3010) + 5(1102) + (0300) + 5(1110)
+ 4(0012) + 4(2001) + 3(0020) + 8(0101) + 3(1000)

Sl 3: (2022) + (1220) + (4011) + 2(2103) + (2030) + (1301) + (0122) + 7(2111) + 2(0203)
+ (0130) + 5(1013) + 7(0211) + 5(3002) + 3(2200) + 11(1021) + 5(3010) + 20(1102) + 3(0300)
+ 20(1110) + 2(0004) + 14(0012) + 16(2001) + 12(0020) + 23(0101) + 9(1000)

Affine A4 ⊃ A4 - Level 3 - Class 1

Sl 0: $(\dot{1}000)$ **Highest Weight** $\mathbf{(\overset{2}{1}000)}$

Sl 1: $(\dot{2}001)$ + $(\dot{0}101)$ + $(\dot{1}000)$

Sl 2: $(\dot{3}002)$ + $(\dot{1}102)$ + $(\dot{1}110)$ + 3$(\dot{2}001)$ + $(\dot{0}020)$ + 3$(\dot{0}101)$ + 3$(\dot{1}000)$

Sl 3: $(\dot{2}103)$ + $(\dot{2}111)$ + $(\dot{0}211)$ + 3$(\dot{3}002)$ + $(\dot{1}021)$ + 2$(\dot{3}010)$ + 5$(\dot{1}102)$ + 5$(\dot{1}110)$ + 2$(\dot{0}012)$
+ 9$(\dot{2}001)$ + 2$(\dot{0}020)$ + 9$(\dot{0}101)$ + 7$(\dot{1}000)$

Affine A4 ⊃ A4 - Level 3 - Class 2

Sl 0: $(\dot{0}003)$ **Highest Weight** $\mathbf{(\overset{0}{0}003)}$

Sl 1: $(\dot{1}012)$ + $(\dot{0}003)$ + $(\dot{0}011)$

Sl 2: $(\dot{2}021)$ + $(\dot{1}004)$ + $(\dot{0}202)$ + 3$(\dot{1}012)$ + $(\dot{1}020)$ + $(\dot{1}101)$ + 3$(\dot{0}003)$ + 3$(\dot{0}011)$

Sl 3: $(\dot{3}030)$ + $(\dot{2}013)$ + $(\dot{1}211)$ + 3$(\dot{2}021)$ + $(\dot{0}113)$ + 2$(\dot{2}102)$ + 2$(\dot{0}121)$ + $(\dot{2}110)$ + 3$(\dot{1}004)$
+ 2$(\dot{0}202)$ + 10$(\dot{1}012)$ + $(\dot{0}210)$ + 4$(\dot{1}020)$ + 5$(\dot{1}101)$ + 7$(\dot{0}003)$ + 8$(\dot{0}011)$ + 2$(\dot{0}100)$

Affine A4 ⊃ A4 - Level 3 - Class 2

Sl 0: $(\dot{0}210)$ **Highest Weight** $\mathbf{(\overset{0}{0}210)}$

Sl 1: $(\dot{1}300)$ + $(\dot{0}121)$ + $(\dot{2}110)$ + $(\dot{0}202)$ + 2$(\dot{0}210)$ + $(\dot{1}020)$ + $(\dot{1}101)$

Sl 2: 2$(\dot{1}211)$ + $(\dot{0}032)$ + $(\dot{3}200)$ + $(\dot{2}021)$ + $(\dot{0}113)$ + $(\dot{4}010)$ + $(\dot{2}102)$ + 3$(\dot{1}300)$ + 4$(\dot{0}121)$
+ 4$(\dot{2}110)$ + 3$(\dot{0}202)$ + 2$(\dot{1}012)$ + 8$(\dot{0}210)$ + $(\dot{3}001)$ + 4$(\dot{1}020)$ + 5$(\dot{1}101)$ + $(\dot{0}011)$ + $(\dot{0}100)$

Affine A4 ⊃ A4 - Level 3 - Class 2

Sl 0: $(\dot{1}020)$ **Highest Weight** $\mathbf{(\overset{0}{1}020)}$

Sl 1: $(\dot{0}121)$ + $(\dot{2}110)$ + $(\dot{1}012)$ + $(\dot{0}210)$ + 2$(\dot{1}020)$ + $(\dot{1}101)$ + $(\dot{0}011)$

Sl 2: $(\dot{1}211)$ + $(\dot{3}200)$ + 2$(\dot{2}021)$ + $(\dot{0}113)$ + $(\dot{0}040)$ + $(\dot{2}102)$ + $(\dot{1}300)$ + 4$(\dot{0}121)$ + 4$(\dot{2}110)$
+ $(\dot{1}004)$ + 2$(\dot{0}202)$ + 4$(\dot{1}012)$ + 4$(\dot{0}210)$ + $(\dot{3}001)$ + 8$(\dot{1}020)$ + 5$(\dot{1}101)$ + $(\dot{0}003)$ + 4$(\dot{0}011)$
+ $(\dot{2}000)$ + $(\dot{0}100)$

Affine A4 ⊃ A4 - Level 3 - Class 2

Sl 0: (1101) **Highest Weight** $\overset{0}{(1101)}$

Sl 1: (2110) + (0202) + (1012) + (0210) + (3001) + (1020) + 3(1101) + (0011) + (2000) + (0100)

Sl 2: (1211) + (2021) + (0113) + (4010) + 3(2102) + (1300) + 2(0121) + 5(2110) + 4(0202) + 5(1012) + 5(0210) + 4(3001) + 5(1020) + 12(1101) + (0003) + 5(0011) + 4(2000) + 4(0100)

Sl 3: (1122) + 2(3111) + 2(1203) + (1130) + (0401) + 2(2013) + 8(1211) + (0032) + 2(4002) + 2(3200) + 8(2021) + 5(0113) + (0040) + 4(4010) + 14(2102) + 5(1300) + 11(0121) + 21(2110) + 2(1004) + 17(0202) + 21(1012) + 19(0210) + 15(3001) + 21(1020) + 41(1101) + 5(0003) + 18(0011) + 13(2000) + 13(0100)

Affine A4 ⊃ A4 - Level 3 - Class 2

Sl 0: (0011) **Highest Weight** $\overset{1}{(0011)}$

Sl 1: (1012) + (1020) + (1101) + (0003) + 2(0011) + (0100)

Sl 2: (2021) + (2102) + (0121) + (2110) + (1004) + (0202) + 5(1012) + (0210) + 4(1020) + 5(1101) + 3(0003) + 7(0011) + (2000) + 3(0100)

Sl 3: (3111) + (1130) + 2(2013) + 2(1211) + 6(2021) + 2(0113) + 6(2102) + (1300) + 6(0121) + 6(2110) + 4(1004) + 6(0202) + 19(1012) + 6(0210) + 3(3001) + 15(1020) + 19(1101) + 9(0003) + 21(0011) + 4(2000) + 9(0100)

Affine A4 ⊃ A4 - Level 3 - Class 2

Sl 0: (2000) **Highest Weight** $\overset{1}{(2000)}$

Sl 1: (3001) + (1101) + (2000) + (0100)

Sl 2: (2102) + (2110) + (0202) + 3(3001) + (1020) + 4(1101) + (0011) + 4(2000) + 2(0100)

Sl 3: (1203) + (1211) + (4002) + (2021) + (4010) + 5(2102) + (0121) + 5(2110) + 3(0202) + 3(1012) + 3(0210) + 9(3001) + 3(1020) + 14(1101) + 4(0011) + 9(2000) + 7(0100)

Affine A4 ⊃ A4 - Level 3 - Class 2

Sl 0: $(\dot{0}100)$ **Highest Weight** $(\overset{2}{0}100)$

Sl 1: $(\dot{1}101) + (\dot{0}011) + (\dot{2}000) + (\dot{0}100)$

Sl 2: $(\dot{2}102) + (\dot{1}012) + (\dot{0}210) + (\dot{3}001) + (\dot{1}020) + 4(\dot{1}101) + 3(\dot{0}011) + 2(\dot{2}000) + 4(\dot{0}100)$

Sl 3: $(\dot{2}013) + (\dot{1}211) + (\dot{4}002) + (\dot{2}021) + 4(\dot{2}102) + (\dot{0}121) + 3(\dot{2}110) + 3(\dot{0}202) + 5(\dot{1}012)$

$+ 3(\dot{0}210) + 4(\dot{3}001) + 5(\dot{1}020) + 14(\dot{1}101) + 2(\dot{0}003) + 9(\dot{0}011) + 7(\dot{2}000) + 9(\dot{0}100)$

Affine A5 Branching Rules

Affine A5 ⊃ A5 - Level 1 - Class 0

Sl 0: (00000) **Highest Weight** $\overset{1}{(00000)}$

Sl 1: (10001)

Sl 2: (01010) + 2(10001) + (00000)

Sl 3: (20010) + (01002) + (00200) + 2(01010) + 4(10001) + 2(00000)

Sl 4: (11100) + (00111) + (20002) + 2(20010) + 2(01002) + (00200) + 6(01010) + 8(10001) + 4(00000)

Sl 5: (11011) + 3(11100) + 3(00111) + 2(20002) + 6(20010) + 6(01002) + 4(00200) + 11(01010) + 16(10001) + 6(00000)

Sl 6: (10201) + 4(11011) + (30100) + (03000) + (00103) + (00030) + 7(11100) + 7(00111) + 6(20002) + 12(20010) + 12(01002) + 8(00200) + 25(01010) + 29(10001) + 12(00000)

Affine A5 ⊃ A5 - Level 1 - Class 1

Sl 0: (10000) **Highest Weight** $\overset{0}{(10000)}$

Sl 1: (01001) + (10000)

Sl 2: (00110) + (20001) + 2(01001) + 2(10000)

Sl 3: (11010) + (00102) + 2(00110) + 2(20001) + 5(01001) + 4(10000)

Sl 4: (11002) + (10200) + 3(11010) + (00021) + 2(00102) + 5(00110) + 5(20001) + 10(01001) + 8(10000)

Sl 5: (10111) + (02100) + (30010) + 3(11002) + 2(10200) + 8(11010) + 2(00021) + 6(00102) + 11(00110) + 10(20001) + 20(01001) + 14(10000)

Sl 6: (02011) + (21100) + 4(10111) + (30002) + 2(02100) + 2(30010) + 8(11002) + 6(10200) + (00013) + 18(11010) + 5(00021) + 12(00102) + 23(00110) + 21(20001) + 38(01001) + 25(10000)

Affine A5 ⊃ A5 - Level 1 - Class 2

Sl 0: (01000) **Highest Weight (01000)**

Sl 1: (00101) + (20000) + (01000)

Sl 2: (11001) + (00020) + 2(00101) + (20000) + 3(01000)

Sl 3: (10110) + 3(11001) + (00012) + (00020) + 5(00101) + 3(20000) + 5(01000)

Sl 4: (10102) + (02010) + 3(10110) + (30001) + 7(11001) + 2(00012) + 4(00020) + 10(00101)
+ 5(20000) + 11(01000)

Sl 5: (21010) + (10021) + (02002) + (01200) + 3(10102) + 2(02010) + 8(10110) + 2(30001)
+ 16(11001) + 5(00012) + 7(00020) + 21(00101) + 11(20000) + 18(01000)

Sl 6: (01111) + (21002) + (20200) + 3(21010) + 3(10021) + 2(02002) + 2(01200) + 9(10102)
+ 7(02010) + 18(10110) + 5(30001) + (00004) + 33(11001) + 10(00012) + 16(00020) + 39(00101)
+ 19(20000) + 35(01000)

Affine A5 ⊃ A5 - Level 1 - Class 3

Sl 0: (00100) **Highest Weight (00100)**

Sl 1: (11000) + (00011) + (00100)

Sl 2: (10101) + 2(11000) + 2(00011) + 3(00100)

Sl 3: (10020) + (02001) + 3(10101) + (30000) + (00003) + 4(11000) + 4(00011) + 6(00100)

Sl 4: (01110) + (21001) + (10012) + 2(10020) + 2(02001) + 8(10101) + (30000) + (00003)
+ 9(11000) + 9(00011) + 11(00100)

Sl 5: (20110) + (01102) + 3(01110) + 3(21001) + 3(10012) + 6(10020) + 6(02001) + 17(10101)
+ 3(30000) + 3(00003) + 18(11000) + 18(00011) + 20(00100)

Sl 6: (20102) + (12010) + (01021) + 3(20110) + 3(01102) + (00300) + 8(01110) + 8(21001)
+ 8(10012) + 13(10020) + 13(02001) + 37(10101) + 6(30000) + 6(00003) + 33(11000) + 33(00011)
+ 38(00100)

Affine A5 ⊃ A5 - Level 2 - Class 0

sl 0: (00200) **Highest Weight** $(00200)^{0}$

sl 1: (11100) + (00111) + (00200) + (01010)

sl 2: (10201) + (22000) + (11011) + (00022) + 3(11100) + 3(00111) + (20010) + (01002)
+ 4(00200) + 3(01010) + (10001)

sl 3: (21101) + (10112) + (10120) + (02101) + 4(10201) + 2(22000) + 5(11011) + 2(00022)
+ 2(30100) + 2(03000) + 2(00103) + 2(00030) + 9(11100) + 9(00111) + (20002) + 4(20010)
+ 4(01002) + 9(00200) + 11(01010) + 3(10001) + (00000)

Affine A5 ⊃ A5 - Level 2 - Class 0

sl 0: (01010) **Highest Weight** $(01010)^{0}$

sl 1: (11100) + (00111) + (20010) + (01002) + (00200) + 2(01010) + (10001)

sl 2: (10201) + 2(11011) + (30100) + (03000) + (00103) + (00030) + 4(11100) + 4(00111)
+ (20002) + 3(20010) + 3(01002) + 3(00200) + 8(01010) + 4(10001) + (00000)

sl 3: (21101) + (10112) + 2(10120) + 2(02101) + (30011) + (11003) + 5(10201) + 2(22000)
+ 10(11011) + 2(00022) + 3(30100) + 2(03000) + 3(00103) + 2(00030) + 14(11100) + 14(00111)
+ 4(20002) + 12(20010) + 12(01002) + 11(00200) + 21(01010) + 13(10001) + 2(00000)

Affine A5 ⊃ A5 - Level 2 - Class 0

sl 0: (10001) **Highest Weight** $(10001)^{0}$

sl 1: (20010) + (01002) + (01010) + 2(10001) + (00000)

sl 2: (11011) + (11100) + (00111) + 2(20002) + 3(20010) + 3(01002) + (00200) + 4(01010)
+ 6(10001) + 2(00000)

sl 3: (10120) + (02101) + (30011) + (11003) + (10201) + 6(11011) + (30100) + (00103)
+ 5(11100) + 5(00111) + 6(20002) + 10(20010) + 10(01002) + 3(00200) + 13(01010) + 16(10001)
+ 4(00000)

Affine A5 ⊃ A5 - Level 2 - Class 0

Sl 0: (00000) **Highest Weight** $\overset{2}{(00000)}$

Sl 1: (10001)

Sl 2: (20002) + (01010) + 2(10001) + (00000)

Sl 3: (11011) + 2(20002) + 2(20010) + 2(01002) + (00200) + 2(01010) + 5(10001) + 2(00000)

Sl 4: (02020) + (30011) + (11003) + (10201) + 4(11011) + 2(11100) + 2(00111) + 7(20002) + 5(20010) + 5(01002) + (00200) + 8(01010) + 12(10001) + 5(00000)

Affine A5 ⊃ A5 - Level 2 - Class 1

Sl 0: (00110) **Highest Weight** $\overset{0}{(00110)}$

Sl 1: (10200) + (11010) + (00021) + (00102) + 2(00110) + (01001)

Sl 2: (21100) + 2(10111) + (02100) + (11002) + 3(10200) + (00013) + 4(11010) + 3(00021) + 3(00102) + 7(00110) + (20001) + 4(01001) + (10000)

Sl 3: (20201) + (21011) + (10022) + (01201) + (13000) + (10103) + (10030) + 2(02011) + 4(21100) + 9(10111) + 4(02100) + 2(30010) + 5(11002) + 10(10200) + 3(00013) + 14(11010) + 9(00021) + 11(00102) + 19(00110) + 5(20001) + 13(01001) + 3(10000)

Affine A5 ⊃ A5 - Level 2 - Class 1

Sl 0: (01001) **Highest Weight** $\overset{0}{(01001)}$

Sl 1: (11010) + (00102) + (00110) + (20001) + 2(01001) + (10000)

Sl 2: (10111) + (02100) + (30010) + 2(11002) + (10200) + 4(11010) + (00021) + 3(00102) + 4(00110) + 3(20001) + 7(01001) + 3(10000)

Sl 3: (21011) + (01201) + (10103) + (10030) + 2(02011) + 2(21100) + 6(10111) + (30002) + 3(02100) + 3(30010) + 8(11002) + 5(10200) + (00013) + 15(11010) + 4(00021) + 10(00102) + 13(00110) + 11(20001) + 19(01001) + 8(10000)

Affine A5 ⊃ A5 - Level 2 - Class 1

sl 0: (10000) **Highest Weight** $\overset{1}{(10000)}$

sl 1: (20001) + (01001) + (10000)

sl 2: (11002) + (11010) + (00110) + 3(20001) + 3(01001) + 3(10000)

sl 3: (02011) + (10111) + (30002) + (30010) + 4(11002) + (10200) + 5(11010) + 2(00102) + 3(00110) + 8(20001) + 9(01001) + 7(10000)

Affine A5 ⊃ A5 - Level 2 - Class 2

sl 0: (00020) **Highest Weight** $\overset{0}{(00020)}$

sl 1: (10110) + (00012) + (00020) + (00101)

sl 2: (20200) + (10021) + (10102) + (02010) + 3(10110) + (00004) + (11001) + 2(00012) + 4(00020) + 3(00101) + (01000)

sl 3: (20111) + (12100) + (01111) + 2(20200) + (10013) + 2(21010) + 4(10021) + (02002) + 2(01200) + 4(10102) + 2(02010) + 10(10110) + (00004) + 5(11001) + 7(00012) + 8(00020) + 10(00101) + (20000) + 2(01000)

Affine A5 ⊃ A5 - Level 2 - Class 2

sl 0: (00101) **Highest Weight** $\overset{0}{(00101)}$

sl 1: (10110) + (11001) + (00012) + (00020) + 2(00101) + (01000)

sl 2: (21010) + (10021) + (01200) + 2(10102) + (02010) + 4(10110) + 4(11001) + 3(00012) + 3(00020) + 7(00101) + (20000) + 3(01000)

sl 3: (20111) + (12100) + 2(01111) + (21002) + 2(20200) + (10013) + 4(21010) + 5(10021) + 2(02002) + 3(01200) + 8(10102) + 5(02010) + 15(10110) + 2(30001) + (00004) + 14(11001) + 9(00012) + 10(00020) + 19(00101) + 4(20000) + 9(01000)

Affine A5 ⊃ A5 - Level 2 - Class 2

Sl 0: $\dot{(20000)}$ **Highest Weight** $\overset{0}{\mathbf{(20000)}}$

Sl 1: $\dot{(11001)} + \dot{(20000)} + \dot{(01000)}$

Sl 2: $\dot{(02002)} + \dot{(10110)} + \dot{(30001)} + 3\dot{(11001)} + \dot{(00101)} + 3\dot{(20000)} + 2\dot{(01000)}$

Sl 3: $\dot{(01111)} + \dot{(21002)} + \dot{(21010)} + 2\dot{(02002)} + 2\dot{(10102)} + 2\dot{(02010)} + 3\dot{(10110)} + 3\dot{(30001)}$
$+ 9\dot{(11001)} + \dot{(00020)} + 4\dot{(00101)} + 6\dot{(20000)} + 6\dot{(01000)}$

Affine A5 ⊃ A5 - Level 2 - Class 2

Sl 0: $\dot{(01000)}$ **Highest Weight** $\overset{1}{\mathbf{(01000)}}$

Sl 1: $\dot{(11001)} + \dot{(00101)} + \dot{(20000)} + \dot{(01000)}$

Sl 2: $\dot{(10102)} + \dot{(02010)} + \dot{(10110)} + \dot{(30001)} + 4\dot{(11001)} + \dot{(00020)} + 3\dot{(00101)} + 2\dot{(20000)}$
$+ 4\dot{(01000)}$

Sl 3: $\dot{(01111)} + \dot{(21002)} + 2\dot{(21010)} + \dot{(10021)} + 2\dot{(02002)} + \dot{(01200)} + 4\dot{(10102)} + 3\dot{(02010)}$
$+ 6\dot{(10110)} + 3\dot{(30001)} + 13\dot{(11001)} + 2\dot{(00012)} + 2\dot{(00020)} + 10\dot{(00101)} + 7\dot{(20000)} + 9\dot{(01000)}$

Affine A5 ⊃ A5 - Level 2 - Class 3

Sl 0: $\dot{(00011)}$ **Highest Weight** $\overset{0}{\mathbf{(00011)}}$

Sl 1: $\dot{(10020)} + \dot{(10101)} + \dot{(00003)} + 2\dot{(00011)} + \dot{(00100)}$

Sl 2: $\dot{(20110)} + \dot{(01110)} + 2\dot{(10012)} + 3\dot{(10020)} + \dot{(02001)} + 4\dot{(10101)} + 2\dot{(00003)} + \dot{(11000)}$
$+ 6\dot{(00011)} + 3\dot{(00100)}$

Sl 3: $\dot{(20021)} + \dot{(11200)} + \dot{(20102)} + \dot{(12010)} + \dot{(01021)} + 4\dot{(20110)} + 2\dot{(01102)} + \dot{(10004)}$
$+ 5\dot{(01110)} + 2\dot{(21001)} + 7\dot{(10012)} + 10\dot{(10020)} + 3\dot{(02001)} + 14\dot{(10101)} + 5\dot{(00003)} + 5\dot{(11000)}$
$+ 16\dot{(00011)} + 8\dot{(00100)}$

Affine A5 ⊃ A5 - Level 2 - Class 3

sl 0: (11000) **Highest Weight** $\overset{0}{(11000)}$

sl 1: (02001) + (10101) + (30000) + 2(11000) + (00100)

sl 2: (01102) + (01110) + 2(21001) + (10020) + 3(02001) + 4(10101) + 2(30000) + 6(11000) + (00011) + 3(00100)

sl 3: (12002) + (00211) + (20102) + (12010) + (01021) + 2(20110) + 4(01102) + (40001) + 5(01110) + 7(21001) + 2(10012) + 3(10020) + 10(02001) + 14(10101) + 5(30000) + 16(11000) + 5(00011) + 8(00100)

Affine A5 ⊃ A5 - Level 2 - Class 3

sl 0: (00100) **Highest Weight** $\overset{1}{(00100)}$

sl 1: (10101) + (11000) + (00011) + (00100)

sl 2: (01110) + (21001) + (10012) + (10020) + (02001) + 4(10101) + 3(11000) + 3(00011) + 4(00100)

sl 3: (20102) + (12010) + (01021) + 2(20110) + 2(01102) + (00300) + 4(01110) + 4(21001) + 4(10012) + 5(10020) + 5(02001) + 14(10101) + 2(30000) + 2(00003) + 9(11000) + 9(00011) + 10(00100)

Affine A6 Branching Rules

Affine A6 ⊃ A6 - Level 1 - Class 0

```
                                           1
Sl 0:   (000000)  Highest Weight  (000000)

Sl 1:   (100001)

Sl 2:   (010010) + 2(100001) +  (000000)

Sl 3:   (001100) +  (200010) +  (010002) + 2(010010) + 4(100001) + 2(000000)

Sl 4:   (110100) +  (001011) +  (200002) + 2(001100) + 2(200010) + 2(010002) + 6(010010)

     + 8(100001) + 4(000000)

Sl 5:   (110011) +  (102000) +  (000201) + 3(110100) + 3(001011) + 2(200002) + 5(001100)

     + 6(200010) + 6(010002) + 12(010010) + 16(100001) + 6(000000)
```

Affine A6 ⊃ A6 - Level 1 - Class 1

```
                                           0
Sl 0:   (100000)  Highest Weight  (100000)

Sl 1:   (010001) +  (100000)

Sl 2:   (001010) +  (200001) + 2(010001) + 2(100000)

Sl 3:   (110010) +  (001002) +  (000200) + 2(001010) + 2(200001) + 5(010001) + 4(100000)

Sl 4:   (101100) +  (110002) +  (000111) + 3(110010) + 2(001002) +  (000200) + 6(001010)

     + 5(200001) + 10(010001) + 8(100000)

Sl 5:   (101011) +  (020100) + 3(101100) +  (300010) + 3(110002) + 3(000111) + 8(110010)

     + 6(001002) + 4(000200) + 12(001010) + 10(200001) + 21(010001) + 14(100000)
```

Affine A6 ⊃ A6 - Level 1 - Class 2

Sl 0: (010000) **Highest Weight** (010000)

Sl 1: (001001) + (200000) + (010000)

Sl 2: (110001) + (000110) + 2(001001) + (200000) + 3(010000)

Sl 3: (101010) + (000102) + 3(110001) + 2(000110) + 5(001001) + 3(200000) + 5(010000)

Sl 4: (101002) + (100200) + (020010) + 3(101010) + (000021) + (300001) + 2(000102) + 7(110001) + 5(000110) + 11(001001) + 5(200000) + 11(010000)

Sl 5: (100111) + (011100) + (210010) + (020002) + 3(101002) + 2(100200) + 2(020010) + 9(101010) + 2(000021) + 2(300001) + 6(000102) + 16(110001) + 11(000110) + 22(001001) + 11(200000) + 19(010000)

Affine A6 ⊃ A6 - Level 1 - Class 3

Sl 0: (001000) **Highest Weight** (001000)

Sl 1: (000101) + (110000) + (001000)

Sl 2: (101001) + (000020) + 2(000101) + 2(110000) + 3(001000)

Sl 3: (100110) + (020001) + 3(101001) + (000012) + (000020) + (300000) + 5(000101) + 4(110000) + 6(001000)

Sl 4: (011010) + (100102) + (210001) + 3(100110) + 2(020001) + 8(101001) + 2(000012) + 4(000020) + (300000) + 10(000101) + 9(110000) + 12(001000)

Sl 5: (201010) + (011002) + (010200) + (100021) + 3(011010) + 3(100102) + 3(210001) + 8(100110) + 6(020001) + 18(101001) + 5(000012) + 7(000020) + 3(300000) + 22(000101) + 18(110000) + 21(001000)

Affine A6 ⊃ A6 - Level 2 - Class 0

Sl 0: (001100) **Highest Weight** (001100)

Sl 1: (102000) + (000201) + (110100) + (001011) + 2(001100) + (010010)

Sl 2: 2(101101) + (211000) + (000112) + (110011) + (021000) + (000120) + 3(102000) + 3(000201) + 4(110100) + 4(001011) + 7(001100) + (200010) + (010002) + 4(010010) + (100001)

Affine A6 ⊃ A6 - Level 2 - Class 0

```
                                    0
Sl 0:   (010010)   Highest Weight  (010010)

Sl 1:   (110100) +   (001011) +   (001100) +   (200010) +   (010002) +  2(010010) +   (100001)

Sl 2:   (101101) +  2(110011) +   (021000) +   (000120) +   (300100) +   (102000) +   (001003)

   +    (000201) +  4(110100) +  4(001011) +   (200002) +  4(001100) +  3(200010) +  3(010002)

   +   8(010010) +  4(100001) +   (000000)

Sl 3:   (210101) +   (101012) +   (100210) +   (012001) +  2(101020) +  2(020101) +  6(101101)

   +    (300011) +  2(211000) +   (110003) +  2(000112) + 10(110011) +  3(021000) +  3(000120)

   +   3(300100) +  5(102000) +  3(001003) +  5(000201) + 15(110100) + 15(001011) +  4(200002)

   +  14(001100) + 12(200010) + 12(010002) + 22(010010) + 13(100001) +  2(000000)
```

Affine A6 ⊃ A6 - Level 2 - Class 0

```
                                    0
Sl 0:   (100001)   Highest Weight  (100001)

Sl 1:   (200010) +   (010002) +   (010010) +  2(100001) +   (000000)

Sl 2:   (110011) +   (110100) +   (001011) +  2(200002) +   (001100) +  3(200010) +  3(010002)

   +   4(010010) +  6(100001) +  2(000000)

Sl 3:   (101020) +   (020101) +   (101101) +   (300011) +   (110003) +  6(110011) +   (300100)

   +    (102000) +   (001003) +   (000201) +  5(110100) +  5(001011) +  6(200002) +  4(001100)

   +  10(200010) + 10(010002) + 13(010010) + 16(100001) +  4(000000)
```

Affine A6 ⊃ A6 - Level 2 - Class 0

```
                                    2
Sl 0:   (000000)   Highest Weight  (000000)

Sl 1:   (100001)

Sl 2:   (200002) +   (010010) +  2(100001) +   (000000)

Sl 3:   (110011) +  2(200002) +   (001100) +  2(200010) +  2(010002) +  2(010010) +  5(100001)

   +   2(000000)

Sl 4:   (020020) +   (101101) +   (300011) +   (110003) +  4(110011) +  2(110100) +  2(001011)

   +   7(200002) +  2(001100) +  5(200010) +  5(010002) +  8(010010) + 12(100001) +  5(000000)
```

Affine A6 ⊃ A6 - Level 2 - Class 1

sl 0: (000200) **Highest Weight** $\overset{0}{(000200)}$

sl 1: (101100) + (000111) + (000200) + (001010)

sl 2: (202000) + (100201) + (101011) + (020100) + (000022) + 3(101100) + 3(000111)
+ (110010) + (001002) + 4(000200) + 3(001010) + (010001)

Affine A6 ⊃ A6 - Level 2 - Class 1

sl 0: (001010) **Highest Weight** $\overset{0}{(001010)}$

sl 1: (101100) + (000111) + (110010) + (001002) + (000200) + 2(001010) + (010001)

sl 2: (100201) + (210100) + 2(101011) + (012000) + (020100) + 4(101100) + (000103)
+ (000030) + (110002) + 4(000111) + 4(110010) + 3(001002) + 3(000200) + 8(001010)
+ (200001) + 4(010001) + (100000)

Affine A6 ⊃ A6 - Level 2 - Class 1

sl 0: (010001) **Highest Weight** $\overset{0}{(010001)}$

sl 1: (110010) + (001002) + (001010) + (200001) + 2(010001) + (100000)

sl 2: (101011) + (020100) + (101100) + (300010) + 2(110002) + (000111) + 4(110010)
+ 3(001002) + (000200) + 4(001010) + 3(200001) + 7(010001) + 3(100000)

Affine A6 ⊃ A6 - Level 2 - Class 1

sl 0: (100000) **Highest Weight** $\overset{1}{(100000)}$

sl 1: (200001) + (010001) + (100000)

sl 2: (110002) + (110010) + (001010) + 3(200001) + 3(010001) + 3(100000)

sl 3: (020011) + (101011) + (300002) + (101100) + (300010) + 4(110002) + 5(110010)
+ 2(001002) + (000200) + 3(001010) + 8(200001) + 9(010001) + 7(100000)

Affine A6 ⊃ A6 - Level 2 - Class 2

Sl 0: (000110) **Highest Weight** **(000110)**

Sl 1: (100200) + (101010) + (000021) + (000102) + 2(000110) + (001001)

Sl 2: (201100) + 2(100111) + (011100) + (101002) + 3(100200) + (020010) + 4(101010)
+ (000013) + 3(000021) + 3(000102) + (110001) + 7(000110) + 4(001001) + (010000)

Affine A6 ⊃ A6 - Level 2 - Class 2

Sl 0: (001001) **Highest Weight** **(001001)**

Sl 1: (101010) + (000102) + (110001) + (000110) + 2(001001) + (010000)

Sl 2: (100111) + (011100) + (210010) + 2(101002) + (100200) + (020010) + 4(101010)
+ (000021) + 3(000102) + 4(110001) + 4(000110) + 7(001001) + (200000) + 3(010000)

Sl 3: (201011) + (010201) + (120100) + 2(011011) + 2(201100) + (100103) + (100030)
+ (003000) + (210002) + 6(100111) + 4(011100) + 4(210010) + 2(020002) + 8(101002)
+ 5(100200) + 5(020010) + 16(101010) + (000013) + 4(000021) + 2(300001) + 10(000102)
+ 14(110001) + 13(000110) + 20(001001) + 4(200000) + 9(010000)

Affine A6 ⊃ A6 - Level 2 - Class 2

Sl 0: (200000) **Highest Weight** **(200000)**

Sl 1: (110001) + (200000) + (010000)

Sl 2: (020002) + (101010) + (300001) + 3(110001) + (001001) + 3(200000) + 2(010000)

Sl 3: (011011) + (210002) + (210010) + 2(020002) + 2(101002) + (100200) + 2(020010)
+ 3(101010) + 3(300001) + 9(110001) + (000110) + 4(001001) + 6(200000) + 6(010000)

Affine A6 ⊃ A6 - Level 2 - Class 2

```
sl 0:  (010000)  Highest Weight  (010000) 1

sl 1:  (110001) + (001001) + (200000) + (010000)

sl 2:  (101002) + (020010) + (101010) + (300001) + 4(110001) + (000110) + 3(001001)
     + 2(200000) + 4(010000)

sl 3:  (011011) + (210002) + (100111) + (011100) + 2(210010) + 2(020002) + 4(101002)
     + (100200) + 3(020010) + 6(101010) + 3(300001) + 2(000102) + 13(110001) + 3(000110)
     + 10(001001) + 7(200000) + 9(010000)
```

Affine A6 ⊃ A6 - Level 2 - Class 3

```
sl 0:  (000020)  Highest Weight  (000020) 0

sl 1:  (100110) + (000012) + (000020) + (000101)

sl 2:  (200200) + (100021) + (011010) + (100102) + 3(100110) + (101001) + (000004)
     + 2(000012) + 4(000020) + 3(000101) + (001000)

sl 3:  (200111) + (111100) + (010111) + 2(200200) + 2(201010) + (100013) + (011002)
     + 2(010200) + 4(100021) + 3(011010) + 4(100102) + 10(100110) + (020001) + 5(101001)
     + (000004) + 7(000012) + 8(000020) + 10(000101) + (110000) + 2(001000)
```

Affine A6 ⊃ A6 - Level 2 - Class 3

```
sl 0:  (000101)  Highest Weight  (000101) 0

sl 1:  (100110) + (101001) + (000012) + (000020) + 2(000101) + (001000)

sl 2:  (201010) + (010200) + (100021) + (011010) + 2(100102) + 4(100110) + (020001)
     + 4(101001) + 3(000012) + 3(000020) + 7(000101) + (110000) + 3(001000)

sl 3:  (200111) + (111100) + 2(010111) + (201002) + 2(200200) + (120010) + (002100)
     + 4(201010) + (100013) + 2(011002) + 3(010200) + 5(100021) + 6(011010) + 8(100102)
     + 2(210001) + 15(100110) + 3(020001) + 15(101001) + (000004) + 9(000012) + 10(000020)
     + 19(000101) + 5(110000) + 9(001000)
```

Affine A6 ⊃ A6 - Level 2 - Class 3

Sl 0: $(\dot{1}10000)$ **Highest Weight** $(\overset{0}{1}10000)$

Sl 1: $(\dot{0}20001)$ + $(\dot{1}01001)$ + $(\dot{3}00000)$ + $2(\dot{1}10000)$ + $(\dot{0}01000)$

Sl 2: $(\dot{0}11002)$ + $(\dot{0}11010)$ + $2(\dot{2}10001)$ + $(\dot{1}00110)$ + $3(\dot{0}20001)$ + $4(\dot{1}01001)$ + $2(\dot{3}00000)$
+ $(\dot{0}00101)$ + $6(\dot{1}10000)$ + $3(\dot{0}01000)$

Sl 3: $(\dot{0}02011)$ + $(\dot{1}20002)$ + $(\dot{0}10111)$ + $(\dot{2}01002)$ + $(\dot{1}20010)$ + $2(\dot{2}01010)$ + $4(\dot{0}11002)$
+ $(\dot{0}10200)$ + $5(\dot{0}11010)$ + $(\dot{4}00001)$ + $2(\dot{1}00102)$ + $7(\dot{2}10001)$ + $4(\dot{1}00110)$ + $10(\dot{0}20001)$
+ $14(\dot{1}01001)$ + $(\dot{0}00020)$ + $5(\dot{3}00000)$ + $5(\dot{0}00101)$ + $16(\dot{1}10000)$ + $8(\dot{0}01000)$

Affine A6 ⊃ A6 - Level 2 - Class 3

Sl 0: $(\dot{0}01000)$ **Highest Weight** $(\overset{1}{0}01000)$

Sl 1: $(\dot{1}01001)$ + $(\dot{0}00101)$ + $(\dot{1}10000)$ + $(\dot{0}01000)$

Sl 2: $(\dot{0}11010)$ + $(\dot{1}00102)$ + $(\dot{2}10001)$ + $(\dot{1}00110)$ + $(\dot{0}20001)$ + $4(\dot{1}01001)$ + $(\dot{0}00020)$
+ $3(\dot{0}00101)$ + $3(\dot{1}10000)$ + $4(\dot{0}01000)$

Sl 3: $(\dot{0}10111)$ + $(\dot{2}01002)$ + $(\dot{1}20010)$ + $(\dot{0}02100)$ + $2(\dot{2}01010)$ + $2(\dot{0}11002)$ + $(\dot{0}10200)$
+ $(\dot{1}00021)$ + $4(\dot{0}11010)$ + $4(\dot{1}00102)$ + $4(\dot{2}10001)$ + $6(\dot{1}00110)$ + $5(\dot{0}20001)$ + $14(\dot{1}01001)$
+ $2(\dot{0}00012)$ + $2(\dot{0}00020)$ + $2(\dot{3}00000)$ + $10(\dot{0}00101)$ + $9(\dot{1}10000)$ + $10(\dot{0}01000)$

Affine A7 Branching Rules

Affine A7 ⊃ A7 - Level 1 - Class 0

Sl 0: (0000000) **Highest Weight** $\overset{1}{(0000000)}$

Sl 1: (1000001)

Sl 2: (0100010) + 2(1000001) + (0000000)

Sl 3: (0010100) + (2000010) + (0100002) + 2(0100010) + 4(1000001) + 2(0000000)

Sl 4: (1100100) + (0010011) + (0002000) + 2(0010100) + (2000002) + 2(2000010) + 2(0100002) + 6(0100010) + 8(1000001) + 4(0000000)

Affine A7 ⊃ A7 - Level 1 - Class 1

Sl 0: (1000000) **Highest Weight** $\overset{0}{(1000000)}$

Sl 1: (0100001) + (1000000)

Sl 2: (0010010) + (2000001) + 2(0100001) + 2(1000000)

Sl 3: (1100010) + (0001100) + (0010002) + 2(0010010) + 2(2000001) + 5(0100001) + 4(1000000)

Sl 4: (1010100) + (0001011) + (1100002) + 3(1100010) + 2(0001100) + 2(0010002) + 6(0010010) + 5(2000001) + 10(0100001) + 8(1000000)

Affine A7 ⊃ A7 - Level 1 - Class 2

Sl 0: (0100000) **Highest Weight** $\overset{0}{(0100000)}$

Sl 1: (0010001) + (2000000) + (0100000)

Sl 2: (0001010) + (1100001) + 2(0010001) + (2000000) + 3(0100000)

Sl 3: (1010010) + (0001002) + (0000200) + 2(0001010) + 3(1100001) + 5(0010001) + 3(2000000) + 5(0100000)

Sl 4: (1001100) + (1010002) + (0200010) + 3(1010010) + (0000111) + 2(0001002) + (0000200) + (3000001) + 6(0001010) + 7(1100001) + 11(0010001) + 5(2000000) + 11(0100000)

Affine A7 ⊃ A7 - Level 1 - Class 3

Sl 0: (0010000) **Highest Weight** **(0010000)**

Sl 1: (0001001) + (1100000) + (0010000)

Sl 2: (1010001) + (0000110) + 2(0001001) + 2(1100000) + 3(0010000)

Sl 3: (1001010) + (0200001) + 3(1010001) + (0000102) + 2(0000110) + 5(0001001) + (3000000) + 4(1100000) + 6(0010000)

Sl 4: (0110010) + (1001002) + (1000200) + 3(1001010) + (2100001) + 2(0200001) + (0000021) + 8(1010001) + 2(0000102) + 5(0000110) + 11(0001001) + (3000000) + 9(1100000) + 12(0010000)

Affine A7 ⊃ A7 - Level 1 - Class 4

Sl 0: (0001000) **Highest Weight** **(0001000)**

Sl 1: (1010000) + (0000101) + (0001000)

Sl 2: (1001001) + (0200000) + (0000020) + 2(1010000) + 2(0000101) + 3(0001000)

Sl 3: (1000110) + (0110001) + 3(1001001) + (2100000) + (0000012) + (0200000) + (0000020) + 5(1010000) + 5(0000101) + 6(0001000)

Sl 4: (0101010) + (2010001) + (1000102) + 3(1000110) + 3(0110001) + 8(1001001) + 2(2100000) + 2(0000012) + 4(0200000) + 4(0000020) + 10(1010000) + 10(0000101) + 13(0001000)

Affine A7 ⊃ A7 - Level 2 - Class 0

Sl 0: (0002000) **Highest Weight** **(0002000)**

Sl 1: (1011000) + (0001101) + (0002000) + (0010100)

Sl 2: (1002001) + (2020000) + (1010101) + (0000202) + (0201000) + (0001020) + 3(1011000) + 3(0001101) + (1100100) + (0010011) + 4(0002000) + 3(0010100) + (0100010)

Affine A7 ⊃ A7 - Level 2 - Class 0

Sl 0: ·(0010100) **Highest Weight** $\overset{0}{(0010100)}$

Sl 1: ·(1011000) + ·(0001101) + ·(1100100) + ·(0010011) + ·(0002000) + 2·(0010100) + ·(0100010)

Sl 2: ·(1002001) + 2·(1010101) + ·(2101000) + ·(0120000) + ·(0001012) + ·(0000210) + ·(0201000) + ·(0001020) + ·(1100011) + 4·(1011000) + 4·(0001101) + 4·(1100100) + 4·(0010011) + 3·(0002000) + 8·(0010100) + ·(2000010) + ·(0100002) + 4·(0100010) + ·(1000001)

Affine A7 ⊃ A7 - Level 2 - Class 0

Sl 0: ·(0100010) **Highest Weight** $\overset{0}{(0100010)}$

Sl 1: ·(1100100) + ·(0010011) + ·(0010100) + ·(2000010) + ·(0100002) + 2·(0100010) + ·(1000001)

Sl 2: ·(1010101) + ·(0201000) + ·(0001020) + 2·(1100011) + ·(1011000) + ·(0001101) + ·(3000100) + ·(0010003) + 4·(1100100) + 4·(0010011) + ·(0002000) + 4·(0010100) + ·(2000002) + 3·(2000010) + 3·(0100002) + 8·(0100010) + 4·(1000001) + ·(0000000)

Affine A7 ⊃ A7 - Level 2 - Class 0

Sl 0: ·(1000001) **Highest Weight** $\overset{0}{(1000001)}$

Sl 1: ·(2000010) + ·(0100002) + ·(0100010) + 2·(1000001) + ·(0000000)

Sl 2: ·(1100011) + ·(1100100) + ·(0010011) + ·(0010100) + 2·(2000002) + 3·(2000010) + 3·(0100002) + 4·(0100010) + 6·(1000001) + 2·(0000000)

Sl 3: ·(1010020) + ·(0200101) + ·(1010101) + ·(3000011) + ·(1100003) + 6·(1100011) + ·(1011000) + ·(0001101) + ·(3000100) + ·(0010003) + 5·(1100100) + 5·(0010011) + ·(0002000) + 4·(0010100) + 6·(2000002) + 10·(2000010) + 10·(0100002) + 13·(0100010) + 16·(1000001) + 4·(0000000)

Affine A7 ⊃ A7 - Level 2 - Class 0

Sl 0: ·(0000000) **Highest Weight** $\overset{2}{(0000000)}$

Sl 1: ·(1000001)

Sl 2: ·(2000002) + ·(0100010) + 2·(1000001) + ·(0000000)

Sl 3: ·(1100011) + ·(0010100) + 2·(2000002) + 2·(2000010) + 2·(0100002) + 2·(0100010) + 5·(1000001) + 2·(0000000)

Affine A7 ⊃ A7 - Level 2 - Class 1

Sl 0: (0001100) **Highest Weight** $\overset{0}{\textbf{(0001100)}}$

Sl 1: (1002000) + (1010100) + (0000201) + (0001011) + 2(0001100) + (0010010)

Sl 2: (2011000) + 2(1001101) + (0111000) + (1010011) + (0000112) + 3(1002000) + (0200100)
+ (0000120) + 4(1010100) + 3(0000201) + 4(0001011) + (1100010) + 7(0001100) + (0010002)
+ 4(0010010) + (0100001)

Affine A7 ⊃ A7 - Level 2 - Class 1

Sl 0: (0010010) **Highest Weight** $\overset{0}{\textbf{(0010010)}}$

Sl 1: (1010100) + (0001011) + (1100010) + (0001100) + (0010002) + 2(0010010) + (0100001)

Sl 2: (1001101) + (0111000) + (2100100) + 2(1010011) + (1002000) + (0200100) + (0000120)
+ 4(1010100) + (0001003) + (0000201) + 4(0001011) + (1100002) + 4(1100010) + 4(0001100)
+ 3(0010002) + 8(0010010) + (2000001) + 4(0100001) + (1000000)

Affine A7 ⊃ A7 - Level 2 - Class 1

Sl 0: (0100001) **Highest Weight** $\overset{0}{\textbf{(0100001)}}$

Sl 1: (1100010) + (0010002) + (0010010) + (2000001) + 2(0100001) + (1000000)

Sl 2: (1010011) + (0200100) + (1010100) + (0001011) + (3000010) + 2(1100002) + 4(1100010)
+ (0001100) + 3(0010002) + 4(0010010) + 3(2000001) + 7(0100001) + 3(1000000)

Affine A7 ⊃ A7 - Level 2 - Class 1

Sl 0: (1000000) **Highest Weight** $\overset{1}{\textbf{(1000000)}}$

Sl 1: (2000001) + (0100001) + (1000000)

Sl 2: (1100002) + (1100010) + (0010010) + 3(2000001) + 3(0100001) + 3(1000000)

Sl 3: (0200011) + (1010011) + (1010100) + (3000002) + (3000010) + 4(1100002) + 5(1100010)
+ (0001100) + 2(0010002) + 3(0010010) + 8(2000001) + 9(0100001) + 7(1000000)

Affine A7 ⊃ A7 - Level 2 - Class 2

Sl 0: (0000200) **Highest Weight (0000200)**

Sl 1: (1001100) + (0000111) + (0000200) + (0001010)

Sl 2: (2002000) + (1000201) + (1001011) + (0110100) + 3(1001100) + (0000022) + (1010010)
+ 3(0000111) + (0001002) + 4(0000200) + 3(0001010) + (0010001)

Affine A7 ⊃ A7 - Level 2 - Class 2

Sl 0: (0001010) **Highest Weight (0001010)**

Sl 1: (1001100) + (1010010) + (0000111) + (0001002) + (0000200) + 2(0001010) + (0010001)

Sl 2: (2010100) + (1000201) + (0102000) + 2(1001011) + (0110100) + 4(1001100) + (1010002)
+ (0200010) + (0000103) + (0000030) + 4(1010010) + 4(0000111) + 3(0001002) + 3(0000200)
+ 8(0001010) + (1100001) + 4(0010001) + (0100000)

Affine A7 ⊃ A7 - Level 2 - Class 2

Sl 0: (0010001) **Highest Weight (0010001)**

Sl 1: (1010010) + (0001002) + (0001010) + (1100001) + 2(0010001) + (0100000)

Sl 2: (1001011) + (0110100) + (2100010) + (1001100) + 2(1010002) + (0200010) + 4(1010010)
+ (0000111) + 3(0001002) + (0000200) + 4(0001010) + 4(1100001) + 7(0010001) + (2000000)
+ 3(0100000)

Affine A7 ⊃ A7 - Level 2 - Class 2

Sl 0: (2000000) **Highest Weight (2000000)**

Sl 1: (1100001) + (2000000) + (0100000)

Sl 2: (0200002) + (1010010) + (3000001) + 3(1100001) + (0010001) + 3(2000000) + 2(0100000)

Sl 3: (0110011) + (2100002) + (2100010) + (1001100) + 2(0200002) + 2(1010002) + 2(0200010)
+ 3(1010010) + 3(3000001) + (0001010) + 9(1100001) + 4(0010001) + 6(2000000) + 6(0100000)

Affine A7 ⊃ A7 - Level 2 - Class 2

```
        .                          1
Sl 0:   (0100000)  Highest Weight  (0100000)

Sl 1:   (1100001) + (0010001) + (2000000) + (0100000)

Sl 2:   (1010002) + (0200010) + (1010010) + (3000001) + (0001010) + 4(1100001) + 3(0010001)
      + 2(2000000) + 4(0100000)

Sl 3:   (0110011) + (1001011) + (0110100) + (2100002) + 2(2100010) + (1001100) + 2(0200002)
      + 4(1010002) + 3(0200010) + 6(1010010) + 2(0001002) + (0000200) + 3(3000001) + 3(0001010)
      + 13(1100001) + 10(0010001) + 7(2000000) + 9(0100000)
```

Affine A7 ⊃ A7 - Level 2 - Class 3

```
        .                          0
Sl 0:   (0000110)  Highest Weight  (0000110)

Sl 1:   (1000200) + (1001010) + (0000021) + (0000102) + 2(0000110) + (0001001)

Sl 2:   (2001100) + (0101100) + 2(1000111) + (0110010) + (1001002) + 3(1000200) + 4(1001010)
      + (0000013) + 3(0000021) + (1010001) + 3(0000102) + 7(0000110) + 4(0001001) + (0010000)
```

Affine A7 ⊃ A7 - Level 2 - Class 3

```
        .                          0
Sl 0:   (0001001)  Highest Weight  (0001001)

Sl 1:   (1001010) + (1010001) + (0000102) + (0000110) + 2(0001001) + (0010000)

Sl 2:   (0101100) + (2010010) + (1000111) + (0110010) + 2(1001002) + (1000200) + 4(1001010)
      + (0200001) + (0000021) + 4(1010001) + 3(0000102) + 4(0000110) + 7(0001001) + (1100000)
      + 3(0010000)
```

Affine A7 ⊃ A7 - Level 2 - Class 3

```
        .                          0
Sl 0:   (1100000)  Highest Weight  (0001001)

Sl 1:   (0200001) + (1010001) + (3000000) + 2(1100000) + (0010000)

Sl 2:   (0110002) + (0110010) + (1001010) + 2(2100001) + 3(0200001) + 4(1010001) + (0001001)
      + 2(3000000) + 6(1100000) + 3(0010000)
```

Affine A7 ⊃ A7 - Level 2 - Class 3

sl 0: (0010000) **Highest Weight** $\overset{1}{(0010000)}$

sl 1: (1010001) + (0001001) + (1100000) + (0010000)

sl 2: (0110010) + (1001002) + (1001010) + (2100001) + (0200001) + 4(1010001) + (0000110)
+ 3(0001001) + 3(1100000) + 4(0010000)

sl 3: (0101011) + (0020100) + (2010002) + (1200010) + (0101100) + 2(2010010) + (1000111)
+ 2(0110002) + 4(0110010) + 4(1001002) + (1000200) + 6(1001010) + 4(2100001) + 5(0200001)
+ 14(1010001) + 2(0000102) + 3(0000110) + 10(0001001) + 2(3000000) + 9(1100000) + 10(0010000)

Affine A7 ⊃ A7 - Level 2 - Class 4

sl 0: (0000020) **Highest Weight** $\overset{0}{(0000020)}$

sl 1: (1000110) + (0000012) + (0000020) + (0000101)

sl 2: (2000200) + (0101010) + (1000021) + (1000102) + 3(1000110) + (1001001) + (0000004)
+ 2(0000012) + 4(0000020) + 3(0000101) + (0001000)

Affine A7 ⊃ A7 - Level 2 - Class 4

sl 0: (0000101) **Highest Weight** $\overset{0}{(0000101)}$

sl 1: (1000110) + (1001001) + (0000012) + (0000020) + 2(0000101) + (0001000)

sl 2: (2001010) + (0100200) + (0101010) + (1000021) + 2(1000102) + 4(1000110) + (0110001)
+ 4(1001001) + 3(0000012) + 3(0000020) + (1010000) + 7(0000101) + 3(0001000)

Affine A7 ⊃ A7 - Level 2 - Class 4

sl 0: (0200000) **Highest Weight** $\overset{0}{(0200000)}$

sl 1: (0110001) + (2100000) + (0200000) + (1010000)

sl 2: (0020002) + (0101010) + (1200001) + (2010001) + 3(0110001) + (1001001) + (4000000)
+ 2(2100000) + 4(0200000) + 3(1010000) + (0001000)

Affine A7 ⊃ A7 - Level 2 - Class 4

```
        •                          0
Sl 0:  (1010000)  Highest Weight  (1010000)

Sl 1:  (0110001) + (1001001) + (2100000) + (0200000) + 2(1010000) + (0001000)

Sl 2:  (0101002) + (0020010) + (0101010) + (1200001) + 2(2010001) + (1000110) + 4(0110001)

     + 4(1001001) + 3(2100000) + 3(0200000) + 7(1010000) + (0000101) + 3(0001000)
```

Affine A7 ⊃ A7 - Level 2 - Class 4

```
        •                          1
Sl 0:  (0001000)  Highest Weight  (0001000)

Sl 1:  (1001001) + (1010000) + (0000101) + (0001000)

Sl 2:  (0101010) + (2010001) + (1000102) + (1000110) + (0110001) + 4(1001001) + (0200000)

     + (0000020) + 3(1010000) + 3(0000101) + 4(0001000)

Sl 3:  (1110010) + (0100111) + (0011100) + (2001002) + 2(2001010) + 2(0101002) + (0100200)

     + (0020010) + 4(0101010) + (1200001) + (1000021) + 4(2010001) + 4(1000102) + 6(1000110)

     + 6(0110001) + 14(1001001) + 2(2100000) + 2(0000012) + 2(0200000) + 2(0000020) + 10(1010000)

     + 10(0000101) + 10(0001000)
```

(All weights in the Sl 1 – Sl 3 rows carry a dot over the first digit.)

Affine A8 Branching Rules

Affine A8 ⊃ A8 - Level 1 - Class 0

sl 0: (00000000) **Highest Weight** $\overset{1}{(00000000)}$

sl 1: (10000001)

sl 2: (01000010) + 2(10000001) + (00000000)

sl 3: (00100100) + (20000010) + (01000002) + 2(01000010) + 4(10000001) + 2(00000000)

sl 4: (11000100) + (00100011) + (00011000) + 2(00100100) + (20000002) + 2(20000010)
+ 2(01000002) + 6(01000010) + 8(10000001) + 4(00000000)

sl 5: (10101000) + (00010101) + (11000011) + 3(11000100) + 3(00100011) + 2(00011000)
+ 6(00100100) + 2(20000002) + 6(20000010) + 6(01000002) + 12(01000010) + 16(10000001)
+ 6(00000000)

sl 6: (10100101) + (10020000) + (02001000) + (00010020) + (00002001) + 3(10101000)
+ 3(00010101) + 4(11000011) + (30000100) + (00100003) + 8(11000100) + 8(00100011)
+ 5(00011000) + 14(00100100) + 6(20000002) + 12(20000010) + 12(01000002) + 27(01000010)
+ 30(10000001) + 12(00000000)

Affine A8 ⊃ A8 - Level 1 - Class 1

sl 0: (10000000) **Highest Weight** $\overset{0}{(10000000)}$

sl 1: (01000001) + (10000000)

sl 2: (00100010) + (20000001) + 2(01000001) + 2(10000000)

sl 3: (00010100) + (11000010) + (00100002) + 2(00100010) + 2(20000001) + 5(01000001)
+ 4(10000000)

sl 4: (10100100) + (00010011) + (00002000) + (11000002) + 2(00010100) + 3(11000010)
+ 2(00100002) + 6(00100010) + 5(20000001) + 10(01000001) + 8(10000000)

sl 5: (10100011) + (10011000) + (02000100) + (00001101) + 3(10100100) + 3(00010011)
+ (00002000) + (30000010) + 3(11000002) + 6(00010100) + 8(11000010) + 6(00100002)
+ 13(00100010) + 10(20000001) + 21(01000001) + 14(10000000)

Affine A8 ⊃ A8 - Level 1 - Class 2

Sl 0: (01000000) **Highest Weight (01000000)**

Sl 1: (00100001) + (20000000) + (01000000)

Sl 2: (00010010) + (11000001) + 2(00100001) + (20000000) + 3(01000000)

Sl 3: (10100010) + (00001100) + (00010002) + 2(00010010) + 3(11000001) + 5(00100001) + 3(20000000) + 5(01000000)

Sl 4: (10010100) + (10100002) + (02000010) + (00001011) + 3(10100010) + 2(00001100) + 2(00010002) + 6(00010010) + (30000001) + 7(11000001) + 11(00100001) + 5(20000000) + 11(01000000)

Sl 5: (10010011) + (01100100) + (10002000) + 3(10010100) + (21000010) + (02000002) + (00000201) + 3(10100002) + 2(02000010) + 3(00001011) + 9(10100010) + 5(00001100) + 6(00010002) + 13(00010010) + 2(30000001) + 16(11000001) + 23(00100001) + 11(20000000) + 19(01000000)

Affine A8 ⊃ A8 - Level 1 - Class 3

Sl 0: (00100000) **Highest Weight (00100000)**

Sl 1: (00010001) + (11000000) + (00100000)

Sl 2: (10100001) + (00001010) + 2(00010001) + 2(11000000) + 3(00100000)

Sl 3: (10010010) + (02000001) + (00001002) + (00000200) + 3(10100001) + 2(00001010) + 5(00010001) + (30000000) + 4(11000000) + 6(00100000)

Sl 4: (10001100) + (01100010) + (10010002) + 3(10010010) + (00000111) + (21000001) + 2(02000001) + 2(00001002) + (00000200) + 8(10100001) + 6(00001010) + 11(00010001) + (30000000) + 9(11000000) + 12(00100000)

Sl 5: (01010100) + (10001011) + (20100010) + (01100002) + 3(10001100) + 3(01100010) + 3(10010002) + 9(10010010) + 3(00000111) + 3(21000001) + 6(02000001) + 6(00001002) + 4(00000200) + 18(10100001) + 12(00001010) + 24(00010001) + 3(30000000) + 18(11000000) + 22(00100000)

Affine A8 ⊃ A8 - Level 1 - Class 4

sl 0: (00010000) **Highest Weight** $\overset{0}{(}$**00010000)**

sl 1: (00001001) + (10100000) + (00010000)

sl 2: (10010001) + (00000110) + (02000000) + 2(00001001) + 2(10100000) + 3(00010000)

sl 3: (10001010) + (01100001) + 3(10010001) + (00000102) + 2(00000110) + (21000000)
+ (02000000) + 5(00001001) + 5(10100000) + 6(00010000)

sl 4: (01010010) + (10001002) + (10000200) + (20100001) + 3(10001010) + 3(01100001)
+ 8(10010001) + (00000021) + 2(00000102) + 5(00000110) + 2(21000000) + 4(02000000)
+ 11(00001001) + 10(10100000) + 13(00010000)

sl 5: (01001100) + (20010010) + (01010002) + (00200010) + (10000111) + 3(01010010)
+ (12000001) + 3(10001002) + 2(10000200) + 3(20100001) + 9(10001010) + 8(01100001)
+ 19(10010001) + 2(00000021) + 6(00000102) + 11(00000110) + 5(21000000) + 7(02000000)
+ 23(00001001) + 22(10100000) + 23(00010000)

Affine A8 ⊃ A8 - Level 2 - Class 0

sl 0: (00011000) **Highest Weight** $\overset{0}{(}$**00011000)**

sl 1: (10020000) + (00002001) + (10101000) + (00010101) + 2(00011000) + (00100100)

Affine A8 ⊃ A8 - Level 2 - Class 0

sl 0: (00100100) **Highest Weight** $\overset{0}{(}$**00100100)**

sl 1: (10101000) + (00010101) + (11000100) + (00100011) + (00011000) + 2(00100100)
+ (01000010)

Affine A8 ⊃ A8 - Level 2 - Class 0

sl 0: (01000010) **Highest Weight** $\overset{0}{(}$**01000010)**

sl 1: (11000100) + (00100011) + (00100100) + (20000010) + (01000002) + 2(01000010)
+ (10000001)

sl 2: (10100101) + (02001000) + (00010020) + (10101000) + (00010101) + 2(11000011)
+ (30000100) + (00100003) + 4(11000100) + 4(00100011) + (00011000) + 4(00100100)
+ (20000002) + 3(20000010) + 3(01000002) + 8(01000010) + 4(10000001) + (00000000)

Affine A8 ⊃ A8 - Level 2 - Class 0

Sl 0: $(\dot{1}0000001)$ **Highest Weight** $(\overset{0}{1}0000001)$

Sl 1: $(\dot{2}0000010)$ + $(\dot{0}1000002)$ + $(\dot{0}1000010)$ + $2(\dot{1}0000001)$ + $(\dot{0}0000000)$

Sl 2: $(\dot{1}1000011)$ + $(\dot{1}1000100)$ + $(\dot{0}0100011)$ + $(\dot{0}0100100)$ + $2(\dot{2}0000002)$ + $3(\dot{2}0000010)$
+ $3(\dot{0}1000002)$ + $4(\dot{0}1000010)$ + $6(\dot{1}0000001)$ + $2(\dot{0}0000000)$

Sl 3: $(\dot{1}0100020)$ + $(\dot{0}2000101)$ + $(\dot{1}0100101)$ + $(\dot{3}0000011)$ + $(\dot{1}1000003)$ + $(\dot{1}0101000)$
+ $(\dot{0}0010101)$ + $6(\dot{1}1000011)$ + $(\dot{3}0000100)$ + $(\dot{0}0100003)$ + $5(\dot{1}1000100)$ + $5(\dot{0}0100011)$
+ $(\dot{0}0011000)$ + $4(\dot{0}0100100)$ + $6(\dot{2}0000002)$ + $10(\dot{2}0000010)$ + $10(\dot{0}1000002)$ + $13(\dot{0}1000010)$
+ $16(\dot{1}0000001)$ + $4(\dot{0}0000000)$

Affine A8 ⊃ A8 - Level 2 - Class 0

Sl 0: $(\dot{0}0000000)$ **Highest Weight** $(\overset{2}{0}0000000)$

Sl 1: $(\dot{1}0000001)$

Sl 2: $(\dot{2}0000002)$ + $(\dot{0}1000010)$ + $2(\dot{1}0000001)$ + $(\dot{0}0000000)$

Sl 3: $(\dot{1}1000011)$ + $(\dot{0}0100100)$ + $2(\dot{2}0000002)$ + $2(\dot{2}0000010)$ + $2(\dot{0}1000002)$ + $2(\dot{0}1000010)$
+ $5(\dot{1}0000001)$ + $2(\dot{0}0000000)$

Affine A8 ⊃ A8 - Level 2 - Class 1

Sl 0: $(\dot{0}0002000)$ **Highest Weight** $(\overset{0}{0}0002000)$

Sl 1: $(\dot{1}0011000)$ + $(\dot{0}0001101)$ + $(\dot{0}0002000)$ + $(\dot{0}0010100)$

Affine A8 ⊃ A8 - Level 2 - Class 1

Sl 0: $(\dot{0}0010100)$ **Highest Weight** $(\overset{0}{0}0010100)$

Sl 1: $(\dot{1}0011000)$ + $(\dot{0}0001101)$ + $(\dot{1}0100100)$ + $(\dot{0}0010011)$ + $(\dot{0}0002000)$ + $2(\dot{0}0010100)$
+ $(\dot{0}0100010)$

Affine A8 ⊃ A8 - Level 2 - Class 1

Sl 0: (00100010) **Highest Weight** ^{0}**(00100010)**

Sl 1: (10100100) + (00010011) + (00010100) + (11000010) + (00100002) + 2(00100010)

+ (01000001)

Affine A8 ⊃ A8 - Level 2 - Class 1

Sl 0: (01000001) **Highest Weight** ^{0}**(01000001)**

Sl 1: (11000010) + (00100002) + (00100010) + (20000001) + 2(01000001) + (10000000)

Sl 2: (10100011) + (02000100) + (10100100) + (00010011) + (30000010) + 2(11000002)

+ (00010100) + 4(11000010) + 3(00100002) + 4(00100010) + 3(20000001) + 7(01000001)

+ 3(10000000)

Affine A8 ⊃ A8 - Level 2 - Class 1

Sl 0: (10000000) **Highest Weight** ^{1}**(10000000)**

Sl 1: (20000001) + (01000001) + (10000000)

Sl 2: (11000002) + (11000010) + (00100010) + 3(20000001) + 3(01000001) + 3(10000000)

Sl 3: (02000011) + (10100011) + (10100100) + (30000002) + (30000010) + 4(11000002)

+ (00010100) + 5(11000010) + 2(00100002) + 3(00100010) + 8(20000001) + 9(01000001)

+ 7(10000000)

Affine A8 ⊃ A8 - Level 2 - Class 2

Sl 0: (00001100) **Highest Weight** ^{0}**(00001100)**

Sl 1: (10002000) + (10010100) + (00000201) + (00001011) + 2(00001100) + (00010010)

Affine A8 ⊃ A8 - Level 2 - Class 2

Sl 0: (00010010) **Highest Weight** ^{0}**(00010010)**

Sl 1: (10010100) + (00001011) + (10100010) + (00001100) + (00010002) + 2(00010010)

+ (00100001)

Affine A8 ⊃ A8 - Level 2 - Class 2

sl 0: $(\dot{0}0100001)$ **Highest Weight** $(\overset{0}{0}0100001)$

sl 1: $(\dot{1}0100010)$ + $(\dot{0}0010002)$ + $(\dot{0}0010010)$ + $(\dot{1}1000001)$ + $2(\dot{0}0100001)$ + $(\dot{0}1000000)$

sl 2: $(\dot{1}0010011)$ + $(\dot{0}1100100)$ + $(\dot{1}0010100)$ + $(\dot{2}1000010)$ + $2(\dot{1}0100002)$ + $(\dot{0}2000010)$
+ $(\dot{0}0001011)$ + $4(\dot{1}0100010)$ + $(\dot{0}0001100)$ + $3(\dot{0}0010002)$ + $4(\dot{0}0010010)$ + $4(\dot{1}1000001)$
+ $7(\dot{0}0100001)$ + $(\dot{2}0000000)$ + $3(\dot{0}1000000)$

Affine A8 ⊃ A8 - Level 2 - Class 2

sl 0: $(\dot{2}0000000)$ **Highest Weight** $(\overset{0}{2}0000000)$

sl 1: $(\dot{1}1000001)$ + $(\dot{2}0000000)$ + $(\dot{0}1000000)$

sl 2: $(\dot{0}2000002)$ + $(\dot{1}0100010)$ + $(\dot{3}0000001)$ + $3(\dot{1}1000001)$ + $(\dot{0}0100001)$ + $3(\dot{2}0000000)$
+ $2(\dot{0}1000000)$

Affine A8 ⊃ A8 - Level 2 - Class 2

sl 0: $(\dot{0}1000000)$ **Highest Weight** $(\overset{1}{0}1000000)$

sl 1: $(\dot{1}1000001)$ + $(\dot{0}0100001)$ + $(\dot{2}0000000)$ + $(\dot{0}1000000)$

sl 2: $(\dot{1}0100002)$ + $(\dot{0}2000010)$ + $(\dot{1}0100010)$ + $(\dot{0}0010010)$ + $(\dot{3}0000001)$ + $4(\dot{1}1000001)$
+ $3(\dot{0}0100001)$ + $2(\dot{2}0000000)$ + $4(\dot{0}1000000)$

Affine A8 ⊃ A8 - Level 2 - Class 3

sl 0: $(\dot{0}0000200)$ **Highest Weight** (00000200)

sl 1: $(\dot{1}0001100)$ + $(\dot{0}0000111)$ + $(\dot{0}0000200)$ + $(\dot{0}0001010)$

Affine A8 ⊃ A8 - Level 2 - Class 3

sl 0: $(\dot{0}0001010)$ **Highest Weight** $(\overset{0}{0}0001010)$

sl 1: $(\dot{1}0001100)$ + $(\dot{1}0010010)$ + $(\dot{0}0000111)$ + $(\dot{0}0001002)$ + $(\dot{0}0000200)$ + $2(\dot{0}0001010)$
+ $(\dot{0}0010001)$

sl 2: $(\dot{2}0010100)$ + $(\dot{0}1002000)$ + $(\dot{1}0000201)$ + $(\dot{0}1010100)$ + $2(\dot{1}0001011)$ + $4(\dot{1}0001100)$
+ $(\dot{0}1100010)$ + $(\dot{1}0010002)$ + $4(\dot{1}0010010)$ + $(\dot{0}0000103)$ + $(\dot{0}0000030)$ + $4(\dot{0}0000111)$
+ $3(\dot{0}0001002)$ + $3(\dot{0}0000200)$ + $(\dot{1}0100001)$ + $8(\dot{0}0001010)$ + $4(\dot{0}0010001)$ + $(\dot{0}0100000)$

Affine A8 ⊃ A8 - Level 2 - Class 3

Sl 0: (00010001) **Highest Weight** $\overset{0}{(00010001)}$

Sl 1: (10010010) + (00001002) + (10100001) + (00001010) + 2(00010001) + (00100000)

Sl 2: (01010100) + (10001011) + (20100010) + (10001100) + (01100010) + 2(10010002)
+ 4(10010010) + (00000111) + (02000001) + 3(00001002) + (00000200) + 4(10100001)
+ 4(00001010) + 7(00010001) + (11000000) + 3(00100000)

Affine A8 ⊃ A8 - Level 2 - Class 3

Sl 0: (11000000) **Highest Weight** $\overset{0}{(11000000)}$

Sl 1: (02000001) + (10100001) + (30000000) + 2(11000000) + (00100000)

Sl 2: (01100002) + (01100010) + (10010010) + 2(21000001) + 3(02000001) + 4(10100001)
+ (00010001) + 2(30000000) + 6(11000000) + 3(00100000)

Affine A8 ⊃ A8 - Level 2 - Class 3

Sl 0: (00100000) **Highest Weight** $\overset{1}{(00100000)}$

Sl 1: (10100001) + (00010001) + (11000000) + (00100000)

Sl 2: (01100010) + (10010002) + (10010010) + (21000001) + (02000001) + 4(10100001)
+ (00001010) + 3(00010001) + 3(11000000) + 4(00100000)

Affine A8 ⊃ A8 - Level 2 - Class 4

Sl 0: (00000110) **Highest Weight** $\overset{0}{(00000110)}$

Sl 1: (10000200) + (10001010) + (00000021) + (00000102) + 2(00000110) + (00001001)

Sl 2: (20001100) + (01001100) + 2(10000111) + (01010010) + (10001002) + 3(10000200)
+ 4(10001010) + (00000013) + (10010001) + 3(00000021) + 3(00000102) + 7(00000110)
+ 4(00001001) + (00010000)

Affine A8 ⊃ A8 - Level 2 - Class 4

Sl 0: (00001001) **Highest Weight** $\overset{0}{(00001001)}$

Sl 1: (10001010) + (10010001) + (00000102) + (00000110) + 2(00001001) + (00010000)

Sl 2: (01001100) + (20010010) + (10000111) + (01010010) + 2(10001002) + (10000200)
+ 4(10001010) + (01100001) + 4(10010001) + (00000021) + 3(00000102) + 4(00000110)
+ 7(00001001) + (10100000) + 3(00010000)

Affine A8 ⊃ A8 - Level 2 - Class 4

Sl 0: (02000000) **Highest Weight** $\overset{0}{(02000000)}$

Sl 1: (01100001) + (21000000) + (02000000) + (10100000)

Sl 2: (00200002) + (01010010) + (12000001) + (20100001) + 3(01100001) + (10010001)
+ (40000000) + 2(21000000) + 4(02000000) + 3(10100000) + (00010000)

Affine A8 ⊃ A8 - Level 2 - Class 4

Sl 0: (10100000) **Highest Weight** $\overset{0}{(10100000)}$

Sl 1: (01100001) + (10010001) + (21000000) + (02000000) + 2(10100000) + (00010000)

Sl 2: (01010002) + (00200010) + (01010010) + (12000001) + 2(20100001) + (10001010)
+ 4(01100001) + 4(10010001) + 3(21000000) + 3(02000000) + (00001001) + 7(10100000)
+ 3(00010000)

Affine A8 ⊃ A8 - Level 2 - Class 4

Sl 0: (00010000) **Highest Weight** $\overset{1}{(00010000)}$

Sl 1: (10010001) + (00001001) + (10100000) + (00010000)

Sl 2: (01010010) + (10001002) + (20100001) + (10001010) + (01100001) + 4(10010001)
+ (00000110) + (02000000) + 3(00001001) + 3(10100000) + 4(00010000)

Affine B3 Branching Rules

Affine B3 ⊃ B3 - Level 1 - Class 0

sl 0: $\overset{\bullet}{(}00)_{1}$ **Highest Weight** $\overset{0}{(}00)_{1}$

sl 1: $\overset{\bullet}{(}02)_{0} + \overset{\bullet}{(}00)_{1}$

sl 2: $\overset{\bullet}{(}10)_{1} + \overset{\bullet}{(}02)_{0} + \overset{\bullet}{(}10)_{0} + 2\overset{\bullet}{(}00)_{1}$

sl 3: $\overset{\bullet}{(}02)_{1} + 2\overset{\bullet}{(}10)_{1} + 3\overset{\bullet}{(}02)_{0} + \overset{\bullet}{(}10)_{0} + 3\overset{\bullet}{(}00)_{1} + \overset{\bullet}{(}00)_{0}$

sl 4: $\overset{\bullet}{(}12)_{0} + \overset{\bullet}{(}00)_{3} + 2\overset{\bullet}{(}02)_{1} + 4\overset{\bullet}{(}10)_{1} + \overset{\bullet}{(}00)_{2} + 5\overset{\bullet}{(}02)_{0} + 3\overset{\bullet}{(}10)_{0} + 6\overset{\bullet}{(}00)_{1} + \overset{\bullet}{(}00)_{0}$

sl 5: $\overset{\bullet}{(}02)_{2} + 2\overset{\bullet}{(}12)_{0} + \overset{\bullet}{(}20)_{0} + \overset{\bullet}{(}00)_{3} + 5\overset{\bullet}{(}02)_{1} + 8\overset{\bullet}{(}10)_{1} + 2\overset{\bullet}{(}00)_{2} + 10\overset{\bullet}{(}02)_{0} + 5\overset{\bullet}{(}10)_{0} + 9\overset{\bullet}{(}00)_{1}$
$+ 2\overset{\bullet}{(}00)_{0}$

sl 6: $\overset{\bullet}{(}20)_{1} + 2\overset{\bullet}{(}02)_{2} + \overset{\bullet}{(}04)_{0} + \overset{\bullet}{(}10)_{2} + 5\overset{\bullet}{(}12)_{0} + 2\overset{\bullet}{(}20)_{0} + 3\overset{\bullet}{(}00)_{3} + 10\overset{\bullet}{(}02)_{1} + 15\overset{\bullet}{(}10)_{1} + 4\overset{\bullet}{(}00)_{2}$
$+ 16\overset{\bullet}{(}02)_{0} + 11\overset{\bullet}{(}10)_{0} + 16\overset{\bullet}{(}00)_{1} + 3\overset{\bullet}{(}00)_{0}$

sl 7: $\overset{\bullet}{(}12)_{1} + 2\overset{\bullet}{(}20)_{1} + 6\overset{\bullet}{(}02)_{2} + 2\overset{\bullet}{(}04)_{0} + 2\overset{\bullet}{(}10)_{2} + 10\overset{\bullet}{(}12)_{0} + 5\overset{\bullet}{(}20)_{0} + 5\overset{\bullet}{(}00)_{3} + 20\overset{\bullet}{(}02)_{1} + 26\overset{\bullet}{(}10)_{1}$
$+ 8\overset{\bullet}{(}00)_{2} + 30\overset{\bullet}{(}02)_{0} + 18\overset{\bullet}{(}10)_{0} + 24\overset{\bullet}{(}00)_{1} + 6\overset{\bullet}{(}00)_{0}$

Affine B3 ⊃ B3 - Level 1 - Class 0

sl 0: $\overset{\bullet}{(}00)_{0}$ **Highest Weight** $\overset{1}{(}00)_{0}$

sl 1: $\overset{\bullet}{(}10)_{0}$

sl 2: $\overset{\bullet}{(}00)_{2} + \overset{\bullet}{(}02)_{0} + \overset{\bullet}{(}10)_{0} + \overset{\bullet}{(}00)_{0}$

sl 3: $\overset{\bullet}{(}02)_{1} + \overset{\bullet}{(}00)_{2} + \overset{\bullet}{(}02)_{0} + 3\overset{\bullet}{(}10)_{0} + \overset{\bullet}{(}00)_{1} + \overset{\bullet}{(}00)_{0}$

sl 4: $\overset{\bullet}{(}20)_{0} + 2\overset{\bullet}{(}02)_{1} + \overset{\bullet}{(}10)_{1} + 3\overset{\bullet}{(}00)_{2} + 3\overset{\bullet}{(}02)_{0} + 4\overset{\bullet}{(}10)_{0} + \overset{\bullet}{(}00)_{1} + 3\overset{\bullet}{(}00)_{0}$

sl 5: $\overset{\bullet}{(}10)_{2} + \overset{\bullet}{(}12)_{0} + \overset{\bullet}{(}20)_{0} + 5\overset{\bullet}{(}02)_{1} + 2\overset{\bullet}{(}10)_{1} + 4\overset{\bullet}{(}00)_{2} + 5\overset{\bullet}{(}02)_{0} + 9\overset{\bullet}{(}10)_{0} + 3\overset{\bullet}{(}00)_{1} + 3\overset{\bullet}{(}00)_{0}$

sl 6: $\overset{\bullet}{(}02)_{2} + \overset{\bullet}{(}04)_{0} + 2\overset{\bullet}{(}10)_{2} + 2\overset{\bullet}{(}12)_{0} + 4\overset{\bullet}{(}20)_{0} + 9\overset{\bullet}{(}02)_{1} + 5\overset{\bullet}{(}10)_{1} + 9\overset{\bullet}{(}00)_{2} + 11\overset{\bullet}{(}02)_{0} + 13\overset{\bullet}{(}10)_{0}$
$+ 5\overset{\bullet}{(}00)_{1} + 7\overset{\bullet}{(}00)_{0}$ (continued next page)

Sl 7: $\dot{(}12)_{1}$ + $2\dot{(}02)_{2}$ + $\dot{(}04)_{0}$ + $5\dot{(}10)_{2}$ + $6\dot{(}12)_{0}$ + $6\dot{(}20)_{0}$ + $\dot{(}00)_{3}$ + $19\dot{(}02)_{1}$ + $10\dot{(}10)_{1}$ + $13\dot{(}00)_{2}$

+ $17\dot{(}02)_{0}$ + $25\dot{(}10)_{0}$ + $10\dot{(}00)_{1}$ + $8\dot{(}00)_{0}$

Affine B3 ⊃ B3 - Level 1 - Class 1

Sl 0: $\dot{(}01)_{0}$ **Highest Weight** $\overset{0}{(}01)_{0}$

Sl 1: $\dot{(}01)_{1}$ + $\dot{(}01)_{0}$

Sl 2: $\dot{(}11)_{0}$ + $2\dot{(}01)_{1}$ + $2\dot{(}01)_{0}$

Sl 3: $\dot{(}01)_{2}$ + $\dot{(}03)_{0}$ + $2\dot{(}11)_{0}$ + $4\dot{(}01)_{1}$ + $4\dot{(}01)_{0}$

Sl 4: $\dot{(}11)_{1}$ + $2\dot{(}01)_{2}$ + $2\dot{(}03)_{0}$ + $5\dot{(}11)_{0}$ + $8\dot{(}01)_{1}$ + $7\dot{(}01)_{0}$

Sl 5: $\dot{(}03)_{1}$ + $3\dot{(}11)_{1}$ + $5\dot{(}01)_{2}$ + $4\dot{(}03)_{0}$ + $10\dot{(}11)_{0}$ + $15\dot{(}01)_{1}$ + $12\dot{(}01)_{0}$

Sl 6: $\dot{(}21)_{0}$ + $\dot{(}01)_{3}$ + $3\dot{(}03)_{1}$ + $7\dot{(}11)_{1}$ + $10\dot{(}01)_{2}$ + $9\dot{(}03)_{0}$ + $19\dot{(}11)_{0}$ + $27\dot{(}01)_{1}$ + $21\dot{(}01)_{0}$

Sl 7: $\dot{(}11)_{2}$ + $\dot{(}13)_{0}$ + $2\dot{(}21)_{0}$ + $2\dot{(}01)_{3}$ + $7\dot{(}03)_{1}$ + $16\dot{(}11)_{1}$ + $20\dot{(}01)_{2}$ + $17\dot{(}03)_{0}$ + $35\dot{(}11)_{0}$ + $47\dot{(}01)_{1}$

+ $34\dot{(}01)_{0}$

Affine B3 ⊃ B3 - Level 2 - Class 0

Sl 0: $\dot{(}02)_{0}$ **Highest Weight** $\overset{0}{(}02)_{0}$

Sl 1: $\dot{(}02)_{1}$ + $\dot{(}10)_{1}$ + $\dot{(}02)_{0}$ + $\dot{(}10)_{0}$ + $\dot{(}00)_{1}$

Sl 2: $\dot{(}02)_{2}$ + $\dot{(}12)_{0}$ + $\dot{(}20)_{0}$ + $3\dot{(}02)_{1}$ + $3\dot{(}10)_{1}$ + $\dot{(}00)_{2}$ + $4\dot{(}02)_{0}$ + $2\dot{(}10)_{0}$ + $2\dot{(}00)_{1}$ + $\dot{(}00)_{0}$

Sl 3: $\dot{(}12)_{1}$ + $\dot{(}20)_{1}$ + $3\dot{(}02)_{2}$ + $\dot{(}04)_{0}$ + $2\dot{(}10)_{2}$ + $5\dot{(}12)_{0}$ + $2\dot{(}20)_{0}$ + $2\dot{(}00)_{3}$ + $9\dot{(}02)_{1}$ + $8\dot{(}10)_{1}$

+ $3\dot{(}00)_{2}$ + $9\dot{(}02)_{0}$ + $7\dot{(}10)_{0}$ + $6\dot{(}00)_{1}$ + $\dot{(}00)_{0}$

Sl 4: $\dot{(}22)_{0}$ + $\dot{(}02)_{3}$ + $\dot{(}04)_{1}$ + $\dot{(}10)_{3}$ + $5\dot{(}12)_{1}$ + $4\dot{(}20)_{1}$ + $11\dot{(}02)_{2}$ + $4\dot{(}04)_{0}$ + $6\dot{(}10)_{2}$ + $14\dot{(}12)_{0}$

+ $9\dot{(}20)_{0}$ + $4\dot{(}00)_{3}$ + $24\dot{(}02)_{1}$ + $22\dot{(}10)_{1}$ + $9\dot{(}00)_{2}$ + $23\dot{(}02)_{0}$ + $15\dot{(}10)_{0}$ + $12\dot{(}00)_{1}$ + $4\dot{(}00)_{0}$

Affine B3 ⊃ B3 - Level 2 - Class 0

sl 0: $\overset{\bullet}{(}10)_{0}$ **Highest Weight** $\overset{0}{(}10)_{0}$

sl 1: $\overset{\bullet}{(}02)_{1} + \overset{\bullet}{(}00)_{2} + \overset{\bullet}{(}02)_{0} + \overset{\bullet}{(}10)_{0} + \overset{\bullet}{(}00)_{0}$

sl 2: $\overset{\bullet}{(}10)_{2} + \overset{\bullet}{(}12)_{0} + \overset{\bullet}{(}20)_{0} + 3\overset{\bullet}{(}02)_{1} + \overset{\bullet}{(}10)_{1} + 2\overset{\bullet}{(}00)_{2} + 2\overset{\bullet}{(}02)_{0} + 4\overset{\bullet}{(}10)_{0} + \overset{\bullet}{(}00)_{1} + \overset{\bullet}{(}00)_{0}$

sl 3: $\overset{\bullet}{(}12)_{1} + 2\overset{\bullet}{(}02)_{2} + \overset{\bullet}{(}04)_{0} + 3\overset{\bullet}{(}10)_{2} + 3\overset{\bullet}{(}12)_{0} + 4\overset{\bullet}{(}20)_{0} + 9\overset{\bullet}{(}02)_{1} + 4\overset{\bullet}{(}10)_{1} + 6\overset{\bullet}{(}00)_{2} + 7\overset{\bullet}{(}02)_{0}$
$+ 8\overset{\bullet}{(}10)_{0} + 2\overset{\bullet}{(}00)_{1} + 3\overset{\bullet}{(}00)_{0}$

sl 4: $\overset{\bullet}{(}02)_{3} + \overset{\bullet}{(}04)_{1} + \overset{\bullet}{(}30)_{0} + 5\overset{\bullet}{(}12)_{1} + 2\overset{\bullet}{(}20)_{1} + \overset{\bullet}{(}00)_{4} + 6\overset{\bullet}{(}02)_{2} + 4\overset{\bullet}{(}04)_{0} + 10\overset{\bullet}{(}10)_{2} + 11\overset{\bullet}{(}12)_{0}$
$+ 10\overset{\bullet}{(}20)_{0} + 2\overset{\bullet}{(}00)_{3} + 24\overset{\bullet}{(}02)_{1} + 12\overset{\bullet}{(}10)_{1} + 13\overset{\bullet}{(}00)_{2} + 16\overset{\bullet}{(}02)_{0} + 21\overset{\bullet}{(}10)_{0} + 7\overset{\bullet}{(}00)_{1} + 5\overset{\bullet}{(}00)_{0}$

Affine B3 ⊃ B3 - Level 2 - Class 0

sl 0: $\overset{\bullet}{(}00)_{2}$ **Highest Weight** $\overset{0}{(}00)_{2}$

sl 1: $\overset{\bullet}{(}02)_{1} + \overset{\bullet}{(}00)_{2} + \overset{\bullet}{(}10)_{0}$

sl 2: $\overset{\bullet}{(}04)_{0} + \overset{\bullet}{(}10)_{2} + \overset{\bullet}{(}20)_{0} + 2\overset{\bullet}{(}02)_{1} + \overset{\bullet}{(}10)_{1} + 3\overset{\bullet}{(}00)_{2} + \overset{\bullet}{(}02)_{0} + 2\overset{\bullet}{(}10)_{0} + \overset{\bullet}{(}00)_{0}$

sl 3: $\overset{\bullet}{(}12)_{1} + \overset{\bullet}{(}02)_{2} + \overset{\bullet}{(}04)_{0} + 3\overset{\bullet}{(}10)_{2} + 2\overset{\bullet}{(}12)_{0} + 2\overset{\bullet}{(}20)_{0} + 7\overset{\bullet}{(}02)_{1} + 2\overset{\bullet}{(}10)_{1} + 5\overset{\bullet}{(}00)_{2} + 3\overset{\bullet}{(}02)_{0}$
$+ 6\overset{\bullet}{(}10)_{0} + \overset{\bullet}{(}00)_{1} + \overset{\bullet}{(}00)_{0}$

sl 4: $\overset{\bullet}{(}20)_{2} + \overset{\bullet}{(}04)_{1} + 4\overset{\bullet}{(}12)_{1} + \overset{\bullet}{(}20)_{1} + \overset{\bullet}{(}00)_{4} + 4\overset{\bullet}{(}02)_{2} + 4\overset{\bullet}{(}04)_{0} + 8\overset{\bullet}{(}10)_{2} + 6\overset{\bullet}{(}12)_{0} + 8\overset{\bullet}{(}20)_{0}$
$+ \overset{\bullet}{(}00)_{3} + 16\overset{\bullet}{(}02)_{1} + 7\overset{\bullet}{(}10)_{1} + 13\overset{\bullet}{(}00)_{2} + 9\overset{\bullet}{(}02)_{0} + 12\overset{\bullet}{(}10)_{0} + 3\overset{\bullet}{(}00)_{1} + 4\overset{\bullet}{(}00)_{0}$

Affine B3 ⊃ B3 - Level 2 - Class 0

sl 0: $\overset{\bullet}{(}00)_{1}$ **Highest Weight** $\overset{1}{(}00)_{1}$

sl 1: $\overset{\bullet}{(}10)_{1} + \overset{\bullet}{(}02)_{0} + \overset{\bullet}{(}00)_{1}$

sl 2: $\overset{\bullet}{(}12)_{0} + \overset{\bullet}{(}00)_{3} + \overset{\bullet}{(}02)_{1} + 3\overset{\bullet}{(}10)_{1} + 2\overset{\bullet}{(}02)_{0} + \overset{\bullet}{(}10)_{0} + 3\overset{\bullet}{(}00)_{1}$ (continued next page)

Sl 3: $(^{\bullet}_{1}\,20) + 2(^{\bullet}_{2}\,02) + (^{\bullet}_{0}\,04) + 3(^{\bullet}_{0}\,12) + (^{\bullet}_{0}\,20) + 2(^{\bullet}_{3}\,00) + 4(^{\bullet}_{1}\,02) + 8(^{\bullet}_{1}\,10) + (^{\bullet}_{2}\,00) + 7(^{\bullet}_{0}\,02)$
$+ 2(^{\bullet}_{0}\,10) + 6(^{\bullet}_{1}\,00) + (^{\bullet}_{0}\,00)$

Sl 4: $(^{\bullet}_{1}\,04) + (^{\bullet}_{3}\,10) + 2(^{\bullet}_{1}\,12) + 4(^{\bullet}_{1}\,20) + 6(^{\bullet}_{2}\,02) + 2(^{\bullet}_{0}\,04) + 2(^{\bullet}_{2}\,10) + 11(^{\bullet}_{0}\,12) + 3(^{\bullet}_{0}\,20) + 6(^{\bullet}_{3}\,00)$
$+ 13(^{\bullet}_{1}\,02) + 20(^{\bullet}_{1}\,10) + 3(^{\bullet}_{2}\,00) + 15(^{\bullet}_{0}\,02) + 8(^{\bullet}_{0}\,10) + 14(^{\bullet}_{1}\,00) + (^{\bullet}_{0}\,00)$

Affine B3 ⊃ B3 - Level 2 - Class 0

Sl 0: $(^{\bullet}_{0}\,00)$ **Highest Weight** $(^{2}_{0}\,00)$

Sl 1: $(^{\bullet}_{0}\,10)$

Sl 2: $(^{\bullet}_{0}\,20) + (^{\bullet}_{2}\,00) + (^{\bullet}_{0}\,02) + (^{\bullet}_{0}\,10) + (^{\bullet}_{0}\,00)$

Sl 3: $(^{\bullet}_{2}\,10) + (^{\bullet}_{0}\,12) + (^{\bullet}_{0}\,20) + 2(^{\bullet}_{1}\,02) + (^{\bullet}_{2}\,00) + (^{\bullet}_{0}\,02) + 4(^{\bullet}_{0}\,10) + (^{\bullet}_{1}\,00) + (^{\bullet}_{0}\,00)$

Sl 4: $(^{\bullet}_{1}\,12) + (^{\bullet}_{4}\,00) + (^{\bullet}_{2}\,02) + (^{\bullet}_{0}\,04) + 2(^{\bullet}_{2}\,10) + 2(^{\bullet}_{0}\,12) + 5(^{\bullet}_{0}\,20) + 5(^{\bullet}_{1}\,02) + 2(^{\bullet}_{1}\,10) + 5(^{\bullet}_{2}\,00)$
$+ 5(^{\bullet}_{0}\,02) + 6(^{\bullet}_{0}\,10) + (^{\bullet}_{1}\,00) + 4(^{\bullet}_{0}\,00)$

Affine B3 ⊃ B3 - Level 2 - Class 1

Sl 0: $(^{\bullet}_{1}\,01)$ **Highest Weight** $(^{0}_{1}\,01)$

Sl 1: $(^{\bullet}_{2}\,01) + (^{\bullet}_{0}\,03) + (^{\bullet}_{0}\,11) + 2(^{\bullet}_{1}\,01) + (^{\bullet}_{0}\,01)$

Sl 2: $(^{\bullet}_{1}\,03) + 2(^{\bullet}_{1}\,11) + 3(^{\bullet}_{2}\,01) + 2(^{\bullet}_{0}\,03) + 4(^{\bullet}_{0}\,11) + 6(^{\bullet}_{1}\,01) + 3(^{\bullet}_{0}\,01)$

Sl 3: $(^{\bullet}_{2}\,11) + (^{\bullet}_{0}\,13) + (^{\bullet}_{0}\,21) + (^{\bullet}_{3}\,01) + 4(^{\bullet}_{1}\,03) + 8(^{\bullet}_{1}\,11) + 9(^{\bullet}_{2}\,01) + 7(^{\bullet}_{0}\,03) + 12(^{\bullet}_{0}\,11) + 15(^{\bullet}_{1}\,01)$
$+ 8(^{\bullet}_{0}\,01)$

Sl 4: $(^{\bullet}_{1}\,21) + 2(^{\bullet}_{2}\,03) + (^{\bullet}_{0}\,05) + 5(^{\bullet}_{2}\,11) + 4(^{\bullet}_{0}\,13) + 6(^{\bullet}_{0}\,21) + 5(^{\bullet}_{3}\,01) + 14(^{\bullet}_{1}\,03) + 24(^{\bullet}_{1}\,11) + 24(^{\bullet}_{2}\,01)$
$+ 19(^{\bullet}_{0}\,03) + 32(^{\bullet}_{0}\,11) + 37(^{\bullet}_{1}\,01) + 19(^{\bullet}_{0}\,01)$

Sl 5: $(^{\bullet}_{3}\,11) + 3(^{\bullet}_{1}\,13) + 5(^{\bullet}_{1}\,21) + (^{\bullet}_{4}\,01) + 9(^{\bullet}_{2}\,03) + 3(^{\bullet}_{0}\,05) + 18(^{\bullet}_{2}\,11) + 15(^{\bullet}_{0}\,13) + 20(^{\bullet}_{0}\,21) + 16(^{\bullet}_{3}\,01)$
$+ 40(^{\bullet}_{1}\,03) + 67(^{\bullet}_{1}\,11) + 60(^{\bullet}_{2}\,01) + 47(^{\bullet}_{0}\,03) + 79(^{\bullet}_{0}\,11) + 84(^{\bullet}_{1}\,01) + 42(^{\bullet}_{0}\,01)$

Affine B3 ⊃ B3 - Level 2 - Class 1

Sl 0: ${}^{\bullet}_{0}(01)$ **Highest Weight** ${}^{1}_{0}(01)$

Sl 1: ${}^{\bullet}_{0}(11) + {}^{\bullet}_{1}(01) + {}^{\bullet}_{0}(01)$

Sl 2: ${}^{\bullet}_{1}(11) + {}^{\bullet}_{2}(01) + {}^{\bullet}_{0}(03) + 3\,{}^{\bullet}_{0}(11) + 3\,{}^{\bullet}_{1}(01) + 3\,{}^{\bullet}_{0}(01)$

Sl 3: ${}^{\bullet}_{0}(21) + {}^{\bullet}_{3}(01) + 2\,{}^{\bullet}_{1}(03) + 4\,{}^{\bullet}_{1}(11) + 4\,{}^{\bullet}_{2}(01) + 4\,{}^{\bullet}_{0}(03) + 8\,{}^{\bullet}_{0}(11) + 9\,{}^{\bullet}_{1}(01) + 7\,{}^{\bullet}_{0}(01)$

Sl 4: ${}^{\bullet}_{2}(03) + 2\,{}^{\bullet}_{2}(11) + 2\,{}^{\bullet}_{0}(13) + 4\,{}^{\bullet}_{0}(21) + 3\,{}^{\bullet}_{3}(01) + 7\,{}^{\bullet}_{1}(03) + 14\,{}^{\bullet}_{1}(11) + 13\,{}^{\bullet}_{2}(01) + 11\,{}^{\bullet}_{0}(03) + 22\,{}^{\bullet}_{0}(11)$
$+ 23\,{}^{\bullet}_{1}(01) + 15\,{}^{\bullet}_{0}(01)$

Sl 5: ${}^{\bullet}_{1}(13) + 2\,{}^{\bullet}_{1}(21) + {}^{\bullet}_{4}(01) + 5\,{}^{\bullet}_{2}(03) + {}^{\bullet}_{0}(05) + 9\,{}^{\bullet}_{2}(11) + 8\,{}^{\bullet}_{0}(13) + 13\,{}^{\bullet}_{0}(21) + 10\,{}^{\bullet}_{3}(01) + 23\,{}^{\bullet}_{1}(03)$
$+ 41\,{}^{\bullet}_{1}(11) + 35\,{}^{\bullet}_{2}(01) + 29\,{}^{\bullet}_{0}(03) + 54\,{}^{\bullet}_{0}(11) + 55\,{}^{\bullet}_{1}(01) + 32\,{}^{\bullet}_{0}(01)$

Affine B3 ⊃ A1 + A1 + A1 - Level 1 - Class 0

Sl 0: ${}^{0}_{1}(\bullet 0)$ **Highest Weight** ${}^{0}_{1}(00)$

Sl 1: ${}^{1}_{0}(\bullet 2)$

Sl 2: ${}^{2}_{1}(\bullet 0) + {}^{0}_{1}(\bullet 2) + {}^{0}_{1}(\bullet 0)$

Sl 3: ${}^{1}_{2}(\bullet 2) + 2\,{}^{1}_{0}(\bullet 2) + {}^{1}_{0}(\bullet 0)$

Sl 4: ${}^{2}_{1}(\bullet 2) + {}^{0}_{1}(\bullet 4) + 2\,{}^{2}_{1}(\bullet 0) + {}^{0}_{3}(\bullet 0) + 2\,{}^{0}_{1}(\bullet 2) + 3\,{}^{0}_{1}(\bullet 0)$

Sl 5: ${}^{3}_{0}(\bullet 2) + 3\,{}^{1}_{2}(\bullet 2) + {}^{1}_{0}(\bullet 4) + {}^{1}_{2}(\bullet 0) + 5\,{}^{1}_{0}(\bullet 2) + 2\,{}^{1}_{0}(\bullet 0)$

Sl 6: ${}^{2}_{1}(\bullet 4) + {}^{2}_{3}(\bullet 0) + 3\,{}^{2}_{1}(\bullet 2) + {}^{0}_{3}(\bullet 2) + 2\,{}^{0}_{1}(\bullet 4) + 5\,{}^{2}_{1}(\bullet 0) + {}^{0}_{3}(\bullet 0) + 6\,{}^{0}_{1}(\bullet 2) + 6\,{}^{0}_{1}(\bullet 0)$

Sl 7: ${}^{3}_{2}(\bullet 2) + {}^{1}_{2}(\bullet 4) + 2\,{}^{3}_{0}(\bullet 2) + 8\,{}^{1}_{2}(\bullet 2) + 3\,{}^{1}_{0}(\bullet 4) + {}^{3}_{0}(\bullet 0) + 3\,{}^{1}_{2}(\bullet 0) + 11\,{}^{1}_{0}(\bullet 2) + 5\,{}^{1}_{0}(\bullet 0)$

Affine B3 ⊃ A1 + A1 + A1 - Level 1 - Class 1

```
          0                        0
Sl 0:    ( •1)  Highest Weight   ( 01)
          0                        0

          1
Sl 1:    ( •1)
          1

          2          0          0          0
Sl 2:    ( •1) +    ( •1) +    ( •3) +    ( •1)
          0          2          0          0

          1          1
Sl 3:    ( •3) +   3( •1)
          1          1

          2          2          0          2          0          0          0
Sl 4:    ( •1) +    ( •3) +    ( •3) +   2( •1) +   2( •1) +   2( •3) +   4( •1)
          2          0          2          0          2          0          0

          3          1          1          1
Sl 5:    ( •1) +    ( •1) +   4( •3) +   7( •1)
          1          3          1          1

          2          2          2          0          0          2          0          0          0
Sl 6:    ( •3) +   3( •1) +   3( •3) +   3( •3) +    ( •5) +   6( •1) +   6( •1) +   5( •3) +   8( •1)
          2          2          0          2          0          0          2          0          0

          3          1          1          3          1          1          1
Sl 7:    ( •3) +    ( •3) +    ( •5) +   3( •1) +   3( •1) +  10( •3) +  17( •1)
          1          3          1          1          3          1          1
```

Affine B3 ⊃ A1 + A1 + A1 - Level 2 - Class 0

```
          0                        0
Sl 0:    ( •2)  Highest Weight   ( 02)
          0                        0

          1          1
Sl 1:    ( •2) +    ( •0)
          1          1

          2          2          0          0          2          0          0          0
Sl 2:    ( •2) +    ( •2) +    ( •2) +    ( •4) +    ( •0) +    ( •0) +   2( •2) +    ( •0)
          2          0          2          0          0          2          0          0

          3          1          1          3          1          1          1
Sl 3:    ( •2) +    ( •2) +   2( •4) +    ( •0) +    ( •0) +   5( •2) +   4( •0)
          1          3          1          1          3          1          1

          2          4          2          2          0          0          0          2          2          0
Sl 4:    ( •4) +    ( •2) +   5( •2) +   2( •4) +    ( •2) +   2( •4) +    ( •6) +   3( •0) +   6( •2) +   6( •2)
          2          0          2          0          4          2          0          2          0          2

          0          2          0          0          0
    +    3( •4) +   3( •0) +   3( •0) +   8( •2) +   4( •0)
          0          0          2          0          0

          3          3          1          1          3          3          1          1          3          1
Sl 5:    ( •2) +   2( •4) +   2( •4) +    ( •6) +    ( •0) +   6( •2) +   6( •2) +  10( •4) +   5( •0) +   5( •0)
          3          1          3          1          3          1          3          1          1          3

          1          1
    +   21( •2) +  14( •0)
          1          1
```

Affine B3 ⊃ A1 + A1 + A1 - Level 2 - Class 0

```
          0                        0
Sl 0:    ( •0)  Highest Weight   ( 10)
          0                        0

          1
Sl 1:    ( •2)
          1

          2          2          0          0          2          0          0          0
Sl 2:    ( •0) +    ( •2) +    ( •2) +    ( •4) +    ( •0) +    ( •0) +    ( •2) +    ( •0)
          2          0          2          0          0          2          0          0
```

(continued next page)

sl 3: $({}^{3}_{1}\cdot 2) + ({}^{1}_{3}\cdot 2) + ({}^{1}_{1}\cdot 4) + 5({}^{1}_{1}\cdot 2) + 2({}^{1}_{1}\cdot 0)$

sl 4: $({}^{2}_{2}\cdot 4) + 3({}^{2}_{2}\cdot 2) + 2({}^{2}_{0}\cdot 4) + 2({}^{0}_{2}\cdot 4) + ({}^{4}_{0}\cdot 0) + 4({}^{2}_{2}\cdot 0) + 4({}^{2}_{0}\cdot 2) + ({}^{0}_{4}\cdot 0) + 4({}^{0}_{2}\cdot 2) + 3({}^{0}_{0}\cdot 4)$

$+ 4({}^{2}_{0}\cdot 0) + 4({}^{0}_{2}\cdot 0) + 5({}^{0}_{0}\cdot 2) + 5({}^{0}_{0}\cdot 0)$

sl 5: $({}^{3}_{3}\cdot 2) + ({}^{3}_{1}\cdot 4) + ({}^{1}_{3}\cdot 4) + ({}^{1}_{1}\cdot 6) + 6({}^{3}_{1}\cdot 2) + 6({}^{1}_{3}\cdot 2) + 7({}^{1}_{1}\cdot 4) + 2({}^{3}_{1}\cdot 0) + 2({}^{1}_{3}\cdot 0) + 20({}^{1}_{1}\cdot 2)$

$+ 8({}^{1}_{1}\cdot 0)$

Affine B3 ⊃ A1 + A1 + A1 - Level 2 - Class 0

sl 0: $({}^{0}_{2}\cdot 0)$ **Highest Weight** $({}^{0}_{2}\,00)$

sl 1: $({}^{1}_{1}\cdot 2)$

sl 2: $({}^{2}_{0}\cdot 4) + ({}^{2}_{2}\cdot 0) + ({}^{0}_{2}\cdot 2) + ({}^{2}_{0}\cdot 0) + ({}^{0}_{2}\cdot 0) + ({}^{0}_{0}\cdot 2) + ({}^{0}_{0}\cdot 0)$

sl 3: $({}^{3}_{1}\cdot 2) + ({}^{1}_{3}\cdot 2) + ({}^{1}_{1}\cdot 4) + 4({}^{1}_{1}\cdot 2) + ({}^{1}_{1}\cdot 0)$

sl 4: $({}^{2}_{2}\cdot 4) + ({}^{4}_{2}\cdot 0) + 2({}^{2}_{2}\cdot 2) + 2({}^{2}_{0}\cdot 4) + 2({}^{0}_{2}\cdot 4) + 3({}^{2}_{2}\cdot 0) + 3({}^{2}_{0}\cdot 2) + ({}^{0}_{4}\cdot 0) + 3({}^{0}_{2}\cdot 2) + 2({}^{0}_{0}\cdot 4)$

$+ 3({}^{2}_{0}\cdot 0) + 5({}^{0}_{2}\cdot 0) + 3({}^{0}_{0}\cdot 2) + 3({}^{0}_{0}\cdot 0)$

sl 5: $({}^{3}_{3}\cdot 2) + ({}^{3}_{1}\cdot 4) + ({}^{1}_{3}\cdot 4) + ({}^{1}_{1}\cdot 6) + 5({}^{3}_{1}\cdot 2) + 5({}^{1}_{3}\cdot 2) + 5({}^{1}_{1}\cdot 4) + ({}^{3}_{1}\cdot 0) + ({}^{1}_{3}\cdot 0) + 15({}^{1}_{1}\cdot 2)$

$+ 5({}^{1}_{1}\cdot 0)$

Affine B3 ⊃ A1 + A1 + A1 - Level 2 - Class 0

sl 0: $({}^{1}_{1}\cdot 0)$ **Highest Weight** $({}^{1}_{1}\,00)$

sl 1: $({}^{2}_{0}\cdot 2) + ({}^{0}_{2}\cdot 2) + ({}^{0}_{0}\cdot 2)$

sl 2: $({}^{1}_{1}\cdot 4) + ({}^{3}_{1}\cdot 0) + ({}^{1}_{3}\cdot 0) + 2({}^{1}_{1}\cdot 2) + 3({}^{1}_{1}\cdot 0)$

sl 3: $3({}^{2}_{2}\cdot 2) + ({}^{2}_{0}\cdot 4) + ({}^{0}_{2}\cdot 4) + 4({}^{2}_{0}\cdot 2) + 4({}^{0}_{2}\cdot 2) + ({}^{0}_{0}\cdot 4) + ({}^{2}_{0}\cdot 0) + ({}^{0}_{2}\cdot 0) + 4({}^{0}_{0}\cdot 2) + 2({}^{0}_{0}\cdot 0)$

sl 4: $({}^{3}_{1}\cdot 4) + ({}^{1}_{3}\cdot 4) + ({}^{3}_{3}\cdot 0) + 2({}^{3}_{1}\cdot 2) + 2({}^{1}_{3}\cdot 2) + 6({}^{1}_{1}\cdot 4) + 4({}^{3}_{1}\cdot 0) + 4({}^{1}_{3}\cdot 0) + 10({}^{1}_{1}\cdot 2) + 11({}^{1}_{1}\cdot 0)$

(continued next page)

```
              4           2           2           2           0           4           2           2           0           0
Sl 5:        ( •2) +     ( •2) +    3( •4) +     ( •6) +     ( •6) +    2( •2) +   13( •2) +    5( •4) +    2( •2) +    5( •4)
              2           4           2           0           2           0           2           0           4           2

              0           2           2           0           0           2           0           0           0
    +        ( •6) +    4( •0) +   16( •2) +   16( •2) +    6( •4) +    5( •0) +    5( •0) +   15( •2) +    5( •0)
              0           2           0           2           0           0           2           0           0

              3           3           3           1           1           5           3           3           1           1
Sl 6:        ( •4) +    2( •2) +    7( •4) +    7( •4) +    3( •6) +     ( •0) +    5( •0) +   12( •2) +     ( •0) +   12( •2)
              3           3           1           3           1           1           3           1           5           3

              1           3           1           1           1
    +       24( •4) +   14( •0) +   14( •0) +   39( •2) +   34( •0)
              1           1           3           1           1
```

Affine B3 $\supset$ A1 + A1 + A1 - Level 2 - Class 1

```
              0                           0
Sl 0:        ( •1)  Highest Weight ( 01)
              1                           1

              1           1           1
Sl 1:        ( •1) +     ( •3) +     ( •1)
              2           0           0

              2           2           0           0           0
Sl 2:        ( •3) +    2( •1) +     ( •1) +    2( •3) +    3( •1)
              1           1           3           1           1

              3           3           1           1           3           1           1           1
Sl 3:        ( •1) +     ( •3) +    3( •3) +     ( •5) +     ( •1) +    5( •1) +    4( •3) +    6( •1)
              2           0           2           0           0           2           0           0

              2           2           4           2           2           0           0           2           0           0
Sl 4:        ( •3) +     ( •5) +     ( •1) +    2( •1) +    7( •3) +    2( •3) +    2( •5) +   10( •1) +    4( •1) +    9( •3)
              3           1           1           3           1           3           1           1           3           1

              0
    +       12( •1)
              1

              3           3           1           1           3           3           1           1           1           3
Sl 5:       3( •3) +     ( •5) +     ( •3) +    3( •5) +    6( •1) +    5( •3) +    2( •1) +   15( •3) +    5( •5) +    7( •1)
              2           0           4           2           2           0           4           2           0           0

              1           1           1
    +       21( •1) +   18( •3) +   21( •1)
              2           0           0
```

Affine B3 $\supset$ A3 - Level 1 - Class 0

```
              0                           0
Sl 0:        ( 0•)  Highest Weight ( 00)
              1                           1

              1
Sl 1:        ( 0•)
              0

              1           0
Sl 2:        ( 1•) +     ( 0•)
              0           1

              0           1
Sl 3:        ( 1•) +    2( 0•)
              1           0

              1           1           0
Sl 4:        ( 0•) +    2( 1•) +    3( 0•)
              2           0           1

              2           0           1
Sl 5:        ( 0•) +    3( 1•) +    4( 0•)
              1           1           0

              0           3           1           1           0
Sl 6:        ( 2•) +     ( 0•) +    2( 0•) +    5( 1•) +    6( 0•)
              1           0           2           0           1
```

(continued next page)

```
            1          2          0          0          1
sl 7:     ( 2•) +   3( 0•) +    ( 0•) +   7( 1•) +   8( 0•)
            0          1          3          1          0

            2          0          3          1          1          0
sl 8:     ( 1•) +   2( 2•) +    ( 0•) +   6( 0•) +  11( 1•) +  12( 0•)
            1          1          0          2          0          1
```

Affine B3 ⊃ A3 - Level 1 - Class 1

```
            0                          0
sl 0:     ( 0•)   Highest Weight ( 01)
            0                          0

            0
sl 1:     ( 1•)
            0

            1          0
sl 2:     ( 0•) +    ( 0•)
            1          0

            2          0          0
sl 3:     ( 0•) +    ( 0•) +   2( 1•)
            0          2          0

            0          1          0
sl 4:     ( 2•) +   3( 0•) +   2( 0•)
            0          1          0

            1          2          0          0
sl 5:     ( 1•) +   2( 0•) +   2( 0•) +   5( 1•)
            1          0          2          0

            2          0          0          1          0
sl 6:     ( 1•) +    ( 1•) +   2( 2•) +   7( 0•) +   5( 0•)
            0          2          0          1          0

            1          2          0          0
sl 7:    4( 1•) +   5( 0•) +   5( 0•) +  10( 1•)
            1          0          2          0

            2          2          0          0          1          0
sl 8:     ( 0•) +   3( 1•) +   3( 1•) +   6( 2•) +  15( 0•) +   9( 0•)
            2          0          2          0          1          0
```

Affine B3 ⊃ A3 - Level 2 - Class 0

```
            0                          0
sl 0:     ( 0•)   Highest Weight ( 02)
            0                          0

            0
sl 1:     ( 1•)
            0

            0          1          0
sl 2:     ( 2•) +    ( 0•) +    ( 0•)
            0          1          0

            1          2          0          0
sl 3:     ( 1•) +    ( 0•) +    ( 0•) +   3( 1•)
            1          0          2          0

            2          2          0          0          1          0
sl 4:     ( 0•) +    ( 1•) +    ( 1•) +   3( 2•) +   5( 0•) +   3( 0•)
            2          0          2          0          1          0

            3          1          0          1          2          0          0
sl 5:     ( 0•) +    ( 0•) +    ( 3•) +   6( 1•) +   4( 0•) +   4( 0•) +   9( 1•)
            1          3          0          1          0          2          0

            1          4          2          0          2          0          0          1          0
sl 6:    2( 2•) +    ( 0•) +   4( 0•) +    ( 0•) +   6( 1•) +   6( 1•) +  10( 2•) +  16( 0•) +   8( 0•)
            1          0          2          4          0          2          0          1          0

            2          2          0          3          1          0          1          2          0          0
sl 7:     ( 1•) +   2( 2•) +   2( 2•) +   5( 0•) +   5( 0•) +   4( 3•) +  23( 1•) +  14( 0•) +  14( 0•) +  24( 1•)
            2          0          2          1          3          0          1          0          2          0
```

Affine B3 ⊃ A3 - Level 2 - Class 0

```
            0                       0
Sl 0:     ( 1•)  Highest Weight   ( 10)
            0                       0

            1          0
Sl 1:     ( 0•) +    ( 0•)
            1          0

            1          2          0          0
Sl 2:     ( 1•) +    ( 0•) +    ( 0•) +   2( 1•)
            1          0          2          0

            2          0          0          1          0
Sl 3:     ( 1•) +    ( 1•) +   2( 2•) +   4( 0•) +   2( 0•)
            0          2          0          1          0

            3          1          0          1          2          0          0
Sl 4:     ( 0•) +    ( 0•) +    ( 3•) +   5( 1•) +   4( 0•) +   4( 0•) +   7( 1•)
            1          3          0          1          0          2          0

            1          2          2          0          0          1          0
Sl 5:     ( 2•) +   3( 0•) +   5( 1•) +   5( 1•) +   7( 2•) +  14( 0•) +   6( 0•)
            1          2          0          2          0          1          0

            2          2          0          3          1          0          1          2          0          0
Sl 6:     ( 1•) +   2( 2•) +   2( 2•) +   4( 0•) +   4( 0•) +   3( 3•) +  20( 1•) +  13( 0•) +  13( 0•) +  20( 1•)
            2          0          2          1          3          0          1          0          2          0

            3          1          1          4          2          0          2          0          0          1
Sl 7:     ( 1•) +    ( 1•) +   8( 2•) +   2( 0•) +  12( 0•) +   2( 0•) +  19( 1•) +  19( 1•) +  23( 2•) +  41( 0•)
            1          3          1          0          2          4          0          2          0          1

            0
   +  15( 0•)
            0
```

Affine B3 ⊃ A3 - Level 2 - Class 0

```
            0                       0
Sl 0:     ( 0•)  Highest Weight   ( 00)
            2                       2

            1
Sl 1:     ( 0•)
            1

            1          2          0          0
Sl 2:     ( 1•) +    ( 0•) +    ( 0•) +    ( 1•)
            1          0          2          0

            2          0          0          1          0
Sl 3:     ( 1•) +    ( 1•) +    ( 2•) +   3( 0•) +    ( 0•)
            0          2          0          1          0

            2          1          1          2          0          0
Sl 4:     ( 2•) +    ( 0•) +   4( 1•) +   3( 0•) +   4( 0•) +   4( 1•)
            0          3          1          0          2          0

            1          2          2          0          0          1          0
Sl 5:     ( 2•) +   2( 0•) +   4( 1•) +   4( 1•) +   4( 2•) +  10( 0•) +   3( 0•)
            1          2          0          2          0          1          0

            2          2          0          3          1          0          1          2          0          0
Sl 6:     ( 1•) +   2( 2•) +   2( 2•) +   3( 0•) +   3( 0•) +   2( 3•) +  14( 1•) +  10( 0•) +  10( 0•) +  12( 1•)
            2          0          2          1          3          0          1          0          2          0

            3          1          1          4          2          0          2          0          0          1
Sl 7:     ( 1•) +    ( 1•) +   6( 2•) +    ( 0•) +   8( 0•) +    ( 0•) +  14( 1•) +  14( 1•) +  14( 2•) +  28( 0•)
            1          3          1          0          2          4          0          2          0          1

            0
   +   8( 0•)
            0
```

Affine B3 ⊃ A3 - Level 2 - Class 0

```
          1                       1
sl 0:    ( 0•)   Highest Weight ( 00)
          1                       1

          2           0           0
sl 1:    ( 0•) +     ( 0•) +     ( 1•)
          0           2           0

          2           0           0           1           0
sl 2:    ( 1•) +     ( 1•) +     ( 2•) +    3( 0•) +     ( 0•)
          0           2           0           1           0

          1           2           0           0
sl 3:   3( 1•) +    3( 0•) +    3( 0•) +    4( 1•)
          1           0           2           0

          1           2           2           0           0           1           0
sl 4:    ( 2•) +    2( 0•) +    4( 1•) +    4( 1•) +    4( 2•) +   11( 0•) +    4( 0•)
          1           2           0           2           0           1           0

          2           0           3           1           0           1           2           0           0
sl 5:    ( 2•) +     ( 2•) +    2( 0•) +    2( 0•) +     ( 3•) +   13( 1•) +   10( 0•) +   10( 0•) +   13( 1•)
          0           2           1           3           0           1           0           2           0

          3           1           1           4           2           0           2           0           0           1
sl 6:    ( 1•) +     ( 1•) +    6( 2•) +     ( 0•) +    8( 0•) +     ( 0•) +   15( 1•) +   15( 1•) +   16( 2•) +   32( 0•)
          1           3           1           0           2           4           0           2           0           1

        0
   + 10( 0•)
        0

          2           2           0           3           1           0           1           2           0           0
sl 7:   3( 1•) +    6( 2•) +    6( 2•) +   10( 0•) +   10( 0•) +    7( 3•) +   44( 1•) +   29( 0•) +   29( 0•) +   37( 1•)
          2           0           2           1           3           0           1           0           2           0
```

Affine B3 ⊃ A3 - Level 2 - Class 1

```
          0                       0
sl 0:    ( 0•)   Highest Weight ( 01)
          1                       1

          0           1
sl 1:    ( 1•) +     ( 0•)
          1           0

          1           1           0
sl 2:    ( 0•) +    2( 1•) +    2( 0•)
          2           0           1

          1           2           0           0           1
sl 3:    ( 2•) +    2( 0•) +     ( 0•) +    4( 1•) +    4( 0•)
          0           1           3           1           0

          2           0           3           1           1           0
sl 4:    ( 1•) +    2( 2•) +     ( 0•) +    5( 0•) +    8( 1•) +    7( 0•)
          1           1           0           2           0           1

          3           1           1           2           0           0           1
sl 5:    ( 1•) +    3( 1•) +    5( 2•) +    9( 0•) +    3( 0•) +   15( 1•) +   12( 0•)
          0           2           0           1           3           1           0

          2           1           2           0           0           3           1           1           0
sl 6:    ( 0•) +     ( 3•) +    7( 1•) +    2( 1•) +   10( 2•) +    6( 0•) +   18( 0•) +   26( 1•) +   21( 0•)
          3           0           1           3           1           0           2           0           1

          2           3           1           0           3           1           1           2           0           0
sl 7:    ( 2•) +    2( 0•) +     ( 0•) +    2( 3•) +    5( 1•) +   15( 1•) +   20( 2•) +   32( 0•) +   11( 0•) +   46( 1•)
          1           2           4           1           0           2           0           1           3           1

        1
   + 33( 0•)
        0
```

Affine B4 Branching Rules

Affine B4 ⊃ B4 - Level 1 - Class 0

Sl 0: $({}^{\bullet}000)_{1}$ **Highest Weight** $({}^{0}000)_{1}$

Sl 1: $({}^{\bullet}010)_{0} + ({}^{\bullet}000)_{1}$

Sl 2: $({}^{\bullet}100)_{1} + ({}^{\bullet}002)_{0} + ({}^{\bullet}010)_{0} + 2({}^{\bullet}000)_{1}$

Sl 3: $({}^{\bullet}002)_{1} + 2({}^{\bullet}100)_{1} + ({}^{\bullet}002)_{0} + 3({}^{\bullet}010)_{0} + ({}^{\bullet}100)_{0} + 3({}^{\bullet}000)_{1}$

Sl 4: $({}^{\bullet}110)_{0} + 2({}^{\bullet}002)_{1} + ({}^{\bullet}010)_{1} + ({}^{\bullet}000)_{3} + 4({}^{\bullet}100)_{1} + 3({}^{\bullet}002)_{0} + 5({}^{\bullet}010)_{0} + ({}^{\bullet}100)_{0} + 6({}^{\bullet}000)_{1}$
$+ ({}^{\bullet}000)_{0}$

Sl 5: $({}^{\bullet}010)_{2} + ({}^{\bullet}102)_{0} + 2({}^{\bullet}110)_{0} + 5({}^{\bullet}002)_{1} + 2({}^{\bullet}010)_{1} + ({}^{\bullet}000)_{3} + 8({}^{\bullet}100)_{1} + 5({}^{\bullet}002)_{0} + 10({}^{\bullet}010)_{0}$
$+ ({}^{\bullet}000)_{2} + 3({}^{\bullet}100)_{0} + 9({}^{\bullet}000)_{1} + ({}^{\bullet}000)_{0}$

Sl 6: $({}^{\bullet}012)_{0} + ({}^{\bullet}002)_{2} + ({}^{\bullet}200)_{1} + 2({}^{\bullet}010)_{2} + 2({}^{\bullet}102)_{0} + 5({}^{\bullet}110)_{0} + 10({}^{\bullet}002)_{1} + ({}^{\bullet}200)_{0} + 5({}^{\bullet}010)_{1}$
$+ 3({}^{\bullet}000)_{3} + 15({}^{\bullet}100)_{1} + 11({}^{\bullet}002)_{0} + 16({}^{\bullet}010)_{0} + 2({}^{\bullet}000)_{2} + 5({}^{\bullet}100)_{0} + 16({}^{\bullet}000)_{1} + 2({}^{\bullet}000)_{0}$

Sl 7: $({}^{\bullet}102)_{1} + 2({}^{\bullet}012)_{0} + 2({}^{\bullet}002)_{2} + 2({}^{\bullet}200)_{1} + ({}^{\bullet}020)_{0} + 6({}^{\bullet}010)_{2} + 6({}^{\bullet}102)_{0} + 10({}^{\bullet}110)_{0} + ({}^{\bullet}100)_{2}$
$+ 20({}^{\bullet}002)_{1} + 2({}^{\bullet}200)_{0} + 10({}^{\bullet}010)_{1} + 5({}^{\bullet}000)_{3} + 26({}^{\bullet}100)_{1} + 18({}^{\bullet}002)_{0} + 30({}^{\bullet}010)_{0} + 4({}^{\bullet}000)_{2} + 11({}^{\bullet}100)_{0}$
$+ 24({}^{\bullet}000)_{1} + 3({}^{\bullet}000)_{0}$

Affine B4 ⊃ B4 - Level 1 - Class 0

Sl 0: $({}^{\bullet}000)_{0}$ **Highest Weight** $({}^{1}000)_{0}$

Sl 1: $({}^{\bullet}100)_{0}$

Sl 2: $({}^{\bullet}002)_{0} + ({}^{\bullet}000)_{2} + ({}^{\bullet}100)_{0} + ({}^{\bullet}000)_{0}$

Sl 3: $({}^{\bullet}010)_{1} + ({}^{\bullet}002)_{0} + ({}^{\bullet}010)_{0} + ({}^{\bullet}000)_{2} + 3({}^{\bullet}100)_{0} + ({}^{\bullet}000)_{0}$ (continued next page)

Sl 4: $(002)^{\bullet}_{1} + (200)^{\bullet}_{0} + 2(010)^{\bullet}_{1} + 3(002)^{\bullet}_{0} + (010)^{\bullet}_{0} + 3(000)^{\bullet}_{2} + 4(100)^{\bullet}_{0} + (000)^{\bullet}_{1} + 3(000)^{\bullet}_{0}$

Sl 5: $(102)^{\bullet}_{0} + (100)^{\bullet}_{2} + 2(002)^{\bullet}_{1} + (200)^{\bullet}_{0} + 5(010)^{\bullet}_{1} + (100)^{\bullet}_{1} + 5(002)^{\bullet}_{0} + 3(010)^{\bullet}_{0} + 4(000)^{\bullet}_{2}$

$+ 9(100)^{\bullet}_{0} + (000)^{\bullet}_{1} + 3(000)^{\bullet}_{0}$

Sl 6: $(002)^{\bullet}_{2} + (020)^{\bullet}_{0} + 2(102)^{\bullet}_{0} + (110)^{\bullet}_{0} + 2(100)^{\bullet}_{2} + 5(002)^{\bullet}_{1} + 4(200)^{\bullet}_{0} + 9(010)^{\bullet}_{1} + 2(100)^{\bullet}_{1}$

$+ 11(002)^{\bullet}_{0} + 5(010)^{\bullet}_{0} + 9(000)^{\bullet}_{2} + 13(100)^{\bullet}_{0} + 3(000)^{\bullet}_{1} + 7(000)^{\bullet}_{0}$

Sl 7: $(110)^{\bullet}_{1} + (012)^{\bullet}_{0} + 2(002)^{\bullet}_{2} + (020)^{\bullet}_{0} + (010)^{\bullet}_{2} + 6(102)^{\bullet}_{0} + 2(110)^{\bullet}_{0} + 5(100)^{\bullet}_{2} + 10(002)^{\bullet}_{1}$

$+ 6(200)^{\bullet}_{0} + 19(010)^{\bullet}_{1} + 5(100)^{\bullet}_{1} + 17(002)^{\bullet}_{0} + 11(010)^{\bullet}_{0} + 13(000)^{\bullet}_{2} + 25(100)^{\bullet}_{0} + 5(000)^{\bullet}_{1} + 8(000)^{\bullet}_{0}$

Affine B4 ⊃ B4 - Level 1 - Class 1

Sl 0: $(001)^{\bullet}_{0}$ **Highest Weight** $(001)^{0}_{0}$

Sl 1: $(001)^{\bullet}_{1} + (001)^{\bullet}_{0}$

Sl 2: $(101)^{\bullet}_{0} + 2(001)^{\bullet}_{1} + 2(001)^{\bullet}_{0}$

Sl 3: $(011)^{\bullet}_{0} + (001)^{\bullet}_{2} + 2(101)^{\bullet}_{0} + 4(001)^{\bullet}_{1} + 4(001)^{\bullet}_{0}$

Sl 4: $(101)^{\bullet}_{1} + (003)^{\bullet}_{0} + 2(011)^{\bullet}_{0} + 2(001)^{\bullet}_{2} + 5(101)^{\bullet}_{0} + 8(001)^{\bullet}_{1} + 7(001)^{\bullet}_{0}$

Sl 5: $(011)^{\bullet}_{1} + 3(101)^{\bullet}_{1} + 2(003)^{\bullet}_{0} + 5(011)^{\bullet}_{0} + 5(001)^{\bullet}_{2} + 10(101)^{\bullet}_{0} + 15(001)^{\bullet}_{1} + 12(001)^{\bullet}_{0}$

Sl 6: $(003)^{\bullet}_{1} + (201)^{\bullet}_{0} + 3(011)^{\bullet}_{1} + (001)^{\bullet}_{3} + 7(101)^{\bullet}_{1} + 4(003)^{\bullet}_{0} + 11(011)^{\bullet}_{0} + 10(001)^{\bullet}_{2} + 20(101)^{\bullet}_{0}$

$+ 27(001)^{\bullet}_{1} + 21(001)^{\bullet}_{0}$

Sl 7: $(111)^{\bullet}_{0} + (101)^{\bullet}_{2} + 3(003)^{\bullet}_{1} + 2(201)^{\bullet}_{0} + 8(011)^{\bullet}_{1} + 2(001)^{\bullet}_{3} + 16(101)^{\bullet}_{1} + 9(003)^{\bullet}_{0} + 21(011)^{\bullet}_{0}$

$+ 20(001)^{\bullet}_{2} + 37(101)^{\bullet}_{0} + 48(001)^{\bullet}_{1} + 34(001)^{\bullet}_{0}$

Sl 8: $(011)^{\bullet}_{2} + (103)^{\bullet}_{0} + 3(111)^{\bullet}_{0} + 3(101)^{\bullet}_{2} + 7(003)^{\bullet}_{1} + 6(201)^{\bullet}_{0} + 18(011)^{\bullet}_{1} + 5(001)^{\bullet}_{3} + 33(101)^{\bullet}_{1}$

$+ 18(003)^{\bullet}_{0} + 40(011)^{\bullet}_{0} + 37(001)^{\bullet}_{2} + 67(101)^{\bullet}_{0} + 82(001)^{\bullet}_{1} + 56(001)^{\bullet}_{0}$

Affine B4 ⊃ A1 + A1 + C2 - Level 1 - Class 0

Sl 0: $({}^{1}_{0}\cdot 00)$ **Highest Weight** $({}^{1}_{0}000)$

Sl 1: $({}^{0}_{1}\cdot 10)$

Sl 2: $({}^{1}_{0}\cdot 02) + ({}^{1}_{2}\cdot 00) + ({}^{1}_{0}\cdot 00)$

Sl 3: $({}^{2}_{1}\cdot 10) + ({}^{0}_{1}\cdot 02) + 2({}^{0}_{1}\cdot 10)$

Sl 4: $({}^{1}_{2}\cdot 02) + ({}^{1}_{0}\cdot 20) + 2({}^{1}_{0}\cdot 02) + ({}^{1}_{0}\cdot 10) + ({}^{3}_{0}\cdot 00) + 2({}^{1}_{2}\cdot 00) + 3({}^{1}_{0}\cdot 00)$

Sl 5: $({}^{0}_{1}\cdot 12) + ({}^{2}_{1}\cdot 02) + 3({}^{2}_{1}\cdot 10) + ({}^{0}_{3}\cdot 10) + 2({}^{0}_{1}\cdot 02) + 5({}^{0}_{1}\cdot 10) + ({}^{0}_{1}\cdot 00)$

Sl 6: $({}^{1}_{2}\cdot 20) + ({}^{1}_{0}\cdot 12) + ({}^{3}_{0}\cdot 02) + 3({}^{1}_{2}\cdot 02) + 2({}^{1}_{0}\cdot 20) + ({}^{1}_{2}\cdot 10) + 6({}^{1}_{0}\cdot 02) + ({}^{3}_{2}\cdot 00) + 2({}^{1}_{0}\cdot 10)$
$+ ({}^{3}_{0}\cdot 00) + 5({}^{1}_{2}\cdot 00) + 6({}^{1}_{0}\cdot 00)$

Sl 7: $({}^{2}_{1}\cdot 12) + 3({}^{0}_{1}\cdot 12) + ({}^{2}_{3}\cdot 10) + 3({}^{2}_{1}\cdot 02) + ({}^{0}_{3}\cdot 02) + ({}^{0}_{1}\cdot 20) + 8({}^{2}_{1}\cdot 10) + 2({}^{0}_{3}\cdot 10) + 6({}^{0}_{1}\cdot 02)$
$+ 11({}^{0}_{1}\cdot 10) + ({}^{2}_{1}\cdot 00) + 2({}^{0}_{1}\cdot 00)$

Affine B4 ⊃ A1 + A1 + C2 - Level 1 - Class 1

Sl 0: $({}^{0}_{0}\cdot 01)$ **Highest Weight** $({}^{0}_{0}001)$

Sl 1: $({}^{1}_{1}\cdot 01)$

Sl 2: $({}^{0}_{0}\cdot 11) + ({}^{2}_{0}\cdot 01) + ({}^{0}_{2}\cdot 01) + ({}^{0}_{0}\cdot 01)$

Sl 3: $({}^{1}_{1}\cdot 11) + 3({}^{1}_{1}\cdot 01)$

Sl 4: $({}^{2}_{0}\cdot 11) + ({}^{0}_{2}\cdot 11) + ({}^{0}_{0}\cdot 03) + ({}^{2}_{2}\cdot 01) + 2({}^{0}_{0}\cdot 11) + 2({}^{2}_{0}\cdot 01) + 2({}^{0}_{2}\cdot 01) + 4({}^{0}_{0}\cdot 01)$

Sl 5: $({}^{1}_{1}\cdot 03) + 4({}^{1}_{1}\cdot 11) + ({}^{3}_{1}\cdot 01) + ({}^{1}_{3}\cdot 01) + 7({}^{1}_{1}\cdot 01)$

Sl 6: $({}^{2}_{2}\cdot 11) + ({}^{2}_{0}\cdot 03) + ({}^{0}_{2}\cdot 03) + ({}^{0}_{0}\cdot 21) + 3({}^{2}_{0}\cdot 11) + 3({}^{0}_{2}\cdot 11) + 2({}^{0}_{0}\cdot 03) + 3({}^{2}_{2}\cdot 01) + 6({}^{0}_{0}\cdot 11)$
$+ 6({}^{2}_{0}\cdot 01) + 6({}^{0}_{2}\cdot 01) + 8({}^{0}_{0}\cdot 01)$

Sl 7: $({}^{1}_{1}\cdot 21) + ({}^{3}_{1}\cdot 11) + ({}^{1}_{3}\cdot 11) + 4({}^{1}_{1}\cdot 03) + 11({}^{1}_{1}\cdot 11) + 3({}^{3}_{1}\cdot 01) + 3({}^{1}_{3}\cdot 01) + 17({}^{1}_{1}\cdot 01)$

Affine B4 ⊃ A3 + A1 - Level 1 - Class 0

sl 0: $\left({}^{1}_{0}\,0\cdot0\right)$ **Highest Weight** $\left({}^{1}_{0}\,000\right)$

sl 1: $\left({}^{0}_{1}\,0\cdot2\right)$

sl 2: $\left({}^{0}_{1}\,1\cdot0\right) + \left({}^{1}_{0}\,0\cdot2\right) + \left({}^{1}_{0}\,0\cdot0\right)$

sl 3: $\left({}^{1}_{0}\,1\cdot2\right) + 2\left({}^{0}_{1}\,0\cdot2\right) + \left({}^{0}_{1}\,0\cdot0\right)$

sl 4: $\left({}^{2}_{1}\,0\cdot0\right) + \left({}^{0}_{1}\,1\cdot2\right) + \left({}^{1}_{0}\,0\cdot4\right) + 2\left({}^{0}_{1}\,1\cdot0\right) + 2\left({}^{1}_{0}\,0\cdot2\right) + 3\left({}^{1}_{0}\,0\cdot0\right)$

sl 5: $\left({}^{1}_{2}\,0\cdot2\right) + 3\left({}^{1}_{0}\,1\cdot2\right) + \left({}^{1}_{0}\,1\cdot0\right) + \left({}^{0}_{1}\,0\cdot4\right) + 5\left({}^{0}_{1}\,0\cdot2\right) + 2\left({}^{0}_{1}\,0\cdot0\right)$

sl 6: $\left({}^{2}_{1}\,0\cdot2\right) + \left({}^{1}_{0}\,2\cdot0\right) + \left({}^{0}_{1}\,1\cdot4\right) + 2\left({}^{2}_{1}\,0\cdot0\right) + 3\left({}^{0}_{1}\,1\cdot2\right) + \left({}^{0}_{3}\,0\cdot0\right) + 2\left({}^{1}_{0}\,0\cdot4\right) + 5\left({}^{0}_{1}\,1\cdot0\right) + 6\left({}^{1}_{0}\,0\cdot2\right)$
$+ 6\left({}^{1}_{0}\,0\cdot0\right)$

sl 7: $\left({}^{0}_{1}\,2\cdot2\right) + \left({}^{3}_{0}\,0\cdot2\right) + \left({}^{1}_{0}\,1\cdot4\right) + 3\left({}^{1}_{2}\,0\cdot2\right) + 8\left({}^{1}_{0}\,1\cdot2\right) + \left({}^{1}_{2}\,0\cdot0\right) + 3\left({}^{1}_{0}\,1\cdot0\right) + 3\left({}^{0}_{1}\,0\cdot4\right) + 11\left({}^{0}_{1}\,0\cdot2\right)$
$+ 5\left({}^{0}_{1}\,0\cdot0\right)$

Affine B4 ⊃ A3 + A1 - Level 1 - Class 1

sl 0: $\left({}^{0}_{0}\,0\cdot1\right)$ **Highest Weight** $\left({}^{0}_{0}\,001\right)$

sl 1: $\left({}^{0}_{0}\,1\cdot1\right)$

sl 2: $\left({}^{1}_{1}\,0\cdot1\right) + \left({}^{0}_{0}\,0\cdot3\right) + \left({}^{0}_{0}\,0\cdot1\right)$

sl 3: $\left({}^{2}_{0}\,0\cdot1\right) + \left({}^{0}_{0}\,1\cdot3\right) + \left({}^{0}_{2}\,0\cdot1\right) + 2\left({}^{0}_{0}\,1\cdot1\right)$

sl 4: $\left({}^{1}_{1}\,0\cdot3\right) + \left({}^{0}_{0}\,2\cdot1\right) + 3\left({}^{1}_{1}\,0\cdot1\right) + 2\left({}^{0}_{0}\,0\cdot3\right) + 3\left({}^{0}_{0}\,0\cdot1\right)$

sl 5: $\left({}^{1}_{1}\,1\cdot1\right) + \left({}^{2}_{0}\,0\cdot3\right) + \left({}^{0}_{2}\,0\cdot3\right) + 2\left({}^{2}_{0}\,0\cdot1\right) + 3\left({}^{0}_{0}\,1\cdot3\right) + 2\left({}^{0}_{2}\,0\cdot1\right) + 6\left({}^{0}_{0}\,1\cdot1\right)$

sl 6: $\left({}^{2}_{0}\,1\cdot1\right) + \left({}^{0}_{0}\,2\cdot3\right) + \left({}^{0}_{2}\,1\cdot1\right) + 4\left({}^{1}_{1}\,0\cdot3\right) + 2\left({}^{0}_{0}\,2\cdot1\right) + 8\left({}^{1}_{1}\,0\cdot1\right) + \left({}^{0}_{0}\,0\cdot5\right) + 4\left({}^{0}_{0}\,0\cdot3\right) + 7\left({}^{0}_{0}\,0\cdot1\right)$

sl 7: $\left({}^{1}_{1}\,1\cdot3\right) + 4\left({}^{1}_{1}\,1\cdot1\right) + 3\left({}^{2}_{0}\,0\cdot3\right) + \left({}^{0}_{0}\,1\cdot5\right) + 3\left({}^{0}_{2}\,0\cdot3\right) + 6\left({}^{2}_{0}\,0\cdot1\right) + 8\left({}^{0}_{0}\,1\cdot3\right) + 6\left({}^{0}_{2}\,0\cdot1\right) + 13\left({}^{0}_{0}\,1\cdot1\right)$

Affine B4 ⊃ D4 - Level 1 - Class 0

```
         1                       1
Sl 0:   ( 00•)  Highest Weight  ( 000)
         0                       0

         0
Sl 1:   ( 00•)
         1

         0          1
Sl 2:   ( 01•) +   ( 00•)
         1          0

         1          0
Sl 3:   ( 01•) +  2( 00•)
         0          1

         1          0          1
Sl 4:   ( 10•) +  2( 01•) +  3( 00•)
         0          1          0

         0          1          0
Sl 5:   ( 10•) +  3( 01•) +  4( 00•)
         1          0          1

         1          1          1          0          1
Sl 6:   ( 02•) +   ( 00•) +  2( 10•) +  5( 01•) +  6( 00•)
         0          2          0          1          0

         0          2          0          1          0
Sl 7:   ( 02•) +   ( 00•) +  3( 10•) +  7( 01•) +  8( 00•)
         1          1          1          0          1

         0          1          1          3          1           0           1
Sl 8:   ( 11•) +  2( 02•) +  2( 00•) +   ( 00•) +  6( 10•) + 11( 01•) + 12( 00•)
         1          0          2          0          0           1           0
```

Affine B4 ⊃ D4 - Level 1 - Class 1

```
         0                       0
Sl 0:   ( 00•)  Highest Weight  ( 001)
         0                       0

         0
Sl 1:   ( 01•)
         0

         0          0
Sl 2:   ( 10•) +   ( 00•)
         0          0

         1          0
Sl 3:   ( 00•) +  2( 01•)
         1          0

         0          0          2          0          0
Sl 4:   ( 02•) +   ( 00•) +   ( 00•) +  2( 10•) +  2( 00•)
         0          2          0          0          0

         0          1          0
Sl 5:   ( 11•) +  3( 00•) +  4( 01•)
         0          1          0

         1          0          0          2          0          0
Sl 6:   ( 01•) +  2( 02•) +  2( 00•) +  2( 00•) +  6( 10•) +  4( 00•)
         1          0          2          0          0          0

         0          2          0          1          0
Sl 7:   ( 01•) +   ( 01•) +  3( 11•) +  7( 00•) +  9( 01•)
         2          0          0          1          0

         0          1          0          0          2           0          0
Sl 8:   ( 20•) +  4( 01•) +  5( 02•) +  5( 00•) +  5( 00•) + 12( 10•) +  8( 00•)
         0          1          0          2          0           0          0
```

Affine B5 Branching Rules

Affine B5 ⊃ B5 - Level 1 - Class 0

sl 0: $\overset{\bullet}{(}0000)_{1}$ **Highest Weight** $\overset{0}{(}0000)_{1}$

sl 1: $\overset{\bullet}{(}0100)_{0}$ + $\overset{\bullet}{(}0000)_{1}$

sl 2: $\overset{\bullet}{(}0002)_{0}$ + $\overset{\bullet}{(}1000)_{1}$ + $\overset{\bullet}{(}0100)_{0}$ + $2\overset{\bullet}{(}0000)_{1}$

sl 3: $\overset{\bullet}{(}0010)_{1}$ + $\overset{\bullet}{(}0002)_{0}$ + $2\overset{\bullet}{(}1000)_{1}$ + $\overset{\bullet}{(}0010)_{0}$ + $3\overset{\bullet}{(}0100)_{0}$ + $3\overset{\bullet}{(}0000)_{1}$

sl 4: $\overset{\bullet}{(}1100)_{0}$ + $\overset{\bullet}{(}0002)_{1}$ + $2\overset{\bullet}{(}0010)_{1}$ + $\overset{\bullet}{(}0000)_{3}$ + $3\overset{\bullet}{(}0002)_{0}$ + $4\overset{\bullet}{(}1000)_{1}$ + $\overset{\bullet}{(}0010)_{0}$ + $5\overset{\bullet}{(}0100)_{0}$
\+ $\overset{\bullet}{(}1000)_{0}$ + $6\overset{\bullet}{(}0000)_{1}$

sl 5: $\overset{\bullet}{(}1002)_{0}$ + $\overset{\bullet}{(}0100)_{2}$ + $2\overset{\bullet}{(}1100)_{0}$ + $2\overset{\bullet}{(}0002)_{1}$ + $5\overset{\bullet}{(}0010)_{1}$ + $\overset{\bullet}{(}0100)_{1}$ + $\overset{\bullet}{(}0000)_{3}$ + $5\overset{\bullet}{(}0002)_{0}$
\+ $8\overset{\bullet}{(}1000)_{1}$ + $3\overset{\bullet}{(}0010)_{0}$ + $10\overset{\bullet}{(}0100)_{0}$ + $\overset{\bullet}{(}1000)_{0}$ + $9\overset{\bullet}{(}0000)_{1}$ + $\overset{\bullet}{(}0000)_{0}$

sl 6: $\overset{\bullet}{(}0110)_{0}$ + $\overset{\bullet}{(}0002)_{2}$ + $2\overset{\bullet}{(}1002)_{0}$ + $\overset{\bullet}{(}2000)_{1}$ + $\overset{\bullet}{(}1010)_{0}$ + $2\overset{\bullet}{(}0100)_{2}$ + $5\overset{\bullet}{(}1100)_{0}$ + $5\overset{\bullet}{(}0002)_{1}$
\+ $10\overset{\bullet}{(}0010)_{1}$ + $2\overset{\bullet}{(}0100)_{1}$ + $3\overset{\bullet}{(}0000)_{3}$ + $11\overset{\bullet}{(}0002)_{0}$ + $15\overset{\bullet}{(}1000)_{1}$ + $5\overset{\bullet}{(}0010)_{0}$ + $16\overset{\bullet}{(}0100)_{0}$ + $\overset{\bullet}{(}0000)_{2}$
\+ $3\overset{\bullet}{(}1000)_{0}$ + $16\overset{\bullet}{(}0000)_{1}$ + $\overset{\bullet}{(}0000)_{0}$

sl 7: $\overset{\bullet}{(}1010)_{1}$ + $\overset{\bullet}{(}0102)_{0}$ + $2\overset{\bullet}{(}0110)_{0}$ + $2\overset{\bullet}{(}0002)_{2}$ + $\overset{\bullet}{(}0010)_{2}$ + $6\overset{\bullet}{(}1002)_{0}$ + $2\overset{\bullet}{(}2000)_{1}$ + $2\overset{\bullet}{(}1010)_{0}$
\+ $6\overset{\bullet}{(}0100)_{2}$ + $10\overset{\bullet}{(}1100)_{0}$ + $10\overset{\bullet}{(}0002)_{1}$ + $20\overset{\bullet}{(}0010)_{1}$ + $\overset{\bullet}{(}2000)_{0}$ + $5\overset{\bullet}{(}0100)_{1}$ + $5\overset{\bullet}{(}0000)_{3}$ + $18\overset{\bullet}{(}0002)_{0}$
\+ $26\overset{\bullet}{(}1000)_{1}$ + $11\overset{\bullet}{(}0010)_{0}$ + $30\overset{\bullet}{(}0100)_{0}$ + $2\overset{\bullet}{(}0000)_{2}$ + $5\overset{\bullet}{(}1000)_{0}$ + $24\overset{\bullet}{(}0000)_{1}$ + $2\overset{\bullet}{(}0000)_{0}$

Affine B5 ⊃ B5 - Level 1 - Class 0

sl 0: $\overset{\bullet}{(}0000)_{0}$ **Highest Weight** $\overset{1}{(}0000)_{0}$

sl 1: $\overset{\bullet}{(}1000)_{0}$

sl 2: $\overset{\bullet}{(}0010)_{0}$ + $\overset{\bullet}{(}0000)_{2}$ + $\overset{\bullet}{(}1000)_{0}$ + $\overset{\bullet}{(}0000)_{0}$

sl 3: $\overset{\bullet}{(}0100)_{1}$ + $\overset{\bullet}{(}0002)_{0}$ + $\overset{\bullet}{(}0010)_{0}$ + $\overset{\bullet}{(}0000)_{2}$ + $3\overset{\bullet}{(}1000)_{0}$ + $\overset{\bullet}{(}0000)_{0}$ (continued next page)

Sl 4: $({}^{\bullet}_{1}\,0002)$ + $({}^{\bullet}_{0}\,2000)$ + 2$({}^{\bullet}_{1}\,0100)$ + $({}^{\bullet}_{0}\,0002)$ + 3$({}^{\bullet}_{0}\,0010)$ + $({}^{\bullet}_{0}\,0100)$ + 3$({}^{\bullet}_{2}\,0000)$ + 4$({}^{\bullet}_{0}\,1000)$
+ 3$({}^{\bullet}_{0}\,0000)$

Sl 5: $({}^{\bullet}_{0}\,1010)$ + 2$({}^{\bullet}_{1}\,0002)$ + $({}^{\bullet}_{2}\,1000)$ + $({}^{\bullet}_{1}\,0010)$ + $({}^{\bullet}_{0}\,2000)$ + 5$({}^{\bullet}_{1}\,0100)$ + 3$({}^{\bullet}_{0}\,0002)$ + 5$({}^{\bullet}_{0}\,0010)$
+ $({}^{\bullet}_{0}\,0100)$ + 4$({}^{\bullet}_{2}\,0000)$ + 9$({}^{\bullet}_{0}\,1000)$ + $({}^{\bullet}_{1}\,0000)$ + 3$({}^{\bullet}_{0}\,0000)$

Sl 6: $({}^{\bullet}_{2}\,0010)$ + $({}^{\bullet}_{0}\,1002)$ + $({}^{\bullet}_{0}\,0200)$ + 2$({}^{\bullet}_{0}\,1010)$ + 5$({}^{\bullet}_{1}\,0002)$ + 2$({}^{\bullet}_{2}\,1000)$ + 2$({}^{\bullet}_{1}\,0010)$ + 4$({}^{\bullet}_{0}\,2000)$
+ 9$({}^{\bullet}_{1}\,0100)$ + 5$({}^{\bullet}_{0}\,0002)$ + $({}^{\bullet}_{1}\,1000)$ + 11$({}^{\bullet}_{0}\,0010)$ + 3$({}^{\bullet}_{0}\,0100)$ + 9$({}^{\bullet}_{2}\,0000)$ + 13$({}^{\bullet}_{0}\,1000)$ + $({}^{\bullet}_{1}\,0000)$
+ 7$({}^{\bullet}_{0}\,0000)$

Affine B5 ⊃ B5 - Level 1 - Class 1

Sl 0: $({}^{\bullet}_{0}\,0001)$ **Highest Weight** $({}^{0}_{0}\,0001)$

Sl 1: $({}^{\bullet}_{1}\,0001)$ + $({}^{\bullet}_{0}\,0001)$

Sl 2: $({}^{\bullet}_{0}\,1001)$ + 2$({}^{\bullet}_{1}\,0001)$ + 2$({}^{\bullet}_{0}\,0001)$

Sl 3: $({}^{\bullet}_{0}\,0101)$ + $({}^{\bullet}_{2}\,0001)$ + 2$({}^{\bullet}_{0}\,1001)$ + 4$({}^{\bullet}_{1}\,0001)$ + 4$({}^{\bullet}_{0}\,0001)$

Sl 4: $({}^{\bullet}_{1}\,1001)$ + $({}^{\bullet}_{0}\,0011)$ + 2$({}^{\bullet}_{0}\,0101)$ + 2$({}^{\bullet}_{2}\,0001)$ + 5$({}^{\bullet}_{0}\,1001)$ + 8$({}^{\bullet}_{1}\,0001)$ + 7$({}^{\bullet}_{0}\,0001)$

Sl 5: $({}^{\bullet}_{1}\,0101)$ + $({}^{\bullet}_{0}\,0003)$ + 3$({}^{\bullet}_{1}\,1001)$ + 2$({}^{\bullet}_{0}\,0011)$ + 5$({}^{\bullet}_{0}\,0101)$ + 5$({}^{\bullet}_{2}\,0001)$ + 10$({}^{\bullet}_{0}\,1001)$ + 15$({}^{\bullet}_{1}\,0001)$
+ 12$({}^{\bullet}_{0}\,0001)$

Sl 6: $({}^{\bullet}_{1}\,0011)$ + $({}^{\bullet}_{0}\,2001)$ + 3$({}^{\bullet}_{1}\,0101)$ + $({}^{\bullet}_{3}\,0001)$ + 2$({}^{\bullet}_{0}\,0003)$ + 7$({}^{\bullet}_{1}\,1001)$ + 5$({}^{\bullet}_{0}\,0011)$ + 11$({}^{\bullet}_{0}\,0101)$
+ 10$({}^{\bullet}_{2}\,0001)$ + 20$({}^{\bullet}_{0}\,1001)$ + 27$({}^{\bullet}_{1}\,0001)$ + 21$({}^{\bullet}_{0}\,0001)$

Sl 7: $({}^{\bullet}_{0}\,1101)$ + $({}^{\bullet}_{1}\,0003)$ + $({}^{\bullet}_{2}\,1001)$ + 3$({}^{\bullet}_{1}\,0011)$ + 2$({}^{\bullet}_{0}\,2001)$ + 8$({}^{\bullet}_{1}\,0101)$ + 2$({}^{\bullet}_{3}\,0001)$ + 4$({}^{\bullet}_{0}\,0003)$
+ 16$({}^{\bullet}_{1}\,1001)$ + 11$({}^{\bullet}_{0}\,0011)$ + 22$({}^{\bullet}_{0}\,0101)$ + 20$({}^{\bullet}_{2}\,0001)$ + 37$({}^{\bullet}_{0}\,1001)$ + 48$({}^{\bullet}_{1}\,0001)$ + 34$({}^{\bullet}_{0}\,0001)$

Affine B5 ⊃ A1 + A1 + B3 - Level 1 - Class 0

sl 0: $({}^{0}_{1}\cdot 000)$ **Highest Weight** $({}^{0}_{1}\,0000)$

sl 1: $({}^{1}_{0}\cdot 100)$

sl 2: $({}^{0}_{1}\cdot 010) + ({}^{2}_{1}\cdot 000) + ({}^{0}_{1}\cdot 000)$

sl 3: $({}^{1}_{0}\cdot 002) + ({}^{1}_{2}\cdot 100) + 2({}^{1}_{0}\cdot 100)$

sl 4: $({}^{2}_{1}\cdot 010) + ({}^{0}_{1}\cdot 200) + ({}^{0}_{1}\cdot 002) + 2({}^{0}_{1}\cdot 010) + 2({}^{2}_{1}\cdot 000) + ({}^{0}_{3}\cdot 000) + 3({}^{0}_{1}\cdot 000)$

sl 5: $({}^{1}_{0}\cdot 110) + ({}^{1}_{2}\cdot 002) + 2({}^{1}_{0}\cdot 002) + ({}^{1}_{0}\cdot 010) + ({}^{3}_{0}\cdot 100) + 3({}^{1}_{2}\cdot 100) + 5({}^{1}_{0}\cdot 100)$

sl 6: $({}^{0}_{1}\cdot 102) + ({}^{2}_{1}\cdot 200) + ({}^{2}_{1}\cdot 002) + 3({}^{2}_{1}\cdot 010) + ({}^{0}_{3}\cdot 010) + 2({}^{0}_{1}\cdot 200) + 2({}^{0}_{1}\cdot 002) + 6({}^{0}_{1}\cdot 010)$

$+ ({}^{0}_{1}\cdot 100) + ({}^{2}_{3}\cdot 000) + 5({}^{2}_{1}\cdot 000) + ({}^{0}_{3}\cdot 000) + 6({}^{0}_{1}\cdot 000)$

Affine B5 ⊃ A1 + A1 + B3 - Level 1 - Class 1

sl 0: $({}^{0}_{0}\cdot 001)$ **Highest Weight** $({}^{0}_{0}\,0001)$

sl 1: $({}^{1}_{1}\cdot 001)$

sl 2: $({}^{0}_{0}\cdot 101) + ({}^{2}_{0}\cdot 001) + ({}^{0}_{2}\cdot 001) + ({}^{0}_{0}\cdot 001)$

sl 3: $({}^{1}_{1}\cdot 101) + 3({}^{1}_{1}\cdot 001)$

sl 4: $({}^{0}_{0}\cdot 011) + ({}^{2}_{0}\cdot 101) + ({}^{0}_{2}\cdot 101) + 2({}^{0}_{0}\cdot 101) + ({}^{2}_{2}\cdot 001) + 2({}^{2}_{0}\cdot 001) + 2({}^{0}_{2}\cdot 001) + 4({}^{0}_{0}\cdot 001)$

sl 5: $({}^{1}_{1}\cdot 011) + 4({}^{1}_{1}\cdot 101) + ({}^{3}_{1}\cdot 001) + ({}^{1}_{3}\cdot 001) + 7({}^{1}_{1}\cdot 001)$

sl 6: $({}^{2}_{0}\cdot 011) + ({}^{0}_{2}\cdot 011) + ({}^{0}_{0}\cdot 201) + ({}^{0}_{0}\cdot 003) + ({}^{2}_{2}\cdot 101) + 2({}^{0}_{0}\cdot 011) + 3({}^{2}_{0}\cdot 101) + 3({}^{0}_{2}\cdot 101)$

$+ 6({}^{0}_{0}\cdot 101) + 3({}^{2}_{2}\cdot 001) + 6({}^{2}_{0}\cdot 001) + 6({}^{0}_{2}\cdot 001) + 8({}^{0}_{0}\cdot 001)$

Affine B5 $\supset$ A3 + C2 - Level 1 - Class 0

Sl 0: $({}^{0}_{1}0\cdot00)$ **Highest Weight** $({}^{0}_{1}0000)$

Sl 1: $({}^{1}_{0}0\cdot10)$

Sl 2: $({}^{0}_{1}0\cdot02) + ({}^{1}_{0}1\cdot00) + ({}^{0}_{1}0\cdot00)$

Sl 3: $({}^{0}_{1}1\cdot10) + ({}^{1}_{0}0\cdot02) + 2({}^{1}_{0}0\cdot10)$

Sl 4: $({}^{1}_{0}1\cdot02) + ({}^{0}_{1}0\cdot20) + ({}^{1}_{2}0\cdot00) + 2({}^{0}_{1}0\cdot02) + 2({}^{1}_{0}1\cdot00) + ({}^{0}_{1}0\cdot10) + 3({}^{0}_{1}0\cdot00)$

Sl 5: $({}^{2}_{1}0\cdot10) + ({}^{1}_{0}0\cdot12) + ({}^{0}_{1}1\cdot02) + 3({}^{0}_{1}1\cdot10) + 2({}^{1}_{0}0\cdot02) + 5({}^{1}_{0}0\cdot10) + ({}^{1}_{0}0\cdot00)$

Sl 6: $({}^{1}_{0}1\cdot20) + ({}^{1}_{2}0\cdot02) + 3({}^{1}_{0}1\cdot02) + ({}^{0}_{1}0\cdot12) + ({}^{1}_{0}1\cdot10) + ({}^{0}_{1}2\cdot00) + 2({}^{0}_{1}0\cdot20) + ({}^{3}_{0}0\cdot00)$

$+ 2({}^{1}_{2}0\cdot00) + 6({}^{0}_{1}0\cdot02) + 5({}^{1}_{0}1\cdot00) + 2({}^{0}_{1}0\cdot10) + 6({}^{0}_{1}0\cdot00)$

Sl 7: $({}^{0}_{1}1\cdot12) + ({}^{2}_{1}0\cdot02) + ({}^{1}_{0}2\cdot10) + 3({}^{2}_{1}0\cdot10) + 3({}^{1}_{0}0\cdot12) + 3({}^{0}_{1}1\cdot02) + ({}^{0}_{3}0\cdot10) + ({}^{1}_{0}0\cdot20)$

$+ 8({}^{0}_{1}1\cdot10) + 6({}^{1}_{0}0\cdot02) + 11({}^{1}_{0}0\cdot10) + ({}^{0}_{1}1\cdot00) + 2({}^{1}_{0}0\cdot00)$

Affine B5 $\supset$ A3 + C2 - Level 1 - Class 1

Sl 0: $({}^{0}_{0}0\cdot01)$ **Highest Weight** $({}^{0}_{0}0001)$

Sl 1: $({}^{0}_{0}1\cdot01)$

Sl 2: $({}^{1}_{1}0\cdot01) + ({}^{0}_{0}0\cdot11) + ({}^{0}_{0}0\cdot01)$

Sl 3: $({}^{0}_{0}1\cdot11) + ({}^{2}_{0}0\cdot01) + ({}^{0}_{2}0\cdot01) + 2({}^{0}_{0}1\cdot01)$

Sl 4: $({}^{1}_{1}0\cdot11) + ({}^{0}_{0}2\cdot01) + 3({}^{1}_{1}0\cdot01) + ({}^{0}_{0}0\cdot03) + 2({}^{0}_{0}0\cdot11) + 3({}^{0}_{0}0\cdot01)$

Sl 5: $({}^{2}_{0}0\cdot11) + ({}^{1}_{1}1\cdot01) + ({}^{0}_{0}1\cdot03) + ({}^{0}_{2}0\cdot11) + 3({}^{0}_{0}1\cdot11) + 2({}^{2}_{0}0\cdot01) + 2({}^{0}_{2}0\cdot01) + 6({}^{0}_{0}1\cdot01)$

Sl 6: $({}^{1}_{1}0\cdot03) + ({}^{0}_{0}2\cdot11) + ({}^{2}_{0}1\cdot01) + 4({}^{1}_{1}0\cdot11) + ({}^{0}_{2}1\cdot01) + 2({}^{0}_{0}2\cdot01) + ({}^{0}_{0}0\cdot21) + 8({}^{1}_{1}0\cdot01)$

Affine B5 ⊃ D4 + A1 - Level 1 - Class 0

```
          0                          0
sl 0:    ( 00•0)  Highest Weight  ( 0000)
          1                          1

          1
sl 1:    ( 00•2)
          0

          1            0             0
sl 2:    ( 01•0) +    ( 00•2) +     ( 00•0)
          0            1             1

          0            1             1
sl 3:    ( 01•2) +   2( 00•2) +     ( 00•0)
          1            0             0

          0            1             1             0             0             0
sl 4:    ( 10•0) +    ( 01•2) +    2( 01•0) +    ( 00•4) +    2( 00•2) +    3( 00•0)
          1            0             0             1             1             1

          1            0             0             1             1             1
sl 5:    ( 10•2) +   3( 01•2) +     ( 01•0) +    ( 00•4) +    5( 00•2) +    2( 00•0)
          0            1             1             0             0             0

          0            0             2             1             0             1             1             0
sl 6:    ( 02•0) +    ( 10•2) +     ( 00•0) +    ( 01•4) +    2( 10•0) +    3( 01•2) +    5( 01•0) +    2( 00•4)
          1            1             1             0             1             0             0             1

          0            0
     +  6( 00•2) +   6( 00•0)
          1            1

          1            1             1             0             1             0             0             1
sl 7:    ( 02•2) +    ( 00•2) +    3( 10•2) +    ( 01•4) +     ( 10•0) +    8( 01•2) +    3( 01•0) +    3( 00•4)
          0            2             0             1             0             1             1             0

          1            1
     + 11( 00•2) +   5( 00•0)
          0            0
```

Affine B5 ⊃ D4 + A1 - Level 1 - Class 1

```
          0                          0
sl 0:    ( 00•1)  Highest Weight  ( 0001)
          0                          0

          0
sl 1:    ( 01•1)
          0

          0            0             0
sl 2:    ( 10•1) +    ( 00•3) +     ( 00•1)
          0            0             0

          1            0             0
sl 3:    ( 00•1) +    ( 01•3) +    2( 01•1)
          1            0             0

          0            0             0             2             0             0             0
sl 4:    ( 02•1) +    ( 10•3) +     ( 00•1) +    ( 00•1) +    2( 10•1) +    2( 00•3) +    3( 00•1)
          0            0             2             0             0             0             0

          0            1             1             0             0
sl 5:    ( 11•1) +    ( 00•3) +    3( 00•1) +   3( 01•3) +    5( 01•1)
          0            1             1             0             0

          1            0             0             2             0             0             0             2
sl 6:    ( 01•1) +    ( 02•3) +     ( 00•3) +    ( 00•3) +    2( 02•1) +    3( 10•3) +    2( 00•1) +    2( 00•1)
          1            0             2             0             0             0             2             0

          0            0             0             0
     +  7( 10•1) +    ( 00•5) +    4( 00•3) +   6( 00•1)
          0            0             0             0
```

Affine B5 ⊃ D5 - Level 1 - Class 0

Sl 0: $({}^{0}_{1}\,000\bullet)$ **Highest Weight** $({}^{0}_{1}\,0000)$

Sl 1: $({}^{1}_{0}\,000\bullet)$

Sl 2: $({}^{1}_{0}\,001\bullet) + ({}^{0}_{1}\,000\bullet)$

Sl 3: $({}^{0}_{1}\,001\bullet) + 2({}^{1}_{0}\,000\bullet)$

Sl 4: $({}^{0}_{1}\,010\bullet) + 2({}^{1}_{0}\,001\bullet) + 3({}^{0}_{1}\,000\bullet)$

Sl 5: $({}^{1}_{0}\,010\bullet) + 3({}^{0}_{1}\,001\bullet) + 4({}^{1}_{0}\,000\bullet)$

Sl 6: $({}^{1}_{0}\,100\bullet) + ({}^{0}_{1}\,002\bullet) + 2({}^{0}_{1}\,010\bullet) + 5({}^{1}_{0}\,001\bullet) + 6({}^{0}_{1}\,000\bullet)$

Sl 7: $({}^{0}_{1}\,100\bullet) + ({}^{1}_{0}\,002\bullet) + 3({}^{1}_{0}\,010\bullet) + 7({}^{0}_{1}\,001\bullet) + 8({}^{1}_{0}\,000\bullet)$

Affine B5 ⊃ D5 - Level 1 - Class 1

Sl 0: $({}^{0}_{0}\,000\bullet)$ **Highest Weight** $({}^{0}_{0}\,0001)$

Sl 1: $({}^{0}_{0}\,001\bullet)$

Sl 2: $({}^{0}_{0}\,010\bullet) + ({}^{0}_{0}\,000\bullet)$

Sl 3: $({}^{0}_{0}\,100\bullet) + 2({}^{0}_{0}\,001\bullet)$

Sl 4: $({}^{1}_{1}\,000\bullet) + ({}^{0}_{0}\,002\bullet) + 2({}^{0}_{0}\,010\bullet) + 2({}^{0}_{0}\,000\bullet)$

Sl 5: $({}^{0}_{0}\,011\bullet) + ({}^{0}_{2}\,000\bullet) + ({}^{2}_{0}\,000\bullet) + 2({}^{0}_{0}\,100\bullet) + 4({}^{0}_{0}\,001\bullet)$

Sl 6: $({}^{0}_{0}\,101\bullet) + 3({}^{1}_{1}\,000\bullet) + 2({}^{0}_{0}\,002\bullet) + 5({}^{0}_{0}\,010\bullet) + 4({}^{0}_{0}\,000\bullet)$

Affine B6 Branching Rules

Affine B6 ⊃ B6 - Level 1 - Class 0

sl 0: $(^{\cdot}_{1}\,00000)$ **Highest Weight** $(^{0}_{1}\,00000)$

sl 1: $(^{\cdot}_{0}\,01000) + (^{\cdot}_{1}\,00000)$

sl 2: $(^{\cdot}_{0}\,00010) + (^{\cdot}_{1}\,10000) + (^{\cdot}_{0}\,01000) + 2(^{\cdot}_{1}\,00000)$

sl 3: $(^{\cdot}_{1}\,00100) + (^{\cdot}_{0}\,00002) + (^{\cdot}_{0}\,00010) + 2(^{\cdot}_{1}\,10000) + 3(^{\cdot}_{0}\,01000) + 3(^{\cdot}_{1}\,00000)$

sl 4: $(^{\cdot}_{1}\,00002) + (^{\cdot}_{0}\,11000) + 2(^{\cdot}_{1}\,00100) + (^{\cdot}_{0}\,00002) + 3(^{\cdot}_{0}\,00010) + (^{\cdot}_{3}\,00000) + (^{\cdot}_{0}\,00100)$
$+ 4(^{\cdot}_{1}\,10000) + 5(^{\cdot}_{0}\,01000) + 6(^{\cdot}_{1}\,00000)$

sl 5: $(^{\cdot}_{0}\,10010) + (^{\cdot}_{2}\,01000) + 2(^{\cdot}_{1}\,00002) + (^{\cdot}_{1}\,00010) + 2(^{\cdot}_{0}\,11000) + 5(^{\cdot}_{1}\,00100) + 3(^{\cdot}_{0}\,00002)$
$+ 5(^{\cdot}_{0}\,00010) + (^{\cdot}_{3}\,00000) + (^{\cdot}_{0}\,00100) + 8(^{\cdot}_{1}\,10000) + 10(^{\cdot}_{0}\,01000) + (^{\cdot}_{0}\,10000) + 9(^{\cdot}_{1}\,00000)$

Affine B6 ⊃ B6 - Level 1 - Class 0

sl 0: $(^{\cdot}_{0}\,00000)$ **Highest Weight** $(^{1}_{0}\,00000)$

sl 1: $(^{\cdot}_{0}\,10000)$

sl 2: $(^{\cdot}_{0}\,00100) + (^{\cdot}_{2}\,00000) + (^{\cdot}_{0}\,10000) + (^{\cdot}_{0}\,00000)$

sl 3: $(^{\cdot}_{1}\,01000) + (^{\cdot}_{0}\,00002) + (^{\cdot}_{0}\,00100) + (^{\cdot}_{2}\,00000) + 3(^{\cdot}_{0}\,10000) + (^{\cdot}_{0}\,00000)$

sl 4: $(^{\cdot}_{1}\,00010) + (^{\cdot}_{0}\,20000) + 2(^{\cdot}_{1}\,01000) + (^{\cdot}_{0}\,00002) + (^{\cdot}_{0}\,00010) + 3(^{\cdot}_{0}\,00100) + 3(^{\cdot}_{2}\,00000)$
$+ 4(^{\cdot}_{0}\,10000) + 3(^{\cdot}_{0}\,00000)$

sl 5: $(^{\cdot}_{0}\,10100) + (^{\cdot}_{1}\,00002) + 2(^{\cdot}_{1}\,00010) + (^{\cdot}_{2}\,10000) + (^{\cdot}_{0}\,20000) + 5(^{\cdot}_{1}\,01000) + 3(^{\cdot}_{0}\,00002)$
$+ (^{\cdot}_{0}\,00010) + 5(^{\cdot}_{0}\,00100) + (^{\cdot}_{0}\,01000) + 4(^{\cdot}_{2}\,00000) + 9(^{\cdot}_{0}\,10000) + 3(^{\cdot}_{0}\,00000)$

Affine B6 ⊃ B6 - Level 1 - Class 1

Sl 0: $({}^{\bullet}_{0}\,00001)$ **Highest Weight** $({}^{0}_{0}\,00001)$

Sl 1: $({}^{\bullet}_{1}\,00001) + ({}^{\bullet}_{0}\,00001)$

Sl 2: $({}^{\bullet}_{0}\,10001) + 2({}^{\bullet}_{1}\,00001) + 2({}^{\bullet}_{0}\,00001)$

Sl 3: $({}^{\bullet}_{0}\,01001) + ({}^{\bullet}_{2}\,00001) + 2({}^{\bullet}_{0}\,10001) + 4({}^{\bullet}_{1}\,00001) + 4({}^{\bullet}_{0}\,00001)$

Sl 4: $({}^{\bullet}_{0}\,00101) + ({}^{\bullet}_{1}\,10001) + 2({}^{\bullet}_{0}\,01001) + 2({}^{\bullet}_{2}\,00001) + 5({}^{\bullet}_{0}\,10001) + 8({}^{\bullet}_{1}\,00001) + 7({}^{\bullet}_{0}\,00001)$

Sl 5: $({}^{\bullet}_{1}\,01001) + ({}^{\bullet}_{0}\,00011) + 2({}^{\bullet}_{0}\,00101) + 3({}^{\bullet}_{1}\,10001) + 5({}^{\bullet}_{0}\,01001) + 5({}^{\bullet}_{2}\,00001) + 10({}^{\bullet}_{0}\,10001)$
$+ 15({}^{\bullet}_{1}\,00001) + 12({}^{\bullet}_{0}\,00001)$

Sl 6: $({}^{\bullet}_{1}\,00101) + ({}^{\bullet}_{0}\,20001) + 3({}^{\bullet}_{1}\,01001) + ({}^{\bullet}_{0}\,00003) + 2({}^{\bullet}_{0}\,00011) + ({}^{\bullet}_{3}\,00001) + 5({}^{\bullet}_{0}\,00101)$
$+ 7({}^{\bullet}_{1}\,10001) + 11({}^{\bullet}_{0}\,01001) + 10({}^{\bullet}_{2}\,00001) + 20({}^{\bullet}_{0}\,10001) + 27({}^{\bullet}_{1}\,00001) + 21({}^{\bullet}_{0}\,00001)$

Affine B6 ⊃ A1 + A1 + B4 - Level 1 - Class 0

Sl 0: $({}^{1}_{0}\,\bullet 0000)$ **Highest Weight** $({}^{1}_{0}\,00000)$

Sl 1: $({}^{0}_{1}\,\bullet 1000)$

Sl 2: $({}^{1}_{0}\,\bullet 0100) + ({}^{1}_{2}\,\bullet 0000) + ({}^{1}_{0}\,\bullet 0000)$

Sl 3: $({}^{0}_{1}\,\bullet 0010) + ({}^{2}_{1}\,\bullet 1000) + 2({}^{0}_{1}\,\bullet 1000)$

Sl 4: $({}^{1}_{0}\,\bullet 0002) + ({}^{1}_{2}\,\bullet 0100) + ({}^{1}_{0}\,\bullet 2000) + 2({}^{1}_{0}\,\bullet 0100) + ({}^{3}_{0}\,\bullet 0000) + 2({}^{1}_{2}\,\bullet 0000) + 3({}^{1}_{0}\,\bullet 0000)$

Sl 5: $({}^{0}_{1}\,\bullet 1100) + ({}^{2}_{1}\,\bullet 0010) + ({}^{0}_{1}\,\bullet 0002) + 2({}^{0}_{1}\,\bullet 0010) + 3({}^{2}_{1}\,\bullet 1000) + ({}^{0}_{3}\,\bullet 1000) + 5({}^{0}_{1}\,\bullet 1000)$

Sl 6: $({}^{1}_{0}\,\bullet 1010) + ({}^{1}_{2}\,\bullet 0002) + 2({}^{1}_{0}\,\bullet 0002) + ({}^{1}_{2}\,\bullet 2000) + ({}^{1}_{0}\,\bullet 0010) + ({}^{3}_{0}\,\bullet 0100) + 3({}^{1}_{2}\,\bullet 0100)$
$+ 2({}^{1}_{0}\,\bullet 2000) + 6({}^{1}_{0}\,\bullet 0100) + ({}^{3}_{2}\,\bullet 0000) + ({}^{3}_{0}\,\bullet 0000) + 5({}^{1}_{2}\,\bullet 0000) + 6({}^{1}_{0}\,\bullet 0000)$

Affine B6 ⊃ A1 + A1 + B4 - Level 1 - Class 1

sl 0: $({}^{0}_{0}\cdot 0001)$ **Highest Weight** $({}^{0}_{0}00001)$

sl 1: $({}^{1}_{1}\cdot 0001)$

sl 2: $({}^{0}_{0}\cdot 1001) + ({}^{2}_{0}\cdot 0001) + ({}^{0}_{2}\cdot 0001) + ({}^{0}_{0}\cdot 0001)$

sl 3: $({}^{1}_{1}\cdot 1001) + 3({}^{1}_{1}\cdot 0001)$

sl 4: $({}^{0}_{0}\cdot 0101) + ({}^{2}_{0}\cdot 1001) + ({}^{0}_{2}\cdot 1001) + 2({}^{0}_{0}\cdot 1001) + ({}^{2}_{2}\cdot 0001) + 2({}^{2}_{0}\cdot 0001) + 2({}^{0}_{2}\cdot 0001)$
$+ 4({}^{0}_{0}\cdot 0001)$

sl 5: $({}^{1}_{1}\cdot 0101) + 4({}^{1}_{1}\cdot 1001) + ({}^{3}_{1}\cdot 0001) + ({}^{1}_{3}\cdot 0001) + 7({}^{1}_{1}\cdot 0001)$

sl 6: $({}^{0}_{0}\cdot 0011) + ({}^{2}_{0}\cdot 0101) + ({}^{0}_{2}\cdot 0101) + ({}^{0}_{0}\cdot 2001) + 2({}^{0}_{0}\cdot 0101) + ({}^{2}_{2}\cdot 1001) + 3({}^{2}_{0}\cdot 1001)$
$+ 3({}^{0}_{2}\cdot 1001) + 6({}^{0}_{0}\cdot 1001) + 3({}^{2}_{2}\cdot 0001) + 6({}^{2}_{0}\cdot 0001) + 6({}^{0}_{2}\cdot 0001) + 8({}^{0}_{0}\cdot 0001)$

Affine B6 ⊃ A3 + B3 - Level 1 - Class 0

sl 0: $({}^{1}_{0}0\cdot 000)$ **Highest Weight** $({}^{1}_{0}00000)$

sl 1: $({}^{0}_{1}0\cdot 100)$

sl 2: $({}^{1}_{0}0\cdot 010) + ({}^{0}_{1}1\cdot 000) + ({}^{1}_{0}0\cdot 000)$

sl 3: $({}^{0}_{1}0\cdot 002) + ({}^{1}_{0}1\cdot 100) + 2({}^{0}_{1}0\cdot 100)$

sl 4: $({}^{0}_{1}1\cdot 010) + ({}^{1}_{0}0\cdot 200) + ({}^{1}_{0}0\cdot 002) + 2({}^{1}_{0}0\cdot 010) + ({}^{2}_{1}0\cdot 000) + 2({}^{0}_{1}1\cdot 000) + 3({}^{1}_{0}0\cdot 000)$

sl 5: $({}^{1}_{0}1\cdot 002) + ({}^{0}_{1}0\cdot 110) + ({}^{1}_{2}0\cdot 100) + 2({}^{0}_{1}0\cdot 002) + 3({}^{1}_{0}1\cdot 100) + ({}^{0}_{1}0\cdot 010) + 5({}^{0}_{1}0\cdot 100)$

sl 6: $({}^{1}_{0}0\cdot 102) + ({}^{2}_{1}0\cdot 010) + ({}^{0}_{1}1\cdot 200) + ({}^{0}_{1}1\cdot 002) + 3({}^{0}_{1}1\cdot 010) + 2({}^{1}_{0}0\cdot 200) + 2({}^{1}_{0}0\cdot 002)$
$+ 6({}^{1}_{0}0\cdot 010) + ({}^{1}_{0}2\cdot 000) + 2({}^{2}_{1}0\cdot 000) + ({}^{1}_{0}0\cdot 100) + ({}^{0}_{3}0\cdot 000) + 5({}^{0}_{1}1\cdot 000) + 6({}^{1}_{0}0\cdot 000)$

Affine B6 ⊃ A3 + B3 - Level 1 - Class 1

Sl 0: $(^{0}_{0}\,0\cdot001)$ **Highest Weight** $(^{0}_{0}\,00001)$

Sl 1: $(^{0}_{0}\,1\cdot001)$

Sl 2: $(^{1}_{1}\,0\cdot001) + (^{0}_{0}\,0\cdot101) + (^{0}_{0}\,0\cdot001)$

Sl 3: $(^{0}_{0}\,1\cdot101) + (^{2}_{0}\,0\cdot001) + (^{0}_{2}\,0\cdot001) + 2(^{0}_{0}\,1\cdot001)$

Sl 4: $(^{1}_{1}\,0\cdot101) + (^{0}_{0}\,0\cdot011) + (^{0}_{0}\,2\cdot001) + 3(^{1}_{1}\,0\cdot001) + 2(^{0}_{0}\,0\cdot101) + 3(^{0}_{0}\,0\cdot001)$

Sl 5: $(^{0}_{0}\,1\cdot011) + (^{2}_{0}\,0\cdot101) + (^{0}_{2}\,0\cdot101) + (^{1}_{1}\,1\cdot001) + 3(^{0}_{0}\,1\cdot101) + 2(^{2}_{0}\,0\cdot001) + 2(^{0}_{2}\,0\cdot001)$
$+ 6(^{0}_{0}\,1\cdot001)$

Sl 6: $(^{1}_{1}\,0\cdot011) + (^{0}_{0}\,2\cdot101) + 4(^{1}_{1}\,0\cdot101) + (^{0}_{0}\,0\cdot201) + (^{0}_{0}\,0\cdot003) + (^{2}_{0}\,1\cdot001) + (^{0}_{2}\,1\cdot001)$
$+ 2(^{0}_{0}\,0\cdot011) + 2(^{0}_{0}\,2\cdot001) + 8(^{1}_{1}\,0\cdot001) + 5(^{0}_{0}\,0\cdot101) + 7(^{0}_{0}\,0\cdot001)$

Affine B6 ⊃ D4 + C2 - Level 1 - Class 0

Sl 0: $(^{1}_{0}\,00\cdot00)$ **Highest Weight** $(^{1}_{0}\,00000)$

Sl 1: $(^{0}_{1}\,00\cdot10)$

Sl 2: $(^{0}_{1}\,01\cdot00) + (^{1}_{0}\,00\cdot02) + (^{1}_{0}\,00\cdot00)$

Sl 3: $(^{1}_{0}\,01\cdot10) + (^{0}_{1}\,00\cdot02) + 2(^{0}_{1}\,00\cdot10)$

Sl 4: $(^{0}_{1}\,01\cdot02) + (^{1}_{0}\,10\cdot00) + (^{1}_{0}\,00\cdot20) + 2(^{0}_{1}\,01\cdot00) + 2(^{1}_{0}\,00\cdot02) + (^{1}_{0}\,00\cdot10) + 3(^{1}_{0}\,00\cdot00)$

Sl 5: $(^{0}_{1}\,10\cdot10) + (^{1}_{0}\,01\cdot02) + 3(^{1}_{0}\,01\cdot10) + (^{0}_{1}\,00\cdot12) + 2(^{0}_{1}\,00\cdot02) + 5(^{0}_{1}\,00\cdot10) + (^{0}_{1}\,00\cdot00)$

Sl 6: $(^{1}_{0}\,10\cdot02) + (^{0}_{1}\,01\cdot20) + (^{1}_{0}\,02\cdot00) + 3(^{0}_{1}\,01\cdot02) + (^{1}_{2}\,00\cdot00) + (^{0}_{1}\,01\cdot10) + 2(^{1}_{0}\,10\cdot00)$
$+ (^{1}_{0}\,00\cdot12) + 2(^{1}_{0}\,00\cdot20) + 5(^{0}_{1}\,01\cdot00) + 6(^{1}_{0}\,00\cdot02) + 2(^{1}_{0}\,00\cdot10) + 6(^{1}_{0}\,00\cdot00)$

Affine B6 ⊃ D4 + C2 - Level 1 - Class 1

sl 0: $({}^{0}_{0}\,00\cdot 01)$ **Highest Weight** $({}^{0}_{0}\,00001)$

sl 1: $({}^{0}_{0}\,01\cdot 01)$

sl 2: $({}^{0}_{0}\,10\cdot 01) + ({}^{0}_{0}\,00\cdot 11) + ({}^{0}_{0}\,00\cdot 01)$

sl 3: $({}^{1}_{1}\,00\cdot 01) + ({}^{0}_{0}\,01\cdot 11) + 2({}^{0}_{0}\,01\cdot 01)$

sl 4: $({}^{0}_{0}\,10\cdot 11) + ({}^{0}_{0}\,02\cdot 01) + ({}^{0}_{2}\,00\cdot 01) + ({}^{2}_{0}\,00\cdot 01) + 2({}^{0}_{0}\,10\cdot 01) + ({}^{0}_{0}\,00\cdot 03) + 2({}^{0}_{0}\,00\cdot 11)$
$+ 3({}^{0}_{0}\,00\cdot 01)$

sl 5: $({}^{0}_{0}\,11\cdot 01) + ({}^{1}_{1}\,00\cdot 11) + ({}^{0}_{0}\,01\cdot 03) + 3({}^{1}_{1}\,00\cdot 01) + 3({}^{0}_{0}\,01\cdot 11) + 5({}^{0}_{0}\,01\cdot 01)$

sl 6: $({}^{1}_{1}\,01\cdot 01) + ({}^{0}_{0}\,02\cdot 11) + ({}^{0}_{0}\,10\cdot 03) + ({}^{0}_{2}\,00\cdot 11) + ({}^{2}_{0}\,00\cdot 11) + 3({}^{0}_{0}\,10\cdot 11) + 2({}^{0}_{0}\,02\cdot 01)$
$+ 2({}^{0}_{2}\,00\cdot 01) + 2({}^{2}_{0}\,00\cdot 01) + 7({}^{0}_{0}\,10\cdot 01) + ({}^{0}_{0}\,00\cdot 21) + 2({}^{0}_{0}\,00\cdot 03) + 5({}^{0}_{0}\,00\cdot 11) + 6({}^{0}_{0}\,00\cdot 01)$

Affine B6 ⊃ D5 + A1 - Level 1 - Class 0

sl 0: $({}^{1}_{0}\,000\cdot 0)$ **Highest Weight** $({}^{1}_{0}\,00000)$

sl 1: $({}^{0}_{1}\,000\cdot 2)$

sl 2: $({}^{0}_{1}\,001\cdot 0) + ({}^{1}_{0}\,000\cdot 2) + ({}^{1}_{0}\,000\cdot 0)$

sl 3: $({}^{1}_{0}\,001\cdot 2) + 2({}^{0}_{1}\,000\cdot 2) + ({}^{0}_{1}\,000\cdot 0)$

sl 4: $({}^{1}_{0}\,010\cdot 0) + ({}^{0}_{1}\,001\cdot 2) + 2({}^{0}_{1}\,001\cdot 0) + ({}^{1}_{0}\,000\cdot 4) + 2({}^{1}_{0}\,000\cdot 2) + 3({}^{1}_{0}\,000\cdot 0)$

sl 5: $({}^{0}_{1}\,010\cdot 2) + 3({}^{1}_{0}\,001\cdot 2) + ({}^{1}_{0}\,001\cdot 0) + ({}^{0}_{1}\,000\cdot 4) + 5({}^{0}_{1}\,000\cdot 2) + 2({}^{0}_{1}\,000\cdot 0)$

sl 6: $({}^{0}_{1}\,100\cdot 0) + ({}^{1}_{0}\,002\cdot 0) + ({}^{1}_{0}\,010\cdot 2) + 2({}^{1}_{0}\,010\cdot 0) + ({}^{0}_{1}\,001\cdot 4) + 3({}^{0}_{1}\,001\cdot 2) + 5({}^{0}_{1}\,001\cdot 0)$
$+ 2({}^{1}_{0}\,000\cdot 4) + 6({}^{1}_{0}\,000\cdot 2) + 6({}^{1}_{0}\,000\cdot 0)$

Affine B6 ⊃ D5 + A1 - Level 1 - Class 1

Sl 0: $({}^{0}_{0}000\cdot1)$ **Highest Weight** $({}^{0}_{0}00001)$

Sl 1: $({}^{0}_{0}001\cdot1)$

Sl 2: $({}^{0}_{0}010\cdot1) + ({}^{0}_{0}000\cdot3) + ({}^{0}_{0}000\cdot1)$

Sl 3: $({}^{0}_{0}100\cdot1) + ({}^{0}_{0}001\cdot3) + 2({}^{0}_{0}001\cdot1)$

Sl 4: $({}^{1}_{1}000\cdot1) + ({}^{0}_{0}002\cdot1) + ({}^{0}_{0}010\cdot3) + 2({}^{0}_{0}010\cdot1) + 2({}^{0}_{0}000\cdot3) + 3({}^{0}_{0}000\cdot1)$

Sl 5: $({}^{0}_{0}011\cdot1) + ({}^{0}_{0}100\cdot3) + ({}^{0}_{2}000\cdot1) + ({}^{2}_{0}000\cdot1) + 2({}^{0}_{0}100\cdot1) + 3({}^{0}_{0}001\cdot3) + 5({}^{0}_{0}001\cdot1)$

Sl 6: $({}^{0}_{0}101\cdot1) + ({}^{1}_{1}000\cdot3) + 3({}^{1}_{1}000\cdot1) + ({}^{0}_{0}002\cdot3) + 2({}^{0}_{0}002\cdot1) + 3({}^{0}_{0}010\cdot3) + 6({}^{0}_{0}010\cdot1)$

$+ ({}^{0}_{0}000\cdot5) + 4({}^{0}_{0}000\cdot3) + 6({}^{0}_{0}000\cdot1)$

Affine B6 ⊃ D6 - Level 1 - Class 0

Sl 0: $({}^{1}_{0}0000\cdot)$ **Highest Weight** $({}^{1}_{0}00000)$

Sl 1: $({}^{0}_{1}0000\cdot)$

Sl 2: $({}^{0}_{1}0001\cdot) + ({}^{1}_{0}0000\cdot)$

Sl 3: $({}^{1}_{0}0001\cdot) + 2({}^{0}_{1}0000\cdot)$

Sl 4: $({}^{1}_{0}0010\cdot) + 2({}^{0}_{1}0001\cdot) + 3({}^{1}_{0}0000\cdot)$

Sl 5: $({}^{0}_{1}0010\cdot) + 3({}^{1}_{0}0001\cdot) + 4({}^{0}_{1}0000\cdot)$

Sl 6: $({}^{0}_{1}0100\cdot) + ({}^{1}_{0}0002\cdot) + 2({}^{1}_{0}0010\cdot) + 5({}^{0}_{1}0001\cdot) + 6({}^{1}_{0}0000\cdot)$

Affine B6 ⊃ D6 - Level 1 - Class 0

```
          0                         0
Sl 0:   ( 0000•)  Highest Weight  ( 00001)
          0                         0

          0
Sl 1:   ( 0001•)
          0

          0            0
Sl 2:   ( 0010•) +   ( 0000•)
          0            0

          0            0
Sl 3:   ( 0100•) +  2( 0001•)
          0            0

          0            0            0            0
Sl 4:   ( 1000•) +   ( 0002•) +  2( 0010•) +  2( 0000•)
          0            0            0            0

          1            0            0            0
Sl 5:   ( 0000•) +   ( 0011•) +  2( 0100•) +  4( 0001•)
          1            0            0            0

          0            0            2            0            0            0            0
Sl 6:   ( 0101•) +   ( 0000•) +   ( 0000•) +  2( 1000•) +  2( 0002•) +  5( 0010•) +  4( 0000•)
          0            2            0            0            0            0            0
```

Affine B7 Branching Rules

Affine B7 ⊃ B7 - Level 1 - Class 0

Sl 0: $(^{\bullet}000000)_{1}$ **Highest Weight** $(^{0}000000)_{1}$

Sl 1: $(^{\bullet}010000)_{0}$ + $(^{\bullet}000000)_{1}$

Sl 2: $(^{\bullet}000100)_{0}$ + $(^{\bullet}100000)_{1}$ + $(^{\bullet}010000)_{0}$ + 2$(^{\bullet}000000)_{1}$

Sl 3: $(^{\bullet}001000)_{1}$ + $(^{\bullet}000002)_{0}$ + $(^{\bullet}000100)_{0}$ + 2$(^{\bullet}100000)_{1}$ + 3$(^{\bullet}010000)_{0}$ + 3$(^{\bullet}000000)_{1}$

Sl 4: $(^{\bullet}000010)_{1}$ + $(^{\bullet}110000)_{0}$ + 2$(^{\bullet}001000)_{1}$ + $(^{\bullet}000002)_{0}$ + $(^{\bullet}000010)_{0}$ + 3$(^{\bullet}000100)_{0}$ + $(^{\bullet}000000)_{3}$
+ 4$(^{\bullet}100000)_{1}$ + 5$(^{\bullet}010000)_{0}$ + 6$(^{\bullet}000000)_{1}$

Sl 5: $(^{\bullet}100100)_{0}$ + $(^{\bullet}000002)_{1}$ + 2$(^{\bullet}000010)_{1}$ + $(^{\bullet}010000)_{2}$ + 2$(^{\bullet}110000)_{0}$ + 5$(^{\bullet}001000)_{1}$ + 3$(^{\bullet}000002)_{0}$
+ $(^{\bullet}000010)_{0}$ + 5$(^{\bullet}000100)_{0}$ + $(^{\bullet}001000)_{0}$ + $(^{\bullet}000000)_{3}$ + 8$(^{\bullet}100000)_{1}$ + 10$(^{\bullet}010000)_{0}$ + 9$(^{\bullet}000000)_{1}$

Sl 6: $(^{\bullet}100002)_{0}$ + $(^{\bullet}011000)_{0}$ + $(^{\bullet}000100)_{2}$ + 2$(^{\bullet}100100)_{0}$ + 2$(^{\bullet}000002)_{1}$ + 5$(^{\bullet}000010)_{1}$ + $(^{\bullet}200000)_{1}$
+ 2$(^{\bullet}010000)_{2}$ + $(^{\bullet}000100)_{1}$ + 5$(^{\bullet}110000)_{0}$ + 10$(^{\bullet}001000)_{1}$ + 5$(^{\bullet}000002)_{0}$ + 3$(^{\bullet}000010)_{0}$ + 11$(^{\bullet}000100)_{0}$
+ $(^{\bullet}001000)_{0}$ + 3$(^{\bullet}000000)_{3}$ + 15$(^{\bullet}100000)_{1}$ + 16$(^{\bullet}010000)_{0}$ + $(^{\bullet}100000)_{0}$ + 16$(^{\bullet}000000)_{1}$

Affine B7 ⊃ B7 - Level 1 - Class 0

Sl 0: $(^{\bullet}000000)_{0}$ **Highest Weight** $(^{1}000000)_{0}$

Sl 1: $(^{\bullet}100000)_{0}$

Sl 2: $(^{\bullet}001000)_{0}$ + $(^{\bullet}000000)_{2}$ + $(^{\bullet}100000)_{0}$ + $(^{\bullet}000000)_{0}$

Sl 3: $(^{\bullet}000010)_{0}$ + $(^{\bullet}010000)_{1}$ + $(^{\bullet}001000)_{0}$ + $(^{\bullet}000000)_{2}$ + 3$(^{\bullet}100000)_{0}$ + $(^{\bullet}000000)_{0}$

Sl 4: $(^{\bullet}000100)_{1}$ + $(^{\bullet}000002)_{0}$ + $(^{\bullet}000010)_{0}$ + $(^{\bullet}200000)_{0}$ + 2$(^{\bullet}010000)_{1}$ + 3$(^{\bullet}001000)_{0}$ + 3$(^{\bullet}000000)_{2}$
+ 4$(^{\bullet}100000)_{0}$ + 3$(^{\bullet}000000)_{0}$

Sl 5: $(^{\bullet}000002)_{1}$ + $(^{\bullet}101000)_{0}$ + 2$(^{\bullet}000100)_{1}$ + $(^{\bullet}000002)_{0}$ + $(^{\bullet}100000)_{2}$ + 3$(^{\bullet}000010)_{0}$ + $(^{\bullet}200000)_{0}$
+ 5$(^{\bullet}010000)_{1}$ + $(^{\bullet}000100)_{0}$ + 5$(^{\bullet}001000)_{0}$ + 4$(^{\bullet}000000)_{2}$ + 9$(^{\bullet}100000)_{0}$ + 3$(^{\bullet}000000)_{0}$

(continued next page)

Sl 6: $(\dot{1}00010)_{0}$ + $(\dot{0}01000)_{2}$ + $(\dot{0}20000)_{0}$ + $2(\dot{0}00002)_{1}$ + $2(\dot{1}01000)_{0}$ + $(\dot{0}00010)_{1}$ + $5(\dot{0}00100)_{1}$

+ $3(\dot{0}00002)_{0}$ + $2(\dot{1}00000)_{2}$ + $5(\dot{0}00010)_{0}$ + $4(\dot{2}00000)_{0}$ + $9(\dot{0}10000)_{1}$ + $(\dot{0}00100)_{0}$ + $11(\dot{0}01000)_{0}$

+ $(\dot{0}10000)_{0}$ + $9(\dot{0}00000)_{2}$ + $13(\dot{1}00000)_{0}$ + $7(\dot{0}00000)_{0}$

Affine B7 ⊃ B7 - Level 1 - Class 1

Sl 0: $(\dot{0}00001)_{0}$ **Highest Weight** $(\overset{0}{0}00001)_{0}$

Sl 1: $(\dot{0}00001)_{1}$ + $(\dot{0}00001)_{0}$

Sl 2: $(\dot{1}00001)_{0}$ + $2(\dot{0}00001)_{1}$ + $2(\dot{0}00001)_{0}$

Sl 3: $(\dot{0}10001)_{0}$ + $(\dot{0}00001)_{2}$ + $2(\dot{1}00001)_{0}$ + $4(\dot{0}00001)_{1}$ + $4(\dot{0}00001)_{0}$

Sl 4: $(\dot{0}01001)_{0}$ + $(\dot{1}00001)_{1}$ + $2(\dot{0}10001)_{0}$ + $2(\dot{0}00001)_{2}$ + $5(\dot{1}00001)_{0}$ + $8(\dot{0}00001)_{1}$ + $7(\dot{0}00001)_{0}$

Sl 5: $(\dot{0}10001)_{1}$ + $(\dot{0}00101)_{0}$ + $2(\dot{0}01001)_{0}$ + $3(\dot{1}00001)_{1}$ + $5(\dot{0}10001)_{0}$ + $5(\dot{0}00001)_{2}$ + $10(\dot{1}00001)_{0}$

+ $15(\dot{0}00001)_{1}$ + $12(\dot{0}00001)_{0}$

Sl 6: $(\dot{0}01001)_{1}$ + $(\dot{0}00011)_{0}$ + $(\dot{2}00001)_{0}$ + $3(\dot{0}10001)_{1}$ + $2(\dot{0}00101)_{0}$ + $5(\dot{0}01001)_{0}$ + $(\dot{0}00001)_{3}$

+ $7(\dot{1}00001)_{1}$ + $11(\dot{0}10001)_{0}$ + $10(\dot{0}00001)_{2}$ + $20(\dot{1}00001)_{0}$ + $27(\dot{0}00001)_{1}$ + $21(\dot{0}00001)_{0}$

Sl 7: $(\dot{0}00101)_{1}$ + $(\dot{1}10001)_{0}$ + $3(\dot{0}01001)_{1}$ + $(\dot{0}00003)_{0}$ + $(\dot{1}00001)_{2}$ + $2(\dot{0}00011)_{0}$ + $2(\dot{2}00001)_{0}$

+ $8(\dot{0}10001)_{1}$ + $5(\dot{0}00101)_{0}$ + $11(\dot{0}01001)_{0}$ + $2(\dot{0}00001)_{3}$ + $16(\dot{1}00001)_{1}$ + $22(\dot{0}10001)_{0}$ + $20(\dot{0}00001)_{2}$

+ $37(\dot{1}00001)_{0}$ + $48(\dot{0}00001)_{1}$ + $34(\dot{0}00001)_{0}$

Sl 8: $(\dot{1}01001)_{0}$ + $(\dot{0}00011)_{1}$ + $(\dot{0}10001)_{2}$ + $3(\dot{0}00101)_{1}$ + $3(\dot{1}10001)_{0}$ + $8(\dot{0}01001)_{1}$ + $2(\dot{0}00003)_{0}$

+ $3(\dot{1}00001)_{2}$ + $5(\dot{0}00011)_{0}$ + $6(\dot{2}00001)_{0}$ + $18(\dot{0}10001)_{1}$ + $11(\dot{0}00101)_{0}$ + $23(\dot{0}01001)_{0}$ + $5(\dot{0}00001)_{3}$

+ $33(\dot{1}00001)_{1}$ + $42(\dot{0}10001)_{0}$ + $37(\dot{0}00001)_{2}$ + $68(\dot{1}00001)_{0}$ + $82(\dot{0}00001)_{1}$ + $56(\dot{0}00001)_{0}$

Affine B7 ⊃ A1 + A1 + B5 - Level 1 - Class 0

Sl 0: $(^{1}_{0}\cdot 00000)$ **Highest Weight** $(^{1}_{0}\,000000)$

Sl 1: $(^{0}_{1}\cdot 10000)$

Sl 2: $(^{1}_{0}\cdot 01000) + (^{1}_{2}\cdot 00000) + (^{1}_{0}\cdot 00000)$

Sl 3: $(^{0}_{1}\cdot 00100) + (^{2}_{1}\cdot 10000) + 2(^{0}_{1}\cdot 10000)$

Sl 4: $(^{1}_{0}\cdot 00010) + (^{1}_{2}\cdot 01000) + (^{1}_{0}\cdot 20000) + 2(^{1}_{0}\cdot 01000) + (^{3}_{0}\cdot 00000) + 2(^{1}_{2}\cdot 00000) + 3(^{1}_{0}\cdot 00000)$

Sl 5: $(^{0}_{1}\cdot 00002) + (^{0}_{1}\cdot 11000) + (^{2}_{1}\cdot 00100) + 2(^{0}_{1}\cdot 00100) + 3(^{2}_{1}\cdot 10000) + (^{0}_{3}\cdot 10000) + 5(^{0}_{1}\cdot 10000)$

Sl 6: $(^{1}_{0}\cdot 10100) + (^{1}_{2}\cdot 00010) + (^{1}_{0}\cdot 00002) + 2(^{1}_{0}\cdot 00010) + (^{1}_{2}\cdot 20000) + (^{3}_{0}\cdot 01000) + 3(^{1}_{2}\cdot 01000)$
$+ 2(^{1}_{0}\cdot 20000) + 6(^{1}_{0}\cdot 01000) + (^{3}_{2}\cdot 00000) + (^{3}_{0}\cdot 00000) + 5(^{1}_{2}\cdot 00000) + 6(^{1}_{0}\cdot 00000)$

Affine B7 ⊃ A1 + A1 + B5 - Level 1 - Class 1

Sl 0: $(^{0}_{0}\cdot 00001)$ **Highest Weight** $(^{0}_{0}\,000001)$

Sl 1: $(^{1}_{1}\cdot 00001)$

Sl 2: $(^{0}_{0}\cdot 10001) + (^{2}_{0}\cdot 00001) + (^{0}_{2}\cdot 00001) + (^{0}_{0}\cdot 00001)$

Sl 3: $(^{1}_{1}\cdot 10001) + 3(^{1}_{1}\cdot 00001)$

Sl 4: $(^{0}_{0}\cdot 01001) + (^{2}_{0}\cdot 10001) + (^{0}_{2}\cdot 10001) + 2(^{0}_{0}\cdot 10001) + (^{2}_{2}\cdot 00001) + 2(^{2}_{0}\cdot 00001) + 2(^{0}_{2}\cdot 00001)$
$+ 4(^{0}_{0}\cdot 00001)$

Sl 5: $(^{1}_{1}\cdot 01001) + 4(^{1}_{1}\cdot 10001) + (^{3}_{1}\cdot 00001) + (^{1}_{3}\cdot 00001) + 7(^{1}_{1}\cdot 00001)$

Sl 6: $(^{0}_{0}\cdot 00101) + (^{2}_{0}\cdot 01001) + (^{0}_{2}\cdot 01001) + (^{0}_{0}\cdot 20001) + 2(^{0}_{0}\cdot 01001) + (^{2}_{2}\cdot 10001) + 3(^{2}_{0}\cdot 10001)$
$+ 3(^{0}_{2}\cdot 10001) + 6(^{0}_{0}\cdot 10001) + 3(^{2}_{2}\cdot 00001) + 6(^{2}_{0}\cdot 00001) + 6(^{0}_{2}\cdot 00001) + 8(^{0}_{0}\cdot 00001)$

Affine B7 ⊃ A3 + B4 - Level 1 - Class 0

```
          1                         1
Sl 0:   ( 0•0000)  Highest Weight ( 000000)
          0                         0

          0
Sl 1:   ( 0•1000)
          1

          1             0             1
Sl 2:   ( 0•0100) +   ( 1•0000) +   ( 0•0000)
          0             1             0

          0             1             0
Sl 3:   ( 0•0010) +   ( 1•1000) +  2( 0•1000)
          1             0             1

          1             0             1             1             2             0             1
Sl 4:   ( 0•0002) +   ( 1•0100) +   ( 0•2000) +  2( 0•0100) +   ( 0•0000) +  2( 1•0000) +  3( 0•0000)
          0             1             0             0             1             1             0

          1             0             0             0             1             1             0
Sl 5:   ( 1•0010) +   ( 0•1100) +   ( 0•0002) +  2( 0•0010) +   ( 0•1000) +  3( 1•1000) +  5( 0•1000)
          0             1             1             1             2             0             1

          1             0             2             1             0             1             0
Sl 6:   ( 0•1010) +   ( 1•0002) +   ( 0•0100) +  2( 0•0002) +   ( 1•2000) +   ( 0•0010) +  3( 1•0100)
          0             1             1             0             1             0             1

          1             1             1             2             0             0             1
    +  2( 0•2000) +  6( 0•0100) +   ( 2•0000) +  2( 0•0000) +   ( 0•0000) +  5( 1•0000) +  6( 0•0000)
          0             0             0             1             3             1             0
```

Affine B7 ⊃ A3 + B4 - Level 1 - Class 1

```
          0                         0
Sl 0:   ( 0•0001)  Highest Weight ( 000001)
          0                         0

          0
Sl 1:   ( 1•0001)
          0

          0             1             0
Sl 2:   ( 0•1001) +   ( 0•0001) +   ( 0•0001)
          0             1             0

          0             2             0             0
Sl 3:   ( 1•1001) +   ( 0•0001) +   ( 0•0001) +  2( 1•0001)
          0             0             2             0

          1             0             0             0             1             0
Sl 4:   ( 0•1001) +   ( 0•0101) +   ( 2•0001) +  2( 0•1001) +  3( 0•0001) +  3( 0•0001)
          1             0             0             0             1             0

          0             2             0             0             1             2             0
Sl 5:   ( 1•0101) +   ( 0•1001) +   ( 0•1001) +  3( 1•1001) +   ( 1•0001) +  2( 0•0001) +  2( 0•0001)
          0             0             2             0             1             0             2

          0
    +  6( 1•0001)
          0

          1             0             0             0             1             0             2
Sl 6:   ( 0•0101) +   ( 0•0011) +   ( 2•1001) +   ( 0•2001) +  4( 0•1001) +  2( 0•0101) +   ( 1•0001)
          1             0             0             0             1             0             0

          0             0             0             1             0
    +   ( 1•0001) +  2( 2•0001) +  5( 0•1001) +  8( 0•0001) +  7( 0•0001)
          2             0             0             1             0
```

Affine B7 ⊃ D4 + B3 - Level 1 - Class 0

Sl 0: $({}^{1}_{0}\,00\cdot000)$ **Highest Weight** $({}^{1}_{0}\,000000)$

Sl 1: $({}^{0}_{1}\,00\cdot100)$

Sl 2: $({}^{1}_{0}\,00\cdot010) + ({}^{0}_{1}\,01\cdot000) + ({}^{1}_{0}\,00\cdot000)$

Sl 3: $({}^{1}_{0}\,01\cdot100) + ({}^{0}_{1}\,00\cdot002) + 2({}^{0}_{1}\,00\cdot100)$

Sl 4: $({}^{0}_{1}\,01\cdot010) + ({}^{1}_{0}\,00\cdot200) + ({}^{1}_{0}\,00\cdot002) + ({}^{1}_{0}\,10\cdot000) + 2({}^{1}_{0}\,00\cdot010) + 2({}^{0}_{1}\,01\cdot000) + 3({}^{1}_{0}\,00\cdot000)$

Sl 5: $({}^{1}_{0}\,01\cdot002) + ({}^{0}_{1}\,10\cdot100) + ({}^{0}_{1}\,00\cdot110) + 3({}^{1}_{0}\,01\cdot100) + 2({}^{0}_{1}\,00\cdot002) + ({}^{0}_{1}\,00\cdot010) + 5({}^{0}_{1}\,00\cdot100)$

Sl 6: $({}^{1}_{0}\,10\cdot010) + ({}^{0}_{1}\,01\cdot200) + ({}^{0}_{1}\,01\cdot002) + ({}^{1}_{0}\,00\cdot102) + 3({}^{0}_{1}\,01\cdot010) + ({}^{1}_{0}\,02\cdot000) + ({}^{1}_{2}\,00\cdot000)$

$+ 2({}^{1}_{0}\,00\cdot200) + 2({}^{1}_{0}\,00\cdot002) + 2({}^{1}_{0}\,10\cdot000) + 6({}^{1}_{0}\,00\cdot010) + 5({}^{0}_{1}\,01\cdot000) + ({}^{1}_{0}\,00\cdot100) + 6({}^{1}_{0}\,00\cdot000)$

Affine B7 ⊃ D4 + B3 - Level 1 - Class 1

Sl 0: $({}^{0}_{0}\,00\cdot001)$ **Highest Weight** $({}^{0}_{0}\,000001)$

Sl 1: $({}^{0}_{0}\,01\cdot001)$

Sl 2: $({}^{0}_{0}\,10\cdot001) + ({}^{0}_{0}\,00\cdot101) + ({}^{0}_{0}\,00\cdot001)$

Sl 3: $({}^{0}_{0}\,01\cdot101) + ({}^{1}_{1}\,00\cdot001) + 2({}^{0}_{0}\,01\cdot001)$

Sl 4: $({}^{0}_{0}\,10\cdot101) + ({}^{0}_{0}\,02\cdot001) + ({}^{0}_{2}\,00\cdot001) + ({}^{2}_{0}\,00\cdot001) + 2({}^{0}_{0}\,10\cdot001) + ({}^{0}_{0}\,00\cdot011) + 2({}^{0}_{0}\,00\cdot101)$

$+ 3({}^{0}_{0}\,00\cdot001)$

Sl 5: $({}^{1}_{1}\,00\cdot101) + ({}^{0}_{0}\,11\cdot001) + ({}^{0}_{0}\,01\cdot011) + 3({}^{0}_{0}\,01\cdot101) + 3({}^{1}_{1}\,00\cdot001) + 5({}^{0}_{0}\,01\cdot001)$

Sl 6: $({}^{0}_{0}\,10\cdot011) + ({}^{0}_{0}\,02\cdot101) + ({}^{1}_{1}\,01\cdot001) + ({}^{0}_{2}\,00\cdot101) + ({}^{2}_{0}\,00\cdot101) + 3({}^{0}_{0}\,10\cdot101) + 2({}^{0}_{0}\,02\cdot001)$

$+ 2({}^{0}_{2}\,00\cdot001) + 2({}^{2}_{0}\,00\cdot001) + ({}^{0}_{0}\,00\cdot201) + ({}^{0}_{0}\,00\cdot003) + 7({}^{0}_{0}\,10\cdot001) + 2({}^{0}_{0}\,00\cdot011) + 5({}^{0}_{0}\,00\cdot101)$

$+ 6({}^{0}_{0}\,00\cdot001)$

Affine B7 ⊃ D5 + C2 - Level 1 - Class 0

```
            1                         1
Sl 0:     ( 000•00)  Highest Weight ( 000000)
            0                         0

            0
Sl 1:     ( 000•10)
            1

            0             1             1
Sl 2:     ( 001•00) +   ( 000•02) +   ( 000•00)
            1             0             0

            1             0             0
Sl 3:     ( 001•10) +   ( 000•02) +  2( 000•10)
            0             1             1

            0             1             0             1             1             1             1
Sl 4:     ( 001•02) +   ( 010•00) +  2( 001•00) +   ( 000•20) +  2( 000•02) +   ( 000•10) +  3( 000•00)
            1             0             1             0             0             0             0

            0             1             1             0             0             0             0
Sl 5:     ( 010•10) +   ( 001•02) +  3( 001•10) +   ( 000•12) +  2( 000•02) +  5( 000•10) +   ( 000•00)
            1             0             0             1             1             1             1

            1             0             1             0             0             1             0
Sl 6:     ( 010•02) +   ( 100•00) +   ( 002•00) +   ( 001•20) +  3( 001•02) +  2( 010•00) +   ( 001•10)
            0             1             0             1             1             0             1

            1             0             1             1             1             1
    +     ( 000•12) +  5( 001•00) +  2( 000•20) +  6( 000•02) +  2( 000•10) +  6( 000•00)
            0             1             0             0             0             0
```

Affine B7 ⊃ D5 + C2 - Level 1 - Class 1

```
            0                         0
Sl 0:     ( 000•01)  Highest Weight ( 000001)
            0                         0

            0
Sl 1:     ( 001•01)
            0

            0             0             0
Sl 2:     ( 010•01) +   ( 000•11) +   ( 000•01)
            0             0             0

            0             0             0
Sl 3:     ( 100•01) +   ( 001•11) +  2( 001•01)
            0             0             0

            1             0             0             0             0             0             0
Sl 4:     ( 000•01) +   ( 010•11) +   ( 002•01) +  2( 010•01) +   ( 000•03) +  2( 000•11) +  3( 000•01)
            1             0             0             0             0             0             0

            0             0             0             2             0             0             0
Sl 5:     ( 011•01) +   ( 100•11) +   ( 000•01) +   ( 000•01) +  2( 100•01) +   ( 001•03) +  3( 001•11)
            0             0             2             0             0             0             0

            0
    +    5( 001•01)
            0

            0             1             0             0             1             0             0
Sl 6:     ( 101•01) +   ( 000•11) +   ( 002•11) +   ( 010•03) +  3( 000•01) +  3( 010•11) +  2( 002•01)
            0             1             0             0             1             0             0

            0             0             0             0             0
    +    6( 010•01) +   ( 000•21) +  2( 000•03) +  5( 000•11) +  6( 000•01)
            0             0             0             0             0
```

Affine B7 ⊃ D6 + A1 - Level 1 - Class 0

Sl 0: $(^{1}_{0}\,0000\bullet 0)$ **Highest Weight** $(^{1}_{0}\,000000)$

Sl 1: $(^{0}_{1}\,0000\bullet 2)$

Sl 2: $(^{0}_{1}\,0001\bullet 0) + (^{1}_{0}\,0000\bullet 2) + (^{1}_{0}\,0000\bullet 0)$

Sl 3: $(^{1}_{0}\,0001\bullet 2) + 2(^{0}_{1}\,0000\bullet 2) + (^{0}_{1}\,0000\bullet 0)$

Sl 4: $(^{1}_{0}\,0010\bullet 0) + (^{0}_{1}\,0001\bullet 2) + 2(^{0}_{1}\,0001\bullet 0) + (^{1}_{0}\,0000\bullet 4) + 2(^{1}_{0}\,0000\bullet 2) + 3(^{1}_{0}\,0000\bullet 0)$

Sl 5: $(^{0}_{1}\,0010\bullet 2) + 3(^{1}_{0}\,0001\bullet 2) + (^{1}_{0}\,0001\bullet 0) + (^{0}_{1}\,0000\bullet 4) + 5(^{0}_{1}\,0000\bullet 2) + 2(^{0}_{1}\,0000\bullet 0)$

Sl 6: $(^{0}_{1}\,0100\bullet 0) + (^{1}_{0}\,0002\bullet 0) + (^{1}_{0}\,0010\bullet 2) + 2(^{1}_{0}\,0010\bullet 0) + (^{0}_{1}\,0001\bullet 4) + 3(^{0}_{1}\,0001\bullet 2) + 5(^{0}_{1}\,0001\bullet 0)$
$+ 2(^{1}_{0}\,0000\bullet 4) + 6(^{1}_{0}\,0000\bullet 2) + 6(^{1}_{0}\,0000\bullet 0)$

Affine B7 ⊃ D6 + A1 - Level 1 - Class 1

Sl 0: $(^{0}_{0}\,0000\bullet 1)$ **Highest Weight** $(^{0}_{0}\,000001)$

Sl 1: $(^{0}_{0}\,0001\bullet 1)$

Sl 2: $(^{0}_{0}\,0010\bullet 1) + (^{0}_{0}\,0000\bullet 3) + (^{0}_{0}\,0000\bullet 1)$

Sl 3: $(^{0}_{0}\,0100\bullet 1) + (^{0}_{0}\,0001\bullet 3) + 2(^{0}_{0}\,0001\bullet 1)$

Sl 4: $(^{0}_{0}\,1000\bullet 1) + (^{0}_{0}\,0002\bullet 1) + (^{0}_{0}\,0010\bullet 3) + 2(^{0}_{0}\,0010\bullet 1) + 2(^{0}_{0}\,0000\bullet 3) + 3(^{0}_{0}\,0000\bullet 1)$

Sl 5: $(^{1}_{1}\,0000\bullet 1) + (^{0}_{0}\,0011\bullet 1) + (^{0}_{0}\,0100\bullet 3) + 2(^{0}_{0}\,0100\bullet 1) + 3(^{0}_{0}\,0001\bullet 3) + 5(^{0}_{0}\,0001\bullet 1)$

Sl 6: $(^{0}_{0}\,0101\bullet 1) + (^{0}_{0}\,1000\bullet 3) + (^{0}_{2}\,0000\bullet 1) + (^{2}_{0}\,0000\bullet 1) + 2(^{0}_{0}\,1000\bullet 1) + (^{0}_{0}\,0002\bullet 3) + 2(^{0}_{0}\,0002\bullet 1)$
$+ 3(^{0}_{0}\,0010\bullet 3) + 6(^{0}_{0}\,0010\bullet 1) + (^{0}_{0}\,0000\bullet 5) + 4(^{0}_{0}\,0000\bullet 3) + 6(^{0}_{0}\,0000\bullet 1)$

Affine B7 ⊃ D7 - Level 1 - Class 0

```
         1                    1
Sl 0:   ( 00000•)  Highest Weight ( 000000)
         0                    0

         0
Sl 1:   ( 00000•)
         1

         0             1
Sl 2:   ( 00001•) +   ( 00000•)
         1             0

         1             0
Sl 3:   ( 00001•) +  2( 00000•)
         0             1

         1             0             1
Sl 4:   ( 00010•) +  2( 00001•) +  3( 00000•)
         0             1             0

         0             1             0
Sl 5:   ( 00010•) +  3( 00001•) +  4( 00000•)
         1             0             1

         0             1             1             0             1
Sl 6:   ( 00100•) +   ( 00002•) +  2( 00010•) +  5( 00001•) +  6( 00000•)
         1             0             0             1             0
```

Affine B7 ⊃ D7 - Level 1 - Class 1

```
         0                    0
Sl 0:   ( 00000•)  Highest Weight ( 000001)
         0                    0

         0
Sl 1:   ( 00001•)
         0

         0             0
Sl 2:   ( 00010•) +   ( 00000•)
         0             0

         0             0
Sl 3:   ( 00100•) +  2( 00001•)
         0             0

         0             0             0             0
Sl 4:   ( 01000•) +   ( 00002•) +  2( 00010•) +  2( 00000•)
         0             0             0             0

         0             0             0             0
Sl 5:   ( 10000•) +   ( 00011•) +  2( 00100•) +  4( 00001•)
         0             0             0             0

         0             1             0             0             0             0
Sl 6:   ( 00101•) +   ( 00000•) +  2( 01000•) +  2( 00002•) +  5( 00010•) +  4( 00000•)
         0             1             0             0             0             0
```

Affine B8 Branching Rules

Affine B8 ⊃ B8 - Level 1 - Class 0

```
          .                          0
Sl 0:    ( 0000000)  Highest Weight  ( 0000000)
           1                          1

          .              .
Sl 1:    ( 0100000) +   ( 0000000)
           0              1

          .              .              .               .
Sl 2:    ( 0001000) +   ( 1000000) +   ( 0100000) +  2( 0000000)
           0              1              0               1

          .              .              .               .               .               .
Sl 3:    ( 0000010) +   ( 0010000) +   ( 0001000) +  2( 1000000) +  3( 0100000) +  3( 0000000)
           0              1              0               1               0               1

          .              .              .               .               .               .
Sl 4:    ( 0000100) +   ( 1100000) +   ( 0000002) +   ( 0000010) +  2( 0010000) +  3( 0001000)
           1              0              0               0               1               0

          .              .              .               .
      +  ( 0000000) +  4( 1000000) +  5( 0100000) +  6( 0000000)
           3              1              0               1

          .              .              .               .               .               .
Sl 5:    ( 1001000) +   ( 0000002) +  2( 0000100) +   ( 0100000) +  2( 1100000) +   ( 0000002)
           0              1              1               2               0               0

          .              .              .               .               .               .
      + 3( 0000010) +  5( 0010000) +   ( 0000100) +  5( 0001000) +   ( 0000000) +  8( 1000000)
           0              1              0               0               3               1

          .              .
      + 10( 0100000) + 9( 0000000)
           0              1
```

Affine B8 ⊃ B8 - Level 1 - Class 1

```
                                     1
Sl 0:    ( 0000000)  Highest Weight  ( 0000000)
           0                          0

          .
Sl 1:    ( 1000000)
           0

          .              .              .               .
Sl 2:    ( 0010000) +   ( 0000000) +   ( 1000000) +   ( 0000000)
           0              2              0               0

          .              .              .               .               .               .
Sl 3:    ( 0000100) +   ( 0100000) +   ( 0010000) +   ( 0000000) +  3( 1000000) +   ( 0000000)
           0              1              0               2               0               0

          .              .              .               .               .               .
Sl 4:    ( 0001000) +   ( 0000002) +   ( 0000100) +   ( 2000000) +  2( 0100000) +  3( 0010000)
           1              0              0               0               1               0

          .              .              .
      + 3( 0000000) +  4( 1000000) +  3( 0000000)
           2              0              0

          .              .              .               .               .               .
Sl 5:    ( 0000010) +   ( 1010000) +  2( 0001000) +   ( 0000002) +   ( 0000010) +  3( 0000100)
           1              0              1               0               0               0

          .              .              .               .               .               .
      +  ( 1000000) +   ( 2000000) +  5( 0100000) +  5( 0010000) +  4( 0000000) +  9( 1000000)
           2              0              1               0               2               0

          .
      + 3( 0000000)
           0
```

Affine B8 ⊃ B8 - Level 1 - Class 1

```
          .                            0
Sl 0:   ( 0000001)  Highest Weight  ( 0000001)
          0                            0

          .            .
Sl 1:   ( 0000001) +  ( 0000001)
          1            0

          .            .              .
Sl 2:   ( 1000001) + 2( 0000001) + 2( 0000001)
          0            1              0

          .            .              .              .              .
Sl 3:   ( 0100001) +  ( 0000001) + 2( 1000001) + 4( 0000001) + 4( 0000001)
          0            2              0              1              0

          .            .              .              .              .              .
Sl 4:   ( 0010001) +  ( 1000001) + 2( 0100001) + 2( 0000001) + 5( 1000001) + 8( 0000001)
          0            1              0              2              0              1

          .
   +  7( 0000001)
          0

          .            .              .              .              .              .
Sl 5:   ( 0001001) +  ( 0100001) + 2( 0010001) + 3( 1000001) + 5( 0100001) + 5( 0000001)
          0            1              0              1              0              2

          .               .              .
   + 10( 1000001) + 15( 0000001) + 12( 0000001)
          0               1              0
```

Affine B8 ⊃ A1 + A1 + B6 - Level 1 - Class 0

```
          1                            1
Sl 0:   ( •000000)  Highest Weight  ( 0000000)
          0                            0

          0
Sl 1:   ( •100000)
          1

          1            1              1
Sl 2:   ( •010000) +  ( •000000) +  ( •000000)
          0            2              0

          0            2              0
Sl 3:   ( •001000) +  ( •100000) + 2( •100000)
          1            1              1

          1            1              1              1              3              1
Sl 4:   ( •000100) +  ( •010000) +  ( •200000) + 2( •010000) +  ( •000000) + 2( •000000)
          0            2              0              0              0              2

          1
   +  3( •000000)
          0

          0            0              2              0              2              0
Sl 5:   ( •000010) +  ( •110000) +  ( •001000) + 2( •001000) + 3( •100000) +  ( •100000)
          1            1              1              1              1              3

          0
   +  5( •100000)
          1
```

Affine B8 ⊃ A1 + A1 + B6 - Level 1 - Class 1

```
         0                        0
Sl 0:   ( •000001)  Highest Weight ( 0000001)
         0                        0

         1
Sl 1:   ( •000001)
         1

         0              2              0              0
Sl 2:   ( •100001) +   ( •000001) +   ( •000001) +   ( •000001)
         0              0              2              0

         1              1
Sl 3:   ( •100001) +  3( •000001)
         1              1

         0              2              0              0              2              2
Sl 4:   ( •010001) +   ( •100001) +   ( •100001) +  2( •100001) +   ( •000001) +  2( •000001)
         0              0              2              0              2              0

         0              0
   +   2( •000001) +  4( •000001)
         2              0

         1              1              3              1              1
Sl 5:   ( •010001) +  4( •100001) +   ( •000001) +   ( •000001) +  7( •000001)
         1              1              1              3              1

         0              2              0              0              0              2
Sl 6:   ( •001001) +   ( •010001) +   ( •010001) +   ( •200001) +  2( •010001) +   ( •100001)
         0              0              2              0              0              2

         2              0              0              2              2              0
   +   3( •100001) +  3( •100001) +  6( •100001) +  3( •000001) +  6( •000001) +  6( •000001)
         0              2              0              2              0              2

         0
   +   8( •000001)
         0
```

Affine B8 ⊃ A3 + B5 - Level 1 - Class 0

```
         1                         1
Sl 0:   ( 0•00000)  Highest Weight ( 0000000)
         0                         0

         0
Sl 1:   ( 0•10000)
         1

         1              0              1
Sl 2:   ( 0•01000) +   ( 1•00000) +   ( 0•00000)
         0              1              0

         0              1              0
Sl 3:   ( 0•00100) +   ( 1•10000) +  2( 0•10000)
         1              0              1

         1              0              1              1              2              0
Sl 4:   ( 0•00010) +   ( 1•01000) +   ( 0•20000) +  2( 0•01000) +   ( 0•00000) +  2( 1•00000)
         0              1              0              0              1              1

         1
   +   3( 0•00000)
         0

         0              1              0              0              1              1
Sl 5:   ( 0•00002) +   ( 1•00100) +   ( 0•11000) +  2( 0•00100) +   ( 0•10000) +  3( 1•10000)
         1              0              1              1              2              0

         0
   +   5( 0•10000)
         1
```

Affine B8 ⊃ A3 + B5 - Level 1 - Class 1

```
         0                                0
Sl 0:   ( 0•00001)   Highest Weight  ( 0000001)
         0                                0

         0
Sl 1:   ( 1•00001)
         0

         0             1            0
Sl 2:   ( 0•10001) +  ( 0•00001) +  ( 0•00001)
         0             1            0

         0             2            0            0
Sl 3:   ( 1•10001) +  ( 0•00001) +  ( 0•00001) + 2( 1•00001)
         0             0            2            0

         0             1             0             0             1             0
Sl 4:   ( 0•01001) +  ( 0•10001) + 2( 0•10001) +  ( 2•00001) + 3( 0•00001) + 3( 0•00001)
         0             1             0             0             1             0

         0             2             0             0             1             2
Sl 5:   ( 1•01001) +  ( 0•10001) +  ( 0•10001) + 3( 1•10001) +  ( 1•00001) + 2( 0•00001)
         0             0             2             0             1             0

         0             0
    +  2( 0•00001) + 6( 1•00001)
         2             0

         1             0             0             0             0             1
Sl 6:   ( 0•01001) +  ( 0•00101) +  ( 0•20001) +  ( 2•10001) + 2( 0•01001) + 4( 0•10001)
         1             0             0             0             0             1

         2             0             0             0             1             0
    +   ( 1•00001) +  ( 1•00001) + 5( 0•10001) + 2( 2•00001) + 8( 0•00001) + 7( 0•00001)
         0             2             0             0             1             0
```

Affine B8 ⊃ D4 + B4 - Level 1 - Class 0

```
         1                                1
Sl 0:   ( 00•0000)   Highest Weight  ( 0000000)
         0                                0

         0
Sl 1:   ( 00•1000)
         1

         1             0             1
Sl 2:   ( 00•0100) +  ( 01•0000) +  ( 00•0000)
         0             1             0

         0             1             0
Sl 3:   ( 00•0010) +  ( 01•1000) + 2( 00•1000)
         1             0             1

         0             1             1             1             1             0
Sl 4:   ( 01•0100) +  ( 00•0002) +  ( 00•2000) + 2( 00•0100) +  ( 10•0000) + 2( 01•0000)
         1             0             0             0             0             1

         1
    +  3( 00•0000)
         0

         1             0             0             0             0             1
Sl 5:   ( 01•0010) +  ( 00•1100) +  ( 00•0002) +  ( 10•1000) + 2( 00•0010) + 3( 01•1000)
         0             1             1             1             1             0

         0
    +  5( 00•1000)
         1
```

Affine B8 ⊃ D4 + B4 - Level 1 - Class 1

```
          0                        0
Sl 0:    ( 00•0001)   Highest Weight ( 0000001)
          0                        0

          0
Sl 1:    ( 01•0001)
          0

          0                0                0
Sl 2:    ( 10•0001) +    ( 00•1001) +    ( 00•0001)
          0                0                0

          0                1                0
Sl 3:    ( 01•1001) +    ( 00•0001) +  2( 01•0001)
          0                1                0

          0                0                0                0                2                0
Sl 4:    ( 10•1001) +    ( 00•0101) +    ( 02•0001) +    ( 00•0001) +    ( 00•0001) +  2( 10•0001)
          0                0                0                2                0                0

          0                0
      + 2( 00•1001) +  3( 00•0001)
          0                0

          0                1                0                0                1                0
Sl 5:    ( 01•0101) +    ( 00•1001) +    ( 11•0001) +  3( 01•1001) +  3( 00•0001) +  5( 01•0001)
          0                1                0                0                1                0
```

Affine B8 ⊃ D5 + B3 - Level 1 - Class 0

```
          1                        1
Sl 0:    ( 000•000)   Highest Weight ( 0000000)
          0                        0

          0
Sl 1:    ( 000•100)
          1

          1                0                1
Sl 2:    ( 000•010) +    ( 001•000) +    ( 000•000)
          0                1                0

          1                0                0
Sl 3:    ( 001•100) +    ( 000•002) +  2( 000•100)
          0                1                1

          0                1                1                1                1                0
Sl 4:    ( 001•010) +    ( 010•000) +    ( 000•200) +    ( 000•002) +  2( 000•010) +  2( 001•000)
          1                0                0                0                0                1

          1
      + 3( 000•000)
          0

          1                0                0                1                0                0
Sl 5:    ( 001•002) +    ( 010•100) +    ( 000•110) +  3( 001•100) +  2( 000•002) +    ( 000•010)
          0                1                1                0                1                1

          0
      + 5( 000•100)
          1

          1                0                0                0                0                1
Sl 6:    ( 010•010) +    ( 001•200) +    ( 001•002) +  3( 001•010) +    ( 100•000) +    ( 000•102)
          0                1                1                1                1                0

          1                1                1                1                1                0
      +  ( 002•000) +  2( 010•000) +  2( 000•200) +  2( 000•002) +  6( 000•010) +  5( 001•000)
          0                0                0                0                0                1

          1                1
      +  ( 000•100) +  6( 000•000)
          0                0
```

Affine B8 ⊃ D5 + B3 - Level 1 - Class 1

```
         0                                0
Sl 0:   ( 000•001)  Highest Weight ( 0000001)
         0                                0

         0
Sl 1:   ( 001•001)
         0

         0             0             0
Sl 2:   ( 010•001) +  ( 000•101) +  ( 000•001)
         0             0             0

         0             0             0
Sl 3:   ( 100•001) +  ( 001•101) + 2( 001•001)
         0             0             0

         0             1             0             0             0             0
Sl 4:   ( 010•101) +  ( 000•001) +  ( 002•001) + 2( 010•001) +  ( 000•011) + 2( 000•101)
         0             1             0             0             0             0

         0
     + 3( 000•001)
         0

         0             0             0             2             0             0
Sl 5:   ( 100•101) +  ( 011•001) +  ( 000•001) +  ( 000•001) +  ( 001•011) + 2( 100•001)
         0             0             2             0             0             0

         0             0
     + 3( 001•101) + 5( 001•001)
         0             0

         0             1             0             0             0             1
Sl 6:   ( 101•001) +  ( 000•101) +  ( 010•011) +  ( 002•101) + 3( 010•101) + 3( 000•001)
         0             1             0             0             0             1

         0             0             0             0             0             0
     + 2( 002•001) + 6( 010•001) +  ( 000•201) +  ( 000•003) + 2( 000•011) + 5( 000•101)
         0             0             0             0             0             0

         0
     + 6( 000•001)
         0
```

Affine B8 ⊃ D6 + C2 - Level 1 - Class 0

```
         1                                1
Sl 0:   ( 0000•00)  Highest Weight ( 0000000)
         0                                0

         0
Sl 1:   ( 0000•10)
         1

         0             1             1
Sl 2:   ( 0001•00) +  ( 0000•02) +  ( 0000•00)
         1             0             0

         1             0             0
Sl 3:   ( 0001•10) +  ( 0000•02) + 2( 0000•10)
         0             1             1

         1             0             0             1             1             1
Sl 4:   ( 0010•00) +  ( 0001•02) + 2( 0001•00) +  ( 0000•20) + 2( 0000•02) +  ( 0000•10)
         0             1             1             0             0             0

         1
     + 3( 0000•00)
         0

         0             1             1             0             0             0
Sl 5:   ( 0010•10) +  ( 0001•02) + 3( 0001•10) +  ( 0000•12) + 2( 0000•02) + 5( 0000•10)
         1             0             0             1             1             1

         0
     +  ( 0000•00)
         1
```

Affine B8 ⊃ D6 + C2 - Level 1 - Class 1

```
          0                        0
Sl 0:    ( 0000•01)  Highest Weight ( 0000001)
          0                        0

          0
Sl 1:    ( 0001•01)
          0

          0              0              0
Sl 2:    ( 0010•01) +   ( 0000•11) +   ( 0000•01)
          0              0              0

          0              0              0
Sl 3:    ( 0100•01) +   ( 0001•11) +  2( 0001•01)
          0              0              0

          0              0              0              0              0              0
Sl 4:    ( 1000•01) +   ( 0010•11) +   ( 0002•01) +  2( 0010•01) +   ( 0000•03) +  2( 0000•11)
          0              0              0              0              0              0

          0
     +  3( 0000•01)
          0

          1              0              0              0              0              0
Sl 5:    ( 0000•01) +   ( 0011•01) +   ( 0100•11) +  2( 0100•01) +   ( 0001•03) +  3( 0001•11)
          1              0              0              0              0              0

          0
     +  5( 0001•01)
          0

          0              0              0              2              0              0
Sl 6:    ( 0101•01) +   ( 1000•11) +   ( 0000•01) +   ( 0000•01) +  2( 1000•01) +   ( 0002•11)
          0              0              2              0              0              0

          0              0              0              0              0              0
     +   ( 0010•03) +  3( 0010•11) +  2( 0002•01) +  6( 0010•01) +   ( 0000•21) +  2( 0000•03)
          0              0              0              0              0              0

          0              0
     +  5( 0000•11) +  6( 0000•01)
          0              0
```

Affine B8 ⊃ D7 + A1 - Level 1 - Class 0

```
          1                         1
Sl 0:    ( 00000•0)  Highest Weight ( 0000000)
          0                         0

          0
Sl 1:    ( 00000•2)
          1

          0              1              1
Sl 2:    ( 00001•0) +   ( 00000•2) +   ( 00000•0)
          1              0              0

          1              0              0
Sl 3:    ( 00001•2) +  2( 00000•2) +   ( 00000•0)
          0              1              1

          1              0              0              1              1              1
Sl 4:    ( 00010•0) +   ( 00001•2) +  2( 00001•0) +   ( 00000•4) +  2( 00000•2) +  3( 00000•0)
          0              1              1              0              0              0

          0              1              1              0              0              0
Sl 5:    ( 00010•2) +  3( 00001•2) +   ( 00001•0) +   ( 00000•4) +  5( 00000•2) +  2( 00000•0)
          1              0              0              1              1              1

          0              1              1              1              0              0
Sl 6:    ( 00100•0) +   ( 00002•0) +   ( 00010•2) +  2( 00010•0) +   ( 00001•4) +  3( 00001•2)
          1              0              0              0              1              1

          0              1              1              1
     +  5( 00001•0) +  2( 00000•4) +  6( 00000•2) +  6( 00000•0)
          1              0              0              0
```

Affine B8 ⊃ D7 + A1 - Level 1 - Class 1

```
          0                        0
Sl 0:    ( 00000•1)  Highest Weight ( 0000001)
          0                        0

          0
Sl 1:    ( 00001•1)
          0

          0           0           0
Sl 2:    ( 00010•1) + ( 00000•3) + ( 00000•1)
          0           0           0

          0           0            0
Sl 3:    ( 00100•1) + ( 00001•3) + 2( 00001•1)
          0           0            0

          0           0           0            0            0            0
Sl 4:    ( 01000•1) + ( 00002•1) + ( 00010•3) + 2( 00010•1) + 2( 00000•3) + 3( 00000•1)
          0           0           0            0            0            0

          0           0           0            0            0            0
Sl 5:    ( 10000•1) + ( 00011•1) + ( 00100•3) + 2( 00100•1) + 3( 00001•3) + 5( 00001•1)
          0           0           0            0            0            0

          0           1           0            0           0            0
Sl 6:    ( 00101•1) + ( 00000•1) + ( 01000•3) + 2( 01000•1) + ( 00002•3) + 2( 00002•1)
          0           1           0            0           0            0

           0            0           0            0            0
    +  3( 00010•3) + 6( 00010•1) + ( 00000•5) + 4( 00000•3) + 6( 00000•1)
           0            0           0            0            0
```

Affine B8 ⊃ D8 - Level 1 - Class 0

```
          0                        0
Sl 0:    ( 000000•)  Highest Weight ( 0000000)
          1                        1

          1
Sl 1:    ( 000000•)
          0

          1           0
Sl 2:    ( 000001•) + ( 000000•)
          0           1

          0            1
Sl 3:    ( 000001•) + 2( 000000•)
          1            0

          0            1            0
Sl 4:    ( 000010•) + 2( 000001•) + 3( 000000•)
          1            0            1

          1            0            1
Sl 5:    ( 000010•) + 3( 000001•) + 4( 000000•)
          0            1            0
```

Affine B8 ⊃ D8 - Level 1 - Class 1

```
          0                        0
Sl 0:    ( 000000•)  Highest Weight ( 0000001)
          0                        0

          0
Sl 1:    ( 000001•)
          0

          0           0
Sl 2:    ( 000010•) + ( 000000•)
          0           0

          0            0
Sl 3:    ( 000100•) + 2( 000001•)
          0            0

          0           0            0            0
Sl 4:    ( 001000•) + ( 000002•) + 2( 000010•) + 2( 000000•)
          0           0            0            0

          0           0            0            0
Sl 5:    ( 010000•) + ( 000011•) + 2( 000100•) + 4( 000001•)
          0           0            0            0
```

Affine C2 Branching Rules

Affine C2 ⊃ C2 - Level 1 - Class 0

Sl 0: (•01) **Highest Weight (001)**

Sl 1: (•20) + (•01)

Sl 2: (•21) + (•20) + 2(•01) + (•00)

Sl 3: 2(•21) + (•02) + 3(•20) + 3(•01) + (•00)

Sl 4: (•40) + (•03) + 4(•21) + 2(•02) + 5(•20) + 6(•01) + 2(•00)

Sl 5: (•22) + 2(•40) + (•03) + 8(•21) + 4(•02) + 10(•20) + 9(•01) + 3(•00)

Sl 6: (•41) + 2(•22) + 4(•40) + 3(•03) + 15(•21) + 8(•02) + 16(•20) + 16(•01) + 6(•00)

Sl 7: 2(•41) + 6(•22) + 8(•40) + 5(•03) + 26(•21) + 14(•02) + 29(•20) + 24(•01) + 9(•00)

Sl 8: (•23) + 5(•41) + 11(•22) + 16(•40) + 10(•03) + 45(•21) + 25(•02) + 45(•20) + 40(•01) + 15(•00)

Sl 9: (•60) + 2(•23) + 10(•41) + (•04) + 23(•22) + 27(•40) + 16(•03) + 75(•21) + 42(•02) + 75(•20) + 60(•01) + 22(•00)

Affine C2 ⊃ C2 - Level 1 - Class 0

Sl 0: (•00) **Highest Weight (100)**

Sl 1: (•20)

Sl 2: (•02) + (•20) + (•01) + (•00)

Sl 3: (•21) + (•02) + 3(•20) + (•01) + (•00)

Sl 4: (•40) + 2(•21) + 3(•02) + 4(•20) + 3(•01) + 3(•00)

Sl 5: (•22) + (•40) + 5(•21) + 4(•02) + 9(•20) + 5(•01) + 3(•00)

Sl 6: 2(•22) + 4(•40) + (•03) + 9(•21) + 9(•02) + 13(•20) + 10(•01) + 7(•00)

Sl 7: (•41) + 5(•22) + 6(•40) + 2(•03) + 18(•21) + 13(•02) + 25(•20) + 15(•01) + 8(•00)

Sl 8: 3(•41) + (•04) + 9(•22) + 13(•40) + 5(•03) + 31(•21) + 25(•02) + 37(•20) + 27(•01) + 16(•00)

Sl 9: (•60) + (•23) + 6(•41) + (•04) + 19(•22) + 21(•40) + 9(•03) + 55(•21) + 37(•02) + 64(•20) + 41(•01) + 20(•00)

Affine C2 ⊃ C2 - Level 1 - Class 1

Sl 0: (•10) **Highest Weight (010)**

Sl 1: (•11) + (•10)

Sl 2: (•30) + 2(•11) + 2(•10)

Sl 3: (•12) + 2(•30) + 4(•11) + 4(•10)

Sl 4: (•31) + 2(•12) + 4(•30) + 8(•11) + 7(•10)

Sl 5: 3(•31) + 5(•12) + 8(•30) + 14(•11) + 12(•10)

Sl 6: (•50) + (•13) + 6(•31) + 10(•12) + 15(•30) + 25(•11) + 20(•10)

Sl 7: (•32) + 2(•50) + 2(•13) + 13(•31) + 19(•12) + 26(•30) + 43(•11) + 32(•10)

Sl 8: 3(•32) + 5(•50) + 5(•13) + 25(•31) + 34(•12) + 45(•30) + 71(•11) + 51(•10)

Sl 9: (•51) + 7(•32) + 10(•50) + 11(•13) + 45(•31) + 60(•12) + 75(•30) +115(•11) + 80(•10)

Affine C2 ⊃ C2 - Level 2 - Class 0

Sl 0: (•02) **Highest Weight (002)**

Sl 1: (•21) + (•02) + (•20)

Sl 2: (•22) + (•40) + 2(•21) + 3(•02) + 2(•20) + (•01) + (•00)

Sl 3: (•41) + 3(•22) + 2(•40) + (•03) + 6(•21) + 5(•02) + 6(•20) + 3(•01) + (•00)

Sl 4: (•42) + 3(•41) + (•04) + 8(•22) + 7(•40) + 3(•03) + 14(•21) + 12(•02) + 12(•20) + 7(•01) + 4(•00)

Affine C2 ⊃ C2 - Level 2 - Class 0

Sl 0: (•20) **Highest Weight (020)**

Sl 1: (•21) + (•02) + (•20) + (•01) + (•00)

Sl 2: (•22) + (•40) + 3(•21) + 2(•02) + 4(•20) + 2(•01) + (•00)

Sl 3: (•41) + 3(•22) + 4(•40) + 2(•03) + 8(•21) + 6(•02) + 8(•20) + 6(•01) + 3(•00)

Sl 4: (•60) + (•23) + 4(•41) + (•04) + 10(•22) + 9(•40) + 4(•03) + 21(•21) + 13(•02) + 20(•20) + 12(•01) + 5(•00)

Sl 5: 2(•42) + 2(•60) + 4(•23) + 13(•41) + 3(•04) + 24(•22) + 24(•40) + 13(•03) + 48(•21) + 30(•02) + 39(•20) + 27(•01) + 12(•00)

Affine C2 ⊃ C2 - Level 2 - Class 0

Sl 0: (•01) **Highest Weight (101)**

Sl 1: (•21) + (•20) + (•01)

Sl 2: (•40) + (•03) + 3(•21) + (•02) + 2(•20) + 3(•01) + (•00)

Sl 3: (•41) + 2(•22) + 2(•40) + 2(•03) + 8(•21) + 3(•02) + 7(•20) + 6(•01) + (•00)

Sl 4: (•23) + 4(•41) + 6(•22) + 8(•40) + 6(•03) + 19(•21) + 9(•02) + 14(•20) + 14(•01) + 4(•00)

Sl 5: (•42) + 2(•60) + 4(•23) + 11(•41) + (•04) + 19(•22) + 18(•40) + 13(•03) + 44(•21) + 20(•02) + 34(•20) + 26(•01) + 7(•00)

Affine C2 ⊃ C2 - Level 2 - Class 0

Sl 0: (•00) **Highest Weight (200)**

Sl 1: (•20)

Sl 2: (•40) + (•02) + (•20) + (•01) + (•00)

Sl 3: (•22) + (•40) + 2(•21) + (•02) + 4(•20) + (•01) + (•00)

Sl 4: (•41) + (•04) + 2(•22) + 4(•40) + (•03) + 5(•21) + 5(•02) + 6(•20) + 4(•01) + 4(•00)

Sl 5: (•60) + (•23) + 3(•41) + (•04) + 8(•22) + 7(•40) + 2(•03) + 14(•21) + 8(•02) + 16(•20) + 7(•01) + 4(•00)

Affine C2 ⊃ C2 - Level 2 - Class 1

Sl 0: (•11) **Highest Weight (011)**

Sl 1: (•12) + (•30) + 2(•11) + (•10)

Sl 2: 2(•31) + 3(•12) + 3(•30) + 5(•11) + 3(•10)

Sl 3: (•32) + (•50) + (•13) + 6(•31) + 8(•12) + 8(•30) + 13(•11) + 7(•10)

Sl 4: (•51) + 4(•32) + 4(•50) + 5(•13) + 16(•31) + 20(•12) + 20(•30) + 29(•11) + 16(•10)

Sl 5: (•33) + 4(•51) + (•14) + 13(•32) + 12(•50) + 14(•13) + 40(•31) + 46(•12) + 45(•30) + 62(•11) + 33(•10)

Sl 6: (•52) + (•70) + 4(•33) + 13(•51) + 5(•14) + 35(•32) + 30(•50) + 36(•13) + 92(•31) +100(•12) + 95(•30) +127(•11) + 66(•10)

Sl 7: 5(•52) + (•15) + 5(•70) + 14(•33) + 35(•51) + 15(•14) + 85(•32) + 70(•50) + 85(•13) +198(•31) +208(•12) +193(•30) +249(•11) +128(•10)

Sl 8: 2(•71) + 2(•34) + 16(•52) + 3(•15) + 14(•70) + 40(•33) + 86(•51) + 40(•14) +193(•32) +153(•50) +186(•13) +408(•31) +415(•12) +377(•30) +475(•11) +238(•10)

Affine C2 ⊃ C2 - Level 2 - Class 1

Sl 0: (•10) **Highest Weight (110)**

Sl 1: (•30) + (•11) + (•10)

Sl 2: (•31) + (•12) + 2(•30) + 3(•11) + 3(•10)

Sl 3: (•50) + (•13) + 3(•31) + 4(•12) + 6(•30) + 8(•11) + 6(•10)

Sl 4: 2(•32) + 3(•50) + 3(•13) + 10(•31) + 11(•12) + 14(•30) + 19(•11) + 13(•10)

Sl 5: 2(•51) + (•14) + 7(•32) + 8(•50) + 9(•13) + 26(•31) + 28(•12) + 32(•30) + 42(•11) + 26(•10)

Sl 6: (•70) + 2(•33) + 7(•51) + 3(•14) + 21(•32) + 21(•50) + 24(•13) + 61(•31) + 65(•12) + 68(•30) + 88(•11) + 52(•10)

Sl 7: 2(•52) + 3(•70) + 8(•33) + 21(•51) + 10(•14) + 54(•32) + 49(•50) + 57(•13) +136(•31) +140(•12) +140(•30) +177(•11) + 98(•10)

Sl 8: (•71) + (•34) + 8(•52) + (•15) + 9(•70) + 24(•33) + 55(•51) + 27(•14) +128(•32) +108(•50) +128(•13) +286(•31) +288(•12) +275(•30) +344(•11) +183(•10)

Affine C2 ⊃ C2 - Level 3 - Class 0

Sl 0: (•03) **Highest Weight (003)**

Sl 1: (•22) + (•03) + (•21)

Sl 2: (•23) + (•41) + 2(•22) + (•40) + 3(•03) + 3(•21) + (•02) + (•01)

Sl 3: (•42) + (•60) + 3(•23) + 3(•41) + (•04) + 7(•22) + 3(•40) + 6(•03) + 8(•21) + 3(•02) + 3(•20) + 2(•01)

Affine C2 ⊃ C2 - Level 3 - Class 0

Sl 0: (•21) **Highest Weight (021)**

Sl 1: (•22) + (•40) + (•03) + 2(•21) + (•02) + (•20) + (•01)

Sl 2: (•23) + 2(•41) + 4(•22) + 3(•40) + 3(•03) + 7(•21) + 3(•02) + 4(•20) + 3(•01)

Sl 3: 2(•42) + (•60) + 4(•23) + 8(•41) + 2(•04) + 13(•22) + 10(•40) + 8(•03) + 20(•21) + 10(•02) + 11(•20) + 8(•01) + 2(•00)

Affine C2 ⊃ C2 - Level 3 - Class 0

Sl 0: (•02) **Highest Weight (102)**

Sl 1: (•22) + (•21) + (•02) + (•20)

Sl 2: (•41) + (•04) + 3(•22) + 2(•40) + (•03) + 3(•21) + 4(•02) + 2(•20) + (•01) + (•00)

Sl 3: (•42) + (•60) + 2(•23) + 4(•41) + 2(•04) + 10(•22) + 5(•40) + 3(•03) + 11(•21) + 8(•02) + 8(•20) + 3(•01) + (•00)

Affine C2 ⊃ C2 - Level 3 - Class 0

Sl 0: (•20) **Highest Weight (120)**

Sl 1: (•40) + (•21) + (•02) + (•20) + (•01) + (•00)

Sl 2: (•41) + 2(•22) + 2(•40) + 4(•21) + 2(•02) + 5(•20) + 2(•01) + (•00)

Sl 3: (•42) + (•60) + (•23) + 4(•41) + (•04) + 6(•22) + 8(•40) + 3(•03) + 12(•21) + 8(•02) + 10(•20) + 7(•01) + 4(•00)

Affine C2 ⊃ C2 - Level 3 - Class 0

Sl 0: (•01) **Highest Weight (201)**

Sl 1: (•21) + (•20) + (•01)

Sl 2: (•41) + (•40) + (•03) + 3(•21) + (•02) + 2(•20) + 3(•01) + (•00)

Sl 3: (•60) + (•23) + 3(•41) + 3(•22) + 3(•40) + 2(•03) + 9(•21) + 3(•02) + 7(•20) + 6(•01) + (•00)

Sl 4: (•61) + 2(•42) + (•05) + 2(•60) + 4(•23) + 10(•41) + (•04) + 10(•22) + 12(•40) + 8(•03) + 23(•21) + 10(•02) + 15(•20) + 15(•01) + 4(•00)

Affine C2 ⊃ C2 - Level 3 - Class 0

Sl 0: (•00) **Highest Weight (300)**

Sl 1: (•20)

Sl 2: (•40) + (•02) + (•20) + (•01) + (•00)

Sl 3: (•60) + (•22) + (•40) + 2(•21) + (•02) + 4(•20) + (•01) + (•00)

Sl 4: (•42) + (•60) + 2(•41) + (•04) + 2(•22) + 5(•40) + (•03) + 5(•21) + 5(•02) + 6(•20) + 4(•01) + 4(•00)

Affine C2 ⊃ C2 - Level 3 - Class 1

Sl 0: (•12) **Highest Weight (012)**

Sl 1: (•13) + (•31) + 2(•12) + (•30) + (•11)

Sl 2: 2(•32) + (•50) + 3(•13) + 4(•31) + 6(•12) + 3(•30) + 4(•11) + (•10)

Sl 3: (•33) + 2(•51) + (•14) + 7(•32) + 4(•50) + 9(•13) + 13(•31) + 16(•12) + 10(•30) + 12(•11) + 4(•10)

Sl 4: 2(•52) + (•70) + 5(•33) + 8(•51) + 5(•14) + 22(•32) + 13(•50) + 24(•13) + 37(•31) + 41(•12) + 27(•30) + 32(•11) + 12(•10)

Affine C2 ⊃ C2 - Level 3 - Class 1

Sl 0: (•30) **Highest Weight (030)**

Sl 1: (•31) + (•12) + (•30) + (•11) + (•10)

Sl 2: (•32) + (•50) + (•13) + 3(•31) + 3(•12) + 4(•30) + 4(•11) + 2(•10)

Sl 3: (•33) + (•51) + 4(•32) + 4(•50) + 4(•13) + 10(•31) + 10(•12) + 10(•30) + 11(•11) + 6(•10)

Sl 4: (•52) + (•70) + 3(•33) + 6(•51) + 3(•14) + 14(•32) + 11(•50) + 13(•13) + 28(•31) + 27(•12) + 26(•30) + 29(•11) + 14(•10)

Affine C2 ⊃ C2 - Level 3 - Class 1

Sl 0: (•11) **Highest Weight (111)**

Sl 1: (•31) + (•12) + (•30) + 2(•11) + (•10)

Sl 2: (•32) + (•50) + (•13) + 4(•31) + 4(•12) + 4(•30) + 6(•11) + 3(•10)

Sl 3: 2(•51) + (•14) + 5(•32) + 4(•50) + 5(•13) + 13(•31) + 13(•12) + 12(•30) + 17(•11) + 8(•10)

Sl 4: (•52) + (•70) + 3(•33) + 8(•51) + 4(•14) + 18(•32) + 14(•50) + 17(•13) + 38(•31) + 37(•12) + 33(•30) + 42(•11) + 20(•10)

Sl 5: (•71) + (•34) + 6(•52) + (•15) + 5(•70) + 14(•33) + 27(•51) + 14(•14) + 55(•32) + 41(•50) + 50(•13) +100(•31) + 95(•12) + 82(•30) +100(•11) + 46(•10)

Affine C2 ⊃ C2 - Level 3 - Class 1

Sl 0: (•10) **Highest Weight (210)**

Sl 1: (•30) + (•11) + (•10)

Sl 2: (•50) + (•31) + (•12) + 2(•30) + 3(•11) + 3(•10)

Sl 3: (•51) + (•32) + 2(•50) + (•13) + 4(•31) + 4(•12) + 7(•30) + 8(•11) + 6(•10)

Sl 4: (•70) + (•33) + 3(•51) + (•14) + 5(•32) + 7(•50) + 4(•13) + 14(•31) + 13(•12) + 16(•30) + 20(•11) + 14(•10)

Sl 5: 2(•52) + (•15) + 3(•70) + 4(•33) + 11(•51) + 4(•14) + 18(•32) + 18(•50) + 15(•13) + 39(•31) + 35(•12) + 40(•30) + 47(•11) + 28(•10)

Affine C2 ⊃ A1 + A1 - Level 1 - Class 0

```
Sl 0:    (0•1)  Highest Weight (001)
Sl 1:    (1•0)
Sl 2:    (2•1) +    (0•1)
Sl 3:    (1•2) +  2(1•0)
Sl 4:  2(2•1) +    (0•3) +  3(0•1)
Sl 5:    (3•0) +  3(1•2) +  4(1•0)
Sl 6:    (2•3) +  5(2•1) +    (0•3) +  6(0•1)
Sl 7:    (3•2) +  2(3•0) +  7(1•2) +  8(1•0)
Sl 8:    (4•1) +  2(2•3) + 10(2•1) +  4(0•3) + 12(0•1)
Sl 9:  3(3•2) +    (1•4) +  5(3•0) + 14(1•2) + 16(1•0)
```

Affine C2 ⊃ A1 + A1 - Level 1 - Class 1

```
Sl 0:    (0•0)  Highest Weight (010)
Sl 1:    (1•1)
Sl 2:    (2•0) +    (0•2) +    (0•0)
Sl 3:  3(1•1)
Sl 4:    (2•2) +  2(2•0) +  2(0•2) +  3(0•0)
Sl 5:    (3•1) +    (1•3) +  6(1•1)
Sl 6:  3(2•2) +  5(2•0) +  5(0•2) +  6(0•0)
Sl 7:  3(3•1) +  3(1•3) + 13(1•1)
Sl 8:    (4•0) +  8(2•2) +    (0•4) + 10(2•0) + 10(0•2) + 11(0•0)
```

Affine C2 ⊃ A1 + A1 - Level 2 - Class 0

```
Sl 0:    (0•2)  Highest Weight (002)
Sl 1:    (1•1)
Sl 2:    (2•2) +    (2•0) +    (0•2) +    (0•0)
Sl 3:    (3•1) +    (1•3) +  3(1•1)
Sl 4:    (4•2) +  3(2•2) +    (0•4) +  3(2•0) +  4(0•2) +  3(0•0)
Sl 5:    (3•3) +  4(3•1) +  4(1•3) +  9(1•1)
Sl 6:  2(4•2) +  2(2•4) +  2(4•0) + 11(2•2) +  2(0•4) +  9(2•0) +  9(0•2) +  7(0•0)
Sl 7:    (5•1) +  5(3•3) +    (1•5) + 12(3•1) + 12(1•3) + 24(1•1)
```

Affine C2 ⊃ A1 + A1 - Level 2 - Class 0

Sl 0: (0•0) **Highest Weight (020)**

Sl 1: (1•1)

Sl 2: (2•2) + (2•0) + (0•2) + (0•0)

Sl 3: (3•1) + (1•3) + 4(1•1)

Sl 4: (4•0) + 4(2•2) + (0•4) + 4(2•0) + 4(0•2) + 4(0•0)

Sl 5: (3•3) + 5(3•1) + 5(1•3) + 12(1•1)

Sl 6: 2(4•2) + 2(2•4) + 2(4•0) + 14(2•2) + 2(0•4) + 12(2•0) + 12(0•2) + 10(0•0)

Sl 7: (5•1) + 6(3•3) + (1•5) + 16(3•1) + 16(1•3) + 33(1•1)

Affine C2 ⊃ A1 + A1 - Level 2 - Class 0

Sl 0: (1•1) **Highest Weight (101)**

Sl 1: (2•0) + (0•2) + (0•0)

Sl 2: (3•1) + (1•3) + 3(1•1)

Sl 3: 3(2•2) + 3(2•0) + 3(0•2) + 3(0•0)

Sl 4: (3•3) + 4(3•1) + 4(1•3) + 10(1•1)

Sl 5: (4•2) + (2•4) + 2(4•0) + 10(2•2) + 2(0•4) + 10(2•0) + 10(0•2) + 8(0•0)

Sl 6: (5•1) + 5(3•3) + (1•5) + 13(3•1) + 13(1•3) + 28(1•1)

Sl 7: 6(4•2) + 6(2•4) + 6(4•0) + 32(2•2) + 6(0•4) + 26(2•0) + 26(0•2) + 20(0•0)

Affine C2 ⊃ A1 + A1 - Level 2 - Class 1

Sl 0: (0•1) **Highest Weight (011)**

Sl 1: (1•2) + (1•0)

Sl 2: 2(2•1) + (0•3) + 2(0•1)

Sl 3: (3•2) + (3•0) + 4(1•2) + 4(1•0)

Sl 4: (4•1) + 2(2•3) + 7(2•1) + 3(0•3) + 7(0•1)

Sl 5: 5(3•2) + 2(1•4) + 5(3•0) + 13(1•2) + 11(1•0)

Sl 6: (4•3) + 4(4•1) + 9(2•3) + (0•5) + 22(2•1) + 9(0•3) + 19(0•1)

Sl 7: (5•2) + 2(3•4) + (5•0) + 17(3•2) + 8(1•4) + 15(3•0) + 37(1•2) + 29(1•0)

Sl 8: 5(4•3) + 2(2•5) + 15(4•1) + 29(2•3) + 3(0•5) + 59(2•1) + 26(0•3) + 46(0•1)

Affine C2 ⊃ A1 + A1 - Level 3 - Class 0

Sl 0: (0•3) **Highest Weight (003)**

Sl 1: (1•2)

Sl 2: (2•3) + (2•1) + (0•3) + (0•1)

Sl 3: (3•2) + (1•4) + (3•0) + 3(1•2) + (1•0)

Sl 4: (4•3) + (4•1) + 3(2•3) + (0•5) + 4(2•1) + 4(0•3) + 4(0•1)

Sl 5: (5•2) + (3•4) + 5(3•2) + 4(1•4) + 3(3•0) + 10(1•2) + 5(1•0)

Sl 6: (6•3) + 3(4•3) + 2(2•5) + 5(4•1) + 12(2•3) + 2(0•5) + 15(2•1) + 10(0•3) + 10(0•1)

Sl 7: (5•4) + 4(5•2) + 6(3•4) + (1•6) + 2(5•0) + 18(3•2) + 13(1•4) + 11(3•0) + 29(1•2) + 16(1•0)

Affine C2 ⊃ A1 + A1 - Level 3 - Class 0

Sl 0: (0•1) **Highest Weight (021)**
Sl 1: (1•2) + (1•0)
Sl 2: (2•3) + 2(2•1) + (0•3) + 2(0•1)
Sl 3: 2(3•2) + (1•4) + (3•0) + 5(1•2) + 4(1•0)
Sl 4: (4•3) + 2(4•1) + 5(2•3) + (0•5) + 9(2•1) + 5(0•3) + 8(0•1)
Sl 5: (5•2) + 2(3•4) + (5•0) + 10(3•2) + 6(1•4) + 7(3•0) + 19(1•2) + 13(1•0)
Sl 6: (6•1) + 6(4•3) + 3(2•5) + 10(4•1) + 22(2•3) + 3(0•5) + 33(2•1) + 16(0•3) + 24(0•1)
Sl 7: (5•4) + 7(5•2) + 12(3•4) + 2(1•6) + 5(5•0) + 39(3•2) + 23(1•4) + 25(3•0) + 61(1•2) + 39(1•0)

Affine C2 ⊃ A1 + A1 - Level 3 - Class 0

Sl 0: (1•2) **Highest Weight (102)**
Sl 1: (2•1) + (0•3) + (0•1)
Sl 2: (3•2) + (1•4) + (3•0) + 3(1•2) + 2(1•0)
Sl 3: (4•1) + 3(2•3) + 5(2•1) + 3(0•3) + 4(0•1)
Sl 4: (5•2) + (3•4) + 6(3•2) + 4(1•4) + 4(3•0) + 12(1•2) + 7(1•0)
Sl 5: 3(4•3) + (2•5) + 6(4•1) + 12(2•3) + 2(0•5) + 19(2•1) + 11(0•3) + 14(0•1)
Sl 6: (5•4) + 4(5•2) + 7(3•4) + (1•6) + 3(5•0) + 23(3•2) + 15(1•4) + 16(3•0) + 38(1•2) + 23(1•0)

Affine C2 ⊃ A1 + A1 - Level 3 - Class 1

Sl 0: (0•2) **Highest Weight (012)**
Sl 1: (1•3) + (1•1)
Sl 2: 2(2•2) + (0•4) + (2•0) + 2(0•2) + (0•0)
Sl 3: (3•3) + 2(3•1) + 4(1•3) + 5(1•1)
Sl 4: 2(4•2) + 2(2•4) + (4•0) + 8(2•2) + 3(0•4) + 5(2•0) + 8(0•2) + 4(0•0)
Sl 5: (5•3) + (5•1) + 6(3•3) + 2(1•5) + 10(3•1) + 14(1•3) + 18(1•1)
Sl 6: (6•2) + 2(4•4) + 9(4•2) + 10(2•4) + (0•6) + 6(4•0) + 30(2•2) + 10(0•4) + 18(2•0) + 23(0•2) + 13(0•0)
Sl 7: 5(5•3) + 3(3•5) + 7(5•1) + 25(3•3) + 9(1•5) + 36(3•1) + 44(1•3) + 55(1•1)

Affine C2 ⊃ A1 + A1 - Level 3 - Class 1

Sl 0: (0•0) **Highest Weight (030)**
Sl 1: (1•1)
Sl 2: (2•2) + (2•0) + (0•2) + (0•0)
Sl 3: (3•3) + (3•1) + (1•3) + 4(1•1)
Sl 4: (4•2) + (2•4) + (4•0) + 5(2•2) + (0•4) + 4(2•0) + 4(0•2) + 4(0•0)
Sl 5: (5•1) + 4(3•3) + (1•5) + 7(3•1) + 7(1•3) + 13(1•1)
Sl 6: (4•4) + (6•0) + 6(4•2) + 6(2•4) + (0•6) + 4(4•0) + 20(2•2) + 4(0•4) + 14(2•0) + 14(0•2) + 11(0•0)
Sl 7: 2(5•3) + 2(3•5) + 5(5•1) + 17(3•3) + 5(1•5) + 26(3•1) + 26(1•3) + 40(1•1)

Affine C2 ⊃ A1 + A1 - Level 3 - Class 1

Sl 0: (1•1) **Highest Weight (111)**

Sl 1: (2•2) + (2•0) + (0•2) + (0•0)

Sl 2: 2(3•1) + 2(1•3) + 4(1•1)

Sl 3: (4•2) + (2•4) + (4•0) + 6(2•2) + (0•4) + 5(2•0) + 5(0•2) + 4(0•0)

Sl 4: (5•1) + 4(3•3) + (1•5) + 9(3•1) + 9(1•3) + 16(1•1)

Sl 5: (4•4) + 7(4•2) + 7(2•4) + 6(4•0) + 25(2•2) + 6(0•4) + 18(2•0) + 18(0•2) + 12(0•0)

Sl 6: 2(5•3) + 2(3•5) + 6(5•1) + 20(3•3) + 6(1•5) + 34(3•1) + 34(1•3) + 52(1•1)

Sl 7: 2(6•2) + 8(4•4) + 2(2•6) + 2(6•0) + 30(4•2) + 30(2•4) + 2(0•6) + 22(4•0) + 84(2•2) + 22(0•4) + 56(2•0) + 56(0•2) + 36(0•0)

Affine C3 Branching Rules

Affine C3 ⊃ C3 - Level 1 - Class 0

sl 0: (•010) **Highest Weight (0010)**
sl 1: (•101) + (•200) + (•010)
sl 2: (•210) + (•020) + 2(•101) + (•200) + 3(•010) + (•000)
sl 3: (•111) + 3(•210) + 2(•020) + 5(•101) + 4(•200) + 5(•010) + (•000)
sl 4: (•301) + (•030) + 3(•111) + (•400) + 7(•210) + (•002) + 6(•020) + 10(•101) + 7(•200) + 11(•010) + 3(•000)
sl 5: 2(•220) + (•012) + 3(•301) + 2(•030) + 8(•111) + 2(•400) + 16(•210) + 3(•002) + 12(•020) + 21(•101) + 16(•200) + 19(•010) + 4(•000)
sl 6: (•121) + (•410) + (•202) + 5(•220) + 2(•012) + 8(•301) + 6(•030) + 20(•111) + 6(•400) + 33(•210) + 6(•002) + 27(•020) + 39(•101) + 28(•200) + 37(•010) + 9(•000)
sl 7: (•311) + 4(•121) + 3(•410) + 3(•202) + 15(•220) + 6(•012) + 18(•301) + 13(•030) + 43(•111) + 12(•400) + 67(•210) + 15(•002) + 50(•020) + 75(•101) + 55(•200) + 62(•010) + 14(•000)

Affine C3 ⊃ C3 - Level 1 - Class 0

sl 0: (•000) **Highest Weight (1000)**
sl 1: (•200)
sl 2: (•020) + (•200) + (•010) + (•000)
sl 3: (•210) + (•002) + (•020) + (•101) + 3(•200) + (•010) + (•000)
sl 4: (•111) + (•400) + 2(•210) + (•002) + 4(•020) + 2(•101) + 4(•200) + 4(•010) + 3(•000)
sl 5: (•220) + 3(•111) + (•400) + 6(•210) + 3(•002) + 6(•020) + 6(•101) + 10(•200) + 6(•010) + 3(•000)
sl 6: (•202) + 2(•220) + 2(•301) + 2(•030) + 7(•111) + 4(•400) + 12(•210) + 5(•002) + 15(•020) + 12(•101) + 15(•200) + 14(•010) + 8(•000)
sl 7: (•121) + (•410) + 2(•202) + 7(•220) + (•012) + 4(•301) + 3(•030) + 17(•111) + 6(•400) + 27(•210) + 11(•002) + 24(•020) + 26(•101) + 31(•200) + 23(•010) + 9(•000)

Affine C3 ⊃ C3 - Level 1 - Class 1

sl 0: (•001) **Highest Weight (0001)**
sl 1: (•110) + (•001)
sl 2: (•201) + 2(•110) + 2(•001) + (•100)
sl 3: (•120) + 2(•201) + (•011) + (•300) + 4(•110) + 4(•001) + 2(•100)
sl 4: (•021) + (•310) + 2(•120) + 5(•201) + 3(•011) + 2(•300) + 9(•110) + 7(•001) + 4(•100)
sl 5: (•211) + 2(•021) + 3(•310) + (•102) + 6(•120) + 10(•201) + 7(•011) + 5(•300) + 18(•110) + 12(•001) + 8(•100)
sl 6: (•401) + (•130) + 3(•211) + (•003) + 5(•021) + 7(•310) + 3(•102) + 14(•120) + 21(•201) + 15(•011) + 11(•300) + 34(•110) + 22(•001) + 15(•100)
sl 7: (•320) + (•112) + 2(•401) + 3(•130) + 9(•211) + (•003) + (•500) + 11(•021) + 16(•310) + 8(•102) + 30(•120) + 40(•201) + 31(•011) + 22(•300) + 64(•110) + 36(•001) + 27(•100)

Affine C3 ⊃ C3 - Level 1 - Class 1

Sl 0: (•100) **Highest Weight (0100)**

Sl 1: (•110) + (•100)

Sl 2: (•011) + (•300) + 2(•110) + (•001) + 2(•100)

Sl 3: (•120) + (•201) + 2(•011) + 2(•300) + 5(•110) + 2(•001) + 4(•100)

Sl 4: (•310) + (•102) + 3(•120) + 3(•201) + 5(•011) + 4(•300) + 11(•110) + 4(•001) + 8(•100)

Sl 5: (•211) + (•021) + 3(•310) + 2(•102) + 8(•120) + 8(•201) + 11(•011) + 9(•300) + 22(•110) + 9(•001) + 14(•100)

Sl 6: (•130) + 4(•211) + (•500) + 3(•021) + 8(•310) + 6(•102) + 18(•120) + 18(•201) + 23(•011) + 18(•300) + 43(•110) + 18(•001) + 25(•100)

Sl 7: (•320) + (•112) + (•401) + 3(•130) + 11(•211) + 2(•500) + 9(•021) + 19(•310) + 13(•102) + 39(•120) + 39(•201) + 45(•011) + 34(•300) + 81(•110) + 33(•001) + 43(•100)

Affine C3 ⊃ C3 - Level 2 - Class 0

Sl 0: (•002) **Highest Weight (0002)**

Sl 1: (•111) + (•002) + (•020)

Sl 2: (•202) + (•220) + 3(•111) + (•210) + 3(•002) + 2(•020) + (•101) + (•200)

Sl 3: (•311) + (•121) + 3(•202) + 3(•220) + (•012) + 2(•301) + 2(•030) + 8(•111) + (•400) + 4(•210) + 6(•002) + 7(•020) + 4(•101) + 2(•200) + (•010) + (•000)

Affine C3 ⊃ C3 - Level 2 - Class 0

Sl 0: (•020) **Highest Weight (0020)**

Sl 1: (•111) + (•210) + (•002) + (•020) + (•101) + (•200)

Sl 2: (•202) + (•220) + (•301) + (•030) + 4(•111) + (•400) + 3(•210) + 2(•002) + 5(•020) + 3(•101) + 2(•200) + 2(•010) + (•000)

Sl 3: (•311) + 2(•121) + (•410) + 3(•202) + 6(•220) + 2(•012) + 4(•301) + 3(•030) + 13(•111) + 2(•400) + 12(•210) + 7(•002) + 11(•020) + 11(•101) + 8(•200) + 5(•010) + (•000)

Affine C3 ⊃ C3 - Level 2 - Class 0

Sl 0: (•101) **Highest Weight (0101)**

Sl 1: (•111) + (•210) + (•020) + 2(•101) + (•010)

Sl 2: (•220) + (•012) + 2(•301) + (•030) + 4(•111) + 4(•210) + (•002) + 3(•020) + 6(•101) + 2(•200) + 3(•010)

Sl 3: (•311) + 2(•121) + (•410) + 2(•202) + 5(•220) + 3(•012) + 7(•301) + 5(•030) + 14(•111) + 2(•400) + 14(•210) + 4(•002) + 11(•020) + 16(•101) + 6(•200) + 9(•010) + (•000)

Affine C3 ⊃ C3 - Level 2 - Class 0

Sl 0: (•200) **Highest Weight (0200)**

Sl 1: (•210) + (•020) + (•200) + (•010) + (•000)

Sl 2: (•220) + (•111) + (•400) + 3(•210) + (•002) + 2(•020) + 2(•101) + 4(•200) + 2(•010) + (•000)

Sl 3: (•121) + (•410) + (•202) + 3(•220) + 2(•301) + 2(•030) + 5(•111) + 4(•400) + 9(•210) + 2(•002) + 8(•020) + 6(•101) + 8(•200) + 7(•010) + 3(•000)

Affine C3 ⊃ C3 - Level 2 - Class 0

Sl 0: (•010) **Highest Weight (1010)**

Sl 1: (•210) + (•101) + (•200) + (•010)

Sl 2: (•301) + (•030) + (•111) + (•400) + 3(•210) + 2(•020) + 3(•101) + 2(•200) + 4(•010) + (•000)

Sl 3: (•121) + (•410) + 3(•220) + (•012) + 3(•301) + 2(•030) + 6(•111) + 2(•400) + 11(•210) + (•002) + 5(•020) + 10(•101) + 8(•200) + 8(•010) + (•000)

Affine C3 ⊃ C3 - Level 2 - Class 0

Sl 0: (•000) **Highest Weight (2000)**

Sl 1: (•200)

Sl 2: (•400) + (•020) + (•200) + (•010) + (•000)

Sl 3: (•220) + (•400) + 2(•210) + (•002) + (•020) + (•101) + 4(•200) + (•010) + (•000)

Sl 4: (•040) + (•410) + (•202) + 2(•220) + (•301) + (•030) + 2(•111) + 4(•400) + 5(•210) + (•002) + 6(•020) + 2(•101) + 6(•200) + 5(•010) + 4(•000)

Affine C3 ⊃ C3 - Level 2 - Class 1

Sl 0: (•011) **Highest Weight (0011)**

Sl 1: (•102) + (•120) + (•201) + 2(•011) + (•110)

Sl 2: 2(•211) + (•021) + (•310) + 3(•102) + 4(•120) + 3(•201) + 6(•011) + (•300) + 4(•110) + (•001) + (•100)

Sl 3: (•302) + (•320) + (•112) + (•401) + 2(•130) + 8(•211) + 5(•021) + 5(•310) + 9(•102) + 13(•120) + 11(•201) + 16(•011) + 5(•300) + 13(•110) + 4(•001) + 3(•100)

Affine C3 ⊃ C3 - Level 2 - Class 1

Sl 0: (•110) **Highest Weight (0110)**

Sl 1: (•120) + (•201) + (•011) + (•300) + 2(•110) + (•001) + (•100)

Sl 2: (•211) + (•021) + 2(•310) + (•102) + 4(•120) + 4(•201) + 4(•011) + 3(•300) + 7(•110) + 3(•001) + 3(•100)

Sl 3: (•320) + (•112) + (•401) + 2(•130) + 6(•211) + (•500) + 5(•021) + 8(•310) + 5(•102) + 14(•120) + 14(•201) + 13(•011) + 9(•300) + 21(•110) + 8(•001) + 8(•100)

Affine C3 ⊃ C3 - Level 2 - Class 1

Sl 0: (•001) **Highest Weight (1001)**

Sl 1: (•201) + (•110) + (•001)

Sl 2: (•021) + (•310) + (•120) + 3(•201) + (•011) + 3(•110) + 3(•001) + (•100)

Sl 3: (•401) + (•130) + 2(•211) + (•003) + 3(•021) + 3(•310) + (•102) + 5(•120) + 9(•201) + 4(•011) + 3(•300) + 9(•110) + 7(•001) + 2(•100)

Sl 4: (•221) + 2(•320) + 2(•112) + 4(•401) + 4(•130) + 9(•211) + 2(•003) + (•500) + 11(•021) + 12(•310) + 5(•102) + 17(•120) + 24(•201) + 14(•011) + 8(•300) + 25(•110) + 16(•001) + 7(•100)

Affine C3 ⊃ C3 - Level 2 - Class 1

Sl 0: (•100) **Highest Weight (1100)**

Sl 1: (•300) + (•110) + (•100)

Sl 2: (•310) + (•120) + (•011) + 2(•300) + 3(•110) + (•001) + 3(•100)

Sl 3: (•130) + (•211) + (•500) + 3(•310) + (•102) + 4(•120) + 3(•201) + 3(•011) + 6(•300) + 9(•110) + 2(•001) + 6(•100)

Sl 4: (•031) + 2(•320) + (•112) + (•401) + 3(•130) + 5(•211) + 3(•500) + 3(•021) + 11(•310) + 4(•102) + 14(•120) + 10(•201) + 11(•011) + 14(•300) + 23(•110) + 7(•001) + 14(•100)

Affine C3 ⊃ A1 + C2 - Level 1 - Class 0

Sl 0: (0•10) **Highest Weight (0010)**

Sl 1: (1•01) + (1•00)

Sl 2: (0•11) + (2•10) + 2(0•10)

Sl 3: 2(1•20) + 3(1•01) + 2(1•00)

Sl 4: (2•11) + (0•30) + 3(0•11) + 3(2•10) + 6(0•10)

Sl 5: (1•21) + (1•02) + (3•01) + 6(1•20) + 8(1•01) + (3•00) + 6(1•00)

Sl 6: (2•30) + (0•12) + 4(2•11) + 3(0•30) + 9(0•11) + 9(2•10) + 14(0•10)

Sl 7: 5(1•21) + 2(3•20) + 4(1•02) + 3(3•01) + 17(1•20) + 19(1•01) + 2(3•00) + 13(1•00)

Sl 8: (2•12) + (0•31) + 4(2•30) + 3(0•12) + 13(2•11) + 9(0•30) + (4•10) + 23(0•11) + 22(2•10) + 31(0•10)

Affine C3 ⊃ A1 + C2 - Level 1 - Class 0

Sl 0: (1•00) **Highest Weight (1000)**

Sl 1: (0•10)

Sl 2: (1•20) + (1•00)

Sl 3: (0•11) + (2•10) + 2(0•10)

Sl 4: (1•02) + 2(1•20) + 2(1•01) + (3•00) + 3(1•00)

Sl 5: (2•11) + (0•30) + 3(0•11) + 3(2•10) + 5(0•10)

Sl 6: (1•21) + (3•20) + 2(1•02) + 7(1•20) + 5(1•01) + (3•00) + 6(1•00)

Sl 7: (2•30) + (0•12) + 4(2•11) + 3(0•30) + 8(0•11) + 8(2•10) + 12(0•10)

Sl 8: (1•40) + (3•02) + 4(1•21) + 2(3•20) + 6(1•02) + 2(3•01) + 16(1•20) + 14(1•01) + 4(3•00) + 14(1•00)

Affine C3 ⊃ A1 + C2 - Level 1 - Class 1

```
Sl 0:   (0•01)  Highest Weight (0001)
Sl 1:   (1•10)
Sl 2:   (2•01) +   (0•20) +   (0•01) +   (0•00)
Sl 3:   (1•11) +  3(1•10)
Sl 4:   (0•21) +   (2•20) +  2(2•01) +  2(0•20) +  4(0•01) +   (2•00) +  2(0•00)
Sl 5:   (1•30) +  4(1•11) +   (3•10) +  7(1•10)
Sl 6:   (2•21) +  2(0•21) +  3(2•20) +  2(0•02) +  6(2•01) +  7(0•20) +  8(0•01) +  3(2•00) +  5(0•00)
Sl 7:   (1•12) +   (3•11) +  4(1•30) + 11(1•11) +  3(3•10) + 17(1•10)
Sl 8:  3(2•21) +   (0•40) +   (0•03) +  2(2•02) +  7(0•21) +   (4•01) + 10(2•20) +  4(0•02) + 13(2•01)
     + 16(0•20) + 19(0•01) +  8(2•00) + 10(0•00)
```

Affine C3 ⊃ A1 + C2 - Level 1 - Class 1

```
Sl 0:   (0•00)  Highest Weight (0100)
Sl 1:   (1•10)
Sl 2:   (0•20) +   (0•01) +   (2•00) +   (0•00)
Sl 3:   (1•11) +  3(1•10)
Sl 4:   (2•20) +   (0•02) +   (2•01) +  3(0•20) +  3(0•01) +  2(2•00) +  3(0•00)
Sl 5:   (1•30) +  4(1•11) +   (3•10) +  8(1•10)
Sl 6:   (2•02) +  2(0•21) +  4(2•20) +  2(0•02) +  4(2•01) +  8(0•20) +  8(0•01) +  5(2•00) +  7(0•00)
Sl 7:   (1•12) +   (3•11) +  4(1•30) + 12(1•11) +  3(3•10) + 20(1•10)
Sl 8:  2(2•21) +   (0•40) +  3(2•02) +  6(0•21) + 12(2•20) +  7(0•02) + 12(2•01) + 20(0•20) +   (4•00)
     + 18(0•01) + 11(2•00) + 15(0•00)
```

Affine C3 ⊃ A1 + C2 - Level 2 - Class 0

```
Sl 0:   (0•02)  Highest Weight (0002)
Sl 1:   (1•11)
Sl 2:   (2•02) +   (0•21) +   (2•20) +   (0•02) +   (0•20) +   (0•01)
Sl 3:   (1•12) +   (3•11) +   (1•30) +  4(1•11) +  2(1•10)
Sl 4:   (0•22) +   (4•02) +  2(2•21) +   (0•40) +  3(2•02) +  3(0•21) +  4(2•20) +  5(0•02) +  3(2•01)
     +  5(0•20) +  3(0•01) +   (2•00) +  2(0•00)
Sl 5:   (3•12) +  2(1•31) +   (3•30) +  5(1•12) +  5(3•11) +  6(1•30) + 15(1•11) +  2(3•10) +  9(1•10)
```

Affine C3 ⊃ A1 + C2 - Level 2 - Class 0

```
Sl 0:   (0•20)  Highest Weight (0020)
Sl 1:   (1•11) +   (1•10)
Sl 2:   (2•02) +   (0•21) +   (2•20) +   (0•02) +   (2•01) +  2(0•20) +  2(0•01) +   (2•00) +   (0•00)
Sl 3:   (1•12) +   (3•11) +  2(1•30) +  6(1•11) +   (3•10) +  5(1•10)
Sl 4:   (0•22) +  3(2•21) +   (0•40) +   (4•20) +  4(2•02) +  5(0•21) +  7(2•20) +  5(0•02) +  6(2•01)
     + 10(0•20) +  7(0•01) +  3(2•00) +  4(0•00)
Sl 5:   (3•12) +  3(1•31) +  2(3•30) +  7(1•12) +  7(3•11) + 11(1•30) + 25(1•11) +  6(3•10) + 21(1•10)
```

Affine C3 ⊃ A1 + C2 - Level 2 - Class 0

Sl 0: (0•01) **Highest Weight (0101)**
Sl 1: (1•11) + (1•10)
Sl 2: (0•21) + (2•20) + (0•02) + 2(2•01) + 2(0•20) + 2(0•01) + (0•00)
Sl 3: (1•12) + (3•11) + 2(1•30) + 6(1•11) + (3•10) + 6(1•10)
Sl 4: 3(2•21) + (0•40) + (0•03) + 3(2•02) + 6(0•21) + (4•01) + 7(2•20) + 4(0•02) + 8(2•01)
+ 9(0•20) + 10(0•01) + 3(2•00) + 4(0•00)
Sl 5: (3•12) + 3(1•31) + 2(3•30) + 7(1•12) + 7(3•11) + 12(1•30) + 27(1•11) + 7(3•10) + 24(1•10)

Affine C3 ⊃ A1 + C2 - Level 2 - Class 0

Sl 0: (0•00) **Highest Weight (0200)**
Sl 1: (1•10)
Sl 2: (2•20) + (0•20) + (0•01) + (2•00) + (0•00)
Sl 3: (1•30) + 2(1•11) + (3•10) + 4(1•10)
Sl 4: (2•21) + (0•40) + (2•02) + (0•21) + 4(2•20) + 2(0•02) + 3(2•01) + 5(0•20) + (4•00)
+ 4(0•01) + 4(2•00) + 4(0•00)
Sl 5: (1•31) + (3•30) + 2(1•12) + 2(3•11) + 6(1•30) + 11(1•11) + 5(3•10) + 15(1•10)

Affine C3 ⊃ A1 + C2 - Level 2 - Class 0

Sl 0: (1•10) **Highest Weight (1010)**
Sl 1: (2•01) + (0•20) + (0•01) + (2•00) + (0•00)
Sl 2: (1•30) + 2(1•11) + (3•10) + 4(1•10)
Sl 3: (2•21) + 2(0•21) + 4(2•20) + (0•02) + 4(2•01) + 5(0•20) + 5(0•01) + 3(2•00) + 3(0•00)
Sl 4: (1•31) + (3•30) + 2(1•12) + 2(3•11) + 6(1•30) + 12(1•11) + 5(3•10) + 16(1•10)
Sl 5: 2(2•40) + (2•03) + (0•22) + 8(2•21) + 3(0•40) + (0•03) + (4•20) + 5(2•02) + 11(0•21)
+ 2(4•01) + 17(2•20) + 7(0•02) + 18(2•01) + 20(0•20) + 2(4•00) + 18(0•01) + 12(2•00) + 11(0•00)
Sl 6: (3•31) + (1•50) + (1•13) + 2(3•12) + 8(1•31) + 7(3•30) + 14(1•12) + 14(3•11) + 28(1•30)
+ (5•10) + 50(1•11) + 20(3•10) + 54(1•10)

Affine C3 ⊃ A1 + C2 - Level 2 - Class 0

Sl 0: (2•00) **Highest Weight (2000)**
Sl 1: (1•10)
Sl 2: (2•20) + (0•20) + (2•00) + (0•00)
Sl 3: (1•30) + (1•11) + (3•10) + 3(1•10)
Sl 4: (2•40) + (2•02) + (0•21) + 3(2•20) + (0•02) + 2(2•01) + 3(0•20) + (4•00) + 2(0•01)
+ 4(2•00) + 3(0•00)
Sl 5: (1•31) + (3•30) + (1•12) + (3•11) + 4(1•30) + 6(1•11) + 4(3•10) + 10(1•10)
Sl 6: (2•22) + 2(2•40) + (0•22) + 4(2•21) + 2(0•40) + 2(4•20) + 3(2•02) + 5(0•21) + 13(2•20)
+ 4(0•02) + 7(2•01) + 12(0•20) + 2(4•00) + 7(0•01) + 9(2•00) + 7(0•00)

Affine C3 ⊃ A1 + C2 - Level 2 - Class 1

Sl 0: (0•11) **Highest Weight (0011)**

Sl 1: (1•02) + (1•20) + (1•01)

Sl 2: (0•12) + 2(2•11) + (0•30) + 3(0•11) + (2•10) + 2(0•10)

Sl 3: (3•02) + 3(1•21) + (3•20) + 4(1•02) + (3•01) + 6(1•20) + 5(1•01) + 2(1•00)

Sl 4: 2(2•12) + 2(0•31) + (4•11) + 3(2•30) + 4(0•12) + 10(2•11) + 5(0•30) + 13(0•11) + 7(2•10) + 9(0•10)

Sl 5: 2(1•22) + 3(3•21) + 3(1•40) + (1•03) + 5(3•02) + 16(1•21) + 7(3•20) + 16(1•02) + 6(3•01) + 25(1•20) + 21(1•01) + 2(3•00) + 9(1•00)

Affine C3 ⊃ A1 + C2 - Level 2 - Class 1

Sl 0: (0•10) **Highest Weight (0110)**

Sl 1: (1•20) + (1•01) + (1•00)

Sl 2: (2•11) + (0•30) + 2(0•11) + 2(2•10) + 3(0•10)

Sl 3: 2(1•21) + (3•20) + 2(1•02) + (3•01) + 6(1•20) + 6(1•01) + (3•00) + 4(1•00)

Sl 4: (2•12) + (0•31) + 3(2•30) + 2(0•12) + 8(2•11) + 5(0•30) + (4•10) + 11(0•11) + 10(2•10) + 12(0•10)

Sl 5: (1•22) + 2(3•21) + 3(1•40) + (1•03) + 2(3•02) + 14(1•21) + 7(3•20) + 11(1•02) + 7(3•01) + 27(1•20) + 24(1•01) + 5(3•00) + 14(1•00)

Affine C3 ⊃ A1 + C2 - Level 2 - Class 1

Sl 0: (1•01) **Highest Weight (1001)**

Sl 1: (0•11) + (2•10) + (0•10)

Sl 2: (1•21) + (3•01) + 2(1•20) + 3(1•01) + (1•00)

Sl 3: (2•30) + (0•12) + 3(2•11) + 2(0•30) + 4(0•11) + 4(2•10) + 5(0•10)

Sl 4: (3•21) + (1•40) + (1•03) + 6(1•21) + 2(3•20) + 3(1•02) + 4(3•01) + 10(1•20) + 12(1•01) + (3•00) + 5(1•00)

Sl 5: (2•31) + 3(2•12) + 3(0•31) + (4•11) + 7(2•30) + 5(0•12) + 15(2•11) + 9(0•30) + 2(4•10) + 18(0•11) + 17(2•10) + 17(0•10)

Sl 6: (1•41) + (3•40) + (3•03) + 3(1•22) + 7(3•21) + 6(1•40) + 4(1•03) + 3(3•02) + 28(1•21) + (5•01) + 12(3•20) + 16(1•02) + 15(3•01) + 40(1•20) + 40(1•01) + 6(3•00) + 17(1•00)

Affine C3 ⊃ A1 + C2 - Level 2 - Class 1

Sl 0: (1•00) **Highest Weight (1100)**

Sl 1: (2•10) + (0•10)

Sl 2: 2(1•20) + (1•01) + (3•00) + 2(1•00)

Sl 3: (2•30) + (2•11) + (0•30) + 2(0•11) + 4(2•10) + 4(0•10)

Sl 4: (1•40) + 2(1•21) + 2(3•20) + 2(1•02) + (3•01) + 8(1•20) + 6(1•01) + 3(3•00) + 7(1•00)

Sl 5: (2•31) + (2•12) + (0•31) + 5(2•30) + 2(0•12) + 8(2•11) + 6(0•30) + 2(4•10) + 10(0•11) + 15(2•10) + 14(0•10)

Affine C4 Branching Rules

Affine C4 ⊃ C4 - Level 1 - Class 0

Sl 0: (•0001) **Highest Weight (00001)**

Sl 1: (•1010) + (•0001)

Sl 2: (•2001) + (•0200) + 2(•1010) + 2(•0001) + (•0100)

Sl 3: (•1110) + 2(•2001) + (•0101) + (•2100) + (•0200) + 5(•1010) + 4(•0001) + (•2000) + 2(•0100)

Sl 4: (•0201) + (•3010) + 3(•1110) + 5(•2001) + (•0020) + 3(•0101) + 3(•2100) + 4(•0200) + 10(•1010) + 8(•0001) + 2(•2000) + 5(•0100) + (•0000)

Sl 5: (•2101) + (•0120) + 2(•0201) + (•2200) + (•1011) + 3(•3010) + (•0300) + 8(•1110) + 11(•2001) + 3(•0020) + 8(•0101) + 8(•2100) + 8(•0200) + 22(•1010) + 13(•0001) + 6(•2000) + 10(•0100) + (•0000)

Affine C4 ⊃ C4 - Level 1 - Class 0

Sl 0: (•0100) **Highest Weight (00100)**

Sl 1: (•1010) + (•2000) + (•0100)

Sl 2: (•0101) + (•2100) + (•0200) + 2(•1010) + (•0001) + (•2000) + 3(•0100) + (•0000)

Sl 3: (•1110) + (•2001) + (•0020) + 2(•0101) + 3(•2100) + 2(•0200) + 6(•1010) + 2(•0001) + 4(•2000) + 5(•0100) + (•0000)

Sl 4: (•1011) + (•3010) + (•0300) + 4(•1110) + 3(•2001) + 2(•0020) + (•4000) + 6(•0101) + 7(•2100) + 7(•0200) + 13(•1010) + 5(•0001) + 7(•2000) + 12(•0100) + 3(•0000)

Sl 5: (•2101) + (•0120) + (•0201) + 2(•2200) + 3(•1011) + 3(•3010) + 2(•0300) + 12(•1110) + (•0002) + 9(•2001) + 7(•0020) + 2(•4000) + 13(•0101) + 17(•2100) + 14(•0200) + 30(•1010) + 10(•0001) + 17(•2000) + 21(•0100) + 4(•0000)

Affine C4 ⊃ C4 - Level 1 - Class 0

Sl 0: (•0000) **Highest Weight (10000)**

Sl 1: (•2000)

Sl 2: (•0200) + (•2000) + (•0100) + (•0000)

Sl 3: (•0020) + (•2100) + (•0200) + (•1010) + 3(•2000) + (•0100) + (•0000)

Sl 4: (•1110) + (•0002) + (•0020) + (•4000) + (•0101) + 2(•2100) + 4(•0200) + 2(•1010) + (•0001) + 4(•2000) + 4(•0100) + 3(•0000)

Sl 5: (•2200) + (•1011) + 3(•1110) + (•0002) + (•2001) + 4(•0020) + (•4000) + 2(•0101) + 6(•2100) + 6(•0200) + 7(•1010) + (•0001) + 10(•2000) + 6(•0100) + 3(•0000)

Affine C4 ⊃ C4 - Level 1 - Class 1

Sl 0: (•0010) **Highest Weight (00010)**

Sl 1: (•1001) + (•1100) + (•0010)

Sl 2: (•2010) + (•0110) + 2(•1001) + 2(•1100) + 3(•0010) + (•1000)

Sl 3: (•1101) + (•1200) + 3(•2010) + 3(•0110) + 5(•1001) + (•3000) + 5(•1100) + 6(•0010) + 2(•1000)

Sl 4: (•0210) + (•3001) + (•1020) + 3(•1101) + (•3100) + 3(•1200) + (•0011) + 8(•2010) + 8(•0110) + 11(•1001) + 2(•3000) + 12(•1100) + 12(•0010) + 5(•1000)

Sl 5: (•0111) + 2(•2110) + 3(•0210) + 3(•3001) + 3(•1020) + 9(•1101) + 3(•3100) + 9(•1200) + 4(•0011) + 19(•2010) + 19(•0110) + 23(•1001) + 6(•3000) + 26(•1100) + 23(•0010) + 10(•1000)

Sl 6: (•1201) + (•2011) + (•0030) + (•4010) + (•1300) + 3(•0111) + 7(•2110) + (•1002) + 9(•0210) + 8(•3001) + 10(•1020) + 23(•1101) + 9(•3100) + 22(•1200) + 10(•0011) + 42(•2010) + 43(•0110) + 46(•1001) + 14(•3000) + 53(•1100) + 45(•0010) + 20(•1000)

Affine C4 ⊃ C4 - Level 1 - Class 1

Sl 0: (•1000) **Highest Weight (01000)**

Sl 1: (•1100) + (•1000)

Sl 2: (•0110) + (•3000) + 2(•1100) + (•0010) + 2(•1000)

Sl 3: (•1200) + (•0011) + (•2010) + 2(•0110) + (•1001) + 2(•3000) + 5(•1100) + 2(•0010) + 4(•1000)

Sl 4: (•1020) + (•1101) + (•3100) + 3(•1200) + 2(•0011) + 3(•2010) + 6(•0110) + 3(•1001) + 4(•3000) + 11(•1100) + 5(•0010) + 8(•1000)

Sl 5: (•2110) + (•1002) + (•0210) + 3(•1020) + 4(•1101) + 3(•3100) + 8(•1200) + 5(•0011) + 9(•2010) + 14(•0110) + 8(•1001) + 9(•3000) + 23(•1100) + 11(•0010) + 14(•1000)

Sl 6: (•2011) + (•1300) + (•0111) + 4(•2110) + 2(•1002) + 4(•0210) + 2(•3001) + 9(•1020) + (•5000) + 11(•1101) + 8(•3100) + 19(•1200) + 12(•0011) + 22(•2010) + 32(•0110) + 19(•1001) + 18(•3000) + 46(•1100) + 23(•0010) + 26(•1000)

Affine C4 ⊃ C4 - Level 2 - Class 0

Sl 0: (•0002) **Highest Weight (00002)**

Sl 1: (•1011) + (•0002) + (•0020)

Sl 2: (•2002) + (•2020) + (•0201) + 3(•1011) + (•1110) + 3(•0002) + 2(•0020) + (•0101) + (•0200)

Affine C4 ⊃ C4 - Level 2 - Class 0

Sl 0: (•0020) **Highest Weight (00020)**

Sl 1: (•1011) + (•1110) + (•0002) + (•0020) + (•0101) + (•0200)

Sl 2: (•2002) + (•2020) + (•2101) + (•0120) + (•0201) + (•2200) + 4(•1011) + 4(•1110) + 2(•0002) + (•2001) + 5(•0020) + 3(•0101) + (•2100) + 2(•0200) + 2(•1010) + (•2000)

Affine C4 ⊃ C4 - Level 2 - Class 0

Sl 0: (•0101) **Highest Weight (00101)**

Sl 1: (•1011) + (•1110) + (•2001) + (•0020) + 2(•0101) + (•1010)

Sl 2: (•2020) + (•0102) + 2(•2101) + (•0120) + (•0201) + 4(•1011) + (•3010) + (•0300) + 5(•1110) + (•0002) + 3(•2001) + 3(•0020) + 7(•0101) + (•2100) + 2(•0200) + 4(•1010) + (•0001) + (•0100)

Affine C4 ⊃ C4 - Level 2 - Class 0

Sl 0: (•0200) **Highest Weight (00200)**

Sl 1: (•1110) + (•0020) + (•2100) + (•0200) + (•1010) + (•2000)

Sl 2: (•2020) + (•0201) + (•2200) + (•1011) + (•3010) + (•0300) + 4(•1110) + (•0002) + (•2001) + 2(•0020) + (•4000) + 2(•0101) + 3(•2100) + 5(•0200) + 3(•1010) + (•0001) + 2(•2000) + 2(•0100) + (•0000)

Affine C4 ⊃ C4 - Level 2 - Class 0

Sl 0: (•1010) **Highest Weight (01010)**

Sl 1: (•1110) + (•2001) + (•0101) + (•2100) + (•0200) + 2(•1010) + (•0001) + (•0100)

Sl 2: (•2101) + (•0120) + (•0201) + (•2200) + (•1011) + 2(•3010) + (•0300) + 5(•1110) + 4(•2001) + 2(•0020) + 4(•0101) + 4(•2100) + 3(•0200) + 8(•1010) + 3(•0001) + 2(•2000) + 3(•0100)

Affine C4 ⊃ C4 - Level 2 - Class 0

Sl 0: (•2000) **Highest Weight (02000)**

Sl 1: (•2100) + (•0200) + (•2000) + (•0100) + (•0000)

Sl 2: (•2200) + (•1110) + (•0020) + (•4000) + 3(•2100) + 2(•0200) + 2(•1010) + 4(•2000) + 2(•0100) + (•0000)

Sl 3: (•1210) + (•2020) + (•4100) + 3(•2200) + (•1011) + 2(•3010) + 2(•0300) + 5(•1110) + (•0002) + (•2001) + 2(•0020) + 4(•4000) + 2(•0101) + 9(•2100) + 8(•0200) + 6(•1010) + (•0001) + 8(•2000) + 7(•0100) + 3(•0000)

Affine C4 ⊃ C4 - Level 2 - Class 0

Sl 0: (•0001) **Highest Weight (10001)**

Sl 1: (•2001) + (•1010) + (•0001)

Sl 2: (•0201) + (•3010) + (•1110) + 3(•2001) + (•0101) + (•0200) + 3(•1010) + 3(•0001) + (•0100)

Sl 3: (•1210) + (•0021) + (•4001) + 2(•2101) + (•0120) + 3(•0201) + (•2200) + (•1011) + 3(•3010) + 6(•1110) + 9(•2001) + (•0020) + 4(•0101) + 3(•2100) + 2(•0200) + 10(•1010) + 7(•0001) + (•2000) + 2(•0100)

Affine C4 ⊃ C4 - Level 2 - Class 0

```
Sl 0:  (•0100)  Highest Weight (10100)
Sl 1:  (•2100) +  (•1010) +  (•2000) +  (•0100)
Sl 2:  (•3010) +  (•0300) +  (•1110) +  (•4000) +  (•0101) + 3(•2100) + 2(•0200) + 3(•1010)
     + (•0001) + 2(•2000) + 4(•0100) +  (•0000)
Sl 3:  (•1210) +  (•2101) +  (•0120) +  (•4100) + 3(•2200) +  (•1011) + 3(•3010) + 2(•0300)
     + 6(•1110) + 3(•2001) + 2(•0020) + 2(•4000) + 3(•0101) + 11(•2100) + 5(•0200) + 11(•1010)
     + 2(•0001) + 8(•2000) + 8(•0100) +  (•0000)
```

Affine C4 ⊃ C4 - Level 2 - Class 0

```
Sl 0:  (•0000)  Highest Weight (20000)
Sl 1:  (•2000)
Sl 2:  (•4000) +  (•0200) +  (•2000) +  (•0100) +  (•0000)
Sl 3:  (•2200) +  (•0020) +  (•4000) + 2(•2100) +  (•0200) +  (•1010) + 4(•2000) +  (•0100)
     + (•0000)
```

Affine C4 ⊃ C4 - Level 2 - Class 1

```
Sl 0:  (•0011)  Highest Weight (00011)
Sl 1:  (•1002) +  (•1020) +  (•1101) + 2(•0011) +  (•0110)
Sl 2: 2(•2011) +  (•0111) +  (•2110) + 3(•1002) +  (•0210) + 4(•1020) + 4(•1101) +  (•1200)
     + 6(•0011) +  (•2010) + 4(•0110) +  (•1001) +  (•1100)
```

Affine C4 ⊃ C4 - Level 2 - Class 1

```
Sl 0:  (•0110)  Highest Weight (00110)
Sl 1:  (•1020) +  (•1101) +  (•1200) +  (•0011) +  (•2010) + 2(•0110) +  (•1001) +  (•1100)
Sl 2:  (•2011) +  (•0111) + 2(•2110) +  (•1002) + 2(•0210) +  (•3001) + 4(•1020) + 5(•1101)
     + (•3100) + 4(•1200) + 4(•0011) + 4(•2010) + 8(•0110) + 4(•1001) +  (•3000) + 4(•1100)
     + 2(•0010) +  (•1000)
Sl 3:  (•3020) +  (•1102) +  (•3101) + 2(•1120) + 3(•1201) +  (•3200) + 6(•2011) +  (•0030)
     + (•4010) + 2(•1300) + 6(•0111) + 11(•2110) + 5(•1002) + 9(•0210) + 5(•3001) + 16(•1020)
     + 20(•1101) + 5(•3100) + 15(•1200) + 14(•0011) + 18(•2010) + 25(•0110) + 14(•1001) + 5(•3000)
     + 15(•1100) + 7(•0010) + 3(•1000)
```

Affine C4 ⊃ C4 - Level 2 - Class 1

```
Sl 0:  (•1001)  Highest Weight (01001)
Sl 1:  (•1101) +  (•2010) +  (•0110) + 2(•1001) +  (•0010)
Sl 2:  (•0111) +  (•2110) +  (•0210) + 2(•3001) +  (•1020) + 4(•1101) +  (•1200) +  (•0011)
     + 4(•2010) + 4(•0110) + 6(•1001) + 2(•1100) + 3(•0010)
```

Affine C4 ⊃ C4 - Level 2 - Class 1

Sl 0: (•1100) **Highest Weight (01100)**

Sl 1: (•1200) + (•2010) + (•0110) + (•3000) + 2(•1100) + (•0010) + (•1000)

Sl 2: (•2110) + (•0210) + (•1020) + (•1101) + 2(•3100) + 4(•1200) + (•0011) + 4(•2010) + 4(•0110) + 2(•1001) + 3(•3000) + 7(•1100) + 3(•0010) + 3(•1000)

Sl 3: (•1120) + (•1201) + (•3200) + (•2011) + (•4010) + 2(•1300) + (•0111) + 6(•2110) + (•1002) + 5(•0210) + 2(•3001) + 6(•1020) + (•5000) + 7(•1101) + 8(•3100) + 14(•1200) + 4(•0011) + 15(•2010) + 15(•0110) + 8(•1001) + 9(•3000) + 21(•1100) + 9(•0010) + 8(•1000)

Affine C4 ⊃ C4 - Level 2 - Class 1

Sl 0: (•0010) **Highest Weight (10010)**

Sl 1: (•2010) + (•1001) + (•1100) + (•0010)

Sl 2: (•0210) + (•3001) + (•1101) + (•3100) + (•1200) + 3(•2010) + 2(•0110) + 3(•1001) + 3(•1100) + 4(•0010) + (•1000)

Sl 3: (•1201) + (•0030) + (•4010) + (•1300) + (•0111) + 3(•2110) + 3(•0210) + 3(•3001) + 2(•1020) + 6(•1101) + 3(•3100) + 5(•1200) + (•0011) + 12(•2010) + 7(•0110) + 10(•1001) + 3(•3000) + 10(•1100) + 9(•0010) + 2(•1000)

Affine C4 ⊃ C4 - Level 2 - Class 1

Sl 0: (•1000) **Highest Weight (11000)**

Sl 1: (•3000) + (•1100) + (•1000)

Sl 2: (•3100) + (•1200) + (•0110) + 2(•3000) + 3(•1100) + (•0010) + 3(•1000)

Sl 3: (•1300) + (•2110) + (•1020) + (•5000) + 3(•3100) + 4(•1200) + (•0011) + 3(•2010) + 3(•0110) + (•1001) + 6(•3000) + 9(•1100) + 2(•0010) + 6(•1000)

Affine C4 ⊃ A1 + C3 - Level 1 - Class 0

Sl 0: (0•001) **Highest Weight (00001)**

Sl 1: (1•010)

Sl 2: (0•110) + (2•001) + (0•001) + (0•100)

Sl 3: (1•101) + (1•200) + 3(1•010)

Sl 4: (0•201) + (2•110) + 3(0•110) + 2(2•001) + 4(0•001) + (2•100) + 3(0•100)

Sl 5: (1•210) + (1•020) + 4(1•101) + (3•010) + 3(1•200) + 8(1•010) + 2(1•000)

Sl 6: (2•201) + (0•120) + 2(0•201) + 2(0•011) + 4(2•110) + 2(0•300) + 9(0•110) + 6(2•001) + 9(0•001) + 4(2•100) + 8(0•100)

Sl 7: (1•111) + 5(1•210) + (3•101) + 4(1•020) + 12(1•101) + (3•200) + 3(3•010) + 10(1•200) + 20(1•010) + 5(1•000)

Affine C4 ⊃ A1 + C3 - Level 1 - Class 0

Sl 0: (0•100) **Highest Weight (00100)**

Sl 1: (1•010) + (1•000)

Sl 2: (0•110) + (0•001) + (2•100) + 2(0•100)

Sl 3: (1•101) + 2(1•200) + 3(1•010) + 2(1•000) (continued next page) sl 4: (0•011) + (2•110) + (0•300) + 4(0•110) + (2•001) + 3(0•001) + 3(2•100) + 6(0•100)

sl 5: (1•210) + 2(1•020) + 4(1•101) + (3•010) + 6(1•200) + 10(1•010) + (3•000) + 6(1•000)

sl 6: (0•120) + (2•011) + 2(0•201) + (2•300) + 3(0•011) + 5(2•110) + 3(0•300) + 13(0•110) + 4(2•001) + 9(0•001) + 9(2•100) + 15(0•100)

sl 7: (1•111) + 6(1•210) + (1•002) + (3•101) + 7(1•020) + 14(1•101) + 2(3•200) + 3(3•010) + 19(1•200) + 26(1•010) + 2(3•000) + 13(1•000)

Affine C4 ⊃ A1 + C3 - Level 1 - Class 0

sl 0: (1•000) **Highest Weight (10000)**

sl 1: (0•100)

sl 2: (1•200) + (1•000)

sl 3: (0•110) + (2•100) + 2(0•100)

sl 4: (1•020) + 2(1•200) + 2(1•010) + (3•000) + 3(1•000)

sl 5: (0•011) + (2•110) + (0•300) + 3(0•110) + (0•001) + 3(2•100) + 5(0•100)

sl 6: (1•210) + (1•002) + 2(1•020) + 2(1•101) + (3•200) + 7(1•200) + 5(1•010) + (3•000) + 6(1•000)

sl 7: (0•120) + (2•011) + (0•201) + (2•300) + 3(0•011) + 4(2•110) + 3(0•300) + 9(0•110) + (2•001) + 4(0•001) + 8(2•100) + 12(0•100)

Affine C4 ⊃ A1 + C3 - Level 1 - Class 1

sl 0: (0•010) **Highest Weight (00010)**

sl 1: (1•001) + (1•100)

sl 2: (0•101) + (2•010) + (0•200) + 2(0•010) + (0•000)

sl 3: 2(1•110) + 3(1•001) + 3(1•100)

sl 4: (0•210) + (2•101) + (0•020) + 3(0•101) + (2•200) + 3(2•010) + 3(0•200) + 7(0•010) + (2•000) + 2(0•000)

sl 5: (1•201) + (1•011) + (1•300) + 8(1•110) + (3•001) + 8(1•001) + (3•100) + 9(1•100)

sl 6: (0•111) + (2•210) + (2•020) + 4(0•210) + 4(2•101) + 4(0•020) + 10(0•101) + 4(2•200) + 10(2•010) + 10(0•200) + 17(0•010) + 3(2•000) + 6(0•000)

sl 7: 2(1•120) + 5(1•201) + 5(1•011) + 2(3•110) + 5(1•300) + 24(1•110) + 3(3•001) + 21(1•001) + 3(3•100) + 24(1•100)

Affine C4 ⊃ A1 + C3 - Level 1 - Class 1

sl 0: (0•000) **Highest Weight (01000)**

sl 1: (1•100)

sl 2: (0•200) + (0•010) + (2•000) + (0•000)

sl 3: (1•110) + 3(1•100)

sl 4: (0•020) + (0•101) + (2•200) + (2•010) + 3(0•200) + 3(0•010) + 2(2•000) + 3(0•000)

sl 5: (1•011) + (1•300) + 4(1•110) + 2(1•001) + (3•100) + 8(1•100)

sl 6: (2•020) + 2(0•210) + (0•002) + (2•101) + 3(0•020) + 4(0•101) + 4(2•200) + 4(2•010) + 8(0•200) + 9(0•010) + 5(2•000) + 7(0•000)

sl 7: (1•120) + (1•201) + 4(1•011) + (3•110) + 4(1•300) + 14(1•110) + 7(1•001) + 3(3•100) + 20(1•100)

Affine C4 ⊃ A1 + C3 - Level 2 - Class 0

Sl 0: (0•002) **Highest Weight (00002)**
Sl 1: (1•011)
Sl 2: (0•111) + (2•002) + (2•020) + (0•002) + (0•020) + (0•101)
Sl 3: (1•102) + (1•120) + (3•011) + (1•201) + 4(1•011) + 2(1•110)
Sl 4: (0•202) + (0•220) + 2(2•111) + (4•002) + 4(0•111) + (2•210) + 3(2•002) + 4(2•020) + 2(0•210) + 5(0•002) + 3(2•101) + 5(0•020) + 4(0•101) + (2•200) + 2(0•200) + (0•010)

Affine C4 ⊃ A1 + C3 - Level 2 - Class 0

Sl 0: (0•020) **Highest Weight (00020)**
Sl 1: (1•011) + (1•110)
Sl 2: (0•111) + (2•002) + (2•020) + (0•210) + (0•002) + (2•101) + 2(0•020) + 2(0•101) + (2•200) + (0•200) + (0•010)
Sl 3: (1•102) + 2(1•120) + (3•011) + 2(1•201) + 6(1•011) + (3•110) + (1•300) + 6(1•110) + 2(1•001) + 2(1•100)
Sl 4: (0•202) + (0•220) + 3(2•111) + (0•301) + (0•030) + 6(0•111) + (4•020) + 3(2•210) + 4(2•002) + (0•400) + 7(2•020) + 6(0•210) + 5(0•002) + 7(2•101) + 11(0•020) + 10(0•101) + 4(2•200) + 4(2•010) + 6(0•200) + 6(0•010) + (2•000) + 2(0•000)

Affine C4 ⊃ A1 + C3 - Level 2 - Class 0

Sl 0: (0•101) **Highest Weight (00101)**
Sl 1: (1•011) + (1•110) + (1•001)
Sl 2: (0•111) + (2•020) + (0•210) + (0•002) + 2(2•101) + 2(0•020) + 3(0•101) + (2•010) + (0•200) + 2(0•010)
Sl 3: (1•102) + 2(1•120) + (3•011) + 3(1•201) + 6(1•011) + (3•110) + (1•300) + 8(1•110) + (3•001) + 5(1•001) + 2(1•100)
Sl 4: (0•220) + (0•012) + 3(2•111) + 2(0•301) + (0•030) + 7(0•111) + 4(2•210) + 3(2•002) + (4•101) + 7(2•020) + 8(0•210) + 4(0•002) + 11(2•101) + 10(0•020) + 16(0•101) + 4(2•200) + 7(2•010) + 7(0•200) + 10(0•010) + (0•000)

Affine C4 ⊃ A1 + C3 - Level 2 - Class 0

Sl 0: (0•200) **Highest Weight (00200)**
Sl 1: (1•110) + (1•100)
Sl 2: (2•020) + (0•210) + (0•020) + (0•101) + (2•200) + (2•010) + 2(0•200) + 2(0•010) + (2•000) + (0•000)
Sl 3: (1•120) + (1•201) + 2(1•011) + (3•110) + 2(1•300) + 6(1•110) + 2(1•001) + (3•100) + 5(1•100)
Sl 4: (0•220) + (2•111) + 2(0•111) + 3(2•210) + (2•002) + (0•400) + 4(2•020) + 6(0•210) + 2(0•002) + 4(2•101) + 6(0•020) + (4•200) + 7(0•101) + 7(2•200) + 6(2•010) + 10(0•200) + 8(0•010) + 3(2•000) + 4(0•000)

Affine C4 ⊃ A1 + C3 - Level 2 - Class 0

Sl 0: (0•010) **Highest Weight (01010)**

Sl 1: (1•110) + (1•001) + (1•100)

Sl 2: (0•210) + (2•101) + (0•020) + 2(0•101) + (2•200) + 2(2•010) + 2(0•200) + 3(0•010) + (0•000)

Sl 3: (1•120) + 2(1•201) + 2(1•011) + (3•110) + 2(1•300) + 8(1•110) + (3•001) + 6(1•001) + (3•100) + 6(1•100)

Sl 4: (2•111) + (0•301) + (0•030) + 3(0•111) + 4(2•210) + (0•400) + 4(2•020) + 8(0•210) + (0•002) + 8(2•101) + 7(0•020) + 12(0•101) + (4•010) + 7(2•200) + 11(2•010) + 10(0•200) + 15(0•010) + 3(2•000) + 4(0•000)

Affine C4 ⊃ A1 + C3 - Level 2 - Class 0

Sl 0: (0•000) **Highest Weight (02000)**

Sl 1: (1•100)

Sl 2: (2•200) + (0•200) + (0•010) + (2•000) + (0•000)

Sl 3: (1•300) + 2(1•110) + (3•100) + 4(1•100)

Sl 4: (2•210) + (0•400) + (2•020) + (0•210) + 2(0•020) + (0•101) + 4(2•200) + 3(2•010) + 5(0•200) + 4(0•010) + (4•000) + 4(2•000) + 4(0•000)

Affine C4 ⊃ A1 + C3 - Level 2 - Class 0

Sl 0: (1•001) **Highest Weight (10001)**

Sl 1: (0•101) + (2•010) + (0•010)

Sl 2: (1•201) + 2(1•110) + (3•001) + 3(1•001) + (1•100)

Sl 3: (0•111) + (2•210) + 2(0•210) + 3(2•101) + (0•020) + 4(0•101) + (2•200) + 4(2•010) + 2(0•200) + 5(0•010)

Sl 4: (1•021) + (1•310) + (3•201) + 2(1•120) + 6(1•201) + 3(1•011) + 2(3•110) + 2(1•300) + 12(1•110) + 4(3•001) + 12(1•001) + (3•100) + 6(1•100)

Affine C4 ⊃ A1 + C3 - Level 2 - Class 0

Sl 0: (1•100) **Highest Weight (10100)**

Sl 1: (2•010) + (0•200) + (0•010) + (2•000) + (0•000)

Sl 2: (1•300) + 2(1•110) + (1•001) + (3•100) + 4(1•100)

Sl 3: (2•210) + 2(0•210) + (2•101) + (0•020) + 2(0•101) + 4(2•200) + 4(2•010) + 5(0•200) + 5(0•010) + 3(2•000) + 3(0•000)

Sl 4: (1•310) + 2(1•120) + 2(1•201) + (3•300) + 2(1•011) + 2(3•110) + 6(1•300) + 13(1•110) + (3•001) + 6(1•001) + 5(3•100) + 16(1•100)

Affine C4 ⊃ A1 + C3 - Level 2 - Class 0

Sl 0: (2•000) **Highest Weight (20000)**

Sl 1: (1•100)

Sl 2: (2•200) + (0•200) + (2•000) + (0•000)

Sl 3: (1•300) + (1•110) + (3•100) + 3(1•100)

Sl 4: (2•400) + (2•020) + (0•210) + (0•020) + 3(2•200) + 2(2•010) + 3(0•200) + 2(0•010) + (4•000) + 4(2•000) + 3(0•000)

Affine C4 ⊃ A1 + C3 - Level 2 - Class 1

Sl 0: (0•011) **Highest Weight (00011)**

Sl 1: (1•002) + (1•020) + (1•101)

Sl 2: (0•102) + (0•120) + 2(2•011) + (0•201) + 3(0•011) + (2•110) + 2(0•110) + (0•001)

Sl 3: 3(1•111) + (3•002) + (3•020) + 2(1•210) + 4(1•002) + (3•101) + 6(1•020) + 6(1•101) + 2(1•200) + 2(1•010)

Sl 4: 2(0•211) + 2(2•102) + (0•021) + 3(2•120) + (0•310) + 4(0•102) + (4•011) + 3(2•201) + 6(0•120) + 10(2•011) + 6(0•201) + (2•300) + 14(0•011) + 8(2•110) + 2(0•300) + 12(0•110) + 3(2•001) + 4(0•001) + (2•100) + 3(0•100)

Affine C4 ⊃ A1 + C3 - Level 2 - Class 1

Sl 0: (0•110) **Highest Weight (00110)**

Sl 1: (1•020) + (1•101) + (1•200) + (1•010)

Sl 2: (0•120) + (2•011) + (0•201) + 2(0•011) + 2(2•110) + (0•300) + 4(0•110) + (2•001) + 2(0•001) + (2•100) + 2(0•100)

Sl 3: 2(1•111) + (3•020) + 4(1•210) + 2(1•002) + (3•101) + 6(1•020) + 8(1•101) + (3•200) + (3•010) + 6(1•200) + 7(1•010) + 2(1•000)

Sl 4: (0•211) + (2•102) + (0•021) + 3(2•120) + 2(0•310) + 2(0•102) + 4(2•201) + 7(0•120) + 8(2•011) + 8(0•201) + (4•110) + 3(2•300) + 12(0•011) + 14(2•110) + 6(0•300) + 21(0•110) + 7(2•001) + 10(0•001) + 7(2•100) + 10(0•100)

Affine C4 ⊃ A1 + C3 - Level 2 - Class 1

Sl 0: (0•001) **Highest Weight (01001)**

Sl 1: (1•101) + (1•010)

Sl 2: (0•201) + (0•011) + (2•110) + 2(0•110) + 2(2•001) + 2(0•001) + (0•100)

Sl 3: (1•111) + 2(1•210) + (3•101) + 2(1•020) + 6(1•101) + (3•010) + 2(1•200) + 6(1•010)

Sl 4: (0•021) + (2•120) + (0•310) + (0•102) + 3(2•201) + 3(0•120) + 3(2•011) + 6(0•201) + (2•300) + 5(0•011) + 8(2•110) + 2(0•300) + (4•001) + 12(0•110) + 8(2•001) + 10(0•001) + 3(2•100) + 5(0•100)

Affine C4 ⊃ A1 + C3 - Level 2 - Class 1

sl 0: (0•100) **Highest Weight (01100)**

sl 1: (1•200) + (1•010) + (1•000)

sl 2: (2•110) + (0•300) + 2(0•110) + (0•001) + 2(2•100) + 3(0•100)

sl 3: 2(1•210) + 2(1•020) + 2(1•101) + (3•200) + (3•010) + 6(1•200) + 6(1•010) + (3•000) + 4(1•000)

sl 4: (2•120) + (0•310) + (2•201) + 2(0•120) + (2•011) + 2(0•201) + 3(2•300) + 3(0•011) + 8(2•110) + 5(0•300) + 12(0•110) + 3(2•001) + (4•100) + 5(0•001) + 10(2•100) + 12(0•100)

Affine C4 ⊃ A1 + C3 - Level 2 - Class 1

sl 0: (1•010) **Highest Weight (10010)**

sl 1: (0•110) + (2•001) + (0•001) + (2•100) + (0•100)

sl 2: (1•210) + 2(1•101) + (3•010) + 2(1•200) + 4(1•010) + (1•000)

sl 3: (2•201) + (0•120) + 2(0•201) + (2•300) + (0•011) + 4(2•110) + 2(0•300) + 6(0•110) + 4(2•001) + 5(0•001) + 4(2•100) + 5(0•100)

sl 4: (1•301) + (1•030) + 2(1•111) + (3•210) + (1•400) + 8(1•210) + 2(3•101) + 5(1•020) + 12(1•101) + 2(3•200) + 5(3•010) + 11(1•200) + 18(1•010) + (3•000) + 5(1•000)

Affine C4 ⊃ A1 + C3 - Level 2 - Class 1

sl 0: (1•000) **Highest Weight (11000)**

sl 1: (2•100) + (0•100)

sl 2: 2(1•200) + (1•010) + (3•000) + 2(1•000)

sl 3: (2•300) + (2•110) + (0•300) + 2(0•110) + 4(2•100) + 4(0•100)

sl 4: (1•400) + 2(1•210) + 2(1•020) + (1•101) + 2(3•200) + (3•010) + 8(1•200) + 6(1•010) + 3(3•000) + 7(1•000)

Affine C4 ⊃ C2 + C2 - Level 1 - Class 0

sl 0: (00•01) **Highest Weight (00001)**

sl 1: (01•10)

sl 2: (02•01) + (00•20) + (10•00) + (00•01) + (00•00)

sl 3: (11•10) + (01•11) + 3(01•10)

sl 4: (02•20) + (20•01) + 2(02•01) + (10•20) + (00•21) + 2(10•01) + 2(02•00) + 2(00•20) + 2(10•00) + 4(00•01) + 2(00•00)

sl 5: (11•11) + (03•10) + (01•30) + 4(11•10) + 4(01•11) + 9(01•10)

sl 6: (02•21) + (12•01) + (20•20) + 4(02•20) + 2(20•01) + (10•02) + (12•00) + 8(02•01) + 5(10•20) + 2(00•21) + (20•00) + 6(10•01) + 2(00•02) + 5(02•00) + 7(00•20) + 7(10•00) + 8(00•01) + 6(00•00)

Affine C4 ⊃ C2 + C2 - Level 1 - Class 0

Sl 0: (00•00) **Highest Weight (00100)**
Sl 1: (01•10)
Sl 2: (10•01) + (02•00) + (00•20) + (10•00) + (00•01) + (00•00)
Sl 3: (11•10) + (01•11) + 4(01•10)
Sl 4: (02•20) + 2(02•01) + 2(10•20) + (20•00) + 3(10•01) + (00•02) + 3(02•00) + 3(00•20) + 3(10•00) + 3(00•01) + 4(00•00)
Sl 5: (11•11) + (03•10) + (01•30) + 5(11•10) + 5(01•11) + 13(01•10)
Sl 6: (12•01) + (02•02) + (20•20) + (10•21) + 6(02•20) + (20•01) + (10•02) + 2(12•00) + 7(02•01) + 7(10•20) + 2(00•21) + 2(20•00) + 11(10•01) + 2(00•02) + 9(02•00) + 9(00•20) + 10(10•00) + 10(00•01) + 9(00•00)

Affine C4 ⊃ C2 + C2 - Level 1 - Class 1

Sl 0: (00•10) **Highest Weight (00010)**
Sl 1: (01•01) + (01•00)
Sl 2: (02•10) + (10•10) + (00•11) + 2(00•10)
Sl 3: (11•01) + 2(01•20) + (11•00) + 3(01•01) + 3(01•00)
Sl 4: (02•11) + (20•10) + (10•11) + 4(02•10) + (00•30) + 5(10•10) + 3(00•11) + 6(00•10)
Sl 5: (03•01) + 2(11•20) + (01•21) + 4(11•01) + (01•02) + (03•00) + 7(01•20) + 4(11•00) + 10(01•01) + 9(01•00)
Sl 6: (02•30) + (20•11) + 2(12•10) + 5(02•11) + (10•30) + 3(20•10) + 6(10•11) + (00•12) + 14(02•10) + 3(00•30) + 15(10•10) + 9(00•11) + 16(00•10)

Affine C4 ⊃ C2 + C2 - Level 2 - Class 0

Sl 0: (00•02) **Highest Weight (00002)**
Sl 1: (01•11)
Sl 2: (02•02) + (02•20) + (00•21) + (10•01) + (00•02) + (00•20) + (00•01)
Sl 3: (03•11) + (11•11) + (01•12) + (01•30) + (11•10) + 4(01•11) + 2(01•10)
Sl 4: (04•02) + (12•20) + 2(02•21) + (20•02) + (12•01) + 3(02•02) + (20•20) + (10•21) + (00•22) + 4(02•20) + 2(10•02) + (00•40) + 4(02•01) + 3(10•20) + 3(00•21) + (20•00) + 3(10•01) + 5(00•02) + (02•00) + 5(00•20) + (10•00) + 3(00•01) + 2(00•00)

Affine C4 ⊃ C2 + C2 - Level 2 - Class 0

Sl 0: (00•20) **Highest Weight (00020)**
Sl 1: (01•11) + (01•10)
Sl 2: (02•02) + (02•20) + (02•01) + (10•20) + (00•21) + (10•01) + (00•02) + (02•00) + 2(00•20) + 2(00•01) + (00•00)
Sl 3: (03•11) + 2(11•11) + (01•12) + (03•10) + 2(01•30) + 2(11•10) + 6(01•11) + 6(01•10)

Affine C4 ⊃ C2 + C2 - Level 2 - Class 0

sl 0: (00•01) **Highest Weight (00101)**

sl 1: (01•11) + (01•10)

sl 2: (02•20) + (10•02) + 2(02•01) + (10•20) + (00•21) + (10•01) + (00•02) + 2(00•20) + (10•00) + 2(00•01) + (00•00)

sl 3: (03•11) + 2(11•11) + (01•12) + (03•10) + 2(01•30) + 3(11•10) + 7(01•11) + 7(01•10)

Affine C4 ⊃ C2 + C2 - Level 2 - Class 0

sl 0: (00•00) **Highest Weight (00200)**

sl 1: (01•10)

sl 2: (02•20) + (10•01) + (02•00) + (00•20) + (10•00) + (00•01) + (00•00)

sl 3: (11•11) + (03•10) + (01•30) + 2(11•10) + 2(01•11) + 5(01•10)

sl 4: (12•20) + (02•21) + (20•02) + (12•01) + (02•02) + (20•20) + (10•21) + (04•00) + 5(02•20) + (20•01) + (10•02) + (00•40) + (12•00) + 5(02•01) + 5(10•20) + (00•21) + 2(20•00) + 5(10•01) + 2(00•02) + 5(02•00) + 5(00•20) + 4(10•00) + 4(00•01) + 5(00•00)

Affine C4 ⊃ C2 + C2 - Level 2 - Class 0

sl 0: (01•10) **Highest Weight (01010)**

sl 1: (02•01) + (10•20) + (10•01) + (02•00) + (00•20) + (10•00) + (00•01) + (00•00)

sl 2: (11•11) + (03•10) + (01•30) + 3(11•10) + 3(01•11) + 6(01•10)

sl 3: (12•20) + (02•21) + 2(12•01) + (02•02) + (20•20) + 2(10•21) + 6(02•20) + 2(20•01) + 2(10•02) + 2(12•00) + 7(02•01) + 7(10•20) + 2(00•21) + (20•00) + 8(10•01) + (00•02) + 6(02•00) + 6(00•20) + 6(10•00) + 6(00•01) + 4(00•00)

sl 4: (03•30) + (21•11) + (11•12) + (13•10) + 4(03•11) + 4(11•30) + (01•31) + 3(21•10) + 13(11•11) + 3(01•12) + 9(03•10) + 9(01•30) + 22(11•10) + 22(01•11) + 31(01•10)

Affine C4 ⊃ C2 + C2 - Level 2 - Class 0

sl 0: (10•01) **Highest Weight (10001)**

sl 1: (11•10) + (01•11) + (01•10)

sl 2: (12•01) + (10•21) + (02•20) + 2(02•01) + 2(10•20) + (20•00) + 3(10•01) + (00•02) + (02•00) + (00•20) + (10•00) + (00•01) + (00•00)

sl 3: (03•11) + (11•30) + (21•10) + 3(11•11) + (01•12) + 2(03•10) + 2(01•30) + 6(11•10) + 6(01•11) + 8(01•10)

Affine C4 ⊃ C2 + C2 - Level 2 - Class 1

sl 0: (00•11) **Highest Weight (00011)**

sl 1: (01•02) + (01•20) + (01•01)

sl 2: 2(02•11) + (10•11) + (00•12) + (02•10) + (00•30) + (10•10) + 3(00•11) + 2(00•10)

sl 3: (03•02) + (03•20) + (11•02) + (03•01) + 2(11•20) + 3(01•21) + 2(11•01) + 4(01•02) + 6(01•20) + (11•00) + 6(01•01) + 2(01•00)

sl 4: (04•11) + 2(12•11) + 2(02•12) + 3(02•30) + 2(20•11) + (10•12) + 2(12•10) + 11(02•11) + 2(10•30) + 2(00•31) + 2(20•10) + 9(10•11) + 4(00•12) + 9(02•10) + 5(00•30) + 8(10•10) + 13(00•11) + 9(00•10)

Affine C4 ⊃ C2 + C2 - Level 2 - Class 1

Sl 0: (00•10) **Highest Weight (00110)**

Sl 1: (01•20) + (01•01) + (01•00)

Sl 2: (02•11) + (10•11) + 2(02•10) + (00•30) + 2(10•10) + 2(00•11) + 3(00•10)

Sl 3: (03•20) + (11•02) + (03•01) + 2(11•20) + 2(01•21) + 3(11•01) + 2(01•02) + (03•00) + 7(01•20) + 2(11•00) + 7(01•01) + 5(01•00)

Sl 4: 2(12•11) + (02•12) + (04•10) + 3(02•30) + 2(20•11) + (10•12) + 3(12•10) + 10(02•11) + 3(10•30) + (00•31) + 3(20•10) + 10(10•11) + 2(00•12) + 14(02•10) + 5(00•30) + 13(10•10) + 11(00•11) + 13(00•10)

Affine C4 ⊃ C2 + C2 - Level 2 - Class 1

Sl 0: (01•01) **Highest Weight (01001)**

Sl 1: (10•11) + (02•10) + (10•10) + (00•11) + (00•10)

Sl 2: (03•01) + (11•20) + (01•21) + 2(11•01) + (01•02) + 3(01•20) + (11•00) + 4(01•01) + 2(01•00)

Sl 3: (12•11) + (02•30) + (10•12) + 2(12•10) + 5(02•11) + 2(10•30) + 2(20•10) + 6(10•11) + (00•12) + 7(02•10) + 2(00•30) + 8(10•10) + 5(00•11) + 6(00•10)

Sl 4: (03•21) + (13•01) + (03•02) + (21•20) + 3(11•21) + 4(03•20) + 2(21•01) + 3(11•02) + (01•40) + (01•03) + 6(03•01) + 11(11•20) + 9(01•21) + (21•00) + 14(11•01) + 7(01•02) + 3(03•00) + 19(01•20) + 8(11•00) + 21(01•01) + 11(01•00)

Affine C5 Branching Rules

Affine C5 ⊃ C5 - Level 1 - Class 0

```
Sl 0:   (•00010)  Highest Weight (000010)
Sl 1:   (•10001) +   (•10100) +   (•00010)
Sl 2:   (•20010) +   (•01010) + 2(•10001) +   (•02000) + 2(•10100) + 3(•00010) +   (•01000)
Sl 3:   (•11001) +   (•11100) + 3(•20010) +   (•00200) + 3(•01010) +   (•21000) + 5(•10001)
    +   (•02000) + 6(•10100) + 6(•00010) +   (•20000) + 2(•01000)
Sl 4:   (•02010) +   (•10110) +   (•30001) + 3(•11001) +   (•30100) + 4(•11100) +   (•00101)
    + 8(•20010) + 2(•00200) + 9(•01010) + 3(•21000) + 11(•10001) + 5(•02000) + 13(•10100)
    + 13(•00010) + 2(•20000) + 6(•01000) +   (•00000)
Sl 5:   (•01101) + 2(•21010) +   (•01200) + 3(•02010) + 4(•10110) + 3(•30001) +   (•22000)
    + 9(•11001) + 3(•30100) +   (•03000) +   (•00020) + 12(•11100) + 4(•00101) + 20(•20010)
    + 7(•00200) + 21(•01010) + 9(•21000) + 24(•10001) + 10(•02000) + 31(•10100) + 24(•00010)
    + 7(•20000) + 12(•01000) +   (•00000)
```

Affine C5 ⊃ C5 - Level 1 - Class 0

```
Sl 0:   (•01000)  Highest Weight (001000)
Sl 1:   (•10100) +   (•20000) +   (•01000)
Sl 2:   (•01010) +   (•21000) +   (•02000) + 2(•10100) +   (•00010) +   (•20000) + 3(•01000)
    +   (•00000)
Sl 3:   (•11100) +   (•00101) +   (•20010) +   (•00200) + 2(•01010) + 3(•21000) +   (•10001)
    + 2(•02000) + 6(•10100) + 2(•00010) + 4(•20000) + 5(•01000) +   (•00000)
Sl 4:   (•10110) +   (•11001) +   (•30100) +   (•03000) +   (•00020) + 4(•11100) + 2(•00101)
    + 3(•20010) + 2(•00200) + 7(•01010) +   (•40000) + 7(•21000) + 3(•10001) + 7(•02000)
    + 13(•10100) + 6(•00010) + 7(•20000) + 12(•01000) + 3(•00000)
Sl 5:   (•21010) +   (•10011) +   (•01200) +   (•02010) + 4(•10110) + 2(•22000) + 4(•11001)
    + 3(•30100) + 2(•03000) + 2(•00020) + 12(•11100) + 6(•00101) + 10(•20010) + 8(•00200)
    + 16(•01010) + 2(•40000) + 17(•21000) + 9(•10001) + 14(•02000) + 31(•10100) + 12(•00010)
    + 17(•20000) + 21(•01000) + 4(•00000)
```

Affine C5 ⊃ C5 - Level 1 - Class 0

```
Sl 0:   (•00000)  Highest Weight (100000)
Sl 1:   (•20000)
Sl 2:   (•02000) +   (•20000) +   (•01000) +   (•00000)
Sl 3:   (•00200) +   (•21000) +   (•02000) +   (•10100) + 3(•20000) +   (•01000) +   (•00000)
Sl 4:   (•00020) +   (•11100) +   (•00200) +   (•01010) +   (•40000) + 2(•21000) + 4(•02000)
    + 2(•10100) +   (•00010) + 4(•20000) + 4(•01000) + 3(•00000)
Sl 5:   (•10110) +   (•22000) +   (•00002) +   (•00020) + 3(•11100) +   (•00101) +   (•20010)
    + 4(•00200) + 2(•01010) +   (•40000) + 6(•21000) +   (•10001) + 6(•02000) + 7(•10100)
    +   (•00010) + 10(•20000) + 6(•01000) + 3(•00000)
```

Affine C5 ⊃ C5 - Level 1 - Class 1

Sl 0: (•00001) **Highest Weight (000001)**

Sl 1: (•10010) + (•00001)

Sl 2: (•20001) + (•01100) + 2(•10010) + 2(•00001) + (•00100)

Sl 3: (•11010) + 2(•20001) + (•01001) + (•20100) + 2(•01100) + 5(•10010) + (•11000) + 4(•00001) + 2(•00100)

Sl 4: (•02001) + (•30010) + (•10200) + 3(•11010) + (•00110) + 5(•20001) + (•12000) + 3(•01001) + 3(•20100) + 5(•01100) + 11(•10010) + 3(•11000) + 8(•00001) + 5(•00100) + (•10000)

Sl 5: (•01110) + (•21001) + 2(•02001) + (•21100) + (•10101) + (•02100) + 3(•30010) + 2(•10200) + 9(•11010) + 4(•00110) + 11(•20001) + 3(•12000) + 8(•01001) + 9(•20100) + 12(•01100) + 23(•10010) + (•30000) + 8(•11000) + 14(•00001) + 11(•00100) + 2(•10000)

Affine C5 ⊃ C5 - Level 1 - Class 1

Sl 0: (•00100) **Highest Weight (000100)**

Sl 1: (•10010) + (•11000) + (•00100)

Sl 2: (•01001) + (•20100) + (•01100) + 2(•10010) + 2(•11000) + (•00001) + 3(•00100) + (•10000)

Sl 3: (•11010) + (•00110) + (•20001) + (•12000) + 2(•01001) + 3(•20100) + 3(•01100) + 6(•10010) + (•30000) + 5(•11000) + 2(•00001) + 6(•00100) + 2(•10000)

Sl 4: (•10101) + (•02100) + (•30010) + (•10200) + 4(•11010) + 3(•00110) + (•31000) + 3(•20001) + 3(•12000) + 6(•01001) + 8(•20100) + 9(•01100) + 14(•10010) + 2(•30000) + 12(•11000) + 5(•00001) + 13(•00100) + 5(•10000)

Sl 5: (•01110) + (•21001) + (•10020) + (•02001) + 2(•21100) + 3(•10101) + 3(•02100) + 3(•30010) + 4(•10200) + 13(•11010) + (•00011) + 9(•00110) + 3(•31000) + 9(•20001) + 9(•12000) + 14(•01001) + 20(•20100) + 22(•01100) + 32(•10010) + 6(•30000) + 27(•11000) + 11(•00001) + 25(•00100) + 10(•10000)

Affine C5 ⊃ C5 - Level 1 - Class 1

Sl 0: (•10000) **Highest Weight (010000)**

Sl 1: (•11000) + (•10000)

Sl 2: (•01100) + (•30000) + 2(•11000) + (•00100) + 2(•10000)

Sl 3: (•00110) + (•12000) + (•20100) + 2(•01100) + (•10010) + 2(•30000) + 5(•11000) + 2(•00100) + 4(•10000)

Sl 4: (•10200) + (•11010) + (•00011) + 2(•00110) + (•31000) + 3(•12000) + (•01001) + 3(•20100) + 6(•01100) + 3(•10010) + 4(•30000) + 11(•11000) + (•00001) + 5(•00100) + 8(•10000)

Sl 5: (•10020) + (•21100) + (•10101) + (•02100) + 3(•10200) + 4(•11010) + 2(•00011) + 6(•00110) + 3(•31000) + (•20001) + 8(•12000) + 3(•01001) + 9(•20100) + 14(•01100) + 9(•10010) + 9(•30000) + 23(•11000) + 2(•00001) + 11(•00100) + 14(•10000)

Affine C5 ⊃ A1 + C4 - Level 1 - Class 0

```
Sl 0:   (0•0010)  Highest Weight (000010)
Sl 1:   (1•0001) +   (1•0100)
Sl 2:   (0•1001) +   (0•1100) +   (2•0010) +  2(0•0010) +   (0•1000)
Sl 3:  2(1•1010) +  3(1•0001) +   (1•2000) +  3(1•0100)
Sl 4:   (0•2010) +   (0•0110) +   (2•1001) +  3(0•1001) +   (2•1100) +  4(0•1100) +  3(2•0010)
    +  7(0•0010) +   (2•1000) +  3(0•1000)
Sl 5:   (1•2001) +   (1•0101) +   (1•2100) +  2(1•0200) +  8(1•1010) +   (3•0001) +  8(1•0001)
    +   (3•0100) +  3(1•2000) + 10(1•0100) +  2(1•0000)
Sl 6:   (0•1101) +   (2•2010) +   (0•1200) +   (2•0110) +  4(0•2010) +  5(0•0110) +  4(2•1001)
    + 10(0•1001) +  5(2•1100) +  2(0•3000) + 13(0•1100) + 10(2•0010) + 18(0•0010) +  4(2•1000)
    +  9(0•1000)
```

Affine C5 ⊃ A1 + C4 - Level 1 - Class 0

```
Sl 0:   (0•1000)  Highest Weight (001000)
Sl 1:   (1•0100) +   (1•0000)
Sl 2:   (0•1100) +   (0•0010) +   (2•1000) +  2(0•1000)
Sl 3:   (1•1010) +  2(1•2000) +  3(1•0100) +  2(1•0000)
Sl 4:   (0•0110) +   (0•1001) +   (2•1100) +   (0•3000) +  4(0•1100) +   (2•0010) +  3(0•0010)
    +  3(2•1000) +  6(0•1000)
Sl 5:   (1•0101) +   (1•2100) +  2(1•0200) +  4(1•1010) +  2(1•0001) +   (3•0100) +  6(1•2000)
    + 10(1•0100) +   (3•0000) +  6(1•0000)
Sl 6:   (0•1200) +   (0•0011) +   (2•0110) +  2(0•2010) +  4(0•0110) +   (2•1001) +   (2•3000)
    +  4(0•1001) +  5(2•1100) +  3(0•3000) + 13(0•1100) +  4(2•0010) + 10(0•0010) +  9(2•1000)
    + 15(0•1000)
```

Affine C5 ⊃ A1 + C4 - Level 1 - Class 0

```
Sl 0:   (1•0000)  Highest Weight (100000)
Sl 1:   (0•1000)
Sl 2:   (1•2000) +   (1•0000)
Sl 3:   (0•1100) +   (2•1000) +  2(0•1000)
Sl 4:   (1•0200) +  2(1•2000) +  2(1•0100) +   (3•0000) +  3(1•0000)
Sl 5:   (0•0110) +   (2•1100) +   (0•3000) +  3(0•1100) +   (0•0010) +  3(2•1000) +  5(0•1000)
Sl 6:   (1•0020) +   (1•2100) +  2(1•0200) +  2(1•1010) +   (3•2000) +  7(1•2000) +  5(1•0100)
    +   (3•0000) +  6(1•0000)
Sl 7:   (0•1200) +   (0•0011) +   (2•0110) +   (0•2010) +  3(0•0110) +   (2•3000) +   (0•1001)
    +  4(2•1100) +  3(0•3000) +  9(0•1100) +   (2•0010) +  4(0•0010) +  8(2•1000) + 12(0•1000)
```

Affine C5 ⊃ A1 + C4 - Level 1 - Class 1

Sl 0: (0•0001) **Highest Weight (000001)**

Sl 1: (1•0010)

Sl 2: (0•1010) + (2•0001) + (0•0001) + (0•0100)

Sl 3: (1•1001) + (1•1100) + 3(1•0010)

Sl 4: (0•2001) + (2•1010) + (0•0200) + 3(0•1010) + 2(2•0001) + 4(0•0001) + (2•0100) + (0•2000) + 3(0•0100)

Sl 5: (1•2010) + (1•0110) + 4(1•1001) + 4(1•1100) + (3•0010) + 8(1•0010) + 2(1•1000)

Sl 6: (0•1110) + (2•2001) + 2(0•2001) + 2(0•0101) + (2•0200) + 2(0•2100) + 4(2•1010) + 2(0•0200) + 10(0•1010) + 6(2•0001) + (2•2000) + 9(0•0001) + 4(2•0100) + 3(0•2000) + 9(0•0100) + (0•0000)

Affine C5 ⊃ A1 + C4 - Level 1 - Class 1

Sl 0: (0•0100) **Highest Weight (000100)**

Sl 1: (1•0010) + (1•1000)

Sl 2: (0•1010) + (0•0001) + (2•0100) + (0•2000) + 2(0•0100) + (0•0000)

Sl 3: (1•1001) + 2(1•1100) + 3(1•0010) + 3(1•1000)

Sl 4: (0•0101) + (0•2100) + (2•1010) + (0•0200) + 4(0•1010) + (2•0001) + (2•2000) + 3(0•0001) + 3(2•0100) + 3(0•2000) + 7(0•0100) + (2•0000) + 2(0•0000)

Sl 5: (1•2010) + 2(1•0110) + 4(1•1001) + (1•3000) + 8(1•1100) + (3•0010) + 10(1•0010) + (3•1000) + 9(1•1000)

Sl 6: (0•1110) + (2•0101) + 2(0•2001) + (0•0020) + (2•2100) + 3(0•0101) + (2•0200) + 4(0•2100) + 5(2•1010) + 5(0•0200) + 14(0•1010) + 4(2•0001) + 4(2•2000) + 9(0•0001) + 10(2•0100) + 10(0•2000) + 18(0•0100) + 3(2•0000) + 6(0•0000)

Affine C5 ⊃ A1 + C4 - Level 1 - Class 1

Sl 0: (0•0000) **Highest Weight (010000)**

Sl 1: (1•1000)

Sl 2: (0•2000) + (0•0100) + (2•0000) + (0•0000)

Sl 3: (1•1100) + 3(1•1000)

Sl 4: (0•0200) + (0•1010) + (2•2000) + (2•0100) + 3(0•2000) + 3(0•0100) + 2(2•0000) + 3(0•0000)

Sl 5: (1•0110) + (1•3000) + 4(1•1100) + 2(1•0010) + (3•1000) + 8(1•1000)

Sl 6: (0•0020) + (0•0101) + (2•0200) + 2(0•2100) + (2•1010) + 3(0•0200) + 4(0•1010) + 4(2•2000) + (0•0001) + 4(2•0100) + 8(0•2000) + 9(0•0100) + 5(2•0000) + 7(0•0000)

Affine C5 ⊃ C2 + C3 - Level 1 - Class 0

Sl 0: (00•010) **Highest Weight (000010)**
Sl 1: (01•001) + (01•100)
Sl 2: (02•010) + (00•101) + (10•010) + (00•200) + 2(00•010) + (10•000) + (00•000)
Sl 3: (11•001) + 2(01•110) + (11•100) + 3(01•001) + 4(01•100)
Sl 4: (02•101) + (10•101) + (00•210) + (02•200) + (20•010) + (00•020) + 4(02•010)
+ 2(10•200) + 3(00•101) + 5(10•010) + 3(00•200) + 7(00•010) + 2(02•000) + 2(10•000)
+ 3(00•000)
Sl 5: (01•201) + 2(11•110) + (01•011) + (03•001) + (01•300) + 4(11•001) + 9(01•110)
+ (03•100) + 5(11•100) + 10(01•001) + 14(01•100)

Affine C5 ⊃ C2 + C3 - Level 1 - Class 0

Sl 0: (00•000) **Highest Weight (001000)**
Sl 1: (01•100)
Sl 2: (10•010) + (00•200) + (00•010) + (02•000) + (10•000) + (00•000)
Sl 3: (01•110) + (11•100) + (01•001) + 4(01•100)
Sl 4: (10•101) + (02•200) + (00•020) + 2(02•010) + 2(10•200) + (00•101) + 3(10•010)
+ 3(00•200) + (20•000) + 4(00•010) + 3(02•000) + 3(10•000) + 4(00•000)
Sl 5: (11•110) + (01•011) + (01•300) + (11•001) + 6(01•110) + (03•100) + 5(11•100)
+ 5(01•001) + 13(01•100)

Affine C5 ⊃ C2 + C3 - Level 1 - Class 0

Sl 0: (10•000) **Highest Weight (100000)**
Sl 1: (01•100)
Sl 2: (10•200) + (00•010) + (02•000) + (10•000) + (00•000)
Sl 3: (01•110) + (11•100) + 3(01•100)
Sl 4: (10•020) + (02•200) + (02•010) + 2(10•200) + (00•101) + 2(10•010) + (12•000)
+ 2(00•200) + 2(00•010) + 2(02•000) + 4(10•000) + 2(00•000)
Sl 5: (11•110) + (01•011) + (01•300) + 4(01•110) + (03•100) + 4(11•100) + 2(01•001)
+ 9(01•100)

Affine C5 ⊃ C2 + C3 - Level 1 - Class 1

Sl 0: (00•001) **Highest Weight (000001)**
Sl 1: (01•010)
Sl 2: (02•001) + (00•110) + (10•100) + (00•001) + (00•100)
Sl 3: (01•101) + (11•010) + (01•200) + 3(01•010) + (01•000)
Sl 4: (02•110) + (00•201) + (20•001) + (10•110) + 2(02•001) + 2(10•001) + 3(00•110)
+ 2(02•100) + 3(10•100) + 4(00•001) + 3(00•100)
Sl 5: (11•101) + (01•210) + (01•020) + (03•010) + (11•200) + 4(01•101) + 4(11•010)
+ 4(01•200) + 10(01•010) + (11•000) + 4(01•000)

Affine C5 $\supset$ C2 + C3 - Level 1 - Class 1

Sl 0: (00•100) **Highest Weight (000100)**

Sl 1: (01•010) + (01•000)

Sl 2: (10•001) + (00•110) + (02•100) + (10•100) + (00•001) + 2(00•100)

Sl 3: (01•101) + (11•010) + 2(01•200) + 4(01•010) + (11•000) + 3(01•000)

Sl 4: (02•110) + 2(10•110) + (00•011) + 2(02•001) + (00•300) + (20•100) + 3(10•001)
+ 4(00•110) + 4(02•100) + 5(10•100) + 3(00•001) + 7(00•100)

Sl 5: (11•101) + (01•210) + 2(01•020) + (03•010) + 2(11•200) + 5(01•101) + 5(11•010)
+ 8(01•200) + 15(01•010) + (03•000) + 4(11•000) + 9(01•000)

Sl 6: (02•011) + (10•201) + (02•300) + (20•110) + (10•011) + (00•120) + (12•001)
+ 7(02•110) + (10•300) + 2(00•201) + (20•001) + 10(10•110) + 3(00•011) + 2(12•100)
+ 7(02•001) + 3(00•300) + 3(20•100) + 11(10•001) + 14(00•110) + 15(02•100) + 17(10•100)
+ 11(00•001) + 19(00•100)

Sl 7: (11•210) + (01•111) + (03•101) + 2(11•020) + 6(11•101) + 7(01•210) + (01•002)
+ 2(03•200) + (21•010) + 9(01•020) + 5(03•010) + 10(11•200) + 20(01•101) + 20(11•010)
+ 28(01•200) + (21•000) + 45(01•010) + 4(03•000) + 13(11•000) + 25(01•000)

Affine C5 $\supset$ C2 + C3 - Level 1 - Class 1

Sl 0: (01•000) **Highest Weight (010000)**

Sl 1: (10•100) + (00•100)

Sl 2: (01•200) + (01•010) + (11•000) + 2(01•000)

Sl 3: (10•110) + (00•110) + 2(02•100) + 3(10•100) + (00•001) + 3(00•100)

Sl 4: (01•020) + (11•200) + (01•101) + (11•010) + 4(01•200) + 5(01•010) + (03•000)
+ 3(11•000) + 6(01•000)

Sl 5: (10•011) + 2(02•110) + (10•300) + 4(10•110) + (00•011) + (12•100) + (02•001)
+ (00•300) + (20•100) + 2(10•001) + 5(00•110) + 7(02•100) + 10(10•100) + 3(00•001)
+ 9(00•100)

Sl 6: (11•020) + (11•101) + 2(01•210) + (01•002) + (03•200) + 4(01•020) + (03•010)
+ 5(11•200) + 6(01•101) + 6(11•010) + 14(01•200) + (21•000) + 17(01•010) + 3(03•000)
+ 9(11•000) + 16(01•000)

Sl 7: (10•120) + (12•110) + 2(02•011) + (10•201) + 2(02•300) + (20•110) + 4(10•011)
+ (00•120) + 10(02•110) + 4(10•300) + 2(00•201) + (20•001) + 16(10•110) + 4(00•011)
+ 5(12•100) + 6(02•001) + 4(00•300) + 5(20•100) + 9(10•001) + 17(00•110) + 23(02•100)
+ 28(10•100) + 11(00•001) + 24(00•100)

Affine C6 Branching Rules

Affine C6 ⊃ C6 - Level 1 - Class 0

Sl 0: (•000001) **Highest Weight (0000001)**

Sl 1: (•100010) + (•000001)

Sl 2: (•200001) + (•010100) + 2(•100010) + 2(•000001) + (•000100)

Sl 3: (•110010) + 2(•200001) + (•010001) + (•200100) + (•002000) + 2(•010100) + 5(•100010) + (•101000) + 4(•000001) + 2(•000100)

Sl 4: (•020001) + (•101100) + (•300010) + 3(•110010) + (•001010) + 5(•200001) + (•111000) + 3(•010001) + 3(•200100) + (•002000) + 6(•010100) + 11(•100010) + (•020000) + 3(•101000) + 8(•000001) + 5(•000100) + (•010000)

Affine C6 ⊃ C6 - Level 1 - Class 0

Sl 0: (•000100) **Highest Weight (0000100)**

Sl 1: (•100010) + (•101000) + (•000100)

Sl 2: (•010001) + (•200100) + (•010100) + 2(•100010) + (•020000) + 2(•101000) + (•000001) + 3(•000100) + (•010000)

Sl 3: (•110010) + (•001010) + (•200001) + (•111000) + 2(•010001) + 3(•200100) + (•002000) + 3(•010100) + 6(•100010) + (•210000) + (•020000) + 6(•101000) + 2(•000001) + 6(•000100) + (•200000) + 2(•010000)

Affine C6 ⊃ C6 - Level 1 - Class 0

Sl 0: (•010000) **Highest Weight (0010000)**

Sl 1: (•101000) + (•200000) + (•010000)

Sl 2: (•010100) + (•210000) + (•020000) + 2(•101000) + (•000100) + (•200000) + 3(•010000) + (•000000)

Sl 3: (•001010) + (•111000) + (•200100) + (•002000) + 2(•010100) + (•100010) + 3(•210000) + 2(•020000) + 6(•101000) + 2(•000100) + 4(•200000) + 5(•010000) + (•000000)

Affine C6 ⊃ C6 - Level 1 - Class 0

Sl 0: (•000000) **Highest Weight (1000000)**

Sl 1: (•200000)

Sl 2: (•020000) + (•200000) + (•010000) + (•000000)

Sl 3: (•002000) + (•210000) + (•020000) + (•101000) + 3(•200000) + (•010000) + (•000000)

Sl 4: (•000200) + (•111000) + (•002000) + (•010100) + (•400000) + 2(•210000) + 4(•020000) + 2(•101000) + (•000100) + 4(•200000) + 4(•010000) + 3(•000000)

Sl 5: (•101100) + (•000020) + (•000200) + (•220000) + (•001010) + 3(•111000) + (•200100) + 4(•002000) + 2(•010100) + (•100010) + (•400000) + 6(•210000) + 6(•020000) + 7(•101000) + (•000100) + 10(•200000) + 6(•010000) + 3(•000000)

Affine C6 ⊃ C6 - Level 1 - Class 1

Sl 0: (•000010) **Highest Weight (0000010)**

Sl 1: (•100001) + (•100100) + (•000010)

Sl 2: (•200010) + (•010010) + 2(•100001) + (•011000) + 2(•100100) + 3(•000010) + (•001000)

Sl 3: (•110001) + (•110100) + (•001100) + 3(•200010) + 3(•010010) + (•201000) + 5(•100001) + 2(•011000) + 6(•100100) + 6(•000010) + (•110000) + 2(•001000)

Sl 4: (•020010) + (•101010) + (•300001) + 3(•110001) + (•300100) + (•102000) + 4(•110100) + (•001001) + 3(•001100) + 8(•200010) + 9(•010010) + (•120000) + 3(•201000) + 11(•100001) + 6(•011000) + 14(•100100) + 13(•000010) + 3(•110000) + 6(•001000) + (•100000)

Affine C6 ⊃ C6 - Level 1 - Class 0

Sl 0: (•001000) **Highest Weight (0001000)**

Sl 1: (•100100) + (•110000) + (•001000)

Sl 2: (•010010) + (•201000) + (•011000) + 2(•100100) + (•000010) + 2(•110000) + 3(•001000) + (•100000)

Sl 3: (•110100) + (•001001) + (•001100) + (•200010) + 2(•010010) + (•120000) + 3(•201000) + (•100001) + 3(•011000) + 6(•100100) + 2(•000010) + (•300000) + 5(•110000) + 6(•001000) + 2(•100000)

Affine C6 ⊃ C6 - Level 1 - Class 0

Sl 0: (•100000) **Highest Weight (0100000)**

Sl 1: (•110000) + (•100000)

Sl 2: (•011000) + (•300000) + 2(•110000) + (•001000) + 2(•100000)

Sl 3: (•001100) + (•120000) + (•201000) + 2(•011000) + (•100100) + 2(•300000) + 5(•110000) + 2(•001000) + 4(•100000)

Sl 4: (•000110) + (•102000) + (•110100) + 2(•001100) + (•010010) + (•310000) + 3(•120000) + 3(•201000) + 6(•011000) + 3(•100100) + (•000010) + 4(•300000) + 11(•110000) + 5(•001000) + 8(•100000)

Affine C6 ⊃ A1 + C5 - Level 1 - Class 0

Sl 0: (0•00001) **Highest Weight (0000001)**

Sl 1: (1•00010)

Sl 2: (0•10010) + (2•00001) + (0•00001) + (0•00100)

Sl 3: (1•10001) + (1•10100) + 3(1•00010)

Sl 4: (0•20001) + (0•01100) + (2•10010) + 3(0•10010) + 2(2•00001) + (0•11000) + 4(0•00001) + (2•00100) + 3(0•00100)

Sl 5: (1•20010) + (1•01010) + 4(1•10001) + (1•02000) + 4(1•10100) + (3•00010) + 8(1•00010) + 2(1•01000)

Affine C6 ⊃ A1 + C5 - Level 1 - Class 0

Sl 0: (0•00100) **Highest Weight (0000100)**

Sl 1: (1•00010) + (1•01000)

Sl 2: (0•10010) + (0•11000) + (0•00001) + (2•00100) + 2(0•00100) + (0•10000)

Sl 3: (1•10001) + 2(1•10100) + 3(1•00010) + (1•20000) + 3(1•01000)

Sl 4: (0•01001) + (0•20100) + (0•01100) + (2•10010) + 4(0•10010) + (2•11000) + (2•00001) + 4(0•11000) + 3(0•00001) + 3(2•00100) + 7(0•00100) + (2•10000) + 3(0•10000)

Sl 5: (1•20010) + 2(1•01010) + (1•21000) + 4(1•10001) + 2(1•02000) + 8(1•10100) + (3•00010) + 10(1•00010) + (3•01000) + 3(1•20000) + 10(1•01000) + 2(1•00000)

Affine C6 ⊃ A1 + C5 - Level 1 - Class 0

Sl 0: (0•10000) **Highest Weight (0010000)**

Sl 1: (1•01000) + (1•00000)

Sl 2: (0•11000) + (0•00100) + (2•10000) + 2(0•10000)

Sl 3: (1•10100) + 2(1•20000) + 3(1•01000) + 2(1•00000)

Sl 4: (0•01100) + (0•10010) + (2•11000) + (0•30000) + 4(0•11000) + (2•00100) + 3(0•00100) + 3(2•10000) + 6(0•10000)

Sl 5: (1•01010) + (1•21000) + 2(1•02000) + 4(1•10100) + 2(1•00010) + (3•01000) + 6(1•20000) + 10(1•01000) + (3•00000) + 6(1•00000)

Affine C6 ⊃ A1 + C5 - Level 1 - Class 0

Sl 0: (1•00000) **Highest Weight (1000000)**

Sl 1: (0•10000)

Sl 2: (1•20000) + (1•00000)

Sl 3: (0•11000) + (2•10000) + 2(0•10000)

Sl 4: (1•02000) + 2(1•20000) + 2(1•01000) + (3•00000) + 3(1•00000)

Sl 5: (0•01100) + (2•11000) + (0•30000) + 3(0•11000) + (0•00100) + 3(2•10000) + 5(0•10000)

Sl 6: (1•00200) + (1•21000) + 2(1•02000) + 2(1•10100) + (3•20000) + 7(1•20000) + 5(1•01000) + (3•00000) + 6(1•00000)

Affine C6 ⊃ A1 + C5 - Level 1 - Class 1

Sl 0: (0•00010) **Highest Weight (0000010)**

Sl 1: (1•00001) + (1•00100)

Sl 2: (0•10001) + (0•10100) + (2•00010) + 2(0•00010) + (0•01000)

Sl 3: 2(1•10010) + (1•11000) + 3(1•00001) + 3(1•00100)

Sl 4: (0•20010) + (0•01010) + (2•10001) + 3(0•10001) + (2•10100) + (0•02000) + 4(0•10100) + 3(2•00010) + 7(0•00010) + (2•01000) + (0•20000) + 3(0•01000)

Sl 5: (1•20001) + (1•01001) + (1•20100) + 2(1•01100) + 8(1•10010) + (3•00001) + 4(1•11000) + 8(1•00001) + (3•00100) + 10(1•00100) + 2(1•10000)

Affine C6 ⊃ A1 + C5 - Level 1 - Class 1

```
Sl 0:   (0•01000)  Highest Weight (0001000)
Sl 1:   (1•00100) +   (1•10000)
Sl 2:   (0•10100) +   (0•00010) +   (2•01000) +   (0•20000) +  2(0•01000) +   (0•00000)
Sl 3:   (1•10010) +  2(1•11000) +  3(1•00100) +  3(1•10000)
Sl 4:   (0•01010) +   (0•21000) +   (0•10001) +   (2•10100) +   (0•02000) +  4(0•10100) +   (2•00010)
     + 3(0•00010) +   (2•20000) +  3(2•01000) +  3(0•20000) +  7(0•01000) +   (2•00000) +  2(0•00000)
Sl 5:   (1•01001) +   (1•20100) +  2(1•01100) +  4(1•10010) +   (1•30000) +  8(1•11000) +  2(1•00001)
     +  (3•00100) + 10(1•00100) +   (3•10000) +  9(1•10000)
```

Affine C6 ⊃ A1 + C5 - Level 1 - Class 1

```
Sl 0:   (0•00000)  Highest Weight (0100000)
Sl 1:   (1•10000)
Sl 2:   (0•20000) +   (0•01000) +   (2•00000) +   (0•00000)
Sl 3:   (1•11000) +  3(1•10000)
Sl 4:   (0•02000) +   (0•10100) +   (2•20000) +   (2•01000) +  3(0•20000) +  3(0•01000) +  2(2•00000)
     + 3(0•00000)
Sl 5:   (1•01100) +   (1•30000) +  4(1•11000) +  2(1•00100) +   (3•10000) +  8(1•10000)
```

Affine C6 ⊃ C2 + C4 - Level 1 - Class 0

```
Sl 0:   (00•0001)  Highest Weight (0000001)
Sl 1:   (01•0010)
Sl 2:   (02•0001) +   (00•1010) +   (10•0100) +   (00•0001) +   (00•0100)
Sl 3:   (01•1001) +   (11•0010) +   (01•1100) +  3(01•0010) +   (01•1000)
Sl 4:   (00•2001) +   (02•1010) +   (10•1010) +   (20•0001) +   (00•0200) +  2(02•0001) +  3(00•1010)
     + 2(10•0001) +  2(02•0100) +   (10•2000) +  3(10•0100) +  4(00•0001) +   (00•2000) +  3(00•0100)
     +  (00•0000)
Sl 5:   (01•2010) +   (11•1001) +   (01•0110) +   (11•1100) +  4(01•1001) +   (03•0010) +  4(11•0010)
     + 5(01•1100) + 10(01•0010) +   (11•1000) +  5(01•1000)
```

Affine C6 ⊃ C2 + C4 - Level 1 - Class 0

```
Sl 0:   (00•0100)  Highest Weight (0000100)
Sl 1:   (01•0010) +   (01•1000)
Sl 2:   (00•1010) +   (10•0001) +   (02•0100) +   (10•0100) +   (00•0001) +   (00•2000) +  2(00•0100)
     +  (10•0000) +   (00•0000)
Sl 3:   (01•1001) +   (11•0010) +  2(01•1100) +  4(01•0010) +   (11•1000) +  4(01•1000)
Sl 4:   (02•1010) +   (00•0101) +  2(10•1010) +   (00•2100) +   (00•0200) +  2(02•0001) +  4(00•1010)
     +  (02•2000) +   (20•0100) +  3(10•0001) +  4(02•0100) +  2(10•2000) +  5(10•0100) +  3(00•0001)
     + 3(00•2000) +  8(00•0100) +  2(02•0000) +  2(10•0000) +  3(00•0000)
Sl 5:   (01•2010) +   (11•1001) +  2(01•0110) +  2(11•1100) +  5(01•1001) +   (03•0010) +   (01•3000)
     + 5(11•0010) + 10(01•1100) + 15(01•0010) +   (03•1000) +  5(11•1000) + 14(01•1000)
```

Affine C6 ⊃ C2 + C4 - Level 1 - Class 0

```
Sl 0:   (00•0000)  Highest Weight (0010000)
Sl 1:   (01•1000)
Sl 2:   (10•0100) +   (00•2000) +   (00•0100) +   (02•0000) +   (10•0000) +   (00•0000)
Sl 3:   (01•1100) +   (01•0010) +   (11•1000) + 4(01•1000)
Sl 4:   (10•1010) +   (00•0200) +   (00•1010) +   (02•2000) + 2(02•0100) + 2(10•2000) + 3(10•0100)
    +   (00•0001) + 3(00•2000) + 4(00•0100) +   (20•0000) + 3(02•0000) + 3(10•0000) + 4(00•0000)
Sl 5:   (01•0110) +   (11•1100) +   (01•1001) +   (01•3000) +   (11•0010) + 6(01•1100) + 5(01•0010)
    +   (03•1000) + 5(11•1000) + 13(01•1000)
```

Affine C6 ⊃ C2 + C4 - Level 1 - Class 0

```
Sl 0:   (10•0000)  Highest Weight (1000000)
Sl 1:   (01•1000)
Sl 2:   (10•2000) +   (00•0100) +   (02•0000) +   (10•0000) +   (00•0000)
Sl 3:   (01•1100) +   (11•1000) + 3(01•1000)
Sl 4:   (10•0200) +   (00•1010) +   (02•2000) +   (02•0100) + 2(10•2000) + 2(10•0100) + 2(00•2000)
    + 2(00•0100) +   (12•0000) + 2(02•0000) + 4(10•0000) + 2(00•0000)
Sl 5:   (01•0110) +   (11•1100) +   (01•3000) + 4(01•1100) + 2(01•0010) +   (03•1000) + 4(11•1000)
    + 9(01•1000)
```

Affine C6 ⊃ C2 + C4 - Level 1 - Class 1

```
Sl 0:   (00•0010)  Highest Weight (0000010)
Sl 1:   (01•0001) +   (01•0100)
Sl 2:   (00•1001) +   (02•0010) +   (10•0010) +   (00•1100) + 2(00•0010) +   (10•1000) +   (00•1000)
Sl 3:  2(01•1010) +   (11•0001) +   (11•0100) + 3(01•0001) +   (01•2000) + 4(01•0100) +   (01•0000)
Sl 4:   (02•1001) +   (00•2010) +   (10•1001) +   (00•0110) +   (02•1100) +   (20•0010) + 2(10•1100)
    + 3(00•1001) + 4(02•0010) + 5(10•0010) + 4(00•1100) + 7(00•0010) + 2(02•1000) + 3(10•1000)
    + 4(00•1000)
Sl 5:   (01•2001) +   (01•0101) + 2(11•1010) +   (01•2100) + 2(01•0200) +   (03•0001) + 9(01•1010)
    + 4(11•0001) +   (03•0100) +   (11•2000) + 5(11•0100) + 10(01•0001) + 5(01•2000) + 15(01•0100)
    +   (11•0000) + 4(01•0000)
```

Affine C6 ⊃ C2 + C4 - Level 1 - Class 1

```
Sl 0:   (00•1000)  Highest Weight (0001000)
Sl 1:   (01•0100) +   (01•0000)
Sl 2:   (10•0010) +   (00•1100) +   (00•0010) +   (02•1000) +   (10•1000) + 2(00•1000)
Sl 3:   (01•1010) +   (11•0100) +   (01•0001) + 2(01•2000) + 4(01•0100) +   (11•0000) + 3(01•0000)
Sl 4:   (10•1001) +   (00•0110) +   (02•1100) + 2(10•1100) +   (00•1001) + 2(02•0010) +   (00•3000)
    + 3(10•0010) + 4(00•1100) +   (20•1000) + 4(00•0010) + 4(02•1000) + 5(10•1000) + 7(00•1000)
Sl 5:   (01•0101) +   (11•1010) +   (01•2100) + 2(01•0200) + 6(01•1010) +   (11•0001) +   (03•0100)
    + 2(11•2000) + 5(11•0100) + 5(01•0001) + 8(01•2000) + 15(01•0100) +   (03•0000) + 4(11•0000)
    + 9(01•0000)
```

Affine C6 ⊃ C2 + C4 - Level 1 - Class 1

Sl 0: (01•0000) **Highest Weight (0100000)**

Sl 1: (10•1000) + (00•1000)

Sl 2: (01•2000) + (01•0100) + (11•0000) + 2(01•0000)

Sl 3: (10•1100) + (00•1100) + (00•0010) + 2(02•1000) + 3(10•1000) + 3(00•1000)

Sl 4: (01•0200) + (01•1010) + (11•2000) + (11•0100) + 4(01•2000) + 5(01•0100) + (03•0000) + 3(11•0000) + 6(01•0000)

Sl 5: (10•0110) + (00•0110) + 2(02•1100) + (10•3000) + 4(10•1100) + (00•1001) + (02•0010) + (00•3000) + 2(10•0010) + 5(00•1100) + (12•1000) + (20•1000) + 3(00•0010) + 7(02•1000) + 10(10•1000) + 9(00•1000)

Affine C6 ⊃ C3 + C3 - Level 1 - Class 0

Sl 0: (000•001) **Highest Weight (0000001)**

Sl 1: (001•010)

Sl 2: (002•001) + (010•100) + (000•110) + (000•001) + (000•100)

Sl 3: (011•010) + (001•101) + (001•200) + 3(001•010) + (100•000) + (001•000)

Sl 4: (020•001) + (002•110) + (010•110) + 2(002•001) + (101•100) + (000•201) + 2(010•001) + 2(002•100) + 3(010•100) + 3(000•110) + 4(000•001) + 3(000•100)

Sl 5: (011•101) + (110•010) + (003•010) + (011•200) + (001•210) + 4(011•010) + (001•020) + 4(001•101) + (100•200) + 2(100•010) + 4(001•200) + 2(011•000) + 10(001•010) + 2(100•000) + 4(001•000)

Affine C6 ⊃ C3 + C3 - Level 1 - Class 0

Sl 0: (000•100) **Highest Weight (0000100)**

Sl 1: (001•010) + (001•000)

Sl 2: (010•001) + (002•100) + (010•100) + (000•110) + (000•001) + 2(000•100)

Sl 3: (011•010) + (001•101) + (100•010) + 2(001•200) + (011•000) + 4(001•010) + (100•000) + 3(001•000)

Sl 4: (002•110) + (101•001) + (020•100) + 2(010•110) + 2(002•001) + (101•100) + 3(010•001) + (000•011) + 4(002•100) + (000•300) + 6(010•100) + 4(000•110) + 3(000•001) + 7(000•100)

Sl 5: (011•101) + (110•010) + (003•010) + 2(011•200) + (001•210) + (100•101) + 6(011•010) + 2(001•020) + 5(001•101) + 2(100•200) + (110•000) + 5(100•010) + (003•000) + 8(001•200) + 5(011•000) + 15(001•010) + 3(100•000) + 10(001•000)

Affine C6 ⊃ C3 + C3 - Level 1 - Class 1

```
Sl 0:   (000•010)  Highest Weight (0000010)
Sl 1:   (001•001) +   (001•100)
Sl 2:   (002•010) +   (010•010) +   (000•101) +   (000•200) +   (010•000) +  2(000•010) +   (000•000)
Sl 3:   (011•001) +   (011•100) +  2(001•110) +  3(001•001) +   (100•100) +  4(001•100)
Sl 4:   (002•101) +   (020•010) +   (101•010) +   (010•101) +   (002•200) +  4(002•010) +  2(010•200)
    +   (000•210) +  5(010•010) +   (000•020) +   (101•000) +  3(000•101) +  2(002•000) +  3(000•200)
    +  3(010•000) +  7(000•010) +  3(000•000)
Sl 5:  2(011•110) +   (110•001) +   (003•001) +   (001•201) +  4(011•001) +   (001•011) +   (110•100)
    +   (100•110) +   (003•100) +   (001•300) +  6(011•100) +  9(001•110) +  2(100•001) + 10(001•001)
    +  5(100•100) + 14(001•100)
```

Affine C6 ⊃ C3 + C3 - Level 1 - Class 1

```
Sl 0:   (000•000)  Highest Weight (0001000)
Sl 1:   (001•100)
Sl 2:   (010•010) +   (002•000) +   (000•200) +   (010•000) +   (000•010) +   (000•000)
Sl 3:   (011•100) +   (001•110) +   (100•001) +   (001•001) +   (100•100) +  4(001•100)
Sl 4:   (101•010) +   (010•101) +   (002•200) +  2(002•010) +  2(010•200) +   (020•000) +  4(010•010)
    +   (000•020) +   (101•000) +   (000•101) +  3(002•000) +  3(000•200) +  4(010•000) +  4(000•010)
    +  4(000•000)
Sl 5:   (011•110) +  2(011•001) +   (001•011) +   (110•100) +  2(100•110) +   (003•100) +   (001•300)
    +  6(011•100) +  6(001•110) +  3(100•001) +  5(001•001) +  5(100•100) + 14(001•100)
```

Affine C7 Branching Rules

Affine C7 ⊃ C7 - Level 1 - Class 0

```
Sl 0:   (•0000010)  Highest weight (00000010)
Sl 1:   (•1000001) +   (•1000100) +   (•0000010)
Sl 2:   (•2000010) +   (•0100010) +   (•0101000) +  2(•1000001) +  2(•1000100) +  3(•0000010)
    +   (•0001000)
Sl 3:   (•1100001) +   (•1100100) +   (•0010100) +  3(•2000010) +  3(•0100010) +   (•2001000)
    +   (•0020000) +  2(•0101000) +  5(•1000001) +  6(•1000100) +  6(•0000010) +   (•1010000)
    +  2(•0001000)
```

Affine C7 ⊃ C7 - Level 1 - Class 0

```
Sl 0:   (•0001000)  Highest weight (00001000)
Sl 1:   (•1000100) +   (•1010000) +   (•0001000)
Sl 2:   (•0100010) +   (•2001000) +   (•0101000) +  2(•1000100) +   (•0200000) +   (•0000010)
    +  2(•1010000) +  3(•0001000) +   (•0100000)
Sl 3:   (•1100100) +   (•0010001) +   (•0010100) +   (•2000010) +  2(•0100010) +   (•1110000)
    +  3(•2001000) +   (•0020000) +  3(•0101000) +   (•1000001) +  6(•1000100) +   (•2100000)
    +   (•0200000) +  2(•0000010) +  6(•1010000) +  6(•0001000) +   (•2000000) +  2(•0100000)
```

Affine C7 ⊃ C7 - Level 1 - Class 0

```
Sl 0:   (•0100000)  Highest weight (00100000)
Sl 1:   (•1010000) +   (•2000000) +   (•0100000)
Sl 2:   (•0101000) +   (•2100000) +   (•0200000) +  2(•1010000) +   (•0001000) +   (•2000000)
    +  3(•0100000) +   (•0000000)
Sl 3:   (•0010100) +   (•1110000) +   (•2001000) +   (•0020000) +  2(•0101000) +   (•1000100)
    +  3(•2100000) +  2(•0200000) +  6(•1010000) +  2(•0001000) +  4(•2000000) +  5(•0100000)
    +   (•0000000)
```

Affine C7 ⊃ C7 - Level 1 - Class 0

```
Sl 0:   (•0000000)  Highest weight (10000000)
Sl 1:   (•2000000)
Sl 2:   (•0200000) +   (•2000000) +   (•0100000) +   (•0000000)
Sl 3:   (•0020000) +   (•2100000) +   (•0200000) +   (•1010000) +  3(•2000000) +   (•0100000)
    +   (•0000000)
Sl 4:   (•0002000) +   (•1110000) +   (•0020000) +   (•0101000) +   (•4000000) +  2(•2100000)
    +  4(•0200000) +  2(•1010000) +   (•0001000) +  4(•2000000) +  4(•0100000) +  3(•0000000)
```

Affine C7 ⊃ C7 - Level 1 - Class 1

```
Sl 0:   (•0000001)  Highest weight (00000001)
Sl 1:   (•1000010) +   (•0000001)
Sl 2:   (•2000001) +   (•0100100) +  2(•1000010) +  2(•0000001) +   (•0000100)
Sl 3:   (•1100010) +  2(•2000001) +   (•0100001) +   (•0011000) +   (•2000100) +  2(•0100100)
     + 5(•1000010) +   (•1001000) +  4(•0000001) +  2(•0000100)
Sl 4:   (•0200001) +   (•1010100) +   (•3000010) +  3(•1100010) +   (•0010010) +   (•1101000)
     + 5(•2000001) +  3(•0100001) +  2(•0011000) +  3(•2000100) +  6(•0100100) + 11(•1000010)
     +  (•0110000) +  3(•1001000) +  8(•0000001) +  5(•0000100) +   (•0010000)
```

Affine C7 ⊃ C7 - Level 1 - Class 1

```
Sl 0:   (•0000100)  Highest weight (00000100)
Sl 1:   (•1000010) +   (•1001000) +   (•0000100)
Sl 2:   (•0100001) +   (•2000100) +   (•0100100) +  2(•1000010) +   (•0110000) +  2(•1001000)
     +  (•0000001) +  3(•0000100) +   (•0010000)
Sl 3:   (•1100010) +   (•0010010) +   (•1101000) +   (•2000001) +  2(•0100001) +   (•0011000)
     + 3(•2000100) +  3(•0100100) +  6(•1000010) +   (•2010000) +  2(•0110000) +  6(•1001000)
     + 2(•0000001) +  6(•0000100) +   (•1100000) +  2(•0010000)
```

Affine C7 ⊃ C7 - Level 1 - Class 1

```
Sl 0:   (•0010000)  Highest weight (00010000)
Sl 1:   (•1001000) +   (•1100000) +   (•0010000)
Sl 2:   (•0100100) +   (•2010000) +   (•0110000) +  2(•1001000) +   (•0000100) +  2(•1100000)
     + 3(•0010000) +   (•1000000)
Sl 3:   (•0010010) +   (•1101000) +   (•0011000) +   (•2000100) +  2(•0100100) +   (•1200000)
     +  (•1000010) +  3(•2010000) +  3(•0110000) +  6(•1001000) +  2(•0000100) +   (•3000000)
     + 5(•1100000) +  6(•0010000) +  2(•1000000)
```

Affine C7 ⊃ C7 - Level 1 - Class 1

```
Sl 0:   (•1000000)  Highest weight (01000000)
Sl 1:   (•1100000) +   (•1000000)
Sl 2:   (•0110000) +   (•3000000) +  2(•1100000) +   (•0010000) +  2(•1000000)
Sl 3:   (•0011000) +   (•1200000) +   (•2010000) +  2(•0110000) +   (•1001000) +  2(•3000000)
     + 5(•1100000) +  2(•0010000) +  4(•1000000)
Sl 4:   (•0001100) +   (•1020000) +   (•1101000) +  2(•0011000) +   (•0100100) +   (•3100000)
     + 3(•1200000) +  3(•2010000) +  6(•0110000) +  3(•1001000) +   (•0000100) +  4(•3000000)
     + 11(•1100000) + 5(•0010000) +  8(•1000000)
```

Affine C7 ⊃ A1 + C6 - Level 1 - Class 0

Sl 0: (0•000010) **Highest weight (00000010)**

Sl 1: (1•000001) + (1•000100)

Sl 2: (0•100001) + (0•100100) + (2•000010) + 2(0•000010) + (0•001000)

Sl 3: 2(1•100010) + (1•101000) + 3(1•000001) + 3(1•000100)

Sl 4: (0•200010) + (0•010010) + (2•100001) + 3(0•100001) + (0•011000) + (2•100100)
+ 4(0•100100) + 3(2•000010) + 7(0•000010) + (0•110000) + (2•001000) + 3(0•001000)

Sl 5: (1•200001) + (1•010001) + (1•200100) + 2(1•010100) + 8(1•100010) + (1•020000)
+ (3•000001) + 4(1•101000) + 8(1•000001) + (3•000100) + 10(1•000100) + 2(1•010000)

Affine C7 ⊃ A1 + C6 - Level 1 - Class 0

Sl 0: (0•001000) **Highest weight (00001000)**

Sl 1: (1•000100) + (1•010000)

Sl 2: (0•100100) + (0•000010) + (0•110000) + (2•001000) + 2(0•001000) + (0•100000)

Sl 3: (1•100010) + 2(1•101000) + 3(1•000100) + (1•200000) + 3(1•010000)

Sl 4: (0•010010) + (0•201000) + (0•100001) + (0•011000) + (2•100100) + 4(0•100100)
+ (2•000010) + 3(0•000010) + (2•110000) + 4(0•110000) + 3(2•001000) + 7(0•001000)
+ (2•100000) + 3(0•100000)

Affine C7 ⊃ A1 + C6 - Level 1 - Class 0

Sl 0: (0•100000) **Highest weight (00100000)**

Sl 1: (1•010000) + (1•000000)

Sl 2: (0•110000) + (0•001000) + (2•100000) + 2(0•100000)

Sl 3: (1•101000) + 2(1•200000) + 3(1•010000) + 2(1•000000)

Sl 4: (0•011000) + (0•100100) + (2•110000) + (0•300000) + 4(0•110000) + (2•001000)
+ 3(0•001000) + 3(2•100000) + 6(0•100000)

Sl 5: (1•010100) + (1•210000) + 2(1•020000) + 4(1•101000) + 2(1•000100) + (3•010000)
+ 6(1•200000) + 10(1•010000) + (3•000000) + 6(1•000000)

Affine C7 ⊃ A1 + C6 - Level 1 - Class 0

Sl 0: (1•000000) **Highest weight (10000000)**

Sl 1: (0•100000)

Sl 2: (1•200000) + (1•000000)

Sl 3: (0•110000) + (2•100000) + 2(0•100000)

Sl 4: (1•020000) + 2(1•200000) + 2(1•010000) + (3•000000) + 3(1•000000)

Sl 5: (0•011000) + (2•110000) + (0•300000) + 3(0•110000) + (0•001000) + 3(2•100000)
+ 5(0•100000)

Affine C7 ⊃ A1 + C6 - Level 1 - Class 1

```
Sl 0:   (0•000001)  Highest weight (00000001)
Sl 1:   (1•000010)
Sl 2:   (0•100010) +   (2•000001) +   (0•000001) +   (0•000100)
Sl 3:   (1•100001) +   (1•100100) +  3(1•000010)
Sl 4:   (0•200001) +   (0•010100) +   (2•100010) +  3(0•100010) +  2(2•000001) +   (0•101000)
     +  4(0•000001) +   (2•000100) +  3(0•000100)
```

Affine C7 ⊃ A1 + C6 - Level 1 - Class 1

```
Sl 0:   (0•000100)  Highest weight (00000100)
Sl 1:   (1•000010) +   (1•001000)
Sl 2:   (0•100010) +   (0•101000) +   (0•000001) +   (2•000100) +  2(0•000100) +   (0•010000)
Sl 3:   (1•100001) +  2(1•100100) +  3(1•000010) +   (1•110000) +  3(1•001000)
Sl 4:   (0•010001) +   (0•200100) +   (0•010100) +   (2•100010) +  4(0•100010) +   (2•101000)
     +  (0•020000) +   (2•000001) +  4(0•101000) +  3(0•000001) +  3(2•000100) +  7(0•000100)
     +  (2•010000) +   (0•200000) +  3(0•010000)
```

Affine C7 ⊃ A1 + C6 - Level 1 - Class 1

```
Sl 0:   (0•010000)  Highest weight (00010000)
Sl 1:   (1•001000) +   (1•100000)
Sl 2:   (0•101000) +   (0•000100) +   (2•010000) +   (0•200000) +  2(0•010000) +   (0•000000)
Sl 3:   (1•100100) +  2(1•110000) +  3(1•001000) +  3(1•100000)
Sl 4:   (0•010100) +   (0•100010) +   (0•210000) +   (2•101000) +   (0•020000) +  4(0•101000)
     +  (2•000100) +  3(0•000100) +   (2•200000) +  3(2•010000) +  3(0•200000) +  7(0•010000)
     +  (2•000000) +  2(0•000000)
```

Affine C7 ⊃ A1 + C6 - Level 1 - Class 1

```
Sl 0:   (0•000000)  Highest weight (01000000)
Sl 1:   (1•100000)
Sl 2:   (0•200000) +   (0•010000) +   (2•000000) +   (0•000000)
Sl 3:   (1•110000) +  3(1•100000)
Sl 4:   (0•020000) +   (0•101000) +   (2•200000) +   (2•010000) +  3(0•200000) +  3(0•010000)
     + 2(2•000000) +  3(0•000000)
```

Affine C7 ⊃ C2 + C5 - Level 1 - Class 0

```
Sl 0:   (00•00010)  Highest weight (00000010)
Sl 1:   (01•00001) +   (01•00100)
Sl 2:   (00•10001) +   (02•00010) +   (00•10100) +   (10•00010) +  2(00•00010) +   (10•01000)
     +  (00•01000)
Sl 3:  2(01•10010) +   (11•00001) +   (11•00100) +   (01•11000) +  3(01•00001) +  4(01•00100)
     +  (01•10000)
```

Affine C7 ⊃ C2 + C5 - Level 1 - Class 0

```
Sl 0:   (00•01000)  Highest weight (00001000)
Sl 1:   (01•00100) +  (01•10000)
Sl 2:   (00•10100) +  (10•00010) +  (00•00010) +  (02•01000) +  (10•01000) +  (00•20000)
      + 2(00•01000) +  (10•00000) +  (00•00000)
Sl 3:   (01•10010) +  (11•00100) + 2(01•11000) +  (01•00001) + 4(01•00100) +  (11•10000)
      + 4(01•10000)
```

Affine C7 ⊃ C2 + C5 - Level 1 - Class 0

```
Sl 0:   (00•00000)  Highest weight (00100000)
Sl 1:   (01•10000)
Sl 2:   (10•01000) +  (00•20000) +  (00•01000) +  (02•00000) +  (10•00000) +  (00•00000)
Sl 3:   (01•11000) +  (01•00100) +  (11•10000) + 4(01•10000)
Sl 4:   (10•10100) +  (00•02000) +  (00•10100) +  (02•20000) +  (00•00010) + 2(02•01000)
      + 2(10•20000) + 3(10•01000) + 3(00•20000) + 4(00•01000) +  (20•00000) + 3(02•00000)
      + 3(10•00000) + 4(00•00000)
```

Affine C7 ⊃ C2 + C5 - Level 1 - Class 0

```
Sl 0:   (10•00000)  Highest weight (10000000)
Sl 1:   (01•10000)
Sl 2:   (10•20000) +  (00•01000) +  (02•00000) +  (10•00000) +  (00•00000)
Sl 3:   (01•11000) +  (11•10000) + 3(01•10000)
Sl 4:   (10•02000) +  (00•10100) +  (02•20000) +  (02•01000) + 2(10•20000) + 2(10•01000)
      + 2(00•20000) + 2(00•01000) +  (12•00000) + 2(02•00000) + 4(10•00000) + 2(00•00000)
```

Affine C7 ⊃ C2 + C5 - Level 1 - Class 1

```
Sl 0:   (00•00001)  Highest weight (00000001)
Sl 1:   (01•00010)
Sl 2:   (00•10010) +  (02•00001) +  (10•00100) +  (00•00001) +  (00•00100)
Sl 3:   (01•10001) +  (01•10100) +  (11•00010) + 3(01•00010) +  (01•01000)
Sl 4:   (00•20001) +  (02•10010) +  (10•10010) +  (00•01100) +  (20•00001) + 3(00•10010)
      + 2(02•00001) +  (10•11000) + 2(10•00001) + 2(02•00100) + 3(10•00100) +  (00•11000)
      + 4(00•00001) + 3(00•00100) +  (00•10000)
Sl 5:   (01•20010) +  (01•01010) +  (11•10001) +  (11•10100) + 4(01•10001) +  (01•02000)
      +  (03•00010) + 5(01•10100) + 4(11•00010) +10(01•00010) +  (11•01000) +  (01•20000)
      + 5(01•01000)
```

Affine C7 ⊃ C2 + C5 - Level 1 - Class 1

Sl 0: (00•00100) **Highest weight (00000100)**

Sl 1: (01•00010) + (01•01000)

Sl 2: (00•10010) + (10•00001) + (02•00100) + (10•00100) + (00•11000) + (00•00001) + 2(00•00100) + (10•10000) + (00•10000)

Sl 3: (01•10001) + 2(01•10100) + (11•00010) + 4(01•00010) + (11•01000) + (01•20000) + 4(01•01000) + (01•00000)

Affine C7 ⊃ C2 + C5 - Level 1 - Class 1

Sl 0: (00•10000) **Highest weight (00010000)**

Sl 1: (01•01000) + (01•00000)

Sl 2: (10•00100) + (00•11000) + (00•00100) + (02•10000) + (10•10000) + 2(00•10000)

Sl 3: (01•10100) + (01•00010) + (11•01000) + 2(01•20000) + 4(01•01000) + (11•00000) + 3(01•00000)

Sl 4: (10•10010) + (00•01100) + (00•10010) + (02•11000) + 2(10•11000) + 2(02•00100) + (00•30000) + 3(10•00100) + 4(00•11000) + (00•00001) + 4(00•00100) + (20•10000) + 4(02•10000) + 5(10•10000) + 7(00•10000)

Sl 5: (01•01010) + (11•10100) + (01•21000) + (01•10001) + 2(01•02000) + 6(01•10100) + (11•00010) + 5(01•00010) + (03•01000) + 2(11•20000) + 5(11•01000) + 8(01•20000) + 15(01•01000) + (03•00000) + 4(11•00000) + 9(01•00000)

Affine C7 ⊃ C2 + C5 - Level 1 - Class 1

Sl 0: (01•00000) **Highest weight (01000000)**

Sl 1: (10•10000) + (00•10000)

Sl 2: (01•20000) + (01•01000) + (11•00000) + 2(01•00000)

Sl 3: (10•11000) + (00•11000) + (00•00100) + 2(02•10000) + 3(10•10000) + 3(00•10000)

Sl 4: (01•02000) + (01•10100) + (11•20000) + (11•01000) + 4(01•20000) + 5(01•01000) + (03•00000) + 3(11•00000) + 6(01•00000)

Sl 5: (10•01100) + (00•01100) + (00•10010) + 2(02•11000) + (10•30000) + 4(10•11000) + (02•00100) + (00•30000) + 2(10•00100) + 5(00•11000) + 3(00•00100) + (12•10000) + (20•10000) + 7(02•10000) + 10(10•10000) + 9(00•10000)

Affine C7 ⊃ C3 + C4 - Level 1 - Class 0

Sl 0: (000•0010) **Highest weight (00000010)**

Sl 1: (001•0001) + (001•0100)

Sl 2: (002•0010) + (010•0010) + (000•1001) + (000•1100) + (010•1000) + 2(000•0010) + (000•1000)

Sl 3: (011•0001) + 2(001•1010) + (011•0100) + 3(001•0001) + (100•0100) + (001•2000) + 4(001•0100) + (100•0000) + (001•0000) (continued next page)

Sl 4: (002•1001) + (020•0010) + (010•1001) + (002•1100) + (101•0010) + (000•2010)
+ 2(010•1100) + (000•0110) + 4(002•0010) + 5(010•0010) + 3(000•1001) + (101•1000)
+ 4(000•1100) + 2(002•1000) + 4(010•1000) + 7(000•0010) + 4(000•1000)

Affine C7 $\supset$ C3 + C4 - Level 1 - Class 0

Sl 0: (000•1000) **Highest weight (00001000)**
Sl 1: (001•0100) + (001•0000)
Sl 2: (010•0010) + (000•1100) + (002•1000) + (010•1000) + (000•0010) + 2(000•1000)
Sl 3: (001•1010) + (011•0100) + (100•0001) + (001•0001) + (100•0100) + 2(001•2000)
+ 4(001•0100) + (011•0000) + (100•0000) + 3(001•0000)
Sl 4: (010•1001) + (002•1100) + (101•0010) + 2(010•1100) + (000•0110) + 2(002•0010)
+ (020•1000) + 4(010•0010) + (000•1001) + (101•1000) + (000•3000) + 4(000•1100)
+ 4(002•1000) + 6(010•1000) + 4(000•0010) + 7(000•1000)

Affine C7 $\supset$ C3 + C4 - Level 1 - Class 0

Sl 0: (001•0000) **Highest weight (00100000)**
Sl 1: (010•1000) + (000•1000)
Sl 2: (100•0100) + (001•2000) + (001•0100) + (011•0000) + (100•0000) + 2(001•0000)
Sl 3: (010•1100) + (010•0010) + (101•1000) + (000•1100) + 2(002•1000) + 4(010•1000)
+ (000•0010) + 3(000•1000)
Sl 4: (100•1010) + (001•0200) + (011•2000) + (001•1010) + 2(011•0100) + 2(100•2000)
+ (001•0001) + 3(100•0100) + 4(001•2000) + 6(001•0100) + (110•0000) + (003•0000)
+ 4(011•0000) + 3(100•0000) + 7(001•0000)

Affine C7 $\supset$ C3 + C4 - Level 1 - Class 0

Sl 0: (100•0000) **Highest weight (10000000)**
Sl 1: (010•1000)
Sl 2: (100•2000) + (001•0100) + (011•0000) + (100•0000) + (001•0000)
Sl 3: (010•1100) + (101•1000) + (002•1000) + 3(010•1000) + (000•0010) + (000•1000)
Sl 4: (100•0200) + (011•2000) + (001•1010) + (011•0100) + 2(100•2000) + 2(100•0100)
+ (102•0000) + 2(001•2000) + 3(001•0100) + 3(011•0000) + 4(100•0000) + 3(001•0000)
Sl 5: (010•0110) + (101•1100) + (002•1100) + (010•3000) + 4(010•1100) + (012•1000)
+ (002•0010) + (020•1000) + 2(010•0010) + (000•1001) + 4(101•1000) + 2(000•1100)
+ 4(002•1000) + 10(010•1000) + 2(000•0010) + 4(000•1000)

Affine C7 ⊃ C3 + C4 - Level 1 - Class 1

Sl 0: (000•0001) **Highest weight (00000001)**

Sl 1: (001•0010)

Sl 2: (002•0001) + (000•1010) + (010•0100) + (000•0001) + (000•0100)

Sl 3: (011•0010) + (001•1001) + (001•1100) + 3(001•0010) + (100•1000) + (001•1000)

Sl 4: (002•1010) + (020•0001) + (010•1010) + (000•2001) + 2(002•0001) + (101•0100)

+ 2(010•0001) + (000•0200) + 2(002•0100) + (010•2000) + 3(000•1010) + 3(010•0100)

+ 4(000•0001) + (000•2000) + 3(000•0100) + (010•0000) + (000•0000)

Affine C7 ⊃ C3 + C4 - Level 1 - Class 1

Sl 0: (000•0100) **Highest weight (00000100)**

Sl 1: (001•0010) + (001•1000)

Sl 2: (010•0001) + (002•0100) + (000•1010) + (010•0100) + (000•0001) + (000•2000)

+ 2(000•0100) + (010•0000) + (000•0000)

Sl 3: (011•0010) + (001•1001) + 2(001•1100) + (100•0010) + (011•1000) + 4(001•0010)

+ (100•1000) + 4(001•1000)

Sl 4: (002•1010) + (101•0001) + 2(010•1010) + (020•0100) + (000•0101) + 2(002•0001)

+ (101•0100) + (000•2100) + (002•2000) + 3(010•0001) + (000•0200) + 4(002•0100)

+ 2(010•2000) + 4(000•1010) + 6(010•0100) + 3(000•0001) + (101•0000) + 3(000•2000)

+ 8(000•0100) + 2(002•0000) + 3(010•0000) + 3(000•0000)

Affine C7 ⊃ C3 + C4 - Level 1 - Class 1

Sl 0: (000•0000) **Highest weight (00010000)**

Sl 1: (001•1000)

Sl 2: (010•0100) + (000•2000) + (000•0100) + (002•0000) + (010•0000) + (000•0000)

Sl 3: (001•1100) + (100•0010) + (011•1000) + (001•0010) + (100•1000) + 4(001•1000)

Sl 4: (010•1010) + (101•0100) + (002•2000) + (010•0001) + (000•0200) + 2(002•0100)

+ 2(010•2000) + (000•1010) + 4(010•0100) + (020•0000) + (000•0001) + (101•0000)

+ 3(000•2000) + 4(000•0100) + 3(002•0000) + 4(010•0000) + 4(000•0000)

Affine C7 ⊃ C3 + C4 - Level 1 - Class 1

Sl 0: (010•0000) **Highest weight (01000000)**

Sl 1: (100•1000) + (001•1000)

Sl 2: (010•2000) + (010•0100) + (101•0000) + (000•0100) + (002•0000) + 2(010•0000)

+ (000•0000)

Sl 3: (100•1100) + (001•1100) + 2(011•1000) + (001•0010) + 3(100•1000) + 4(001•1000)

Sl 4: (010•0200) + (010•1010) + (101•2000) + (101•0100) + (002•2000) + 2(002•0100)

+ 4(010•2000) + (000•1010) + 5(010•0100) + (012•0000) + (020•0000) + (000•0001)

+ 3(101•0000) + 2(000•2000) + 3(000•0100) + 3(002•0000) + 7(010•0000) + 3(000•0000)

Affine C8 Branching Rules

Affine C8 ⊃ C8 - Level 1 - Class 0

Sl 0: (•00000001) **Highest Weight (000000001)**

Sl 1: (•10000010) + (•00000001)

Sl 2: (•20000001) + (•01000100) + 2(•10000010) + 2(•00000001) + (•00000100)

Sl 3: (•11000010) + 2(•20000001) + (•00101000) + (•01000001) + (•20000100) + 2(•01000100)
+ 5(•10000010) + (•10001000) + 4(•00000001) + 2(•00000100)

Affine C8 ⊃ C8 - Level 1 - Class 0

Sl 0: (•00000100) **Highest Weight (000000100)**

Sl 1: (•10000010) + (•10001000) + (•00000100)

Sl 2: (•01000001) + (•20000100) + (•01000100) + 2(•10000010) + (•01010000) + 2(•10001000)
+ (•00000001) + 3(•00000100) + (•00010000)

Sl 3: (•11000010) + (•00100010) + (•11001000) + (•20000001) + (•00101000) + 2(•01000001)
+ 3(•20000100) + 3(•01000100) + (•20010000) + 6(•10000010) + (•00200000) + 2(•01010000)
+ 6(•10001000) + 2(•00000001) + 6(•00000100) + (•10100000) + 2(•00010000)

Affine C8 ⊃ C8 - Level 1 - Class 0

Sl 0: (•00010000) **Highest Weight (000010000)**

Sl 1: (•10001000) + (•10100000) + (•00010000)

Sl 2: (•01000100) + (•20010000) + (•01010000) + 2(•10001000) + (•00000100) + (•02000000)
+ 2(•10100000) + 3(•00010000) + (•01000000)

Sl 3: (•00100010) + (•11001000) + (•00101000) + (•20000100) + 2(•01000100) + (•11100000)
+ 3(•20010000) + (•10000010) + (•00200000) + 3(•01010000) + 6(•10001000) + 2(•00000100)
+ (•21000000) + (•02000000) + 6(•10100000) + 6(•00010000) + (•20000000) + 2(•01000000)

Affine C8 ⊃ C8 - Level 1 - Class 0

Sl 0: (•01000000) **Highest Weight (001000000)**

Sl 1: (•10100000) + (•20000000) + (•01000000)

Sl 2: (•01010000) + (•21000000) + (•02000000) + 2(•10100000) + (•00010000) + (•20000000)
+ 3(•01000000) + (•00000000)

Sl 3: (•00101000) + (•11100000) + (•20010000) + (•00200000) + 2(•01010000) + (•10001000)
+ 3(•21000000) + 2(•02000000) + 6(•10100000) + 2(•00010000) + 4(•20000000) + 5(•01000000)
+ (•00000000)

Sl 4: (•00010100) + (•10110000) + (•11001000) + (•00020000) + 2(•00101000) + (•01000100)
+ (•30100000) + (•03000000) + 4(•11100000) + 3(•20010000) + 2(•00200000) + 7(•01010000)
+ 3(•10001000) + (•40000000) + (•00000100) + 7(•21000000) + 7(•02000000) + 13(•10100000)
+ 6(•00010000) + 7(•20000000) + 12(•01000000) + 3(•00000000)

Affine C8 ⊃ C8 - Level 1 - Class 0

Sl 0: (•00000000) **Highest Weight (100000000)**

Sl 1: (•20000000)

Sl 2: (•02000000) + (•20000000) + (•01000000) + (•00000000)

Sl 3: (•00200000) + (•21000000) + (•02000000) + (•10100000) + 3(•20000000) + (•01000000)
+ (•00000000)

Sl 4: (•00020000) + (•11100000) + (•00200000) + (•01010000) + (•40000000) + 2(•21000000)
+ 4(•02000000) + 2(•10100000) + (•00010000) + 4(•20000000) + 4(•01000000) + 3(•00000000)

Affine C8 ⊃ C8 - Level 1 - Class 1

Sl 0: (•00000010) **Highest Weight (000000010)**

Sl 1: (•10000001) + (•10000100) + (•00000010)

Sl 2: (•20000010) + (•01000010) + (•01001000) + 2(•10000001) + 2(•10000100) + 3(•00000010)
+ (•00001000)

Sl 3: (•11000001) + (•11000100) + (•00100100) + 3(•20000010) + 3(•01000010) + (•00110000)
+ (•20001000) + 2(•01001000) + 5(•10000001) + 6(•10000100) + (•10010000) + 6(•00000010)
+ 2(•00001000)

Affine C8 ⊃ C8 - Level 1 - Class 1

Sl 0: (•00001000) **Highest Weight (000001000)**

Sl 1: (•10000100) + (•10010000) + (•00001000)

Sl 2: (•01000010) + (•20001000) + (•01001000) + 2(•10000100) + (•01100000) + 2(•10010000)
+ (•00000010) + 3(•00001000) + (•00100000)

Sl 3: (•11000100) + (•00100001) + (•00100100) + (•20000010) + (•11010000) + 2(•01000010)
+ (•00110000) + 3(•20001000) + 3(•01001000) + (•10000001) + 6(•10000100) + (•20100000)
+ 2(•01100000) + 6(•10010000) + 2(•00000010) + 6(•00001000) + (•11000000) + 2(•00100000)

Sl 4: (•10100010) + (•02001000) + (•00010010) + (•10101000) + (•11000001) + (•30000100)
+ 4(•11000100) + 2(•00100001) + (•00011000) + 3(•00100100) + (•30010000) + 3(•20000010)
+ (•10200000) + 4(•11010000) + 7(•01000010) + 3(•00110000) + 8(•20001000) + 10(•01001000)
+ 3(•10000001) + 14(•10000100) + (•12000000) + 3(•20100000) + 6(•01100000) + 14(•10010000)
+ 6(•00000010) + 14(•00001000) + 3(•11000000) + 6(•00100000) + (•10000000)

Affine C8 ⊃ C8 - Level 1 - Class 1

Sl 0: (•00100000) **Highest Weight (000100000)**

Sl 1: (•10010000) + (•11000000) + (•00100000)

Sl 2: (•01001000) + (•20100000) + (•01100000) + 2(•10010000) + (•00001000) + 2(•11000000)
+ 3(•00100000) + (•10000000)

Sl 3: (•00100100) + (•11010000) + (•00110000) + (•20001000) + 2(•01001000) + (•10000100)
+ (•12000000) + 3(•20100000) + 3(•01100000) + 6(•10010000) + 2(•00001000) + (•30000000)
+ 5(•11000000) + 6(•00100000) + 2(•10000000)

Affine C8 ⊃ C8 - Level 1 - Class 1

```
Sl 0:  (•10000000)  Highest Weight (010000000)
Sl 1:  (•11000000) + (•10000000)
Sl 2:  (•01100000) + (•30000000) + 2(•11000000) + (•00100000) + 2(•10000000)
Sl 3:  (•00110000) + (•12000000) + (•20100000) + 2(•01100000) + (•10010000) + 2(•30000000)
     + 5(•11000000) + 2(•00100000) + 4(•10000000)
Sl 4:  (•00011000) + (•10200000) + (•11010000) + 2(•00110000) + (•01001000) + (•31000000)
     + 3(•12000000) + 3(•20100000) + 6(•01100000) + 3(•10010000) + (•00001000) + 4(•30000000)
     + 11(•11000000) + 5(•00100000) + 8(•10000000)
```

Affine C8 ⊃ A1 + C7 - Level 1 - Class 0

```
Sl 0:  (0•0000001)  Highest Weight (000000001)
Sl 1:  (1•0000010)
Sl 2:  (0•1000010) + (2•0000001) + (0•0000001) + (0•0000100)
Sl 3:  (1•1000001) + (1•1000100) + 3(1•0000010)
Sl 4:  (0•2000001) + (0•0100100) + (2•1000010) + 3(0•1000010) + (0•1001000) + 2(2•0000001)
     + 4(0•0000001) + (2•0000100) + 3(0•0000100)
Sl 5:  (1•2000010) + (1•0100010) + (1•0101000) + 4(1•1000001) + 4(1•1000100) + (3•0000010)
     + 8(1•0000010) + 2(1•0001000)
```

Affine C8 ⊃ A1 + C7 - Level 1 - Class 0

```
Sl 0:  (0•0000100)  Highest Weight (000000100)
Sl 1:  (1•0000010) + (1•0001000)
Sl 2:  (0•1000010) + (0•1001000) + (0•0000001) + (2•0000100) + 2(0•0000100) + (0•0010000)
Sl 3:  (1•1000001) + 2(1•1000100) + 3(1•0000010) + (1•1010000) + 3(1•0001000)
Sl 4:  (0•0100001) + (0•2000100) + (0•0100100) + (2•1000010) + 4(0•1000010) + (0•0110000)
     + (2•1001000) + 4(0•1001000) + (2•0000001) + 3(0•0000001) + 3(2•0000100) + 7(0•0000100)
     + (0•1100000) + (2•0010000) + 3(0•0010000)
```

Affine C8 ⊃ A1 + C7 - Level 1 - Class 0

```
Sl 0:  (0•0010000)  Highest Weight (000010000)
Sl 1:  (1•0001000) + (1•0100000)
Sl 2:  (0•1001000) + (0•0000100) + (0•1100000) + (2•0010000) + 2(0•0010000) + (0•1000000)
Sl 3:  (1•1000100) + 2(1•1010000) + 3(1•0001000) + (1•2000000) + 3(1•0100000)
Sl 4:  (0•0100100) + (0•1000010) + (0•2010000) + (0•0110000) + (2•1001000) + 4(0•1001000)
     + (2•0000100) + 3(0•0000100) + (2•1100000) + 4(0•1100000) + 3(2•0010000) + 7(0•0010000)
     + (2•1000000) + 3(0•1000000)
Sl 5:  (1•0100010) + (1•2001000) + 2(1•0101000) + 4(1•1000100) + (1•2100000) + 2(1•0200000)
     + 2(1•0000010) + 8(1•1010000) + (3•0001000) + 10(1•0001000) + (3•0100000) + 3(1•2000000)
     + 10(1•0100000) + 2(1•0000000)
```

Affine C8 ⊃ A1 + C7 - Level 1 - Class 0

sl 0: (0•1000000) **Highest Weight (001000000)**
sl 1: (1•0100000) + (1•0000000)
sl 2: (0•1100000) + (0•0010000) + (2•1000000) + 2(0•1000000)
sl 3: (1•1010000) + 2(1•2000000) + 3(1•0100000) + 2(1•0000000)
sl 4: (0•0110000) + (0•1001000) + (2•1100000) + (0•3000000) + 4(0•1100000) + (2•0010000)
+ 3(0•0010000) + 3(2•1000000) + 6(0•1000000)

Affine C8 ⊃ A1 + C7 - Level 1 - Class 0

sl 0: (1•0000000) **Highest Weight (100000000)**
sl 1: (0•1000000)
sl 2: (1•2000000) + (1•0000000)
sl 3: (0•1100000) + (2•1000000) + 2(0•1000000)
sl 4: (1•0200000) + 2(1•2000000) + 2(1•0100000) + (3•0000000) + 3(1•0000000)

Affine C8 ⊃ A1 + C7 - Level 1 - Class 1

sl 0: (0•0000010) **Highest Weight (000000010)**
sl 1: (1•0000001) + (1•0000100)
sl 2: (0•1000001) + (0•1000100) + (2•0000010) + 2(0•0000010) + (0•0001000)
sl 3: 2(1•1000010) + (1•1001000) + 3(1•0000001) + 3(1•0000100)
sl 4: (0•2000010) + (0•0100010) + (2•1000001) + (0•0101000) + 3(0•1000001) + (2•1000100)
+ 4(0•1000100) + 3(2•0000010) + 7(0•0000010) + (0•1010000) + (2•0001000) + 3(0•0001000)

Affine C8 ⊃ A1 + C7 - Level 1 - Class 1

sl 0: (0•0001000) **Highest Weight (000001000)**
sl 1: (1•0000100) + (1•0010000)
sl 2: (0•1000100) + (0•0000010) + (0•1010000) + (2•0001000) + 2(0•0001000) + (0•0100000)
sl 3: (1•1000010) + 2(1•1001000) + 3(1•0000100) + (1•1100000) + 3(1•0010000)
sl 4: (0•0100010) + (0•2001000) + (0•0101000) + (0•1000001) + (2•1000100) + 4(0•1000100)
+ (2•0000010) + (2•1010000) + (0•0200000) + 3(0•0000010) + 4(0•1010000) + 3(2•0001000)
+ 7(0•0001000) + (2•0100000) + (0•2000000) + 3(0•0100000)

Affine C8 ⊃ A1 + C7 - Level 1 - Class 1

sl 0: (0•0100000) **Highest Weight (000100000)**
sl 1: (1•0010000) + (1•1000000)
sl 2: (0•1010000) + (0•0001000) + (2•0100000) + (0•2000000) + 2(0•0100000) + (0•0000000)
sl 3: (1•1001000) + 2(1•1100000) + 3(1•0010000) + 3(1•1000000)
sl 4: (0•0101000) + (0•1000100) + (0•2100000) + (2•1010000) + (0•0200000) + 4(0•1010000)
+ (2•0001000) + 3(0•0001000) + (2•2000000) + 3(2•0100000) + 3(0•2000000) + 7(0•0100000)
+ (2•0000000) + 2(0•0000000)

Affine C8 ⊃ A1 + C7 - Level 1 - Class 1

Sl 0: (0•0000000) **Highest Weight (010000000)**

Sl 1: (1•1000000)

Sl 2: (0•2000000) + (0•0100000) + (2•0000000) + (0•0000000)

Sl 3: (1•1100000) + 3(1•1000000)

Sl 4: (0•0200000) + (0•1010000) + (2•2000000) + (2•0100000) + 3(0•2000000) + 3(0•0100000) + 2(2•0000000) + 3(0•0000000)

Affine C8 ⊃ C2 + C6 - Level 1 - Class 0

Sl 0: (00•000001) **Highest Weight (000000001)**

Sl 1: (01•000010)

Sl 2: (00•100010) + (02•000001) + (10•000100) + (00•000001) + (00•000100)

Sl 3: (01•100001) + (01•100100) + (11•000010) + 3(01•000010) + (01•001000)

Sl 4: (00•200001) + (02•100010) + (00•010100) + (10•100010) + 3(00•100010) + (20•000001) + 2(02•000001) + (10•101000) + 2(10•000001) + 2(02•000100) + (00•101000) + 3(10•000100) + 4(00•000001) + 3(00•000100) + (00•010000)

Affine C8 ⊃ C2 + C6 - Level 1 - Class 0

Sl 0: (00•000100) **Highest Weight (000000100)**

Sl 1: (01•000010) + (01•001000)

Sl 2: (00•100010) + (10•000001) + (02•000100) + (00•101000) + (10•000100) + (00•000001) + 2(00•000100) + (10•010000) + (00•010000)

Sl 3: (01•100001) + 2(01•100100) + (11•000010) + 4(01•000010) + (11•001000) + (01•110000) + 4(01•001000) + (01•100000)

Affine C8 ⊃ C2 + C6 - Level 1 - Class 0

Sl 0: (00•010000) **Highest Weight (000010000)**

Sl 1: (01•001000) + (01•100000)

Sl 2: (00•101000) + (10•000100) + (00•000100) + (02•010000) + (10•010000) + (00•200000) + 2(00•010000) + (10•000000) + (00•000000)

Sl 3: (01•100100) + (01•000010) + (11•001000) + 2(01•110000) + 4(01•001000) + (11•100000) + 4(01•100000)

Affine C8 ⊃ C2 + C6 - Level 1 - Class 0

Sl 0: (00•000000) **Highest Weight (001000000)**

Sl 1: (01•100000)

Sl 2: (10•010000) + (00•200000) + (00•010000) + (02•000000) + (10•000000) + (00•000000)

Sl 3: (01•110000) + (01•001000) + (11•100000) + 4(01•100000)

Sl 4: (10•101000) + (00•020000) + (00•101000) + (00•000100) + (02•200000) + 2(02•010000) + 2(10•200000) + 3(10•010000) + 3(00•200000) + 4(00•010000) + (20•000000) + 3(02•000000) + 3(10•000000) + 4(00•000000)

Affine C8 ⊃ C2 + C6 - Level 1 - Class 0

Sl 0: (10•000000) **Highest Weight (100000000)**

Sl 1: (01•100000)

Sl 2: (10•200000) + (00•010000) + (02•000000) + (10•000000) + (00•000000)

Sl 3: (01•110000) + (11•100000) + 3(01•100000)

Sl 4: (10•020000) + (00•101000) + (02•200000) + (02•010000) + 2(10•200000) + 2(10•010000)
+ 2(00•200000) + 2(00•010000) + (12•000000) + 2(02•000000) + 4(10•000000) + 2(00•000000)

Affine C8 ⊃ C2 + C6 - Level 1 - Class 1

Sl 0: (00•000010) **Highest Weight (000000010)**

Sl 1: (01•000001) + (01•000100)

Sl 2: (00•100001) + (00•100100) + (02•000010) + (10•000010) + 2(00•000010) + (10•001000)
+ (00•001000)

Sl 3: 2(01•100010) + (11•000001) + (01•101000) + (11•000100) + 3(01•000001) + 4(01•000100)
+ (01•010000)

Affine C8 ⊃ C2 + C6 - Level 1 - Class 1

Sl 0: (00•001000) **Highest Weight (000001000)**

Sl 1: (01•000100) + (01•010000)

Sl 2: (00•100100) + (10•000010) + (00•000010) + (02•001000) + (10•001000) + (00•110000)
+ 2(00•001000) + (10•100000) + (00•100000)

Sl 3: (01•100010) + 2(01•101000) + (11•000100) + (01•000001) + 4(01•000100) + (11•010000)
+ (01•200000) + 4(01•010000) + (01•000000)

Affine C8 ⊃ C2 + C6 - Level 1 - Class 1

Sl 0: (00•100000) **Highest Weight (000100000)**

Sl 1: (01•010000) + (01•000000)

Sl 2: (10•001000) + (00•110000) + (00•001000) + (02•100000) + (10•100000) + 2(00•100000)

Sl 3: (01•101000) + (01•000100) + (11•010000) + 2(01•200000) + 4(01•010000) + (11•000000)
+ 3(01•000000)

Sl 4: (10•100100) + (00•011000) + (00•100100) + (02•110000) + 2(10•110000) + (00•000010)
+ 2(02•001000) + (00•300000) + 3(10•001000) + 4(00•110000) + 4(00•001000) + (20•100000)
+ 4(02•100000) + 5(10•100000) + 7(00•100000)

Affine C8 ⊃ C2 + C6 - Level 1 - Class 1

Sl 0: (01•000000) **Highest Weight (010000000)**

Sl 1: (10•100000) + (00•100000)

Sl 2: (01•200000) + (01•010000) + (11•000000) + 2(01•000000)

Sl 3: (10•110000) + (00•110000) + (00•001000) + 2(02•100000) + 3(10•100000) + 3(00•100000)

Sl 4: (01•020000) + (01•101000) + (11•200000) + (11•010000) + 4(01•200000) + 5(01•010000)
+ (03•000000) + 3(11•000000) + 6(01•000000)

Affine C8 ⊃ C3 + C5 - Level 1 - Class 0

Sl 0: (000•00001) **Highest Weight (000000001)**
Sl 1: (001•00010)
Sl 2: (002•00001) + (000•10010) + (010•00100) + (000•00001) + (000•00100)
Sl 3: (001•10001) + (011•00010) + (001•10100) + 3(001•00010) + (100•01000) + (001•01000)
Sl 4: (002•10010) + (000•20001) + (020•00001) + (010•10010) + (000•01100) + 2(002•00001)
+ (101•00100) + (010•11000) + 2(010•00001) + 3(000•10010) + 2(002•00100) + 3(010•00100)
+ (000•11000) + 4(000•00001) + 3(000•00100) + (010•10000) + (000•10000)

Affine C8 ⊃ C3 + C5 - Level 1 - Class 0

Sl 0: (000•00100) **Highest Weight (000000100)**
Sl 1: (001•00010) + (001•01000)
Sl 2: (010•00001) + (000•10010) + (002•00100) + (010•00100) + (000•11000) + (000•00001)
+ 2(000•00100) + (010•10000) + (000•10000)
Sl 3: (001•10001) + (011•00010) + 2(001•10100) + (100•00010) + (011•01000) + 4(001•00010)
+ (100•01000) + (001•20000) + 4(001•01000) + (100•00000) + (001•00000)

Affine C8 ⊃ C3 + C5 - Level 1 - Class 0

Sl 0: (000•10000) **Highest Weight (000010000)**
Sl 1: (001•01000) + (001•00000)
Sl 2: (010•00100) + (000•11000) + (000•00100) + (002•10000) + (010•10000) + 2(000•10000)
Sl 3: (001•10100) + (100•00010) + (011•01000) + (001•00010) + (100•01000) + 2(001•20000)
+ 4(001•01000) + (011•00000) + (100•00000) + 3(001•00000)
Sl 4: (010•10010) + (000•01100) + (002•11000) + (101•00100) + 2(010•11000) + (010•00001)
+ (000•10010) + 2(002•00100) + 4(010•00100) + (000•30000) + (020•10000) + 4(000•11000)
+ (000•00001) + (101•10000) + 4(000•00100) + 4(002•10000) + 6(010•10000) + 7(000•10000)

Affine C8 ⊃ C3 + C5 - Level 1 - Class 0

Sl 0: (001•00000) **Highest Weight (001000000)**
Sl 1: (010•10000) + (000•10000)
Sl 2: (100•01000) + (001•20000) + (001•01000) + (011•00000) + (100•00000) + 2(001•00000)
Sl 3: (010•11000) + (010•00100) + (000•11000) + (101•10000) + (000•00100) + 2(002•10000)
+ 4(010•10000) + 3(000•10000)
Sl 4: (100•10100) + (001•02000) + (001•10100) + (011•20000) + 2(011•01000) + (001•00010)
+ 2(100•20000) + 3(100•01000) + 4(001•20000) + 6(001•01000) + (110•00000) + (003•00000)
+ 4(011•00000) + 3(100•00000) + 7(001•00000)

Affine C8 ⊃ C3 + C5 - Level 1 - Class 0

sl 0: (100•00000) **Highest Weight (100000000)**

sl 1: (010•10000)

sl 2: (100•20000) + (001•01000) + (011•00000) + (100•00000) + (001•00000)

sl 3: (010•11000) + (101•10000) + (000•00100) + (002•10000) + 3(010•10000) + (000•10000)

sl 4: (100•02000) + (001•10100) + (011•20000) + (011•01000) + 2(100•20000) + 2(100•01000) + 2(001•20000) + 3(001•01000) + (102•00000) + 3(011•00000) + 4(100•00000) + 3(001•00000)

sl 5: (010•01100) + (101•11000) + (002•11000) + (010•30000) + 4(010•11000) + (000•10010) + (002•00100) + 2(010•00100) + (012•10000) + (020•10000) + 2(000•11000) + 4(101•10000) + 2(000•00100) + 4(002•10000) + 10(010•10000) + 4(000•10000)

Affine C8 ⊃ C3 + C5 - Level 1 - Class 1

sl 0: (000•00010) **Highest Weight (000000010)**

sl 1: (001•00001) + (001•00100)

sl 2: (002•00010) + (000•10001) + (010•00010) + (000•10100) + (010•01000) + 2(000•00010) + (000•01000)

sl 3: (011•00001) + 2(001•10010) + (011•00100) + (001•11000) + 3(001•00001) + (100•00100) + 4(001•00100) + (100•10000) + (001•10000)

Affine C8 ⊃ C3 + C5 - Level 1 - Class 1

sl 0: (000•01000) **Highest Weight (000001000)**

sl 1: (001•00100) + (001•10000)

sl 2: (010•00010) + (000•10100) + (002•01000) + (010•01000) + (000•00010) + (000•20000) + 2(000•01000) + (010•00000) + (000•00000)

sl 3: (001•10010) + (011•00100) + (100•00001) + 2(001•11000) + (001•00001) + (100•00100) + 4(001•00100) + (011•10000) + (100•10000) + 4(001•10000)

Affine C8 ⊃ C3 + C5 - Level 1 - Class 1

sl 0: (000•00000) **Highest Weight (000100000)**

sl 1: (001•10000)

sl 2: (010•01000) + (000•20000) + (000•01000) + (002•00000) + (010•00000) + (000•00000)

sl 3: (001•11000) + (100•00100) + (001•00100) + (011•10000) + (100•10000) + 4(001•10000)

sl 4: (010•10100) + (010•00010) + (000•02000) + (101•01000) + (000•10100) + (002•20000) + 2(002•01000) + 2(010•20000) + 4(010•01000) + (000•00010) + 3(000•20000) + (020•00000) + 4(000•01000) + (101•00000) + 3(002•00000) + 4(010•00000) + 4(000•00000)

Affine C8 ⊃ C3 + C5 - Level 1 - Class 1

Sl 0: (010•00000) **Highest Weight (010000000)**

Sl 1: (100•10000) + (001•10000)

Sl 2: (010•20000) + (010•01000) + (000•01000) + (101•00000) + (002•00000) + 2(010•00000) + (000•00000)

Sl 3: (100•11000) + (001•11000) + (001•00100) + 2(011•10000) + 3(100•10000) + 4(001•10000)

Sl 4: (010•02000) + (010•10100) + (101•20000) + (101•01000) + (000•10100) + (002•20000) + 2(002•01000) + 4(010•20000) + 5(010•01000) + (000•00010) + (012•00000) + 2(000•20000) + (020•00000) + 3(000•01000) + 3(101•00000) + 3(002•00000) + 7(010•00000) + 3(000•00000)

Affine C8 ⊃ C4 + C4 - Level 1 - Class 0

Sl 0: (0000•0001) **Highest Weight (000000001)**

Sl 1: (0001•0010)

Sl 2: (0002•0001) + (0010•0100) + (0000•1010) + (0000•0001) + (0000•0100)

Sl 3: (0011•0010) + (0001•1001) + (0001•1100) + 3(0001•0010) + (0100•1000) + (0001•1000)

Sl 4: (0020•0001) + (0002•1010) + (0101•0100) + (0010•1010) + 2(0002•0001) + (0000•2001) + 2(0010•0001) + 2(0002•0100) + (0010•2000) + 3(0010•0100) + (0000•0200) + 3(0000•1010) + (1000•0000) + 4(0000•0001) + (0000•2000) + (0010•0000) + 3(0000•0100) + (0000•0000)

Affine C8 ⊃ C4 + C4 - Level 1 - Class 0

Sl 0: (0000•0100) **Highest Weight (000000100)**

Sl 1: (0001•0010) + (0001•1000)

Sl 2: (0010•0001) + (0002•0100) + (0010•0100) + (0000•1010) + (0000•0001) + (0000•2000) + (0010•0000) + 2(0000•0100) + (0000•0000)

Sl 3: (0011•0010) + (0001•1001) + (0100•0010) + (0011•1000) + 2(0001•1100) + 4(0001•0010) + (0100•1000) + 4(0001•1000)

Affine C8 ⊃ C4 + C4 - Level 1 - Class 0

Sl 0: (0000•0000) **Highest Weight (000010000)**

Sl 1: (0001•1000)

Sl 2: (0010•0100) + (0002•0000) + (0000•2000) + (0010•0000) + (0000•0100) + (0000•0000)

Sl 3: (0100•0010) + (0011•1000) + (0001•1100) + (0001•0010) + (0100•1000) + 4(0001•1000)

Affine C8 ⊃ C4 + C4 - Level 1 - Class 1

```
sl 0:  (0000•0010)  Highest Weight (000000010)
sl 1:  (0001•0001) +  (0001•0100)
sl 2:  (0002•0010) +  (0010•0010) +  (0000•1001) +  (0010•1000) +  (0000•1100) + 2(0000•0010)
    +  (0000•1000)
sl 3:  (0011•0001) +  (0011•0100) + 2(0001•1010) +  (0100•0100) + 3(0001•0001) +  (0001•2000)
    + 4(0001•0100) +  (0100•0000) +  (0001•0000)
sl 4:  (0020•0010) +  (0002•1001) +  (0101•0010) +  (0010•1001) +  (0002•1100) + 2(0010•1100)
    + 4(0002•0010) +  (0101•1000) +  (0000•2010) + 5(0010•0010) +  (0000•0110) +  (1000•1000)
    + 3(0000•1001) + 2(0002•1000) + 4(0010•1000) + 4(0000•1100) + 7(0000•0010) + 4(0000•1000)
```

Affine C8 ⊃ C4 + C4 - Level 1 - Class 1

```
sl 0:  (0000•1000)  Highest Weight (000001000)
sl 1:  (0001•0100) +  (0001•0000)
sl 2:  (0010•0010) +  (0002•1000) +  (0010•1000) +  (0000•1100) +  (0000•0010) + 2(0000•1000)
sl 3:  (0011•0100) +  (0100•0001) +  (0001•1010) +  (0100•0100) +  (0001•0001) + 2(0001•2000)
    +  (0011•0000) + 4(0001•0100) +  (0100•0000) + 3(0001•0000)
sl 4:  (0101•0010) +  (0010•1001) +  (0002•1100) +  (0020•1000) + 2(0010•1100) +  (1000•0010)
    + 2(0002•0010) +  (0101•1000) + 4(0010•0010) +  (0000•0110) +  (1000•1000) +  (0000•1001)
    + 4(0002•1000) +  (0000•3000) + 6(0010•1000) + 4(0000•1100) + 4(0000•0010) + 7(0000•1000)
```

Affine D4 Branching Rules

Affine D4 ⊃ D4 - Level 1

```
         • 0                        1 0
sl 0:   ( 0 )   Highest Weight     ( 0 )
         0 0                        0 0

         • 0
sl 1:   ( 1 )
         0 0

         • 0        • 2        • 0        • 0        • 0
sl 2:   ( 0 ) +    ( 0 ) +    ( 0 ) +    ( 1 ) +    ( 0 )
         2 0        0 0        0 2        0 0        0 0

         • 1        • 0        • 2        • 0        • 0        • 0
sl 3:   ( 0 ) +    ( 0 ) +    ( 0 ) +    ( 0 ) +   4( 1 ) +    ( 0 )
         1 1        2 0        0 0        0 2        0 0        0 0

         • 0        • 1        • 0        • 2        • 0        • 0        • 0
sl 4:   ( 2 ) +   3( 0 ) +   3( 0 ) +   3( 0 ) +   3( 0 ) +   5( 1 ) +   4( 0 )
         0 0        1 1        2 0        0 0        0 2        0 0        0 0

         • 0        • 2        • 0        • 0        • 1        • 0        • 2        • 0        • 0        • 0
sl 5:   ( 1 ) +    ( 1 ) +    ( 1 ) +    ( 2 ) +   7( 0 ) +   5( 0 ) +   5( 0 ) +   5( 0 ) +  12( 1 ) +   4( 0 )
         2 0        0 0        0 2        0 0        1 1        2 0        0 0        0 2        0 0        0 0

         • 2        • 0        • 2        • 0        • 2        • 0        • 0        • 1        • 0        • 2
sl 6:   ( 0 ) +    ( 0 ) +    ( 0 ) +   2( 1 ) +   2( 1 ) +   2( 1 ) +   5( 2 ) +  14( 0 ) +  11( 0 ) +  11( 0 )
         2 0        2 2        0 2        2 0        0 0        0 2        0 0        1 1        2 0        0 0

         • 0        • 0        • 0
   +  11( 0 ) +  18( 1 ) +   9( 0 )
         0 2        0 0        0 0

         • 1        • 2        • 0        • 2        • 0        • 2        • 0        • 0        • 1        • 0
sl 7:   ( 1 ) +   2( 0 ) +   2( 0 ) +   2( 0 ) +   6( 1 ) +   6( 1 ) +   6( 1 ) +   8( 2 ) +  29( 0 ) +  17( 0 )
         1 1        2 0        2 2        0 2        2 0        0 0        0 2        0 0        1 1        2 0

         • 2        • 0        • 0        • 0
   +  17( 0 ) +  17( 0 ) +  36( 1 ) +  11( 0 )
         0 0        0 2        0 0        0 0
```

Affine D4 ⊃ D4 - Level 1

```
         • 0                        0 0
sl 0:   ( 0 )   Highest Weight     ( 0 )
         0 1                        0 1

         • 1        • 0
sl 1:   ( 0 ) +    ( 0 )
         1 0        0 1

         • 0        • 1        • 0
sl 2:   ( 1 ) +   2( 0 ) +   2( 0 )
         0 1        1 0        0 1

         • 0        • 2        • 0        • 1        • 0
sl 3:   ( 0 ) +    ( 0 ) +   2( 1 ) +   4( 0 ) +   4( 0 )
         2 1        0 1        0 1        1 0        0 1

         • 1        • 0        • 2        • 0        • 0        • 1        • 0
sl 4:   ( 1 ) +   2( 0 ) +   2( 0 ) +    ( 0 ) +   5( 1 ) +   8( 0 ) +   7( 0 )
         1 0        2 1        0 1        0 3        0 1        1 0        0 1

         • 1        • 1        • 0        • 2        • 0        • 0        • 1        • 0
sl 5:   ( 0 ) +   3( 1 ) +   5( 0 ) +   5( 0 ) +    ( 0 ) +  10( 1 ) +  15( 0 ) +  12( 0 )
         1 2        1 0        2 1        0 1        0 3        0 1        1 0        0 1

         • 0        • 1        • 3        • 1        • 1        • 0        • 2        • 0        • 0        • 1
sl 6:   ( 2 ) +    ( 0 ) +    ( 0 ) +   3( 0 ) +   7( 1 ) +  10( 0 ) +  10( 0 ) +   3( 0 ) +  20( 1 ) +  27( 0 )
         0 1        3 0        1 0        1 2        1 0        2 1        0 1        0 3        0 1        1 0

         • 0
   +  21( 0 )
         0 1
```

(continued next page)

```
           • 0        • 2          • 0          • 1          • 3          • 1           • 1           • 0           • 2          • 0
sl 7:     ( 1 ) +    ( 1 ) +    2( 2 ) +     2( 0 ) +     2( 0 ) +     8( 0 ) +    16( 1 ) +     20( 0 ) +     20( 0 ) +     6( 0 )
           2 1        0 1          0 1          3 0          1 0          1 2           1 0           2 1           0 1          0 3

            • 0         • 1         • 0
   + 36( 1 ) + 48( 0 ) + 34( 0 )
            0 1         1 0         0 1
```

Affine D4 ⊃ D4 - Level 2

```
          • 0                        0 0
sl 0:    ( 0 )   Highest Weight     ( 0 )
          0 2                        0 2

          • 1         • 0         • 0
sl 1:    ( 0 ) +     ( 0 ) +     ( 1 )
          1 1         0 2         0 0

          • 2         • 0         • 0          • 1          • 0         • 2          • 0          • 0          • 0
sl 2:    ( 0 ) +     ( 1 ) +     ( 2 ) +    3( 0 ) +     ( 0 ) +     ( 0 ) +    3( 0 ) +     2( 1 ) +     ( 0 )
          2 0         0 2         0 0          1 1          2 0         0 0          0 2          0 0          0 0

          • 1          • 2        • 0         • 2          • 0          • 2          • 0          • 0          • 1          • 0
sl 3:    ( 1 ) +    2( 0 ) +     ( 0 ) +     ( 0 ) +    2( 1 ) +     2( 1 ) +     3( 1 ) +     2( 2 ) +     9( 0 ) +     3( 0 )
          1 1          2 0        2 2         0 2          2 0          0 0          0 2          0 0          1 1          2 0

          • 2          • 0          • 0         • 0
   +    3( 0 ) +    6( 0 ) +    7( 1 ) +     ( 0 )
          0 0          0 2          0 0         0 0

          • 0         • 1         • 3          • 1          • 2          • 0          • 2          • 0          • 0          • 2
sl 4:    ( 2 ) +     ( 0 ) +     ( 0 ) +    5( 1 ) +     7( 0 ) +     4( 0 ) +     4( 0 ) +     ( 0 ) +     6( 1 ) +     6( 1 )
          0 2         3 1         1 1          1 1          2 0          2 2          0 2          0 4          2 0          0 0

          • 0          • 0           • 1          • 0          • 2           • 0           • 0          • 0
   +    9( 1 ) +    9( 2 ) +    24( 0 ) +    9( 0 ) +     9( 0 ) +    15( 0 ) +    15( 1 ) +     4( 0 )
          0 2          0 0           1 1          2 0          0 0           0 2           0 0          0 0
```

Affine D4 ⊃ D4 - Level 2

```
          • 0                        0 0
sl 0:    ( 1 )   Highest Weight     ( 1 )
          0 0                        0 0

          • 1         • 0         • 2         • 0         • 0         • 0
sl 1:    ( 0 ) +     ( 0 ) +     ( 0 ) +     ( 0 ) +     ( 1 ) +     ( 0 )
          1 1         2 0         0 0         0 2         0 0         0 0

          • 0         • 2         • 0         • 0          • 1          • 0          • 2          • 0          • 0         • 0
sl 2:    ( 1 ) +     ( 1 ) +     ( 1 ) +     ( 2 ) +    4( 0 ) +     2( 0 ) +     2( 0 ) +     2( 0 ) +     5( 1 ) +     ( 0 )
          2 0         0 0         0 2         0 0          1 1          2 0          0 0          0 2          0 0         0 0

          • 1          • 2          • 0          • 2          • 0          • 2          • 0          • 0           • 1          • 0
sl 3:    ( 1 ) +    2( 0 ) +     2( 0 ) +     2( 0 ) +     3( 1 ) +     3( 1 ) +     3( 1 ) +     5( 2 ) +    13( 0 ) +     7( 0 )
          1 1          2 0          2 2          0 2          2 0          0 0          0 2          0 0           1 1          2 0

          • 2          • 0           • 0          • 0
   +    7( 0 ) +    7( 0 ) +    10( 1 ) +    4( 0 )
          0 0          0 2           0 0          0 0

          • 1         • 3         • 1         • 0          • 1         • 0          • 2          • 0         • 4          • 2
sl 4:    ( 0 ) +     ( 0 ) +     ( 0 ) +     ( 3 ) +    7( 1 ) +     ( 0 ) +    6( 0 ) +     6( 0 ) +     ( 0 ) +    6( 0 )
          3 1         1 1         1 3         0 0          1 1         4 0          2 0          2 2         0 0          0 2

          • 0           • 0           • 2           • 0           • 0           • 1           • 0           • 2           • 0           • 0
   +     ( 0 ) +   12( 1 ) +    12( 1 ) +    12( 1 ) +    13( 2 ) +    37( 0 ) +    16( 0 ) +    16( 0 ) +    16( 0 ) +    29( 1 )
          0 4           2 0           0 0           0 2           0 0           1 1           2 0           0 0           0 2           0 0

          • 0
   +    6( 0 )
          0 0
```

Affine D4 ⊃ D4 - Level 2

```
           • 0                   2 0
Sl 0:      ( 0 )  Highest Weight ( 0 )
           0 0                   0 0

           • 0
Sl 1:      ( 1 )
           0 0

           • 0         • 0         • 2         • 0         • 0         • 0
Sl 2:      ( 2 ) +     ( 0 ) +     ( 0 ) +     ( 0 ) +     ( 1 ) +     ( 0 )
           0 0         2 0         0 0         0 2         0 0         0 0

           • 0         • 2         • 0         • 0         • 1         • 0         • 2         • 0         • 0         • 0
Sl 3:      ( 1 ) +     ( 1 ) +     ( 1 ) +     ( 2 ) +   2( 0 ) +     ( 0 ) +     ( 0 ) +     ( 0 ) +   5( 1 ) +     ( 0 )
           2 0         0 0         0 2         0 0         1 1         2 0         0 0         0 2         0 0         0 0

           • 1         • 0         • 2         • 0         • 4         • 2         • 0         • 0         • 2         • 0
Sl 4:      ( 1 ) +     ( 0 ) +     ( 0 ) +     ( 0 ) +     ( 0 ) +     ( 0 ) +     ( 0 ) +   2( 1 ) +   2( 1 ) +   2( 1 )
           1 1         4 0         2 0         2 2         0 0         0 2         0 4         2 0         0 0         0 2

           • 0         • 1         • 0         • 2         • 0         • 0         • 0
    +    6( 2 ) +   7( 0 ) +   5( 0 ) +   5( 0 ) +   5( 0 ) +   7( 1 ) +   5( 0 )
           0 0         1 1         2 0         0 0         0 2         0 0         0 0

           • 1         • 3         • 1         • 0         • 1         • 0         • 2         • 0         • 4         • 2
Sl 5:      ( 0 ) +     ( 0 ) +     ( 0 ) +     ( 3 ) +   5( 1 ) +     ( 0 ) +   3( 0 ) +   3( 0 ) +     ( 0 ) +   3( 0 )
           3 1         1 1         1 3         0 0         1 1         4 0         2 0         2 2         0 0         0 2

           • 0         • 0         • 2         • 0         • 0         • 1         • 0         • 2         • 0         • 0
    +      ( 0 ) +   9( 1 ) +   9( 1 ) +   9( 1 ) +  10( 2 ) +  22( 0 ) +   9( 0 ) +   9( 0 ) +   9( 0 ) +  22( 1 )
           0 4         2 0         0 0         0 2         0 0         1 1         2 0         0 0         0 2         0 0

           • 0
    +    5( 0 )
           0 0
```

Affine D4 ⊃ D4 - Level 2

```
           • 1                   0 1
Sl 0:      ( 0 )  Highest Weight ( 0 )
           1 0                   1 0

           • 0         • 2         • 0         • 1         • 0
Sl 1:      ( 0 ) +     ( 0 ) +     ( 1 ) +   2( 0 ) +     ( 0 )
           2 1         0 1         0 1         1 0         0 1

           • 1         • 1         • 0         • 2         • 0         • 1         • 0
Sl 2:      ( 0 ) +   2( 1 ) +   3( 0 ) +   3( 0 ) +   4( 1 ) +   6( 0 ) +   3( 0 )
           1 2         1 0         2 1         0 1         0 1         1 0         0 1

           • 0         • 2         • 0         • 1         • 3         • 1         • 1         • 0         • 2         • 0
Sl 3:      ( 1 ) +     ( 1 ) +     ( 2 ) +     ( 0 ) +     ( 0 ) +   5( 0 ) +   8( 1 ) +   9( 0 ) +   9( 0 ) +   2( 0 )
           2 1         0 1         0 1         3 0         1 0         1 2         1 0         2 1         0 1         0 3

           • 0         • 1         • 0
    +   12( 1 ) +  16( 0 ) +   8( 0 )
           0 1         1 0         0 1

           • 1         • 2         • 0         • 2         • 0         • 2         • 0         • 0         • 1         • 3
Sl 4:      ( 2 ) +   2( 0 ) +     ( 0 ) +     ( 0 ) +   5( 1 ) +   5( 1 ) +     ( 1 ) +   6( 2 ) +   5( 0 ) +   5( 0 )
           1 0         2 1         2 3         0 3         2 1         0 1         0 3         0 1         3 0         1 0

           • 1         • 1         • 0         • 2         • 0         • 0         • 1         • 0
    +   17( 0 ) +  25( 1 ) +  25( 0 ) +  25( 0 ) +   6( 0 ) +  34( 1 ) +  39( 0 ) +  19( 0 )
           1 2         1 0         2 1         0 1         0 3         0 1         1 0         0 1

           • 1         • 3         • 1         • 1         • 0         • 2         • 0         • 4         • 2         • 0
Sl 5:      ( 1 ) +     ( 1 ) +   3( 1 ) +   5( 2 ) +     ( 0 ) +  10( 0 ) +   4( 0 ) +     ( 0 ) +   4( 0 ) +  19( 1 )
           3 0         1 0         1 2         1 0         4 1         2 1         2 3         0 1         0 3         2 1

           • 2         • 0         • 0         • 1         • 3         • 1         • 1         • 0         • 2         • 0
    +   19( 1 ) +   5( 1 ) +  21( 2 ) +  16( 0 ) +  16( 0 ) +  51( 0 ) +  72( 1 ) +  63( 0 ) +  63( 0 ) +  18( 0 )
           0 1         0 3         0 1         3 0         1 0         1 2         1 0         2 1         0 1         0 3

           • 0         • 1         • 0
    +   86( 1 ) +  90( 0 ) +  43( 0 )
           0 1         1 0         0 1
```

Affine D4 ⊃ D4 - Level 2

```
         • 0                       1 0
Sl 0:   ( 0 )   Highest Weight   ( 0 )
         0 1                       0 1

         • 0       • 1       • 0
Sl 1:   ( 1 ) +   ( 0 ) +   ( 0 )
         0 1       1 0       0 1

         • 1       • 0       • 2       • 0       • 0       • 1       • 0
Sl 2:   ( 1 ) +   ( 0 ) +   ( 0 ) +   ( 0 ) +  3( 1 ) +  3( 0 ) +  3( 0 )
         1 0       2 1       0 1       0 3       0 1       1 0       0 1

         • 0       • 1       • 3       • 1       • 1       • 0       • 2       • 0       • 0       • 1
Sl 3:   ( 2 ) +   ( 0 ) +   ( 0 ) +  2( 0 ) +  4( 1 ) +  4( 0 ) +  4( 0 ) +  2( 0 ) +  9( 1 ) +  9( 0 )
         0 1       3 0       1 0       1 2       1 0       2 1       0 1       0 3       0 1       1 0

         • 0
    +  7( 0 )
         0 1

         • 2       • 0       • 2       • 0       • 0       • 1       • 3       • 1       • 1       • 0
Sl 4:   ( 0 ) +  2( 1 ) +  2( 1 ) +   ( 1 ) +  4( 2 ) +  3( 0 ) +  3( 0 ) +  8( 0 ) + 15( 1 ) + 13( 0 )
         2 1       2 1       0 1       0 3       0 1       3 0       1 0       1 2       1 0       2 1

         • 2       • 0       • 0       • 1       • 0
    + 13( 0 ) +  6( 0 ) + 24( 1 ) + 23( 0 ) + 16( 0 )
         0 1       0 3       0 1       1 0       0 1

         • 1       • 1       • 0       • 2       • 0       • 4       • 2       • 0       • 2       • 0
Sl 5:   ( 1 ) +  2( 2 ) +   ( 0 ) +  6( 0 ) +   ( 0 ) +   ( 0 ) +   ( 0 ) +  9( 1 ) +  9( 1 ) +  4( 1 )
         1 2       1 0       4 1       2 1       2 3       0 1       0 3       2 1       0 1       0 3

         • 0       • 1       • 3       • 1       • 1       • 0       • 2       • 0       • 0       • 1
    + 14( 2 ) + 10( 0 ) + 10( 0 ) + 28( 0 ) + 44( 1 ) + 36( 0 ) + 36( 0 ) + 15( 0 ) + 60( 1 ) + 57( 0 )
         0 1       3 0       1 0       1 2       1 0       2 1       0 1       0 3       0 1       1 0

         • 0
    + 34( 0 )
         0 1

         • 1       • 3       • 0       • 1       • 3       • 1       • 1       • 0       • 2       • 0
Sl 6:   ( 0 ) +   ( 0 ) +   ( 3 ) +  2( 1 ) +  2( 1 ) +  7( 1 ) + 10( 2 ) +  3( 0 ) + 22( 0 ) +  6( 0 )
         3 2       1 2       0 1       3 0       1 0       1 2       1 0       4 1       2 1       2 3

         • 4       • 2       • 0       • 0       • 2       • 0       • 0       • 1       • 3       • 1
    +  3( 0 ) +  6( 0 ) +   ( 0 ) + 32( 1 ) + 32( 1 ) + 13( 1 ) + 42( 2 ) + 30( 0 ) + 30( 0 ) + 81( 0 )
         0 1       0 3       0 5       2 1       0 1       0 3       0 1       3 0       1 0       1 2

         • 1       • 0       • 2       • 0       • 0       • 1       • 0
    +118( 1 ) + 92( 0 ) + 92( 0 ) + 37( 0 ) +142( 1 ) +130( 0 ) + 73( 0 )
         1 0       2 1       0 1       0 3       0 1       1 0       0 1
```

Affine D4 ⊃ D4 - Level 3

```
         • 1                       0 1
Sl 0:   ( 0 )   Highest Weight   ( 0 )
         1 1                       1 1

         • 2       • 0       • 2       • 0       • 2       • 0       • 0       • 1       • 0       • 2
Sl 1:   ( 0 ) +   ( 0 ) +   ( 0 ) +   ( 1 ) +   ( 1 ) +   ( 1 ) +   ( 2 ) +  3( 0 ) +   ( 0 ) +   ( 0 )
         2 0       2 2       0 2       2 0       0 0       0 2       0 0       1 1       2 0       0 0

         • 0       • 0
    +   ( 0 ) +   ( 1 )
         0 2       0 0

         • 1       • 3       • 1       • 1       • 2       • 0       • 2       • 0       • 2       • 0
Sl 2:   ( 0 ) +   ( 0 ) +   ( 0 ) +  4( 1 ) +  4( 0 ) +  4( 0 ) +  4( 0 ) +  5( 1 ) +  5( 1 ) +  5( 1 )
         3 1       1 1       1 3       1 1       2 0       2 2       0 2       2 0       0 0       0 2

         • 0       • 1       • 0       • 2       • 0       • 0
    +  4( 2 ) + 13( 0 ) +  4( 0 ) +  4( 0 ) +  4( 0 ) +  5( 1 )
         0 0       1 1       2 0       0 0       0 2       0 0   (continued next page)
```

Sl 3: (•2; 0; 2 2) + 2(•2; 1; 2 0) + 2(•0; 1; 2 2) + 2(•2; 1; 0 2) + 2(•0; 2; 2 0) + 2(•2; 2; 0 0) + 2(•0; 2; 0 2) + 6(•1; 0; 3 1) + 6(•3; 0; 1 1) + 6(•1; 0; 1 3)
+ (•0; 3; 0 0) + 22(•1; 1; 1 1) + 2(•0; 0; 4 0) + 16(•2; 0; 2 0) + 16(•0; 0; 2 2) + 2(•4; 0; 0 0) + 16(•2; 0; 0 2) + 2(•0; 0; 0 4) + 19(•0; 1; 2 0) + 19(•2; 1; 0 0)
+ 19(•0; 1; 0 2) + 18(•0; 2; 0 0) + 45(•1; 0; 1 1) + 14(•0; 0; 2 0) + 14(•2; 0; 0 0) + 14(•0; 0; 0 2) + 17(•0; 1; 0 0) + 2(•0; 0; 0 0)

Affine D4 ⊃ D4 - Level 3

Sl 0: (•0; 0; 0 2) **Highest Weight** (1 0; 0; 0 2)

Sl 1: (•0; 1; 0 2) + (•1; 0; 1 1) + (•0; 0; 0 2) + (•0; 1; 0 0)

Sl 2: (•1; 1; 1 1) + (•2; 0; 2 0) + (•0; 0; 2 2) + (•2; 0; 0 2) + (•0; 0; 0 4) + 3(•0; 1; 0 2) + 2(•0; 2; 0 0) + 4(•1; 0; 1 1) + (•0; 0; 2 0) + (•2; 0; 0 0)
+ 4(•0; 0; 0 2) + 2(•0; 1; 0 0) + (•0; 0; 0 0)

Sl 3: (•2; 1; 2 0) + (•0; 2; 0 2) + (•1; 0; 3 1) + (•3; 0; 1 1) + 2(•1; 0; 1 3) + (•0; 3; 0 0) + 6(•1; 1; 1 1) + 3(•2; 0; 2 0) + 4(•0; 0; 2 2) + 4(•2; 0; 0 2)
+ 2(•0; 0; 0 4) + 4(•0; 1; 2 0) + 4(•2; 1; 0 0) + 11(•0; 1; 0 2) + 5(•0; 2; 0 0) + 15(•1; 0; 1 1) + 3(•0; 0; 2 0) + 3(•2; 0; 0 0) + 9(•0; 0; 0 2) + 9(•0; 1; 0 0)
+ (•0; 0; 0 0)

Affine D4 ⊃ D4 - Level 3

Sl 0: (•0; 1; 0 0) **Highest Weight** (1 0; 1; 0 0)

Sl 1: (•0; 2; 0 0) + (•1; 0; 1 1) + (•0; 0; 2 0) + (•2; 0; 0 0) + (•0; 0; 0 2) + (•0; 1; 0 0) + (•0; 0; 0 0)

Sl 2: (•1; 1; 1 1) + 2(•0; 1; 2 0) + 2(•2; 1; 0 0) + 2(•0; 1; 0 2) + 2(•0; 2; 0 0) + 5(•1; 0; 1 1) + 2(•0; 0; 2 0) + 2(•2; 0; 0 0) + 2(•0; 0; 0 2) + 6(•0; 1; 0 0)
+ (•0; 0; 0 0)

Sl 3: (•0; 2; 2 0) + (•2; 2; 0 0) + (•0; 2; 0 2) + (•1; 0; 3 1) + (•3; 0; 1 1) + (•1; 0; 1 3) + (•0; 3; 0 0) + 6(•1; 1; 1 1) + (•0; 0; 4 0) + 4(•2; 0; 2 0)
+ 4(•0; 0; 2 2) + (•4; 0; 0 0) + 4(•2; 0; 0 2) + (•0; 0; 0 4) + 6(•0; 1; 2 0) + 6(•2; 1; 0 0) + 6(•0; 1; 0 2) + 11(•0; 2; 0 0) + 18(•1; 0; 1 1) + 9(•0; 0; 2 0)
+ 9(•2; 0; 0 0) + 9(•0; 0; 0 2) + 12(•0; 1; 0 0) + 5(•0; 0; 0 0)

Affine D4 $\supset$ D4 - Level 3

```
         • 0                     3 0
sl 0:   ( 0 )   Highest Weight  ( 0 )
         0 0                     0 0

         • 0
sl 1:   ( 1 )
         0 0

         • 0        • 0        • 2        • 0        • 0        • 0
sl 2:   ( 2 ) +    ( 0 ) +    ( 0 ) +    ( 0 ) +    ( 1 ) +    ( 0 )
         0 0        2 0        0 0        0 2        0 0        0 0

         • 0        • 0        • 2        • 0        • 0        • 1        • 0        • 2        • 0        • 0
sl 3:   ( 3 ) +    ( 1 ) +    ( 1 ) +    ( 1 ) +    ( 2 ) +   2( 0 ) +    ( 0 ) +    ( 0 ) +    ( 0 ) +   5( 1 )
         0 0        2 0        0 0        0 2        0 0        1 1        2 0        0 0        0 2        0 0

         • 0
     +  ( 0 )
         0 0
```

Affine D4 $\supset$ D4 - Level 3

```
         • 0                     0 0
sl 0:   ( 0 )   Highest Weight  ( 0 )
         0 3                     0 3

         • 1        • 0        • 0
sl 1:   ( 0 ) +    ( 0 ) +    ( 1 )
         1 2        0 3        0 1

         • 2        • 0        • 0        • 1        • 1        • 0        • 2        • 0        • 0        • 0
sl 2:   ( 0 ) +    ( 1 ) +    ( 2 ) +   3( 0 ) +    ( 1 ) +    ( 0 ) +    ( 0 ) +   3( 0 ) +   3( 1 ) +    ( 0 )
         2 1        0 3        0 1        1 2        1 0        2 1        0 1        0 3        0 1        0 1

         • 3        • 1        • 1        • 2        • 0        • 2        • 0        • 2        • 0        • 0
sl 3:   ( 0 ) +    ( 1 ) +    ( 2 ) +   3( 0 ) +    ( 0 ) +    ( 0 ) +   2( 1 ) +   2( 1 ) +   3( 1 ) +   3( 2 )
         3 0        1 2        1 0        2 1        2 3        0 3        2 1        0 1        0 3        0 1

         • 1        • 3        • 1        • 1        • 0        • 2        • 0        • 0        • 1        • 0
     +  ( 0 ) +    ( 0 ) +  10( 0 ) +   5( 1 ) +   4( 0 ) +   4( 0 ) +   7( 0 ) +   9( 1 ) +   3( 0 ) +   2( 0 )
         3 0        1 0        1 2        1 0        2 1        0 1        0 3        0 1        1 0        0 1
```

Affine D4 $\supset$ D4 - Level 3

```
         • 2                     0 2
sl 0:   ( 0 )   Highest Weight  ( 0 )
         0 1                     0 1

         • 3        • 1        • 1        • 2        • 0        • 1
sl 1:   ( 0 ) +    ( 0 ) +    ( 1 ) +   2( 0 ) +    ( 1 ) +    ( 0 )
         1 0        1 2        1 0        0 1        0 1        1 0

         • 2        • 0        • 0        • 2        • 0        • 3        • 1        • 1        • 0        • 2
sl 2:   ( 0 ) +    ( 0 ) +    ( 1 ) +   2( 1 ) +    ( 2 ) +   3( 0 ) +   4( 0 ) +   5( 1 ) +   2( 0 ) +   7( 0 )
         2 1        2 3        2 1        0 1        0 1        1 0        1 2        1 0        2 1        0 1

         • 0        • 0        • 1        • 0
     +  ( 0 ) +   4( 1 ) +   4( 0 ) +    ( 0 )
         0 3        0 1        1 0        0 1

         • 1        • 3        • 1        • 1        • 2        • 0        • 4        • 2        • 0        • 2
sl 3:   ( 0 ) +    ( 1 ) +   2( 1 ) +   2( 2 ) +   6( 0 ) +   3( 0 ) +    ( 0 ) +    ( 0 ) +   6( 1 ) +   9( 1 )
         3 2        1 0        1 2        1 0        2 1        2 3        0 1        0 3        2 1        0 1

         • 0        • 0        • 1        • 3        • 1        • 1        • 0        • 2        • 0        • 0
     + 2( 1 ) +   6( 2 ) +   3( 0 ) +  10( 0 ) +  16( 0 ) +  19( 1 ) +   9( 0 ) +  20( 0 ) +   4( 0 ) +  16( 1 )
         0 3        0 1        3 0        1 0        1 2        1 0        2 1        0 1        0 3        0 1

         • 1        • 0
     +14( 0 ) +   4( 0 )
         1 0        0 1
```

Affine D4 ⊃ D4 - Level 3

```
          • 0                    0 0
Sl 0:    ( 1 )  Highest Weight  ( 1 )
          0 1                    0 1

          • 1       • 1       • 0       • 2       • 0       • 0       • 1       • 0
Sl 1:    ( 0 ) +   ( 1 ) +   ( 0 ) +   ( 0 ) +   ( 0 ) +  2( 1 ) +   ( 0 ) +   ( 0 )
          1 2       1 0       2 1       0 1       0 3       0 1       1 0       0 1

          • 2       • 0       • 2       • 0       • 0       • 1       • 3       • 1       • 1       • 0
Sl 2:    ( 0 ) +   ( 1 ) +   ( 1 ) +   ( 1 ) +  2( 2 ) +   ( 0 ) +   ( 0 ) +  5( 0 ) +  5( 1 ) +  4( 0 )
          2 1       2 1       0 1       0 3       0 1       3 0       1 0       1 2       1 0       2 1

          • 2       • 0       • 0       • 1       • 0
    +   4( 0 ) +  3( 0 ) +  8( 1 ) +  5( 0 ) +  3( 0 )
          0 1       0 3       0 1       1 0       0 1

          • 1       • 3       • 1       • 1       • 2       • 0       • 2       • 0       • 2       • 0
Sl 3:    ( 1 ) +   ( 1 ) +  2( 1 ) +  2( 2 ) +  6( 0 ) +  2( 0 ) +  2( 0 ) +  7( 1 ) +  7( 1 ) +  4( 1 )
          3 0       1 0       1 2       1 0       2 1       2 3       0 3       2 1       0 1       0 3

          • 0       • 1       • 3       • 1       • 1       • 0       • 2       • 0       • 0       • 1
    +   9( 2 ) +  5( 0 ) +  5( 0 ) + 20( 0 ) + 21( 1 ) + 16( 0 ) + 16( 0 ) +  9( 0 ) + 25( 1 ) + 17( 0 )
          0 1       3 0       1 0       1 2       1 0       2 1       0 1       0 3       0 1       1 0

          • 0
    +   9( 0 )
          0 1
```

Affine D4 ⊃ D4 - Level 3

```
          • 1                    1 1
Sl 0:    ( 0 )  Highest Weight  ( 0 )
          1 0                    1 0

          • 1       • 0       • 2       • 0       • 1       • 0
Sl 1:    ( 1 ) +   ( 0 ) +   ( 0 ) +   ( 1 ) +  2( 0 ) +   ( 0 )
          1 0       2 1       0 1       0 1       1 0       0 1

          • 0       • 2       • 0       • 1       • 3       • 1       • 1       • 0       • 2       • 0
Sl 2:    ( 1 ) +   ( 1 ) +   ( 2 ) +   ( 0 ) +   ( 0 ) +  2( 0 ) +  5( 1 ) +  4( 0 ) +  4( 0 ) +  5( 1 )
          2 1       0 1       0 1       3 0       1 0       1 2       1 0       2 1       0 1       0 1

          • 1       • 0
    +   7( 0 ) +  3( 0 )
          1 0       0 1

          • 1       • 1       • 0       • 2       • 0       • 4       • 2       • 0       • 2       • 0
Sl 3:    ( 1 ) +  2( 2 ) +   ( 0 ) +  3( 0 ) +   ( 0 ) +   ( 0 ) +   ( 0 ) +  6( 1 ) +  6( 1 ) +   ( 1 )
          1 2       1 0       4 1       2 1       2 3       0 1       0 3       2 1       0 1       0 3

          • 0       • 1       • 3       • 1       • 1       • 0       • 2       • 0       • 0       • 1
    +   6( 2 ) +  5( 0 ) +  5( 0 ) + 11( 0 ) + 20( 1 ) + 15( 0 ) + 15( 0 ) +  3( 0 ) + 18( 1 ) + 21( 0 )
          0 1       3 0       1 0       1 2       1 0       2 1       0 1       0 3       0 1       1 0

          • 0
    +   9( 0 )
          0 1
```

Affine D4 ⊃ D4 - Level 3

```
          • 0                    2 0
Sl 0:    ( 0 )  Highest Weight  ( 0 )
          0 1                    0 1

          • 0       • 1       • 0
Sl 1:    ( 1 ) +   ( 0 ) +   ( 0 )
          0 1       1 0       0 1

          • 0       • 1       • 0       • 2       • 0       • 0       • 1       • 0
Sl 2:    ( 2 ) +   ( 1 ) +   ( 0 ) +   ( 0 ) +   ( 0 ) +  3( 1 ) +  3( 0 ) +  3( 0 )
          0 1       1 0       2 1       0 1       0 3       0 1       1 0       0 1
```

(continued next page)

```
         • 1         • 0         • 2         • 0         • 0         • 1         • 3         • 1         • 1         • 0
sl 3:    ( 2 ) +    ( 1 ) +    ( 1 ) +    ( 1 ) +   3( 2 ) +    ( 0 ) +    ( 0 ) +   3( 0 ) +   5( 1 ) +   4( 0 )
          1 0         2 1         0 1         0 3         0 1         3 0         1 0         1 2         1 0         2 1

         • 2         • 0         • 0         • 1         • 0
    +   4( 0 ) +   2( 0 ) +  10( 1 ) +   9( 0 ) +   7( 0 )
          0 1         0 3         0 1         1 0         0 1
```

Affine D4 ⊃ A1 + A1 + A1 + A1 - Level 1

```
          1 0                        1 0
sl 0:    ( • )   Highest Weight     ( 0 )
          0 0                        0 0

          0 1
sl 1:    ( • )
          1 1

          1 0         1 2         1 0         1 0
sl 2:    ( • ) +    ( • ) +    ( • ) +    ( • )
          2 0         0 0         0 2         0 0

          2 1         0 1
sl 3:    ( • ) +   3( • )
          1 1         1 1

          1 2         1 0         1 2         3 0         1 0         1 2         1 0         1 0
sl 4:    ( • ) +    ( • ) +    ( • ) +    ( • ) +   2( • ) +   2( • ) +   2( • ) +   4( • )
          2 0         2 2         0 2         0 0         2 0         0 0         0 2         0 0

          2 1         0 1         0 3         0 1         0 1
sl 5:   4( • ) +    ( • ) +    ( • ) +    ( • ) +   7( • )
          1 1         3 1         1 1         1 3         1 1

          1 2         3 0         3 2         3 0         1 2         1 0         1 2         3 0         1 0         1 2
sl 6:    ( • ) +    ( • ) +    ( • ) +    ( • ) +   3( • ) +   3( • ) +   3( • ) +    ( • ) +   6( • ) +   6( • )
          2 2         2 0         0 0         0 2         2 0         2 2         0 2         0 0         2 0         0 0

          1 0         1 0
    +   6( • ) +   8( • )
          0 2         0 0

          2 1         2 3         2 1         2 1         0 1         0 3         0 1         0 1
sl 7:    ( • ) +    ( • ) +    ( • ) +  11( • ) +   3( • ) +   3( • ) +   3( • ) +  17( • )
          3 1         1 1         1 3         1 1         3 1         1 1         1 3         1 1

          3 2         3 0         3 2         1 2         3 0         3 2         3 0         1 0         1 2         1 0
sl 8:    ( • ) +    ( • ) +    ( • ) +   4( • ) +   2( • ) +   2( • ) +   2( • ) +    ( • ) +   9( • ) +   9( • )
          2 0         2 2         0 2         2 2         2 0         0 0         0 2         4 0         2 0         2 2

          1 4         1 2         1 0         3 0         1 0         1 2         1 0         1 0
    +    ( • ) +   9( • ) +    ( • ) +   5( • ) +  13( • ) +  13( • ) +  13( • ) +  18( • )
          0 0         0 2         0 4         0 0         2 0         0 0         0 2         0 0
```

Affine D4 ⊃ A1 + A1 + A1 + A1 - Level 2

```
          0 0                        0 0
sl 0:    ( • )   Highest Weight     ( 0 )
          0 2                        0 2

          1 1
sl 1:    ( • )
          1 1

          2 2         2 0         0 0         0 2         2 0         0 0         0 2         0 0         0 0
sl 2:    ( • ) +    ( • ) +    ( • ) +    ( • ) +    ( • ) +    ( • ) +    ( • ) +    ( • ) +    ( • )
          2 0         0 2         2 2         0 2         0 0         2 0         0 0         0 2         0 0

          3 1         1 1         1 3         1 1         1 1
sl 3:    ( • ) +    ( • ) +    ( • ) +    ( • ) +   5( • )
          1 1         3 1         1 1         1 3         1 1

          2 2         4 0         2 2         2 0         2 2         0 0         0 2         0 4         2 0         2 2
sl 4:    ( • ) +    ( • ) +   3( • ) +   2( • ) +   2( • ) +    ( • ) +   2( • ) +    ( • ) +   3( • ) +   3( • )
          2 2         0 2         2 0         2 2         0 2         4 2         2 2         0 2         2 0         0 0

          2 0         0 2         0 0         0 2         0 0         2 0         0 0         0 2         0 0         0 0
    +   3( • ) +   3( • ) +   3( • ) +   3( • ) +    ( • ) +   3( • ) +   3( • ) +   3( • ) +   6( • ) +   4( • )
          0 2         2 0         2 2         0 2         0 4         0 0         2 0         0 0         0 2         0 0
```

(continued next page)

```
        3 1        3 3        3 1        1 3        1 1        1 3        3 1        1 1        1 3        1 1
Sl 5:   ( • ) +    ( • ) +    ( • ) +    ( • ) +    ( • ) +    ( • ) +  6( • ) +  6( • ) +  6( • ) +  6( • )
        3 1        1 1        1 3        3 1        3 3        1 3        1 1        3 1        1 1        1 3

        1 1
   + 21( • )
        1 1
```

Affine D4 ⊃ A1 + A1 + A1 + A1 - Level 2

```
        0 0                       0 0
Sl 0:   ( • )  Highest Weight   ( 1 )
        0 0                       0 0

        1 1
Sl 1:   ( • )
        1 1

        2 0        2 2        2 0        0 2        0 0        0 2        2 0        0 0        0 2        0 0
Sl 2:   ( • ) +    ( • ) +    ( • ) +    ( • ) +    ( • ) +    ( • ) +    ( • ) +    ( • ) +    ( • ) +    ( • )
        2 0        0 0        0 2        2 0        2 2        0 2        0 0        2 0        0 0        0 2

        0 0
    +   ( • )
        0 0

        3 1        1 1        1 3        1 1        1 1
Sl 3:   ( • ) +    ( • ) +    ( • ) +    ( • ) +  7( • )
        1 1        3 1        1 1        1 3        1 1

        2 2        2 2        2 0        2 2        0 2        4 0        2 0        2 2        2 0        0 0
Sl 4:   ( • ) +  3( • ) +  3( • ) +  3( • ) +  3( • ) +    ( • ) +  4( • ) +  4( • ) +  4( • ) +    ( • )
        2 2        2 0        2 2        0 2        2 2        0 0        2 0        0 0        0 2        4 0

        0 2        0 0        0 4        0 2        0 0        2 0        0 0        0 2        0 0        0 0
    +  4( • ) +  4( • ) +    ( • ) +  4( • ) +    ( • ) +  5( • ) +  5( • ) +  5( • ) +  5( • ) +  7( • )
        2 0        2 2        0 0        0 2        0 4        0 0        2 0        0 0        0 2        0 0

        3 1        3 3        3 1        1 3        1 1        1 3        3 1        1 1        1 3        1 1
Sl 5:   ( • ) +    ( • ) +    ( • ) +    ( • ) +    ( • ) +    ( • ) +  8( • ) +  8( • ) +  8( • ) +  8( • )
        3 1        1 1        1 3        3 1        3 3        1 3        1 1        3 1        1 1        1 3

        1 1
   + 30( • )
        1 1
```

Affine D4 ⊃ A1 + A1 + A1 + A1 - Level 2

```
        0 1                       0 1
Sl 0:   ( • )  Highest Weight   ( 0 )
        1 0                       1 0

        1 0        1 2        1 0
Sl 1:   ( • ) +    ( • ) +    ( • )
        2 1        0 1        0 1

        2 1        2 1        0 1        0 3        0 1        0 1
Sl 2:   ( • ) +  2( • ) +    ( • ) +    ( • ) +  2( • ) +  3( • )
        1 2        1 0        3 0        1 0        1 2        1 0

        3 0        3 2        1 2        1 0        1 2        3 0        1 0        1 2        1 0        1 0
Sl 3:   ( • ) +    ( • ) +  3( • ) +    ( • ) +    ( • ) +    ( • ) +  5( • ) +  5( • ) +    ( • ) +  6( • )
        2 1        0 1        2 1        2 3        0 3        0 1        2 1        0 1        0 3        0 1

        2 1        2 3        4 1        2 1        2 3        2 1        0 3        0 1        0 3        0 1
Sl 4:   ( • ) +    ( • ) +    ( • ) +  2( • ) +  2( • ) +  8( • ) +    ( • ) +  2( • ) +  2( • ) +    ( • )
        3 2        1 2        1 0        3 0        1 0        1 2        3 0        3 2        1 2        1 4

        2 1        0 1        0 3        0 1        0 1
   + 10( • ) +  4( • ) +  4( • ) + 10( • ) + 13( • )
        1 0        3 0        1 0        1 2        1 0

        3 2        3 0        3 2        1 2        1 4        1 2        3 0        3 2        3 0        1 0
Sl 5:  3( • ) +    ( • ) +    ( • ) +    ( • ) +    ( • ) +  3( • ) +  6( • ) +  6( • ) +    ( • ) +  2( • )
        2 1        2 3        0 3        4 1        2 1        2 3        2 1        0 1        0 3        4 1

        1 2        1 0        1 4        1 2        3 0        1 0        1 2        1 0        1 0
   + 17( • ) +  6( • ) +  2( • ) +  6( • ) +  7( • ) + 22( • ) + 22( • ) +  7( • ) + 22( • )
        2 1        2 3        0 1        0 3        0 1        2 1        0 1        0 3        0 1
```

Affine D4 ⊃ A1 + A1 + A1 + A1 - Level 3

```
          0 1                   0 1
Sl 0:    ( • )  Highest Weight ( 0 )
          1 1                   1 1

          1 2        1 0        1 2        1 0        1 2        1 0        1 0
Sl 1:    ( • ) +    ( • ) +    ( • ) +    ( • ) +    ( • ) +    ( • ) +    ( • )
          2 0        2 2        0 2        2 0        0 0        0 2        0 0

          2 1        2 3        2 1        2 1        0 1        0 3        0 1        0 1
Sl 1:    ( • ) +    ( • ) +    ( • ) +   4( • ) +   2( • ) +   2( • ) +   2( • ) +   6( • )
          3 1        1 1        1 3        1 1        3 1        1 1        1 3        1 1

          3 2        3 2        3 0        3 2        1 2        1 0        1 4        1 2        1 0        1 4
Sl 3:    ( • ) +   2( • ) +   2( • ) +   2( • ) +    ( • ) +    ( • ) +    ( • ) +   6( • ) +    ( • ) +    ( • )
          2 2        2 0        2 2        0 2        4 0        4 2        2 0        2 2        2 4        0 2

          1 2        3 0        3 2        3 0        1 0        1 2        1 0        1 4        1 2        1 0
    +    ( • ) +   2( • ) +   2( • ) +   2( • ) +    ( • ) +   8( • ) +   8( • ) +    ( • ) +   8( • ) +    ( • )
          0 4        2 0        0 0        0 2        4 0        2 0        2 2        0 0        0 2        0 4

          3 0        1 0        1 2        1 0        1 0
    +    ( • ) +   8( • ) +   8( • ) +   8( • ) +   7( • )
          0 0        2 0        0 0        0 2        0 0

          4 1        4 3        4 1        2 3        2 1        2 3        4 1        2 1        2 3        2 1
Sl 4:    ( • ) +    ( • ) +    ( • ) +   3( • ) +   3( • ) +   3( • ) +   4( • ) +  12( • ) +  12( • ) +  12( • )
          3 1        1 1        1 3        3 1        3 3        1 3        1 1        3 1        1 1        1 3

          0 1        0 3        0 1        0 5        0 3        0 1        2 1        0 1        0 3        0 1
    +    ( • ) +   4( • ) +   4( • ) +    ( • ) +   4( • ) +    ( • ) +  32( • ) +  14( • ) +  14( • ) +  14( • )
          5 1        3 1        3 3        1 1        1 3        1 5        1 1        3 1        1 1        1 3

          0 1
    +  34( • )
          1 1

          3 2        3 4        3 2        5 2        5 0        5 2        3 2        3 0        3 4        3 2
Sl 5:    ( • ) +    ( • ) +    ( • ) +    ( • ) +    ( • ) +    ( • ) +   2( • ) +   2( • ) +   2( • ) +  16( • )
          4 2        2 2        2 4        2 0        2 2        0 2        4 0        4 2        2 0        2 2

          3 0        3 4        3 2        1 4        1 2        1 0        1 4        1 2        1 4        5 0
    +   2( • ) +   2( • ) +   2( • ) +    ( • ) +   6( • ) +    ( • ) +   6( • ) +   6( • ) +    ( • ) +    ( • )
          2 4        0 2        0 4        4 0        4 2        4 4        2 2        2 4        0 4        2 0

          5 2        5 0        3 0        3 2        3 0        3 4        3 2        3 0        1 2        1 0
    +    ( • ) +    ( • ) +   2( • ) +  18( • ) +  18( • ) +   2( • ) +  18( • ) +   2( • ) +   9( • ) +   9( • )
          0 0        0 2        4 0        2 0        2 2        0 0        0 2        0 4        4 0        4 2

          1 4        1 2        1 0        1 4        1 2        5 0        3 0        3 2        3 0        1 0
    +   9( • ) +  47( • ) +   9( • ) +   9( • ) +   9( • ) +    ( • ) +  16( • ) +  16( • ) +  16( • ) +   9( • )
          2 0        2 2        2 4        0 2        0 4        0 0        2 0        0 0        0 2        4 0

          1 2        1 0        1 4        1 2        1 0        3 0        1 0        1 2        1 0        1 0
    +  49( • ) +  49( • ) +   9( • ) +  49( • ) +   9( • ) +  14( • ) +  42( • ) +  42( • ) +  42( • ) +  33( • )
          2 0        2 2        0 0        0 2        0 4        0 0        2 0        0 0        0 2        0 0
```

Affine D4 ⊃ A1 + A1 + A1 + A1 - Level 3

```
          1 0                   1 0
Sl 0:    ( • )  Highest Weight ( 0 )
          0 2                   0 2

          2 1        0 1        0 1
Sl 1:    ( • ) +    ( • ) +    ( • )
          1 1        1 3        1 1

          3 2        1 2        3 0        1 2        1 0        1 2        1 0        3 0        1 0        1 2
Sl 2:    ( • ) +    ( • ) +    ( • ) +    ( • ) +   2( • ) +   2( • ) +    ( • ) +    ( • ) +    ( • ) +    ( • )
          2 0        2 2        0 2        2 0        2 2        0 2        0 4        0 0        2 0        0 0

          1 0        1 0
    +   3( • ) +   2( • )
          0 2        0 0  (continued next page)
```

Sl 3: $\begin{pmatrix}2\,3\\ \bullet\\ 3\,1\end{pmatrix} + \begin{pmatrix}4\,1\\ \bullet\\ 1\,1\end{pmatrix} + 2\begin{pmatrix}2\,1\\ \bullet\\ 3\,1\end{pmatrix} + 2\begin{pmatrix}2\,3\\ \bullet\\ 1\,1\end{pmatrix} + 3\begin{pmatrix}2\,1\\ \bullet\\ 1\,3\end{pmatrix} + \begin{pmatrix}0\,1\\ \bullet\\ 3\,3\end{pmatrix} + \begin{pmatrix}0\,3\\ \bullet\\ 1\,3\end{pmatrix} + 8\begin{pmatrix}2\,1\\ \bullet\\ 1\,1\end{pmatrix} + 2\begin{pmatrix}0\,1\\ \bullet\\ 3\,1\end{pmatrix} + 2\begin{pmatrix}0\,3\\ \bullet\\ 1\,1\end{pmatrix}$

$+ 5\begin{pmatrix}0\,1\\ \bullet\\ 1\,3\end{pmatrix} + 8\begin{pmatrix}0\,1\\ \bullet\\ 1\,1\end{pmatrix}$

Sl 4: $\begin{pmatrix}3\,2\\ \bullet\\ 4\,0\end{pmatrix} + \begin{pmatrix}3\,4\\ \bullet\\ 2\,0\end{pmatrix} + 3\begin{pmatrix}3\,2\\ \bullet\\ 2\,2\end{pmatrix} + \begin{pmatrix}1\,2\\ \bullet\\ 4\,2\end{pmatrix} + \begin{pmatrix}1\,4\\ \bullet\\ 2\,2\end{pmatrix} + \begin{pmatrix}1\,2\\ \bullet\\ 2\,4\end{pmatrix} + \begin{pmatrix}5\,0\\ \bullet\\ 0\,2\end{pmatrix} + 5\begin{pmatrix}3\,2\\ \bullet\\ 2\,0\end{pmatrix} + 3\begin{pmatrix}3\,0\\ \bullet\\ 2\,2\end{pmatrix} + 3\begin{pmatrix}3\,2\\ \bullet\\ 0\,2\end{pmatrix}$

$+ \begin{pmatrix}3\,0\\ \bullet\\ 0\,4\end{pmatrix} + \begin{pmatrix}1\,2\\ \bullet\\ 4\,0\end{pmatrix} + 2\begin{pmatrix}1\,0\\ \bullet\\ 4\,2\end{pmatrix} + \begin{pmatrix}1\,4\\ \bullet\\ 2\,0\end{pmatrix} + 11\begin{pmatrix}1\,2\\ \bullet\\ 2\,2\end{pmatrix} + 2\begin{pmatrix}1\,0\\ \bullet\\ 2\,4\end{pmatrix} + 2\begin{pmatrix}1\,4\\ \bullet\\ 0\,2\end{pmatrix} + 2\begin{pmatrix}1\,2\\ \bullet\\ 0\,4\end{pmatrix} + 4\begin{pmatrix}3\,0\\ \bullet\\ 2\,0\end{pmatrix} + 4\begin{pmatrix}3\,2\\ \bullet\\ 0\,0\end{pmatrix}$

$+ 6\begin{pmatrix}3\,0\\ \bullet\\ 0\,2\end{pmatrix} + \begin{pmatrix}1\,0\\ \bullet\\ 4\,0\end{pmatrix} + 10\begin{pmatrix}1\,2\\ \bullet\\ 2\,0\end{pmatrix} + 12\begin{pmatrix}1\,0\\ \bullet\\ 2\,2\end{pmatrix} + \begin{pmatrix}1\,4\\ \bullet\\ 0\,0\end{pmatrix} + 12\begin{pmatrix}1\,2\\ \bullet\\ 0\,2\end{pmatrix} + 4\begin{pmatrix}1\,0\\ \bullet\\ 0\,4\end{pmatrix} + 4\begin{pmatrix}3\,0\\ \bullet\\ 0\,0\end{pmatrix} + 9\begin{pmatrix}1\,0\\ \bullet\\ 2\,0\end{pmatrix} + 9\begin{pmatrix}1\,2\\ \bullet\\ 0\,0\end{pmatrix}$

$+ 16\begin{pmatrix}1\,0\\ \bullet\\ 0\,2\end{pmatrix} + 9\begin{pmatrix}1\,0\\ \bullet\\ 0\,0\end{pmatrix}$

Sl 5: $\begin{pmatrix}4\,3\\ \bullet\\ 3\,1\end{pmatrix} + \begin{pmatrix}2\,3\\ \bullet\\ 3\,3\end{pmatrix} + 2\begin{pmatrix}4\,1\\ \bullet\\ 3\,1\end{pmatrix} + 2\begin{pmatrix}4\,3\\ \bullet\\ 1\,1\end{pmatrix} + 3\begin{pmatrix}4\,1\\ \bullet\\ 1\,3\end{pmatrix} + \begin{pmatrix}2\,1\\ \bullet\\ 5\,1\end{pmatrix} + 8\begin{pmatrix}2\,3\\ \bullet\\ 3\,1\end{pmatrix} + 5\begin{pmatrix}2\,1\\ \bullet\\ 3\,3\end{pmatrix} + \begin{pmatrix}2\,5\\ \bullet\\ 1\,1\end{pmatrix} + 5\begin{pmatrix}2\,3\\ \bullet\\ 1\,3\end{pmatrix}$

$+ \begin{pmatrix}2\,1\\ \bullet\\ 1\,5\end{pmatrix} + \begin{pmatrix}0\,1\\ \bullet\\ 5\,3\end{pmatrix} + 2\begin{pmatrix}0\,3\\ \bullet\\ 3\,3\end{pmatrix} + \begin{pmatrix}0\,5\\ \bullet\\ 1\,3\end{pmatrix} + 9\begin{pmatrix}4\,1\\ \bullet\\ 1\,1\end{pmatrix} + 18\begin{pmatrix}2\,1\\ \bullet\\ 3\,1\end{pmatrix} + 18\begin{pmatrix}2\,3\\ \bullet\\ 1\,1\end{pmatrix} + 21\begin{pmatrix}2\,1\\ \bullet\\ 1\,3\end{pmatrix} + \begin{pmatrix}0\,1\\ \bullet\\ 5\,1\end{pmatrix} + 5\begin{pmatrix}0\,3\\ \bullet\\ 3\,1\end{pmatrix}$

$+ 8\begin{pmatrix}0\,1\\ \bullet\\ 3\,3\end{pmatrix} + \begin{pmatrix}0\,5\\ \bullet\\ 1\,1\end{pmatrix} + 8\begin{pmatrix}0\,3\\ \bullet\\ 1\,3\end{pmatrix} + 2\begin{pmatrix}0\,1\\ \bullet\\ 1\,5\end{pmatrix} + 47\begin{pmatrix}2\,1\\ \bullet\\ 1\,1\end{pmatrix} + 16\begin{pmatrix}0\,1\\ \bullet\\ 3\,1\end{pmatrix} + 16\begin{pmatrix}0\,3\\ \bullet\\ 1\,1\end{pmatrix} + 25\begin{pmatrix}0\,1\\ \bullet\\ 1\,3\end{pmatrix} + 41\begin{pmatrix}0\,1\\ \bullet\\ 1\,1\end{pmatrix}$

Affine D4 ⊃ A1 + A1 + A1 + A1 - Level 3

Sl 0: $\begin{pmatrix}0\,0\\ \bullet\\ 0\,3\end{pmatrix}$ **Highest Weight** $\begin{pmatrix}0\,0\\ 0\\ 0\,3\end{pmatrix}$

Sl 1: $\begin{pmatrix}1\,1\\ \bullet\\ 1\,2\end{pmatrix}$

Sl 2: $\begin{pmatrix}2\,2\\ \bullet\\ 2\,1\end{pmatrix} + \begin{pmatrix}2\,0\\ \bullet\\ 0\,3\end{pmatrix} + \begin{pmatrix}0\,0\\ \bullet\\ 2\,3\end{pmatrix} + \begin{pmatrix}0\,2\\ \bullet\\ 0\,3\end{pmatrix} + \begin{pmatrix}2\,0\\ \bullet\\ 0\,1\end{pmatrix} + \begin{pmatrix}0\,0\\ \bullet\\ 2\,1\end{pmatrix} + \begin{pmatrix}0\,2\\ \bullet\\ 0\,1\end{pmatrix} + \begin{pmatrix}0\,0\\ \bullet\\ 0\,3\end{pmatrix} + \begin{pmatrix}0\,0\\ \bullet\\ 0\,1\end{pmatrix}$

Sl 3: $\begin{pmatrix}3\,3\\ \bullet\\ 3\,0\end{pmatrix} + \begin{pmatrix}3\,1\\ \bullet\\ 1\,2\end{pmatrix} + \begin{pmatrix}1\,1\\ \bullet\\ 3\,2\end{pmatrix} + \begin{pmatrix}1\,3\\ \bullet\\ 1\,2\end{pmatrix} + \begin{pmatrix}1\,1\\ \bullet\\ 1\,4\end{pmatrix} + \begin{pmatrix}3\,1\\ \bullet\\ 1\,0\end{pmatrix} + \begin{pmatrix}1\,1\\ \bullet\\ 3\,0\end{pmatrix} + \begin{pmatrix}1\,3\\ \bullet\\ 1\,0\end{pmatrix} + 5\begin{pmatrix}1\,1\\ \bullet\\ 1\,2\end{pmatrix} + 2\begin{pmatrix}1\,1\\ \bullet\\ 1\,0\end{pmatrix}$

Sl 4: $\begin{pmatrix}4\,2\\ \bullet\\ 2\,1\end{pmatrix} + \begin{pmatrix}2\,2\\ \bullet\\ 4\,1\end{pmatrix} + \begin{pmatrix}2\,4\\ \bullet\\ 2\,1\end{pmatrix} + \begin{pmatrix}2\,2\\ \bullet\\ 2\,3\end{pmatrix} + \begin{pmatrix}4\,0\\ \bullet\\ 0\,3\end{pmatrix} + 5\begin{pmatrix}2\,2\\ \bullet\\ 2\,1\end{pmatrix} + 2\begin{pmatrix}2\,0\\ \bullet\\ 2\,3\end{pmatrix} + 2\begin{pmatrix}2\,2\\ \bullet\\ 0\,3\end{pmatrix} + \begin{pmatrix}0\,0\\ \bullet\\ 4\,3\end{pmatrix} + 2\begin{pmatrix}0\,2\\ \bullet\\ 2\,3\end{pmatrix}$

$+ \begin{pmatrix}0\,4\\ \bullet\\ 0\,3\end{pmatrix} + \begin{pmatrix}4\,0\\ \bullet\\ 0\,1\end{pmatrix} + 4\begin{pmatrix}2\,0\\ \bullet\\ 2\,1\end{pmatrix} + 4\begin{pmatrix}2\,2\\ \bullet\\ 0\,1\end{pmatrix} + 3\begin{pmatrix}2\,0\\ \bullet\\ 0\,3\end{pmatrix} + \begin{pmatrix}0\,0\\ \bullet\\ 4\,1\end{pmatrix} + 4\begin{pmatrix}0\,2\\ \bullet\\ 2\,1\end{pmatrix} + 3\begin{pmatrix}0\,0\\ \bullet\\ 2\,3\end{pmatrix} + \begin{pmatrix}0\,4\\ \bullet\\ 0\,1\end{pmatrix} + 3\begin{pmatrix}0\,2\\ \bullet\\ 0\,3\end{pmatrix}$

$+ \begin{pmatrix}0\,0\\ \bullet\\ 0\,5\end{pmatrix} + 4\begin{pmatrix}2\,0\\ \bullet\\ 0\,1\end{pmatrix} + 4\begin{pmatrix}0\,0\\ \bullet\\ 2\,1\end{pmatrix} + 4\begin{pmatrix}0\,2\\ \bullet\\ 0\,1\end{pmatrix} + 6\begin{pmatrix}0\,0\\ \bullet\\ 0\,3\end{pmatrix} + 6\begin{pmatrix}0\,0\\ \bullet\\ 0\,1\end{pmatrix}$

Sl 5: $\begin{pmatrix}3\,3\\ \bullet\\ 3\,2\end{pmatrix} + \begin{pmatrix}5\,1\\ \bullet\\ 1\,2\end{pmatrix} + 3\begin{pmatrix}3\,3\\ \bullet\\ 3\,0\end{pmatrix} + 2\begin{pmatrix}3\,1\\ \bullet\\ 3\,2\end{pmatrix} + 2\begin{pmatrix}3\,3\\ \bullet\\ 1\,2\end{pmatrix} + \begin{pmatrix}3\,1\\ \bullet\\ 1\,4\end{pmatrix} + \begin{pmatrix}1\,1\\ \bullet\\ 5\,2\end{pmatrix} + 2\begin{pmatrix}1\,3\\ \bullet\\ 3\,2\end{pmatrix} + \begin{pmatrix}1\,1\\ \bullet\\ 3\,4\end{pmatrix} + \begin{pmatrix}1\,5\\ \bullet\\ 1\,2\end{pmatrix}$

$+ \begin{pmatrix}1\,3\\ \bullet\\ 1\,4\end{pmatrix} + 3\begin{pmatrix}3\,1\\ \bullet\\ 3\,0\end{pmatrix} + 3\begin{pmatrix}3\,3\\ \bullet\\ 1\,0\end{pmatrix} + 8\begin{pmatrix}3\,1\\ \bullet\\ 1\,2\end{pmatrix} + 3\begin{pmatrix}1\,3\\ \bullet\\ 3\,0\end{pmatrix} + 8\begin{pmatrix}1\,1\\ \bullet\\ 3\,2\end{pmatrix} + 8\begin{pmatrix}1\,3\\ \bullet\\ 1\,2\end{pmatrix} + 6\begin{pmatrix}1\,1\\ \bullet\\ 1\,4\end{pmatrix} + 6\begin{pmatrix}3\,1\\ \bullet\\ 1\,0\end{pmatrix} + 6\begin{pmatrix}1\,1\\ \bullet\\ 3\,0\end{pmatrix}$

$+ 6\begin{pmatrix}1\,3\\ \bullet\\ 1\,0\end{pmatrix} + 24\begin{pmatrix}1\,1\\ \bullet\\ 1\,2\end{pmatrix} + 14\begin{pmatrix}1\,1\\ \bullet\\ 1\,0\end{pmatrix}$

Affine D4 ⊃ A1 + A1 + A1 + A1 - Level 3

```
              0 0                  0 0
sl 0:     ( • )  Highest Weight  ( 1 )
              0 1                  0 1

              1 1        1 1
sl 1:     ( • ) +    ( • )
              1 2        1 0

              2 2        2 0        2 2        2 0        0 2        0 0        0 2        2 0        0 0        0 2
sl 2:     ( • ) +    ( • ) +    ( • ) +    ( • ) +    ( • ) +    ( • ) +    ( • ) +  2( • ) +  2( • ) +  2( • )
              2 1        2 1        0 1        0 3        2 1        2 3        0 3        0 1        2 1        0 1

              0 0        0 0
   +      ( • ) +  2( • )
              0 3        0 1

              3 1        3 3        3 1        1 3        1 1        1 3        1 1        3 1        1 1        1 3
sl 3:     ( • ) +    ( • ) +  2( • ) +    ( • ) +  2( • ) +  2( • ) +    ( • ) +  2( • ) +  2( • ) +  2( • )
              3 0        1 0        1 2        3 0        3 2        1 2        1 4        1 0        3 0        1 0

              1 1        1 1
   +    9( • ) +  8( • )
              1 2        1 0

              4 2        2 2        2 4        2 2        4 0        4 2        4 0        2 0        2 2        2 0
sl 4:     ( • ) +    ( • ) +    ( • ) +  2( • ) +    ( • ) +    ( • ) +    ( • ) +    ( • ) + 11( • ) +  4( • )
              2 1        4 1        2 1        2 3        2 1        0 1        0 3        4 1        2 1        2 3

              2 4        2 2        0 2        0 0        0 4        0 2        0 4        4 0        2 0        2 2
   +      ( • ) +  4( • ) +    ( • ) +    ( • ) +    ( • ) +  4( • ) +    ( • ) +  2( • ) + 11( • ) + 11( • )
              0 1        0 3        4 1        4 3        2 1        2 3        0 3        0 1        2 1        0 1

              2 0        0 0        0 2        0 0        0 4        0 2        0 0        2 0        0 0        0 2
   +    5( • ) +  2( • ) + 11( • ) +  5( • ) +  2( • ) +  5( • ) +    ( • ) + 11( • ) + 11( • ) + 11( • )
              0 3        4 1        2 1        2 3        0 1        0 3        0 5        0 1        2 1        0 1

              0 0        0 0
   +    7( • ) + 13( • )
              0 3        0 1

              3 3        5 1        3 3        3 1        3 3        3 1        1 1        1 3        1 1        1 5
sl 5:     ( • ) +    ( • ) +  3( • ) +  5( • ) +  5( • ) +  2( • ) +    ( • ) +  5( • ) +  2( • ) +    ( • )
              3 2        1 2        3 0        3 2        1 2        1 4        5 2        3 2        3 4        1 2

              1 3        5 1        3 1        3 3        3 1        1 1        1 3        1 1        1 5        1 3
   +    2( • ) +    ( • ) +  7( • ) +  7( • ) + 19( • ) +    ( • ) +  7( • ) + 19( • ) +    ( • ) + 19( • )
              1 4        1 0        3 0        1 0        1 2        5 0        3 0        3 2        1 0        1 2

              1 1        3 1        1 1        1 3        1 1        1 1
   +   10( • ) + 18( • ) + 18( • ) + 18( • ) + 53( • ) + 43( • )
              1 4        1 0        3 0        1 0        1 2        1 0
```

Affine D5 Branching Rules

Affine D5 ⊃ D5 - Level 1

```
          • 0                    1 0
sl 0:    ( 00 )  Highest Weight ( 00 )
          0 0                    0 0

          • 0
sl 1:    ( 10 )
          0 0

          • 1          • 0          • 0          • 0
sl 2:    ( 00 ) +     ( 00 ) +     ( 10 ) +     ( 00 )
          0 1          2 0          0 0          0 0

          • 0          • 1          • 0          • 0          • 0
sl 3:    ( 01 ) +    2( 00 ) +     ( 00 ) +    3( 10 ) +     ( 00 )
          1 0          0 1          2 0          0 0          0 0

          • 2          • 0          • 0          • 0          • 1          • 0          • 0          • 0
sl 4:    ( 00 ) +     ( 00 ) +     ( 20 ) +    2( 01 ) +    4( 00 ) +    3( 00 ) +    5( 10 ) +    3( 00 )
          1 0          1 2          0 0          1 0          0 1          2 0          0 0          0 0

          • 1          • 0          • 2          • 0          • 0          • 0          • 1          • 0           • 0
sl 5:    ( 10 ) +     ( 10 ) +    2( 00 ) +    2( 00 ) +     ( 20 ) +    6( 01 ) +    8( 00 ) +    4( 00 ) + 10( 10 )
          0 1          2 0          1 0          1 2          0 0          1 0          0 1          2 0           0 0
          • 0
    +   4( 00 )
          0 0

          • 1          • 0          • 1          • 0          • 2          • 0          • 0           • 0           • 1
sl 6:    ( 00 ) +     ( 02 ) +    3( 10 ) +    2( 10 ) +    5( 00 ) +    5( 00 ) +    4( 20 ) + 11( 01 ) + 16( 00 )
          2 1          0 0          0 1          2 0          1 0          1 2          0 0           1 0           0 1
          • 0          • 0          • 0
    + 10( 00 ) + 16( 10 ) +   8( 00 )
          2 0          0 0          0 0
```

Affine D5 ⊃ D5 - Level 1

```
          • 0                    0 0
sl 0:    ( 00 )  Highest Weight ( 00 )
          0 1                    0 1

          • 1          • 0
sl 1:    ( 00 ) +     ( 00 )
          1 0          0 1

          • 0          • 1          • 0
sl 2:    ( 10 ) +    2( 00 ) +    2( 00 )
          0 1          1 0          0 1

          • 1          • 0          • 0          • 1          • 0
sl 3:    ( 01 ) +     ( 00 ) +    2( 10 ) +    4( 00 ) +    4( 00 )
          0 0          2 1          0 1          1 0          0 1

          • 1          • 1          • 1          • 0          • 0          • 1          • 0
sl 4:    ( 10 ) +     ( 00 ) +    2( 01 ) +    2( 00 ) +    5( 10 ) +    8( 00 ) +    7( 00 )
          1 0          0 2          0 0          2 1          0 1          1 0          0 1

          • 0          • 1          • 3          • 1          • 1          • 0           • 0           • 1           • 0
sl 5:    ( 01 ) +    3( 10 ) +     ( 00 ) +    2( 00 ) +    5( 01 ) +    5( 00 ) + 10( 10 ) + 15( 00 ) + 12( 00 )
          1 1          1 0          0 0          0 2          0 0          2 1           0 1           1 0           0 1

          • 2          • 0          • 0          • 1          • 1          • 3          • 1           • 1           • 0
sl 6:    ( 00 ) +     ( 20 ) +    3( 01 ) +     ( 00 ) +    7( 10 ) +     ( 00 ) +    5( 00 ) + 11( 01 ) + 10( 00 )
          1 1          0 1          1 1          3 0          1 0          0 0          0 2           0 0           2 1
          • 0          • 1          • 0
    + 20( 10 ) + 27( 00 ) + 21( 00 )
          0 1          1 0          0 1
```

Affine D5 ⊃ D5 - Level 1

```
           • 0                 0 0
Sl 0:    ( 00 )  Highest Weight ( 00 )
           1 0                 1 0

           • 0        • 0
Sl 1:    ( 01 ) +   ( 00 )
           0 0        1 0

           • 0        • 2        • 0        • 0        • 0
Sl 2:    ( 10 ) +   ( 00 ) +   ( 00 ) +   ( 01 ) +  2( 00 )
           1 0        0 0        0 2        0 0        1 0

           • 1        • 0        • 2        • 0        • 0        • 0
Sl 3:    ( 00 ) +  2( 10 ) +   ( 00 ) +   ( 00 ) +  4( 01 ) +  3( 00 )
           1 1        1 0        0 0        0 2        0 0        1 0

           • 0        • 1        • 0        • 0        • 2        • 0        • 0        • 0
Sl 4:    ( 11 ) +  3( 00 ) +   ( 00 ) +  4( 10 ) +  3( 00 ) +  3( 00 ) +  6( 01 ) +  7( 00 )
           0 0        1 1        3 0        1 0        0 0        0 2        0 0        1 0

           • 0        • 2        • 0        • 0        • 1        • 0        • 0        • 2        • 0
Sl 5:    ( 01 ) +   ( 10 ) +   ( 10 ) +  2( 11 ) +  7( 00 ) +   ( 00 ) +  9( 10 ) +  5( 00 ) +  5( 00 )
           2 0        0 0        0 2        0 0        1 1        3 0        1 0        0 0        0 2

           • 0        • 0
     +  13( 01 ) + 10( 00 )
           0 0        1 0

           • 1        • 2        • 0        • 0        • 0        • 2        • 0        • 0        • 1
Sl 6:    ( 01 ) +   ( 00 ) +   ( 00 ) +   ( 20 ) +  2( 01 ) +  2( 10 ) +  2( 10 ) +  6( 11 ) + 15( 00 )
           0 1        2 0        2 2        1 0        2 0        0 0        0 2        0 0        1 1

           • 0        • 0        • 2        • 0        • 0        • 0
     +   3( 00 ) + 17( 10 ) + 11( 00 ) + 11( 00 ) + 21( 01 ) + 19( 00 )
           3 0        1 0        0 0        0 2        0 0        1 0
```

Affine D5 ⊃ D5 - Level 2

```
           • 1                 0 1
Sl 0:    ( 00 )  Highest Weight ( 00 )
           0 1                 0 1

           • 2        • 0        • 0        • 1        • 0
Sl 1:    ( 00 ) +   ( 00 ) +   ( 01 ) +  2( 00 ) +   ( 10 )
           1 0        1 2        1 0        0 1        0 0

           • 1        • 1        • 2        • 0        • 0        • 0        • 1        • 0        • 0
Sl 2:    ( 00 ) +  2( 10 ) +  3( 00 ) +  3( 00 ) +   ( 20 ) +  4( 01 ) +  6( 00 ) +   ( 00 ) +  3( 10 )
           2 1        0 1        1 0        1 2        0 0        1 0        0 1        2 0        0 0

           • 0
     +   ( 00 )
           0 0

           • 2        • 0        • 0        • 2        • 0        • 1        • 0        • 1        • 0
Sl 3:    ( 10 ) +   ( 10 ) +   ( 11 ) +   ( 01 ) +   ( 01 ) +  5( 00 ) +  2( 02 ) +  8( 10 ) +  2( 10 )
           1 0        1 2        1 0        0 0        0 2        2 1        0 0        0 1        2 0

           • 2        • 0        • 0        • 0        • 1        • 0        • 0        • 0
     +   9( 00 ) +  9( 00 ) +  3( 20 ) + 13( 01 ) + 16( 00 ) +  4( 00 ) +  9( 10 ) +  2( 00 )
           1 0        1 2        0 0        1 0        0 1        2 0        0 0        0 0

           • 1        • 1        • 2        • 0        • 0        • 2        • 0        • 0        • 2
Sl 4:    ( 20 ) +  2( 01 ) +   ( 00 ) +   ( 00 ) +   ( 01 ) +  5( 10 ) +  5( 10 ) +  7( 11 ) +  2( 00 )
           0 1        1 1        3 0        3 2        3 0        1 0        1 2        1 0        0 2

           • 2        • 0        • 1        • 0        • 1        • 0        • 2        • 0        • 0
     +   5( 01 ) +  5( 01 ) + 17( 00 ) +  6( 02 ) + 27( 10 ) +  8( 10 ) + 25( 00 ) + 25( 00 ) + 12( 20 )
           0 0        0 2        2 1        0 0        0 1        2 0        1 0        1 2        0 0

           • 0        • 1        • 0        • 0        • 0
     +  37( 01 ) + 40( 00 ) + 12( 00 ) + 23( 10 ) +  5( 00 )
           1 0        0 1        2 0        0 0        0 0
```

Affine D5 ⊃ D5 - Level 2

```
           • 0                       0 0
Sl 0:     ( 10 )  Highest Weight    ( 10 )
           0 0                       0 0

           • 0         • 1         • 0         • 0         • 0
Sl 1:     ( 01 ) +    ( 00 ) +    ( 00 ) +    ( 10 ) +    ( 00 )
           1 0         0 1         2 0         0 0         0 0

           • 1         • 0         • 2         • 0         • 0         • 0         • 1         • 0         • 0
Sl 2:     ( 10 ) +    ( 10 ) +    ( 00 ) +    ( 00 ) +    ( 20 ) +   3( 01 ) +   3( 00 ) +   2( 00 ) +   4( 10 )
           0 1         2 0         1 0         1 2         0 0         1 0         0 1         2 0         0 0

           • 0
    +     ( 00 )
           0 0

           • 0         • 2         • 0         • 1         • 0         • 1         • 0         • 2         • 0
Sl 3:     ( 11 ) +    ( 01 ) +    ( 01 ) +   2( 00 ) +    ( 02 ) +   4( 10 ) +   3( 10 ) +   4( 00 ) +   4( 00 )
           1 0         0 0         0 2         2 1         0 0         0 1         2 0         1 0         1 2

           • 0         • 0         • 1         • 0         • 0         • 0
    +    4( 20 ) +  10( 01 ) +   9( 00 ) +   6( 00 ) +   9( 10 ) +   3( 00 )
           0 0         1 0         0 1         2 0         0 0         0 0

           • 1         • 0         • 2         • 0         • 0         • 0         • 2         • 2         • 0
Sl 4:     ( 01 ) +    ( 01 ) +   2( 10 ) +   2( 10 ) +    ( 30 ) +   5( 11 ) +   2( 00 ) +   3( 01 ) +   3( 01 )
           1 1         3 0         1 0         1 2         0 0         1 0         0 2         0 0         0 2

           • 1         • 0         • 1         • 0         • 0         • 2         • 0         • 0         • 0
    +    8( 00 ) +   5( 02 ) +  16( 10 ) +    ( 00 ) +  10( 10 ) +  13( 00 ) +  13( 00 ) +  11( 20 ) +  28( 01 )
           2 1         0 0         0 1         4 0         2 0         1 0         1 2         0 0         1 0

           • 1         • 0         • 0         • 0
    +   24( 00 ) +  14( 00 ) +  23( 10 ) +   6( 00 )
           0 1         2 0         0 0         0 0
```

Affine D5 ⊃ D5 - Level 2

```
           • 0                       0 0
Sl 0:     ( 00 )  Highest Weight    ( 00 )
           2 0                       2 0

           • 0         • 0         • 0
Sl 1:     ( 01 ) +    ( 00 ) +    ( 10 )
           1 0         2 0         0 0

           • 0         • 0         • 2         • 0         • 0         • 0         • 1         • 0         • 0
Sl 2:     ( 02 ) +    ( 10 ) +    ( 00 ) +    ( 00 ) +    ( 20 ) +   2( 01 ) +    ( 00 ) +   3( 00 ) +   2( 10 )
           0 0         2 0         1 0         1 2         0 0         1 0         0 1         2 0         0 0

           • 0
    +     ( 00 )
           0 0

           • 0         • 2         • 0         • 1         • 0         • 1         • 0         • 2         • 0
Sl 3:     ( 11 ) +    ( 01 ) +    ( 01 ) +    ( 00 ) +    ( 02 ) +   2( 10 ) +   3( 10 ) +   2( 00 ) +   2( 00 )
           1 0         0 0         0 2         2 1         0 0         0 1         2 0         1 0         1 2

           • 0         • 0         • 1         • 0         • 0         • 0
    +    2( 20 ) +   8( 01 ) +   4( 00 ) +   5( 00 ) +   6( 10 ) +    ( 00 )
           0 0         1 0         0 1         2 0         0 0         0 0

           • 1         • 0         • 2         • 0         • 0         • 4         • 2         • 0         • 2
Sl 4:     ( 01 ) +    ( 20 ) +    ( 10 ) +    ( 10 ) +   4( 11 ) +    ( 00 ) +    ( 00 ) +    ( 00 ) +   2( 01 )
           1 1         2 0         1 0         1 2         1 0         0 0         0 2         0 4         0 0

           • 0         • 1         • 0         • 1         • 0         • 0         • 2         • 0         • 0
    +    2( 01 ) +   5( 00 ) +   6( 02 ) +   8( 10 ) +    ( 00 ) +   8( 10 ) +   8( 00 ) +   8( 00 ) +   8( 20 )
           0 2         2 1         0 0         0 1         4 0         2 0         1 0         1 2         0 0

           • 0         • 1         • 0         • 0         • 0
    +   18( 01 ) +  12( 00 ) +  14( 00 ) +  13( 10 ) +   4( 00 )
           1 0         0 1         2 0         0 0         0 0
```

Affine D5 ⊃ D5 - Level 2

```
           • 0                       2  0
sl 0:     ( 00 )  Highest Weight   ( 00 )
           0  0                      0  0

           • 0
sl 1:     ( 10 )
           0  0

           • 0         • 1         • 0         • 0         • 0
sl 2:     ( 20 ) +    ( 00 ) +    ( 00 ) +    ( 10 ) +    ( 00 )
           0  0        0  1        2  0        0  0        0  0

           • 1         • 0         • 0         • 0         • 1         • 0         • 0         • 0
sl 3:     ( 10 ) +    ( 10 ) +    ( 20 ) +   2( 01 ) +   2( 00 ) +    ( 00 ) +   4( 10 ) +    ( 00 )
           0  1        2  0        0  0        1  0        0  1        2  0        0  0        0  0

           • 0         • 2         • 1         • 0         • 1         • 0         • 0         • 2         • 0
sl 4:     ( 11 ) +    ( 00 ) +    ( 00 ) +    ( 02 ) +   3( 10 ) +    ( 00 ) +   2( 10 ) +   2( 00 ) +   2( 00 )
           1  0        0  2        2  1        0  0        0  1        4  0        2  0        1  0        1  2

           • 0         • 0         • 1         • 0         • 0         • 0
     +   5( 20 ) +   5( 01 ) +   6( 00 ) +   5( 00 ) +   7( 10 ) +   4( 00 )
           0  0        1  0        0  1        2  0        0  0        0  0

           • 1         • 0         • 2         • 0         • 0         • 0         • 2         • 2         • 0
sl 5:     ( 01 ) +    ( 01 ) +    ( 10 ) +    ( 10 ) +    ( 30 ) +   4( 11 ) +   2( 00 ) +   2( 01 ) +   2( 01 )
           1  1        3  0        1  0        1  2        0  0        1  0        0  2        0  0        0  2

           • 1         • 0         • 1         • 0         • 0         • 2         • 0         • 0         • 0
     +   4( 00 ) +   2( 02 ) +  11( 10 ) +    ( 00 ) +   8( 10 ) +   6( 00 ) +   6( 00 ) +   9( 20 ) +  18( 01 )
           2  1        0  0        0  1        4  0        2  0        1  0        1  2        0  0        1  0

           • 1         • 0         • 0         • 0
     +  14( 00 ) +   8( 00 ) +  18( 10 ) +   5( 00 )
           0  1        2  0        0  0        0  0
```

Affine D5 ⊃ D5 - Level 2

```
           • 1                       0  1
sl 0:     ( 00 )  Highest Weight   ( 00 )
           1  0                      1  0

           • 1         • 0         • 0         • 1         • 0
sl 1:     ( 01 ) +    ( 00 ) +    ( 10 ) +   2( 00 ) +    ( 00 )
           0  0        2  1        0  1        1  0        0  1

           • 0         • 1         • 3         • 1         • 1         • 0         • 0         • 1         • 0
sl 2:     ( 01 ) +   2( 10 ) +    ( 00 ) +    ( 00 ) +   3( 01 ) +   3( 00 ) +   4( 10 ) +   6( 00 ) +   3( 00 )
           1  1        1  0        0  0        0  2        0  0        2  1        0  1        1  0        0  1

           • 1         • 0         • 2         • 0         • 0         • 0         • 1         • 1         • 3
sl 3:     ( 11 ) +    ( 10 ) +   2( 00 ) +    ( 00 ) +    ( 20 ) +   5( 01 ) +    ( 00 ) +   8( 10 ) +   2( 00 )
           0  0        2  1        1  1        1  3        0  1        1  1        3  0        1  0        0  0

           • 1         • 1         • 0         • 0         • 1         • 0
     +   4( 00 ) +  10( 01 ) +   9( 00 ) +  13( 10 ) +  16( 00 ) +   8( 00 )
           0  2        0  0        2  1        0  1        1  0        0  1

           • 1         • 1         • 0         • 1         • 3         • 1         • 1         • 0         • 2
sl 4:     ( 00 ) +    ( 20 ) +    ( 02 ) +   2( 01 ) +    ( 10 ) +   2( 10 ) +   5( 11 ) +   5( 10 ) +   9( 00 )
           2  2        1  0        0  1        2  0        0  0        0  2        0  0        2  1        1  1

           • 0         • 0         • 0         • 1         • 1         • 3         • 1         • 1         • 0
     +   3( 00 ) +   6( 20 ) +  19( 01 ) +   5( 00 ) +  26( 10 ) +   6( 00 ) +  14( 00 ) +  29( 01 ) +  25( 00 )
           1  3        0  1        1  1        3  0        1  0        0  0        0  2        0  0        2  1

           • 0         • 1         • 0
     +  36( 10 ) +  40( 00 ) +  20( 00 )
           0  1        1  0        0  1
```

Affine D5 ⊃ D5 - Level 2

```
          • 0                     1 0
Sl 0:    ( 00 )  Highest Weight  ( 00 )
          0 1                     0 1

          • 0        • 1        • 0
Sl 1:    ( 10 ) +   ( 00 ) +   ( 00 )
           0 1        1 0        0 1

          • 1        • 1        • 1        • 0        • 0        • 1        • 0
Sl 2:    ( 10 ) +   ( 00 ) +   ( 01 ) +   ( 00 ) +  3( 10 ) +  3( 00 ) +  3( 00 )
           1 0        0 2        0 0        2 1        0 1        1 0        0 1

          • 2        • 0        • 0        • 1        • 1        • 1        • 1        • 0        • 0
Sl 3:    ( 00 ) +   ( 20 ) +  2( 01 ) +   ( 00 ) +  4( 10 ) +  3( 00 ) +  5( 01 ) +  4( 00 ) +  9( 10 )
           1 1        0 1        1 1        3 0        1 0        0 2        0 0        2 1        0 1

          • 1        • 0
     +  9( 00 ) +  7( 00 )
           1 0        0 1

          • 1        • 1        • 1        • 0        • 2        • 0        • 0        • 0        • 1
Sl 4:    ( 01 ) +   ( 10 ) +  2( 11 ) +  2( 10 ) +  5( 00 ) +   ( 00 ) +  4( 20 ) +  9( 01 ) +  3( 00 )
           2 0        0 2        0 0        2 1        1 1        1 3        0 1        1 1        3 0

          • 1        • 3        • 1        • 1        • 0        • 0        • 1        • 0
     + 15( 10 ) +  2( 00 ) + 10( 00 ) + 15( 01 ) + 13( 00 ) + 25( 10 ) + 24( 00 ) + 16( 00 )
           1 0        0 0        0 2        0 0        2 1        0 1        1 0        0 1

          • 0        • 2        • 3        • 1        • 1        • 0        • 1        • 3        • 1
Sl 5:    ( 11 ) +   ( 01 ) +   ( 00 ) +  2( 00 ) +  2( 20 ) +   ( 02 ) +  6( 01 ) +   ( 10 ) +  6( 10 )
           1 1        0 1        2 0        2 2        1 0        0 1        2 0        0 0        0 2

          • 0        • 1        • 0        • 2        • 0        • 0        • 0        • 1        • 1
     +   ( 00 ) + 10( 11 ) +  9( 10 ) + 18( 00 ) +  5( 00 ) + 14( 20 ) + 32( 01 ) + 10( 00 ) + 46( 10 )
           4 1        0 0        2 1        1 1        1 3        0 1        1 1        3 0        1 0

          • 3        • 1        • 1        • 0        • 0        • 1        • 0
     +  7( 00 ) + 28( 00 ) + 44( 01 ) + 37( 00 ) + 64( 10 ) + 59( 00 ) + 35( 00 )
           0 0        0 2        0 0        2 1        0 1        1 0        0 1
```

Affine D5 ⊃ D5 - Level 2

```
          • 0                     0 0
Sl 0:    ( 00 )  Highest Weight  ( 00 )
          0 2                     0 2

          • 1        • 0        • 0
Sl 1:    ( 00 ) +   ( 00 ) +   ( 01 )
           1 1        0 2        0 0

          • 2        • 0        • 0        • 1        • 0        • 2        • 0        • 0        • 0
Sl 2:    ( 00 ) +   ( 10 ) +   ( 11 ) +  3( 00 ) +   ( 10 ) +   ( 00 ) +  3( 00 ) +  2( 01 ) +   ( 00 )
           2 0        0 2        0 0        1 1        1 0        0 0        0 2        0 0        1 0

          • 1        • 1        • 2        • 0        • 0        • 2        • 0        • 0        • 1
Sl 3:    ( 10 ) +   ( 01 ) +  2( 00 ) +   ( 00 ) +  2( 01 ) +  2( 10 ) +  3( 10 ) +  3( 11 ) +  9( 00 )
           1 1        0 1        2 0        2 2        2 0        0 0        0 2        0 0        1 1

          • 0        • 2        • 0        • 0        • 0
     +  4( 10 ) +  3( 00 ) +  6( 00 ) +  7( 01 ) +  2( 00 )
           1 0        0 0        0 2        0 0        1 0

          • 0        • 2        • 1        • 0        • 1        • 1        • 1        • 2        • 0
Sl 4:    ( 20 ) +   ( 01 ) +   ( 00 ) +   ( 02 ) +  5( 10 ) +   ( 00 ) +  5( 01 ) +  7( 00 ) +  4( 00 )
           0 2        1 0        3 1        1 0        1 1        0 3        0 1        2 0        2 2

          • 0        • 0        • 2        • 0        • 0        • 1        • 0        • 0        • 2
     +  2( 20 ) +  6( 01 ) +  6( 10 ) +  9( 10 ) + 11( 11 ) + 25( 00 ) +  2( 00 ) + 12( 10 ) +  9( 00 )
           1 0        2 0        0 0        0 2        0 0        1 1        3 0        1 0        0 0

          • 0        • 0        • 0
     + 15( 00 ) + 16( 01 ) +  7( 00 )
           0 2        0 0        1 0
```

Affine D5 ⊃ D5 - Level 2

```
          • 0                     0 0
sl 0:    ( 01 )  Highest Weight ( 01 )
          0 0                     0 0

          • 1        • 0        • 2        • 0        • 0        • 0
sl 1:    ( 00 ) +   ( 10 ) +   ( 00 ) +   ( 00 ) +   ( 01 ) +   ( 00 )
          1 1        1 0        0 0        0 2        0 0        1 0

          • 0        • 2        • 0        • 0        • 1        • 0        • 2        • 0        • 0
sl 2:    ( 01 ) +   ( 10 ) +   ( 10 ) +   ( 11 ) +  4( 00 ) +  3( 10 ) +  2( 00 ) +  2( 00 ) +  5( 01 )
          2 0        0 0        0 2        0 0        1 1        1 0        0 0        0 2        0 0

          • 0
   +    2( 00 )
          1 0

          • 1        • 1        • 2        • 0        • 0        • 0        • 2        • 0        • 0
sl 3:    ( 10 ) +  2( 01 ) +  2( 00 ) +  2( 00 ) +   ( 20 ) +  3( 01 ) +  3( 10 ) +  3( 10 ) +  6( 11 )
          1 1        0 1        2 0        2 2        1 0        2 0        0 0        0 2        0 0

          • 1        • 0        • 0        • 2        • 0        • 0        • 0
   +   13( 00 ) +  2( 00 ) +  9( 10 ) +  7( 00 ) +  7( 00 ) + 11( 01 ) +  7( 00 )
          1 1        3 0        1 0        0 0        0 2        0 0        1 0

          • 2        • 0        • 0        • 1        • 0        • 1        • 3        • 1        • 0
sl 4:    ( 01 ) +   ( 01 ) +   ( 21 ) +   ( 00 ) +   ( 02 ) +  7( 10 ) +   ( 00 ) +   ( 00 ) +   ( 10 )
          1 0        1 2        0 0        3 1        1 0        1 1        0 1        0 3        3 0

          • 1        • 2        • 0        • 0        • 0        • 2        • 0        • 0        • 1
   +    8( 01 ) +  6( 00 ) +  6( 00 ) +  4( 20 ) + 13( 01 ) + 12( 10 ) + 12( 10 ) + 18( 11 ) + 38( 00 )
          0 1        2 0        2 2        1 0        2 0        0 0        0 2        0 0        1 1

          • 0        • 0        • 2        • 0        • 0        • 0
   +    4( 00 ) + 26( 10 ) + 16( 00 ) + 16( 00 ) + 31( 01 ) + 14( 00 )
          3 0        1 0        0 0        0 2        0 0        1 0
```

Affine D5 ⊃ D5 - Level 2

```
          • 0                     1 0
sl 0:    ( 00 )  Highest Weight ( 00 )
          1 0                     1 0

          • 0        • 0        • 0
sl 1:    ( 10 ) +   ( 01 ) +   ( 00 )
          1 0        0 0        1 0

          • 0        • 1        • 0        • 0        • 2        • 0        • 0        • 0
sl 2:    ( 11 ) +   ( 00 ) +   ( 00 ) +  3( 10 ) +   ( 00 ) +   ( 00 ) +  2( 01 ) +  3( 00 )
          0 0        1 1        3 0        1 0        0 0        0 2        0 0        1 0

          • 1        • 0        • 0        • 2        • 0        • 0        • 1        • 0        • 0
sl 3:    ( 01 ) +   ( 20 ) +  2( 01 ) +   ( 10 ) +   ( 10 ) +  3( 11 ) +  5( 00 ) +  2( 00 ) +  8( 10 )
          0 1        1 0        2 0        0 0        0 2        0 0        1 1        3 0        1 0

          • 2        • 0        • 0        • 0
   +    2( 00 ) +  2( 00 ) +  8( 01 ) +  6( 00 )
          0 0        0 2        0 0        1 0

          • 0        • 1        • 3        • 1        • 0        • 1        • 2        • 0        • 0
sl 4:    ( 02 ) +  2( 10 ) +   ( 00 ) +   ( 00 ) +   ( 10 ) +  4( 01 ) +  2( 00 ) +  2( 00 ) +  4( 20 )
          1 0        1 1        0 1        0 3        3 0        0 1        2 0        2 2        1 0

          • 0        • 2        • 0        • 0        • 1        • 0        • 0        • 2        • 0
   +    6( 01 ) +  4( 10 ) +  4( 10 ) + 12( 11 ) + 17( 00 ) +  6( 00 ) + 21( 10 ) +  8( 00 ) +  8( 00 )
          2 0        0 0        0 2        0 0        1 1        3 0        1 0        0 0        0 2

          • 0        • 0
   +   17( 01 ) + 15( 00 )
          0 0        1 0
```

Affine D5 ⊃ A1 + A1 + A3 - Level 1

```
        1 0                      1 0
Sl 0:  ( •0 )  Highest Weight  ( 00 )
        0 0                      0 0

        0 0
Sl 1:  ( •1 )
        1 0

        1 1         1 0         1 0
Sl 2:  ( •0 ) +    ( •0 ) +    ( •0 )
        0 1         2 0         0 0

        2 0         0 2         0 0          0 0
Sl 3:  ( •1 ) +    ( •0 ) +    ( •0 ) +   2( •1 )
        1 0         1 0         1 2          1 0

        1 1         1 0          1 1         3 0          1 0          1 0
Sl 4:  ( •0 ) +    ( •2 ) +   3( •0 ) +    ( •0 ) +   2( •0 ) +   3( •0 )
        2 1         0 0          0 1         0 0          2 0          0 0

        0 1         2 2         2 0          2 0         0 0          0 2          0 0          0 0
Sl 5:  ( •1 ) +    ( •0 ) +    ( •0 ) +   3( •1 ) +    ( •1 ) +   2( •0 ) +   2( •0 ) +   6( •1 )
        1 1         1 0         1 2          1 0         3 0          1 0          1 2          1 0

        1 0         1 2         1 0         3 1          1 1          1 0          1 1         3 0         3 0
Sl 6:  ( •2 ) +    ( •1 ) +    ( •1 ) +    ( •0 ) +   4( •0 ) +   2( •2 ) +   8( •0 ) +    ( •0 ) +    ( •0 )
        2 0         0 0         0 2         0 1          2 1          0 0          0 1         2 0         0 0

         1 0          1 0
   +  5( •0 ) +   7( •0 )
         2 0          0 0
```

Affine D5 ⊃ A1 + A1 + A3 - Level 1

```
        0 0                      0 0
Sl 0:  ( •0 )  Highest Weight  ( 00 )
        0 1                      0 1

        1 1
Sl 1:  ( •0 )
        1 0

        0 1         2 0         0 0         0 0
Sl 2:  ( •1 ) +    ( •0 ) +    ( •0 ) +    ( •0 )
        0 0         0 1         2 1         0 1

        1 0          1 1
Sl 3:  ( •1 ) +   3( •0 )
        1 1          1 0

        2 1         0 1         0 1         2 0          0 1          2 0          0 0          0 0
Sl 4:  ( •1 ) +    ( •1 ) +    ( •0 ) +    ( •0 ) +   2( •1 ) +   2( •0 ) +   2( •0 ) +   4( •0 )
        0 0         2 0         0 2         2 1          0 0          0 1          2 1          0 1

        1 2          1 0        3 1         1 1          1 1
Sl 5:  ( •0 ) +   4( •1 ) +    ( •0 ) +    ( •0 ) +   7( •0 )
        1 1          1 1        1 0         3 0          1 0

        2 1         2 1         0 1         0 0          2 1          0 1         0 3          0 1          2 0
Sl 6:  ( •1 ) +    ( •0 ) +    ( •0 ) +    ( •2 ) +   3( •1 ) +   3( •1 ) +    ( •0 ) +   2( •0 ) +   3( •0 )
        2 0         0 2         2 2         0 1          0 0          2 0         0 0          0 2          2 1

         0 1          2 0          0 0          0 0
   +  6( •1 ) +   6( •0 ) +   6( •0 ) +   8( •0 )
         0 0          0 1          2 1          0 1
```

Affine D5 ⊃ A1 + A1 + A3 - Level 2

```
         0  1                      0  1
sl 0:    ( •0 )  Highest Weight  ( 00 )
         0  1                      0  1

          1  2        1  0        1  0
sl 1:    ( •0 ) +    ( •0 ) +    ( •1 )
          1  0        1  2        1  0

          2  1        0  2        0  0        2  1        0  1        0  0        0  1        2  0        0  0
sl 2:    ( •0 ) +    ( •1 ) +    ( •1 ) +   2( •0 ) +   2( •0 ) +    ( •2 ) +   3( •0 ) +    ( •0 ) +    ( •0 )
          2  1        0  0        0  2        0  1        2  1        0  0        0  1        0  0        2  0

          0  0
    +    ( •0 )
          0  0

          1  1        3  2        3  0        1  2        1  0        3  0        1  0        1  2        1  0
sl 3:   3( •1 ) +    ( •0 ) +    ( •0 ) +    ( •0 ) +    ( •0 ) +    ( •1 ) +    ( •1 ) +   5( •0 ) +   5( •0 )
          1  1        1  0        1  2        3  0        3  2        1  0        3  0        1  0        1  2

          1  0
    +   7( •1 )
          1  0

          2  2        2  0        0  1        2  0        2  2        2  0        0  2        0  0        0  2
sl 4:    ( •1 ) +    ( •1 ) +    ( •2 ) +    ( •2 ) +   2( •1 ) +   2( •1 ) +   2( •1 ) +   2( •1 ) +   2( •0 )
          2  0        2  2        0  1        2  0        0  0        0  2        2  0        2  2        0  2

          4  1        2  1        2  0        0  1        0  0        0  2        0  0        2  1        0  1
    +    ( •0 ) +   8( •0 ) +   3( •2 ) +    ( •0 ) +   3( •2 ) +   4( •1 ) +   4( •1 ) +  10( •0 ) +  10( •0 )
          0  1        2  1        0  0        4  1        2  0        0  0        0  2        0  1        2  1

          0  0        0  1        2  0        2  0        0  0        0  0
    +   4( •2 ) +  14( •0 ) +   3( •0 ) +   4( •0 ) +   4( •0 ) +   6( •0 )
          0  0        0  1        2  0        0  0        2  0        0  0
```

Affine D5 ⊃ A1 + A1 + A3 - Level 2

```
         0  0                      0  0
sl 0:    ( •0 )  Highest Weight  ( 10 )
         0  0                      0  0

          1  0
sl 1:    ( •1 )
          1  0

          2  1        0  1        0  0        0  1        2  0        2  0        0  0        0  0
sl 2:    ( •0 ) +    ( •0 ) +    ( •2 ) +    ( •0 ) +    ( •0 ) +    ( •0 ) +    ( •0 ) +    ( •0 )
          0  1        2  1        0  0        0  1        2  0        0  0        2  0        0  0

          1  1        3  0        1  0        1  2        1  0        1  0
sl 3:    ( •1 ) +    ( •1 ) +    ( •1 ) +   2( •0 ) +   2( •0 ) +   5( •1 )
          1  1        1  0        3  0        1  0        1  2        1  0

          2  0        2  2        2  0        0  2        0  0        0  2        2  1        2  0        0  0
sl 4:    ( •2 ) +    ( •1 ) +    ( •1 ) +    ( •1 ) +    ( •1 ) +    ( •0 ) +   3( •0 ) +   2( •2 ) +   2( •2 )
          2  0        0  0        0  2        2  0        2  2        0  2        2  1        0  0        2  0

          0  2        0  0        2  1        0  1        0  0        0  1        4  0        2  0        0  0
    +    ( •1 ) +    ( •1 ) +   5( •0 ) +   5( •0 ) +   3( •2 ) +   7( •0 ) +    ( •0 ) +   4( •0 ) +    ( •0 )
          0  0        0  2        0  1        2  1        0  0        0  1        0  0        2  0        4  0

          2  0        0  0        0  0
    +   4( •0 ) +   4( •0 ) +   5( •0 )
          0  0        2  0        0  0

          3  1        1  1        1  3        1  1        1  0        1  1        3  0        3  2        3  0
sl 5:    ( •1 ) +    ( •1 ) +    ( •0 ) +    ( •0 ) +    ( •3 ) +  10( •1 ) +    ( •1 ) +   2( •0 ) +   2( •0 )
          1  1        3  1        1  1        1  3        1  0        1  1        3  0        1  0        1  2

          1  2        1  0        3  0        1  0        1  2        1  0        1  0
    +   2( •0 ) +   2( •0 ) +   6( •1 ) +   6( •1 ) +  10( •0 ) +  10( •0 ) +  22( •1 )
          3  0        3  2        1  0        3  0        1  0        1  2        1  0
```

Affine D5 ⊃ A1 + A1 + A3 - Level 2

```
          2 0                 2 0
Sl 0:    ( •0 ) Highest Weight ( 00 )
          0 0                 0 0

          1 0
Sl 1:    ( •1 )
          1 0

          0 0         2 1         0 1         2 0         2 0         0 0         0 0
Sl 2:    ( •2 ) +    ( •0 ) +    ( •0 ) +    ( •0 ) +    ( •0 ) +    ( •0 ) +    ( •0 )
          2 0         0 1         0 1         2 0         0 0         2 0         0 0

          1 1         3 0         1 0         1 2         1 0         1 0
Sl 3:    ( •1 ) +    ( •1 ) +    ( •1 ) +    ( •0 ) +    ( •0 ) +  4( •1 )
          1 1         1 0         3 0         1 0         1 2         1 0

          2 2         2 0         0 2         0 0         2 1         2 0         0 0         0 2         0 0
Sl 4:    ( •0 ) +    ( •2 ) +    ( •1 ) +    ( •1 ) +  2( •0 ) +  2( •2 ) +  2( •2 ) +    ( •1 ) +    ( •1 )
          0 2         2 0         2 0         2 2         2 1         0 0         2 0         0 0         0 2

          2 1         0 1         0 0         2 0         0 1         4 0         2 0         2 0         0 0
    +  4( •0 ) +  3( •0 ) +  2( •2 ) +    ( •0 ) +  4( •0 ) +    ( •0 ) +  3( •0 ) +  5( •0 ) +  3( •0 )
          0 1         2 1         0 0         4 0         0 1         0 0         2 0         0 0         2 0

          0 0
    +  3( •0 )
          0 0

          3 1         1 1         1 3         1 1         1 0         1 1         3 0         3 2         3 0
Sl 5:    ( •1 ) +    ( •1 ) +    ( •0 ) +    ( •0 ) +    ( •3 ) +  7( •1 ) +    ( •1 ) +    ( •0 ) +    ( •0 )
          1 1         3 1         1 1         1 3         1 0         1 1         3 0         1 0         1 2

          1 2         1 0         3 0         1 0         1 2         1 0         1 0
    +    ( •0 ) +    ( •0 ) +  5( •1 ) +  5( •1 ) +  6( •0 ) +  6( •0 ) + 16( •1 )
          3 0         3 2         1 0         3 0         1 0         1 2         1 0
```

Affine D5 ⊃ A1 + A1 + A3 - Level 2

```
          0 1                 0 1
Sl 0:    ( •0 ) Highest Weight ( 00 )
          1 0                 1 0

          1 1         1 0         1 0
Sl 1:    ( •1 ) +    ( •0 ) +    ( •0 )
          0 0         2 1         0 1

          2 0         0 2         0 0         2 1         0 1         0 1
Sl 2:    ( •1 ) +    ( •0 ) +  2( •1 ) +  2( •0 ) +    ( •0 ) +  3( •0 )
          1 1         1 1         1 1         1 0         3 0         1 0

          1 1         1 0         3 1         1 1         1 3         1 1         3 0         1 1         3 0
Sl 3:    ( •0 ) +    ( •2 ) +    ( •1 ) +  3( •1 ) +    ( •0 ) +  2( •0 ) +    ( •0 ) +  5( •1 ) +    ( •0 )
          2 2         0 1         0 0         2 0         0 0         0 2         2 1         0 0         0 1

          1 0         1 0
    +  5( •0 ) +  6( •0 )
          2 1         0 1

          2 1         0 1         2 0         2 2         2 0         0 2         0 1         2 0         0 0
Sl 4:    ( •2 ) +    ( •1 ) +    ( •1 ) +  3( •0 ) +    ( •0 ) +    ( •0 ) +  2( •2 ) +  8( •1 ) +  2( •1 )
          1 0         1 2         3 1         1 1         1 3         3 1         1 0         1 1         3 1

          0 2         0 0         4 1         2 1         0 0         2 1         0 1         0 1
    +  6( •0 ) +    ( •0 ) +    ( •0 ) +  2( •0 ) + 11( •1 ) + 10( •0 ) +  4( •0 ) + 13( •0 )
          1 1         1 3         1 0         3 0         1 1         1 0         3 0         1 0

          1 2         3 1         3 0         1 0         1 2         1 0         3 1         3 3         3 1
Sl 5:    ( •1 ) +    ( •0 ) +    ( •2 ) +  3( •2 ) +  3( •1 ) +    ( •1 ) +  3( •1 ) +    ( •0 ) +  2( •0 )
          2 1         2 2         0 1         2 1         0 1         0 3         2 0         0 0         0 2

          1 1         1 3         1 1         1 0         3 1         1 1         1 3         1 1         3 0
    +    ( •1 ) +  2( •0 ) +  9( •0 ) +  6( •2 ) +  6( •1 ) + 18( •1 ) +  4( •0 ) + 12( •0 ) +  6( •0 )
          4 0         2 0         2 2         0 1         0 0         2 0         0 0         0 2         2 1

          1 0         1 1         3 0         1 0         1 0
    +  2( •0 ) + 24( •1 ) +  7( •0 ) + 22( •0 ) + 23( •0 )
          4 1         0 0         0 1         2 1         0 1
```

Affine D5 ⊃ A1 + A1 + A3 - Level 2

```
           0  0                        0  0
sl 0:     ( •0 )    Highest Weight    ( 00 )
           0  2                        0  2

           1  1
sl 1:     ( •0 )
           1  1

           2  2        0  1        2  0        0  0        2  0        0  0        0  2        0  0        0  0
sl 2:     ( •0 ) +    ( •1 ) +    ( •0 ) +    ( •0 ) +    ( •1 ) +    ( •1 ) +    ( •0 ) +    ( •0 ) +    ( •1 )
           2  0        0  1        0  2        2  2        0  0        2  0        0  0        0  2        0  0

           1  2        1  0        3  1        1  1        1  0        1  1        1  0
sl 3:     ( •1 ) +    ( •1 ) +    ( •0 ) +    ( •0 ) +    ( •2 ) +   5( •0 ) +   2( •0 )
           1  0        1  2        1  1        3  1        1  0        1  1        1  0

           2  1        0  2        2  1        0  1        0  1        4  0        2  2        2  0        0  0
sl 4:     ( •1 ) +    ( •2 ) +   2( •1 ) +   2( •1 ) +    ( •0 ) +    ( •0 ) +   3( •0 ) +   2( •0 ) +    ( •0 )
           2  1        0  0        0  1        2  1        0  3        0  2        2  0        2  2        4  2

           0  1        2  0        2  2        2  0        0  2        0  0        2  0        0  0        0  2
    +    4( •1 ) +   3( •1 ) +   3( •0 ) +   3( •0 ) +   3( •0 ) +   3( •0 ) +   4( •1 ) +   4( •1 ) +   3( •0 )
           0  1        2  0        0  0        0  2        2  0        2  2        0  0        2  0        0  0

           0  0        0  0
    +    6( •0 ) +   5( •1 )
           0  2        0  0

           1  1        3  2        3  0        1  2        1  0        1  2        3  1        3  0        1  0
sl 5:     ( •2 ) +    ( •1 ) +    ( •1 ) +    ( •1 ) +    ( •1 ) +   2( •0 ) +    ( •0 ) +    ( •2 ) +    ( •2 )
           1  1        1  0        1  2        3  0        3  2        1  2        3  1        1  0        3  0

           1  2        1  0        3  1        1  1        1  0        1  1        3  0        1  0        1  0
    +    6( •1 ) +   6( •1 ) +   6( •0 ) +   6( •0 ) +   7( •2 ) +  22( •0 ) +   2( •0 ) +   2( •0 ) +   8( •0 )
           1  0        1  2        1  1        3  1        1  0        1  1        1  0        3  0        1  0
```

Affine D5 ⊃ A1 + A1 + A3 - Level 2

```
           0  0                        0  0
sl 0:     ( •1 )    Highest Weight    ( 01 )
           0  0                        0  0

           1  1        1  0
sl 1:     ( •0 ) +    ( •0 )
           1  1        1  0

           0  1        2  0        2  2        2  0        0  2        0  0        2  0        0  0        0  2
sl 2:     ( •1 ) +    ( •1 ) +    ( •0 ) +    ( •0 ) +    ( •0 ) +    ( •0 ) +    ( •1 ) +    ( •1 ) +    ( •0 )
           0  1        2  0        0  0        0  2        2  0        2  2        0  0        2  0        0  0

           0  0        0  0
    +     ( •0 ) +   2( •1 )
           0  2        0  0

           1  2        1  0        3  1        1  1        1  0        1  1        3  0        1  0        1  0
sl 3:     ( •1 ) +    ( •1 ) +    ( •0 ) +    ( •0 ) +   2( •2 ) +   7( •0 ) +    ( •0 ) +    ( •0 ) +   4( •0 )
           1  0        1  2        1  1        3  1        1  0        1  1        1  0        3  0        1  0

           2  1        2  1        0  1        0  3        0  1        0  0        2  2        2  0        0  1
sl 4:     ( •1 ) +   3( •1 ) +   3( •1 ) +    ( •0 ) +    ( •0 ) +    ( •3 ) +   3( •0 ) +   3( •0 ) +   5( •1 )
           2  1        0  1        2  1        0  1        0  3        0  0        2  0        2  2        0  1

           4  0        2  0        2  2        2  0        0  0        0  2        0  0        2  0        0  0
    +     ( •1 ) +   5( •1 ) +   4( •0 ) +   4( •0 ) +    ( •1 ) +   4( •0 ) +   4( •0 ) +   7( •1 ) +   7( •1 )
           0  0        2  0        0  0        0  2        4  0        2  0        2  2        0  0        2  0

           0  2        0  0        0  0
    +    5( •0 ) +   5( •0 ) +  10( •1 )
           0  0        0  2        0  0

           1  1        3  2        3  0        1  2        1  0        1  2        3  1        3  0        1  0
sl 5:     ( •2 ) +    ( •1 ) +    ( •1 ) +    ( •1 ) +    ( •1 ) +   3( •0 ) +    ( •0 ) +   2( •2 ) +   2( •2 )
           1  1        1  0        1  2        3  0        3  2        1  2        3  1        1  0        3  0

           1  2        1  0        3  1        1  1        1  0        1  1        3  0        3  0        1  0
    +    8( •1 ) +   8( •1 ) +   8( •0 ) +   8( •0 ) +  12( •2 ) +  32( •0 ) +    ( •0 ) +   5( •0 ) +   5( •0 )
           1  0        1  2        1  1        3  1        1  0        1  1        3  0        1  0        3  0

           1  0
    +   16( •0 )
           1  0
```

Affine D5 ⊃ A1 + A1 + A3 - Level 2

```
                                  1  0
          1  0                    1  0
Sl 0:    ( •0 )  Highest Weight  ( 00 )
          1  0                    1  0

          2  0        0  0        0  0
Sl 1:    ( •1 ) +    ( •1 ) +    ( •1 )
          0  0        2  0        0  0

          1  0        1  1        3  0        1  0        1  0
Sl 2:    ( •2 ) +   2( •0 ) +    ( •0 ) +    ( •0 ) +   3( •0 )
          1  0        1  1        1  0        3  0        1  0

          2  1        0  1        0  1        2  0        2  2        2  0        0  2        0  0        2  0
Sl 3:    ( •1 ) +    ( •1 ) +    ( •1 ) +   3( •1 ) +    ( •0 ) +    ( •0 ) +    ( •0 ) +    ( •0 ) +   4( •1 )
          0  1        2  1        0  1        2  0        0  0        0  2        2  0        2  2        0  0

          0  0        0  2        0  0        0  0
    +   4( •1 ) +   2( •0 ) +   2( •0 ) +   4( •1 )
          2  0        0  0        0  2        0  0

          1  2        3  0        1  0        1  2        1  0        3  1        1  1        1  0        1  1
Sl 4:    ( •0 ) +    ( •2 ) +    ( •2 ) +   2( •1 ) +   2( •1 ) +   2( •0 ) +   2( •0 ) +   6( •2 ) +  12( •0 )
          1  2        1  0        3  0        1  0        1  2        1  1        3  1        1  0        1  1

          3  0        3  0        1  0        1  0
    +    ( •0 ) +   4( •0 ) +   4( •0 ) +  11( •0 )
          3  0        1  0        3  0        1  0

          2  1        2  3        2  1        2  0        0  3        0  1        0  0        2  1        0  1
Sl 5:   3( •1 ) +    ( •0 ) +    ( •0 ) +    ( •3 ) +    ( •0 ) +    ( •0 ) +    ( •3 ) +   7( •1 ) +   7( •1 )
          2  1        0  1        0  3        0  0        2  1        2  3        2  0        0  1        2  1

          0  3        0  1        0  0        4  0        2  0        2  2        2  0        0  1        4  0
    +    ( •0 ) +    ( •0 ) +    ( •3 ) +    ( •1 ) +    ( •1 ) +   4( •0 ) +   4( •0 ) +   9( •1 ) +   2( •1 )
          0  1        0  3        0  0        2  0        4  0        2  0        2  2        0  1        0  0

          2  0        2  2        2  0        0  0        0  2        0  0        2  0        0  0        0  2
    +  13( •1 ) +   6( •0 ) +   6( •0 ) +   2( •1 ) +   6( •0 ) +   6( •0 ) +  17( •1 ) +  17( •1 ) +   7( •0 )
          2  0        0  0        0  2        4  0        2  0        2  2        0  0        2  0        0  0

          0  0        0  0
    +   7( •0 ) +  17( •1 )
          0  2        0  0
```

Affine D6 Branching Rules

Affine D6 ⊃ D6 - Level 1

```
         •  0                   1  0
sl 0:   ( 000 )  Highest Weight ( 000 )
         0  0                   0  0

         •  0
sl 1:   ( 100 )
         0  0

         •  0        •  0        •  0        •  0
sl 2:   ( 001 ) +   ( 000 ) +   ( 100 ) +   ( 000 )
         0  0        2  0        0  0        0  0

         •  0        •  2        •  0        •  0        •  0         •  0        •  0
sl 3:   ( 010 ) +   ( 000 ) +   ( 000 ) +   ( 001 ) +   ( 000 ) +  3( 100 ) +   ( 000 )
         1  0        0  0        0  2        0  0        2  0         0  0        0  0

         •  1        •  0         •  0       •  2        •  0         •  0         •  0         •  0
sl 4:   ( 000 ) +   ( 200 ) +  2( 010 ) +   ( 000 ) +   ( 000 ) +  4( 001 ) +  3( 000 ) +  4( 100 )
         1  1        0  0         1  0       0  0        0  2         0  0         2  0         0  0

          •  0
   +  3( 000 )
          0  0

         •  0         •  1       •  0        •  0         •  0         •  2         •  0         •  0
sl 5:   ( 101 ) +  3( 000 ) +   ( 100 ) +   ( 200 ) +  5( 010 ) +  3( 000 ) +  3( 000 ) +  6( 001 )
         0  0         1  1       2  0        0  0         1  0         0  0         0  2         0  0

          •  0          •  0         •  0
   +  4( 000 ) + 10( 100 ) +  3( 000 )
          2  0          0  0         0  0

         •  0        •  2        •  0        •  0         •  0         •  1         •  0         •  0
sl 6:   ( 001 ) +   ( 100 ) +   ( 100 ) +   ( 020 ) +  2( 101 ) +  7( 000 ) +  2( 100 ) +  4( 200 )
         2  0        0  0        0  2        0  0         0  0         1  1         2  0         0  0

          •  0         •  2         •  0          •  0         •  0          •  0         •  0
   + 10( 010 ) +  5( 000 ) +  5( 000 ) + 14( 001 ) +  9( 000 ) + 14( 100 ) +  8( 000 )
          1  0         0  0         0  2          0  0         2  0          0  0         0  0
```

Affine D6 ⊃ D6 - Level 1

```
         •  1                   0  1
sl 0:   ( 000 )  Highest Weight ( 000 )
         0  0                   0  0

         •  0        •  1
sl 1:   ( 000 ) +   ( 000 )
         1  1        0  0

         •  1         •  0         •  1
sl 2:   ( 100 ) +  2( 000 ) +  2( 000 )
         0  0         1  1         0  0

         •  0        •  1         •  1         •  0         •  1
sl 3:   ( 010 ) +   ( 000 ) +  2( 100 ) +  4( 000 ) +  4( 000 )
         0  1        2  0         0  0         1  1         0  0

         •  0        •  1         •  0         •  1         •  1         •  0         •  1
sl 4:   ( 100 ) +   ( 001 ) +  2( 010 ) +  2( 000 ) +  5( 100 ) +  8( 000 ) +  7( 000 )
         1  1        0  0         0  1         2  0         0  0         1  1         0  0

         •  1        •  1         •  0         •  1         •  0         •  1          •  1          •  0
sl 5:   ( 010 ) +   ( 000 ) +  3( 100 ) +  2( 001 ) +  5( 010 ) +  5( 000 ) + 10( 100 ) + 15( 000 )
         1  0        0  2         1  1         0  0         0  1         2  0          0  0          1  1

          •  1
   + 12( 000 )
          0  0
```

Affine D6 ⊃ D6 - Level 1

```
           •  0                       0  0
Sl 0:    ( 000 )  Highest Weight    ( 000 )
           1  0                       1  0

           •  0        •  0
Sl 1:    ( 010 ) +   ( 000 )
           0  0        1  0

           •  1        •  0        •  0        •  0
Sl 2:    ( 000 ) +   ( 100 ) +   ( 010 ) +  2( 000 )
           0  1        1  0        0  0        1  0

           •  0        •  1        •  0        •  0        •  0
Sl 3:    ( 001 ) +  2( 000 ) +  2( 100 ) +  3( 010 ) +  3( 000 )
           1  0        0  1        1  0        0  0        1  0

           •  0        •  2        •  0        •  0        •  0        •  1        •  0        •  0
Sl 4:    ( 110 ) +   ( 000 ) +   ( 000 ) +  2( 001 ) +   ( 000 ) +  4( 000 ) +  4( 100 ) +  6( 010 )
           0  0        1  0        1  2        1  0        3  0        0  1        1  0        0  0

           •  0
    +   6( 000 )
           1  0

           •  1        •  0        •  0        •  2        •  0        •  0        •  0        •  1
Sl 5:    ( 100 ) +   ( 010 ) +  2( 110 ) +  2( 000 ) +  2( 000 ) +  6( 001 ) +   ( 000 ) +  8( 000 )
           0  1        2  0        0  0        1  0        1  2        1  0        3  0        0  1

           •  0        •  0        •  0
    +   8( 100 ) + 11( 010 ) + 10( 000 )
           1  0        0  0        1  0

           •  0        •  1        •  1        •  0        •  0        •  0        •  2        •  0
Sl 6:    ( 011 ) +   ( 000 ) +  3( 100 ) +   ( 200 ) +  2( 010 ) +  5( 110 ) +  5( 000 ) +  5( 000 )
           0  0        2  1        0  1        1  0        2  0        0  0        1  0        1  2

           •  0        •  0        •  1        •  0        •  0        •  0
    +  12( 001 ) +  3( 000 ) + 16( 000 ) + 16( 100 ) + 19( 010 ) + 17( 000 )
           1  0        3  0        0  1        1  0        0  0        1  0
```

Affine D6 ⊃ A1 + A1 + D4 - Level 1

```
           1  0                       1  0
Sl 0:    ( •00 )  Highest Weight    ( 000 )
           0  0                       0  0

           0  0
Sl 1:    ( •10 )
           1  0

           1  0        1  0        1  0
Sl 2:    ( •01 ) +   ( •00 ) +   ( •00 )
           0  0        2  0        0  0

           0  1        2  0        0  0
Sl 3:    ( •00 ) +   ( •10 ) +  2( •10 )
           1  1        1  0        1  0

           1  0        1  0        1  2        1  0        1  0        3  0        1  0        1  0
Sl 4:    ( •01 ) +   ( •20 ) +   ( •00 ) +   ( •00 ) +  2( •01 ) +   ( •00 ) +  2( •00 ) +  3( •00 )
           2  0        0  0        0  0        0  2        0  0        0  0        2  0        0  0

           0  0        2  1        0  1        2  0        0  0        0  0
Sl 5:    ( •11 ) +   ( •00 ) +  3( •00 ) +  3( •10 ) +   ( •10 ) +  5( •10 )
           1  0        1  1        1  1        1  0        3  0        1  0   (

           1  1        1  0        1  2        1  0        3  0        1  0        1  0        1  2
Sl 6:    ( •10 ) +   ( •20 ) +   ( •00 ) +   ( •00 ) +   ( •01 ) +  3( •01 ) +  2( •20 ) +  2( •00 )
           0  1        2  0        2  0        2  2        0  0        2  0        0  0        0  0

           1  0        1  0        3  0        3  0        1  0        1  0
    +   2( •00 ) +  7( •01 ) +   ( •00 ) +   ( •00 ) +  5( •00 ) +  6( •00 )
           0  2        0  0        2  0        0  0        2  0        0  0
```

Affine D6 ⊃ A1 + A1 + D4 - Level 1

```
        0   1                        0   1
sl 0:  ( •00 )   Highest Weight    ( 000 )
        0   0                        0   0

        1   0
sl 1:  ( •00 )
        1   1

        0   0          2   1          0   1          0   1
sl 2:  ( •10 )  +     ( •00 )  +     ( •00 )  +     ( •00 )
        0   1          0   0          2   0          0   0

        1   1          1   0
sl 3:  ( •10 )  +    3( •00 )
        1   0          1   1

        0   1          2   0          0   0          0   0          2   1          2   1          0   1          0   1
sl 4:  ( •01 )  +     ( •10 )  +     ( •10 )  +    2( •10 )  +     ( •00 )  +    2( •00 )  +    2( •00 )  +    4( •00 )
        0   0          0   1          2   1          0   1          2   0          0   0          2   0          0   0

        1   0          1   1          3   0          1   0          1   0
sl 5:  ( •01 )  +    4( •10 )  +     ( •00 )  +     ( •00 )  +    7( •00 )
        1   1          1   0          1   1          3   1          1   1

        2   1          0   1          0   1          0   1          2   0          0   1          2   0          0   0
sl 6:  ( •01 )  +     ( •01 )  +     ( •20 )  +     ( •00 )  +     ( •10 )  +    2( •01 )  +    3( •10 )  +    3( •10 )
        0   0          2   0          0   0          0   2          2   1          0   0          0   1          2   1

        0   0          2   1          2   1          0   1          0   1
   +  6( •10 )  +    3( •00 )  +    6( •00 )  +    6( •00 )  +    8( •00 )
        0   1          2   0          0   0          2   0          0   0
```

Affine D6 ⊃ A3 + A3 - Level 1

```
        1   0                        1   0
sl 0:  ( 0•0 )   Highest Weight    ( 000 )
        0   0                        0   0

        0   0
sl 1:  ( 0•1 )
        1   0

        1   1          0   0          1   0
sl 2:  ( 0•0 )  +     ( 1•0 )  +     ( 0•0 )
        0   1          1   0          0   0

        1   0          0   2          0   0          0   0
sl 3:  ( 1•1 )  +     ( 0•0 )  +     ( 0•0 )  +    2( 0•1 )
        0   0          1   0          1   2          1   0

        0   1          1   0          2   0          1   1          0   0          1   0
sl 4:  ( 1•0 )  +     ( 0•2 )  +     ( 0•0 )  +    3( 0•0 )  +    2( 1•0 )  +    3( 0•0 )
        1   1          0   0          1   0          0   1          1   0          0   0

        1   2          1   0          1   0          0   1          1   0          0   2          0   0          0   0
sl 5:  ( 1•0 )  +     ( 1•0 )  +     ( 0•1 )  +     ( 0•1 )  +    3( 1•1 )  +    2( 0•0 )  +    2( 0•0 )  +    6( 0•1 )
        0   0          0   2          2   0          1   1          0   0          1   0          1   2          1   0

        2   1          0   0          1   2          1   0          0   1          1   0          1   0          2   0
sl 6:  ( 0•0 )  +     ( 1•2 )  +     ( 0•1 )  +     ( 0•1 )  +    4( 1•0 )  +     ( 2•0 )  +    2( 0•2 )  +    2( 0•0 )
        1   1          1   0          0   0          0   2          1   1          0   0          0   0          1   0

        1   1          0   0          0   0          1   0
   +  8( 0•0 )  +     ( 0•0 )  +    5( 1•0 )  +    7( 0•0 )
        0   1          3   0          1   0          0   0
```

Affine D7 Branching Rules

Affine D7 ⊃ D7 - Level 1

```
          •  0                       1  0
sl 0:   ( 0000 )  Highest Weight   ( 0000 )
          0  0                       0  0

          •  0
sl 1:   ( 1000 )
          0  0

          •  0          •  0          •  0          •  0
sl 2:   ( 0010 ) +    ( 0000 ) +    ( 1000 ) +    ( 0000 )
          0  0          2  0          0  0          0  0

          •  0          •  1          •  0          •  0           •  0           •  0
sl 3:   ( 0100 ) +    ( 0000 ) +    ( 0010 ) +    ( 0000 ) +   3( 1000 ) +    ( 0000 )
          1  0          0  1          0  0          2  0           0  0           0  0

          •  0          •  0           •  0           •  1           •  0           •  0           •  0
sl 4:   ( 0001 ) +    ( 2000 ) +   2( 0100 ) +   2( 0000 ) +   3( 0010 ) +   3( 0000 ) +   4( 1000 )
          1  0          0  0           1  0           0  1           0  0           2  0           0  0

          •  0
   +   3( 0000 )
          0  0

          •  0          •  2          •  0           •  0          •  0          •  0           •  0
sl 5:   ( 1010 ) +    ( 0000 ) +    ( 0000 ) +   2( 0001 ) +    ( 1000 ) +    ( 2000 ) +   5( 0100 )
          0  0          1  0          1  2           1  0          2  0          0  0           1  0

          •  1           •  0           •  0           •  0           •  0
   +   4( 0000 ) +   6( 0010 ) +   4( 0000 ) +   9( 1000 ) +   3( 0000 )
          0  1           0  0           2  0           0  0           0  0
```

Affine D7 ⊃ D7 - Level 1

```
          •  1                       0  1
sl 0:   ( 0000 )  Highest Weight   ( 0000 )
          0  0                       0  0

          •  0          •  1
sl 1:   ( 0000 ) +    ( 0000 )
          1  1          0  0

          •  1           •  0           •  1
sl 2:   ( 1000 ) +   2( 0000 ) +   2( 0000 )
          0  0           1  1           0  0

          •  0          •  1           •  1           •  0           •  1
sl 3:   ( 0100 ) +    ( 0000 ) +   2( 1000 ) +   4( 0000 ) +   4( 0000 )
          0  1          2  0           0  0           1  1           0  0

          •  1          •  0           •  0           •  1           •  1           •  0           •  1
sl 4:   ( 0010 ) +    ( 1000 ) +   2( 0100 ) +   2( 0000 ) +   5( 1000 ) +   8( 0000 ) +   7( 0000 )
          0  0          1  1           0  1           2  0           0  0           1  1           0  0

          •  1          •  0           •  1           •  0           •  0           •  1            •  1
sl 5:   ( 0100 ) +    ( 0001 ) +   2( 0010 ) +   3( 1000 ) +   5( 0100 ) +   5( 0000 ) +  10( 1000 )
          1  0          0  1           0  0           1  1           0  1           2  0            0  0

           •  0            •  1
   +  15( 0000 ) +  12( 0000 )
           1  1            0  0

          •  0          •  1           •  1          •  2           •  0          •  0           •  1
sl 6:   ( 0010 ) +    ( 2000 ) +   3( 0100 ) +    ( 0000 ) +   2( 0001 ) +    ( 0000 ) +   5( 0010 )
          1  1          0  0           1  0          0  1           0  1          3  1           0  0

          •  0            •  0            •  1            •  1            •  0            •  1
   +   7( 1000 ) +  11( 0100 ) +  10( 0000 ) +  20( 1000 ) +  27( 0000 ) +  21( 0000 )
          1  1            0  1            2  0            0  0            1  1            0  0
```

Affine D7 ⊃ D7 - Level 1

```
         •   0                          0   0
Sl 0:   ( 0000 )   Highest Weight    ( 0000 )
         1   0                          1   0

         •   0           •   0
Sl 1:   ( 0100 )  +     ( 0000 )
         0   0           1   0

         •   0           •   0           •   0           •   0
Sl 2:   ( 0001 )  +     ( 1000 )  +     ( 0100 )  +   2( 0000 )
         0   0           1   0           0   0           1   0

         •   0           •   2           •   0           •   0           •   0           •   0           •   0
Sl 3:   ( 0010 )  +     ( 0000 )  +     ( 0000 )  +     ( 0001 )  +   2( 1000 )  +   3( 0100 )  +   3( 0000 )
         1   0           0   0           0   2           0   0           1   0           0   0           1   0

         •   1           •   0           •   0           •   2           •   0           •   0           •   0
Sl 4:   ( 0000 )  +     ( 1100 )  +   2( 0010 )  +     ( 0000 )  +     ( 0000 )  +   4( 0001 )  +     ( 0000 )
         1   1           0   0           1   0           0   0           0   2           0   0           3   0

         •   0           •   0           •   0
   +   4( 1000 )  +   5( 0100 )  +   6( 0000 )
         1   0           0   0           1   0

         •   0           •   0           •   1           •   0           •   0           •   2           •   0
Sl 5:   ( 1001 )  +     ( 0100 )  +   3( 0000 )  +   2( 1100 )  +   5( 0010 )  +   3( 0000 )  +   3( 0000 )
         0   0           2   0           1   1           0   0           1   0           0   0           0   2

         •   0           •   0           •   0           •   0           •   0
   +   6( 0001 )  +     ( 0000 )  +   8( 1000 )  +  11( 0100 )  +   9( 0000 )
         0   0           3   0           1   0           0   0           1   0
```

Affine D7 ⊃ A1 + A1 + D5 - Level 1

```
         1   0                          1   0
Sl 0:   ( •000 )   Highest Weight    ( 0000 )
         0   0                          0   0

         0   0
Sl 1:   ( •100 )
         1   0

         1   0           1   0           1   0
Sl 2:   ( •010 )  +     ( •000 )  +     ( •000 )
         0   0           2   0           0   0

         0   0           2   0           0   0
Sl 3:   ( •001 )  +     ( •100 )  +   2( •100 )
         1   0           1   0           1   0

         1   1           1   0           1   0           1   0           3   0           1   0           1   0
Sl 4:   ( •000 )  +     ( •010 )  +     ( •200 )  +   2( •010 )  +     ( •000 )  +   2( •000 )  +   3( •000 )
         0   1           2   0           0   0           0   0           0   0           2   0           0   0

         0   0           2   0           0   2           0   0           0   0           2   0           0   0
Sl 5:   ( •110 )  +     ( •001 )  +     ( •000 )  +     ( •000 )  +   2( •001 )  +   3( •100 )  +     ( •100 )
         1   0           1   0           1   0           1   2           1   0           1   0           3   0

         0   0
   +   5( •100 )
         1   0

         1   0           1   1           1   1           1   0           3   0           1   0           1   0
Sl 6:   ( •101 )  +     ( •000 )  +   3( •000 )  +     ( •200 )  +     ( •010 )  +   3( •010 )  +   2( •200 )
         0   0           2   1           0   1           2   0           0   0           2   0           0   0

         1   0           3   0           3   0           1   0           1   0
   +   6( •010 )  +     ( •000 )  +     ( •000 )  +   5( •000 )  +   6( •000 )
         0   0           2   0           0   0           2   0           0   0
```

Affine D7 ⊃ A1 + A1 + D5 - Level 1

```
        0   1                       0   1
Sl 0:  ( •000 )  Highest Weight  ( 0000 )
        0   0                       0   0

        1   0
Sl 1:  ( •000 )
        1   1

        0   0         2   1         0   1         0   1
Sl 2:  ( •100 )  +  ( •000 )  +  ( •000 )  +  ( •000 )
        0   1         0   0         2   0         0   0

        1   1         1   0
Sl 3:  ( •100 )  + 3( •000 )
        1   0         1   1

        0   1         2   0         0   0         0   0         2   1         2   1         0   1
Sl 4:  ( •010 )  +  ( •100 )  +  ( •100 )  + 2( •100 )  +  ( •000 )  + 2( •000 )  + 2( •000 )
        0   0         0   1         2   1         0   1         2   0         0   0         2   0

        0   1
   + 4( •000 )
        0   0

        1   0         1   1         3   0         1   0         1   0
Sl 5:  ( •010 )  + 4( •100 )  +  ( •000 )  +  ( •000 )  + 7( •000 )
        1   1         1   0         1   1         3   1         1   1

        0   0         2   1         0   1         0   1         0   1         2   0         2   0
Sl 6:  ( •001 )  +  ( •010 )  +  ( •010 )  +  ( •200 )  + 2( •010 )  +  ( •100 )  + 3( •100 )
        0   1         0   0         2   0         0   0         0   0         2   1         0   1

        0   0         0   0         2   1         2   1         0   1         0   1
   + 3( •100 )  + 6( •100 )  + 3( •000 )  + 6( •000 )  + 6( •000 )  + 8( •000 )
        2   1         0   1         2   0         0   0         2   0         0   0
```

Affine D7 ⊃ A3 + D4 - Level 1

```
        1   0                       1   0
Sl 0:  ( 0•00 )  Highest Weight  ( 0000 )
        0   0                       0   0

        0   0
Sl 1:  ( 0•10 )
        1   0

        1   0         0   0         1   0
Sl 2:  ( 0•01 )  +  ( 1•00 )  +  ( 0•00 )
        0   0         1   0         0   0

        0   1         1   0         0   0
Sl 3:  ( 0•00 )  +  ( 1•10 )  + 2( 0•10 )
        1   1         0   0         1   0

        0   0         1   0         1   2         1   0         1   0         2   0         0   0
Sl 4:  ( 1•01 )  +  ( 0•20 )  +  ( 0•00 )  +  ( 0•00 )  + 2( 0•01 )  +  ( 0•00 )  + 2( 1•00 )
        1   0         0   0         0   0         0   2         0   0         1   0         1   0

        1   0
   + 3( 0•00 )
        0   0

        1   1         0   0         1   0         0   1         1   0         0   0
Sl 5:  ( 1•00 )  +  ( 0•11 )  +  ( 0•10 )  + 3( 0•00 )  + 3( 1•10 )  + 5( 0•10 )
        0   1         1   0         2   0         1   1         0   0         1   0

        1   1         2   0         0   0         0   2         0   0         0   0         1   0
Sl 6:  ( 0•10 )  +  ( 0•01 )  +  ( 1•20 )  +  ( 1•00 )  +  ( 1•00 )  + 3( 1•01 )  + 2( 0•20 )
        0   1         1   0         1   0         1   0         1   2         1   0         0   0

        1   2         1   0         1   0         1   0         2   0         0   0         0   0
   + 2( 0•00 )  + 2( 0•00 )  + 7( 0•01 )  +  ( 2•00 )  + 2( 0•00 )  +  ( 0•00 )  + 5( 1•00 )
        0   0         0   2         0   0         0   0         1   0         3   0         1   0

        1   0
   + 6( 0•00 )
        0   0
```

Affine D7 ⊃ A3 + D4 - Level 1

```
         0    1                        0    1
sl 0:  ( 0•00 )   Highest Weight    ( 0000 )
         0    0                        0    0

         0    0
sl 1:  ( 1•00 )
         0    1

         1    1          0    0          0    1
sl 2:  ( 0•00 )  +     ( 0•10 )  +     ( 0•00 )
         1    0          0    1          0    0

         0    1          2    0          0    0          0    0
sl 3:  ( 1•10 )  +     ( 0•00 )  +     ( 0•00 )  +   2( 1•00 )
         0    0          0    1          2    1          0    1

         1    0          0    1          0    1          1    1          0    0          0    1
sl 4:  ( 0•10 )  +     ( 0•01 )  +     ( 2•00 )  +   3( 0•00 )  +   2( 0•10 )  +   3( 0•00 )
         1    1          0    0          0    0          1    0          0    1          0    0

         0    0          2    1          0    1          1    0          0    1          2    0          0    0
sl 5:  ( 1•01 )  +     ( 0•10 )  +     ( 0•10 )  +     ( 1•00 )  +   3( 1•10 )  +   2( 0•00 )  +   2( 0•00 )
         0    1          0    0          2    0          1    1          0    0          0    1          2    1

         0    0
   +  6( 1•00 )
         0    1

         1    1          0    0          1    0          0    1          0    1          2    1          0    1
sl 6:  ( 0•01 )  +     ( 2•10 )  +   4( 0•10 )  +     ( 0•20 )  +     ( 0•00 )  +     ( 1•00 )  +     ( 1•00 )
         1    0          0    1          1    1          0    0          0    2          0    0          2    0

         0    1          0    1          1    1          0    0          0    1
   +  2( 0•01 )  +   2( 2•00 )  +   8( 0•00 )  +   5( 0•10 )  +   7( 0•00 )
         0    0          0    0          1    0          0    1          0    0
```

Affine D8 Branching Rules

Affine D8 ⊃ D8 - Level 1

```
         •     0                       1     0
Sl 0:   ( 00000 )  Highest Weight  ( 00000 )
         0     0                       0     0

         •     0
Sl 1:   ( 10000 )
         0     0

         •     0          •     0          •     0          •     0
Sl 2:   ( 00100 )  +    ( 00000 )  +    ( 10000 )  +    ( 00000 )
         0     0          2     0          0     0          0     0

         •     0          •     0          •     0          •     0          •     0          •     0
Sl 3:   ( 00001 )  +    ( 01000 )  +    ( 00100 )  +    ( 00000 )  +  3( 10000 )  +    ( 00000 )
         0     0          1     0          0     0          2     0          0     0          0     0

         •     0 .        •     2          •     0          •     0          •     0          •     0          •     0
Sl 4:   ( 00010 )  +    ( 00000 )  +    ( 00000 )  +    ( 00001 )  +    ( 20000 )  +  2( 01000 )  +  3( 00100 )
         1     0          0     0          0     2          0     0          0     0          1     0          0     0

         •     0          •     0          •     0
   +  3( 00000 )  +  4( 10000 )  +  3( 00000 )
         2     0          0     0          0     0

         •     1          •     0          •     0          •     2          •     0          •     0          •     0
Sl 5:   ( 00000 )  +    ( 10100 )  +  2( 00010 )  +    ( 00000 )  +    ( 00000 )  +    ( 10000 )  +  4( 00001 )
         1     1          0     0          1     0          0     0          0     2          2     0          0     0

         •     0          •     0          •     0          •     0          •     0          •     0
   +    ( 20000 )  +  5( 01000 )  +  5( 00100 )  +  4( 00000 )  +  9( 10000 )  +  3( 00000 )
         0     0          1     0          0     0          2     0          0     0          0     0

         •     0          •     0          •     0          •     1          •     0          •     0          •     2
Sl 6:   ( 10001 )  +    ( 00100 )  +    ( 02000 )  +  3( 00000 )  +  2( 10100 )  +  5( 00010 )  +  3( 00000 )
         0     0          2     0          0     0          1     1          0     0          1     0          0     0

         •     0          •     0          •     0          •     0          •     0          •     0          •     0
   +  3( 00000 )  +  2( 10000 )  +  6( 00001 )  +  4( 20000 )  +  9( 01000 )  + 12( 00100 )  +  9( 00000 )
         0     2          2     0          0     0          0     0          1     0          0     0          2     0

         •     0          •     0
   + 13( 10000 )  +  7( 00000 )
         0     0          0     0
```

Affine D8 ⊃ D8 - Level 1

```
         •     0                       0     0
Sl 0:   ( 00000 )  Highest Weight  ( 00000 )
         0     1                       0     1

         •     1          •     0
Sl 1:   ( 00000 )  +    ( 00000 )
         1     0          0     1

         •     0          •     1          •     0
Sl 2:   ( 10000 )  +  2( 00000 )  +  2( 00000 )
         0     1          1     0          0     1

         •     1          •     0          •     0          •     1          •     0
Sl 3:   ( 01000 )  +    ( 00000 )  +  2( 10000 )  +  4( 00000 )  +  4( 00000 )
         0     0          2     1          0     1          1     0          0     1

         •     0          •     1          •     1          •     0          •     0          •     1          •     0
Sl 4:   ( 00100 )  +    ( 10000 )  +  2( 01000 )  +  2( 00000 )  +  5( 10000 )  +  8( 00000 )  +  7( 00000 )
         0     1          1     0          0     0          2     1          0     1          1     0          0     1

         •     0          •     1          •     0          •     1          •     1          •     0          •     0
Sl 5:   ( 01000 )  +    ( 00010 )  +  2( 00100 )  +  3( 10000 )  +  5( 01000 )  +  5( 00000 )  + 10( 10000 )
         1     1          0     0          0     1          1     0          0     0          2     1          0     1

         •     1          •     0
   + 15( 00000 )  + 12( 00000 )
         1     0          0     1
```

(continued next page)

```
          •    1        •    0        •    0         •    0        •    1         •    0         •    1
Sl 6:    ( 00100 ) +   ( 00001 ) +   ( 20000 ) +  3( 01000 ) +  2( 00010 ) +  5( 00100 ) +    ( 00000 )
          1    0        0    1        0    1         1    1        0    0         0    1         3    0

          •    1        •    1        •    0         •    0        •    1         •    0
  +    7( 10000 ) + 11( 01000 ) + 10( 00000 ) + 20( 10000 ) + 27( 00000 ) + 21( 00000 )
          1    0        0    0        2    1         0    1        1    0         0    1
```

Affine D8 ⊃ D8 - Level 1

```
          •    0                          0    0
Sl 0:    ( 00000 )   Highest Weight     ( 00000 )
          1    0                          1    0

          •    0        •    0
Sl 1:    ( 01000 ) +   ( 00000 )
          0    0        1    0

          •    0        •    0        •    0         •    0
Sl 2:    ( 00010 ) +   ( 10000 ) +   ( 01000 ) +  2( 00000 )
          0    0        1    0        0    0         1    0

          •    0        •    1        •    0         •    0        •    0         •    0
Sl 3:    ( 00100 ) +   ( 00000 ) +   ( 00010 ) +  2( 10000 ) +  3( 01000 ) +  3( 00000 )
          1    0        0    1        0    0         1    0        0    0         1    0

          •    0        •    0        •    0         •    1        •    0         •    0         •    0
Sl 4:    ( 00001 ) +   ( 11000 ) +  2( 00100 ) +  2( 00000 ) +  3( 00010 ) +    ( 00000 ) +  4( 10000 )
          1    0        0    0        1    0         0    1        0    0         3    0         1    0

          •    0        •    0
  +    5( 01000 ) +  6( 00000 )
          0    0        1    0

          •    0        •    2        •    0         •    0        •    0         •    0         •    0
Sl 5:    ( 10010 ) +   ( 00000 ) +   ( 00000 ) +  2( 00001 ) +    ( 01000 ) +  2( 11000 ) +  5( 00100 )
          0    0        1    0        1    2         1    0        2    0         0    0         1    0

          •    1        •    0        •    0         •    0        •    0         •    0
  +    4( 00000 ) +  6( 00010 ) +    ( 00000 ) +  8( 10000 ) + 10( 01000 ) +  9( 00000 )
          0    1        0    0        3    0         1    0        0    0         1    0

          •    1        •    0        •    0         •    0        •    2         •    0         •    0
Sl 6:    ( 10000 ) +   ( 01100 ) +   ( 00010 ) +  2( 10010 ) +  2( 00000 ) +  2( 00000 ) +  6( 00001 )
          0    1        0    0        2    0         0    0        1    0         1    2         1    0

          •    0        •    0        •    0         •    0        •    1         •    0         •    0
  +      ( 20000 ) +  2( 01000 ) +  5( 11000 ) + 10( 00100 ) +  8( 00000 ) + 12( 00010 ) +  3( 00000 )
          1    0        2    0        0    0         1    0        0    1         0    0         3    0

          •    0        •    0        •    0
  +   15( 10000 ) + 17( 01000 ) + 16( 00000 )
          1    0        0    0        1    0
```

Affine D8 ⊃ A1 + A1 + D6 - Level 1

```
          1    0                          1    0
Sl 0:    ( •0000 )   Highest Weight     ( 00000 )
          0    0                          0    0

          0    0
Sl 1:    ( •1000 )
          1    0

          1    0        1    0        1    0
Sl 2:    ( •0100 ) +   ( •0000 ) +   ( •0000 )
          0    0        2    0        0    0

          0    0        2    0        0    0
Sl 3:    ( •0010 ) +   ( •1000 ) +  2( •1000 )
          1    0        1    0        1    0

          1    0        1    0        1    0         1    0        3    0         1    0         1    0
Sl 4:    ( •0001 ) +   ( •0100 ) +   ( •2000 ) +  2( •0100 ) +    ( •0000 ) +  2( •0000 ) +  3( •0000 )
          0    0        2    0        0    0         0    0        0    0         2    0         0    0

          0    1        0    0        2    0         0    0        2    0         0    0         0    0
Sl 5:    ( •0000 ) +   ( •1100 ) +   ( •0010 ) +  2( •0010 ) +  3( •1000 ) +    ( •1000 ) +  5( •1000 )
          1    1        1    0        1    0         1    0        1    0         3    0         1    0
```

(continued next page)

```
          1    0        1    0        1    2        1    0         1    0        1    0        3    0
Sl 6:    ( •1010 ) +  ( •0001 ) +  ( •0000 ) +  ( •0000 ) +  2( •0001 ) +  ( •2000 ) +  ( •0100 )
          0    0        2    0        0    0        0    2         0    0        2    0        0    0

          1    0        1    0        1    0        3    0         3    0        1    0        1    0
   +   3( •0100 ) + 2( •2000 ) + 6( •0100 ) +  ( •0000 ) +   ( •0000 ) + 5( •0000 ) + 6( •0000 )
          2    0        0    0        0    0        2    0         0    0        2    0        0    0
```

Affine D8 ⊃ A1 + A1 + D6 - Level 1

```
          0    0                          0    0
Sl 0:    ( •0000 )   Highest Weight   ( 00000 )
          0    1                          0    1

          1    1
Sl 1:    ( •0000 )
          1    0

          0    1        2    0        0    0        0    0
Sl 2:    ( •1000 ) +  ( •0000 ) +  ( •0000 ) +  ( •0000 )
          0    0        0    1        2    1        0    1

          1    0        1    1
Sl 3:    ( •1000 ) + 3( •0000 )
          1    1        1    0

          0    0        2    1        0    1        0    1         2    0        2    0        0    0
Sl 4:    ( •0100 ) +  ( •1000 ) +  ( •1000 ) + 2( •1000 ) +   ( •0000 ) + 2( •0000 ) + 2( •0000 )
          0    1        0    0        2    0        0    0         2    1        0    1        2    1

          0    0
   +   4( •0000 )
          0    1

          1    1        1    0        3    1        1    1         1    1
Sl 5:    ( •0100 ) + 4( •1000 ) +  ( •0000 ) +  ( •0000 ) +  7( •0000 )
          1    0        1    1        1    0        3    0         1    0

          0    1        2    0        0    0        0    0         0    0        2    1        2    1
Sl 6:    ( •0010 ) +  ( •0100 ) +  ( •0100 ) +  ( •2000 ) +  2( •0100 ) +  ( •1000 ) + 3( •1000 )
          0    0        0    1        2    1        0    1         0    1        2    0        0    0

          0    1        0    1        2    0        2    0         0    0        0    0
   +   3( •1000 ) + 6( •1000 ) + 3( •0000 ) + 6( •0000 ) +  6( •0000 ) + 8( •0000 )
          2    0        0    0        2    1        0    1         2    1        0    1
```

Affine D8 ⊃ A3 + D5 - Level 1

```
          1    0                          1    0
Sl 0:    ( 0•000 )   Highest Weight   ( 00000 )
          0    0                          0    0

          0    0
Sl 1:    ( 0•100 )
          1    0

          1    0        0    0        1    0
Sl 2:    ( 0•010 ) +  ( 1•000 ) +  ( 0•000 )
          0    0        1    0        0    0

          0    0        1    0        0    0
Sl 3:    ( 0•001 ) +  ( 1•100 ) + 2( 0•100 )
          1    0        0    0        1    0

          1    1        0    0        1    0        1    0         2    0        0    0        1    0
Sl 4:    ( 0•000 ) +  ( 1•010 ) +  ( 0•200 ) + 2( 0•010 ) +   ( 0•000 ) + 2( 1•000 ) + 3( 0•000 )
          0    1        1    0        0    0        0    0         1    0        1    0        0    0

          1    0        0    0        0    2        0    0         0    0        1    0        1    0
Sl 5:    ( 1•001 ) +  ( 0•110 ) +  ( 0•000 ) +  ( 0•000 ) +  2( 0•001 ) +  ( 0•100 ) + 3( 1•100 )
          0    0        1    0        1    0        1    2         1    0        2    0        0    0

          0    0
   +   5( 0•100 )
          1    0

          1    0        0    1        2    0        1    1         0    0        0    0        1    0
Sl 6:    ( 0•101 ) +  ( 1•000 ) +  ( 0•010 ) + 3( 0•000 ) +   ( 1•200 ) + 3( 1•010 ) + 2( 0•200 )
          0    0        1    1        1    0        0    1         1    0        1    0        0    0

          1    0        1    0        2    0        0    0         0    0        1    0
   +   6( 0•010 ) +  ( 2•000 ) + 2( 0•000 ) +  ( 0•000 ) +  5( 1•000 ) + 6( 0•000 )
          0    0        0    0        1    0        3    0         1    0        0    0
```

Affine D8 $\supset$ A3 + D5 - Level 1

```
           0   0                     0   0
Sl 0:    ( 0•000 )  Highest Weight ( 00000 )
           0   1                     0   1

           0   1
Sl 1:    ( 1•000 )
           0   0

           0   1         1   0         0   0
Sl 2:    ( 0•100 ) +   ( 0•000 ) +   ( 0•000 )
           0   0         1   1         0   1

           0   0         2   1         0   1         0   1
Sl 3:    ( 1•100 ) +   ( 0•000 ) +   ( 0•000 ) +  2( 1•000 )
           0   1         0   0         2   0         0   0

           1   1         0   0         0   0         0   1         1   0         0   0
Sl 4:    ( 0•100 ) +   ( 0•010 ) +   ( 2•000 ) +  2( 0•100 ) +  3( 0•000 ) +  3( 0•000 )
           1   0         0   1         0   1         0   0         1   1         0   1

           0   1         2   0         0   0         0   0         1   1         2   1         0   1
Sl 5:    ( 1•010 ) +   ( 0•100 ) +   ( 0•100 ) +  3( 1•100 ) +   ( 1•000 ) +  2( 0•000 ) +  2( 0•000 )
           0   0         0   1         2   1         0   1         1   0         0   0         2   0

           0   1
   +    6( 1•000 )
           0   0

           1   0         0   1         0   1         0   0         1   1         0   0         2   0
Sl 6:    ( 0•010 ) +   ( 0•001 ) +   ( 2•100 ) +   ( 0•200 ) +  4( 0•100 ) +  2( 0•010 ) +   ( 1•000 )
           1   1         0   0         0   0         0   1         1   0         0   1         0   1

           0   0         0   0         0   1         1   0         0   0
   +     ( 1•000 ) +  2( 2•000 ) +  5( 0•100 ) +  8( 0•000 ) +  7( 0•000 )
           2   1         0   1         0   0         1   1         0   1
```

Affine D8 $\supset$ D4 + D4 - Level 1

```
           1   0                     1   0
Sl 0:    ( 00•00 )  Highest Weight ( 00000 )
           0   0                     0   0

           0   0
Sl 1:    ( 00•10 )
           1   0

           1   0         0   0         1   0
Sl 2:    ( 00•01 ) +   ( 01•00 ) +   ( 00•00 )
           0   0         1   0         0   0

           1   0         0   1         0   0
Sl 3:    ( 01•10 ) +   ( 00•00 ) +  2( 00•10 )
           0   0         1   1         1   0

           0   0         1   0         1   2         1   0         1   0         1   0         0   0
Sl 4:    ( 01•01 ) +   ( 00•20 ) +   ( 00•00 ) +   ( 00•00 ) +   ( 10•00 ) +  2( 00•01 ) +  2( 01•00 )
           1   0         0   0         0   0         0   2         0   0         0   0         1   0

           1   0
   +    3( 00•00 )
           0   0

           1   1         0   0         0   0         1   0         0   1         0   0
Sl 5:    ( 01•00 ) +   ( 10•10 ) +   ( 00•11 ) +  3( 01•10 ) +  3( 00•00 ) +  5( 00•10 )
           0   1         1   0         1   0         0   0         1   1         1   0

           1   0         0   0         0   2         0   0         1   1         0   0         1   0
Sl 6:    ( 10•01 ) +   ( 01•20 ) +   ( 01•00 ) +   ( 01•00 ) +   ( 00•10 ) +  3( 01•01 ) +   ( 02•00 )
           0   0         1   0         1   0         1   2         0   1         1   0         0   0

           1   0         1   0         1   2         1   0         1   0         1   0         0   0
   +     ( 00•00 ) +  2( 00•20 ) +  2( 00•00 ) +  2( 00•00 ) +  2( 10•00 ) +  7( 00•01 ) +  5( 01•00 )
           2   0         0   0         0   0         0   2         0   0         0   0         1   0

           1   0
   +    6( 00•00 )
           0   0
```

Affine E6 Branching Rules

Affine E6 ⊃ E6 - Level 1

```
          •                      1
Sl 0:   ( 0 )  Highest Weight  ( 0 )
         00000                  00000

          •
Sl 1:   ( 1 )
         00000

          •          •          •
Sl 2:   ( 0 ) +    ( 1 ) +    ( 0 )
         10001      00000      00000

          •          •          •          •
Sl 3:   ( 0 ) +  2( 0 ) +  2( 1 ) +    ( 0 )
         00100      10001      00000      00000

          •          •          •          •          •          •          •
Sl 4:   ( 0 ) +    ( 0 ) +    ( 2 ) +    ( 0 ) +  4( 0 ) +  4( 1 ) +  2( 0 )
         11000      00011      00000      00100      10001      00000      00000

          •          •          •          •          •          •          •          •
Sl 5:   ( 1 ) +  2( 0 ) +  2( 0 ) +    ( 2 ) +  4( 0 ) +  7( 0 ) +  7( 1 ) +  3( 0 )
         10001      11000      00011      00000      00100      10001      00000      00000

          •          •          •          •          •          •          •          •
Sl 6:   ( 0 ) +  3( 1 ) +    ( 0 ) +    ( 0 ) +  4( 0 ) +  4( 0 ) +  3( 2 ) +  7( 0 )
         01010      10001      30000      00003      11000      00011      00000      00100

          •          •          •
   + 14( 0 ) + 11( 1 ) +  6( 0 )
         10001      00000      00000

          •          •          •          •          •          •          •          •
Sl 7:   ( 1 ) +    ( 0 ) +    ( 0 ) +  2( 0 ) +  7( 1 ) +    ( 0 ) +    ( 0 ) +  9( 0 )
         00100      20010      01002      01010      10001      30000      00003      11000

          •          •          •          •          •          •
   +  9( 0 ) +  5( 2 ) + 14( 0 ) + 23( 0 ) + 20( 1 ) +  7( 0 )
         00011      00000      00100      10001      00000      00000
```

Affine E6 ⊃ E6 - Level 1

```
          •                      0
Sl 0:   ( 0 )  Highest Weight  ( 0 )
         10000                  10000

          •          •
Sl 1:   ( 0 ) +    ( 0 )
         00010      10000

          •          •          •          •
Sl 2:   ( 1 ) +    ( 0 ) +    ( 0 ) +  2( 0 )
         10000      00002      00010      10000

          •          •          •          •          •
Sl 3:   ( 0 ) +  2( 1 ) +    ( 0 ) +  3( 0 ) +  3( 0 )
         01001      10000      00002      00010      10000

          •          •          •          •          •          •          •
Sl 4:   ( 1 ) +    ( 0 ) +  2( 0 ) +  4( 1 ) +  3( 0 ) +  5( 0 ) +  6( 0 )
         00010      20001      01001      10000      00002      00010      10000

          •          •          •          •          •          •          •          •
Sl 5:   ( 0 ) +    ( 1 ) +  2( 1 ) +  2( 0 ) +  5( 0 ) +  8( 1 ) +  4( 0 ) + 10( 0 )
         10100      00002      00010      20001      01001      10000      00002      00010

          •
   +  9( 0 )
         10000
```
(continued next page)

```
           •          •          •          •          •          •          •          •      
Sl 6:    ( 0 ) +    ( 2 ) +    ( 0 ) +   2( 0 ) +   2( 1 ) +   5( 1 ) +   5( 0 ) +  10( 0 )
         10011      10000      02000      10100      00002      00010      20001      01001    

           •          •          •          •      
     +  15( 1 ) +   9( 0 ) +  16( 0 ) +  16( 0 )
         10000      00002      00010      10000    

           •          •          •          •          •          •          •          •      
Sl 7:    ( 1 ) +    ( 0 ) +   3( 0 ) +   2( 2 ) +    ( 0 ) +   6( 0 ) +   5( 1 ) +  10( 1 )
         01001      21000      10011      10000      02000      10100      00002      00010    

           •          •          •          •          •          •      
     +   9( 0 ) +  20( 0 ) +  27( 1 ) +  14( 0 ) +  29( 0 ) +  24( 0 )
         20001      01001      10000      00002      00010      10000    
```

Affine E6 ⊃ E6 - Level 2

```
           •                     0
Sl 0:    ( 1 )  Highest Weight ( 1 )
         00000                 00000

           •          •          •          •      
Sl 1:    ( 0 ) +    ( 0 ) +    ( 1 ) +    ( 0 )
         00100      10001      00000      00000    

           •          •          •          •          •          •          •          •      
Sl 2:    ( 1 ) +    ( 0 ) +    ( 0 ) +    ( 2 ) +   2( 0 ) +   3( 0 ) +   3( 1 ) +    ( 0 )
         10001      11000      00011      00000      00100      10001      00000      00000    

           •          •          •          •          •          •          •          •      
Sl 3:    ( 1 ) +    ( 0 ) +    ( 0 ) +    ( 0 ) +   4( 1 ) +   4( 0 ) +   4( 0 ) +   3( 2 )
         00100      20010      01002      01010      10001      11000      00011      00000    

           •          •          •          •      
     +   6( 0 ) +   8( 0 ) +   7( 1 ) +   2( 0 )
         00100      10001      00000      00000    

           •          •          •          •          •          •          •          •      
Sl 4:    ( 0 ) +   2( 1 ) +   2( 1 ) +    ( 3 ) +   2( 0 ) +   3( 1 ) +   3( 0 ) +   3( 0 )
         10101      11000      00011      00000      20002      00100      20010      01002    

           •          •          •          •          •          •          •          •      
     +   6( 0 ) +  14( 1 ) +   2( 0 ) +   2( 0 ) +  11( 0 ) +  11( 0 ) +   8( 2 ) +  17( 0 )
         01010      10001      30000      00003      11000      00011      00000      00100    

           •          •          •      
     +  21( 0 ) +  15( 1 ) +   5( 0 )
         10001      00000      00000    
```

Affine E6 ⊃ E6 - Level 2

```
           •                     0
Sl 0:    ( 0 )  Highest Weight ( 0 )
         10001                 10001

           •          •          •          •          •      
Sl 1:    ( 0 ) +    ( 0 ) +    ( 0 ) +   2( 0 ) +    ( 1 )
         11000      00011      00100      10001      00000    

           •          •          •          •          •          •          •          •      
Sl 2:    ( 0 ) +   2( 1 ) +    ( 0 ) +    ( 0 ) +   3( 0 ) +   3( 0 ) +    ( 2 ) +   3( 0 )
         01010      10001      30000      00003      11000      00011      00000      00100    

           •          •          •      
     +   6( 0 ) +   3( 1 ) +    ( 0 )
         10001      00000      00000    

           •          •          •          •          •          •          •          •      
Sl 3:    ( 1 ) +    ( 1 ) +    ( 1 ) +   2( 0 ) +   2( 0 ) +   4( 0 ) +   8( 1 ) +   2( 0 )
         11000      00011      00100      20010      01002      01010      10001      30000    

           •          •          •          •          •          •          •          •      
     +   2( 0 ) +   9( 0 ) +   9( 0 ) +   3( 2 ) +  11( 0 ) +  15( 0 ) +   8( 1 ) +   2( 0 )
         00003      11000      00011      00000      00100      10001      00000      00000    
```

(continued next page)

```
             .          .          .          .          .          .          .          .
Sl 4:      ( 2 ) +    ( 0 ) +    ( 0 ) +   2( 0 ) +    ( 1 ) +    ( 1 ) +   5( 1 ) +   5( 1 )
           10001      10020      02001      10101      30000      00003      11000      00011

             .          .          .          .          .          .          .          .
      +   3( 0 ) +   6( 1 ) +   7( 0 ) +   7( 0 ) +  15( 0 ) +  26( 1 ) +   6( 0 ) +   6( 0 )
           20002      00100      20010      01002      01010      10001      30000      00003

             .          .          .          .          .          .          .
      +  25( 0 ) +  25( 0 ) +  11( 2 ) +  27( 0 ) +  38( 0 ) +  20( 1 ) +   5( 0 )
           11000      00011      00000      00100      10001      00000      00000
```

Affine E6 ⊃ E6 - Level 2

```
             .                         2
Sl 0:      ( 0 )   Highest Weight    ( 0 )
           00000                     00000

             .
Sl 1:      ( 1 )
           00000

             .          .          .          .
Sl 2:      ( 2 ) +    ( 0 ) +    ( 1 ) +    ( 0 )
           00000      10001      00000      00000

             .          .          .          .          .          .
Sl 3:      ( 1 ) +    ( 2 ) +   2( 0 ) +   2( 0 ) +   3( 1 ) +    ( 0 )
           10001      00000      00100      10001      00000      00000

             .          .          .          .          .          .          .          .
Sl 4:      ( 0 ) +    ( 1 ) +    ( 0 ) +   3( 1 ) +   2( 0 ) +   2( 0 ) +   4( 2 ) +   3( 0 )
           20002      00100      01010      10001      11000      00011      00000      00100

             .          .          .
      +   6( 0 ) +   6( 1 ) +   3( 0 )
           10001      00000      00000

             .          .          .          .          .          .          .          .
Sl 5:      ( 0 ) +    ( 1 ) +    ( 1 ) +    ( 3 ) +   2( 0 ) +   3( 1 ) +   2( 0 ) +   2( 0 )
           10101      11000      00011      00000      20002      00100      20010      01002

             .          .          .          .          .          .          .          .
      +   3( 0 ) +  10( 1 ) +   6( 0 ) +   6( 0 ) +   7( 2 ) +  12( 0 ) +  13( 0 ) +  13( 1 )
           01010      10001      11000      00011      00000      00100      10001      00000

             .
      +   4( 0 )
           00000
```

Affine E6 ⊃ E6 - Level 2

```
             .                         0
Sl 0:      ( 0 )   Highest Weight    ( 0 )
           00002                     00002

             .          .          .
Sl 1:      ( 0 ) +    ( 0 ) +    ( 0 )
           01001      00002      00010

             .          .          .          .          .          .          .          .
Sl 2:      ( 0 ) +    ( 1 ) +    ( 1 ) +    ( 0 ) +   2( 0 ) +    ( 1 ) +   3( 0 ) +   2( 0 )
           02000      00002      00010      20001      01001      10000      00002      00010

             .
      +    ( 0 )
           10000

             .          .          .          .          .          .          .          .
Sl 3:      ( 1 ) +    ( 0 ) +    ( 0 ) +    ( 0 ) +   2( 0 ) +   3( 1 ) +   3( 1 ) +   2( 0 )
           01001      21000      10011      02000      10100      00002      00010      20001

             .          .          .          .          .
      +   7( 0 ) +   4( 1 ) +   5( 0 ) +   6( 0 ) +   2( 0 )
           01001      10000      00002      00010      10000
```

(continued next page)

```
           .          .          .          .          .          .          .          .
Sl 4:    ( 0 ) +    ( 2 ) +    ( 0 ) +    ( 1 ) +  4( 1 ) +    ( 0 ) +    ( 0 ) +  2( 0 )
         11010      00002      00110      20001      01001      40000      10003      21000

           .          .          .          .          .          .          .          .
     +  5( 0 ) +  2( 2 ) +  5( 0 ) +  6( 0 ) +  8( 1 ) + 10( 1 ) +  8( 0 ) + 17( 0 )
         10011      10000      02000      10100      00002      00010      20001      01001

           .          .          .          .
     + 12( 1 ) + 13( 0 ) + 13( 0 ) +  7( 0 )
         10000      00002      00010      10000
```

Affine E6 ⊃ E6 - Level 2

```
           .                          0
Sl 0:    ( 0 )   Highest Weight     ( 0 )
         00010                      00010

           .          .          .          .          .
Sl 1:    ( 0 ) +    ( 1 ) +    ( 0 ) +    ( 0 ) +    ( 0 )
         01001      10000      00002      00010      10000

           .          .          .          .          .          .          .          .
Sl 2:    ( 0 ) +    ( 1 ) +    ( 1 ) +    ( 0 ) +  3( 0 ) +  3( 1 ) +  2( 0 ) +  4( 0 )
         10100      00002      00010      20001      01001      10000      00002      00010

           .
     +  2( 0 )
         10000

           .          .          .          .          .          .          .          .
Sl 3:    ( 1 ) +    ( 0 ) +  2( 0 ) +    ( 2 ) +    ( 0 ) +  3( 0 ) +  3( 1 ) +  5( 1 )
         01001      21000      10011      10000      02000      10100      00002      00010

           .          .          .          .          .          .
     +  4( 0 ) +  9( 0 ) +  9( 1 ) +  6( 0 ) +  9( 0 ) +  6( 0 )
         20001      01001      10000      00002      00010      10000

           .          .          .          .          .          .          .          .
Sl 4:    ( 0 ) +    ( 2 ) +    ( 0 ) +    ( 0 ) +  2( 1 ) +  5( 1 ) +    ( 0 ) +  3( 0 )
         11010      00010      00102      00110      20001      01001      10003      21000

           .          .          .          .          .          .          .          .
     +  8( 0 ) +  4( 2 ) +  4( 0 ) + 12( 0 ) + 10( 1 ) + 15( 1 ) + 12( 0 ) + 25( 0 )
         10011      10000      02000      10100      00002      00010      20001      01001

           .          .          .          .
     + 24( 1 ) + 13( 0 ) + 23( 0 ) + 13( 0 )
         10000      00002      00010      10000
```

Affine E6 ⊃ E6 - Level 2

```
           .                          1
Sl 0:    ( 0 )   Highest Weight     ( 0 )
         10000                      10000

           .          .          .
Sl 1:    ( 1 ) +    ( 0 ) +    ( 0 )
         10000      00010      10000

           .          .          .          .          .          .          .
Sl 2:    ( 1 ) +    ( 0 ) +    ( 0 ) +  3( 1 ) +    ( 0 ) +  2( 0 ) +  3( 0 )
         00010      20001      01001      10000      00002      00010      10000

           .          .          .          .          .          .          .          .
Sl 3:    ( 0 ) +    ( 2 ) +  2( 0 ) +    ( 1 ) +  3( 1 ) +  3( 0 ) +  5( 0 ) +  8( 1 )
         10011      10000      10100      00002      00010      20001      01001      10000

           .          .          .
     +  2( 0 ) +  7( 0 ) +  6( 0 )
         00002      00010      10000
```

(continued next page)

```
            •          •           •          •          •           •           •           •
Sl 4:    ( 0 ) +    ( 1 ) +  2( 1 ) +    ( 0 ) +    ( 0 ) +  5( 0 ) +  4( 2 ) +  2( 0 )
         00110      20001     01001      10003      21000     10011     10000     02000

            •          •           •          •          •           •           •           •
   +  7( 0 ) +  4( 1 ) + 11( 1 ) + 10( 0 ) + 15( 0 ) + 21( 1 ) +  8( 0 ) + 16( 0 )
         10100      00002     00010      20001      01001     10000     00002     00010

            •
   + 14( 0 )
         10000

            •          •           •          •          •           •           •           •
Sl 5:    ( 1 ) +    ( 0 ) +  2( 0 ) +    ( 0 ) +  2( 2 ) +  2( 0 ) +  4( 0 ) +  6( 1 )
         10100      11002     11010      00021      00010     00102     00110     20001

            •          •           •          •          •           •           •           •
   + 10( 1 ) +  3( 0 ) +  5( 0 ) + 18( 0 ) + 12( 2 ) +  6( 0 ) + 25( 0 ) + 15( 1 )
         01001      10003     21000      10011      10000     02000     10100     00002

            •          •           •          •          •           •           •
   + 31( 1 ) + 26( 0 ) + 43( 0 ) + 51( 1 ) + 18( 0 ) + 40( 0 ) + 28( 0 )
         00010      20001     01001      10000      00002     00010     10000
```

Affine E6 ⊃ A1 + A5 - Level 1

```
            1                          1
Sl 0:    ( • )   Highest Weight     ( 0 )
         00000                      00000

            0
Sl 1:    ( • )
         00100

            1          1
Sl 2:    ( • ) +    ( • )
         10001      00000

            0          0           2          0
Sl 3:    ( • ) +    ( • ) +    ( • ) +    ( • )
         11000      00011      00100      00100

            1          1           3          1
Sl 4:    ( • ) +  3( • ) +    ( • ) +  2( • )
         01010      10001      00000      00000

            0          2           2          0          0           2           0
Sl 5:    ( • ) +    ( • ) +    ( • ) +  2( • ) +  2( • ) +  2( • ) +  4( • )
         10101      11000      00011      11000      00011     00100     00100

            1          1           1          1          3           1           3           1
Sl 6:    ( • ) +    ( • ) +    ( • ) +  3( • ) +    ( • ) +  7( • ) +    ( • ) +  5( • )
         20010      01002      00200      01010      10001     10001     00000     00000

            2          0           0          0          2           2           0           0
Sl 7:    ( • ) +    ( • ) +    ( • ) +  3( • ) +  3( • ) +  3( • ) +    ( • ) +    ( • )
         10101      10020      02001      10101      11000     00011     30000     00003

            0          0           2          0
   +  5( • ) +  5( • ) +  6( • ) +  8( • )
         11000      00011      00100      00100

            1          1           1          3          1           1           1           1
Sl 8:    ( • ) +    ( • ) +    ( • ) +    ( • ) +  3( • ) +  3( • ) +  2( • ) +  9( • )
         11100      00111      20002      01010      20010     01002     00200     01010

            3          1           3          1
   +  3( • ) + 16( • ) +  3( • ) + 10( • )
         10001      10001      00000      00000
```

Affine E6 ⊃ A1 + A5 - Level 1

```
            0                          0
Sl 0:    ( • )   Highest Weight     ( 0 )
         10000                      10000

            1
Sl 1:    ( • )
         00010   (continued next page)
```

```
            0           2           0
Sl 2:     ( • ) +     ( • ) +     ( • )
          01001       10000       10000

            1           1           1
Sl 3:     ( • ) +     ( • ) +    2( • )
          10100       00002       00010

            0           2           0           0           2           0
Sl 4:     ( • ) +     ( • ) +     ( • ) +    2( • ) +    2( • ) +    3( • )
          00110       01001       20001       01001       10000       10000

            1           1           1           3           1           1
Sl 5:     ( • ) +     ( • ) +    3( • ) +     ( • ) +    2( • ) +    5( • )
          10011       02000       10100       00010       00002       00010

            0           2           0           2           0           2           0           0
Sl 6:     ( • ) +     ( • ) +     ( • ) +     ( • ) +    2( • ) +    3( • ) +    2( • ) +    6( • )
          11010       00110       00102       20001       00110       01001       20001       01001

            2           0
    +    5( • ) +    6( • )
          10000       10000

            1           1           1           3           1           1           3           3
Sl 7:     ( • ) +     ( • ) +    4( • ) +     ( • ) +    2( • ) +    8( • ) +     ( • ) +    2( • )
          01101       21000       10011       10100       02000       10100       00002       00010

            1           1
    +    5( • ) +   11( • )
          00002       00010

            2           0           0           2           0           0           2           0
Sl 8:     ( • ) +     ( • ) +     ( • ) +     ( • ) +    3( • ) +     ( • ) +    3( • ) +    2( • )
          11010       11002       10200       00102       11010       00021       00110       00102

            2           0           2           0           0           4           2           0
    +    3( • ) +    6( • ) +    9( • ) +    6( • ) +   13( • ) +     ( • ) +   10( • ) +   13( • )
          20001       00110       01001       20001       01001       10000       10000       10000
```

Affine E6 ⊃ A1 + A5 - Level 2

```
            0                        0
Sl 0:     ( • )  Highest Weight    ( 1 )
          00000                    00000

            1
Sl 1:     ( • )
          00100

            0           2           0           2           0
Sl 2:     ( • ) +     ( • ) +     ( • ) +     ( • ) +     ( • )
          01010       10001       10001       00000       00000

            1           1           1           3           1
Sl 3:     ( • ) +    2( • ) +    2( • ) +     ( • ) +    3( • )
          10101       11000       00011       00100       00100

            0           0           2           2           2           0           2           0
Sl 4:     ( • ) +     ( • ) +     ( • ) +     ( • ) +     ( • ) +     ( • ) +    2( • ) +     ( • )
          11100       00111       20010       01002       00200       20002       01010       20010

            0           0           0           2           0           4           2           0
    +     ( • ) +     ( • ) +    4( • ) +    5( • ) +    6( • ) +     ( • ) +    3( • ) +    3( • )
          01002       00200       01010       10001       10001       00000       00000       00000

            1           1           1           3           1           1           1           3
Sl 5:     ( • ) +     ( • ) +     ( • ) +     ( • ) +    2( • ) +    2( • ) +    8( • ) +    2( • )
          01110       21001       10012       10101       10020       02001       10101       11000

            3           1           1           1           1           3           1
    +    2( • ) +     ( • ) +     ( • ) +    9( • ) +    9( • ) +    4( • ) +   13( • )
          00011       30000       00003       11000       00011       00100       00100
```

Affine E6 ⊃ A1 + A5 - Level 2

```
           0                      0
Sl 0:    ( • )  Highest Weight  ( 0 )
         10001                  10001

           1            1            1
Sl 1:    ( • ) +      ( • ) +      ( • )
         11000        00011        00100

           2            0            0            0            2            0            2            0
Sl 2:    ( • ) +      ( • ) +      ( • ) +     2( • ) +     2( • ) +     3( • ) +      ( • ) +      ( • )
         01010        20010        01002        01010        10001        10001        00000        00000

           1            1            1            3            3            1            1            1
Sl 3:    ( • ) +      ( • ) +     3( • ) +      ( • ) +      ( • ) +      ( • ) +      ( • ) +     5( • )
         10020        02001        10101        11000        00011        30000        00003        11000

           1            3            1
     +  5( • ) +      ( • ) +     6( • )
         00011        00100        00100

           0            2            2            0            0            2            2            2
Sl 4:    ( • ) +      ( • ) +      ( • ) +     2( • ) +     2( • ) +     3( • ) +     3( • ) +      ( • )
         11011        11100        00111        11100        00111        20010        01002        00200

           0            2            0            0            0            0            4            2
     +  2( • ) +     7( • ) +     5( • ) +     5( • ) +     3( • ) +     9( • ) +      ( • ) +    10( • )
         20002        01010        20010        01002        00200        01010        10001        10001

           0            2            0
     + 13( • ) +     4( • ) +     5( • )
         10001        00000        00000
```

Affine E6 ⊃ A1 + A5 - Level 2

```
           2                      2
Sl 0:    ( • )  Highest Weight  ( 0 )
         00000                  00000

           1
Sl 1:    ( • )
         00100

           0            2            0            2            0
Sl 2:    ( • ) +      ( • ) +      ( • ) +      ( • ) +      ( • )
         00200        10001        10001        00000        00000

           1            1            1            3            1
Sl 3:    ( • ) +      ( • ) +      ( • ) +      ( • ) +     3( • )
         10101        11000        00011        00100        00100

           2            0            0            2            2            0            0            0
Sl 4:    ( • ) +      ( • ) +      ( • ) +      ( • ) +     2( • ) +      ( • ) +      ( • ) +      ( • )
         20002        11100        00111        00200        01010        20010        01002        00200

           0            2            0            4            2            0
     +  2( • ) +     4( • ) +     4( • ) +      ( • ) +     4( • ) +     2( • )
         01010        10001        10001        00000        00000        00000

           1            1            1            3            1            1            1            3
Sl 5:    ( • ) +      ( • ) +      ( • ) +      ( • ) +      ( • ) +      ( • ) +     6( • ) +      ( • )
         01110        21001        10012        10101        10020        02001        10101        11000

           3            1            1            3            1
     +   ( • ) +     6( • ) +     6( • ) +     4( • ) +    10( • )
         00011        11000        00011        00100        00100
```

Affine E6 ⊃ A1 + A5 - Level 2

```
           0                      0
Sl 0:    ( • )  Highest Weight  ( 0 )
         00002                  00002

           1
Sl 1:    ( • )
         01001  (continued next page)
```

sl 2: (2 • 02000) + (0 • 10011) + (0 • 10100) + (2 • 00002) + (2 • 00010) + (0 • 00002) + (0 • 00010)

sl 3: (1 • 11010) + (1 • 00102) + (1 • 00110) + (3 • 01001) + (1 • 20001) + 4(1 • 01001) + 2(1 • 10000)

sl 4: (0 • 20020) + (2 • 01101) + 2(0 • 01101) + (2 • 21000) + 2(2 • 10011) + (0 • 10003) + 2(2 • 02000) + (0 • 21000)
+ 4(0 • 10011) + 3(2 • 10100) + 2(0 • 02000) + (4 • 00002) + 4(0 • 10100) + 3(2 • 00002) + 4(2 • 00010) + 5(0 • 00002)
+ 4(0 • 00010)

Affine E6 ⊃ A1 + A5 - Level 2

sl 0: (0 • 00010) **Highest Weight (0 0 00010)**

sl 1: (1 • 01001) + (1 • 10000)

sl 2: (0 • 10011) + (2 • 10100) + (0 • 02000) + (0 • 10100) + (2 • 00002) + (2 • 00010) + (0 • 00002) + 2(0 • 00010)

sl 3: (1 • 11010) + (1 • 00102) + 2(1 • 00110) + (3 • 01001) + 2(1 • 20001) + 5(1 • 01001) + (3 • 10000) + 4(1 • 10000)

sl 4: (2 • 01101) + (0 • 20101) + (0 • 01020) + 2(0 • 01101) + (2 • 21000) + 3(2 • 10011) + (0 • 10003) + 2(2 • 02000)
+ (0 • 21000) + 5(0 • 10011) + 5(2 • 10100) + 3(0 • 02000) + 7(0 • 10100) + (4 • 00010) + 4(2 • 00002) + 6(2 • 00010)
+ 4(0 • 00002) + 8(0 • 00010)

sl 5: (1 • 10111) + (1 • 02100) + (3 • 11010) + (1 • 30010) + 3(1 • 11002) + 2(1 • 10200) + (3 • 00102) + 8(1 • 11010)
+ 2(1 • 00021) + 2(3 • 00110) + 6(1 • 00102) + 2(3 • 20001) + 11(1 • 00110) + 6(3 • 01001) + 10(1 • 20001) + 22(1 • 01001)
+ 5(3 • 10000) + 15(1 • 10000)

Affine E6 ⊃ A1 + A5 - Level 2

sl 0: (1 • 10000) **Highest Weight (1 0 10000)**

sl 1: (0 • 10100) + (2 • 00010) + (0 • 00010)

sl 2: (1 • 00110) + (1 • 20001) + 2(1 • 01001) + (3 • 10000) + 3(1 • 10000) (continued next page)

Sl 3: $\underset{01101}{\overset{0}{(\,\bullet\,)}} + \underset{10011}{\overset{2}{(\,\bullet\,)}} + \underset{21000}{\overset{0}{(\,\bullet\,)}} + 2\underset{10011}{\overset{0}{(\,\bullet\,)}} + 3\underset{10100}{\overset{2}{(\,\bullet\,)}} + \underset{02000}{\overset{0}{(\,\bullet\,)}} + 4\underset{10100}{\overset{0}{(\,\bullet\,)}} + \underset{00002}{\overset{2}{(\,\bullet\,)}}$

$+ 4\underset{00010}{\overset{2}{(\,\bullet\,)}} + 2\underset{00002}{\overset{0}{(\,\bullet\,)}} + 4\underset{00010}{\overset{0}{(\,\bullet\,)}}$

Sl 4: $\underset{11002}{\overset{1}{(\,\bullet\,)}} + \underset{10200}{\overset{1}{(\,\bullet\,)}} + 3\underset{11010}{\overset{1}{(\,\bullet\,)}} + \underset{00021}{\overset{1}{(\,\bullet\,)}} + \underset{00110}{\overset{3}{(\,\bullet\,)}} + 2\underset{00102}{\overset{1}{(\,\bullet\,)}} + \underset{20001}{\overset{3}{(\,\bullet\,)}} + 6\underset{00110}{\overset{1}{(\,\bullet\,)}}$

$+ 2\underset{01001}{\overset{3}{(\,\bullet\,)}} + 6\underset{20001}{\overset{1}{(\,\bullet\,)}} + 11\underset{01001}{\overset{1}{(\,\bullet\,)}} + 4\underset{10000}{\overset{3}{(\,\bullet\,)}} + 11\underset{10000}{\overset{1}{(\,\bullet\,)}}$

Affine E6 ⊃ A2 + A2 + A2 - Level 1

Sl 0: $\underset{00\bullet 00}{\overset{1}{(\,0\,)}}$ **Highest Weight** $\underset{00000}{\overset{1}{(\,0\,)}}$

Sl 1: $\underset{01\bullet 10}{\overset{0}{(\,0\,)}}$

Sl 2: $\underset{10\bullet 01}{\overset{0}{(\,1\,)}}$

Sl 3: $\underset{11\bullet 00}{\overset{1}{(\,0\,)}} + \underset{00\bullet 11}{\overset{1}{(\,0\,)}} + \underset{00\bullet 00}{\overset{0}{(\,2\,)}} + \underset{00\bullet 00}{\overset{1}{(\,0\,)}}$

Sl 4: $\underset{01\bullet 10}{\overset{1}{(\,1\,)}} + \underset{20\bullet 10}{\overset{0}{(\,0\,)}} + \underset{01\bullet 02}{\overset{0}{(\,0\,)}} + 2\underset{01\bullet 10}{\overset{0}{(\,0\,)}}$

Sl 5: $\underset{10\bullet 01}{\overset{2}{(\,0\,)}} + \underset{10\bullet 20}{\overset{0}{(\,1\,)}} + \underset{02\bullet 01}{\overset{0}{(\,1\,)}} + 3\underset{10\bullet 01}{\overset{0}{(\,1\,)}}$

Sl 6: $\underset{11\bullet 11}{\overset{1}{(\,0\,)}} + \underset{11\bullet 00}{\overset{0}{(\,2\,)}} + \underset{00\bullet 11}{\overset{0}{(\,2\,)}} + \underset{00\bullet 00}{\overset{2}{(\,1\,)}} + 3\underset{11\bullet 00}{\overset{1}{(\,0\,)}} + 3\underset{00\bullet 11}{\overset{1}{(\,0\,)}} + \underset{00\bullet 00}{\overset{0}{(\,2\,)}} + 4\underset{00\bullet 00}{\overset{1}{(\,0\,)}}$

Affine E6 ⊃ A2 + A2 + A2 - Level 2

Sl 0: $\underset{00\bullet 00}{\overset{0}{(\,1\,)}}$ **Highest Weight** $\underset{00000}{\overset{0}{(\,1\,)}}$

Sl 1: $\underset{01\bullet 10}{\overset{1}{(\,0\,)}}$

Sl 2: $\underset{10\bullet 01}{\overset{1}{(\,1\,)}} + \underset{10\bullet 20}{\overset{0}{(\,0\,)}} + \underset{02\bullet 01}{\overset{0}{(\,0\,)}} + \underset{10\bullet 01}{\overset{0}{(\,0\,)}}$

Sl 3: $\underset{11\bullet 11}{\overset{0}{(\,1\,)}} + \underset{11\bullet 00}{\overset{2}{(\,0\,)}} + \underset{00\bullet 11}{\overset{2}{(\,0\,)}} + \underset{00\bullet 00}{\overset{1}{(\,2\,)}} + 2\underset{11\bullet 00}{\overset{0}{(\,1\,)}} + 2\underset{00\bullet 11}{\overset{0}{(\,1\,)}} + \underset{00\bullet 00}{\overset{2}{(\,0\,)}} + 2\underset{00\bullet 00}{\overset{0}{(\,1\,)}}$

Sl 4: $\underset{01\bullet 10}{\overset{2}{(\,1\,)}} + \underset{20\bullet 02}{\overset{1}{(\,0\,)}} + \underset{12\bullet 10}{\overset{1}{(\,0\,)}} + \underset{01\bullet 21}{\overset{1}{(\,0\,)}} + \underset{20\bullet 10}{\overset{0}{(\,2\,)}} + \underset{01\bullet 02}{\overset{0}{(\,2\,)}} + 2\underset{20\bullet 10}{\overset{1}{(\,0\,)}} + 2\underset{01\bullet 02}{\overset{1}{(\,0\,)}}$

$+ 2\underset{01\bullet 10}{\overset{0}{(\,2\,)}} + 6\underset{01\bullet 10}{\overset{1}{(\,0\,)}}$

Sl 5: $\underset{21\bullet 01}{\overset{1}{(\,1\,)}} + \underset{10\bullet 12}{\overset{1}{(\,1\,)}} + \underset{02\bullet 20}{\overset{1}{(\,1\,)}} + \underset{10\bullet 01}{\overset{3}{(\,0\,)}} + 3\underset{10\bullet 20}{\overset{1}{(\,1\,)}} + 3\underset{02\bullet 01}{\overset{1}{(\,1\,)}} + \underset{10\bullet 01}{\overset{0}{(\,3\,)}} + \underset{21\bullet 20}{\overset{0}{(\,0\,)}}$

$+ \underset{02\bullet 12}{\overset{0}{(\,0\,)}} + 8\underset{10\bullet 01}{\overset{1}{(\,1\,)}} + 2\underset{21\bullet 01}{\overset{0}{(\,0\,)}} + 2\underset{10\bullet 12}{\overset{0}{(\,0\,)}} + 2\underset{02\bullet 20}{\overset{0}{(\,0\,)}} + 4\underset{10\bullet 20}{\overset{0}{(\,0\,)}} + 4\underset{02\bullet 01}{\overset{0}{(\,0\,)}} + 8\underset{10\bullet 01}{\overset{0}{(\,0\,)}}$

Affine E6 ⊃ A2 + A2 + A2 - Level 2

```
           0                    0
sl 0:    ( 0 )  Highest Weight ( 0 )
         10•01                10001

           0         0         0
sl 1:    ( 1 ) +   ( 1 ) +   ( 1 )
         11•00     00•11     00•00

           1         1         0         1
sl 2:    ( 0 ) +   ( 0 ) +   ( 2 ) +  2( 0 )
         20•10     01•02     01•10     01•10

           1         1         1         0         0         0         0         0
sl 3:    ( 1 ) +   ( 1 ) +  3( 1 ) +   ( 0 ) +   ( 0 ) +  2( 0 ) +  2( 0 ) +  4( 0 )
         10•20     02•01     10•01     21•01     10•12     10•20     02•01     10•01

           2         1         1         0         2         2         0         0
sl 4:    ( 0 ) +   ( 2 ) +   ( 2 ) +  4( 1 ) +  2( 0 ) +  2( 0 ) +   ( 1 ) +   ( 1 )
         11•11     11•00     00•11     11•11     11•00     00•11     30•00     03•00

           0         0         1         0         0         2         0
   +     ( 1 ) +   ( 1 ) +   ( 2 ) +  6( 1 ) +  6( 1 ) +  3( 0 ) +  5( 1 )
         00•30     00•03     00•00     11•00     00•11     00•00     00•00

           2         2         1         1         0         0         2         1
sl 5:    ( 1 ) +   ( 1 ) +   ( 0 ) +   ( 0 ) +   ( 2 ) +   ( 2 ) +  3( 1 ) +  3( 0 )
         20•10     01•02     20•21     12•02     12•10     01•21     01•10     20•02

           1         1         0         0         1         1         0         1
   +    3( 0 ) +  3( 0 ) +  3( 2 ) +  3( 2 ) +  7( 0 ) +  7( 0 ) +  7( 2 ) + 14( 0 )
         12•10     01•21     20•10     01•02     20•10     01•02     01•10     01•10
```

Affine E6 ⊃ A2 + A2 + A2 - Level 2

```
           2                    2
sl 0:    ( 0 )  Highest Weight ( 0 )
         00•00                00000

           1
sl 1:    ( 0 )
         01•10

           1         0         0
sl 2:    ( 1 ) +   ( 0 ) +   ( 0 )
         10•01     02•20     10•01

           0         2         2         1         0         0         2         0
sl 3:    ( 1 ) +   ( 0 ) +   ( 0 ) +   ( 2 ) +   ( 1 ) +   ( 1 ) +   ( 0 ) +  2( 1 )
         11•11     11•00     00•11     00•00     11•00     00•11     00•00     00•00

           0         2         1         1         1         1         0         1
sl 4:    ( 2 ) +   ( 1 ) +   ( 0 ) +   ( 0 ) +  2( 0 ) +  2( 0 ) +  2( 2 ) +  4( 0 )
         20•02     01•10     12•10     01•21     20•10     01•02     01•10     01•10

           1         1         1         3         1         1         0         0
sl 5:    ( 1 ) +   ( 1 ) +   ( 1 ) +   ( 0 ) +  2( 1 ) +  2( 1 ) +   ( 3 ) +   ( 0 )
         21•01     10•12     02•20     10•01     10•20     02•01     10•01     21•20

           0         1         0         0         0         0         0         0
   +     ( 0 ) +  6( 1 ) +   ( 0 ) +   ( 0 ) +  2( 0 ) +  3( 0 ) +  3( 0 ) +  5( 0 )
         02•12     10•01     21•01     10•12     02•20     10•20     02•01     10•01

           1         2         2         2         0         0         0         0
sl 6:    ( 2 ) +   ( 0 ) +  3( 0 ) +   ( 0 ) +   ( 1 ) +   ( 1 ) +   ( 1 ) +   ( 1 )
         11•11     22•00     11•11     00•22     30•11     11•30     11•03     03•11

           1         1         0         3         2         2         0         0
   +    2( 2 ) +  2( 2 ) +  8( 1 ) +   ( 1 ) +  5( 0 ) +  5( 0 ) +   ( 4 ) +   ( 1 )
         11•00     00•11     11•11     00•00     11•00     00•11     00•00     30•00

           0         0         0         1         0         0         2         0
   +     ( 1 ) +   ( 1 ) +   ( 1 ) +  3( 2 ) +  8( 1 ) +  8( 1 ) +  7( 0 ) +  6( 1 )
         03•00     00•30     00•03     00•00     11•00     00•11     00•00     00•00
```

Affine E7 Branching Rules

Affine E7 ⊃ E7 - Level 1

Sl 0: $(\cdot 00\overset{0}{0}000)$ **Highest Weight** $(\mathbf{100\overset{0}{0}000})$

Sl 1: $(\cdot 10\overset{0}{0}000)$

Sl 2: $(\cdot 00\overset{0}{0}010) + (\cdot 10\overset{0}{0}000) + (\cdot 00\overset{0}{0}000)$

Sl 3: $(\cdot 01\overset{0}{0}000) + (\cdot 00\overset{0}{0}002) + (\cdot 00\overset{0}{0}010) + 2(\cdot 10\overset{0}{0}000) + (\cdot 00\overset{0}{0}000)$

Sl 4: $(\cdot 00\overset{1}{0}001) + (\cdot 20\overset{0}{0}000) + (\cdot 01\overset{0}{0}000) + (\cdot 00\overset{0}{0}002) + 3(\cdot 00\overset{0}{0}010) + 3(\cdot 10\overset{0}{0}000) + 2(\cdot 00\overset{0}{0}000)$

Sl 5: $(\cdot 10\overset{0}{0}010) + 2(\cdot 00\overset{1}{0}001) + (\cdot 20\overset{0}{0}000) + 3(\cdot 01\overset{0}{0}000) + 3(\cdot 00\overset{0}{0}002) + 4(\cdot 00\overset{0}{0}010) + 6(\cdot 10\overset{0}{0}000)$
$+ 2(\cdot 00\overset{0}{0}000)$

Sl 6: $(\cdot 00\overset{0}{1}000) + (\cdot 10\overset{0}{0}002) + 2(\cdot 10\overset{0}{0}010) + 4(\cdot 00\overset{1}{0}001) + 3(\cdot 20\overset{0}{0}000) + 5(\cdot 01\overset{0}{0}000) + 4(\cdot 00\overset{0}{0}002)$
$+ 9(\cdot 00\overset{0}{0}010) + 8(\cdot 10\overset{0}{0}000) + 5(\cdot 00\overset{0}{0}000)$

Sl 7: $(\cdot 00\overset{0}{0}101) + (\cdot 11\overset{0}{0}000) + (\cdot 00\overset{2}{0}000) + (\cdot 00\overset{0}{1}000) + 2(\cdot 10\overset{0}{0}002) + 5(\cdot 10\overset{0}{0}010) + 8(\cdot 00\overset{1}{0}001)$
$+ 4(\cdot 20\overset{0}{0}000) + 10(\cdot 01\overset{0}{0}000) + 8(\cdot 00\overset{0}{0}002) + 13(\cdot 00\overset{0}{0}010) + 15(\cdot 10\overset{0}{0}000) + 5(\cdot 00\overset{0}{0}000)$

Affine E7 ⊃ E7 - Level 1

Sl 0: $(\cdot 00\overset{0}{0}001)$ **Highest Weight** $(\mathbf{000\overset{0}{0}001})$

Sl 1: $(\cdot 00\overset{1}{0}000) + (\cdot 00\overset{0}{0}001)$

Sl 2: $(\cdot 10\overset{0}{0}001) + (\cdot 00\overset{1}{0}000) + 2(\cdot 00\overset{0}{0}001)$

Sl 3: $(\cdot 00\overset{0}{0}100) + 2(\cdot 10\overset{0}{0}001) + 2(\cdot 00\overset{1}{0}000) + 3(\cdot 00\overset{0}{0}001)$

Sl 4: $(\cdot 10\overset{1}{0}000) + (\cdot 00\overset{0}{0}011) + (\cdot 00\overset{0}{0}100) + 4(\cdot 10\overset{0}{0}001) + 4(\cdot 00\overset{1}{0}000) + 5(\cdot 00\overset{0}{0}001)$

Sl 5: $(\cdot 01\overset{0}{0}001) + 2(\cdot 10\overset{1}{0}000) + 2(\cdot 00\overset{0}{0}011) + 3(\cdot 00\overset{0}{0}100) + 7(\cdot 10\overset{0}{0}001) + 7(\cdot 00\overset{1}{0}000) + 8(\cdot 00\overset{0}{0}001)$

Sl 6: $(\cdot 00\overset{1}{0}010) + (\cdot 20\overset{0}{0}001) + 2(\cdot 01\overset{0}{0}001) + 4(\cdot 10\overset{1}{0}000) + (\cdot 00\overset{0}{0}003) + 4(\cdot 00\overset{0}{0}011) + 6(\cdot 00\overset{0}{0}100)$
$+ 13(\cdot 10\overset{0}{0}001) + 11(\cdot 00\overset{1}{0}000) + 13(\cdot 00\overset{0}{0}001)$

Sl 7: $(\cdot 10\overset{0}{0}100) + (\cdot 00\overset{1}{0}002) + 2(\cdot 00\overset{1}{0}010) + 2(\cdot 20\overset{0}{0}001) + 5(\cdot 01\overset{0}{0}001) + 8(\cdot 10\overset{1}{0}000) + (\cdot 00\overset{0}{0}003)$
$+ 8(\cdot 00\overset{0}{0}011) + 11(\cdot 00\overset{0}{0}100) + 22(\cdot 10\overset{0}{0}001) + 19(\cdot 00\overset{1}{0}000) + 19(\cdot 00\overset{0}{0}001)$

Affine E7 ⊃ E7 - Level 2

Sl 0: $(\cdot 00\overset{0}{0}002)$ **Highest Weight** $(\mathbf{000\overset{0}{0}002})$

Sl 1: $(\cdot 00\overset{1}{0}001) + (\cdot 00\overset{0}{0}002) + (\cdot 00\overset{0}{0}010)$ (continued next page)

```
          2              0              0              1              0              0              0
sl 2:  (•000000) +   (•100002) +   (•100010) +  2(•000001) +   (•010000) +  3(•000002) +  2(•000010)

          0
   +   (•100000)

          1              0              2              0              0              0              1
sl 3:  (•100001) +   (•000101) +   (•000000) +  2(•001000) +  3(•100002) +  3(•100010) +  6(•000001)

          0              0              0              0              0              0
   +   (•200000) +  3(•010000) +  5(•000002) +  6(•000010) +  2(•100000) +   (•000000)

          1              0              0              1              0              0              0
sl 4:  (•000100) +   (•200002) +   (•010010) +  4(•100001) +   (•000012) +   (•000020) +  4(•000101)

          0              2              0              0              0              1              0
   +  2(•110000) +  4(•000000) +  4(•001000) +  8(•100002) + 10(•100010) + 15(•000001) +  3(•200000)

          0              0              0              0              0
   +  9(•010000) + 12(•000002) + 12(•000010) +  7(•100000) +   (•000000)
```

Affine E7 ⊃ E7 - Level 2

```
          0                         0
sl 0:  (•000010)  **Highest Weight (0000010)**

          1              0              0              0              0
sl 1:  (•000001) +   (•010000) +   (•000002) +   (•000010) +   (•100000)

          0              0              0              1              0              0              0
sl 2:  (•001000) +   (•100002) +   (•100010) +  3(•000001) +   (•200000) +  2(•010000) +  2(•000002)

          0              0              0
   +  4(•000010) +  2(•100000) +   (•000000)

          1              0              0              2              0              0              0
sl 3:  (•100001) +  2(•000101) +   (•110000) +   (•000000) +  2(•001000) +  3(•100002) +  5(•100010)

          1              0              0              0              0              0              0
   +  8(•000001) +  2(•200000) +  7(•010000) +  6(•000002) +  8(•000010) +  6(•100000) +   (•000000)

          1              0              0              0              1              0              0
sl 4:  (•000100) +   (•200010) +   (•010002) +   (•010010) +  5(•100001) +   (•000012) +  2(•000020)

          0              0              2              0              0              0              1
   +  6(•000101) +  4(•110000) +  3(•000000) +  8(•001000) + 10(•100002) + 14(•100010) + 21(•000001)

          0              0              0              0              0              0
   +  8(•200000) + 16(•010000) + 12(•000002) + 21(•000010) + 12(•100000) +  4(•000000)
```

Affine E7 ⊃ E7 - Level 2

```
          0                         0
sl 0:  (•100000)  **Highest Weight (0100000)**

          0              0              0              0
sl 1:  (•010000) +   (•000010) +   (•100000) +   (•000000)

          0              1              0              0              0              0              0
sl 2:  (•100010) +   (•000001) +   (•200000) +  2(•010000) +   (•000002) +  2(•000010) +  3(•100000)

          0
   +   (•000000)

          0              0              0              0              0              1              0
sl 3:  (•000101) +   (•110000) +   (•001000) +   (•100002) +  3(•100010) +  4(•000001) +  3(•200000)

          0              0              0              0              0
   +  5(•010000) +  2(•000002) +  6(•000010) +  6(•100000) +  2(•000000)

          0              1              0              0              0              0              0
sl 4:  (•010010) +  2(•100001) +   (•000012) +   (•000020) +   (•300000) +  3(•000101) +  3(•110000)

          2              0              0              0              1              0              0
   +   (•000000) +  4(•001000) +  4(•100002) + 10(•100010) + 11(•000001) +  7(•200000) + 13(•010000)

          0              0              0              0
   +  7(•000002) + 13(•000010) + 13(•100000) +  4(•000000)
```

Affine E7 ⊃ E7 - Level 2

Sl 0: $\overset{0}{(\cdot 000000)}$ **Highest Weight** $\overset{0}{(2000000)}$

Sl 1: $\overset{0}{(\cdot 100000)}$

Sl 2: $\overset{0}{(\cdot 200000)} + \overset{0}{(\cdot 000010)} + \overset{0}{(\cdot 100000)} + \overset{0}{(\cdot 000000)}$

Sl 3: $\overset{0}{(\cdot 100010)} + \overset{0}{(\cdot 200000)} + 2\overset{0}{(\cdot 010000)} + \overset{0}{(\cdot 000002)} + \overset{0}{(\cdot 000010)} + 3\overset{0}{(\cdot 100000)} + \overset{0}{(\cdot 000000)}$

Sl 4: $\overset{0}{(\cdot 000020)} + \overset{0}{(\cdot 110000)} + \overset{0}{(\cdot 001000)} + \overset{0}{(\cdot 100002)} + 2\overset{0}{(\cdot 100010)} + 2\overset{1}{(\cdot 000001)} + 4\overset{0}{(\cdot 200000)}$
$+ 3\overset{0}{(\cdot 010000)} + \overset{0}{(\cdot 000002)} + 5\overset{0}{(\cdot 000010)} + 5\overset{0}{(\cdot 100000)} + 3\overset{0}{(\cdot 000000)}$

Sl 5: $\overset{0}{(\cdot 010010)} + \overset{1}{(\cdot 100001)} + \overset{0}{(\cdot 000012)} + \overset{0}{(\cdot 000020)} + \overset{0}{(\cdot 300000)} + 2\overset{0}{(\cdot 000101)} + 3\overset{0}{(\cdot 110000)}$
$+ \overset{2}{(\cdot 000000)} + 2\overset{0}{(\cdot 001000)} + 2\overset{0}{(\cdot 100002)} + 8\overset{0}{(\cdot 100010)} + 6\overset{1}{(\cdot 000001)} + 6\overset{0}{(\cdot 200000)} + 10\overset{0}{(\cdot 010000)}$
$+ 5\overset{0}{(\cdot 000002)} + 8\overset{0}{(\cdot 000010)} + 12\overset{0}{(\cdot 100000)} + 3\overset{0}{(\cdot 000000)}$

Affine E7 ⊃ E7 - Level 2

Sl 0: $\overset{1}{(\cdot 000000)}$ **Highest Weight** $\overset{1}{(0000000)}$

Sl 1: $\overset{0}{(\cdot 000100)} + \overset{0}{(\cdot 100001)} + \overset{1}{(\cdot 000000)} + \overset{0}{(\cdot 000001)}$

Sl 2: $\overset{0}{(\cdot 010001)} + \overset{1}{(\cdot 100000)} + \overset{0}{(\cdot 000011)} + 2\overset{0}{(\cdot 000100)} + 3\overset{0}{(\cdot 100001)} + 3\overset{1}{(\cdot 000000)} + 2\overset{0}{(\cdot 000001)}$

Sl 3: $\overset{0}{(\cdot 100100)} + \overset{1}{(\cdot 000002)} + \overset{1}{(\cdot 000010)} + \overset{0}{(\cdot 200001)} + 3\overset{0}{(\cdot 010001)} + 4\overset{1}{(\cdot 100000)} + 4\overset{0}{(\cdot 000011)}$
$+ 5\overset{0}{(\cdot 000100)} + 8\overset{0}{(\cdot 100001)} + 7\overset{1}{(\cdot 000000)} + 5\overset{0}{(\cdot 000001)}$

Sl 4: $\overset{0}{(\cdot 001001)} + \overset{1}{(\cdot 200000)} + \overset{1}{(\cdot 010000)} + 2\overset{0}{(\cdot 100011)} + 3\overset{0}{(\cdot 100100)} + 3\overset{1}{(\cdot 000002)} + 5\overset{1}{(\cdot 000010)}$
$+ 4\overset{0}{(\cdot 200001)} + 10\overset{0}{(\cdot 010001)} + 11\overset{1}{(\cdot 100000)} + 2\overset{0}{(\cdot 000003)} + 10\overset{0}{(\cdot 000011)} + 14\overset{0}{(\cdot 000100)} + 20\overset{0}{(\cdot 100001)}$
$+ 15\overset{1}{(\cdot 000000)} + 11\overset{0}{(\cdot 000001)}$

Affine E7 ⊃ E7 - Level 2

Sl 0: $\overset{0}{(\cdot 000001)}$ **Highest Weight** $\overset{0}{(1000001)}$

Sl 1: $\overset{0}{(\cdot 100001)} + \overset{1}{(\cdot 000000)} + \overset{0}{(\cdot 000001)}$

Sl 2: $\overset{1}{(\cdot 100000)} + \overset{0}{(\cdot 000011)} + \overset{0}{(\cdot 000100)} + 3\overset{0}{(\cdot 100001)} + 2\overset{1}{(\cdot 000000)} + 3\overset{0}{(\cdot 000001)}$

Sl 3: $\overset{1}{(\cdot 000010)} + \overset{0}{(\cdot 200001)} + 2\overset{0}{(\cdot 010001)} + 3\overset{1}{(\cdot 100000)} + \overset{0}{(\cdot 000003)} + 3\overset{0}{(\cdot 000011)} + 4\overset{0}{(\cdot 000100)}$
$+ 8\overset{0}{(\cdot 100001)} + 6\overset{1}{(\cdot 000000)} + 6\overset{0}{(\cdot 000001)}$

Sl 4: $\overset{1}{(\cdot 010000)} + \overset{0}{(\cdot 100011)} + 2\overset{0}{(\cdot 100100)} + 2\overset{1}{(\cdot 000002)} + 4\overset{1}{(\cdot 000010)} + 4\overset{0}{(\cdot 200001)} + 7\overset{0}{(\cdot 010001)}$
$+ 10\overset{1}{(\cdot 100000)} + 2\overset{0}{(\cdot 000003)} + 10\overset{0}{(\cdot 000011)} + 11\overset{0}{(\cdot 000100)} + 20\overset{0}{(\cdot 100001)} + 14\overset{1}{(\cdot 000000)} + 13\overset{0}{(\cdot 000001)}$

(continued next page)

```
              0            0             2            0             1              0              1
Sl 5:   (•110001) +  (•000110) +  (•000001) + 2(•001001) + 2(•200000) +  (•100003) + 4(•010000)

             0            0             1            1             0              0              1
   +  6(•100011) +  8(•100100) +  7(•000002) + 14(•000010) + 12(•200001) + 23(•010001) + 27(•100000)

             0            0             0            0             1              0
   +  6(•000003) + 26(•000011) + 29(•000100) + 47(•100001) + 32(•000000) + 26(•000001)
```

Affine E7 ⊃ A7 - Level 1

```
            •                             0
Sl 0:   (1000000)  Highest Weight  (1000000)

            •
Sl 1:   (0000100)

            •            •
Sl 2:   (0100001) +  (1000000)

            •            •            •
Sl 3:   (1001000) +  (0000011) +  (0000100)

            •            •            •             •
Sl 4:   (0010010) +  (2000001) + 2(0100001) + 2(1000000)

            •            •            •             •             •
Sl 5:   (1000101) +  (0110000) + 2(1001000) + 2(0000011) + 3(0000100)

            •            •            •             •             •             •             •
Sl 6:   (1100010) +  (0001100) +  (0010002) + 2(0010010) + 2(2000001) + 5(0100001) + 4(1000000)

            •            •            •             •             •             •             •
Sl 7:   (0101001) +  (1000020) +  (2010000) + 3(1000101) + 2(0110000) + 5(1001000) +  (0000003)

            •            •
   +   4(0000011) + 6(0000100)
```

Affine E7 ⊃ A7 - Level 2

```
            •                             0
Sl 0:   (0000002)  Highest Weight  (0000002)

            •
Sl 1:   (0010001)

            •            •            •             •             •
Sl 2:   (0020000) +  (1000011) +  (1000100) +  (0000002) +  (0000010)

            •            •            •             •             •             •             •
Sl 3:   (1010010) +  (0001002) +  (0001010) +  (1100001) + 3(0010001) +  (2000000) +  (0100000)

            •            •            •             •             •             •             •
Sl 4:   (2000020) +  (0011001) +  (1110000) + 2(0100101) +  (2001000) +  (0020000) +  (1000003)

            •            •            •             •             •
   +   2(0101000) + 4(1000011) + 4(1000100) + 4(0000002) + 3(0000010)
```

Affine E7 ⊃ A7 - Level 2

```
            •                             0
Sl 0:   (0000010)  Highest Weight  (0000010)

            •            •
Sl 1:   (0010001) +  (0100000)

            •            •            •             •             •
Sl 2:   (0101000) +  (1000011) +  (1000100) +  (0000002) + 2(0000010)

            •            •            •             •             •             •             •
Sl 3:   (1010010) +  (0001002) + 2(0001010) + 2(1100001) + 3(0010001) +  (2000000) + 3(0100000)

            •            •            •             •             •             •             •
Sl 4:   (0011001) +  (2000101) +  (0100020) +  (1110000) + 2(0100101) +  (2001000) +  (0020000)

            •            •            •             •             •             •
   +    (1000003) + 4(0101000) + 5(1000011) + 6(1000100) + 3(0000002) + 6(0000010)
```

Affine E7 ⊃ A7 - Level 2

Sl 0: $\overset{\bullet}{(0000000)}$ **Highest Weight** $\overset{1}{\mathbf{(0000000)}}$

Sl 1: $\overset{\bullet}{(0001000)}$

Sl 2: $\overset{\bullet}{(0100010)} + \overset{\bullet}{(1000001)} + \overset{\bullet}{(0000000)}$

Sl 3: $\overset{\bullet}{(1001001)} + \overset{\bullet}{(0200000)} + \overset{\bullet}{(0000020)} + \overset{\bullet}{(1010000)} + \overset{\bullet}{(0000101)} + 2\overset{\bullet}{(0001000)}$

Sl 4: $\overset{\bullet}{(1100100)} + \overset{\bullet}{(0010011)} + \overset{\bullet}{(0002000)} + \overset{\bullet}{(0010100)} + \overset{\bullet}{(2000002)} + \overset{\bullet}{(2000010)} + \overset{\bullet}{(0100002)}$
$+ 4\overset{\bullet}{(0100010)} + 4\overset{\bullet}{(1000001)} + 2\overset{\bullet}{(0000000)}$

Sl 5: $\overset{\bullet}{(0101010)} + \overset{\bullet}{(2010001)} + \overset{\bullet}{(1000102)} + 2\overset{\bullet}{(1000110)} + 2\overset{\bullet}{(0110001)} + 5\overset{\bullet}{(1001001)} + \overset{\bullet}{(2100000)}$
$+ \overset{\bullet}{(0000012)} + 3\overset{\bullet}{(0200000)} + 3\overset{\bullet}{(0000020)} + 5\overset{\bullet}{(1010000)} + 5\overset{\bullet}{(0000101)} + 7\overset{\bullet}{(0001000)}$

Affine E7 ⊃ A7 - Level 2

Sl 0: $\overset{\bullet}{(1000001)}$ **Highest Weight** $\overset{0}{\mathbf{(1000001)}}$

Sl 1: $\overset{\bullet}{(1010000)} + \overset{\bullet}{(0000101)} + \overset{\bullet}{(0001000)}$

Sl 2: $\overset{\bullet}{(0010100)} + \overset{\bullet}{(2000010)} + \overset{\bullet}{(0100002)} + 2\overset{\bullet}{(0100010)} + 3\overset{\bullet}{(1000001)} + \overset{\bullet}{(0000000)}$

Sl 3: $\overset{\bullet}{(1000110)} + \overset{\bullet}{(0110001)} + 3\overset{\bullet}{(1001001)} + \overset{\bullet}{(2100000)} + \overset{\bullet}{(0000012)} + \overset{\bullet}{(0200000)} + \overset{\bullet}{(0000020)}$
$+ 4\overset{\bullet}{(1010000)} + 4\overset{\bullet}{(0000101)} + 4\overset{\bullet}{(0001000)}$

Sl 4: $\overset{\bullet}{(1100011)} + \overset{\bullet}{(1011000)} + \overset{\bullet}{(0001101)} + 3\overset{\bullet}{(1100100)} + 3\overset{\bullet}{(0010011)} + \overset{\bullet}{(0002000)} + 5\overset{\bullet}{(0010100)}$
$+ 2\overset{\bullet}{(2000002)} + 5\overset{\bullet}{(2000010)} + 5\overset{\bullet}{(0100002)} + 9\overset{\bullet}{(0100010)} + 11\overset{\bullet}{(1000001)} + 3\overset{\bullet}{(0000000)}$

Affine E7 ⊃ A1 + D6 - Level 1

Sl 0: $\overset{0}{(1\cdot00000)}$ **Highest Weight** $\overset{0}{\mathbf{(1000000)}}$

Sl 1: $\overset{0}{(0\cdot10000)}$

Sl 2: $\overset{0}{(1\cdot00010)} + \overset{0}{(1\cdot00000)}$

Sl 3: $\overset{1}{(0\cdot00001)} + \overset{0}{(2\cdot10000)} + \overset{0}{(0\cdot10000)}$

Sl 4: $\overset{0}{(1\cdot01000)} + \overset{0}{(1\cdot00002)} + 2\overset{0}{(1\cdot00010)} + \overset{0}{(3\cdot00000)} + 2\overset{0}{(1\cdot00000)}$

Sl 5: $\overset{0}{(0\cdot10010)} + \overset{1}{(2\cdot00001)} + 2\overset{1}{(0\cdot00001)} + 2\overset{0}{(2\cdot10000)} + 3\overset{0}{(0\cdot10000)}$

Sl 6: $\overset{0}{(1\cdot00101)} + \overset{0}{(1\cdot20000)} + \overset{2}{(1\cdot00000)} + 2\overset{0}{(1\cdot01000)} + \overset{0}{(3\cdot00010)} + 2\overset{0}{(1\cdot00002)} + 5\overset{0}{(1\cdot00010)}$
$+ \overset{0}{(3\cdot00000)} + 4\overset{0}{(1\cdot00000)}$

Sl 7: $\overset{1}{(0\cdot00100)} + \overset{0}{(2\cdot10010)} + \overset{0}{(0\cdot10002)} + 2\overset{0}{(0\cdot10010)} + 3\overset{1}{(2\cdot00001)} + 5\overset{1}{(0\cdot00001)} + 5\overset{0}{(2\cdot10000)}$
$+ 6\overset{0}{(0\cdot10000)}$

Affine E7 ⊃ A1 + D6 - Level 1

sl 0: $\overset{0}{(0\cdot00001)}$ **Highest Weight** $\overset{0}{(0000001)}$

sl 1: $\overset{1}{(1\cdot00000)}$

sl 2: $\overset{0}{(0\cdot00100)} + \overset{0}{(2\cdot00001)} + \overset{0}{(0\cdot00001)}$

sl 3: $\overset{0}{(1\cdot10001)} + 2\overset{1}{(1\cdot00000)}$

sl 4: $\overset{1}{(0\cdot10000)} + \overset{0}{(0\cdot00011)} + \overset{0}{(2\cdot00100)} + \overset{0}{(0\cdot00100)} + 2\overset{0}{(2\cdot00001)} + 3\overset{0}{(0\cdot00001)}$

sl 5: $\overset{1}{(1\cdot00010)} + 3\overset{0}{(1\cdot10001)} + \overset{1}{(3\cdot00000)} + 4\overset{1}{(1\cdot00000)}$

sl 6: $\overset{0}{(0\cdot01001)} + \overset{1}{(2\cdot10000)} + \overset{0}{(2\cdot00011)} + 2\overset{1}{(0\cdot10000)} + 2\overset{0}{(0\cdot00011)} + 2\overset{0}{(2\cdot00100)} + 4\overset{0}{(0\cdot00100)}$

$+ 5\overset{0}{(2\cdot00001)} + 5\overset{0}{(0\cdot00001)}$

sl 7: $\overset{0}{(1\cdot10100)} + \overset{1}{(1\cdot00002)} + 3\overset{1}{(1\cdot00010)} + \overset{0}{(3\cdot10001)} + 7\overset{0}{(1\cdot10001)} + 2\overset{1}{(3\cdot00000)} + 9\overset{1}{(1\cdot00000)}$

Affine E7 ⊃ A1 + D6 - Level 2

sl 0: $\overset{0}{(0\cdot00002)}$ **Highest Weight** $\overset{0}{(0000002)}$

sl 1: $\overset{1}{(1\cdot00001)}$

sl 2: $\overset{0}{(0\cdot00101)} + \overset{2}{(2\cdot00000)} + \overset{0}{(0\cdot01000)} + \overset{0}{(2\cdot00002)} + \overset{0}{(2\cdot00010)} + \overset{0}{(0\cdot00002)} + \overset{0}{(0\cdot00010)}$

sl 3: $\overset{1}{(1\cdot00100)} + \overset{0}{(1\cdot10002)} + \overset{0}{(1\cdot10010)} + \overset{1}{(3\cdot00001)} + 4\overset{1}{(1\cdot00001)} + 2\overset{0}{(1\cdot10000)}$

sl 4: $\overset{0}{(0\cdot00200)} + \overset{1}{(2\cdot10001)} + 2\overset{1}{(0\cdot10001)} + \overset{0}{(0\cdot00012)} + 2\overset{0}{(2\cdot00101)} + \overset{0}{(0\cdot00020)} + 3\overset{0}{(0\cdot00101)}$

$+ 2\overset{2}{(2\cdot00000)} + 3\overset{0}{(2\cdot01000)} + \overset{0}{(0\cdot20000)} + 2\overset{2}{(0\cdot00000)} + 3\overset{0}{(0\cdot01000)} + \overset{0}{(4\cdot00002)} + 3\overset{0}{(2\cdot00002)}$

$+ 4\overset{0}{(2\cdot00010)} + 5\overset{0}{(0\cdot00002)} + 4\overset{0}{(0\cdot00010)} + \overset{0}{(2\cdot00000)} + 2\overset{0}{(0\cdot00000)}$

Affine E7 ⊃ A1 + D6 - Level 2

sl 0: $\overset{0}{(0\cdot00010)}$ **Highest Weight** $\overset{0}{(0000010)}$

sl 1: $\overset{1}{(1\cdot00001)} + \overset{0}{(1\cdot10000)}$

sl 2: $\overset{0}{(0\cdot00101)} + \overset{0}{(2\cdot01000)} + \overset{2}{(0\cdot00000)} + \overset{0}{(0\cdot01000)} + \overset{0}{(2\cdot00002)} + \overset{0}{(2\cdot00010)} + \overset{0}{(0\cdot00002)}$

$+ 2\overset{0}{(0\cdot00010)} + \overset{0}{(2\cdot00000)} + \overset{0}{(0\cdot00000)}$

sl 3: $\overset{1}{(1\cdot00100)} + \overset{0}{(1\cdot10002)} + 2\overset{0}{(1\cdot10010)} + \overset{1}{(3\cdot00001)} + 5\overset{1}{(1\cdot00001)} + \overset{0}{(3\cdot10000)} + 4\overset{0}{(1\cdot10000)}$

sl 4: $\overset{0}{(0\cdot01010)} + \overset{1}{(2\cdot10001)} + 2\overset{1}{(0\cdot10001)} + \overset{0}{(0\cdot00012)} + 3\overset{0}{(2\cdot00101)} + \overset{0}{(0\cdot00020)} + 4\overset{0}{(0\cdot00101)}$

$+ \overset{0}{(2\cdot20000)} + 2\overset{2}{(2\cdot00000)} + 4\overset{0}{(2\cdot01000)} + 2\overset{0}{(0\cdot20000)} + 3\overset{2}{(0\cdot00000)} + 5\overset{0}{(0\cdot01000)} + \overset{0}{(4\cdot00010)}$

$+ 4\overset{0}{(2\cdot00002)} + 6\overset{0}{(2\cdot00010)} + 4\overset{0}{(0\cdot00002)} + 8\overset{0}{(0\cdot00010)} + 3\overset{0}{(2\cdot00000)} + 3\overset{0}{(0\cdot00000)}$

Affine E7 ⊃ A1 + D6 - Level 2

```
               0                          0
Sl 0:      (0•00000)   Highest Weight (0100000)

               0
Sl 1:      (1•10000)

               0             0             0             0             0
Sl 2:      (0•01000) +   (2•00010) +   (0•00010) +   (2•00000) +   (0•00000)

               0              1            0              0
Sl 3:      (1•10010) +  2(1•00001) +   (3•10000) +  3(1•10000)

               1             0             0             0             0              0              0
Sl 4:      (0•10001) +   (2•00101) +   (0•00020) +   (0•00101) +   (2•20000) +  2(2•01000) +   (0•20000)

               2              0            0              0             0              0             0
      +    (0•00000) +  3(0•01000) +   (2•00002) +  4(2•00010) +  2(0•00002) +  4(0•00010) +   (4•00000)

                0             0
      +   3(2•00000) +  3(0•00000)
```

Affine E7 ⊃ A1 + D6 - Level 2

```
               0                          0
Sl 0:      (2•00000)   Highest Weight (2000000)

               0
Sl 1:      (1•10000)

               0             0             0             0             0
Sl 2:      (0•20000) +   (2•00010) +   (0•00010) +   (2•00000) +   (0•00000)

               0             1             0              0
Sl 3:      (1•10010) +   (1•00001) +   (3•10000) +  3(1•10000)

               1             0             0             0              0             0              0
Sl 4:      (0•10001) +   (2•00020) +   (0•00101) +   (2•20000) +  2(2•01000) +   (0•20000) +  2(0•01000)

               0              0            0              0             0              0              0
      +    (2•00002) +  3(2•00010) +   (0•00002) +  3(0•00010) +   (4•00000) +  4(2•00000) +  2(0•00000)
```

Affine E7 ⊃ A1 + D6 - Level 2

```
               1                          1
Sl 0:      (0•00000)   Highest Weight (0000000)

               0             0
Sl 1:      (1•00100) +   (1•00001)

               1             0             0             1              1
Sl 2:      (0•00010) +   (2•10001) +   (0•10001) +   (2•00000) +  2(0•00000)

               0              1             0            0              0             0              0
Sl 3:      (1•01001) +  2(1•10000) +  2(1•00011) +   (3•00100) +  3(1•00100) +   (3•00001) +  4(1•00001)

               0             1             0             0             1              1              1
Sl 4:      (0•10011) +   (0•01000) +   (2•10100) +   (0•10100) +   (2•00002) +  2(2•00010) +   (0•00002)

                1             0             0            1              1              1
      +   4(0•00010) +  5(2•10001) +  6(0•10001) +   (4•00000) +  5(2•00000) +  6(0•00000)
```

Affine E7 ⊃ A1 + D6 - Level 2

```
               0                          0
Sl 0:      (1•00001)   Highest Weight (1000001)

               0             1             1
Sl 1:      (0•10001) +   (2•00000) +   (0•00000)

               1             0              0            0              0
Sl 2:      (1•10000) +   (1•00011) +  2(1•00100) +   (3•00001) +  3(1•00001)

               0             1             1              1             0              0              1
Sl 3:      (0•10100) +   (2•00010) +   (0•00002) +  2(0•00010) +  3(2•10001) +  4(0•10001) +  4(2•00000)

                1
      +   4(0•00000)   (continued next page)
```

```
          0              0              2              0              1              0              0
Sl 4:  (1•00110) +    (1•20001) +    (1•00001) + 3(1•01001) +    (3•10000) +    (3•00011) +    (1•00003)

          1              0              0              0              0              0
   + 6(1•10000) +   6(1•00011) +   2(3•00100) +   9(1•00100) +   4(3•00001) + 11(1•00001)
```

Affine E7 ⊃ A2 + A5 - Level 1

```
          0                           0
Sl 0:  (10•0000)   Highest Weight (1000000)

          0
Sl 1:  (00•1000)

          0
Sl 2:  (01•0010)

          1              0              0
Sl 3:  (10•0001) +    (02•0000) +    (10•0000)

          0              0              2              0
Sl 4:  (00•0101) +    (11•1000) +    (00•0000) +    (00•1000)

          1              0              0              0
Sl 5:  (01•0100) +    (20•0010) +    (01•0002) +  2(01•0010)

          0              1              1              0              0              0
Sl 6:  (10•1010) +    (02•0001) +  3(10•0001) +    (21•0000) +    (02•0000) +  3(10•0000)

          0              1              0              2              0              0              2
Sl 7:  (11•0101) +    (00•1001) +    (00•0020) +    (11•0000) +  2(00•0101) +  3(11•1000) +    (00•0000)

          0
   +  4(00•1000)
```

Affine E7 ⊃ A2 + A5 - Level 1

```
          0                           0
Sl 0:  (00•0001)   Highest Weight (0000001)

          1
Sl 1:  (01•0000)

          0
Sl 2:  (10•0100)

          1              0              0
Sl 3:  (00•0010) +    (11•0001) +    (00•0001)

          0              1              1
Sl 4:  (01•1001) +    (20•0000) +  2(01•0000)

          1              0              0              0
Sl 5:  (10•1000) +    (10•0011) +    (02•0100) +  2(10•0100)

          1              0              1              1              0              0
Sl 6:  (11•0010) +    (00•1100) +    (00•0002) +  2(00•0010) +  3(11•0001) +  3(00•0001)

          0              0              2              0              1              1              1
Sl 7:  (01•0110) +    (20•1001) +    (01•0001) +  3(01•1001) +    (12•0000) +  2(20•0000) +  5(01•0000)
```

Affine E7 ⊃ A2 + A5 - Level 2

```
          0                           0
Sl 0:  (00•0002)   Highest Weight (0000002)

          1
Sl 1:  (01•0001)

          0              2              0
Sl 2:  (10•0101) +    (02•0000) +    (10•1000)

          1              1              0              1              0              0              0
Sl 3:  (11•0100) +    (00•0011) +    (11•0002) +    (00•0100) +    (11•0010) +    (00•0002) +  2(00•0010)

          0              2              0              0              1              1              1
Sl 4:  (20•0200) +    (01•0010) +    (01•1002) +  2(01•1010) +    (12•0001) +  2(20•0001) +  4(01•0001)

          0              0
   +   (20•0000) +  2(01•0000)   (continued next page)
```

```
           1            1            0            1            0            0            0
Sl 5:   (10•0110) +  (02•1001) +  (21•0101) + 2(10•1001) +  (10•0012) +  (10•0020) + 2(02•0101)
           2            0            0            2            2            0            0
     +  (21•0000) + 5(10•0101) +  (21•1000) + 2(02•0000) + 3(10•0000) + 3(02•1000) + 5(10•1000)
```

Affine E7 ⊃ A2 + A5 - Level 2

```
           0                       0
Sl 0:   (00•0010) Highest Weight (0000010)
           1            0
Sl 1:   (01•0001) +  (01•0000)
           0            2            0            0
Sl 2:   (10•0101) +  (10•0000) +  (02•1000) +  (10•1000)
           1            1            0            1            0            0            0
Sl 3:   (11•0100) +  (00•0011) +  (11•0002) + 2(00•0100) + 2(11•0010) +  (00•0002) + 2(00•0010)
           0            2            0            0            0            1            1
Sl 4:   (20•1010) +  (01•0010) +  (01•1002) +  (01•0200) + 2(01•1010) +  (12•0001) + 2(20•0001)
           1            0            0            0
     + 6(01•0001) +  (12•0000) + 2(20•0000) + 3(01•0000)
           1            1            0            1            0            0            0
Sl 5:   (10•0110) +  (02•1001) +  (21•0101) + 3(10•1001) +  (10•0012) + 2(10•0020) + 3(02•0101)
           2            0            0            2            2            0            0
     +  (21•0000) + 6(10•0101) + 2(21•1000) + 2(02•0000) + 4(10•0000) + 4(02•1000) + 8(10•1000)
```

Affine E7 ⊃ A2 + A5 - Level 2

```
           0                       0
Sl 0:   (01•0000) Highest Weight (0100000)
           0
Sl 1:   (10•1000)
           1            0            0
Sl 2:   (00•0100) +  (11•0010) +  (00•0010)
           0            1            1            0            0            0
Sl 3:   (01•1010) +  (20•0001) + 2(01•0001) +  (12•0000) +  (20•0000) + 2(01•0000)
           1            0            0            0            0            2            0
Sl 4:   (10•1001) +  (10•0020) +  (02•0101) + 2(10•0101) +  (21•1000) + 2(10•0000) + 2(02•1000)
           0
     + 4(10•1000)
           1            0            0            1            2            1            0
Sl 5:   (11•0011) +  (00•1101) +  (11•2000) + 3(11•0100) +  (00•1000) + 2(00•0011) +  (00•2000)
           0            0            0            1            0            0            0
     +  (30•0010) + 2(11•0002) +  (03•0010) + 4(00•0100) + 6(11•0010) + 2(00•0002) + 5(00•0010)
```

Affine E7 ⊃ A2 + A5 - Level 2

```
           0                       0
Sl 0:   (20•0000) Highest Weight (2000000)
           0
Sl 1:   (10•1000)
           0            0            0
Sl 2:   (00•2000) +  (11•0010) +  (00•0010)
           0            1            1            0            0            0
Sl 3:   (01•1010) +  (20•0001) +  (01•0001) +  (12•0000) +  (20•0000) + 2(01•0000)
           1            0            0            0            2            0            0
Sl 4:   (10•1001) +  (02•0020) + 2(10•0101) +  (21•1000) +  (10•0000) + 2(02•1000) + 3(10•1000)
           1            0            0            1            2            1            0
Sl 5:   (11•0011) +  (00•1101) +  (11•2000) + 2(11•0100) +  (00•1000) +  (00•0011) +  (00•2000)
           0            0            0            1            0            0            0
     +  (30•0010) +  (11•0002) +  (03•0010) + 3(00•0100) + 5(11•0010) + 2(00•0002) + 3(00•0010)
```

Affine E7 ⊃ A2 + A5 - Level 2

```
              1                            1
Sl 0:     (00•0000)  Highest Weight  (0000000)

              0
Sl 1:     (01•0100)

              1              0              0
Sl 2:     (10•0010) +    (02•0001) +    (10•0001)

              0              0              2              0              1              1
Sl 3:     (11•1001) +    (00•0110) +    (00•0001) +    (00•1001) +  2(11•0000) +  2(00•0000)

              1              1              0              0              1              0              0
Sl 4:     (01•0101) +    (20•1000) +    (20•0011) +    (12•0100) +  2(01•1000) +  2(01•0011) +    (20•0100)

              0
      +  4(01•0100)

              0              2              0              1              0              1              1
Sl 5:     (10•1011) +    (10•0100) +    (02•1100) +    (21•0010) +  2(10•1100) +    (02•0002) +  2(10•0002)

              1              1              0              0              0
      +  2(02•0010) +  6(10•0010) +  2(21•0001) +  4(02•0001) +  6(10•0001)
```

Affine E7 ⊃ A2 + A5 - Level 2

```
Sl 0:     (10•0001)  Highest Weight  (1000001)

              0              1              1
Sl 1:     (00•1001) +    (11•0000) +    (00•0000)

              1              0              0              0
Sl 2:     (01•1000) +    (01•0011) +    (20•0100) +  2(01•0100)

              0              1              1              1              0              0              0
Sl 3:     (10•1100) +    (10•0002) +    (02•0010) +  3(10•0010) +    (21•0001) +  2(02•0001) +  4(10•0001)

              0              1              2              0              0              0              2
Sl 4:     (11•0110) +    (00•1010) +    (11•0001) +    (00•0102) +  4(11•1001) +  2(00•0110) +  2(00•0001)

              0              1              1              1              1
      +  5(00•1001) +    (30•0000) +    (03•0000) +  6(11•0000) +  5(00•0000)
```

Affine E7 ⊃ A1 + A3 + A3 - Level 1

```
              0                            0
Sl 0:     (100•000)  Highest Weight  (1000000)

              1
Sl 1:     (000•100)

              0
Sl 2:     (001•010)

              1
Sl 3:     (010•001)

              0              2              0              0
Sl 4:     (100•101) +    (100•000) +    (011•000) +  3(100•000)

              1              1              1
Sl 5:     (000•011) +    (101•100) +  5(000•100)

              0              2              0              0              0
Sl 6:     (110•010) +    (001•010) +    (001•200) +    (001•002) +  4(001•010)

              1              1              1              1
Sl 7:     (010•110) +    (200•001) +  4(010•001) +    (002•001)
```

Affine E7 ⊃ A1 + A3 + A3 - Level 2

Sl 0: $(000\overset{0}{\cdot}002)$ **Highest Weight** $(000\overset{0}{0}002)$

Sl 1: $(001\overset{1}{\cdot}001)$

Sl 2: $(010\overset{0}{\cdot}101) + (010\overset{2}{\cdot}000) + (010\overset{0}{\cdot}000) + (002\overset{2}{\cdot}000)$

Sl 3: $(100\overset{1}{\cdot}011) + (011\overset{1}{\cdot}100) + 3(100\overset{1}{\cdot}100)$

Sl 4: $(020\overset{0}{\cdot}200) + (000\overset{0}{\cdot}111) + (101\overset{2}{\cdot}010) + (101\overset{0}{\cdot}200) + (101\overset{0}{\cdot}002) + (000\overset{2}{\cdot}002) + 2(101\overset{0}{\cdot}010)$
$+ 4(000\overset{2}{\cdot}010) + (000\overset{0}{\cdot}200) + 4(000\overset{0}{\cdot}002) + 8(000\overset{0}{\cdot}010)$

Sl 5: $(110\overset{1}{\cdot}110) + 3(110\overset{1}{\cdot}001) + (001\overset{1}{\cdot}102) + 4(001\overset{1}{\cdot}110) + (102\overset{1}{\cdot}001) + (001\overset{3}{\cdot}001) + 12(001\overset{1}{\cdot}001)$

Affine E7 ⊃ A1 + A3 + A3 - Level 2

Sl 0: $(000\overset{0}{\cdot}010)$ **Highest Weight** $(000\overset{0}{0}010)$

Sl 1: $(001\overset{1}{\cdot}001)$

Sl 2: $(010\overset{0}{\cdot}101) + (010\overset{2}{\cdot}000) + 2(010\overset{0}{\cdot}000) + (002\overset{0}{\cdot}000)$

Sl 3: $(100\overset{1}{\cdot}011) + (011\overset{1}{\cdot}100) + 4(100\overset{1}{\cdot}100)$

Sl 4: $(020\overset{0}{\cdot}010) + (000\overset{0}{\cdot}111) + (101\overset{2}{\cdot}010) + (101\overset{0}{\cdot}200) + (101\overset{0}{\cdot}002) + (000\overset{2}{\cdot}002) + 3(101\overset{0}{\cdot}010)$
$+ 5(000\overset{2}{\cdot}010) + 4(000\overset{0}{\cdot}200) + 4(000\overset{0}{\cdot}002) + 9(000\overset{0}{\cdot}010)$

Sl 5: $(110\overset{1}{\cdot}110) + 4(110\overset{1}{\cdot}001) + (001\overset{1}{\cdot}102) + 5(001\overset{1}{\cdot}110) + (102\overset{1}{\cdot}001) + (001\overset{3}{\cdot}001) + 15(001\overset{1}{\cdot}001)$

Affine E7 ⊃ A1 + A3 + A3 - Level 2

Sl 0: $(000\overset{1}{\cdot}000)$ **Highest Weight** $(000\overset{1}{0}000)$

Sl 1: $(001\overset{0}{\cdot}100)$

Sl 2: $(010\overset{1}{\cdot}010)$

Sl 3: $(100\overset{0}{\cdot}110) + (100\overset{2}{\cdot}001) + (011\overset{0}{\cdot}001) + 3(100\overset{0}{\cdot}001)$

Sl 4: $(101\overset{1}{\cdot}101) + (020\overset{1}{\cdot}000) + (000\overset{1}{\cdot}020) + 5(000\overset{1}{\cdot}101) + 3(101\overset{1}{\cdot}000) + (000\overset{3}{\cdot}000) + 8(000\overset{1}{\cdot}000)$

Sl 5: $(110\overset{0}{\cdot}011) + (110\overset{2}{\cdot}100) + 3(110\overset{0}{\cdot}100) + (001\overset{2}{\cdot}011) + (001\overset{0}{\cdot}201) + 4(001\overset{0}{\cdot}011) + (102\overset{0}{\cdot}100)$
$+ 4(001\overset{2}{\cdot}100) + 11(001\overset{0}{\cdot}100)$

Affine E7 ⊃ A1 + A3 + A3 - Level 2

Sl 0: $(100\overset{0}{\bullet}001)$ **Highest Weight** $(100\overset{0}{0}001)$

Sl 1: $(000\overset{1}{\bullet}101) + (101\overset{1}{\bullet}000) + 4(000\overset{1}{\bullet}000)$

Sl 2: $(110\overset{0}{\bullet}100) + (001\overset{0}{\bullet}011) + (001\overset{2}{\bullet}100) + 4(001\overset{0}{\bullet}100)$

Sl 3: $(200\overset{1}{\bullet}010) + (010\overset{1}{\bullet}200) + (010\overset{1}{\bullet}002) + 4(010\overset{1}{\bullet}010) + (002\overset{1}{\bullet}010)$

Sl 4: $(100\overset{2}{\bullet}110) + (100\overset{0}{\bullet}102) + (011\overset{0}{\bullet}110) + 5(100\overset{0}{\bullet}110) + (011\overset{2}{\bullet}001) + (201\overset{0}{\bullet}001) + 5(100\overset{2}{\bullet}001)$
$+ 3(011\overset{0}{\bullet}001) + 14(100\overset{0}{\bullet}001)$

Sl 5: $(020\overset{1}{\bullet}101) + (101\overset{1}{\bullet}020) + (210\overset{1}{\bullet}000) + (000\overset{1}{\bullet}210) + (000\overset{1}{\bullet}012) + 6(101\overset{1}{\bullet}101) + (000\overset{3}{\bullet}101)$
$+ 3(020\overset{1}{\bullet}000) + 5(000\overset{1}{\bullet}020) + 22(000\overset{1}{\bullet}101) + (101\overset{3}{\bullet}000) + (012\overset{1}{\bullet}000) + 13(101\overset{1}{\bullet}000) + 4(000\overset{3}{\bullet}000)$
$+ 28(000\overset{1}{\bullet}000)$

Affine E8 Branching Rules

Affine E8 ⊃ E8 - Level 1

```
                  0                               0
Sl 0:    (•0000000)  Highest Weight (10000000)

                  0
Sl 1:    (•1000000)

                  0                0                0
Sl 2:    (•0000001)  +    (•1000000)  +    (•0000000)

                  0                0                0                0
Sl 3:    (•0100000)  +    (•0000001)  +  2(•1000000)  +    (•0000000)

                  1                0                0                0                0                0
Sl 4:    (•0000000)  +    (•2000000)  +    (•0100000)  +  2(•0000001)  +  3(•1000000)  +  2(•0000000)

                  0                1                0                0                0                0
Sl 5:    (•1000001)  +    (•0000000)  +    (•2000000)  +  3(•0100000)  +  3(•0000001)  +  5(•1000000)

                  0
    +  2(•0000000)

                  0                0                1                0                0                0
Sl 6:    (•0010000)  +  2(•1000001)  +  2(•0000000)  +  3(•2000000)  +  4(•0100000)  +  6(•0000001)

                  0                0
    +  7(•1000000)  +  4(•0000000)

                  0                0                0                0                1                0
Sl 7:    (•0000010)  +    (•1100000)  +    (•0010000)  +  4(•1000001)  +  4(•0000000)  +  4(•2000000)

                  0                0                0                0
    +  8(•0100000)  +  8(•0000001)  + 12(•1000000)  +  4(•0000000)
```

Affine E8 ⊃ E8 - Level 2

```
                  0                               0
Sl 0:    (•1000000)  Highest Weight (01000000)

                  0                0                0                0
Sl 1:    (•0100000)  +    (•0000001)  +    (•1000000)  +    (•0000000)

                  0                1                0                0                0                0
Sl 2:    (•1000001)  +    (•0000000)  +    (•2000000)  +  2(•0100000)  +  2(•0000001)  +  3(•1000000)

                  0
    +    (•0000000)

                  0                0                0                0                1                0
Sl 3:    (•0000010)  +    (•1100000)  +    (•0010000)  +  3(•1000001)  +  3(•0000000)  +  3(•2000000)

                  0                0                0                0
    +  5(•0100000)  +  5(•0000001)  +  6(•1000000)  +  2(•0000000)

                  0                1                0                0                0                0
Sl 4:    (•0100001)  +  2(•1000000)  +    (•0000002)  +  2(•0000010)  +    (•3000000)  +  3(•1100000)

                  0                0                1                0                0                0
    +  4(•0010000)  +  9(•1000001)  +  7(•0000000)  +  7(•2000000)  + 12(•0100000)  + 11(•0000001)

                  0                0
    + 12(•1000000)  +  4(•0000000)
```

Affine E8 ⊃ E8 - Level 2

```
                  0                               0
Sl 0:    (•0000001)  Highest Weight (00000001)

                  1                0                0                0
Sl 1:    (•0000000)  +    (•0100000)  +    (•0000001)  +    (•1000000)

                  0                0                1                0                0                0
Sl 2:    (•0010000)  +    (•1000001)  +  2(•0000000)  +    (•2000000)  +  2(•0100000)  +  3(•0000001)

                  0                0
    +  2(•1000000)  +    (•0000000)   (continued next page)
```

```
                1                  0                  0                  0                  0                  1
Sl 3:     (•1000000)  +    (•0000010)  +    (•1100000)  +  2(•0010000)  +  4(•1000001)  +  4(•0000000)

                0                  0                  0                  0                  0
      +  2(•2000000)  +  6(•0100000)  +  6(•0000001)  +  5(•1000000)  +    (•0000000)

                0                  0                  0                  1                  0                  0
Sl 4:     (•0001000)  +    (•2000001)  +    (•0100001)  +  3(•1000000)  +    (•0000002)  +  3(•0000010)

                0                  0                  0                  1                  0                  0
      +  4(•1100000)  +  6(•0010000)  + 10(•1000001)  + 10(•0000000)  +  7(•2000000)  + 12(•0100000)

                0                  0                  0
      + 13(•0000001)  + 10(•1000000)  +  3(•0000000)
```

Affine E8 ⊃ E8 - Level 2

```
                0                                      0
Sl 0:     (•0000000)   Highest Weight  (20000000)

                0
Sl 1:     (•1000000)

                0                  0                  0                  0
Sl 2:     (•2000000)  +    (•0000001)  +    (•1000000)  +    (•0000000)

                0                  0                  0                  0                  0                  0
Sl 3:     (•1000001)  +    (•2000000)  +  2(•0100000)  +    (•0000001)  +  3(•1000000)  +    (•0000000)

                0                  0                  0                  0                  1                  0
Sl 4:     (•0000002)  +    (•1100000)  +    (•0010000)  +  2(•1000001)  +  2(•0000000)  +  4(•2000000)

                0                  0                  0                  0
      +  3(•0100000)  +  4(•0000001)  +  5(•1000000)  +  3(•0000000)

                0                  1                  0                  0                  0                  0
Sl 5:     (•0100001)  +    (•1000000)  +    (•0000002)  +  2(•0000010)  +    (•3000000)  +  3(•1100000)

                0                  0                  1                  0                  0                  0
      +  2(•0010000)  +  7(•1000001)  +  4(•0000000)  +  6(•2000000)  + 10(•0100000)  +  7(•0000001)

                0                  0
      + 11(•1000000)  +  3(•0000000)
```

Affine E8 ⊃ A8 - Level 1

```
                •                                      0
Sl 0:     (10000000)   Highest Weight  (10000000)

                •
Sl 1:     (00000010)

                •
Sl 2:     (00010000)

                •                  •
Sl 3:     (01000001)  +    (10000000)

                •                  •                  •
Sl 4:     (10000100)  +    (00000002)  +    (00000010)

                •                  •                  •
Sl 5:     (00001001)  +    (10100000)  +    (00010000)

                •                  •                  •                  •
Sl 6:     (00100010)  +    (20000001)  +  2(01000001)  +  2(10000000)

                •                  •                  •                  •                  •
Sl 7:     (01001000)  +    (10000011)  +  2(10000100)  +    (00000002)  +  3(00000010)
```

Affine E8 ⊃ A8 - Level 2

```
                •                                      0
Sl 0:     (01000000)   Highest Weight  (01000000)

                •                  •
Sl 1:     (10000010)  +    (00000001)

                •                  •                  •
Sl 2:     (10010000)  +    (00000101)  +    (00001000)
```

(continued next page)

Sl 3: $(00010\dot{0}10)$ + $(11000\dot{0}01)$ + $2(00100\dot{0}01)$ + $(20000\dot{0}00)$ + $2(01000\dot{0}00)$

Sl 4: $(00101\dot{0}00)$ + $(01000\dot{0}11)$ + $(20000\dot{1}00)$ + $2(01000\dot{1}00)$ + $2(10000\dot{0}02)$ + $4(10000\dot{0}10)$
+ $3(00000\dot{0}01)$

Sl 5: $(01010\dot{0}01)$ + $(10000\dot{1}10)$ + $3(10001\dot{0}01)$ + $(20100\dot{0}00)$ + $2(01100\dot{0}00)$ + $(00000\dot{0}12)$
+ $(00000\dot{0}20)$ + $4(10010\dot{0}00)$ + $4(00000\dot{1}01)$ + $5(00001\dot{0}00)$

Affine E8 ⊃ A8 - Level 2

Sl 0: $(00000\dot{0}01)$ **Highest Weight** $(00000\overset{0}{0}01)$

Sl 1: $(00001\dot{0}00)$

Sl 2: $(00100\dot{0}01)$ + $(01000\dot{0}00)$

Sl 3: $(01000\dot{1}00)$ + $(10000\dot{0}02)$ + $(10000\dot{0}10)$ + $2(00000\dot{0}01)$

Sl 4: $(10001\dot{0}01)$ + $(01100\dot{0}00)$ + $(10010\dot{0}00)$ + $2(00000\dot{1}01)$ + $2(00001\dot{0}00)$

Sl 5: $(10100\dot{0}10)$ + $(00001\dot{1}00)$ + $(00010\dot{0}02)$ + $(00010\dot{0}10)$ + $2(11000\dot{0}01)$ + $4(00100\dot{0}01)$
+ $(20000\dot{0}00)$ + $3(01000\dot{0}00)$

Affine E8 ⊃ A8 - Level 2

Sl 0: $(20000\dot{0}00)$ **Highest Weight** $(20000\overset{0}{0}00)$

Sl 1: $(10000\dot{0}10)$

Sl 2: $(00000\dot{0}20)$ + $(10010\dot{0}00)$ + $(00001\dot{0}00)$

Sl 3: $(00010\dot{0}10)$ + $(11000\dot{0}01)$ + $(00100\dot{0}01)$ + $(20000\dot{0}00)$ + $2(01000\dot{0}00)$

Sl 4: $(00020\dot{0}00)$ + $(01000\dot{0}11)$ + $(20000\dot{1}00)$ + $2(01000\dot{1}00)$ + $(10000\dot{0}02)$ + $3(10000\dot{0}10)$
+ $2(00000\dot{0}01)$

Sl 5: $(01010\dot{0}01)$ + $(10000\dot{1}10)$ + $2(10001\dot{0}01)$ + $(20100\dot{0}00)$ + $(01100\dot{0}00)$ + $(00000\dot{0}12)$
+ $(00000\dot{0}20)$ + $4(10010\dot{0}00)$ + $3(00000\dot{1}01)$ + $3(00001\dot{0}00)$

Affine E8 ⊃ D8 - Level 1

Sl 0: $(100000\overset{0}{0}\bullet)$ **Highest Weight** $(10000\overset{0}{0}00)$

Sl 1: $(000000\overset{1}{0}\bullet)$

Sl 2: $(001000\overset{0}{0}\bullet)$ + $(100000\overset{0}{0}\bullet)$

Sl 3: $(100000\overset{0}{1}\bullet)$ + $(000000\overset{1}{0}\bullet)$

Sl 4: $(000010\overset{0}{0}\bullet)$ + $(110000\overset{0}{0}\bullet)$ + $(001000\overset{0}{0}\bullet)$ + $2(100000\overset{0}{0}\bullet)$ (continued next page)

Sl 5: $(0100000\bullet)^{1}$ + $2(1000001\bullet)^{0}$ + $2(0000000\bullet)^{1}$

Sl 6: $(1001000\bullet)^{0}$ + $(0000001\bullet)^{1}$ + $(0000100\bullet)^{0}$ + $2(1100000\bullet)^{0}$ + $3(0010000\bullet)^{0}$ + $3(1000000\bullet)^{0}$

Sl 7: $(0010001\bullet)^{0}$ + $(2000000\bullet)^{1}$ + $2(0100000\bullet)^{1}$ + $4(1000001\bullet)^{0}$ + $4(0000000\bullet)^{1}$

Affine E8 ⊃ D8 - Level 2

Sl 0: $(0100000\bullet)^{0}$ **Highest Weight** $\mathbf{(01000000)^{0}}$

Sl 1: $(1000000\bullet)^{1}$ + $(0000001\bullet)^{0}$

Sl 2: $(0000010\bullet)^{0}$ + $(1010000\bullet)^{0}$ + $(0001000\bullet)^{0}$ + $(2000000\bullet)^{0}$ + $2(0100000\bullet)^{0}$ + $(0000000\bullet)^{0}$

Sl 3: $(0010000\bullet)^{1}$ + $(2000001\bullet)^{0}$ + $2(0100001\bullet)^{0}$ + $3(1000000\bullet)^{1}$ + $3(0000001\bullet)^{0}$

Sl 4: $(1000001\bullet)^{1}$ + $(0101000\bullet)^{0}$ + $2(1000100\bullet)^{0}$ + $(0000002\bullet)^{0}$ + $(0000000\bullet)^{2}$ + $(2100000\bullet)^{0}$
+ $3(0000010\bullet)^{0}$ + $(0200000\bullet)^{0}$ + $4(1010000\bullet)^{0}$ + $4(0001000\bullet)^{0}$ + $3(2000000\bullet)^{0}$ + $6(0100000\bullet)^{0}$
+ $2(0000000\bullet)^{0}$

Sl 5: $(1010001\bullet)^{0}$ + $(0000100\bullet)^{1}$ + $2(0001001\bullet)^{0}$ + $3(1100000\bullet)^{1}$ + $4(0010000\bullet)^{1}$ + $4(2000001\bullet)^{0}$
+ $8(0100001\bullet)^{0}$ + $11(1000000\bullet)^{1}$ + $8(0000001\bullet)^{0}$

Affine E8 ⊃ D8 - Level 2

Sl 0: $(0000000\bullet)^{0}$ **Highest Weight** $\mathbf{(00000001)^{0}}$

Sl 1: $(0000001\bullet)^{0}$

Sl 2: $(0001000\bullet)^{0}$ + $(0100000\bullet)^{0}$ + $(0000000\bullet)^{0}$

Sl 3: $(0100001\bullet)^{0}$ + $(1000000\bullet)^{1}$ + $2(0000001\bullet)^{0}$

Sl 4: $(1000100\bullet)^{0}$ + $(0000002\bullet)^{0}$ + $(0000010\bullet)^{0}$ + $(0200000\bullet)^{0}$ + $(1010000\bullet)^{0}$ + $2(0001000\bullet)^{0}$
+ $(2000000\bullet)^{0}$ + $3(0100000\bullet)^{0}$ + $2(0000000\bullet)^{0}$

Sl 5: $(0001001\bullet)^{0}$ + $(1100000\bullet)^{1}$ + $(0010000\bullet)^{1}$ + $(2000001\bullet)^{0}$ + $4(0100001\bullet)^{0}$ + $4(1000000\bullet)^{1}$
+ $5(0000001\bullet)^{0}$

Affine E8 ⊃ D8 - Level 2

Sl 0: $(2000000\bullet)^{0}$ **Highest Weight** $\mathbf{(20000000)^{0}}$

Sl 1: $(1000000\bullet)^{1}$

Sl 2: $(0000000\bullet)^{2}$ + $(1010000\bullet)^{0}$ + $(0001000\bullet)^{0}$ + $(2000000\bullet)^{0}$ + $(0100000\bullet)^{0}$

Sl 3: $(0010000\bullet)^{1}$ + $(2000001\bullet)^{0}$ + $(0100001\bullet)^{0}$ + $3(1000000\bullet)^{1}$ + $(0000001\bullet)^{0}$

Sl 4: $(0020000\bullet)^{0}$ + $(1000001\bullet)^{1}$ + $2(1000100\bullet)^{0}$ + $(0000000\bullet)^{2}$ + $(2100000\bullet)^{0}$ + $2(0000010\bullet)^{0}$
+ $(0200000\bullet)^{0}$ + $3(1010000\bullet)^{0}$ + $3(0001000\bullet)^{0}$ + $4(2000000\bullet)^{0}$ + $3(0100000\bullet)^{0}$ + $(0000000\bullet)^{0}$

(continued next page)

Sl 5: $\overset{0}{(1010001\bullet)} + \overset{1}{(0000100\bullet)} + \overset{0}{(0001001\bullet)} + 2\overset{1}{(1100000\bullet)} + 4\overset{1}{(0010000\bullet)} + 4\overset{0}{(2000001\bullet)}$

$+ 5\overset{0}{(0100001\bullet)} + 9\overset{1}{(1000000\bullet)} + 4\overset{0}{(0000001\bullet)}$

Affine E8 ⊃ A1 + E7 - Level 1

Sl 0: $\overset{0}{(1\bullet000000)}$ **Highest Weight** $\overset{\mathbf{0}}{\mathbf{(10000000)}}$

Sl 1: $\overset{0}{(0\bullet100000)}$

Sl 2: $\overset{0}{(1\bullet000001)} + \overset{0}{(1\bullet000000)}$

Sl 3: $\overset{1}{(0\bullet000000)} + \overset{0}{(2\bullet100000)} + \overset{0}{(0\bullet100000)}$

Sl 4: $\overset{0}{(1\bullet010000)} + 2\overset{0}{(1\bullet000001)} + \overset{0}{(3\bullet000000)} + 2\overset{0}{(1\bullet000000)}$

Sl 5: $\overset{0}{(0\bullet100001)} + \overset{1}{(2\bullet000000)} + \overset{1}{(0\bullet000000)} + 2\overset{0}{(2\bullet100000)} + 3\overset{0}{(0\bullet100000)}$

Sl 6: $\overset{0}{(1\bullet000010)} + \overset{0}{(1\bullet200000)} + 2\overset{0}{(1\bullet010000)} + \overset{0}{(3\bullet000001)} + 4\overset{0}{(1\bullet000001)} + \overset{0}{(3\bullet000000)}$

$+ 4\overset{0}{(1\bullet000000)}$

Sl 7: $\overset{0}{(0\bullet001000)} + \overset{0}{(2\bullet100001)} + 2\overset{0}{(0\bullet100001)} + 2\overset{1}{(2\bullet000000)} + 3\overset{1}{(0\bullet000000)} + 5\overset{0}{(2\bullet100000)}$

$+ 5\overset{0}{(0\bullet100000)}$

Affine E8 ⊃ A1 + E7 - Level 2

Sl 0: $\overset{0}{(0\bullet000000)}$ **Highest Weight** $\overset{\mathbf{0}}{\mathbf{(01000000)}}$

Sl 1: $\overset{0}{(1\bullet100000)}$

Sl 2: $\overset{0}{(0\bullet010000)} + \overset{0}{(2\bullet000001)} + \overset{0}{(0\bullet000001)} + \overset{0}{(2\bullet000000)} + \overset{0}{(0\bullet000000)}$

Sl 3: $\overset{0}{(1\bullet100001)} + 2\overset{1}{(1\bullet000000)} + \overset{0}{(3\bullet100000)} + 3\overset{0}{(1\bullet100000)}$

Sl 4: $\overset{1}{(0\bullet100000)} + \overset{0}{(0\bullet000002)} + \overset{0}{(2\bullet000010)} + \overset{0}{(0\bullet000010)} + \overset{0}{(2\bullet200000)} + 2\overset{0}{(2\bullet010000)}$

$+ \overset{0}{(0\bullet200000)} + 3\overset{0}{(0\bullet010000)} + 4\overset{0}{(2\bullet000001)} + 4\overset{0}{(0\bullet000001)} + \overset{0}{(4\bullet000000)} + 3\overset{0}{(2\bullet000000)}$

$+ 3\overset{0}{(0\bullet000000)}$

Sl 5: $\overset{1}{(1\bullet000001)} + \overset{0}{(1\bullet110000)} + 2\overset{0}{(1\bullet001000)} + \overset{0}{(3\bullet100001)} + 6\overset{0}{(1\bullet100001)} + 2\overset{1}{(3\bullet000000)}$

$+ 7\overset{1}{(1\bullet000000)} + 4\overset{0}{(3\bullet100000)} + 11\overset{0}{(1\bullet100000)}$

Affine E8 ⊃ A1 + E7 - Level 2

Sl 0: $\overset{0}{(0\bullet000001)}$ **Highest Weight** $\overset{\mathbf{0}}{\mathbf{(00000001)}}$

Sl 1: $\overset{1}{(1\bullet000000)} + \overset{0}{(1\bullet100000)}$

Sl 2: $\overset{0}{(0\bullet000010)} + \overset{0}{(2\bullet010000)} + \overset{0}{(0\bullet010000)} + \overset{0}{(2\bullet000001)} + 2\overset{0}{(0\bullet000001)} + \overset{0}{(2\bullet000000)}$

$+ \overset{0}{(0\bullet000000)}$ (continued next page)

```
             0               0                1                1                0               0
Sl 3:   (1•001000) +  2(1•100001) +    (3•000000) +  3(1•000000) +    (3•100000) +  4(1•100000)

             0               1                1                0                0               0
Sl 4:   (0•010001) +    (2•100000) +    (0•100000) +    (0•000002) +  2(2•000010) +  3(0•000010)

             0               0                0                0                0               0
    +   (2•200000) +  4(2•010000) +  2(0•200000) +  4(0•010000) +    (4•000001) +  5(2•000001)

             0               0                0
    +  6(0•000001) +  3(2•000000) +  3(0•000000)

             0               1                0                0                0               0
Sl 5:   (1•100010) +  2(1•000001) +  2(1•110000) +    (3•001000) +  4(1•001000) +  2(3•100001)

             0               1                1                0                0
    +  9(1•100001) +  4(3•000000) + 11(1•000000) +  5(3•100000) + 13(1•100000)
```

Affine E8 ⊃ A1 + E7 - Level 2

```
             0                                 0
Sl 0:   (2•000000)   Highest Weight (20000000)

             0
Sl 1:   (1•100000)

             0               0                0                0                0
Sl 2:   (0•200000) +    (2•000001) +    (0•000001) +    (2•000000) +    (0•000000)

             0               1                0                0
Sl 3:   (1•100001) +    (1•000000) +    (3•100000) +  3(1•100000)

             1               0                0                0                0               0
Sl 4:   (0•100000) +    (2•000002) +    (0•000010) +    (2•200000) +  2(2•010000) +    (0•200000)

             0               0                0                0                0               0
    +  2(0•010000) +  3(2•000001) +  3(0•000001) +    (4•000000) +  4(2•000000) +  2(0•000000)

             1               0                0                0                0               1
Sl 5:   (1•000001) +    (1•110000) +    (1•001000) +    (3•100001) +  5(1•100001) +    (3•000000)

             1               0                0
    +  5(1•000000) +  4(3•100000) +  9(1•100000)
```

Affine E8 ⊃ A2 + E6 - Level 1

```
             0                                 0
Sl 0:   (10•00000)   Highest Weight (10000000)

             0
Sl 1:   (00•10000)

             0
Sl 2:   (01•00001)

             1               0                0
Sl 3:   (10•00000) +    (02•00000) +    (10•00000)

             0               0                0
Sl 4:   (00•00010) +    (11•10000) +    (00•10000)

             0               0                0
Sl 5:   (01•01000) +    (20•00001) +  2(01•00001)

             0               1                1                0                0               0
Sl 6:   (10•10001) +    (02•00000) +  2(10•00000) +    (21•00000) +    (02•00000) +  3(10•00000)

             1               0                0                0                0               0
Sl 7:   (00•10000) +    (11•00010) +    (00•00002) +    (00•00010) +  3(11•10000) +  3(00•10000)
```

Affine E8 ⊃ A2 + E6 - Level 2

```
             0                                 0
Sl 0:   (01•00000)   Highest Weight (01000000)

             0
Sl 1:   (10•10000)

             0               0                0
Sl 2:   (00•01000) +    (11•00001) +    (00•00001)   (continued next page)
```

```
            0              1               1              0              0             0
Sl 3:  (01•10001) +  (20•00000) +  2(01•00000) +  (12•00000) +  (20•00000) +  2(01•00000)
            1              0               0              0              0             0
Sl 4:  (10•10000) +  (10•00002) +  (02•00010) +  2(10•00010) +  (21•10000) +  2(02•10000)
            0
    +  4(10•10000)
            0              1               1              0              0             0
Sl 5:  (00•10010) +  (11•00001) +  2(00•00001) +  (11•20000) +  3(11•01000) +  (00•20000)
            0              0               0              0              0
    +  3(00•01000) +  (03•00001) +  (30•00001) +  6(11•00001) +  5(00•00001)
```

Affine E8 ⊃ A2 + E6 - Level 2

```
            0                                  0
Sl 0:  (00•00001)  Highest Weight  (00000001)
            1              0
Sl 1:  (01•00000) +  (01•00000)
            0              0               0
Sl 2:  (10•00010) +  (02•10000) +  (10•10000)
            1              0               0              0              0
Sl 3:  (00•00001) +  (11•01000) +  (00•01000) +  2(11•00001) +  2(00•00001)
            0              0               0              1              1             1
Sl 4:  (01•00100) +  (20•10001) +  2(01•10001) +  (12•00000) +  (20•00000) +  4(01•00000)
            0              0               0
    +  (12•00000) +  2(20•00000) +  3(01•00000)
            0              1               1              0              0             0
Sl 5:  (10•01001) +  (02•10000) +  2(10•10000) +  (21•00010) +  2(10•00002) +  2(02•00010)
            0              0               0              0
    +  4(10•00010) +  2(21•10000) +  4(02•10000) +  6(10•10000)
```

Affine E8 ⊃ A2 + E6 - Level 2

```
            0                                  0
Sl 0:  (20•00000)  Highest Weight  (20000000)
            0
Sl 1:  (10•10000)
            0              0               0
Sl 2:  (00•20000) +  (11•00001) +  (00•00001)
            0              1               1              0              0             0
Sl 3:  (01•10001) +  (20•00000) +  (01•00000) +  (12•00000) +  (20•00000) +  2(01•00000)
            1              0               0              0              0             0
Sl 4:  (10•10000) +  (02•00002) +  2(10•00010) +  (21•10000) +  2(02•10000) +  3(10•10000)
            0              1               1              0              0             0
Sl 5:  (00•10010) +  (11•00001) +  (00•00001) +  (11•20000) +  2(11•01000) +  (00•20000)
            0              0               0              0              0
    +  3(00•01000) +  (03•00001) +  (30•00001) +  5(11•00001) +  3(00•00001)
```

Affine E8 ⊃ A3 + D5 - Level 1

```
            0                                  0
Sl 0:  (100•0000)  Highest Weight  (10000000)
            0
Sl 1:  (000•1000)
            0
Sl 2:  (001•0001)
            1
Sl 3:  (010•0000)  (continued next page)
```

```
            0                 0                 0
Sl 4:   (100•0010) +   (011•0000) +   (100•0000)

            1                 0                 0
Sl 5:   (000•0001) +   (101•1000) +   (000•1000)

            0                 0                 0
Sl 6:   (001•0100) +   (110•0001) +  2(001•0001)

            0                 1                 1               1
Sl 7:   (010•1001) +   (002•0000) +   (200•0000) +  2(010•0000)
```

Affine E8 ⊃ A3 + D5 - Level 2

```
            0                                  0
Sl 0:   (010•0000)  Highest Weight  (01000000)

            0
Sl 1:   (100•1000)

            0                 0                 0
Sl 2:   (000•0100) +   (101•0001) +   (000•0001)

            0                 1                 1
Sl 3:   (001•1001) +   (110•0000) +  2(001•0000)

            1                 0                 0                 0                 0                 0
Sl 4:   (010•1000) +   (010•0002) +   (002•0010) +   (200•0010) +  2(010•0010) +   (111•0000)

            0                 0                 0
     +  (002•0000) +   (200•0000) +  3(010•0000)

            0                 1                 1                 0                 0                 0
Sl 5:   (100•1010) +   (011•0001) +  3(100•0001) +   (201•1000) +  3(011•1000) +  4(100•1000)
```

Affine E8 ⊃ A3 + D5 - Level 2

```
            0                                  0
Sl 0:   (000•0001)  Highest Weight  (00000001)

            1
Sl 1:   (001•0000)

            0                 0                 0
Sl 2:   (010•0010) +   (002•0000) +   (010•0000)

            1                 0                 0
Sl 3:   (100•0001) +   (011•1000) +   (100•1000)

            0                 0                 2                 0                 0                 0
Sl 4:   (101•0100) +   (000•0011) +   (000•0000) +   (000•0100) +   (020•0001) +  2(101•0001)

            0
     + 2(000•0001)

            1                 0                 0                 1                 1                 1
Sl 5:   (001•0010) +   (110•1001) +  2(001•1001) +   (102•0000) +  2(110•0000) +  4(001•0000)
```

Affine E8 ⊃ A3 + D5 - Level 2

```
            0                                  0
Sl 0:   (200•0000)  Highest Weight  (20000000)

            0
Sl 1:   (100•1000)

            0                 0                 0
Sl 2:   (000•2000) +   (101•0001) +   (000•0001)   (continued next page)

            0                 1                 1
Sl 3:   (001•1001) +   (110•0000) +   (001•0000)

            1                 0                 0                0                 0                 0
Sl 4:   (010•1000) +   (002•0002) +   (200•0010) + 2(010•0010) +   (111•0000) +   (002•0000)

            0                 0
     +  (200•0000) +  2(010•0000)

            0                 1                 1                 0                 0                 0
Sl 5:   (100•1010) +   (011•0001) +  2(100•0001) +   (201•1000) +  2(011•1000) +  4(100•1000)
```

Affine E8 ⊃ A4 + A4 - Level 1

Sl 0: $(1000\overset{0}{\cdot}000)$ **Highest Weight** $\overset{0}{(10000000)}$

Sl 1: $(0000\overset{0}{\cdot}100)$

Sl 2: $(0001\overset{0}{\cdot}001)$

Sl 3: $(0010\overset{1}{\cdot}000)$

Sl 4: $(0100\overset{0}{\cdot}010)$

Sl 5: $(1000\overset{1}{\cdot}001) + (0101\overset{0}{\cdot}000) + (1000\overset{0}{\cdot}000)$

Sl 6: $(1001\overset{0}{\cdot}100) + (0000\overset{0}{\cdot}011) + (0000\overset{2}{\cdot}000) + (0000\overset{0}{\cdot}100)$

Sl 7: $(1010\overset{0}{\cdot}001) + (0001\overset{1}{\cdot}010) + 2(0001\overset{0}{\cdot}001)$

Affine E8 ⊃ A4 + A4 - Level 2

Sl 0: $(0100\overset{0}{\cdot}000)$ **Highest Weight** $\overset{0}{(01000000)}$

Sl 1: $(1000\overset{0}{\cdot}100)$

Sl 2: $(1001\overset{0}{\cdot}001) + (0000\overset{1}{\cdot}010) + (0000\overset{0}{\cdot}001)$

Sl 3: $(0001\overset{0}{\cdot}101) + (1010\overset{1}{\cdot}000) + 2(0001\overset{1}{\cdot}000)$

Sl 4: $(1100\overset{0}{\cdot}010) + (0010\overset{1}{\cdot}100) + (0010\overset{0}{\cdot}002) + (0002\overset{0}{\cdot}010) + 2(0010\overset{0}{\cdot}010)$

Sl 5: $(0011\overset{1}{\cdot}001) + (0100\overset{0}{\cdot}110) + (2000\overset{1}{\cdot}001) + 3(0100\overset{1}{\cdot}001) + (1101\overset{0}{\cdot}000) + 2(0011\overset{0}{\cdot}000)$
$+ (2000\overset{0}{\cdot}000) + 3(0100\overset{0}{\cdot}000)$

Affine E8 ⊃ A4 + A4 - Level 2

Sl 0: $(0000\overset{0}{\cdot}001)$ **Highest Weight** $\overset{0}{(00000001)}$

Sl 1: $(0001\overset{1}{\cdot}000)$

Sl 2: $(0010\overset{0}{\cdot}010)$

Sl 3: $(0100\overset{1}{\cdot}001) + (0011\overset{0}{\cdot}000) + (0100\overset{0}{\cdot}000)$

Sl 4: $(0101\overset{0}{\cdot}100) + (1000\overset{0}{\cdot}011) + (1000\overset{2}{\cdot}000) + (1000\overset{0}{\cdot}100)$

Sl 5: $(1001\overset{1}{\cdot}010) + (0110\overset{0}{\cdot}001) + (0000\overset{1}{\cdot}002) + 2(1001\overset{0}{\cdot}001) + 2(0000\overset{1}{\cdot}010) + 2(0000\overset{0}{\cdot}001)$

Affine E8 ⊃ A4 + A4 - Level 2

Sl 0: $(2000\overset{0}{\cdot}000)$ **Highest Weight** $\overset{0}{(20000000)}$

Sl 1: $(1000\overset{0}{\cdot}100)$

Sl 2: $(1001\overset{0}{\cdot}001) + (0000\overset{0}{\cdot}200) + (0000\overset{0}{\cdot}001)$ (continued next page)

Sl 3: $(0001\overset{0}{\bullet}101) + (1010\overset{1}{\bullet}000) + (0001\overset{1}{\bullet}000)$

Sl 4: $(0002\overset{0}{\bullet}002) + (1100\overset{0}{\bullet}010) + (0010\overset{1}{\bullet}100) + 2(0010\overset{0}{\bullet}010)$

Sl 5: $(0011\overset{1}{\bullet}001) + (0100\overset{0}{\bullet}110) + (2000\overset{1}{\bullet}001) + 2(0100\overset{1}{\bullet}001) + (1101\overset{0}{\bullet}000) + (0011\overset{0}{\bullet}000)$

$+ (2000\overset{0}{\bullet}000) + 3(0100\overset{0}{\bullet}000)$

Affine E8 ⊃ A5 + A1 + A2 - Level 1

Sl 0: $(10000\overset{0}{\bullet}00)$ **Highest Weight** $\overset{0}{(10000000)}$

Sl 1: $(00000\overset{1}{\bullet}10)$

Sl 2: $(00001\overset{0}{\bullet}01)$

Sl 3: $(00010\overset{1}{\bullet}00)$

Sl 4: $(00100\overset{0}{\bullet}10)$

Sl 5: $(01000\overset{1}{\bullet}01)$

Sl 6: $(01001\overset{0}{\bullet}00) + (10000\overset{0}{\bullet}11) + (10000\overset{2}{\bullet}00) + (10000\overset{0}{\bullet}00)$

Sl 7: $(10001\overset{1}{\bullet}10) + (00000\overset{1}{\bullet}02) + 2(00000\overset{1}{\bullet}10)$

Affine E8 ⊃ A5 + A1 + A2 - Level 2

Sl 0: $(01000\overset{0}{\bullet}00)$ **Highest Weight** $\overset{0}{(01000000)}$

Sl 1: $(10000\overset{1}{\bullet}10)$

Sl 2: $(10001\overset{0}{\bullet}01) + (00000\overset{2}{\bullet}01) + (00000\overset{0}{\bullet}20) + (00000\overset{0}{\bullet}01)$

Sl 3: $(10010\overset{1}{\bullet}00) + (00001\overset{1}{\bullet}11) + 2(00001\overset{1}{\bullet}00)$

Sl 4: $(10100\overset{0}{\bullet}10) + (00010\overset{2}{\bullet}10) + (00010\overset{0}{\bullet}02) + (00002\overset{0}{\bullet}10) + 2(00010\overset{0}{\bullet}10)$

Sl 5: $(00011\overset{1}{\bullet}01) + (11000\overset{1}{\bullet}01) + (00100\overset{1}{\bullet}20) + 3(00100\overset{1}{\bullet}01)$

Affine E8 ⊃ A5 + A1 + A2 - Level 2

Sl 0: $(00000\overset{0}{\bullet}01)$ **Highest Weight** $\overset{0}{(00000001)}$

Sl 1: $(00001\overset{1}{\bullet}00)$

Sl 2: $(00010\overset{0}{\bullet}10)$

Sl 3: $(00100\overset{1}{\bullet}01)$

Sl 4: $(00101\overset{0}{\bullet}00) + (01000\overset{0}{\bullet}11) + (01000\overset{2}{\bullet}00) + (01000\overset{0}{\bullet}00)$

Sl 5: $(01001\overset{1}{\bullet}10) + (10000\overset{1}{\bullet}02) + 2(10000\overset{1}{\bullet}10)$

Affine E8 ⊃ A5 + A1 + A2 - Level 2

```
              0                                  0
Sl 0:   (20000•00)  Highest Weight  (20000000)

              1
Sl 1:   (10000•10)

              0             2             0
Sl 2:   (10001•01) +  (00000•20) +  (00000•01)

              1             1             1
Sl 3:   (10010•00) +  (00001•11) +  (00001•00)

              0             0             2              0
Sl 4:   (10100•10) +  (00002•02) +  (00010•10) +  2(00010•10)

              1             1             1              1
Sl 5:   (00011•01) +  (11000•01) +  (00100•20) +  2(00100•01)
```

Affine E8 ⊃ A7 + A1 - Level 1

```
              0                                  0
Sl 0:   (100000•0)  Highest Weight  (10000000)

              1
Sl 1:   (000000•1)

              0
Sl 2:   (000010•0)

              0
Sl 3:   (001000•1)

              1             0             0
Sl 4:   (010000•0) +  (100000•2) +  (100000•0)

              0             1
Sl 5:   (100001•1) + 2(000000•1)

              0             1             0             0
Sl 6:   (100100•0) +  (000001•0) +  (000010•2) +  (000010•0)

              1             0             0
Sl 7:   (000100•1) +  (110000•1) + 2(001000•1)
```

Affine E8 ⊃ A7 + A1 - Level 2

```
              0                                  0
Sl 0:   (010000•0)  Highest Weight  (01000000)

              1             0
Sl 1:   (100000•1) +  (000000•1)

              0             0             2             0
Sl 2:   (100010•0) +  (000001•2) +  (000000•0) +  (000001•0)

              1             0             0
Sl 3:   (000010•1) +  (101000•1) + 2(000100•1)

              0             1             1              1              0              0
Sl 4:   (000101•0) +  (110000•0) +  (001000•2) +  2(001000•0) +  (200000•2) +  2(010000•2)

              0             0
    +   (200000•0) + 3(010000•0)

              0             2             0              0              1              1
Sl 5:   (001010•1) +  (010000•1) +  (200001•1) +  3(010001•1) +  (100000•3) +  6(100000•1)

              0             0
    +   (000000•3) + 4(000000•1)
```

Affine E8 ⊃ A7 + A1 - Level 2

```
               0                    0
sl 0:    (000000•1)  Highest Weight (00000001)

               0
sl 1:    (000001•0)

               0
sl 2:    (000100•1)

               1                0               0
sl 3:    (001000•0) +    (010000•2) +    (010000•0)

               0                1               0                0
sl 4:    (010001•1) +  2(100000•1) +    (000000•3) +  2(000000•1)

               0                1               0                0                0                2
sl 5:    (010100•0) +    (100001•0) +    (100010•2) +    (100010•0) +  2(000001•2) +    (000000•0)

               0
    +   3(000001•0)
```

Affine E8 ⊃ A7 + A1 - Level 2

```
               0                    0
sl 0:    (200000•0)  Highest Weight (20000000)

               1
sl 1:    (100000•1)

               0                2               0
sl 2:    (100010•0) +    (000000•2) +    (000001•0)

               1                0               0
sl 3:    (000010•1) +    (101000•1) +    (000100•1)

               0                1               1                1                0                0
sl 4:    (000020•0) +    (110000•0) +    (001000•2) +  2(001000•0) +    (200000•2) +    (010000•2)

               0                0
    +    (200000•0) +  2(010000•0)

               0                2               0                0                1                1
sl 5:    (001010•1) +    (010000•1) +    (200001•1) +  2(010001•1) +    (100000•3) +  5(100000•1)

               0
    +   2(000000•1)
```

Affine F4 Branching Rules

Affine F4 ⊃ F4 - Level 1

Sl 0: (•0001) **Highest Weight (00001)**
Sl 1: (•0010) + (•0001)
Sl 2: (•1001) + (•0002) + (•0010) + 2(•0001)
Sl 3: (•0011) + 2(•1001) + (•0002) + 3(•0010) + (•1000) + 3(•0001)
Sl 4: (•1010) + (•0003) + 2(•0011) + (•0100) + 4(•1001) + 3(•0002) + 5(•0010) + (•1000)
+ 6(•0001) + (•0000)
Sl 5: (•0101) + (•1002) + 2(•1010) + 2(•0003) + 5(•0011) + 2(•0100) + 8(•1001) + 5(•0002)
+ 10(•0010) + 3(•1000) + 9(•0001) + (•0000)
Sl 6: (•0012) + (•2001) + (•0020) + 2(•0101) + 2(•1002) + 5(•1010) + 4(•0003) + 11(•0011)
+ (•2000) + 4(•0100) + 15(•1001) + 11(•0002) + 16(•0010) + 5(•1000) + 16(•0001) + 2(•0000)
Sl 7: (•1011) + 3(•0012) + 2(•2001) + 2(•0020) + 6(•0101) + 6(•1002) + 10(•1010) + 8(•0003)
+ 21(•0011) + 2(•2000) + 9(•0100) + 27(•1001) + 18(•0002) + 30(•0010) + 10(•1000) + 24(•0001)
+ 3(•0000)
Sl 8: (•0110) + (•1003) + 3(•1011) + (•1100) + (•0004) + 7(•0012) + 5(•2001) + 6(•0020)
+ 12(•0101) + 12(•1002) + 21(•1010) + 16(•0003) + 40(•0011) + 5(•2000) + 17(•0100) + 47(•1001)
+ 34(•0002) + 48(•0010) + 16(•1000) + 40(•0001) + 6(•0000)

Affine F4 ⊃ F4 - Level 1

Sl 0: (•0000) **Highest Weight (10000)**
Sl 1: (•1000)
Sl 2: (•0002) + (•1000) + (•0000)
Sl 3: (•0100) + (•0002) + (•0010) + 2(•1000) + (•0000)
Sl 4: (•0011) + (•2000) + (•0100) + 3(•0002) + (•0010) + 3(•1000) + (•0001) + 2(•0000)
Sl 5: (•1002) + 2(•0011) + (•2000) + 3(•0100) + (•1001) + 4(•0002) + 3(•0010) + 6(•1000)
+ (•0001) + 2(•0000)
Sl 6: (•0020) + 2(•1002) + (•1010) + (•0003) + 4(•0011) + 3(•2000) + 5(•0100) + 2(•1001)
+ 9(•0002) + 5(•0010) + 8(•1000) + 3(•0001) + 5(•0000)
Sl 7: (•1100) + (•0012) + (•0020) + (•0101) + 5(•1002) + 2(•1010) + (•0003) + 9(•0011)
+ 4(•2000) + 10(•0100) + 5(•1001) + 13(•0002) + 10(•0010) + 15(•1000) + 5(•0001) + 5(•0000)
Sl 8: (•1011) + 2(•1100) + (•0004) + 2(•0012) + 4(•0020) + 2(•0101) + 9(•1002) + 6(•1010)
+ 4(•0003) + 17(•0011) + 9(•2000) + 16(•0100) + 10(•1001) + 25(•0002) + 16(•0010) + 21(•1000)
+ 10(•0001) + 10(•0000)

Affine F4 ⊃ F4 - Level 2

sl 0: (•0002) **Highest Weight (00002)**

sl 1: (•0011) + (•0100) + (•0002) + (•0010) + (•1000)

sl 2: (•0020) + (•1002) + (•1010) + (•0003) + 3(•0011) + (•2000) + 2(•0100) + (•1001)
+ 4(•0002) + 2(•0010) + 2(•1000) + (•0001) + (•0000)

sl 3: (•1011) + (•1100) + 2(•0012) + 2(•0020) + 2(•0101) + 5(•1002) + 3(•1010) + 2(•0003)
+ 9(•0011) + 2(•2000) + 7(•0100) + 5(•1001) + 8(•0002) + 7(•0010) + 6(•1000) + 2(•0001)
+ (•0000)

Affine F4 ⊃ F4 - Level 2

sl 0: (•0010) **Highest Weight (00010)**

sl 1: (•0011) + (•1001) + (•0002) + (•0010) + (•0001)

sl 2: (•0101) + (•1002) + (•1010) + (•0003) + 3(•0011) + (•0100) + 3(•1001) + 2(•0002)
+ 4(•0010) + (•1000) + 2(•0001)

sl 3: (•1011) + 2(•0012) + (•2001) + 2(•0020) + 3(•0101) + 3(•1002) + 5(•1010) + 4(•0003)
+ 9(•0011) + (•2000) + 4(•0100) + 9(•1001) + 7(•0002) + 9(•0010) + 2(•1000) + 6(•0001)
+ (•0000)

sl 4: (•0021) + (•2010) + (•0102) + (•0110) + 2(•1003) + 5(•1011) + 2(•1100) + (•0004)
+ 8(•0012) + 4(•2001) + 6(•0020) + 12(•0101) + 12(•1002) + 15(•1010) + 10(•0003) + 27(•0011)
+ 3(•2000) + 11(•0100) + 24(•1001) + 16(•0002) + 23(•0010) + 8(•1000) + 13(•0001) + (•0000)

Affine F4 ⊃ F4 - Level 2

sl 0: (•1000) **Highest Weight (01000)**

sl 1: (•0100) + (•0002) + (•1000) + (•0000)

sl 2: (•1002) + (•0011) + (•2000) + 2(•0100) + 2(•0002) + (•0010) + 3(•1000) + (•0000)

sl 3: (•1100) + (•0012) + (•0020) + 3(•1002) + (•1010) + 4(•0011) + 3(•2000) + 5(•0100)
+ (•1001) + 6(•0002) + 2(•0010) + 6(•1000) + (•0001) + 2(•0000)

sl 4: (•0102) + 2(•1011) + (•3000) + 3(•1100) + (•0004) + 3(•0012) + 4(•0020) + 2(•0101)
+ 10(•1002) + 4(•1010) + 2(•0003) + 11(•0011) + 7(•2000) + 13(•0100) + 4(•1001) + 13(•0002)
+ 8(•0010) + 13(•1000) + 2(•0001) + 4(•0000)

Affine F4 ⊃ F4 - Level 2

sl 0: (•0001) **Highest Weight (10001)**

sl 1: (•1001) + (•0010) + (•0001)

sl 2: (•1010) + (•0003) + (•0011) + 3(•1001) + (•0002) + 2(•0010) + 3(•0001)

sl 3: (•0012) + (•2001) + 2(•0101) + (•1002) + 3(•1010) + 2(•0003) + 5(•0011) + (•0100)
+ 8(•1001) + 2(•0002) + 7(•0010) + (•1000) + 6(•0001)

sl 4: (•0110) + (•1003) + 2(•1011) + (•0004) + 4(•0012) + 4(•2001) + 3(•0020) + 7(•0101)
+ 4(•1002) + 11(•1010) + 8(•0003) + 15(•0011) + (•2000) + 4(•0100) + 20(•1001) + 9(•0002)
+ 16(•0010) + 2(•1000) + 14(•0001) + (•0000)

Affine F4 ⊃ F4 - Level 2

Sl 0: (•0000) **Highest Weight (20000)**

Sl 1: (•1000)

Sl 2: (•2000) + (•0002) + (•1000) + (•0000)

Sl 3: (•1002) + (•2000) + 2(•0100) + (•0002) + (•0010) + 3(•1000) + (•0000)

Sl 4: (•1100) + (•0004) + (•0020) + 2(•1002) + (•1010) + 2(•0011) + 4(•2000) + 3(•0100) + 5(•0002) + (•0010) + 5(•1000) + (•0001) + 3(•0000)

Sl 5: (•0102) + (•1011) + (•3000) + 3(•1100) + (•0004) + 2(•0012) + 2(•0020) + (•0101) + 8(•1002) + 2(•1010) + 6(•0011) + 6(•2000) + 10(•0100) + 2(•1001) + 8(•0002) + 5(•0010) + 12(•1000) + (•0001) + 3(•0000)

Affine F4 ⊃ A1 + C3 - Level 1

Sl 0: (0•001) **Highest Weight (00001)**

Sl 1: (1•010)

Sl 2: (0•011) + (2•001) + (0•001)

Sl 3: (1•101) + (1•002) + 2(1•010)

Sl 4: (0•110) + (2•011) + (0•003) + 2(0•011) + (0•100) + 2(2•001) + 3(0•001)

Sl 5: (1•012) + (1•020) + 3(1•101) + (3•010) + 2(1•002) + 5(1•010) + (1•000)

Sl 6: (0•021) + (2•110) + (0•102) + (2•003) + 2(0•110) + 3(2•011) + 2(0•003) + 6(0•011) + (2•100) + 2(0•100) + 5(2•001) + 6(0•001)

Sl 7: (1•111) + 4(1•012) + (3•101) + 3(1•020) + 8(1•101) + (3•002) + 2(3•010) + 6(1•002) + 11(1•010) + 2(1•000)

Sl 8: (2•021) + (0•013) + (0•201) + (2•102) + 3(0•021) + 3(2•110) + 3(0•102) + 3(2•003) + 6(0•110) + 9(2•011) + 5(0•003) + 14(0•011) + 3(2•100) + (4•001) + 5(0•100) + 10(2•001) + 13(0•001)

Affine F4 ⊃ A1 + C3 - Level 1

Sl 0: (1•000) **Highest Weight (10000)**

Sl 1: (0•100)

Sl 2: (1•002) + (1•000)

Sl 3: (0•011) + (2•100) + (0•100)

Sl 4: (1•020) + 2(1•002) + (1•010) + (3•000) + 2(1•000)

Sl 5: (0•102) + (2•011) + 2(0•011) + 2(2•100) + 3(0•100) + (0•001)

Sl 6: (1•012) + (1•200) + 2(1•020) + (1•101) + (3•002) + 5(1•002) + 2(1•010) + (3•000) + 4(1•000)

Sl 7: (2•102) + (0•021) + 2(0•102) + (0•110) + 3(2•011) + (0•003) + 5(0•011) + 5(2•100) + 6(0•100) + (2•001) + 2(0•001)

Sl 8: (1•111) + (1•004) + (3•020) + 3(1•012) + 2(1•200) + 6(1•020) + 3(1•101) + 2(3•002) + (3•010) + 10(1•002) + 6(1•010) + 3(3•000) + 8(1•000)

Affine F4 ⊃ A1 + C3 - Level 2

Sl 0: (0•002) **Highest Weight (00002)**

Sl 1: (1•011) + (1•100)

Sl 2: (2•020) + (0•012) + (0•020) + (0•101) + (2•002) + (2•010) + 2(0•002) + (0•010) + (2•000) + (0•000)

Sl 3: (1•021) + 2(1•102) + (1•110) + (3•011) + (1•003) + 5(1•011) + (3•100) + 4(1•100) + 2(1•001)

Sl 4: (0•022) + (2•111) + 2(0•111) + 3(2•012) + (2•200) + (0•004) + 4(2•020) + 5(0•012) + 2(0•200) + 3(2•101) + 5(0•020) + (4•002) + 5(0•101) + 6(2•002) + 4(2•010) + 8(0•002) + 5(0•010) + 3(2•000) + 3(0•000)

Sl 5: (1•112) + 2(1•120) + (3•021) + 3(1•013) + (1•201) + 2(3•102) + 7(1•021) + (3•110) + 11(1•102) + (3•003) + 8(1•110) + 6(3•011) + 6(1•003) + 22(1•011) + 5(3•100) + 15(1•100) + 2(3•001) + 9(1•001)

Sl 6: (2•202) + (0•121) + 2(2•022) + (0•014) + (0•202) + 2(2•103) + 2(2•030) + 5(0•022) + (0•210) + 9(2•111) + 4(0•103) + 3(0•030) + (4•012) + 4(2•004) + 13(0•111) + 2(4•020) + 16(2•012) + 6(2•200) + 5(0•004) + (4•101) + 20(2•020) + 22(0•012) + 7(0•200) + 16(2•101) + 21(0•020) + 3(4•002) + 21(0•101) + 2(4•010) + 23(2•002) + 19(2•010) + 25(0•002) + 20(0•010) + 2(4•000) + 11(2•000) + 9(0•000)

Affine F4 ⊃ A1 + C3 - Level 2

Sl 0: (0•010) **Highest Weight (00010)**

Sl 1: (1•011) + (1•001)

Sl 2: (0•012) + (2•101) + (0•020) + (0•101) + (2•002) + (2•010) + (0•002) + 2(0•010)

Sl 3: (1•021) + (1•102) + 2(1•110) + (3•011) + 2(1•003) + 5(1•011) + 2(1•100) + (3•001) + 4(1•001)

Sl 4: (2•111) + (0•103) + (0•030) + 2(0•111) + 3(2•012) + (0•004) + 3(2•020) + 5(0•012) + (0•200) + 5(2•101) + 4(0•020) + 7(0•101) + (4•010) + 4(2•002) + 6(2•010) + 5(0•002) + 8(0•010) + (2•000) + 2(0•000)

Sl 5: (1•112) + (1•120) + (3•021) + 3(1•013) + 2(1•201) + (3•102) + 8(1•021) + 2(3•110) + 8(1•102) + 2(3•003) + 11(1•110) + 6(3•011) + 9(1•003) + 23(1•011) + 2(3•100) + 9(1•100) + 5(3•001) + 15(1•001)

Sl 6: (0•121) + 2(2•022) + (2•210) + (0•014) + (0•202) + 3(2•103) + 2(2•030) + 4(0•022) + (0•210) + 9(2•111) + 5(0•103) + 4(0•030) + (4•012) + 2(2•004) + 14(0•111) + (4•020) + 18(2•012) + 3(2•200) + 4(0•004) + 2(4•101) + 14(2•020) + 24(0•012) + 4(0•200) + 25(2•101) + 19(0•020) + 2(4•002) + 27(0•101) + 3(4•010) + 19(2•002) + 24(2•010) + 20(0•002) + 26(0•010) + 5(2•000) + 5(0•000)

Affine F4 ⊃ A1 + C3 - Level 2

Sl 0: (0•000) **Highest Weight (01000)**

Sl 1: (1•100)

Sl 2: (0•020) + (2•002) + (0•002) + (2•000) + (0•000)

Sl 3: (1•102) + 2(1•011) + (3•100) + 3(1•100)

Sl 4: (0•111) + (2•012) + (2•200) + (0•004) + 2(2•020) + (0•012) + (0•200) + 3(0•020)
+ (0•101) + 4(2•002) + (2•010) + 4(0•002) + 2(0•010) + (4•000) + 3(2•000) + 3(0•000)

Sl 5: (1•120) + (1•013) + (3•102) + 2(1•021) + 6(1•102) + 2(1•110) + 2(3•011) + (1•003)
+ 9(1•011) + 4(3•100) + 11(1•100) + 2(1•001)

Affine F4 ⊃ A1 + C3 - Level 2

Sl 0: (1•001) **Highest Weight (10001)**

Sl 1: (0•101) + (2•010) + (0•010)

Sl 2: (1•110) + (1•003) + 2(1•011) + (3•001) + 3(1•001)

Sl 3: (0•111) + (2•012) + 2(0•012) + 3(2•101) + (0•020) + 4(0•101) + (2•002) + 4(2•010)
+ 2(0•002) + 4(0•010)

Sl 4: (1•013) + (1•201) + 3(1•021) + (3•110) + 2(1•102) + (3•003) + 6(1•110) + 2(3•011)
+ 5(1•003) + 11(1•011) + 2(1•100) + 4(3•001) + 11(1•001)

Sl 5: (2•103) + (2•030) + (0•022) + (0•210) + 3(2•111) + 3(0•103) + 2(0•030) + (2•004)
+ 6(0•111) + 8(2•012) + (0•004) + (4•101) + 5(2•020) + 11(0•012) + (0•200) + 14(2•101)
+ 7(0•020) + 17(0•101) + 2(4•010) + 6(2•002) + 16(2•010) + 8(0•002) + 16(0•010) + (2•000)
+ 2(0•000)

Sl 6: (1•031) + 4(1•112) + (3•013) + (3•201) + (1•005) + 2(1•120) + 3(3•021) + 8(1•013)
+ 7(1•201) + 2(3•102) + 19(1•021) + 7(3•110) + 15(1•102) + 6(3•003) + 27(1•110) + 13(3•011)
+ 22(1•003) + 46(1•011) + 2(3•100) + (5•001) + 12(1•100) + 14(3•001) + 35(1•001)

Affine F4 ⊃ A1 + C3 - Level 2

Sl 0: (2•000) **Highest Weight (20000)**

Sl 1: (1•100)

Sl 2: (0•200) + (2•002) + (0•002) + (2•000) + (0•000)

Sl 3: (1•102) + (1•011) + (3•100) + 3(1•100)

Sl 4: (2•004) + (0•111) + (2•200) + 2(2•020) + (0•012) + (0•200) + 2(0•020) + 3(2•002)
+ (2•010) + 3(0•002) + (0•010) + (4•000) + 4(2•000) + 2(0•000)

Sl 5: (1•120) + (1•013) + (3•102) + (1•021) + 5(1•102) + (1•110) + (3•011) + 6(1•011)
+ 4(3•100) + 9(1•100) + (1•001)

Sl 6: (2•022) + (0•202) + 2(0•022) + 2(2•111) + (4•200) + 2(2•004) + 4(0•111) + 4(2•012)
+ 4(2•200) + 2(0•004) + 7(2•020) + 5(0•012) + 6(0•200) + 2(2•101) + 7(0•020) + 2(4•002)
+ 3(0•101) + 12(2•002) + 3(2•010) + 11(0•002) + 4(0•010) + 2(4•000) + 8(2•000) + 6(0•000)

Affine F4 ⊃ A2 + A2 - Level 1

sl 0: (00•01) **Highest Weight (00001)**

sl 1: (01•10)

sl 2: (10•11)

sl 3: (11•01) + (00•12) + (00•20) + (00•01)

sl 4: (01•21) + (20•10) + (01•02) + 2(01•10)

sl 5: (10•30) + (10•03) + (02•11) + 3(10•11) + (10•00)

sl 6: (11•12) + (11•20) + (00•31) + 3(11•01) + 3(00•12) + 2(00•20) + 4(00•01)

sl 7: (20•21) + (01•13) + (20•02) + (12•10) + 4(01•21) + 2(20•10) + 3(01•02) + 6(01•10)

sl 8: (21•11) + 2(10•22) + (02•30) + (02•03) + 3(10•30) + 3(10•03) + 3(02•11) + 9(10•11)
 + (02•00) + 3(10•00)

Affine F4 ⊃ A2 + A2 - Level 1

sl 0: (10•00) **Highest Weight (10000)**

sl 1: (00•20)

sl 2: (01•02)

sl 3: (10•11) + (02•00) + (10•00)

sl 4: (11•20) + (00•12) + (00•20) + (00•01)

sl 5: (20•02) + (01•21) + 2(01•02) + (01•10)

sl 6: (10•22) + (02•11) + (21•00) + 3(10•11) + (02•00) + 3(10•00)

sl 7: (11•12) + 3(11•20) + (00•31) + (00•04) + (11•01) + 2(00•12) + 4(00•20) + 2(00•01)

Affine F4 ⊃ A2 + A2 - Level 2

sl 0: (00•02) **Highest Weight (00002)**

sl 1: (01•11) + (01•00)

sl 2: (10•12) + (02•20) + (10•20) + (10•01)

sl 3: (11•21) + 2(11•02) + (00•13) + (11•10) + 2(00•21) + 2(00•02) + 2(00•10)

sl 4: (20•22) + (12•11) + 2(01•22) + 2(20•11) + 2(01•30) + 2(01•03) + (12•00) + 6(01•11)
 + 2(20•00) + 3(01•00)

sl 5: (21•12) + (10•23) + (02•31) + 2(21•20) + 3(10•31) + 2(10•04) + 3(02•12) + (21•01)
 + 7(10•12) + 4(02•20) + 8(10•20) + 3(02•01) + 6(10•01)

sl 6: (11•32) + (30•21) + (22•02) + 2(11•40) + 4(11•13) + (03•21) + (00•24) + (30•02)
 + 10(11•21) + (03•02) + 2(00•32) + (30•10) + 12(11•02) + (03•10) + 3(00•40) + 6(00•13)
 + 9(11•10) + 11(00•21) + 12(00•02) + 8(00•10)

Affine F4 ⊃ A2 + A2 - Level 2

l 0: (00•10) **Highest Weight (00010)**

sl 1: (01•11)

sl 2: (10•12) + (10•20) + (02•01) + (10•01) (continued next page)

Sl 3: (11•21) + (11•02) + (00•13) + 2(11•10) + 2(00•21) + 2(00•02) + 2(00•10)

Sl 4: (20•30) + (20•03) + (12•11) + 2(01•22) + 2(20•11) + 2(01•30) + 2(01•03) + 6(01•11) + (20•00) + 2(01•00)

Sl 5: (21•12) + (10•23) + (02•31) + (21•20) + 3(10•31) + (10•04) + 3(02•12) + 2(21•01) + 8(10•12) + 3(02•20) + 6(10•20) + 4(02•01) + 8(10•01)

Sl 6: (11•32) + (30•21) + (11•40) + 4(11•13) + (03•21) + (30•02) + (22•10) + 11(11•21) + (03•02) + 3(00•32) + (00•05) + (30•10) + 9(11•02) + (03•10) + 2(00•40) + 6(00•13) + 12(11•10) + 12(00•21) + 9(00•02) + 11(00•10)

Affine F4 ⊃ A2 + A2 - Level 2

Sl 0: (01•00) **Highest Weight (01000)**

Sl 1: (10•20)

Sl 2: (11•02) + (00•21) + (00•02)

Sl 3: (01•22) + (20•11) + (12•00) + 2(01•11) + (20•00) + 2(01•00)

Sl 4: (21•20) + (10•31) + (10•04) + (02•12) + 2(10•12) + 2(02•20) + 4(10•20) + 2(10•01)

Sl 5: (11•40) + (11•13) + (30•02) + 3(11•21) + (03•02) + (00•32) + 6(11•02) + (00•40) + 2(00•13) + 2(11•10) + 4(00•21) + 5(00•02) + 3(00•10)

Sl 6: (12•22) + 3(20•22) + (01•41) + (01•14) + (20•30) + (20•03) + 3(12•11) + 7(01•22) + (31•00) + 6(20•11) + 3(01•30) + 3(01•03) + 4(12•00) + 12(01•11) + 5(20•00) + 8(01•00)

Affine F4 ⊃ A2 + A2 - Level 2

Sl 0: (10•01) **Highest Weight (10001)**

Sl 1: (11•10) + (00•21) + (00•10)

Sl 2: (20•11) + (01•30) + (01•03) + 2(01•11)

Sl 3: (10•31) + (02•12) + (21•01) + 3(10•12) + 2(10•20) + 2(02•01) + 4(10•01)

Sl 4: (11•13) + 4(11•21) + (00•32) + (30•10) + 2(11•02) + (03•10) + (00•40) + 2(00•13) + 6(11•10) + 5(00•21) + 4(00•02) + 5(00•10)

Sl 5: (20•22) + (12•30) + (12•03) + (01•41) + (01•14) + 3(20•30) + 3(20•03) + 3(12•11) + 5(01•22) + 7(20•11) + 6(01•30) + 6(01•03) + 14(01•11) + (20•00) + 3(01•00)

Sl 6: (21•31) + (02•23) + 4(21•12) + (10•50) + 4(10•23) + 3(02•31) + (02•04) + 2(21•20) + (13•01) + 9(10•31) + 2(10•04) + 9(02•12) + 7(21•01) + 20(10•12) + 6(02•20) + 13(10•20) + 10(02•01) + 20(10•01)

Affine F4 ⊃ A2 + A2 - Level 2

Sl 0: (20•00) **Highest Weight (20000)**

Sl 1: (10•20)

Sl 2: (11•02) + (00•40) + (00•02)

Sl 3: (01•22) + (20•11) + (12•00) + (01•11) + (20•00) + 2(01•00)

Sl 4: (02•04) + (21•20) + (10•31) + 2(10•12) + 2(02•20) + 3(10•20) + (10•01)

Sl 5: (11•40) + (11•13) + (30•02) + 2(11•21) + (03•02) + (00•32) + 5(11•02) + (00•40) + (00•13) + (11•10) + 3(00•21) + 3(00•02) + 2(00•10)

Sl 6: (12•22) + 3(20•22) + (01•41) + (01•14) + 2(12•11) + 5(01•22) + (31•00) + 5(20•11) + (04•00) + 2(01•30) + 2(01•03) + 3(12•00) + 8(01•11) + 6(20•00) + 5(01•00)

Affine F4 ⊃ A3 + A1 - Level 1

```
Sl 0:   (000•1)  Highest Weight (00001)
Sl 1:   (001•0)
Sl 2:   (010•1)
Sl 3:   (100•2) +   (100•0)
Sl 4:   (101•1) +   (000•3) +  2(000•1)
Sl 5:   (110•0) +  2(001•2) +  2(001•0)
Sl 6:   (200•1) +   (010•3) +   (002•1) +  3(010•1)
Sl 7:   (011•2) +   (011•0) +  4(100•2) +  3(100•0)
Sl 8:   (101•3) +   (020•1) +  4(101•1) +  3(000•3) +  6(000•1)
```

Affine F4 ⊃ A3 + A1 - Level 1

```
Sl 0:   (100•0)  Highest Weight (10000)
Sl 1:   (000•1)
Sl 2:   (001•2)
Sl 3:   (010•1)
Sl 4:   (011•0) +   (100•2) +   (100•0)
Sl 5:   (101•1) +   (000•3) +  2(000•1)
Sl 6:   (110•2) +  2(001•2) +  2(001•0)
Sl 7:   (200•1) +   (010•3) +   (002•1) +  3(010•1)
Sl 8:   (201•0) +   (011•2) +   (100•4) +  2(011•0) +  3(100•2) +  4(100•0)
```

Affine F4 ⊃ A3 + A1 - Level 2

```
Sl 0:   (000•2)  Highest Weight (00002)
Sl 1:   (001•1)
Sl 2:   (010•2) +   (002•0) +   (010•0)
Sl 3:   (011•1) +   (100•3) +  2(100•1)
Sl 4:   (020•2) +  2(101•2) +  2(101•0) +   (000•4) +  3(000•2) +  2(000•0)
Sl 5:   (110•3) +   (102•1) +  3(110•1) +  3(001•3) +  6(001•1)
Sl 6:   (111•2) +   (200•4) +  2(111•0) +  2(200•2) +  2(010•4) +  3(002•2) +  3(200•0) +  8(010•2)
    +  3(002•0) +  6(010•0)
```

Affine F4 ⊃ A3 + A1 - Level 2

```
Sl 0:   (000•0)  Highest Weight (00010)
Sl 1:   (001•1)
Sl 2:   (010•2) +   (010•0)
Sl 3:   (011•1) +   (100•3) +  2(100•1)
Sl 4:  2(101•2) +   (020•0) +  2(101•0) +   (000•4) +  3(000•2) +  2(000•0)
Sl 5:   (110•3) +   (102•1) +  3(110•1) +  3(001•3) +  6(001•1)
Sl 6:   (111•2) +   (111•0) +  3(200•2) +  2(010•4) +  3(002•2) +  2(200•0) +  8(010•2) +  3(002•0)
    +  6(010•0)
```

Affine F4 ⊃ A3 + A1 - Level 2

Sl 0: (010•0) **Highest Weight (01000)**

Sl 1: (100•1)

Sl 2: (101•2) + (000•2) + (000•0)

Sl 3: (110•1) + (001•3) + 2(001•1)

Sl 4: (111•0) + (200•2) + (010•4) + (002•2) + (200•0) + 2(010•2) + (002•0) + 3(010•0)

Sl 5: (201•1) + (011•3) + 3(011•1) + 3(100•3) + 5(100•1)

Sl 6: (210•2) + (012•2) + (101•4) + 2(020•2) + 7(101•2) + (020•0) + 4(101•0) + 2(000•4) + 6(000•2) + 4(000•0)

Affine F4 ⊃ A3 + A1 - Level 2

Sl 0: (100•1) **Highest Weight (10001)**

Sl 1: (101•0) + (000•2) + (000•0)

Sl 2: (110•1) + (001•3) + 2(001•1)

Sl 3: (200•2) + (002•2) + (200•0) + 3(010•2) + 2(010•0)

Sl 4: (201•1) + (011•3) + 3(011•1) + 3(100•3) + 6(100•1)

Sl 5: (210•0) + (101•4) + (020•2) + (012•0) + 6(101•2) + 2(020•0) + 6(101•0) + 2(000•4) + 7(000•2) + 5(000•0)

Affine F4 ⊃ A3 + A1 - Level 2

Sl 0: (200•0) **Highest Weight (20000)**

Sl 1: (100•1)

Sl 2: (101•2) + (000•2)

Sl 3: (110•1) + (001•3) + (001•1)

Sl 4: (111•0) + (002•4) + (200•2) + (200•0) + 2(010•2) + (002•0) + 2(010•0)

Sl 5: (201•1) + (011•3) + 2(011•1) + 2(100•3) + 4(100•1)

Sl 6: (210•2) + (012•2) + (101•4) + 2(020•2) + 5(101•2) + 3(101•0) + (000•4) + 4(000•2) + 3(000•0)

Affine F4 ⊃ B4 - Level 1

Sl 0: (0000•) **Highest Weight (00001)**

Sl 1: (0001•)

Sl 2: (0100•) + (1000•) + (0000•)

Sl 3: (1001•) + 2(0001•)

Sl 4: (0002•) + (0010•) + (2000•) + 2(0100•) + 2(1000•) + 2(0000•)

Sl 5: (0101•) + 3(1001•) + 5(0001•)

Sl 6: (1010•) + (1100•) + 3(0002•) + 3(0010•) + 2(2000•) + 5(0100•) + 5(1000•) + 4(0000•)

Sl 7: (0011•) + (2001•) + 3(0101•) + 8(1001•) + 11(0001•)

Sl 8: 2(1002•) + (0200•) + 3(1010•) + 3(1100•) + 7(0002•) + 8(0010•) + 5(2000•) + 11(0100•) + 10(1000•) + 8(0000•)

Affine F4 ⊃ B4 - Level 1

Sl 0: (1000•) **Highest Weight (10000)**

Sl 1: (0001•)

Sl 2: (0010•) + (1000•) + (0000•)

Sl 3: (1001•) + 2(0001•)

Sl 4: (1100•) + (0002•) + (0010•) + (0100•) + 3(1000•) + (0000•)

Sl 5: (0101•) + 3(1001•) + 4(0001•)

Sl 6: (1002•) + 2(1100•) + 2(0002•) + 4(0010•) + (2000•) + 3(0100•) + 5(1000•) + 3(0000•)

Sl 7: (0011•) + (2001•) + 3(0101•) + 7(1001•) + 9(0001•)

Sl 8: (0110•) + 2(1002•) + 2(1010•) + (3000•) + 5(1100•) + 6(0002•) + 8(0010•) + 2(2000•) + 7(0100•) + 11(1000•) + 5(0000•)

Affine F4 ⊃ B4 - Level 2

Sl 0: (0000•) **Highest Weight (00002)**

Sl 1: (0001•)

Sl 2: (0002•) + (0100•) + (1000•) + (0000•)

Sl 3: (0101•) + 2(1001•) + 3(0001•)

Sl 4: (1002•) + (0200•) + (1010•) + (1100•) + 3(0002•) + 3(0010•) + 2(2000•) + 4(0100•) + 3(1000•) + 3(0000•)

Sl 5: (1101•) + (0003•) + 2(0011•) + 2(2001•) + 6(0101•) + 9(1001•) + 10(0001•)

Sl 6: (2002•) + 2(0102•) + 2(0110•) + (2100•) + 6(1002•) + 3(0200•) + 8(1010•) + (3000•) + 7(1100•) + 13(0002•) + 11(0010•) + 6(2000•) + 14(0100•) + 11(1000•) + 7(0000•)

Affine F4 ⊃ B4 - Level 2

Sl 0: (0001•) **Highest Weight (00010)**

Sl 1: (0010•) + (0100•) + (1000•) + (0000•)

Sl 2: (0101•) + 2(1001•) + 3(0001•)

Sl 3: (1002•) + (1010•) + 2(1100•) + 3(0002•) + 3(0010•) + (2000•) + 4(0100•) + 4(1000•) + 2(0000•)

Sl 4: (1101•) + (0003•) + 2(0011•) + 2(2001•) + 6(0101•) + 10(1001•) + 10(0001•)

Sl 5: (2010•) + 2(0102•) + 2(0110•) + (2100•) + 7(1002•) + 2(0200•) + 7(1010•) + (3000•) + 9(1100•) + 12(0002•) + 14(0010•) + 6(2000•) + 14(0100•) + 12(1000•) + 7(0000•)

Sl 6: (1003•) + (0201•) + 3(1011•) + (3001•) + 8(1101•) + 5(0003•) + 13(0011•) + 12(2001•) + 26(0101•) + 37(1001•) + 32(0001•)

Affine F4 ⊃ B4 - Level 2

Sl 0: (0100•) **Highest Weight (01000)**

Sl 1: (1001•) + (0001•)

Sl 2: (1010•) + (0002•) + (0010•) + (2000•) + 2(0100•) + (1000•) + (0000•)

Sl 3: (0011•) + (2001•) + 2(0101•) + 4(1001•) + 4(0001•) (continued next page)

Sl 4: (0102•) + (2100•) + 2(1002•) + (0200•) + 4(1010•) + 2(1100•) + 5(0002•) + 5(0010•)
+ 4(2000•) + 7(0100•) + 4(1000•) + 3(0000•)

Sl 5: (1011•) + 3(1101•) + 2(0003•) + 5(0011•) + 5(2001•) + 10(0101•) + 16(1001•) + 13(0001•)

Sl 6: (1110•) + (0012•) + 2(2002•) + (0020•) + (2010•) + 5(0102•) + 3(0110•) + 4(2100•)
+ 12(1002•) + 5(0200•) + 17(1010•) + (3000•) + 11(1100•) + 19(0002•) + 17(0010•) + 12(2000•)
+ 22(0100•) + 14(1000•) + 8(0000•)

Affine F4 ⊃ B4 - Level 2

Sl 0: (1000•) **Highest Weight (10001)**

Sl 1: (1001•) + (0001•)

Sl 2: (1100•) + (0002•) + 2(0010•) + (2000•) + (0100•) + 2(1000•) + (0000•)

Sl 3: (0011•) + (2001•) + 2(0101•) + 5(1001•) + 5(0001•)

Sl 4: (0110•) + 3(1002•) + 3(1010•) + (3000•) + 5(1100•) + 6(0002•) + 7(0010•) + 3(2000•)
+ 6(0100•) + 8(1000•) + 3(0000•)

Sl 5: (1011•) + 3(1101•) + 2(0003•) + 6(0011•) + 6(2001•) + 12(0101•) + 21(1001•) + 17(0001•)

Sl 6: (0012•) + (2002•) + (1200•) + (0020•) + 3(2010•) + 4(0102•) + 6(0110•) + 2(2100•)
+ 17(1002•) + 4(0200•) + 15(1010•) + 3(3000•) + 20(1100•) + 23(0002•) + 27(0010•) + 13(2000•)
+ 23(0100•) + 23(1000•) + 10(0000•)

Affine F4 ⊃ B4 - Level 2

Sl 0: (2000•) **Highest Weight (20000)**

Sl 1: (1001•)

Sl 2: (1010•) + (0002•) + (2000•) + (0100•) + (1000•)

Sl 3: (0011•) + (2001•) + (0101•) + 3(1001•) + 2(0001•)

Sl 4: (0020•) + (2100•) + 2(1002•) + (0200•) + 3(1010•) + (1100•) + 3(0002•) + 3(0010•)
+ 4(2000•) + 4(0100•) + 2(1000•) + 2(0000•)

Sl 5: (1011•) + 2(1101•) + (0003•) + 4(0011•) + 4(2001•) + 6(0101•) + 11(1001•) + 7(0001•)

Sl 6: (1110•) + (0012•) + 2(2002•) + (0020•) + 3(0102•) + 2(0110•) + 3(2100•) + 8(1002•)
+ 3(0200•) + 13(1010•) + (3000•) + 7(1100•) + 13(0002•) + 9(0010•) + 9(2000•) + 13(0100•)
+ 9(1000•) + 4(0000•)

Affine G2 Branching Rules

Affine G2 ⊃ G2 - Level 1

Sl 0: (•01) **Highest Weight (001)**

Sl 1: (•02) + (•01)

Sl 2: (•11) + (•02) + (•10) + 2(•01)

Sl 3: (•03) + 2(•11) + 3(•02) + (•10) + 3(•01) + (•00)

Sl 4: (•12) + 2(•03) + 4(•11) + 5(•02) + 3(•10) + 6(•01) + (•00)

Sl 5: (•04) + 2(•12) + (•20) + 4(•03) + 8(•11) + 10(•02) + 5(•10) + 9(•01) + 2(•00)

Sl 6: (•21) + 2(•04) + 5(•12) + 2(•20) + 9(•03) + 14(•11) + 16(•02) + 10(•10) + 16(•01) + 3(•00)

Sl 7: (•13) + 2(•21) + 5(•04) + 10(•12) + 5(•20) + 16(•03) + 25(•11) + 29(•02) + 16(•10) + 24(•01) + 6(•00)

Sl 8: (•05) + 3(•13) + 5(•21) + 9(•04) + 20(•12) + 9(•20) + 29(•03) + 43(•11) + 46(•02) + 29(•10) + 39(•01) + 8(•00)

Affine G2 ⊃ G2 - Level 1

Sl 0: (•00) **Highest Weight (100)**

Sl 1: (•10)

Sl 2: (•02) + (•10) + (•00)

Sl 3: (•03) + (•02) + 2(•10) + (•01) + (•00)

Sl 4: (•20) + (•03) + (•11) + 3(•02) + 3(•10) + (•01) + 2(•00)

Sl 5: (•12) + (•20) + 3(•03) + 2(•11) + 4(•02) + 6(•10) + 3(•01) + 2(•00)

Sl 6: (•04) + 2(•12) + 3(•20) + 5(•03) + 4(•11) + 9(•02) + 8(•10) + 5(•01) + 5(•00)

Sl 7: (•13) + (•04) + 5(•12) + 4(•20) + 10(•03) + 9(•11) + 13(•02) + 15(•10) + 9(•01) + 5(•00)

Sl 8: 2(•13) + (•21) + 4(•04) + 9(•12) + 9(•20) + 16(•03) + 16(•11) + 25(•02) + 21(•10) + 14(•01) + 10(•00)

Affine G2 ⊃ G2 - Level 2

Sl 0: (•02) **Highest Weight (002)**

Sl 1: (•03) + (•11) + (•02) + (•10) + (•01)

Sl 2: (•04) + (•12) + (•20) + 2(•03) + 3(•11) + 4(•02) + 2(•10) + 2(•01) + (•00)

Sl 3: (•13) + (•21) + 2(•04) + 5(•12) + 2(•20) + 7(•03) + 8(•11) + 8(•02) + 6(•10) + 6(•01) + (•00)

Sl 4: (•22) + (•05) + 4(•13) + 4(•21) + 8(•04) + 13(•12) + 8(•20) + 16(•03) + 20(•11) + 21(•02) + 12(•10) + 12(•01) + 4(•00)

Affine G2 ⊃ G2 - Level 2

Sl 0: (•10) **Highest Weight (010)**

Sl 1: (•03) + (•02) + (•10) + (•00)

Sl 2: (•12) + (•20) + 2(•03) + (•11) + 2(•02) + 3(•10) + (•01) + (•00)

Sl 3: (•13) + (•04) + 3(•12) + 3(•20) + 5(•03) + 4(•11) + 6(•02) + 6(•10) + 2(•01) + 2(•00)

Sl 4: (•30) + (•05) + 3(•13) + 2(•21) + 4(•04) + 9(•12) + 7(•20) + 13(•03) + 10(•11) + 13(•02) + 13(•10) + 7(•01) + 4(•00)

Sl 5: (•14) + 2(•22) + 2(•30) + 3(•05) + 10(•13) + 6(•21) + 12(•04) + 23(•12) + 17(•20) + 29(•03) + 26(•11) + 30(•02) + 25(•10) + 14(•01) + 8(•00)

Affine G2 ⊃ G2 - Level 2

Sl 0: (•01) **Highest Weight (101)**

Sl 1: (•11) + (•02) + (•01)

Sl 2: (•12) + (•03) + 3(•11) + 2(•02) + (•10) + 3(•01)

Sl 3: (•21) + 2(•04) + 3(•12) + (•20) + 4(•03) + 7(•11) + 7(•02) + 2(•10) + 6(•01) + (•00)

Sl 4: (•05) + 2(•13) + 4(•21) + 5(•04) + 10(•12) + 3(•20) + 11(•03) + 18(•11) + 15(•02) + 8(•10) + 13(•01) + (•00)

Sl 5: (•14) + 2(•22) + 3(•05) + 8(•13) + 11(•21) + 16(•04) + 26(•12) + 11(•20) + 27(•03) + 41(•11) + 36(•02) + 17(•10) + 25(•01) + 4(•00)

Affine G2 ⊃ G2 - Level 2

Sl 0: (•00) **Highest Weight (200)**

Sl 1: (•10)

Sl 2: (•20) + (•02) + (•10) + (•00)

Sl 3: (•12) + (•20) + 2(•03) + (•02) + 3(•10) + (•01) + (•00)

Sl 4: (•13) + (•04) + 2(•12) + 4(•20) + 3(•03) + 2(•11) + 5(•02) + 5(•10) + (•01) + 3(•00)

Sl 5: (•30) + (•05) + 3(•13) + (•21) + 2(•04) + 7(•12) + 6(•20) + 10(•03) + 6(•11) + 8(•02) + 12(•10) + 4(•01) + 3(•00)

Affine G2 ⊃ G2 - Level 3

Sl 0: (•03) **Highest Weight (003)**

Sl 1: (•04) + (•12) + (•20) + (•03) + (•11) + (•02) + (•10)

Sl 2: (•05) + 2(•13) + (•21) + 2(•04) + 4(•12) + 2(•20) + 5(•03) + 4(•11) + 3(•02) + 3(•10) + (•01)

Sl 3: (•06) + (•14) + 2(•22) + (•30) + 3(•05) + 7(•13) + 5(•21) + 9(•04) + 13(•12) + 8(•20) + 13(•03) + 12(•11) + 11(•02) + 7(•10) + 4(•01) + 2(•00)

Affine G2 ⊃ G2 - Level 3

Sl 0: (•11) **Highest Weight (011)**

Sl 1: (•04) + (•12) + (•03) + 2(•11) + (•02) + (•01)

Sl 2: (•05) + (•13) + 2(•21) + 3(•04) + 4(•12) + (•20) + 4(•03) + 6(•11) + 4(•02) + (•10) + 3(•01)

Sl 3: 2(•14) + 2(•22) + 3(•05) + 6(•13) + 7(•21) + 10(•04) + 14(•12) + 5(•20) + 12(•03) + 18(•11) + 12(•02) + 5(•10) + 7(•01)

Affine G2 ⊃ G2 - Level 3

Sl 0: (•02) **Highest Weight (102)**

Sl 1: (•12) + (•03) + (•11) + (•02) + (•10) + (•01)

Sl 2: (•13) + (•21) + 2(•04) + 3(•12) + 2(•20) + 3(•03) + 4(•11) + 5(•02) + 2(•10) + 2(•01) + (•00)

Sl 3: (•14) + (•22) + (•30) + 2(•05) + 5(•13) + 5(•21) + 6(•04) + 12(•12) + 5(•20) + 12(•03) + 13(•11) + 11(•02) + 8(•10) + 7(•01) + (•00)

Affine G2 ⊃ G2 - Level 3

sl 0: (•10) **Highest Weight (110)**

sl 1: (•20) + (•03) + (•02) + (•10) + (•00)

sl 2: (•13) + 2(•12) + 2(•20) + 3(•03) + (•11) + 2(•02) + 4(•10) + (•01) + (•00)

sl 3: (•22) + (•30) + (•05) + 4(•13) + (•21) + 3(•04) + 6(•12) + 7(•20) + 8(•03) + 5(•11) + 8(•02) + 8(•10) + 2(•01) + 3(•00)

Affine G2 ⊃ G2 - Level 3

sl 0: (•01) **Highest Weight (201)**

sl 1: (•11) + (•02) + (•01)

sl 2: (•21) + (•12) + (•03) + 3(•11) + 2(•02) + (•10) + 3(•01)

sl 3: (•22) + (•13) + 3(•21) + 3(•04) + 4(•12) + (•20) + 4(•03) + 8(•11) + 7(•02) + 2(•10) + 6(•01) + (•00)

sl 4: (•31) + 2(•14) + 3(•22) + (•30) + 3(•05) + 6(•13) + 10(•21) + 8(•04) + 15(•12) + 4(•20) + 13(•03) + 22(•11) + 16(•02) + 8(•10) + 14(•01) + (•00)

Affine G2 ⊃ G2 - Level 3

sl 0: (•00) **Highest Weight (300)**

sl 1: (•10)

sl 2: (•20) + (•02) + (•10) + (•00)

sl 3: (•30) + (•12) + (•20) + 2(•03) + (•02) + 3(•10) + (•01) + (•00)

sl 4: (•22) + (•30) + 2(•13) + (•04) + 2(•12) + 5(•20) + 3(•03) + 2(•11) + 5(•02) + 5(•10) + (•01) + 3(•00)

Affine G2 ⊃ G2 - Level 4

sl 0: (•04) **Highest Weight (004)**

sl 1: (•05) + (•13) + (•21) + (•04) + (•12) + (•03) + (•11)

sl 2: (•06) + 2(•14) + 2(•22) + 2(•05) + 4(•13) + 3(•21) + 5(•04) + 5(•12) + 2(•20) + 3(•03) + 4(•11) + 2(•02)

Affine G2 ⊃ G2 - Level 4

sl 0: (•12) **Highest Weight (012)**

sl 1: (•05) + (•13) + (•21) + (•04) + 2(•12) + (•20) + (•03) + (•11) + (•02)

sl 2: (•06) + 2(•14) + 2(•22) + (•30) + 3(•05) + 5(•13) + 4(•21) + 5(•04) + 8(•12) + 3(•20) + 5(•03) + 5(•11) + 3(•02) + 2(•10) + (•01)

Affine G2 ⊃ G2 - Level 4

sl 0: (•20) **Highest Weight (020)**

sl 1: (•13) + (•12) + (•20) + (•03) + (•10)

sl 2: (•06) + (•22) + (•30) + (•05) + 3(•13) + (•21) + 2(•04) + 3(•12) + 4(•20) + 3(•03) + (•11) + 2(•02) + 2(•10) + (•00)

Affine G2 ⊃ G2 - Level 4

Sl 0: (•03) **Highest Weight (103)**

Sl 1: (•13) + (•04) + (•12) + (•20) + (•03) + (•11) + (•02) + (•10)

Sl 2: (•14) + (•22) + (•30) + 2(•05) + 4(•13) + 2(•21) + 3(•04) + 5(•12) + 3(•20) + 6(•03) + 4(•11) + 3(•02) + 3(•10) + (•01)

Affine G2 ⊃ G2 - Level 4

Sl 0: (•11) **Highest Weight (111)**

Sl 1: (•21) + (•04) + (•12) + (•03) + 2(•11) + (•02) + (•01)

Sl 2: (•14) + (•22) + (•05) + 2(•13) + 4(•21) + 4(•04) + 5(•12) + (•20) + 4(•03) + 7(•11) + 4(•02) + (•10) + 3(•01)

Affine G2 ⊃ G2 - Level 4

Sl 0: (•02) **Highest Weight (202)**

Sl 1: (•12) + (•03) + (•11) + (•02) + (•10) + (•01)

Sl 2: (•22) + (•13) + (•21) + 2(•04) + 3(•12) + 2(•20) + 3(•03) + 4(•11) + 5(•02) + 2(•10) + 2(•01) + (•00)

Affine G2 ⊃ G2 - Level 4

Sl 0: (•10) **Highest Weight (210)**

Sl 1: (•20) + (•03) + (•02) + (•10) + (•00)

Sl 2: (•30) + (•13) + 2(•12) + 2(•20) + 3(•03) + (•11) + 2(•02) + 4(•10) + (•01) + (•00)

Sl 3: (•23) + 2(•22) + 2(•30) + (•05) + 5(•13) + (•21) + 3(•04) + 6(•12) + 8(•20) + 8(•03) + 5(•11) + 8(•02) + 8(•10) + 2(•01) + 3(•00)

Affine G2 ⊃ G2 - Level 4

Sl 0: (•01) **Highest Weight (301)**

Sl 1: (•11) + (•02) + (•01)

Sl 2: (•21) + (•12) + (•03) + 3(•11) + 2(•02) + (•10) + 3(•01)

Sl 3: (•31) + (•22) + (•13) + 3(•21) + 3(•04) + 4(•12) + (•20) + 4(•03) + 8(•11) + 7(•02) + 2(•10) + 6(•01) + (•00)

Affine G2 ⊃ G2 - Level 4

Sl 0: (•00) **Highest Weight (400)**

Sl 1: (•10)

Sl 2: (•20) + (•02) + (•10) + (•00)

Sl 3: (•30) + (•12) + (•20) + 2(•03) + (•02) + 3(•10) + (•01) + (•00)

Affine G2 ⊃ A1 + A1 - Level 1

```
Sl 0:    (0•1)  Highest Weight (001)
Sl 1:    (1•2)
Sl 2:    (2•1) +    (0•3) +    (0•1)
Sl 3:    (1•4) +  2(1•2) +    (1•0)
Sl 4:    (2•3) +    (0•5) +  2(2•1) +  2(0•3) +  3(0•1)
Sl 5:    (3•2) +  3(1•4) +  5(1•2) +  2(1•0)
Sl 6:    (2•5) +  3(2•3) +  2(0•5) +  5(2•1) +  5(0•3) +  6(0•1)
Sl 7:    (3•4) +    (1•6) +  2(3•2) +  7(1•4) +    (3•0) + 11(1•2) +  5(1•0)
Sl 8:  3(2•5) +    (0•7) +    (4•1) +  8(2•3) +  5(0•5) + 10(2•1) + 11(0•3) + 12(0•1)
```

Affine G2 ⊃ A1 + A1 - Level 1

```
Sl 0:    (1•0)  Highest Weight (100)
Sl 1:    (0•3)
Sl 2:    (1•2) +    (1•0)
Sl 3:    (2•3) +    (0•3) +    (0•1)
Sl 4:    (1•4) +    (3•0) +  2(1•2) +  2(1•0)
Sl 5:  2(2•3) +    (0•5) +    (2•1) +  3(0•3) +  2(0•1)
Sl 6:    (1•6) +    (3•2) +  2(1•4) +    (3•0) +  5(1•2) +  4(1•0)
Sl 7:    (2•5) +  5(2•3) +  2(0•5) +  3(2•1) +  6(0•3) +  4(0•1)
Sl 8:    (3•4) +  2(1•6) +  2(3•2) +  6(1•4) +  3(3•0) + 10(1•2) +  8(1•0)
```

Affine G2 ⊃ A1 + A1 - Level 2

```
Sl 0:    (0•2)  Highest Weight (002)
Sl 1:    (1•3) +    (1•1)
Sl 2:    (2•4) +    (2•2) +    (0•4) +    (2•0) +  2(0•2) +    (0•0)
Sl 3:    (3•3) +  2(1•5) +    (3•1) +  4(1•3) +  4(1•1)
Sl 4:    (2•6) +    (4•2) +  4(2•4) +  2(0•6) +  6(2•2) +  5(0•4) +  3(2•0) +  7(0•2) +  3(0•0)
Sl 5:  2(3•5) +  2(1•7) +  5(3•3) +  8(1•5) +  5(3•1) + 15(1•3) + 13(1•1)
Sl 6:    (2•8) +  2(4•4) +  6(2•6) +    (0•8) +  3(4•2) + 17(2•4) +  7(0•6) +  2(4•0) + 21(2•2) + 16(0•4)
     + 11(2•0) + 20(0•2) +  9(0•0)
Sl 7:  2(3•7) +    (1•9) +    (5•3) + 10(3•5) +  9(1•7) +    (5•1) + 19(3•3) + 28(1•5) + 17(3•1) + 46(1•3)
     + 38(1•1)
Sl 8:  3(4•6) +  5(2•8) +    (0•A) +  9(4•4) + 24(2•6) +  6(0•8) + 14(4•2) + 53(2•4) + 24(0•6) +  6(4•0)
     + 64(2•2) + 47(0•4) + 29(2•0) + 55(0•2) + 24(0•0)
```

Affine G2 ⊃ A1 + A1 - Level 2

```
Sl 0:    (0•0)  Highest Weight (010)
Sl 1:    (1•3)
Sl 2:    (2•2) +    (0•4) +    (2•0) +    (0•2) +    (0•0)
Sl 3:    (3•3) +    (1•5) +  3(1•3) +  2(1•1)
```
(continued next page)

Sl 4: (2•6) + 2(2•4) + (0•6) + (4•0) + 4(2•2) + 3(0•4) + 3(2•0) + 4(0•2) + 3(0•0)

Sl 5: (3•5) + (1•7) + 4(3•3) + 5(1•5) + 2(3•1) + 11(1•3) + 7(1•1)

Sl 6: (4•4) + 5(2•6) + (0•8) + 2(4•2) + 9(2•4) + 5(0•6) + 2(4•0) + 14(2•2) + 10(0•4) + 9(2•0) + 13(0•2) + 8(0•0)

Sl 7: (3•7) + (1•9) + (5•3) + 6(3•5) + 5(1•7) + 14(3•3) + 18(1•5) + 9(3•1) + 33(1•3) + 22(1•1)

Sl 8: 2(4•6) + 3(2•8) + 5(4•4) + 17(2•6) + 4(0•8) + 8(4•2) + 33(2•4) + 16(0•6) + 7(4•0) + 42(2•2) + 30(0•4) + 24(2•0) + 35(0•2) + 20(0•0)

Affine G2 ⊃ A1 + A1 - Level 2

Sl 0: (1•1) **Highest Weight (101)**

Sl 1: (2•2) + (0•4) + (0•2)

Sl 2: (1•5) + (3•1) + 2(1•3) + 3(1•1)

Sl 3: 3(2•4) + (0•6) + 4(2•2) + 3(0•4) + (2•0) + 4(0•2) + 2(0•0)

Sl 4: (3•5) + (1•7) + 2(3•3) + 5(1•5) + 4(3•1) + 9(1•3) + 10(1•1)

Sl 5: (4•4) + 3(2•6) + (0•8) + 2(4•2) + 11(2•4) + 4(0•6) + 15(2•2) + 12(0•4) + 5(2•0) + 14(0•2) + 5(0•0)

Sl 6: (3•7) + 6(3•5) + 6(1•7) + (5•1) + 11(3•3) + 19(1•5) + 13(3•1) + 30(1•3) + 30(1•1)

Sl 7: (4•6) + 3(2•8) + 6(4•4) + 14(2•6) + 4(0•8) + 8(4•2) + 39(2•4) + 15(0•6) + 3(4•0) + 46(2•2) + 35(0•4) + 18(2•0) + 39(0•2) + 16(0•0)

Sl 8: (5•5) + 7(3•7) + 3(1•9) + 2(5•3) + 24(3•5) + 22(1•7) + 4(5•1) + 39(3•3) + 60(1•5) + 40(3•1) + 89(1•3) + 79(1•1)

Affine G2 ⊃ A1 + A1 - Level 2

Sl 0: (2•0) **Highest Weight (200)**

Sl 1: (1•3)

Sl 2: (0•6) + (2•2) + (2•0) + (0•2) + (0•0)

Sl 3: (3•3) + (1•5) + 3(1•3) + (1•1)

Sl 4: (2•6) + 2(2•4) + (0•6) + (4•0) + 3(2•2) + 2(0•4) + 4(2•0) + 3(0•2) + 2(0•0)

Sl 5: (3•5) + (1•7) + 4(3•3) + 4(1•5) + (3•1) + 9(1•3) + 5(1•1)

Sl 6: (4•6) + 4(2•6) + (0•8) + 2(4•2) + 7(2•4) + 5(0•6) + 2(4•0) + 11(2•2) + 7(0•4) + 8(2•0) + 10(0•2) + 6(0•0)

Sl 7: (3•7) + (1•9) + (5•3) + 5(3•5) + 4(1•7) + 12(3•3) + 14(1•5) + 6(3•1) + 26(1•3) + 15(1•1)

Sl 8: 2(4•6) + 3(2•8) + 4(4•4) + 14(2•6) + 3(0•8) + (6•0) + 6(4•2) + 25(2•4) + 13(0•6) + 6(4•0) + 31(2•2) + 21(0•4) + 22(2•0) + 25(0•2) + 15(0•0)

Affine G2 ⊃ A1 + A1 - Level 3

Sl 0: (0•3) **Highest Weight (003)**

Sl 1: (1•4) + (1•2) + (1•0)

Sl 2: (2•5) + 2(2•3) + (0•5) + (2•1) + 2(0•3) + 2(0•1)

Sl 3: (3•6) + (3•4) + 2(1•6) + 2(3•2) + 5(1•4) + (3•0) + 6(1•2) + 3(1•0)

Sl 4: (4•5) + 2(2•7) + 2(4•3) + 6(2•5) + 2(0•7) + (4•1) + 10(2•3) + 6(0•5) + 7(2•1) + 10(0•3) + 7(0•1) (continued next page)

Sl 5: (3•8) + (5•4) + 5(3•6) + 3(1•8) + (5•2) + 10(3•4) + 12(1•6) + (5•0) + 11(3•2) + 22(1•4)
+ 6(3•0) + 24(1•2) + 12(1•0)

Sl 6: 2(4•7) + 3(2•9) + (6•3) + 7(4•5) + 12(2•7) + 2(0•9) + 11(4•3) + 30(2•5) + 10(0•7) + 8(4•1)
+ 43(2•3) + 24(0•5) + 30(2•1) + 34(0•3) + 24(0•1)

Sl 7: (3•A) + 3(5•6) + 8(3•8) + 3(1•A) + 6(5•4) + 27(3•6) + 17(1•8) + 8(5•2) + 45(3•4) + 50(1•6)
+ 4(5•0) + 49(3•2) + 81(1•4) + 22(3•0) + 84(1•2) + 38(1•0)

Affine G2 ⊃ A1 + A1 - Level 3

Sl 0: (0•1) **Highest Weight (011)**

Sl 1: (1•4) + (1•2)

Sl 2: (2•5) + (2•3) + (0•5) + 2(2•1) + 2(0•3) + 2(0•1)

Sl 3: 2(3•4) + 2(1•6) + 2(3•2) + 5(1•4) + 6(1•2) + 2(1•0)

Sl 4: (4•5) + 2(2•7) + (4•3) + 6(2•5) + 2(0•7) + 2(4•1) + 9(2•3) + 6(0•5) + 9(2•1) + 9(0•3)
+ 8(0•1)

Sl 5: (3•8) + (5•4) + 4(3•6) + 3(1•8) + (5•2) + 11(3•4) + 11(1•6) + 12(3•2) + 24(1•4) + 4(3•0)
+ 25(1•2) + 9(1•0)

Sl 6: 2(4•7) + 2(2•9) + 7(4•5) + 13(2•7) + 2(0•9) + (6•1) + 10(4•3) + 31(2•5) + 11(0•7) + 10(4•1)
+ 41(2•3) + 25(0•5) + 36(2•1) + 33(0•3) + 28(0•1)

Sl 7: (3•A) + 2(5•6) + 8(3•8) + 3(1•A) + 7(5•4) + 25(3•6) + 18(1•8) + 8(5•2) + 50(3•4) + 49(1•6)
+ 2(5•0) + 52(3•2) + 89(1•4) + 19(3•0) + 89(1•2) + 34(1•0)

Affine G2 ⊃ A1 + A1 - Level 3

Sl 0: (1•2) **Highest Weight (102)**

Sl 1: (2•3) + (0•5) + (2•1) + (0•3) + (0•1)

Sl 2: (3•4) + (1•6) + (3•2) + 3(1•4) + (3•0) + 4(1•2) + 2(1•0)

Sl 3: (2•7) + (4•3) + 4(2•5) + (0•7) + (4•1) + 7(2•3) + 4(0•5) + 6(2•1) + 6(0•3) + 5(0•1)

Sl 4: 3(3•6) + 2(1•8) + (5•2) + 7(3•4) + 8(1•6) + 9(3•2) + 16(1•4) + 4(3•0) + 19(1•2) + 8(1•0)

Sl 5: (4•7) + (2•9) + 4(4•5) + 8(2•7) + 2(0•9) + 8(4•3) + 22(2•5) + 8(0•7) + 7(4•1) + 33(2•3)
+ 19(0•5) + 26(2•1) + 26(0•3) + 20(0•1)

Sl 6: (5•6) + 5(3•8) + 2(1•A) + 4(5•4) + 18(3•6) + 13(1•8) + 5(5•2) + 37(3•4) + 38(1•6) + 3(5•0)
+ 40(3•2) + 67(1•4) + 19(3•0) + 70(1•2) + 31(1•0)

Affine G2 ⊃ A1 + A1 - Level 3

Sl 0: (1•0) **Highest Weight (110)**

Sl 1: (2•3) + (0•3)

Sl 2: (1•6) + (3•2) + (1•4) + (3•0) + 2(1•2) + 2(1•0)

Sl 3: (4•3) + 2(2•5) + (0•7) + 5(2•3) + 2(0•5) + 2(2•1) + 4(0•3) + 2(0•1)

Sl 4: 2(3•6) + (1•8) + 3(3•4) + 5(1•6) + (5•0) + 5(3•2) + 8(1•4) + 5(3•0) + 10(1•2) + 7(1•0)

Sl 5: (2•9) + 2(4•5) + 4(2•7) + (0•9) + 6(4•3) + 12(2•5) + 4(0•7) + 2(4•1) + 22(2•3) + 10(0•5)
+ 12(2•1) + 16(0•3) + 9(0•1)

Sl 6: (5•6) + 2(3•8) + (1•A) + (5•4) + 12(3•6) + 7(1•8) + 3(5•2) + 18(3•4) + 24(1•6) + 3(5•0)
+ 24(3•2) + 35(1•4) + 15(3•0) + 40(1•2) + 23(1•0) (continued next page)

Sl 7: (4•9) + 4(4•7) + 7(2•9) + (0•B) + 2(6•3) + 14(4•5) + 22(2•7) + 7(0•9) + 26(4•3) + 53(2•5) + 19(0•7) + 14(4•1) + 80(2•3) + 40(0•5) + 48(2•1) + 56(0•3) + 34(0•1)

Affine G2 ⊃ A1 + A1 - Level 3

Sl 0: (2•1) **Highest Weight (201)**

Sl 1: (3•2) + (1•4) + (1•2)

Sl 2: (2•5) + (0•7) + (4•1) + 2(2•3) + (0•5) + 3(2•1) + (0•3) + 2(0•1)

Sl 3: (1•8) + 3(3•4) + 2(1•6) + 4(3•2) + 6(1•4) + (3•0) + 6(1•2) + 2(1•0)

Sl 4: (4•5) + 3(2•7) + (0•9) + 2(4•3) + 8(2•5) + 3(0•7) + 4(4•1) + 11(2•3) + 6(0•5) + 13(2•1) + 8(0•3) + 7(0•1)

Sl 5: (3•8) + (1•A) + (5•4) + 5(3•6) + 5(1•8) + 2(5•2) + 15(3•4) + 13(1•6) + 18(3•2) + 26(1•4) + 5(3•0) + 27(1•2) + 9(1•0)

Sl 6: 3(4•7) + 3(2•9) + (0•B) + 9(4•5) + 16(2•7) + 4(0•9) + (6•1) + 13(4•3) + 36(2•5) + 15(0•7) + 16(4•1) + 48(2•3) + 27(0•5) + 44(2•1) + 32(0•3) + 28(0•1)

Affine G2 ⊃ A1 + A1 - Level 3

Sl 0: (3•0) **Highest Weight (300)**

Sl 1: (2•3)

Sl 2: (1•6) + (3•2) + (3•0) + (1•2) + (1•0)

Sl 3: (0•9) + (4•3) + (2•5) + 3(2•3) + (0•5) + (2•1) + 2(0•3)

Sl 4: (3•6) + (1•8) + 2(3•4) + 3(1•6) + (5•0) + 3(3•2) + 3(1•4) + 4(3•0) + 4(1•2) + 4(1•0)

Sl 5: (2•9) + (4•5) + 2(2•7) + (0•9) + 4(4•3) + 6(2•5) + 2(0•7) + (4•1) + 12(2•3) + 4(0•5) + 5(2•1) + 6(0•3) + 3(0•1)

Sl 6: (5•6) + (3•8) + (1•A) + 7(3•6) + 4(1•8) + 2(5•2) + 8(3•4) + 12(1•6) + 2(5•0) + 13(3•2) + 14(1•4) + 10(3•0) + 17(1•2) + 10(1•0)

Affine G2 ⊃ A1 + A1 - Level 4

Sl 0: (0•4) **Highest Weight (004)**

Sl 1: (1•5) + (1•3) + (1•1)

Sl 2: (2•6) + 2(2•4) + (0•6) + 2(2•2) + 2(0•4) + 2(0•2) + (0•0)

Sl 3: (3•7) + 2(3•5) + 2(1•7) + 2(3•3) + 5(1•5) + 2(3•1) + 7(1•3) + 5(1•1)

Sl 4: (4•8) + (4•6) + 2(2•8) + 3(4•4) + 7(2•6) + 2(0•8) + 2(4•2) + 12(2•4) + 6(0•6) + (4•0) + 11(2•2) + 11(0•4) + 4(2•0) + 10(0•2) + 4(0•0)

Sl 5: (5•7) + 2(3•9) + 2(5•5) + 7(3•7) + 3(1•9) + 2(5•3) + 14(3•5) + 13(1•7) + 2(5•1) + 16(3•3) + 26(1•5) + 12(3•1) + 31(1•3) + 23(1•1)

Affine G2 ⊃ A1 + A1 - Level 4

Sl 0: (0•2) **Highest Weight (012)**

Sl 1: (1•5) + (1•3) + (1•1)

Sl 2: (2•6) + 2(2•4) + (0•6) + 2(2•2) + 2(0•4) + (2•0) + 3(0•2) + (0•0)

Sl 3: (3•7) + 2(3•5) + 2(1•7) + 3(3•3) + 6(1•5) + 2(3•1) + 8(1•3) + 6(1•1) (continued next page)

Sl 4: 2(4•6) + 3(2•8) + 3(4•4) + 8(2•6) + 2(0•8) + 3(4•2) + 14(2•4) + 7(0•6) + (4•0) + 14(2•2) + 12(0•4) + 6(2•0) + 13(0•2) + 5(0•0)

Sl 5: (5•7) + 2(3•9) + 2(5•5) + 8(3•7) + 4(1•9) + 3(5•3) + 17(3•5) + 15(1•7) + 2(5•1) + 21(3•3) + 32(1•5) + 15(3•1) + 39(1•3) + 28(1•1)

Affine G2 ⊃ A1 + A1 - Level 4

Sl 0: (0•0) **Highest Weight (020)**

Sl 1: (1•3)

Sl 2: (2•6) + (2•2) + (0•4) + (2•0) + (0•2) + (0•0)

Sl 3: (3•5) + (1•7) + 2(3•3) + 2(1•5) + 4(1•3) + 2(1•1)

Sl 4: (4•6) + (2•8) + (4•4) + 4(2•6) + (0•8) + (4•2) + 5(2•4) + 3(0•6) + 2(4•0) + 6(2•2) + 5(0•4) + 4(2•0) + 5(0•2) + 4(0•0)

Sl 5: (3•9) + (5•5) + 3(3•7) + 2(1•9) + 2(5•3) + 7(3•5) + 6(1•7) + 11(3•3) + 13(1•5) + 5(3•1) + 19(1•3) + 10(1•1)

Affine G2 ⊃ A1 + A1 - Level 4

Sl 0: (1•3) **Highest Weight (103)**

Sl 1: (2•4) + (0•6) + (2•2) + (0•4) + (2•0) + (0•2) + (0•0)

Sl 2: (3•5) + (1•7) + 2(3•3) + 3(1•5) + (3•1) + 5(1•3) + 3(1•1)

Sl 3: (4•6) + (2•8) + (4•4) + 5(2•6) + (0•8) + 2(4•2) + 8(2•4) + 4(0•6) + (4•0) + 9(2•2) + 7(0•4) + 5(2•0) + 7(0•2) + 3(0•0)

Sl 4: (3•9) + (5•5) + 4(3•7) + 2(1•9) + 2(5•3) + 10(3•5) + 9(1•7) + (5•1) + 15(3•3) + 19(1•5) + 9(3•1) + 26(1•3) + 17(1•1)

Sl 5: 3(4•8) + 2(2•A) + (6•4) + 9(4•6) + 11(2•8) + 2(0•A) + (6•2) + 14(4•4) + 30(2•6) + 9(0•8) + (6•0) + 15(4•2) + 46(2•4) + 23(0•6) + 8(4•0) + 45(2•2) + 33(0•4) + 22(2•0) + 32(0•2) + 15(0•0)

Affine G2 ⊃ A1 + A1 - Level 4

Sl 0: (1•1) **Highest Weight (111)**

Sl 1: (2•4) + (2•2) + (0•4) + (0•2)

Sl 2: (3•5) + (1•7) + (3•3) + 2(1•5) + 2(3•1) + 3(1•3) + 4(1•1)

Sl 3: (2•8) + 2(4•4) + 3(2•6) + (0•8) + 2(4•2) + 8(2•4) + 3(0•6) + 8(2•2) + 6(0•4) + 2(2•0) + 6(0•2) + 2(0•0)

Sl 4: (5•5) + 4(3•7) + 2(1•9) + (5•3) + 9(3•5) + 8(1•7) + 2(5•1) + 11(3•3) + 16(1•5) + 12(3•1) + 20(1•3) + 18(1•1)

Sl 5: 2(4•8) + 2(2•A) + (6•4) + 6(4•6) + 10(2•8) + 2(0•A) + (6•2) + 15(4•4) + 24(2•6) + 8(0•8) + 15(4•2) + 44(2•4) + 18(0•6) + 4(4•0) + 42(2•2) + 30(0•4) + 14(2•0) + 28(0•2) + 10(0•0)

Affine G2 ⊃ A1 + A1 - Level 4

Sl 0: (2•2) **Highest Weight (202)**

Sl 1: (3•3) + (1•5) + (3•1) + (1•3) + (1•1) (continued next page)

Sl 2: (4•4) + (2•6) + (0•8) + (4•2) + 3(2•4) + (0•6) + (4•0) + 4(2•2) + 2(0•4) + 2(2•0) + 2(0•2) + (0•0)

Sl 3: (3•7) + (1•9) + (5•3) + 4(3•5) + 3(1•7) + (5•1) + 7(3•3) + 7(1•5) + 6(3•1) + 9(1•3) + 7(1•1)

Sl 4: (2•A) + 3(4•6) + 4(2•8) + (0•A) + (6•2) + 7(4•4) + 12(2•6) + 4(0•8) + 9(4•2) + 20(2•4) + 8(0•6) + 4(4•0) + 23(2•2) + 12(0•4) + 9(2•0) + 12(0•2) + 5(0•0)

Affine G2 ⊃ A1 + A1 - Level 4

Sl 0: (2•0) **Highest Weight (210)**

Sl 1: (3•3) + (1•3)

Sl 2: (2•6) + (4•2) + (2•4) + (0•6) + (4•0) + 2(2•2) + 2(2•0) + (0•2) + (0•0)

Sl 3: (1•9) + (5•3) + 2(3•5) + (1•7) + 5(3•3) + 3(1•5) + 2(3•1) + 6(1•3) + 2(1•1)

Sl 4: 2(4•6) + 2(2•8) + (0•A) + 3(4•4) + 7(2•6) + 2(0•8) + (6•0) + 5(4•2) + 9(2•4) + 5(0•6) + 5(4•0) + 11(2•2) + 5(0•4) + 9(2•0) + 6(0•2) + 4(0•0)

Affine G2 ⊃ A1 + A1 - Level 4

Sl 0: (3•1) **Highest Weight (301)**

Sl 1: (4•2) + (2•4) + (2•2)

Sl 2: (3•5) + (1•7) + (5•1) + 2(3•3) + (1•5) + 3(3•1) + (1•3) + 2(1•1)

Sl 3: (2•8) + (0•A) + 3(4•4) + 2(2•6) + (0•8) + 4(4•2) + 6(2•4) + (0•6) + (4•0) + 6(2•2) + 3(0•4) + 2(2•0) + 2(0•2)

Sl 4: (1•B) + (5•5) + 3(3•7) + 2(1•9) + 2(5•3) + 8(3•5) + 6(1•7) + 4(5•1) + 11(3•3) + 9(1•5) + 13(3•1) + 10(1•3) + 10(1•1)

Affine G2 ⊃ A1 + A1 - Level 4

Sl 0: (4•0) **Highest Weight (400)**

Sl 1: (3•3)

Sl 2: (2•6) + (4•2) + (4•0) + (2•2) + (2•0)

Sl 3: (1•9) + (5•3) + (3•5) + 3(3•3) + (1•5) + (3•1) + 2(1•3)

Sl 4: (0•C) + (4•6) + (2•8) + 2(4•4) + 3(2•6) + (0•8) + (6•0) + 3(4•2) + 3(2•4) + 2(0•6) + 4(4•0) + 4(2•2) + (0•4) + 4(2•0) + (0•2) + 2(0•0)

Affine G2 ⊃ A2 - Level 1

Sl 0: (00•) **Highest Weight (001)**

Sl 1: (01•)

Sl 2: (10•)

Sl 3: (11•) + (00•)

Sl 4: (20•) + 2(01•)

Sl 5: (02•) + 3(10•)

Sl 6: 3(11•) + 3(00•)

Sl 7: (12•) + 2(20•) + 5(01•)

Sl 8: (21•) + 3(02•) + 6(10•)

Affine G2 ⊃ A2 - Level 1

```
Sl 0:    (10•)  Highest Weight (100)
Sl 1:    (00•)
Sl 2:    (01•)
Sl 3:    (02•) +   (10•)
Sl 4:    (11•) +   (00•)
Sl 5:    (20•) +  2(01•)
Sl 6:    (21•) +   (02•) +  3(10•)
Sl 7:   3(11•) +  3(00•)
```

Affine G2 ⊃ A2 - Level 2

```
Sl 0:    (00•)  Highest Weight (002)
Sl 1:    (01•)
Sl 2:    (02•) +   (10•)
Sl 3:   2(11•) +   (00•)
Sl 4:    (12•) +  2(20•) +  3(01•)
Sl 5:   2(21•) +  3(02•) +  5(10•)
Sl 6:    (22•) +   (30•) +   (03•) +  8(11•) +  5(00•)
Sl 7:    (31•) +  6(12•) +  7(20•) + 11(01•)
Sl 8:    (40•) +  2(13•) +  9(21•) + 12(02•) + 15(10•)
```

Affine G2 ⊃ A2 - Level 2

```
Sl 0:    (01•)  Highest Weight (010)
Sl 1:    (10•)
Sl 2:    (11•) +   (00•)
Sl 3:    (12•) +   (20•) +  2(01•)
Sl 4:    (21•) +  2(02•) +  3(10•)
Sl 5:    (30•) +   (03•) +  5(11•) +  3(00•)
Sl 6:    (31•) +  4(12•) +  5(20•) +  7(01•)
Sl 7:    (13•) +  6(21•) +  7(02•) + 10(10•)
Sl 8:   3(22•) +  4(30•) +  4(03•) + 17(11•) +  9(00•)
```

Affine G2 ⊃ A2 - Level 2

```
Sl 0:    (10•)  Highest Weight (101)
Sl 1:    (11•) +   (00•)
Sl 2:    (20•) +  2(01•)
Sl 3:    (21•) +  2(02•) +  3(10•)
Sl 4:    (30•) +   (03•) +  5(11•) +  3(00•)
Sl 5:   3(12•) +  5(20•) +  7(01•)
Sl 6:    (13•) +  6(21•) +  7(02•) + 11(10•)
Sl 7:   3(22•) +  4(30•) +  4(03•) + 18(11•) + 10(00•)
Sl 8:   3(31•) +   (04•) + 14(12•) + 16(20•) + 22(01•)
```

Affine G2 ⊃ A2 - Level 2

```
Sl 0:   (20•)  Highest Weight (200)
Sl 1:   (10•)
Sl 2:   (11•) +   (00•)
Sl 3:   (12•) +   (20•) +  2(01•)
Sl 4:   (21•) +  2(02•) +  2(10•)
Sl 5:   (30•) +   (03•) +  4(11•) +  2(00•)
Sl 6:   (31•) +   (04•) +  3(12•) +  5(20•) +  5(01•)
Sl 7:   (13•) +  5(21•) +  5(02•) +  8(10•)
Sl 8:  3(22•) +  3(30•) +  3(03•) + 13(11•) +  7(00•)
```

Affine G2 ⊃ A2 - Level 3

```
Sl 0:   (00•)  Highest Weight (003)
Sl 1:   (01•)
Sl 2:   (02•) +   (10•)
Sl 3:   (03•) +  2(11•) +   (00•)
Sl 4:  2(12•) +  2(20•) +  3(01•)
Sl 5:   (13•) +  3(21•) +  4(02•) +  5(10•)
Sl 6:  3(22•) +  2(30•) +  3(03•) + 10(11•) +  5(00•)
Sl 7:   (23•) +  3(31•) +   (04•) + 11(12•) +  9(20•) + 13(01•)
```

Affine G2 ⊃ A2 - Level 3

```
Sl 0:   (01•)  Highest Weight (011)
Sl 1:   (02•) +   (10•)
Sl 2:  2(11•) +   (00•)
Sl 3:  2(12•) +  2(20•) +  3(01•)
Sl 4:   (13•) +  3(21•) +  4(02•) +  5(10•)
Sl 5:  2(22•) +  2(30•) +  3(03•) + 10(11•) +  5(00•)
Sl 6:   (23•) +  3(31•) +   (04•) + 11(12•) + 10(20•) + 13(01•)
Sl 7:  2(32•) +  2(40•) +  7(13•) + 17(21•) + 18(02•) + 19(10•)
```

Affine G2 ⊃ A2 - Level 3

```
Sl 0:   (10•)  Highest Weight (102)
Sl 1:   (11•) +   (00•)
Sl 2:   (12•) +   (20•) +  2(01•)
Sl 3:  2(21•) +  3(02•) +  3(10•)
Sl 4:   (22•) +  2(30•) +  2(03•) +  7(11•) +  3(00•)
Sl 5:  2(31•) +   (04•) +  7(12•) +  8(20•) +  9(01•)
Sl 6:   (32•) +   (40•) +  4(13•) + 13(21•) + 12(02•) + 15(10•)
Sl 7:   (41•) +   (14•) + 11(22•) + 10(30•) + 10(03•) + 31(11•) + 13(00•)
```

Affine G2 ⊃ A2 - Level 3

Sl 0: (11•) **Highest Weight (110)**
Sl 1: (20•) + (01•)
Sl 2: (21•) + (02•) + 2(10•)
Sl 3: (22•) + (30•) + (03•) + 4(11•) + 2(00•)
Sl 4: (31•) + 4(12•) + 4(20•) + 5(01•)
Sl 5: (40•) + 2(13•) + 7(21•) + 7(02•) + 8(10•)
Sl 6: (41•) + (14•) + 6(22•) + 7(30•) + 6(03•) + 18(11•) + 7(00•)
Sl 7: 2(23•) + 8(31•) + 3(04•) + 19(12•) + 19(20•) + 19(01•)

Affine G2 ⊃ A2 - Level 3

Sl 0: (20•) **Highest Weight (201)**
Sl 1: (21•) + (10•)
Sl 2: (30•) + 2(11•) + (00•)
Sl 3: (31•) + 2(12•) + 3(20•) + 3(01•)
Sl 4: (40•) + (13•) + 5(21•) + 4(02•) + 5(10•)
Sl 5: 3(22•) + 5(30•) + 3(03•) + 11(11•) + 4(00•)
Sl 6: (23•) + 6(31•) + 2(04•) + 12(12•) + 14(20•) + 12(01•)
Sl 7: 3(32•) + 4(40•) + 9(13•) + 23(21•) + 16(02•) + 21(10•)

Affine G2 ⊃ A2 - Level 3

Sl 0: (30•) **Highest Weight (300)**
Sl 1: (20•)
Sl 2: (21•) + (10•)
Sl 3: (22•) + (30•) + 2(11•) + (00•)
Sl 4: (31•) + 2(12•) + 2(20•) + 2(01•)
Sl 5: (40•) + (13•) + 4(21•) + 3(02•) + 3(10•)
Sl 6: (41•) + (14•) + 3(22•) + 5(30•) + 3(03•) + 8(11•) + 2(00•)
Sl 7: (23•) + 5(31•) + 2(04•) + 8(12•) + 10(20•) + 7(01•)

Affine G2 ⊃ A2 - Level 4

Sl 0: (00•) **Highest Weight (004)**
Sl 1: (01•)
Sl 2: (02•) + (10•)
Sl 3: (03•) + 2(11•) + (00•)
Sl 4: (04•) + 2(12•) + 2(20•) + 3(01•)
Sl 5: 2(13•) + 3(21•) + 4(02•) + 5(10•)
Sl 6: (14•) + 4(22•) + 2(30•) + 4(03•) + 10(11•) + 5(00•
Sl 7: 3(23•) + 4(31•) + 3(04•) + 13(12•) + 9(20•) + 13(01•)

Affine G2 ⊃ A2 - Level 4

Sl 0: (01•) **Highest Weight (012)**
Sl 1: (02•) + (10•)
Sl 2: (03•) + 2(11•) + (00•)
Sl 3: 3(12•) + 2(20•) + 3(01•)
Sl 4: 2(13•) + 4(21•) + 5(02•) + 5(10•)
Sl 5: (14•) + 4(22•) + 3(30•) + 5(03•) + 12(11•) + 5(00•)
Sl 6: 3(23•) + 5(31•) + 3(04•) + 16(12•) + 12(20•) + 15(01•)
Sl 7: (24•) + 5(32•) + (05•) + 3(40•) + 14(13•) + 24(21•) + 24(02•) + 22(10•)

Affine G2 ⊃ A2 - Level 4

Sl 0: (02•) **Highest Weight (020)**
Sl 1: (11•)
Sl 2: (12•) + (20•) + (01•)
Sl 3: (13•) + 2(21•) + 2(02•) + 2(10•)
Sl 4: 2(22•) + (30•) + 2(03•) + 5(11•) + 2(00•)
Sl 5: (23•) + 2(31•) + (04•) + 7(12•) + 5(20•) + 6(01•)
Sl 6: (24•) + 2(32•) + 2(40•) + 6(13•) + 11(21•) + 11(02•) + 9(10•)
Sl 7: (33•) + 2(41•) + 3(14•) + 12(22•) + 9(30•) + 11(03•) + 23(11•) + 7(00•)

Affine G2 ⊃ A2 - Level 4

Sl 0: (10•) **Highest Weight (103)**
Sl 1: (11•) + (00•)
Sl 2: (12•) + (20•) + 2(01•)
Sl 3: (13•) + 2(21•) + 3(02•) + 3(10•)
Sl 4: 2(22•) + 2(30•) + 3(03•) + 7(11•) + 3(00•)
Sl 5: (23•) + 3(31•) + 2(04•) + 9(12•) + 8(20•) + 9(01•)
Sl 6: 3(32•) + (05•) + 2(40•) + 8(13•) + 16(21•) + 14(02•) + 15(10•)
Sl 7: (33•) + 3(41•) + 4(14•) + 18(22•) + 13(30•) + 15(03•) + 35(11•) + 13(00•)

Affine G2 ⊃ A2 - Level 4

Sl 0: (11•) **Highest Weight (111)**
Sl 1: (12•) + (20•) + (01•)
Sl 2: 2(21•) + 2(02•) + 2(10•)
Sl 3: 2(22•) + 2(30•) + 2(03•) + 6(11•) + 2(00•)
Sl 4: (23•) + 3(31•) + (04•) + 8(12•) + 7(20•) + 7(01•)
Sl 5: 2(32•) + 2(40•) + 6(13•) + 14(21•) + 12(02•) + 12(10•)
Sl 6: (33•) + 3(41•) + 3(14•) + 16(22•) + 13(30•) + 13(03•) + 31(11•) + 10(00•)
Sl 7: 2(42•) + 2(50•) + 12(23•) + 22(31•) + 10(04•) + 42(12•) + 34(20•) + 32(01•)

Affine G2 ⊃ A2 - Level 4

```
Sl 0:   (20•)  Highest Weight (202)
Sl 1:   (21•) +   (10•)
Sl 2:   (22•) +   (30•) +  2(11•) +   (00•)
Sl 3:  2(31•) +  3(12•) +  3(20•) +  3(01•)
Sl 4:   (32•) +  2(40•) +  2(13•) +  7(21•) +  5(02•) +  5(10•)
Sl 5:  2(41•) +   (14•) +  7(22•) +  8(30•) +  5(03•) + 14(11•) +  4(00•)
Sl 6:   (42•) +   (50•) +  4(23•) + 13(31•) +  4(04•) + 19(12•) + 19(20•) + 14(01•)
Sl 7:   (51•) +   (24•) + 11(32•) +  2(05•) + 10(40•) + 17(13•) + 38(21•) + 24(02•) + 26(10•)
```

Affine G2 ⊃ A2 - Level 4

```
Sl 0:   (21•)  Highest Weight (210)
Sl 1:   (30•) +   (11•)
Sl 2:   (31•) +   (12•) +  2(20•) +   (01•)
Sl 3:   (32•) +   (40•) +   (13•) +  4(21•) +  2(02•) +  3(10•)
Sl 4:   (41•) +  4(22•) +  4(30•) +  2(03•) +  7(11•) +  2(00•)
Sl 5:   (50•) +  2(23•) +  7(31•) +   (04•) + 10(12•) +  9(20•) +  7(01•)
Sl 6:   (51•) +   (24•) +  6(32•) +   (05•) +  7(40•) +  9(13•) + 21(21•) + 13(02•) + 13(10•)
```

Affine G2 ⊃ A2 - Level 4

```
Sl 0:   (30•)  Highest Weight (301)
Sl 1:   (31•) +   (20•)
Sl 2:   (40•) +  2(21•) +   (10•)
Sl 3:   (41•) +  2(22•) +  3(30•) +  3(11•) +   (00•)
Sl 4:   (50•) +   (23•) +  5(31•) +  4(12•) +  5(20•) +  3(01•)
Sl 5:  3(32•) +  5(40•) +  3(13•) + 11(21•) +  5(02•) +  6(10•)
Sl 6:   (33•) +  6(41•) +  2(14•) + 12(22•) + 14(30•) +  6(03•) + 17(11•) +  4(00•)
```

Affine G2 ⊃ A2 - Level 4

```
Sl 0:   (40•)  Highest Weight (400)
Sl 1:   (30•)
Sl 2:   (31•) +   (20•)
Sl 3:   (32•) +   (40•) +  2(21•) +   (10•)
Sl 4:   (41•) +  2(22•) +  2(30•) +  2(11•) +   (00•)
Sl 5:   (50•) +   (23•) +  4(31•) +  3(12•) +  3(20•) +  2(01•)
Sl 6:   (51•) +   (24•) +  3(32•) +  5(40•) +  3(13•) +  8(21•) +  4(02•) +  3(10•)
```